Lecture Notes in Artificial Intelligence   5845

Edited by R. Goebel, J. Siekmann, and W. Wahlster

Subseries of Lecture Notes in Computer Science

T0180291

Arturo Hernández Aguirre
Raúl Monroy Borja
Carlos Alberto Reyes García (Eds.)

MICAI 2009: Advances in Artificial Intelligence

8th Mexican International Conference
on Artificial Intelligence
Guanajuato, México, November 9-13, 2009
Proceedings

 Springer

Series Editors

Randy Goebel, University of Alberta, Edmonton, Canada
Jörg Siekmann, University of Saarland, Saarbrücken, Germany
Wolfgang Wahlster, DFKI and University of Saarland, Saarbrücken, Germany

Volume Editors

Arturo Hernández Aguirre
Centro de Investigación en Matemáticas, AC. (CIMAT)
Area de Computación, Calle Jalisco S/N
Mineral de Valenciana Guanajuato, CP 36250, Guanajuato, México
E-mail: artha@cimat.mx

Raúl Monroy Borja
Ciencias de la Computación, Tecnológico de Monterrey
Campus Estado de México, Col. Margarita Maza de Juárez
Atizapán de Zaragoza, CP 52926, Estado de México, México
E-mail: raulm@itesm.mx

Carlos Alberto Reyes García
Instituto Nacional de Astrofísica, Optica y Electrónica (INAOE)
Ciencias Computacionales, Luis Enrique Erro No. 1
Santa María Tonantzintla, CP 72840, Puebla, México
E-mail: kargaxxi@ccc.inaoep.mx

Library of Congress Control Number: 2009937486

CR Subject Classification (1998): I.2, I.2.9, I.5, F.2.2, I.5.4, I.4

LNCS Sublibrary: SL 7 – Artificial Intelligence

ISSN   1867-8211
ISBN 978-3-642-05257-6 Springer Berlin Heidelberg New York

springer.com

© Springer-Verlag Berlin Heidelberg 2009

Typesetting: Camera-ready by author, data conversion by Scientific Publishing Services, Chennai, India
Printed on acid-free paper   SPIN: 12779635   06/3180   5 4 3 2 1 0

Preface

The Mexican International Conference on Artificial Intelligence (MICAI), a yearly international conference organized by the Mexican Society for Artificial Intelligence (SMIA), is a major international AI forum and the main event in the academic life of the country's growing AI community. The proceedings of the previous MICAI events were published by Springer in its *Lecture Notes in Artificial Intelligence* (LNAI) series, vol. 1793, 2313, 2972, 3787, 4293, 4827, and 5317. Since its foundation the conference has been growing in popularity and improving quality.

This volume contains the papers presented at the oral sessions of the 8th Mexican International Conference on Artificial Intelligence, MICAI 2009, held November 9-13, 2009, in Guanajuato, México. The conference received for evaluation 215 submissions by 646 authors from 21 countries. This volume contains revised versions of 63 articles, which after thorough and careful revision were selected by the international Program Committee. Thus the acceptance rate was 29.3% This book is structured into 18 sections, 17 of which correspond to a conference track and are representative of the main current areas of interest for the AI community; the remaining section comprises invited papers.

The conference featured excellent keynote lectures by leading AI experts:

Patricia Melin, Instituto Tecnológico de Tijuana, México
Dieter Hutter, DFKI GmbH, Germany
Josef Kittler, Surrey University, UK
Ramón López de Mantaras, IIIA—CSIC, Spain
José Luis Marroquín, CIMAT, México

In addition to the oral technical sessions and keynote lectures, the conference program included tutorials, workshops, and a poster session, which were published in separate proceedings volumes.

We thank all the people involved in the organization of this conference. In the first place, we thank the authors of the papers of this book. We thank the Track Chairs, the members of the Program Committee and additional reviewers whose great work made possible the excellent collection of papers for this volume. We would also like to thank the General Director of Consejo de Ciencia y Tecnología del Estado de Guanajuato, Pedro Luis López de Alba for his invaluable support. Also thanks to Ramíro Rico, Project and Programs Director at Consejo de Ciencia y Tecnología de Guanajuato. We appreciate the collaboration of the Office of Art and Culture of Guanajuato, and the assistance of the Municipality of Guanajuato City. We would also like to thank the Local Committee at the Centro de Investigación en Matemáticas for the organization of MICAI 2009.

The submission and reviewing processes, as well as the proceedings assemblage process, were conducted on EasyChair (www.easychair.org), a conference management system. EasyChair was developed by Andrei Voronkov, whom we thank for his support. Finally, we would like to express our deep gratitude to Springer's staff for their help and support in editing this volume.

August 2009   Raúl Monroy Borja
                                                  Arturo Hernández Aguirre
                                                Carlos Alberto Reyes García

Conference Organization

MICAI 2009 was organized by the Mexican Society for Artificial Intelligence (SMIA), in collaboration with Centro de Investigación en Matemáticas, AC. (CIMAT)

Conference Committee

General Chair	Carlos Alberto Reyes García
Program Chairs	Arturo Hernández Aguirre
	Raúl Monroy Borja
Workshop Chair	Alexander Gelbukh
Tutorial Chair	Jesús González
Keynote Speaker Chair	Rafael Murrieta
Finance Chair	Grigori Sidorov
Publicity Chair	Alejandro Peña
Grants Chair	Gustavo Arroyo
Logistics Chair	Rafael Murrieta
Doctoral Consortium Chair	Miguel González

Program Committee

Program Chairs	Arturo Hernández Aguirre
	Raúl Monroy Borja

Each Program Committee member was a Track Chair or Co-track Chair of one or more areas:

Mauricio Osorio	Logic and Reasoning
Alexander Gelbukh	Ontologies, Knowledge Management and Knowledge-Based Systems
Raúl Monroy	Uncertainty and Probabilistic Reasoning
Grigori Sidorov	Natural Language Processing
Christian Lemaitre	Multi-agent Systems and Distributed AI
Alexander Gelbukh	Data Mining
Eduardo Morales	Machine Learning
Johan van Horebeek	Pattern Recognition
Mariano Rivera y Alonso Ramírez	Computer Vision and Image Processing
Rafael Murrieta	Robotics
Juan Frausto Solís	Planning and Scheduling
Oscar Castillo	Fuzzy Logic
Carlos A. Reyes García	Neural Networks

Arturo Hernández Aguirre	Intelligent Tutoring Systems
Olac Fuentes	Bioinformatics and Medical Applications
Alberto Ochoa Ortíz Zezzatti	Hybrid Intelligent Systems
Arturo Hernández Aguirre	Evolutionary Algorithms

Program Committee

Ruth Aguilar Ponce
Alfonso Alba
Arantza Aldea
Moises Alencastre-Miranda
Jose Luis Alonzo Velazquez
Rafal Angryk
Annalisa Appice
Miguel Arias Estrada
Gustavo Arroyo
Victor Ayala-Ramirez
Andrew Bagdanov
Sivaji Bandyopadhyay
Maria Lucia Barron-Estrada
Ildar Batyrshin
Bettina Berendt
Phill Burrell
Pedro Cabalar
David Cabanillas
Felix Calderon
Nicoletta Calzolari
Sergio Daniel Cano Ortiz
Gustavo Carneiro
Jesus Ariel Carrasco-Ochoa
Mario Chacon
Edgar Chavez
Zhe Chen
Simon Colton
Quim Comas
Santiago E. Conant-Pablos
Ulises Cortes
Ivan Cruz
Nicandro Cruz-Ramirez
Alfredo Cuzzocrea
Oscar Dalmau
Mats Danielson
Justin Dauwels
Jorge de la Calleja
Marina De Vos

Anne Denton
Michel Devy
Elva Diaz
Juergen Dix
Susana C. Esquivel
Marc Esteva
Claudia Esteves
Vlad Estivill-Castro
Eugene Ezin
Reynaldo Felix
Francisco Fernandez de Vega
Andres Figueroa
Juan J. Flores
Andrea Formisano
Sofia N. Galicia-Haro
Jean Gao
Rene Arnulfo Garcia-Hernandez
Eduardo Garea
Pilar Gomez-Gil
Jesus Gonzalez
Miguel Gonzalez-Mendoza
Javier Gonzalez Contreras
Nicola Guarino
Hartmut Haehnel
Hyoil Han
Yasunari Harada
Rogelio Hasimoto
Jean-Bernard Hayet
Alejandro Guerra Hernández
Juan Arturo Herrera Ortiz
Hugo Hidalgo
Marc-Philippe Huget
Seth Hutchinson
Pablo H. Ibarguengoytia
Hector Jimenez Salazar
Serrano Rubio Juan Pablo
Vicente Julian
Latifur Khan

Mario Koeppen
Mark Kon
Vladik Kreinovich
Reinhard Langmann
Pedro Larranaga
Yulia Ledeneva
Guillermo Leguizamon
Tao Li
James Little
Giovanni Lizarraga
Rene Mac Kinney
Jacek Malec
Jose Luis Marroquin
Jose Fco. Martinez-Trinidad
Francesco Masulli
Pedro Mayorga
Patricia Melin
Efren Mezura-Montes
Dunja Mladenic
Jaime Mora-Vargas
Guillermo Morales
Brendan Mumey
Lourdes Munoz-Gomez
Masaki Murata
Angel Muñoz
Juan Antonio Navarro
Juan Carlos Nieves
Juan Arturo Nolazco Flores
Constantin Orasan
Magdalena Ortiz
Ted Pedersen
Viktor Pekar
Leif Peterson
Alejandro Peña
David Pinto
Michele Piunti
Eunice E. Ponce de Leon
Naren Ramakrishnan
Zbigniew Ras

Fuji Ren
Orion Fausto Reyes-Galaviz
Antonio Robles-Kelly
Horacio Rodriguez
Katya Rodriguez-Vazquez
Raul Rojas
Leandro F. Rojas Peña
Paolo Rosso
Jianhua Ruan
Salvador Ruiz Correa
Rogelio Salinas
Gildardo Sanchez
Elizabeth Santiago-Delangel
Frank-Michael Schleif
Roberto Sepulveda
Leonid Sheremetov
Gerardo Sierra
Thamar Solorio
Humberto Sossa Azuela
Marta Rosa Soto
Luis Enrique Sucar
Cora Beatriz Excelente Toledo
Aurora Torres
Luz Abril Torres-Mendez
Gregorio Toscano
Ivvan Valdez
Aida Valls
Berend Jan van der Zwaag
Johan Van Horebeek
Karin Verspoor
Francois Vialatte
Javier Vigueras
Ricardo Vilalta
Manuel Vilares Ferro
Thomas Villmann
Toby Walsh
Ramon Zatarain
Claudia Zepeda
Zhi-Hua Zhou

Additional Reviewers

Rita M. Acéves-Pérez
Federico Alonso-Pecina
Matt Brown
Selene Cardenas
Nohe Ramon Cazarez-Castro
Mario Chacon
Justin Dauwels
Eugene Ezin
Mohamed Abdel Fattah
Alexander Gelbukh
Selene Hernández-Rodríguez
Rodolfo Ibarra-Orozco
David Juarez-Romero
Vladik Kreinovich
Pedro Larrañaga
Shuo Li
Gang Li
Michael Maher
Pavel Makagonov
Felix Martinez-Rios
Patricia Melín
Yoel Ledo Mezquita
Oscar Montiel
Jaime Mora-Vargas
Rafael Morales
Juan Carlos Nieves

Alberto Ochoa O.Zezzatti
Ivan Olmos
Arturo Olvera
Juan-Arturo Herrera Ortiz
Alberto Portilla-Flores
Orion Fausto Reyes-Galaviz
Carlos A. Reyes-Garcia
David Romero
Jia Rong
Elizabeth Santiago
Hector Sanvicente-Sanchez
Roberto Sepulveda
Rachitkumar Shah
Grigori Sidorov
María Somodevilla
Satoshi Tojo
Benjamin Tovar
Jorge Adolfo Ramirez Uresti
Francisco Fernández de Vega
Salvador Venegas
Huy Quan Vu
Mats Petter Wallander
Toby Walsh
Christian Lemaitre
Rafael Murrieta

Local Organization

Local Chairs

Planning Chair
Registration Chair
Webmaster

Jean Bernard Hayet
Claudia Estevez
Alonso Ramírez
Norma Cortes
Lourdes Navarro
Odalmira Alvarado

Table of Contents

Natural Language Processing

Data Mining

Machine Learning

Pattern Recognition

Computer Vision and Image Processing

Neural Networks

Intelligent Tutoring Systems

Bioinformatics and Medical Applications

Hybrid Intelligent Systems

Evolutionary Algorithms

Semantic Management of Heterogeneous Documents

Dieter Hutter

DFKI, Cartesium 2.41, Enrique Schmidt Str. 5, D-28359 Bremen, Germany
hutter@dfki.de

Abstract. Software Engineering or other design processes produce a variety of different documents written in different formats and formalizations, and interwoven by various relations and dependencies. Working on such document collections is not a simple progression of developing one document after the other, but it's the art of orchestrating a synchronous development of the entire document collection. Design decisions made in one document influence the appearance of others and modifications made to one document have to be propagated along the various interrelations throughout the document collection. In contrast, tool support typically focuses on individual document types ignoring dependencies between different document types, which likely results in inconsistencies between individual documents of the collection.

In this paper we will advocate a semantic management of heterogeneous documents that orchestrates the evolution of the individual documents with the help of specialized tools. Such a management monitors the design process, takes care of the various dependencies between documents, analyses the consequences of changes made in one document to others, and engineers the synchronization steps necessary to obtain a consistent document collection. The semantic basis of such an approach are ontologies formalizing the structuring mechanisms and interrelations of individual document types.

1 Introduction

The management of complex development tasks — from software developments to the development of manufacturing equipment — typically requires the maintenance of evolving heterogeneous documents. Each of these documents represents the state or achievement of the development from a particular view. For instance, in software engineering requirement specifications, specifications of various design layers, source code and various documentation have to be developed. While in early software development methodologies these documents had to be elaborated sequentially (e.g. waterfall model [17]) recent methodologies (e.g. agile software development [2] advocate for an intertwined approached resulting in a parallel evolution of numerous documents.

Documents arising during the development of such systems are mutually related to each other causing various types of dependencies within and between

A. Hernández Aguirre et al. (Eds.): MICAI 2009, LNAI 5845, pp. 1–14, 2009.

individual documents. They are redundant in a sense that bits of information are distributed into various documents. For example, since each function in the source code has to be sufficiently documented, code fragments might be included into the documentation to illustrate the text. Changing the source code implies the change of the documentation. Documents are also intertwined on a semantic level. For example, a security model underlying a system design has to be satisfied by the implementation. Documents evolve over the time. For instance, new functionality is added to the system, existing functionality is adopted to customers' need or errors in the specification that are revealed during testing or verification cause revisions of various documents and furthermore, adaptations in documents based on the revised ones.

Local changes in documents have often global effects. Consider, for instance, software development. Changing the number of parameters of a function in some program causes necessary adaptations in various development documents. The source code has to be changed at positions where this function is used, the documentation explaining the semantics of the function has to be adjusted, and — in case we are concerned with safety-critical systems — the security model (describing in particular potential vulnerabilities of individual functions) has to be updated. Requirement tracing (see Section 2) is a flourishing area developing commercial tools to maintain such types of dependencies in the software development process.

Similar issues can be observed in all types of (non-trivial) development processes. Consider, for example, the development of teaching material for a course consisting of books, lecture notes and slides. Obviously all documents have to agree in their used notations and signatures. Changing the notation of a concept in one document causes the change of the notation in all documents. Once we translate the course material to another (foreign) language, we have to synchronize the content of both document collections. For instance, the insertion of a new paragraph in the German lecture notes causes the insertion of a corresponding paragraph in the English version.

Typically there are countless relationships and dependencies within and between documents that have to be taken into account if individual documents are changed. Considering specific documents some of these dependencies are covered by existing tools (like compilers) if they are written in a formal language. Then restrictions of the language enforced by parsers (e.g. programming languages, XML-dialects equipped with a DTD or XML-schema) guarantee that documents satisfy some (rather simple) constraints. However, compilers are always specific to individual document languages; complex relations or dependencies between different document types are hardly covered by them.

The advent of the Semantic Web has fostered the progress of encoding different documents in a uniform way in order to provide a uniform (syntactical) processing of documents. Therefore, myriads of different XML-languages emerged to encode domain specific knowledge in a uniform way. Nowadays numerous tools support the processing of XML-documents: the use of XML-editors, XML-parsers, and XML-viewers is common practice to create, interpret and visualize domain specific knowledge. However, up to now there is no uniform way to manage the evolution of individual XML-documents or even collections of heterogeneous XML-documents.

In this paper we advocate a uniform semantic management of heterogeneous (XML-)documents that orchestrates the evolution of collections of documents. Such a management monitors the design process, takes care of the various dependencies between documents, analyses the consequences of changes made in one document to itself and others, and engineers the synchronization steps necessary to obtain a consistent document collection. While the mechanisms to maintain the documents is uniform to all considered document types, the semantics necessary to maintain the documents is encoded in document-type specific and collection-type specific ontologies formalizing the dependencies in individual document types as well as the interrelations between them.

2 Some Approaches on Consistency

The problem of maintaining changing knowledge is not a new one. Numerous approaches have been developed to maintain collections of documents in a consistent way. However, these tools are specialized to specific application areas and thus are tailored to individual types of documents. For lack of space we cannot elaborate on all approaches but we highlight some important areas.

In databases, various techniques have been developed to keep a database in a consistent state. The first technique approaching this problem is to keep the database as redundancy-free as possible. In relational databases [3] normal forms were developed to remove structural complexity and dependencies between database entries as far as possible. Dependencies in a table are restricted by various normal forms (see [4,8,5] for such normal forms). Secondly, triggers are introduced in databases to define small programs that are executed once an entry in a table is inserted, changed, or deleted. However, a trigger is an improper tool to enforce database consistency since first the trigger program has only access to limited part of the environment and second, there is an non-determinism with respect to the execution ordering of individual rows if commands operate on multiple rows.

Most prominent examples are tools in the area of software engineering. Requirements traceability, e.g. [13,9,16,19] is concerned with the question of how to acquire and maintain a trace from requirements along their manifestations in development artifacts at different levels of abstraction. This is a key technique for software maintenance, because the information about relationships between different artifacts that is provided by these traces allows to assess the impact and scope of planned changes, cf. [10,16,20,19,6]. A planned change can be propagated along the links that relate dependent items in the artifacts, and thus a work plan of necessary follow-up changes can be acquired. The dependency relationships and the ways in which changes propagate over them is application dependent. Hence common requirement tracing techniques are either very general and thus do not provide sufficient semantic support for propagating changes or they aim at specific phases in the software development process and assume a fixed semantics for the managed documents. [12] gives an overview of existing (mostly commercial) requirement tracing tools. The CDET [18] system for consistent document engineering allows one to specify a kind of system ontology with consistency rules such that the system is able to capture informal consistency requirements. In case of rule violation the system generates consistency reports, which visualize inconsistencies and suggest repair actions. It supports various approaches to determine semantic differences either by analyzing differences on already parsed (and thus analyzed) documents or by compiling only fractions of the documents localized around syntactical differences.

In Formal Methods we developed a tool called MAYA [1] to maintain structured formal specifications and their resulting verification work. Specifications are written in terms of formal theories each consisting of a set of axioms. A theory may import other theories (in terms of theory inclusions) giving rise to directed (definition) links from the imported theories to the importing one. We end up with a directed graph on theories, the so-called *development graph* [11]. Consistency rules, like the one that a theory (for instance, specifying a security model) is implied by a theory (specifying the functional requirements) are represented analogously as directed (theorem) links but this time these links specify proof obligations (or constraints) rather than given facts. The proof of such obligations is done on two layers. On a top layer the system reasons in terms of reachability of theories. If one theory is reachable from another theory the latter is semantically part of the first. Hence theory links postulate reachability relations and can be proven by finding a witness path between source and target theory. On a bottom layer the system can generate new theorem links (and later on use them to establish witness paths) by proving the theorems denoted by the link with the help of a deduction system. MAYA implements a change management by translating changes of the specification into a sequence of basic graph operations. The sequence is used to transform the original development graph to a development graph fitting the changed specification. It incorporates specific knowledge on whether or how such basic graph operations will affect or change proof obligations and potentially existing proofs encoded in the notions of established and non-established theory links. Since it decomposes proof obligations (i.e. theory links) into smaller proof obligations it allows for reuse of proofs once the decomposition of the old and the new proof obligation share common subproblems.

Summing up, existing tools are either domain independent supporting only simple syntactical relations or they are tailored (and thus restricted) to a specific application domain because they make use of the semantics of the documents. To use such tools for the management of heterogeneous document types requires that the semantic dependencies of the various document types have to be fixed when designing the tool which results in an inflexible approach. Rather than developing a specific change management system for a particular application domain, we aim at a generic change management system, that can be tailored to various application domains by instantiating generic mechanisms to individual document types. There is an obvious need for such a tool. The ever increasing complexity of development processes results in an increasing amount of documents written in various formal, semi-formal and natural languages. These documents have to be maintained in a consistent way. However, for non-formal documents such consistency rules are hardly documented but are often only implicitly specified and (hopefully) known by the document engineers. Hence there is the risk that the engineer will underestimate the effects of his changes to the entire documentation resulting in a potential inconsistency of the document collection.

3   What Is Consistency?

In the following, our general scenario is that we start with a set of documents that are considered (or proven) consistent. During the development, some documents will change and we have to check whether the changed documents are either still consistent or, if not, we have to localize the source of inconsistency or compute repair steps to achieve consistency again. Up to now we talked informally about consistency of document collections without giving a precise notion of consistency. In logic consistency is defined either as the absence of contradictions or the existence of a semantic model satisfying the regulations given by the document collection. However, since we are dealing in particular with documents without a formal semantics, we obviously lack a precise description in which cases a document collection would be inconsistent but also what the formal models of our document collection could be. Rather than trying to define some generic denotational semantics for XML-documents we define consistency on a syntactical level as a property of document collections enriched by annotations provided by a semantic analysis. These annotations encode semantic information gained, for instance, by parsing the individual parts of the documents or by semantic analysis of the documents (e.g. resolution of dependencies). Also a user may annotate document parts, in particular to introduce semantic knowledge on informal documents.

Starting with documents represented as XML-trees, the annotation of "semantic" knowledge will render them to (full-fledged) *document graphs*. Annotations are encoded as additional links (or nodes) in the graph making relations and dependencies explicit. A collection of documents constitutes a graph containing the graphs given by annotated individual XML-documents plus additional annotations specifying the interrelationship between the graphs. For instance, given a source code written in some programming language annotations may represent the relation between the definition of a procedure and its calls. Considering the example of writing lecture notes, annotations can encode the correspondences between German paragraphs and their English counterparts. In this case the annotations have to be provided typically by some author who relates corresponding paragraphs, sections or chapters in the first place. In the future, sophisticated natural language processing tools might be useful to extract the (shallow) semantics of individual paragraphs in order to compute such relationships between German and English versions automatically. Hence, the enrichment of XML-documents by annotations depends on the individual document types and has to be processed by document-type specific plugins (e.g. parsers or pre-compilers) or by the user. Analogously, the type of a document collection determine the annotations specifying the various relations between different documents.

Based on document graphs we define consistency as the satisfaction of a set of *consistency constraints* that are formulated as predicates operating on document graphs. Since annotations, which are computed by plugins or users, reflect semantic or derived knowledge about the documents, we are able to formulate in particular sophisticated semantic constraints on document collections. This is in contrast to schema languages like DTD or XML-schema, which only allow for the formalisation of syntactical restrictions on documents but do neither provide any means to specify any semantic constraint nor sophisticated non-local syntactical properties.

3.1   Notions of Consistency

Typically consistency constraints determine the relations between different parts of a document collection. From a technical point of view such constraints constitute relations between nodes in the graph corresponding to the individual parts in the documents. We call these nodes the *constraint nodes* (of a specific constraint). Many of these constraints are structured in a sense that we can decompose the constraints on graphs to collections of constraints operating on subgraphs. For instance, the constraint that a German chapter of a lecture note has an English counterpart can be decomposed to the constraints that each section in the German chapter has an English counterpart in the corresponding English chapter. We can iterate this decomposition until we end up with individual text chunks which do not possess any further structuring. In this example, we might end up with a German and an English sentence and have to check whether both sentences coincide in their semantics. As said before, an author would typically decide on this property. When maintaining formal developments we have, for instance, to prove that a design specification satisfies a particular security model, which reduces technically to the problem of proving that the theory denoting the security model is included in the theory denoting the design specification. Assuming that the theory for the security model is composed of various theories we can decompose the proof obligation (global constraint) into similar proof obligations (local constraints) for each individual sub theory of the security model. At the end we need domain specific knowledge (i.e. tools) to prove the constraints in the base cases, which can be done by specialized theorem provers. Summing up, consistency constraints on complex documents are typically structured and can be decomposed according the structuring mechanisms given in the related graphs (e.g. document structure of the latex-code, or the composition of the theory denoting the security model in formal developments). The way consistency constraints are decomposed depends on the document types involved. Also the constraints arising as the result of the decomposition are document-type specific and have to tackled by specialized tools (plugins) or the user.

This is an important observation since it helps us to design languages and tools for consistency constraints: on one hand consistency constraints denote proof obligations in order to guarantee the integrity of a document collection. On the other hand, a proven consistency constraint represents a useful lemma in order to prove more complex consistency constraints. Especially when changing a consistent document collection, we like to inherit as much proven constraints as possible to the new situation in order to obtain basic building blocks to reprove endangered complex constraints. For example, consider the German and English lecture notes that are synchronized with respect to their document structure. When changing an individual German chapter we still know that all other German chapters have corresponding English counterparts. Hence the author has only to care about the English chapter corresponding to the changed German one. Consider our Formal Method's example. When changing an included theory of the security model we still know that the proofs on other theories of the security model are still valid. Hence we only have to reprove the constraints arising from the changed theory.

In principle, consistency constraints can be verified either by implementing predicates as executable programs or by specifying them explicitly in some logic in and use a generic inference system (calculus) to find a proof. The first approach is easy to implement and results in efficient code for verifying constraints. However, programs are black boxes and lack any explanation of how constraints are proven or why they failed. In the second approach, verification rules are specified and verified in an explicit way. In this case the decomposition of the constraints is also explicitly available and the proofs of the arising sub constraints can be reused in later consistency checks. However, this approach might be rather slow and inefficient if the consistency has to be proven without the use of decomposition rules.

Hence we split the verification process into a declarative part doing the decomposition of constraints and a procedural part which is concerned with the black-box evaluation of non-structured constrains. For the declarative part, decomposition rules are formulated in terms of graph rewriting [7]. Constraints are explicitly represented as annotations within the graph and graph rewriting rules specify the decomposition of these constraints. The execution of a rewriting rule inserts additional annotations to the graph corresponding to the resulting (sub-) constraints. The rewriting rules may depend themselves on document-type specific side conditions that are formulated as new constraints and eventually evaluated by black-box procedures. Applying the rules exhaustively results in a set of non-structured constraints that have to be checked by black-box functions provided to the generic management tool as document-type specific plugins. Hence we combine the efficiency of programming consistency checks with reuse capabilities gained by an explicitly given decomposition of structured constraints.

4   Management of Change

In the previous sections we analyzed the static property of consistency of document collections and illustrated how corresponding constraints guaranteeing the consistency can be formalized and verified. In this section we will discuss the dynamic aspects of maintaining such document collections in a consistent way. As mentioned above, documents representing some development process are constantly changing. Nevertheless they always have to reflect a consistent view on the state of development. For example, correcting specification errors revealed during a test phase results in adaptations in various documents like the requirement specification, the documentation and the source code. A change management dealing with such issues can be split into various phases. In the first phase the location of changes have to be determined in the document, in the second phase the constraints that are affected by the change have to be computed and in the last phase these constraints have to be reproven which typically causes additional changes inside other parts of the document collection.

We start with the issue of detecting differences in two versions of a document collection. There are two alternatives. Either we monitor the activities of the user inside an editor and propagate this information along the way the input is processed or we compute the differences later on the basis of the graph representation of XML-documents after parsing and annotation. The first alternative leads to the notion of *invasive editors*, which are out-of-the-box editors (like Microsoft's Office or OpenOffice) that are extended by specific hooks monitoring the activities of the user and annotating changed positions in the document[1]. In order to benefit from this monitoring also the preprocessing of the documents (e.g. the parsing) has to be incremental in order to avoid the reanalysis of the entire document. The second approach of comparing resulting XML-documents can be applied without incrementally operating parsers. Using XML-based difference algorithms the differences between the original document graph (ignoring annotations added to the graph in later steps) and the actual graph can be computed. Based on this initial analysis of both graphs (providing a mapping between the nodes and some links of the document graph) the mapping is extended to links and nodes included by annotation processes. The differences are translated into an operational view and encoded as a set of operations that transform the original document graph to a new document graph for the changed document (see Figure 1) that coincides with the translation of the changed document in its XML parts. The operations used to transform the graph are simple graph operations like add, delete, move or change nodes or links.

Given a sequence of basic operations transforming the graph of the original document to a graph for the changed document, we are left with the problem of transforming the annotational part of the graph. Therefore we have to distinguish between different sources of annotations. First, there are annotations corresponding to the semantic analysis of the document (e.g. links between definitions and their usage). These annotations are part of the semantics of a document and justified by the parsing process. Hence these annotations will not be translated to the new document but computed from scratch by reparsing the changed document. Second, there are annotations related to specification and verification of consistency constraints (e.g. that two paragraphs are variants in

---

[1] Many editors have already features to monitor the change process of a document.

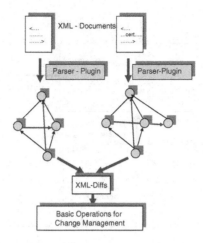

Fig. 1. Analysing the differences

different languages). Typically, these annotations were generated during the verification of consistency constraints and denote either decomposed constraints or proofs thereof. For a management of change we like to inherit this encoded proof information as far as possible. Therefore we have to analyze which annotations under consideration are affected by the graph operations performed.

5 Dependencies of Annotations

In order to transfer an annotation from the original document to the changed one we have to guarantee that the property or constraint denoted by this annotation is still valid. The verification of the property might be based on a node or link of the document that is now subject to the change. One way to reestablish the property is to guarantee that the change of the document does not interfere with those parts of the documents that have been used to establish the annotation in the first place. Consider the example of a constraint denoting that the security model is satisfied by the requirement specification. Such a constraint is typically verified by a formal proof, which relies on axioms specified in the document. If one of these axioms is changed then the proof is no longer valid. Hence we have to inspect the proof in order to guarantee that the constraint is still satisfied. Now suppose, the change is inside the documentation and does not affect any axiom. In this case we do not need to inspect the proof because we know in advance that formal proofs are not affected by changes inside the documentation. If we mark the part of the document collection representing the formalization of the satisfaction property then any change outside this subgraph will not affect the considered constraint. As another example consider the constraint that all citations in a text are listed in the bibliography. Any changes that do not change the list of citations or the bibliography will not endanger this constraint.

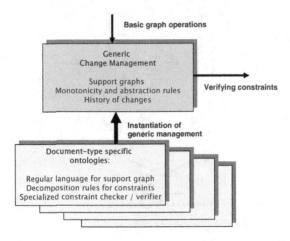

Fig. 2. Instantiation of the generic change management

Suppose, we have no explicit information how a specific constraint was verified. Once a change operation is performed on the document collection, we can only compute those parts of the original collection that may have potentially contributed to the verification of the constraint. If the change does not affect these parts of the collection then the constraint is trivially not affected. Since we aim at a generic management of change, the computation of the relevant parts has to be done in a uniform way for the various constraint types. The idea is to instantiate a generic mechanism by document-type specific rules specified in some kind of *document-type ontology*. The instantiated mechanism will compute these relevant bits of the graph, the so-called support graph. The definition of support graphs is done in terms of reachability properties taking into account the type of links and nodes on the witness path. A simple solution would be to define the support graph as all links and nodes reachable from the constraint nodes via links and nodes of selected types. However, in practice we would lack the ability to deal with intransitive dependencies as defined, for instance, in local definition links in [15]. Intransitive dependencies often occur if constraints argue on relations between nodes instead of relations between subgraph. Hence we restrict the notion of reachability further and allow only specific paths possessing specific sequences of links and nodes types. For instance, the support graph consists of all nodes which are reachable from an constraint node by a sequence of global definition links followed by an individual local link. The restrictions are formalized with the help of regular languages. For each constraint node we allow for an individual regular language specifying the paths that span the support graph. These languages form the *access specifications* stored as document-type specific ontologies and used by the generic mechanism to compute support graphs. Hence, for a given constraint the generic management of change computes its support graph and checks whether the change is located outside of it. If it is outside then the verification of the constraint is not affected by the change and the constraint can be transfered to the new document graph. We are left with the case that the change lies inside the support graph.

The first alternative is to reprove the constraint under the new environment from scratch and therefore submit the arising proof obligation to a document-type specific plugin responsible for dealing with this constraint in the first place. Providing the old (proven) and the changed proof obligation to the plugin allows for a recursive plugin-internal management of change. For instance, a theorem prover could be provided by the new and old proof obligation combined with its proof of the old proof obligations. Then the prover can analyse the difference in the set of axioms and potentially reuse its old proof in order to gain a proof for the new proof obligation.

The second alternative is to use semantic knowledge about monotonicity properties of constraints. Consider the example of citations in a book. Adding new entries to the bibliography will not invalidate the property that all citations are covered by some entry in the bibliography while deleting entries can cause an invalid constraint. Similar arguments hold in Formal Methods. Adding more axioms to a proof obligation does not endanger a given proof if the underlying logic is monotonic. Hence, we incorporate monotonicity arguments also in the generic change management and distinguish the computation of support graphs with respect to the different graph operations like adding, changing, or deleting nodes or links. Consider again the book example and the constraint that all citations have to occur in the bibliography. The corresponding support graph for adding a node or link would cover only the citations but not the bibliography. Vice versa, for deletion of nodes or links the support graph would cover only the bibliography. Another possibility to support the propagation of constraints is to use abstractions to ignore changes in the support graph that are not relevant for verifying the constraint. A typical example would be the elimination of a typo in an entry of the bibliography. An abstraction would hide details of the BibTeX-entries but concentrate only on the BibTeX-key. Abstractions can be also modeled by a more granular representation of knowledge, which allows us to exclude parts of them from the support graph by excluding the corresponding paths in the document-specific access specifications. However, in general the XML-languages under consideration and therefore the granularity of the available structuring mechanisms are fixed.

Another alternative is to consider sequences of operations rather than individual changes. Consider again the Formal Methods example. At some point in the development we prove a constraint. Later on we insert a new axiom into the theory which has no effect on the proven constraint because the constraint is monotonic with respect to enlarging the set of axiom. But for some reason we realize that the new axiom is of no use and delete it in the following step. According to our approach the change management would mark the constraint as broken because the change occurred in its support graph and furthermore, the constraint is not monotonic with respect to deletion of axioms. However, considering the sequence of changes necessary to transform the graph corresponding to the situation in which the constraint was proven to the actual document graph, the insertion and deletion of a new axiom are inverse operation and cancel out each other. This gives rise to the approach of considering the history of changes for each constraint with respect to their support graph. We define rewriting rules on sequences of changes (denoting the histories) that keep the overall transformation invariant. Examples would be cancellation rules allowing one to remove sequences of insertion and deletion of the same object and rules allowing on to permute individual change operations. The goal is to obtain a sequence of operations that are all monotonic such that we can guarantee the satisfiability of the constraint in the resulting document graph.

6   Propagation of Changes

In the last section we formalized sufficient conditions to guarantee that the satisfaction of constraints are not affected by changes. In this section we introduce ways of how a generic change management can support an development if the changes affect consistency constraints. As mentioned before, constraints are considered as predicates on different nodes of a document graph. In Section 3.1 we proposed the use of graph rewriting techniques to decompose constraints along the structuring mechanisms of the graphs below the constraint nodes. We can also use this explicit decomposition technique to speculate about further changes that have to be made in order to guarantee the satisfaction of a constraint after a change. Consider the example of the lecture notes again. Suppose, we insert an additional paragraph in the German version. Decomposing the constraint that the German version of the lecture notes is a translation of the English one results in constraints that all chapters, sections, and paragraphs in both lecture notes correspond to corresponding items in the other document. Adding a new paragraph results in recomputing the decomposition, which delivers a failure in linking the newly inserted German paragraph to any English paragraph. However, the decomposition rules specify the kind of further changes that are necessary to obtain a consistent document collection: we have to add also another English paragraph that forms the counterpart of the added German one. Due to the decomposition rules we also know about the relative position at which the paragraph has to be inserted. However, without any natural language support we obviously do not know how to write this paragraph. Analogously, adding a citation to a text without an existing entry in the bibliography would suggest a new entry with the respective BibTeX-key. Hence, decomposition rules attached to constraints deliver also information on potential repair steps (typically in terms of pattern to be inserted or deleted) necessary to establish the satisfaction of a constraint.

For some non-structured constraints we can specify functional dependencies as they are common in spread-sheet applications (like Excel or OpenCalc). In spread-sheets the content of some cells are calculated with the help of other cells. Changing one of them results in a recomputation of dependent cells. Analogously, we can define programs for individual types of constraints that take some of the constraint nodes as input to compute the graph for the other constraint nodes as output. In a following step this output is compared to the actual graphs of the corresponding constraint nodes, the resulting differences are translated into graph operations which are again executed under the surveillance of the change management.

Typically, objects in the document graph are interconnected by more than one constraint. Hence, there is an issue of how to organize the changes of different related objects in order to satisfy all the constraints. This is an open problem and cannot be solved by a simple generic management. However, the notion of support graphs gives some hints with respect to the sequence in which constraints have to be re-verified. Support graph specify the area of a document graph in which changes might affect a constraint. Once we prioritize constraints then we would like to work on those constraints first the repair of which might affect other constraints. Technically speaking, we sort the repair of constraints according to the overlap of their support graphs with others. Additionally, we delay the repair of decomposed constrained until their child constraints have been repaired or discarded.

7   Conclusion

In this paper we illustrated the issues and proposed solutions for a semantic management of heterogeneous documents. Based on our experiences gained from building a management of change for Formal Methods we generalized and improved the formerly developed techniques to come up with a generic framework for maintaining heterogenous documents. This management provides the generic mechanisms to detect differences between different steps in the development, to transfer the verification of consistency constraints to new situations (if possible), and to suggest additional repair steps in case the consistency could not be verified. Document-type specific knowledge is added to the system in terms of rules specifying the computation of support graphs, rules to decompose constraints, and plugins (or interfaces to external systems) to verify constraints in the first place. This generic change management is currently under development as part of the project FormalSafe (funded by the German Ministry of Technology and Education) and is based on the graph rewriting tool [14].

Acknowlegements

We are grateful to Bernd Krieg-Brückner, Serge Autexier and Christoph Lueth, who read previous versions versions of this article or contributed to this paper by stimulating discussions.

References

1. Autexier, S., Hutter, D.: Formal software development in MAYA. In: Hutter, D., Stephan, W. (eds.) Mechanizing Mathematical Reasoning. LNCS (LNAI), vol. 2605, pp. 407–432. Springer, Heidelberg (2005)

2. Beck, K.: Embracing change with extreme programming. IEEE Computer 32(10) (1999)
3. Codd, E.F.: A Relational Model of Data for Large Shared Data Banks. Communications of the ACM 13(6), 377–387 (1970)
4. Codd, E.F.: Recent Investigations into Relational Data Base Systems. IBM Research Report RJ1385 (April 23, 1974); Republished in Proc. 1974 Congress (Stockholm, Sweden, 1974). North-Holland, New York (1974)
5. Date, C.J., Darwen, H., Lorentzos, N.: Temporal Data and the Relational Model. Morgan Kaufmann, San Francisco (2002)
6. T.AB, Doors XT, http://www.telelogic.com
7. Ehrig, H., Ehrig, K., Prange, U., Taentzer, G.: Fundamentals of Algebraic Graph Transformation. Springer, Heidelberg (2006)
8. Fagin, R.: A Normal Form for Relational Databases That Is Based on Domains and Keys. Communications of the ACM 6, 387–415 (1981)
9. Gotel, O.C.Z., Finkelstein, A.C.W.: An analysis of the requirements traceability problem. In: IEEE International Conference on Requirements Engineering, ICRE 1994 (1994)
10. Grundy, J., Hosking, J., Mugridge, R.: Inconsistency management for multiple-view software development environments. IEEE Transactions on Software Engineering 24(11), 960–981 (1998)
11. Hutter, D.: Management of change in structured verification. In: Proceedings of Automated Software Engineering, ASE 2000. IEEE Computer Society, Los Alamitos (2000)
12. INCOSE Requirements working group, http://www.incose.org
13. Jarke, M.: Requirements tracing. Communication of the ACM 41(12) (1998)
14. Graph Rewrite Generator GrGen.NET, http://www.info.uni-karlsruhe.de/software/grgen/
15. Mossakowski, T., Autexier, S., Hutter, D.: Development graphs – proof management for structured specifications. Journal of Logic and Algebraic Programming 67(1–2), 114–145 (2006)
16. Ramesh, B., Jarke, M.: Towards reference models for requirements traceability. IEEE Transactions on Software Engineering 27(1) (2001)
17. Royce, W.: Managing the development of large software systems: concepts and techniques. In: Proceedings of the 9th International Conference on Software Engineering (1987)
18. Scheffczyk, J., Borghoff, U.M., Rödig, P., Schmitz, L.: Managing inconsistent repositories via prioritized repairs. In: Munson, E., Vion-Dury, J. (eds.) Proceedings of the ACM Symposium on Document Engineering (2004)
19. von Knethen, A.: Automatic change support based on a trace model. In: Proceedings of the Traceability Workshop, Edinburgh, UK (2002)
20. von Knethen, A.: A trace model for system requirements changes on embedded systems. In: 4th International Workshop on Principles of Software, pp. 17–26 (2001)

Possibilistic Well-Founded Semantics

Mauricio Osorio[1] and Juan Carlos Nieves[2]

[1] Universidad de las Américas - Puebla
CENTIA, Sta. Catarina Mártir, Cholula, Puebla, 72820 México
osoriomauri@googlemail.com
[2] Universitat Politècnica de Catalunya
Software Department (LSI)
c/Jordi Girona 1-3, E08034, Barcelona, Spain
jcnieves@lsi.upc.edu

Abstract. Recently, a good set of logic programming semantics has been defined for capturing possibilistic logic program. Practically all of them follow a credulous reasoning approach. This means that given a possibilistic logic program one can infer a set of possibilistic models. However, sometimes it is desirable to associate just one possibilistic model to a given possibilistic logic program. One of the main implications of having just one model associated to a possibilistic logic program is that one can perform queries directly to a possibilistic program and answering these queries in accordance with this model.

In this paper, we introduce an extension of the Well-Founded Semantics, which represents a sceptical reasoning approach, in order to capture possibilistic logic programs. We will show that our new semantics can be considered as an approximation of the possibilistic semantics based on the answer set semantics and the pstable semantic. A relevant feature of the introduced semantics is that it is polynomial time computable.

1 Introduction

In [10], a possibilistic logic programming framework for reasoning under uncertainty was introduced. It is a combination between Answer Set Programming (ASP) [3] and Possibilistic Logic [7]. This framework is able to deal with reasoning that is at the same time non-monotonic and uncertain. Since this framework was defined for normal programs, it was generalized in [11] for capturing possibilistic disjunctive programs and allowing the encoding of uncertain information by using either numerical values or relative likelihoods.

The expressiveness of this approach is rich enough for capturing sophisticated domains such as river basin systems [1, 2]. In fact, one can suggest that the language's expressiveness of the possibilistic logic programs is rich enough for capturing a wide family of problems where one have to confront with incomplete information and uncertain information.

From the logic programming literature [10–13], we can see that all the possibilistic logic program semantics that have defined until now for capturing the semantics of possibilistic logic programs (to the best of our knowledge) follow a credulous reasoning

A. Hernández Aguirre et al. (Eds.): MICAI 2009, LNAI 5845, pp. 15–26, 2009.

approach. This means that given a possibilistic logic program one can infer a set of possibilistic models. However, sometimes it is desirable to associate just one possibilistic model to a given possibilistic. This means to perform a skeptical reasoning approach from a possibilistic knowledge base. It is well-known, in the logic programming community, that a skeptical reasoning approach has several and practical implications in order to apply a logic programming approach into real domain applications. Some of these implications are:

- the process of performing a skeptical reasoning approach usually is polynomial time computable.
- to associate a single model to a logic programs helps to define algorithms for performing top-down queries from a knowledge base.

In the literature of logic programming, we can find several logic programming semantics which perform a skeptical reasoning approach [5, 6, 8]. However, the well-accepted logic programming semantics for performing a skeptical reasoning approach is *the well-founded semantics* introduced by Van Gelder in [8]. There are several results which suggest that the well-founded semantics is a strong logic programming semantic for performing a skeptical reasoning approach. For instance, Dix in [5] showed that the well-founded semantics is a *well-behaved semantics*[1]. It worth mentioning that there are few logic programming semantics which are well-behaved. In fact most of them are variations of the well-founded semantics.

Given that a skeptical reasoning approach has important implications for performing non-monotonic reasoning from a possibilistic knowledge base, in this paper, we introduce a possibilistic version of the well-founded semantics. Our so called possibilistic well-founded semantics will be a combination of some features of possibilistic logic and the standard well-founded semantics. We define the possibilistic well-founded semantics for two classes of possibilistic logic programs: Extended Possibilistic Definite logic programs and Extended Possibilistic Normal Logic programs. We show that our extension of the well-founded semantics preserves the important property of being polynomial time computable. Another important property of our possibilistic well founded semantics is that it can be considered as an approximation of the possibilistic answer set semantics [10, 12] and the possibilistic pstable semantics [13].

The rest of the paper is divided as follows: In §2, some basic concepts *w.r.t.* possibilistic logic are presented. Also the syntax of extended logic programs is defined and a characterization of the well-founded semantics in terms of rewriting systems is presented. In §3, the syntax of the extended possibilistic logic programs is presented. In §4, the definition of the possibilistic well-founded semantics is defined. Finally in the last section our conclusions are presented.

2 Background

In this section, we define some basic concepts of possibilistic logic, logic program syntaxis and a characterization of the well-founded semantics in terms of rewriting systems.

---

[1] A logic programming semantics is called well-behaved if it satisfies the following properties: Cut, Closure, Weak Model-Property, Isomorphy, M_P-extension, Transformation, Relevance, Reduction, PPE and modularity [5].

We assume familiarity with basic concepts in classic logic and in semantics of logic programs *e.g.,* interpretation, model, *etc.* A good introductory treatment of these concepts can be found in [3, 9].

2.1 Possibilistic Logic

A necessity-valued formula is a pair $(\varphi\ \alpha)$ where φ is a classical logic formula and $\alpha \in (0, 1]$ is a positive number. The pair $(\varphi\ \alpha)$ expresses that the formula φ is certain at least to the level α, *i.e.*, $N(\varphi) \geq \alpha$, where N is a necessity measure modeling our possibly incomplete state knowledge [7]. α is not a probability (like it is in probability theory) but it induces a certainty (or confidence) scale. This value is determined by the expert providing the knowledge base. A necessity-valued knowledge base is then defined as a finite set (*i.e.*, a conjunction) of necessity-valued formulae.

Dubois *et al.*[7] introduced a formal system for necessity-valued logic which is based on the following axioms schemata (propositional case):

(A1) $(\varphi \to (\psi \to \varphi)\ 1)$
(A2) $((\varphi \to (\psi \to \xi)) \to ((\varphi \to \psi) \to (\varphi \to \xi))\ 1)$
(A3) $((\sim \varphi \to \sim \psi) \to ((\sim \varphi \to \psi) \to \varphi)\ 1)$

As in classic logic, the symbols $\neg$ and $\to$ are considered primitive connectives, then connectives as $\vee$ and $\wedge$ are defined as abbreviations of $\neg$ and $\to$. Now the inference rules for the axioms are:

(GMP) $(\varphi\ \alpha), (\varphi \to \psi\ \beta) \vdash (\psi\ GLB\{\alpha, \beta\})$
(S) $(\varphi\ \alpha) \vdash (\varphi\ \beta)$ if $\beta \leq \alpha$

According to Dubois *et al.*, basically we need a complete lattice in order to express the levels of uncertainty in Possibilistic Logic. Dubois *et al.*, extended the axioms schemata and the inference rules for considering partially ordered sets. We shall denote by $\vdash_{PL}$ the inference under Possibilistic Logic without paying attention if the necessity-valued formulae are using either a totally ordered set or a partially ordered set for expressing the levels of uncertainty.

The problem of inferring automatically the necessity-value of a classical formula from a possibilistic base was solved by an extended version of *resolution* for possibilistic logic (see [7] for details).

2.2 Syntaxis: Logic Programs

The language of a propositional logic has an alphabet consisting of

(i) proposition symbols: $\perp, \top, p_0, p_1, ...$
(ii) connectives : $\vee, \wedge, \leftarrow, \neg, \textit{not}$
(iii) auxiliary symbols : (,).

where $\vee, \wedge, \leftarrow$ are 2-place connectives, $\neg$, *not* are 1-place connective and $\perp, \top$ are 0-place connective. The proposition symbols, $\perp$, and the propositional symbols of the form $\neg p_i$ ($i \geq 0$) stand for the indecomposable propositions, which we call *atoms*, or *atomic propositions*. Atoms negated by $\neg$ will be called *extended atoms*. We will use the concept of atom without paying attention if it is an extended atom or not. The negation sign $\neg$ is regarded as the so called *strong negation* by the ASP's literature and the negation *not* as the *negation as failure*. A literal is an atom, a (called positive literal), or the negation of an atom *not* a (called negative literal). Given a set of atoms $\{a_1, ..., a_n\}$, we write *not* $\{a_1, ..., a_n\}$ to denote the set of literals $\{not\ a_1, ..., not\ a_n\}$.

An extended normal clause, C, is denoted:

$$a \leftarrow a_1, \ldots, a_j, not\ a_{j+1}, \ldots, not\ a_n$$

where $j + n \geq 0$, a is an atom and each a_i is an atom. When $j + n = 0$ the clause is an abbreviation of $a \leftarrow \top$ such that $\top$ is the proposition symbol that always evaluate to true. An extended normal program P is a finite set of extended normal clauses. When $n = 0$, the clause is called *extended definite clause*. *An extended definite logic program* is a finite set of extended definite clauses. By $\mathcal{L}_P$, we denote the set of atoms in the language of P. Let $Prog_{\mathcal{L}}$ be the set of all normal programs with atoms from $\mathcal{L}$.

We will manage the strong negation ($\neg$), in our logic programs, as it is done in ASP [3]. Basically, it is replaced each atom of the form $\neg a$ by a new atom symbol a' which does not appear in the language of the program. For instance, let P be the extended normal program:

$$a \leftarrow q. \qquad \neg q \leftarrow r. \qquad q \leftarrow \top. \qquad r \leftarrow \top.$$

Then replacing the atom $\neg q$ by a new atom symbol q', we will have:

$$a \leftarrow q. \qquad q' \leftarrow r. \qquad q \leftarrow \top. \qquad r \leftarrow \top.$$

In order not to allow inconsistent models from logic programs, usually it is added a normal clause of the form $f \leftarrow q, q', f$ such that $f \notin \mathcal{L}_P$. We will omit this clause in order to allow an inconsistent level in our possibilistic-WFS. However the user could add this clause without losing generality.

Sometimes we denote an extended normal clause C by $a \leftarrow \mathcal{B}^+, not\ \mathcal{B}^-$, where $\mathcal{B}^+$ contains all the positive body literals and $\mathcal{B}^-$ contains all the negative body literals.

2.3   Well-Founded Semantics

In this section, we present a standard definition of the well-founded semantics in terms of rewriting systems. We start presenting a definition w.r.t. 3-valued logic semantics.

Definition 1 (SEM). *[6] For normal logic program P, we define $HEAD(P) = \{a\,|\,a \leftarrow \mathcal{B}^+, not\ \mathcal{B}^- \in P\}$ — the set of all head-atoms of P. We also define $SEM(P) = \langle P^{true}, P^{false} \rangle$, where $P^{true} := \{p\,|\,p \leftarrow \top \in P\}$ and $P^{false} := \{p\,|\,p \in \mathcal{L}_P \backslash HEAD(P)\}$. $SEM(P)$ is also called model of P.*

In order to present a characterization of the well-funded semantics in terms of rewriting systems, we define some basic transformation rules for normal logic programs.

Definition 2 (Basic Transformation Rules). *[6] A transformation rule is a binary relation on $Prog_{\mathcal{L}}$. The following transformation rules are called* basic. *Let a program $P \in Prog_{\mathcal{L}}$ be given.*

RED$^+$: *This transformation can be applied to P, if there is an atom a which does not occur in HEAD(P). RED$^+$ transforms P to the program where all occurrences of not a are removed.*

RED$^-$: *This transformation can be applied to P, if there is a rule $a \leftarrow \top \in P$. RED$^-$ transforms P to the program where all clauses that contain not a in their bodies are deleted.*

Success: *Suppose that P includes a fact $a \leftarrow \top$ and a clause $q \leftarrow body$ such that $a \in body$. Then we replace the clause $q \leftarrow body$ by $q \leftarrow body \setminus \{a\}$.*

Failure: *Suppose that P contains a clause $q \leftarrow body$ such that $a \in body$ and $a \notin HEAD(P)$. Then we erase the given clause.*

Loop: *We say that P_2 results from P_1 by Loop$_A$ if, by definition, there is a set A of atoms such that 1. for each rule $a \leftarrow body \in P_1$, if $a \in A$, then $body \cap A \neq \emptyset$, 2. $P_2 := \{a \leftarrow body \in P_1 | body \cap A = \emptyset\}$, 3. $P_1 \neq P_2$.*

Let CS_0 be the rewriting system such that contains the transformation rules: RED^+, RED^-, $Success$, $Failure$, and $Loop$. We denote the uniquely determined normal form of a program P with respect to the system CS by $norm_{CS}(P)$. Every system CS induces a semantics SEM$_{CS}$ as follows: SEM$_{CS}(P) := $ SEM$(norm_{CS}(P))$.

In order to illustrate the basic transformation rules, let us consider the following example.

Example 1. Let P be the following normal program:

$$d(b) \leftarrow not \ d(a). \ d(c) \leftarrow not \ d(b). \ d(c) \leftarrow d(a).$$

Now, let us apply CS_0 to P. Since $d(a) \notin HEAD(P)$, then, we can apply **RED$^+$** to P. Thus we get:

$$d(b) \leftarrow \top. \ d(c) \leftarrow not \ d(b). \ d(c) \leftarrow d(a).$$

Notice that now we can apply **RED$^-$** to the new program, thus we get: $d(b) \leftarrow \top$. $d(c) \leftarrow d(a)$.

Finally, we can apply **Failure** to the new program, thus we get: $d(b) \leftarrow \top$. This last program is called the *normal form* of P w.r.t. CS_0, because none of the transformation rules from CS_0 can be applied.

WFS was introduced in [8] and was characterized in terms of rewriting systems in [4]. This characterization is defined as follows:

Lemma 1. *[4] CS_0 is a confluent rewriting system. It induces a 3-valued semantics that it is the Well-founded Semantics.*

3 Possibilistic Logic Programs

In this section, we introduce a standard syntax of extended possibilistic logic programs. In whole paper, we will consider finite lattices. This convention was taken based on the assumption that in real applications rarely we will have an infinite set of labels for expressing the incomplete state of a knowledge base.

3.1 Syntax

First of all, we start defining some relevant concepts[2]. A *possibilistic atom* is a pair $p = (a, q) \in \mathcal{A} \times Q$, where $\mathcal{A}$ is a finite set of atoms and $(Q, \leq)$ is a lattice. We apply the projection $*$ over p as follows: $p^* = a$. Given a set of possibilistic atoms S, we define the generalization of $*$ over S as follows: $S^* = \{p^*|p \in S\}$. Given a lattice $(Q, \leq)$ and $S \subseteq Q$, $LUB(S)$ denotes the least upper bound of S and $GLB(S)$ denotes the greatest lower bound of S.

We define the syntax of a valid extended possibilistic normal logic program as follows: Let $(Q, \leq)$ be a lattice. A extended possibilistic normal clause r is of the form:

$$r := (\alpha : a \leftarrow \mathcal{B}^+, \; not \; \mathcal{B}^-)$$

where $\alpha \in Q$. The projection $*$ over the possibilistic clause r is: $r^* = a \leftarrow \mathcal{B}^+, \; not \; \mathcal{B}^-$. $n(r) = \alpha$ is a necessity degree representing the certainty level of the information described by r.

An extended possibilistic normal logic program P is a tuple of the form $\langle (Q, \leq), N \rangle$, where $(Q, \leq)$ is a lattice and N is a finite set of extended possibilistic normal clauses. The generalization of the projection $*$ over P is as follows: $P^* = \{r^*|r \in N\}$. Notice that P^* is an extended logic normal program. When P^* is an extended definite program, P is called an extended possibilistic definite logic program.

4 Possibilistic Well-Founded Model Semantics

In this section, we introduce the possibilistic version of the well-founded semantics. This semantics will be presented for two classes of possibilistic logic programs: extended possibilistic definite logic programs and extended possibilistic normal logic programs.

4.1 Extended Possibilistic Definite Logic Programs

In this subsection, we are going to deal with the class of extended possibilistic logic programs. For capturing extended possibilistic definite logic program, we are going to consider the basic idea of *possibilistic least model* which was introduced in [10]. For this purpose, we are going to introduce some basic definitions.

The first basic definition that we present is the definition of three basic operators between sets of possibilistic atoms.

Definition 3. *[10] Let $\mathcal{A}$ be a finite set of atoms and $(Q, \leq)$ be a lattice. Consider $\mathcal{PS} = 2^{\mathcal{A} \times Q}$ the finite set of all the possibilistic atom sets induced by $\mathcal{A}$ and Q. $\forall A, B \in \mathcal{PS}$, we define.*
$A \sqcap B = \{(x, GLB\{q_1, q_2\})|(x, q_1) \in A \wedge (x, q_2) \in B\}$
$A \sqcup B = \{(x, q)|(x, q) \in A \; and \; x \notin B^*\} \cup$
$\qquad \{(x, q)|x \notin A^* \; and \; (x, q) \in B\} \cup$
$\qquad \{(x, LUB\{q_1, q_2\})|(x, q_1) \in A \; and \; (x, q_2) \in B\}.$
$A \sqsubseteq B \iff A^* \subseteq B^*, \; and \; \forall x, q_1, q_2,$
$\qquad (x, q_1) \in A \wedge (x, q_2) \in B \; then \; q_1 \leq q_2.$

---

[2] Some concepts presented in this subsection extend some terms presented in [10].

Observe that essentially this definition suggests an extension of the standard operators between sets in order to deal with uncertain values which belong to a partially ordered set. We want to point out that the original version of Definition 3 consider totally ordered sets instead of partially ordered sets.

Like in [10], we are going to introduce a fix-point operator ΠCn. In order to define ΠCn, let us introduce some basic definitions. Given a possibilistic logic program P and $x \in \mathcal{L}_{P^*}$, $H(P, x) = \{r \in P | head(r^*) = x\}$.

Definition 4. *Let $P = \langle (Q, \leq), N \rangle$ be a possibilistic definite logic program, $r \in N$ such that r is of the form $\alpha : a \leftarrow l_1, \ldots, l_n$ and A be a set of possibilistic atoms,*

- *r is β-applicable in A with $\beta = min\{\alpha, \alpha_1, \ldots, \alpha_n\}$ if $\{(l_1, \alpha_1), \ldots, (l_n, \alpha_n)\} \subseteq A$.*
- *r is $\perp_Q$-applicable otherwise.*

And then, for all atom $x \in \mathcal{L}_{P^*}$ we define:

$$App(P, A, x) = \{r \in H(P, x) | r \text{ is } \beta\text{-applicable in } A \text{ and } \beta > \perp_Q\}$$

$\perp_Q$ denotes the bottom element of Q.

Observe that this definition is based on the inferences rules of possibilistic logic. In order to illustrate this definition, let us consider the following example.

Example 2. Let $P = \langle (Q, \leq), N \rangle$ be a possibilistic definite logic program such that $Q = \{0, 1, \ldots, 0.9, 1\}$, $\leq$ be the standard relation between rational number, and N be the following set of possibilistic definite clauses:

$$r_1 = 0.4 : a \leftarrow \top.$$
$$r_2 = 0.3 : b \leftarrow a.$$
$$r_3 = 0.6 : b \leftarrow a.$$
$$r_4 = 0.7 : m \leftarrow n.$$

we can see that if we consider $A = \emptyset$, then r_1 is 0.4-applicable in A. In fact, we can see that $App(P, A, a) = \{r_1\}$. Also, we can see that if $A = \{(a, 0.4)\}$, then r_2 is 0.4-applicable in A and r_3 is 0.6-applicable in A. Observe that $App(P, A, b) = \{r_2, r_3\}$.

Now, we introduce an operator which is based on Definition 4.

Definition 5. *Let P be a possibilistic definite logic program and A be a set of possibilistic atoms. The immediate possibilistic consequence operator ΠT_P maps a set of possibilistic atoms to another one by this way:*

$$\Pi T_P(A) = \{(x, \delta) | x \in HEAD(P^*), App(P, A, x) \neq \emptyset,$$
$$\delta = LUB_{r \in App(P, A, x)}\{\beta | r \text{ is } \beta\text{-applicable in } A\}\}$$

Then the iterated operator ΠT_P^k is defined by

$$\Pi T_P^k = \emptyset \text{ and } \Pi T_P^{n+1} = \Pi T_P(\Pi T_P^n), \forall n \geq 0$$

Observe that ΠT_P is a monotonic operator; therefore, we can insure that ΠT_P always reaches a fix-point.

Proposition 1. *Let P be a possibilistic definite logic program, then ΠT_P has a least fix-point $\bigsqcup_{n \geq 0} \Pi T_P^n$ that we call the set of possibilistic consequences of P and we denote it by $\overline{\Pi} Cn(P)$.*

Example 3. Let P be the extended possibilistic definite logic program introduced in Example 2. It is not difficult to see that $\Pi Cn(P) = \{(a, 0.3), (b, 0.6)\}$.

By considering the operator ΠCn, we define the possibilistic well-founded semantics for extended possibilistic definite logic program as follows: Let $(Q, \leq)$ be a lattice such that $\top_Q$ is the top-element of Q and S be a set of atoms, then $Q_{\top_Q}(S) = \{(a, \top_Q) \mid a \in S\}$.

Definition 6. *Let P be an extended possibilistic definite logic program. S_1 be a set of possibilistic atoms, S_2 be a set of atoms such that $\langle S_1^*, S_2 \rangle$ is the well-founded model of P^*. $\langle S_1, Q_{\top_Q}(S_2) \rangle$ is the possibilistic well-founded model of P if and only if $S_1 = \Pi Cn(P)$.*

Example 4. Let P be the extended possibilistic definite logic program introduced in Example 2, $S_1 = \{(a, 0.3), (b, 0.6)\}$ and $S_2 = \{m, n\}$. As we can see the well-founded model of P^* is $\langle \{a, b\}, \{m, n\} \rangle$, and $\Pi Cn(P) = \{(a, 0.3), (b, 0.6)\}$. This means that the well-founded model of P is:

$$\langle \{(a, 0.3), (b, 0.6)\}, \{(m, 1), (n, 1)\} \rangle$$

4.2 Extended Possibilistic Normal Programs

In this subsection, we are going to deal with the class of extended possibilistic normal programs. In order to define a possibilistic version of the possibilistic well-founded semantics for capturing extended possibilistic normal program, we define a single reduction of an extended possibilistic normal logic program *w.r.t.* a set of atoms. This reduction is defined as follows:

Definition 7. *Let P be an extended possibilistic logic program and S be a set of atoms. We define $R(P, S)$ as the extended possibilistic logic program obtained from P by deleting*

i *all the formulae of the form not a in the bodies of the possibilistic clauses such that $a \in S$, and*
ii *each possibilistic clause that has a formula of the form not a in its body.*

Observe that $R(P, S)$ does not have negative literals. This means that $R(P, S)$ is an extended possibilistic definite logic program. In order to illustrate this definition, let us consider the following example.

Example 5. Let $P = \langle (Q, \leq), N \rangle$ be a possibilistic logic program such that $Q = \{0, 0.1, \ldots, 0.9, 1\}$, $\leq$ be the standard relation between rational number, $S = \{b, d, e\}$ and N be the following set of possibilistic clauses:

$$0.4 : a \leftarrow \ not \ b.$$
$$0.6 : b \leftarrow \ not \ c.$$
$$0.3 : c \leftarrow \ not \ d.$$
$$0.2 : c \leftarrow \ not \ e.$$
$$0.5 : f \leftarrow \ not \ f.$$

As we can see in Definition 7, for inferring $R(P, S)$, we have to apply two steps. The first step is to remove all the negative literals *not a* from P such that a belongs to S. This means that by the first step of Definition 7 we have the following possibilistic program

$$0.4 : a \leftarrow \top.$$
$$0.6 : b \leftarrow \ not \ c.$$
$$0.3 : c \leftarrow \top.$$
$$0.2 : c \leftarrow \top.$$
$$0.5 : f \leftarrow \ not \ f.$$

The next and last step is to remove any possibilistic clause that has a negative literal in its body. This means that $R(P, S)$ is:

$$0.4 : a \leftarrow \top.$$
$$0.3 : c \leftarrow \top.$$
$$0.2 : c \leftarrow \top.$$

As we can see, $R(P, S)$ is a possibilistic definite logic program.

By considering the fix-point operator $\Pi Cn(P)$ and the reduction $R(P, A)$, we define the possibilistic version of the well-founded semantics for extended possibilistic normal logic programs as follows:

Definition 8 (Possibilistic Well-Founded Semantics)
Let $P = \langle (Q, \leq), N \rangle$ be an extended possibilistic logic program, S_1 be a set of possibilistic atoms, S_2 be a set of atoms such that $\langle S_1^*, S_2 \rangle$ is the well-founded model of P^*. $\langle S_1, Q_{\top_Q}(S_2) \rangle$ is the possibilistic well-founded model of P if and only if $S_1 = \Pi Cn(R(P, S_2))$.

In order to illustrate this definition let us consider the following example

Example 6. Let $P = \langle (Q, \leq), N \rangle$ be the possibilistic logic program introduced in Example 5, $S_1 = \{(a\ 0.4), (c\ 0.3)\}$ and $S_2 = \{b, d, e\}$. On order to infer the possibilistic well-founded model of P, we have to infer the well-founded model of the normal program P^*. It is easy to see that $WPF(P^*) = \langle S_1^*, S_2 \rangle$. As we saw in Example 5, $R(P, S_2)$ is

$$0.4 : a \leftarrow \top.$$
$$0.3 : c \leftarrow \top.$$
$$0.2 : c \leftarrow \top.$$

hence, we can see that $Cn(R(P, S_2) = \{(a, 0.4), (c, 0.3)\}$. This suggests that the possibilistic well-founded model of P is

$$\langle\{(a, 0.4), (c, 0.3)\}, \{(b, 1), (d, 1), (e, 1)\}\rangle$$

A first and basic observation that we can see from the definition of the possibilistic well-founded semantics is that the possibilistic well-founded semantics infers the well-founded semantics.

Proposition 2. *Let $P = \langle(Q, \leq), N\rangle$ be an extended possibilistic logic program. If $\langle S_1, s_2\rangle$ is the possibilistic well-founded model of P then $\langle S_1^*, S_2^*)\rangle$ is the well-founded model of P^*.*

Usually, one property that always is desired from a logic program semantics is that it could be polynomial time computable w.r.t. the size of a given logic program. It is well-known that the well-founded semantics is polynomial time computable. For the case of the possibilistic well-founded semantics we can also insure that it is polynomial time computable. For formalizing this property, let us remember that the size of a logic programs is defined as follows: The size of a possibilistic clause r is the number of atom symbols that occurs in r. The size of a possibilistic logic program P is the sum of sizes of the possibilistic clauses that belong to P.

Proposition 3. *Given an extended possibilistic normal program P, there is an algorithm that computes the possibilistic well founded model of P in polynomial time w.r.t. the size of P.*

In the following proposition, a relationship between the possibilistic answer set semantics and the possibilistic well-founded model is formalized. Also a relationship between the possibilistic pstable semantics and the possibilistic well-founded model. In order to formalize these relationships, we define the following functions: $Poss\_ASP(P)$ denotes a function which returns the set of possibilistic answer semantics [10, 12] of a possibilistic logic program P and $Poss\_Pstable(P)$ denotes a function which returns the set of possibilistic pstable models [13] of a possibilistic logic program P.

Proposition 4. *Let P be a possibilistic logic program and $S_1, S_2 \subseteq \mathcal{L}_P$ such that $\langle S_1, S_2\rangle$ is the possibilistic well-founded model of P. Hence, the following conditions holds:*

- *If $Poss\_ASP(P) \neq \emptyset$; hence, if $S = \bigcap_{S' \in Poss\_ASP(P)} S'$, then $S_1 \sqsubseteq S$.*
- *If $S = \bigcap_{S' \in Poss\_Pstable(P)} S'$, then $S_1 \sqsubseteq S$.*

Observe that this proposition essentially suggests that any possibilistic answer set model is an extension of the possibilistic well-founded model. Also that any possibilistic pstable model is an extension of the possibilistic well-founded model. It is worth mentioning that this relationship between the standard answer set semantics and the well-founded semantics was shown by Dix in [5].

5   Conclusions

In this paper have explored the definition of a possibilistic version of the well-founded semantics in order to capture possibilistic logic programs. For this purpose, we first define a possibilistic version of the well-founded semantics for extended possibilistic define logic programs. This definition considers a possibilistic operator for inferring the possibilistic least model of an extended possibilistic definite logic program and the standard definition of the well-founded semantics. In order to define the possibilistic version of the well-founded semantics for extended possibilistic normal logic programs, we introduce a single reduction of a possibilistic logic program in terms of a given set of atoms. We want to point that our construction of the possibilistic well-founded semantics is flexible enough for considering other variants of the well-founded semantics as the explored in [5, 6]. Hence, we can consider our construction of the possibilistic well-founded semantics for exploring different approaches of skeptical reasoning from a possibilistic knowledge base.

We have showed that the actual version of the possibilistic version of the well-founded semantics is polynomial time computable (Proposition 3). This suggests that one can explore efficient algorithms for performing top-down queries from a possibilistic knowledge. In fact, this issue is part of our future work.

Also we have showed that our possibilistic well-founded semantics can be regarded as an approximation of credulous possibilistic semantics as the possibilistic answer set semantics and the possibilistic pstable semantics (Proposition 4). In fact, we can conclude that any possibilistic answer set model is an extension of the possibilistic well-founded model and that any possibilistic pstable model is an extension of the possibilistic well-founded model.

Acknowledgement

We are grateful to anonymous referees for their useful comments. This research has been partially supported by the EC founded project ALIVE (FP7-IST-215890). The views expressed in this paper are not necessarily those of the ALIVE consortium.

References

1. Aulinas, M.: Management of industrial wastewater discharges through agents' argumentation. PhD thesis, PhD on Environmental Sciences, University of Girona, to be presented
2. Aulinas, M., Nieves, J.C., Poch, M., Cortés, U.: Supporting Decision Making in River Basin Systems Using a Declarative Reasoning Approach. In: Finkel, M., Grathwohl, P. (eds.) Proceedings of the AquaTerra Conference (Scientific Fundamentals for River Basic Management), March 2009, p. 75 (2009) ISSN 0935-4948
3. Baral, C.: Knowledge Representation, Reasoning and Declarative Problem Solving. Cambridge University Press, Cambridge (2003)
4. Brass, S., Zukowski, U., Freitag, B.: Transformation-based bottom-up computation of the well-founded model. In: NMELP, pp. 171–201 (1996)
5. Dix, J.: A classification theory of semantics of normal logic programs: II. weak properties. Fundam. Inform. 22(3), 257–288 (1995)

6. Dix, J., Osorio, M., Zepeda, C.: A general theory of confluent rewriting systems for logic programming and its applications. Ann. Pure Appl. Logic 108(1-3), 153–188 (2001)

7. Dubois, D., Lang, J., Prade, H.: Possibilistic logic. In: Gabbay, D., Hogger, C.J., Robinson, J.A. (eds.) Handbook of Logic in Artificial Intelligence and Logic Programming, Nonmonotonic Reasoning and Uncertain Reasoning, vol. 3, pp. 439–513. Oxford University Press, Oxford (1994)

8. Gelder, A.V., Ross, K.A., Schlipf, J.S.: The well-founded semantics for general logic programs. Journal of the ACM 38(3), 620–650 (1991)

9. Mendelson, E.: Introduction to Mathematical Logic, 4th edn. Chapman and Hall/CRC, Boca Raton (1997)

10. Nicolas, P., Garcia, L., Stéphan, I., Lafévre, C.: Possibilistic Uncertainty Handling for Answer Set Programming. Annals of Mathematics and Artificial Intelligence 47(1-2), 139–181 (2006)

11. Nieves, J.C., Osorio, M., Cortés, U.: Semantics for possibilistic disjunctive programs. In: Baral, C., Brewka, G., Schlipf, J. (eds.) LPNMR 2007. LNCS (LNAI), vol. 4483, pp. 315–320. Springer, Heidelberg (2007)

12. Nieves, J.C., Osorio, M., Cortés, U.: Semantics for possibilistic disjunctive programs. In: Costantini, S., Watson, R. (eds.) Answer Set Programming: Advances in Theory and Implementation, pp. 271–284 (2007)

13. Osorio, M., Nieves, J.C.: Pstable semantics for possibilistic logic programs. In: Gelbukh, A., Kuri Morales, Á.F. (eds.) MICAI 2007. LNCS (LNAI), vol. 4827, pp. 294–304. Springer, Heidelberg (2007)

ACO for Solving the Distributed Allocation of a Corporate Semantic Web

Ana B. Rios-Alvarado, Ricardo Marcelín-Jiménez,
and R. Carolina Medina-Ramírez

Department of Electrical Engineering
Universidad Autónoma Metropolitana - Iztapalapa
Atlixco 186, DF, México
{abra,calu,cmed}@xanum.uam.mx

Abstract. This paper outlines a general method for allocating a document collection over a distributed storage system. Documents are organized to make up a corporate semantic web featured by a graph G_1, each of whose nodes represents a set of documents having a common range of semantic indices. There exists a second graph G_2 modelling the distributed storage system. Our approach consists of embedding G_1 into G_2, under size restrictions. We use a meta-heuristic called "Ant Colony Optimization", to solve the corresponding instances of the graph embedding problem, which is known to be a NP problem. Our solution provides an efficient mechanism for information storage and retrieval.

Keywords: Ontologies, ant colony optimization, distributed storage.

1 Introduction

The semantic web is an extension of the current web, intended to provide an improved cooperation between humans and machines [1]. This approach relies on a formal framework where information is given a well-defined meaning. It uses ontologies to support information exchange and search. It is also based on semantic annotations to code content representation. In addition, it works with formal knowledge representation languages in order to describe these ontologies and annotations.

There exist diverse contexts that may profit from the ideas developed by the semantic web. Corporate memories, for instance, share common features with the web, both gather heterogeneous resources and distributed information and have a common interest about information retrieval [2]. Nevertheless, corporate memories have an infrastructure and scope limited to the organization where they are applied. Among the heterogeneous resources belonging to a scientific group or enterprise, for example, documents represent a significant source of collective expertise, requiring an efficient management including storage, handling, querying and propagation. From this meeting between the web and the corporate memories a new solution is born, the Corporate Semantic Web (CSW).

A. Hernández Aguirre et al. (Eds.): MICAI 2009, LNAI 5845, pp. 27–38, 2009.

Formally, a CSW is a collection of resources (either documents or humans) described using semantic annotations (on the document contents or the persons features/competences), which rely on a given ontology [2].

This paper describes a general method to allocate a CSW in a distributed storage system. A system such as this is a collection of interconnected storage devices that contribute with their individual capacities to create an extended system offering improved features. The importance of this emerging technology has been underlined in well-known reference works [3,4]. Although its simplest function is to spread a collection of files across the storage devices attached to a network, desirable attributes of quality must also be incorporated.

Our proposal, that we call a semantic layer, is mainly defined by its two main procedures: information location and query. Location solves document placement and creates the tables supporting query. We consider this approach provides a flexible, scalable and fault-tolerant service. Location is solved using a meta-heuristic called Ant Colony Optimization (ACO). This is a well known technique which has been extensively used to solve hard combinatorial problems. Nevertheless, to the author's best knowledge, we did not find any reference on the use of this method to tackle graph embedding, which is the problem underlying at the location procedure. ACO is based on agents that explore the solution space in order to produce iterative refinements on the output. Due to this agent-based nature, we believe that it might be possible a distributed implementation of the location procedure. This would lead to a self-configurable semantic storage system.

The remaining of this paper is organized as follows. Section 2 is an overview of the previous work which is related to our proposal. Section 3 presents the general structure of our semantic layer. Section 4 is a formal statement of the Graph Embedding problem. Section 5 is about our simulation method. Section 6 describes the experimental agenda and results. Section 7 closes our paper with a summary of our contribution and directions for further work.

2   Related Work

We consider this work at the meeting point between system's engineering and combinatorics. Therefore, our proposal has common links with the leading efforts in the field of semantic storage, as well as the natural connections with optimization problems and, particularly, the graph embedding problem.

Graph embedding instances can be found at the core of many technical problems like circuits design [5], wireless sensor location [6], among many others. Nevertheless, to the author's best knowledge, most of the work has been oriented to study the conditions that render an "easier" version, with a reduced complexity.

Bhatt and Ipsen [7] showed that a complete binary tree with $2n - 1$ nodes can be embedded in a $2n$ node hypercube. Chan and Chin [8] consider embedding $2D$ grids into hypercubes with at least as many nodes as grid points. They developed an embedding scheme for an infinite class of $2D$ grids. In [9] a linear time algorithm is presented for embedding a given graph G in S, where S is an arbitrary fixed surface. Additional works focus on the case when S is a torus. A survey-type paper on torus embeddings is [10].

Erten and Kobourov [11] consider several variations of the simultaneous embedding problem for planar graphs. They begin with a simple proof that not all pairs of planar graphs have simultaneous geometric embeddings. However, using bends, pairs of planar graphs can be simultaneously embedded on the $O(n^2)$ x $O(n^2)$ grid, with at most three bends per edge, where n is the number of vertices. The $O(n)$ time algorithm guarantees that two corresponding vertices in the graphs are mapped to the same location in the final drawing and that both the drawings are without crossings. The special case when both input graphs are trees has several applications, such as contour tree simplification and evolutionary biology.

A different approach was followed by Savage and Wloka, who embedded large random graphs in grids and hypercubes using their Mob Parallel heuristic [12] on a CM-2 Connection Machine.

3   System Description

It is a well-known fact that the IP routing task, at the Internet, is supported by two complementary procedures: the first one, which performs table keeping and the second one, which performs table querying. Following similar principles, we propose the storage of a given CSW based on two procedures. First, we solve document location and build a table, whose entries show the places in charge of a given set of documents. Second, we perform look-up on this table in order to consult the corresponding contents.

This paper addresses our methodology for document location. From our view, a given CSW is a set of documents classified according to an underlying ontology. This ontology itself describes a collection of concepts (categories) which can be mapped to an ordered set of indices. Each document is therefore labelled with an index. Let us assume that the CSW is partitioned in such a way that, all the documents sharing a common index are associated to the same subset. Now, the CSW can be modelled by means of a graph G_1, each of whose nodes represents a subset of the partition. Every documental node $v_i \in G_1$ is featured by 2 parameters, the range $r_{i1} \ldots r_{i2}$ of semantic indices spanning its corresponding documents and the weight $\omega(v_i)$, which is the information these documents amount to. Fig. 1 is an example of a given CSW. There exist 17 different concepts (categories) which, in turn, define 17 indices. Nevertheless, the whole set of documents it is encompassed by a graph having only 5 nodes, each of them featured by its range and weight.

In the complementary part of our description, we model the storage network using a second graph G_2. Each node $v_j \in G_2$, from now on store, has an associated capacity $\kappa(v_j)$ that features the maximal amount of information it is able to contain. Fig. 2 is an example of storage network. Each store shows its capacity. We say that the storage network has homogeneous capacity if, for any store v_j, $\kappa(v_j) = k$. Links in G_2 represent communications channels between stores.

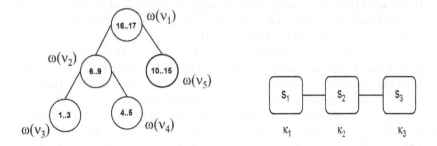

Fig. 1. CSW modelled by G_1 **Fig. 2.** Network Storage modelled by G_2

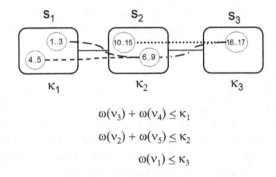

$$\omega(v_3) + \omega(v_4) \leq \kappa_1$$
$$\omega(v_2) + \omega(v_5) \leq \kappa_2$$
$$\omega(v_1) \leq \kappa_3$$

Fig. 3. The embedding G_1 into G_2

Document location implies the embedding of G_1 into G_2. This problem consists of using as few stores as possible in order to place as many documental nodes as possible inside these stores, in such a way that their aggregated weight does not exceed the corresponding capacity. When the particular instance of graph embedding is solved, each store receives a copy of the look-up table. Each row in this table has two parts, the left entry indicates a semantic indices range, while the right entry indicates the store in charge of the documents in this range. Fig. 3 shows how G_1 has been embedded into G_2. Based on this solution we have built table 1.

Any user looking for some content in the CSW may contact any store and ask for a given index. Using its local table, the store will be able to recognize whether

Table 1. Look-up table

Indices range	Store
1...5	s_1
6...15	s_2
16...17	s_3

or not it keeps the corresponding documents. In the first case, it immediately turns in the documents to its client. Otherwise, it works on behalf of its client and contact the proper store to retrieve the documents matching the user's query.

The down side of graph embedding is that it is an NP-complete problem. It is not known an exact algorithm to solve this family of problems in polynomial time. Nevertheless, there exist a vast collection of meta-heuristics developed to tackle this family of problems within bounded accuracy and reasonable complexities. We decided to address our instances of graph embedding using the ant colony optimization method (ACO). It is a probabilistic technique for solving hard computational problems. ACO is inspired by the behaviour of ants in finding efficient paths from their anthill to the places where they collect their food. ACO dates back from the pioneering work of Dorigo [13]. *Due to its cooperative and inherently distributed nature, we consider it is a very promising candidate to develop distributed meta-heuristics, as we propose in this work.*

4   Problem Statement

Let G_1:(V_1, E_1) be a graph representing a CSW and let G_2:(V_2, E_2) be the graph representing a storage network. The embedding of G_1 into G_2 is a couple of maps $S : (\nu, \epsilon)$, where ν assigns each node from G_1 to a store in G_2, while ϵ transforms edges in G_1 to paths in G_2, upon the following restriction:

There exists a function $\omega : V_1 \mapsto \mathbb{R}$, called weight. Also, there is a function $\kappa : V_2 \mapsto \mathbb{R}$, called capacity. Let $N_j = [v_i \mid v_i \in V_1 \; \nu(v_i) = v_j]$ be the set of nodes from V_1 mapped to $v_j \in V_2$.

$$\sum_{v_i \in N_j} \omega(v_i) \leq \kappa(v_j), \forall v_j \in V_2 \tag{1}$$

$$\bigcup_j N_j = V_1 \tag{2}$$

$$N_j \cap N_{j'} = \emptyset, j \neq j' \tag{3}$$

This is, the total weight of nodes in V_1 stored in $v_j \in V_2$, must not exceed the corresponding capacity. Consequently, our optimization problem can be defined.

Problem (GE). Find an embedding S of G_1 into G_2 such that, for a given function $f : S \mapsto \mathbb{R}$ that measures the cost of a given embedding, S has the smallest cost $f(S)$.

For the goals of this work, we will assume that G_2 has homogeneous capacity, this means that each store has the same capacity k. Also, we will assume that there is a linear ordering L on the stores in G_2, such that $succ(v_j)$ is the successor of $v_j \in V_2$, according to L, or null if v_j is the last element in L. GE is an NP-complete problem [12].

5   Ant Colony Optimization - The Method -

Even though it is accepted that the initial work in ACO was due to Dorigo, we will adopt a description wich is slightly different, but fits better with our exposition. According to Gutjhar [14], an ant system is composed of a set $A_1, A_2, \ldots, A_Z$ of agents. A time period in which each agent performs a walk through the construction graph $G = (V, E)$ is called a cycle.

Let $u = (u_0, u_1, \ldots, u_{t-1} = k)$ be the partial walk of a fixed agent in the current cycle m. It means that the agent has travelled along this walk and is currently at node k. As it still has nodes to visit, agent needs to decide its next step. It is written $l \in u$ if a node l is contained in the partial walk u, and $l \notin u$, otherwise. Moreover, let E be the set of arcs in the construction graph. The transition probability $p_{kl}(m, u)$ that the agent moves from node k to node l is

$$p_{kl}(m, u) = \frac{[\tau_{kl}(m)]^\alpha [\eta_{kl}(m)]^\beta}{\sum_{r \notin u, (k,r) \in E} [\tau_{kl}(m)]^\alpha [\eta_{kl}(m)]^\beta} \tag{4}$$

if $l \notin u$ and $(k, l) \in E$, and $p_{kl}(m, u) = 0$, otherwise.

The numbers $\eta_{kl}(m)$ are called *desirability values*, and the numbers $\tau_{kl}(m)$ are called *pheromone values*. The desirability values can be obtained from a greedy heuristic for the problem under consideration. As for the pheromone values, at the beginning of cycle 1, $\tau_{kl}(1) = 1/$(*number of neighbours of k*). At the end of each cycle m the following updating rule is applied. First, for each agent A_i and each arc (k, l), a value $\Delta\tau_{kl}^{(i)}(m)$ is determined as a function of the solution assigned to the walk of A_i in the current cycle. Suppose this solution has a cost value f_i. For each arc (k, l), we set $\Delta\tau_{kl}^{(i)}(m) = \phi(f_i)$, if agent A_i has traversed arc (k, l), and 0 otherwise.

Therein, ϕ is a non-increasing function which depends on the walks of the agents in the cycles $1, \ldots, m-1$ and measures the way that the last solution improves any preceding result. Let the total reward R be

$$R = \sum_{(k,l) \in E} \sum_{i=1}^{Z} \Delta\tau_{kl}^{(i)}(m) \tag{5}$$

Now, if $R = 0$, (no better solution was found during the last cycle), the next value $\tau_{kl}(m+1)$ is $\tau_{kl}(m+1) = \tau_{kl}(m)$. Otherwise, i.e. if $R \geq 0$

$$\tau_{kl}(m+1) = (1-\rho)\tau_{kl}(m) + \rho\Delta\tau_{kl}(m+1) \tag{6}$$

where ρ is called the *evaporation factor*, and the value of $\Delta\tau_{kl}(m+1)$ is

$$\Delta\tau_{kl}(m+1) = \frac{1}{R} \sum_{i=1}^{Z} \Delta\tau_{kl}^{(i)}(m) \tag{7}$$

Our implementation consists of creating Z scout ants. Every ant is charged to perform a random depth first search on G_1. As each ant travels across the graph, it associates the nodes that visits to a given store $v_j \in G_2$. When the aggregated nodes' weight exceeds the capacity of the current store, it reassigns the last node to $succ(v_j)$ and starts this filling process over again, provided that there are still nodes to visit. Every ant exhaustively visits G_1 and reports its solution path to the common nest. We call this procedure *traversal*. Depth first search (DFS) is the building block of this stage. It is implemented using a well known distributed algorithm [15] with time and message complexities $O(n)$ and $O(m)$, respectively. Where n is the order of G_1, and m its size.

In due time, the nest analyzes the cost of each reported solution. Then, it spreads more or less pheromone on each path, depending on its quality, i.e. according to a defined evaluation function, a good path receives more pheromone than a bad one. *Prizing*, as it is also called this stage, is carried out using propagation of information with feedback (PIF) [16]. It is a distributed algorithm with time and message complexities $O(D)$ and $O(n)$, respectively. Where D and n are the diameter and the order of G_1, respectively. As a byproduct, it also builds a spanning tree on G_1. From this moment on, our method profits from this spanning tree, in order to perform successive prizing.

Then, the nest starts the next cycle with Z new scouts that will perform the same procedure: traversal. Nevertheless, eventhough it still is a random search, the prizing phase bias the new resulting paths. Rounds are repeated for a fixed number or until the difference between the best solution of two consecutive rounds does not exceed a given bound.

6 Experiments and Analysis

A key hypothesis that may cause debate is that we assume nodes in G_2, i.e. stores, as static entities or, at least, with lifetimes sufficiently long to validate this assumption. In contrast, many empirical studies on networks' dynamics tend to show that unless storage is supported by fixed capacities, cooperative storage is very volatile. Nevertheless, this very studies consider that high information availability can be achieved, even in P2P environments, by means of data replication. Therefore, we assume a layered architecture and consider our proposal working on a "semantic" layer on top of a "replication" layer. From this perspective, the higher layer works with static logic-stores supported by a lower layer dealing with dynamic, best-effort, storage peers.

Once we had a running simulator, we designed a set of experiments in order to evaluate the quality of our method and its complexity. Let us recall that, as we work with pseudo-random number generators, each individual simulation requires a new seed to grant a new collection of results. So, from now on, when we describe a single experiment, we mean the same simulation repeated under 10 different seeds.

From our point of view, we consider that an instance of the problem is solved when 75% of the agents follow the same trail, which represents the best solution, i.e. the least number of stores from G_1 able to allocate all the nodes from G_2.

Table 2. Variance for the number of stores in the first round

	$n = 100$			$n = 300$			$n = 600$		
Z	μ	σ^2	σ	μ	σ^2	σ	μ	σ^2	σ
5	60.3	1.63	1.27	158.3	1.32	1.14	308.2	1.43	1.19
10	56.4	0.95	1.97	160.1	2.31	1.52	301.3	3.79	1.94
15	58.1	1.23	1.11	158.9	2.55	1.59	308.6	3.65	1.91
20	59.2	0.89	0.94	157.2	2.40	1.55	304.4	3.82	1.95
25	51.4	0.45	0.67	158.3	1.63	1.27	301.2	5.34	2.31
30	55.5	0.56	0.74	154.4	1.20	1.09	305.1	5.45	2.33
35	59.3	0.50	0.70	156.5	1.26	1.12	307.2	5.36	2.32

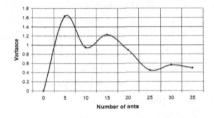

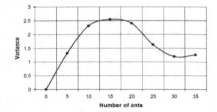

Fig. 4. Variance G_1 size 100   **Fig. 5.** Variance G_1 size 300

In the first set of experiments we investigated the initial number of ants Z that produces, during the first round, the highest variability on the resulting trails. The explanation is that we wanted to bound the resources required to produce the biggest amount of independent random trails (potential solutions) on G_1. For nodes in G_2, we fixed their storage capacity $c=500$. Next, for nodes in G_1, we designed three different initial graphs, with 100, 300 and 600 nodes, respectively. In turn, each of them produced 5 new graphs, with common order but random weights uniformly distributed between 0 and 500. Then, we tested each of the 15 resulting graphs under 7 different values of Z: 5, 10, 15, 20, 25, 30 and 35. This means that we devised 105 experiments.

According to the different instances of the problem, and the 7 levels of agents that we tried, Table 2 shows the mean(μ), variance(σ^2) and standard deviation(σ)

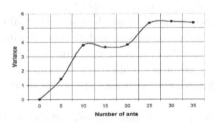

Fig. 6. Variance G_1 size 600

for the number of stores required to contain a given number of documental nodes. This table, as well as, Figs. 4, 5, 6, suggest that there is a minimum number of agents producing the highest variance. The upper bound of this initial value can be roughly featured by the expression $O(\sqrt{n})$.

In the second part of our study we investigated the relationship between the output and the input of the problem, i.e. the most compact achievable embedding for a given set of parameters including the capacity (c) of stores in G_2, as well as, the weight ($w(v_i)$) and the order (n) of nodes in G_1. Table 3 shows the parameters under testing and their corresponding levels. Each individual row represents a different experiment.

A simple analysis will indicate that the least number of stores has, as lower bound, the aggregated weight of all documental nodes, divided by the individual store capacity. In other words,

$$stores \geq \frac{\sum_{i=1}^{n} w(v_i)}{c}, \forall v_i \in V_1 \tag{8}$$

In this particular case, we can approximate the summation in the above formula, since we now that $w(v_i)$ follows a uniform random distribution between $[a, b]$. Therefore, for any i, $w(v_i)$ can be approximated by its mean $\frac{a+b}{2}$. Which, in turns, produces

$$stores \geq \frac{n(\frac{a+b}{2})}{c} \tag{9}$$

The sixth column in table 3 shows the lower bound on the number of stores, for each of the experiments consider in this part of our study. Meanwhile, the fifth column shows the mean value obtained with our simulations. Notice that the problem does not have solution when $w(v_i)$ can be bigger than c.

In the third group of experiments we addressed the influence of the evaporation factor (EF) on the number of rounds. A high evaporation implies a low correlation between the outcomes of consecutive rounds and vice versa. In other words, evaporation features the memory of the previous findings. This time, we tried two evaporation strategies: In the first case, we worked with a fixed factor equal to 0.9. In the second case, we tried with an initial evaporation equal to 0.9 which was decreased by 0.1 on each new round. Table 3 shows the parameters under testing and their corresponding levels. Again, each individual row represents a different experiment. Columns 7 and 8 show the corresponding time complexities for case 1 and 2, respectively. It is quite clear that the second approach is always better than the first one. This new strategy means that we allow a broad initial quest but, as rounds go by, the long term memory prevails and we stick to the best finding to accelerate convergence.

For the second strategy, we evaluated the covariance (S) between n and the number of rounds. The result $S = 399$ indicates that there is a direct dependency between the order of G_1 and the time complexity. In contrast, the covariance between c and the number of rounds is $S = -394.41$, which means an inverse dependency between the storage capacity and the time complexity. We assumed

Table 3. Number of stores and rounds

Z	n	c	$w(v_i)$	Number of stores	Ideal number of stores	Number of rounds EF = 0.9	Number of rounds variable EF	Number of rounds by multiple linear regression
10	100	100	0-20	11.31	10.0	10.37	6.42	7.4598
			0-50	26.15	25.0	12.65	8.03	7.6398
			0-100	53.75	50.0	14.63	9.14	7.9398
		300	0-20	4.46	3.33	8.16	3.81	6.2798
			0-50	9.25	8.33	9.22	4.76	6.4598
			0-100	17.85	16.67	10.74	5.72	6.7598
		900	0-20	2.06	1.11	6.45	2.23	2.7398
			0-50	3.84	2.78	7.32	2.80	2.9198
			0-100	6.35	5.56	8.68	3.38	3.2198
17	300	100	0-60	93.42	90.0	18.18	12.23	8.4998
			0-150	-	-	-	-	-
			0-300	-	-	-	-	-
		300	0-60	32.45	30.0	11.23	6.47	7.3198
			0-150	79.26	75.0	13.24	8.10	7.8598
			0-300	152.89	150.0	16.36	9.74	8.7598
		900	0-60	13.45	10.0	8.42	4.28	3.7798
			0-150	27.69	25.0	9.88	5.36	4.3198
			0-300	53.64	50.0	11.04	6.42	5.2198
30	900	180	-	-	-	-	-	-
			0-450	-	-	-	-	-
			0-900	-	-	-	-	-
		300	0-180	275.18	270.0	14.62	9.63	10.4398
			0-450	-	-	-	-	-
			0-900	-	-	-	-	-
		900	0-180	93.64	90.0	10.76	6.87	6.8998
			0-450	228.59	225.0	14.89	8.60	8.5198
			0-900	455.42	450.0	15.98	10.35	11.2198

there is a linear model that may describe the time complexity as a function of c, n, and $w(v_i)$, then we used the multiple linear regression model [17] and obtained the following expression:

$$rounds = 7.5298 + 0.0040n - 0.0059c + 0.0120\frac{a+b}{2} \qquad (10)$$

The last column in table 3 shows the predicted time complexity for the second strategy, according to this function. We obtained a correlation coefficient equal to 73% between simulation results and the predicted values, which we consider acceptable, for our goals.

7 Conclusions and Further Work

We have presented a general methodology that enables the operation of a Corporate Semantic Web (CSW) on top of a P2P distributed storage system. Our semantic layer proposal, is based on two procedures called location and query. Each peer working as a store, has a look-up table that supports query. Contents are ordered according to a semantic index. Then, the look-up table shows the peer in charge of the contents associated to a given index. Nevertheless, the main contribution of this work is the location procedure that assigns the contents of the CSW to the corresponding store-peers. Location is modelled in terms of the graph embedding of G_1 into G_2. Here, G_1 represents a CSW and G_2 is the P2P system. Graph embedding (GE) is an NP-complete problem that we tackled using the Ant Colony Optimization heuristics (ACO).

We evaluated the quality and complexities of our location method. As ACO consists of ants or agents that explore the solution space and cyclically improve the results, we found the best number of agents that produce the most efficient exploration. Each round consists of two phases, traversal and prizing. Next, we devised a prizing mechanism that accelerates convergence. For the instances of GE here addressed, we were able to propose a model that predicts the total number of rounds as a linear function of each of the parameters under study.

Some important issues remain as directions for further work. For the sake of simplicity we assumed that each document in the CSW is labelled with a single index. What should we do with multi-labelled contents? How should we deal with the CSW growing? Also, we assumed that the storage system has homogeneous capacities, will it worth studying heterogeneous capacities too?

Distributed storage is becoming an emerging technology driving many R-D efforts. From the users point of view, it may turn into the basic mechanism able to unleash the potential benefits of knowledge management. Health sciences, agriculture, geomatics, are only a few examples of the many domains that may dramatically improve their operations with the adoption of this new trend.

References

1. Berners-Lee, T., Hendler, J., Lassila, O.: The semantic web. Scientific American, 28–37 (2001)
2. Dieng-Kuntz, R.: Corporate Semantic Webs. ERCIM News 51, 19–21 (2002)
3. Yanilos, P., Sobti, S.: The evolving field of distributed storage. IEEE Internet Computing 5(5), 35–39 (2001)
4. Kubiatowicz, J., et al.: OceanStore: An architecture for global-scale persistent storage. SIGARCH Comput. Archit. News. 28(5), 190–201 (2000)
5. Kuntz, P., Layzell, P., Snyers, D.: A colony of ant-like agents for partitioning in VLSI technology. In: Husbands, P., Harvey, I. (eds.) Proceedings of the Fourth European Conference on Artificial Life, pp. 417–424. MIT Press, Cambridge (1997)
6. Newsome, J., Song, D.: GEM: Graph EMbedding for routing and data-centric storage in sensor networks without geographic information. In: Proceedings of the 1st international conference on Embedded networked sensor systems, SenSys 2003, pp. 76–88. ACM, New York (2003)

7. Bhatt, S.N., Ipsen, I.C.F.: How to embed trees in hypercubes. Yale University, Dept. of Computer Science. Research Report YALEU/DCS/RR-443 (1985)
8. Chan, M.Y., Chin, F.Y.L.: On embedding rectangular grids in hypercubes. IEEE Transactions on Computers 37(10), 1285–1288 (1988)
9. Mohar, B.: A linear time algorithm for embedding graphs in an arbitrary surface. SIAM Journal on Discrete Mathematics 12(1), 6–26 (1999)
10. Gagarin, A., Kocay, W., Neilson, D.: Embeddings of Small Graphs on the Torus. CUBO 5, 171–251 (2003)
11. Erten, C., Kobourov, S.G.: Simultaneous Embedding of Planar Graphs with Few Bends. Journal of Graph Algorithms and Applications 9(3), 347–364 (2005)
12. Savage, J.E., Wloka, M.G.: MOB a parallel heuristic for graph embedding. Technical Report CS-93-01, Brown University, Dep. Computer Science (1993)
13. Dorigo, M.: Optimization, Learning and Natural Algorithms. Ph.D. Thesis, Dept. of Electronics, Politecnico di Milano, Italy (1992)
14. Gutjahr, W.: A generalized convergence result for the graph-based ant system metaheuristic. University of Vienna, Technical Report 99-09 (1999)
15. Cidon, I.: Yet Another Distributed Depth-First-Search Algorithm. Information Processing Letters 26(6), 301–305 (1988)
16. Segall, A.: Distributed network protocols. IEEE Transaction on Information Theory 29(1), 23–25 (1983)
17. Johnson Norman, L., Leone Fred, C.: Statistics and Experimental Design in Engineering and the Physical Sciences, 2nd edn. vol. 1 (1976)

Implementing *PS-Merge* Operator

Verónica Borja Macías[1] and Pilar Pozos Parra[2]

[1] University of the Mixteca
Carretera a Acatlima Km 2.5
Huajuapan de León Oaxaca, Mexico
vero0304@mixteco.utm.mx
[2] Department of Informatics and Systems
University of Tabasco
Carretera Cunduacán - Jalpa Km. 1
Cunduacán Tabasco, Mexico
pilar.pozos@dais.ujat.mx

Abstract. When information comes from different sources inconsistent
beliefs may appear. To handle inconsistency, several model-based belief
merging operators have been proposed. Starting from the beliefs of a
group of agents which might conflict, these operators return a unique
consistent belief base which represents the beliefs of the group. The
operators, parameterized by a distance between interpretations and
aggregation function, usually only take into account consistent bases.
Consequently, the information in the base, which is not responsible for
conflicts, may be ignored. This paper presents an algorithm for implementing the *PS-Merge* operator, an alternative method of merging that
uses the notion of Partial Satisfiability instead of distance measures. This
operator allows us to take into account inconsistent bases. Also in order
to use the *PS-Merge* operator to solve ITC problems a pre-processing
transformation was proposed.

1   Introduction

Belief merging is concerned with the process of combining the information contained in a set of (possibly inconsistent) belief bases obtained from different
sources to produce a single consistent belief base. Belief merging is an important issue in artificial intelligence and databases, and its applications are many
and diverse [2]. For example, multi-sensor fusion, database integration and expert systems development. Belief merging has applications in multiagent systems
that aim at aggregating the distributed agent-based knowledge into an (ideally)
unique set of consistent propositions. A multiagent system can be considered a
group of autonomous agents who individually hold a set of propositions. Each
set of propositions represents an agent's beliefs on issues on which the group
has to make a collective decision. The merging operator defines the beliefs of
a group, the collective decision or the "true" state of the world for the group.
Though we consider only belief bases, merging operators can typically be used
for merging either beliefs or goals.

A. Hernández Aguirre et al. (Eds.): MICAI 2009, LNAI 5845, pp. 39–50, 2009.
© Springer-Verlag Berlin Heidelberg 2009

Several merging operators have been defined and characterized in a logical way. Among them, model-based merging operators [8,6,14,9] obtain a belief base from a set of interpretations with the help of a distance measure on interpretations and an aggregation function. Usually, model-based merging operators only take into account consistent belief bases: if a base is inconsistent, it will not be considered in the merging process, thus discarding also the information that is not responsible for conflicts. Other merging operators, using a syntax-based approach [1], are based on the selection of some consistent subsets of the set-theoretic union of the belief bases. This allows for taking inconsistent belief bases into account, but such operators usually do not take into account the frequency of each explicit item of belief. For example, the fact that a formula ψ is believed in a base or in n bases is not considered relevant, which is counter-intuitive.

An alternative method of merging uses the notion of Partial Satisfiability to define *PS-Merge*, a model-based merging operator which depends on the syntax of the belief bases [3]. The proposal produces similar results to other merging approaches, but while other approaches require many merging operators in order to achieve satisfactory results for different scenarios the proposal obtains similar results for all these different scenarios with a unique operator. It is worth noticing that *PS-Merge* is not based on distance measures on interpretations, and it takes into account inconsistent bases and the frequency of each explicit item of belief. We propose an implementation of the *PS-Merge* operator.

The rest of the paper is organized as follows. After providing some technical preliminaries, Section 3 describes the notion of Partial Satisfiability and the associated merging operator. In Section 4 we mention some properties that are satisfied by *PS-Merge* operator and compare the operator with other approaches. Section 5 presents an algorithm for implementing the *PS-Merge* operator and Section 6 proposes a pre-processing transformation that transforms timetabling problems proposed for the First International Timetabling Competition into belief merging problems solvable by the algorithm presented if the memory and time space needed are available. Finally, we conclude with a discussion of future work.

2   Preliminaries

We consider a language $\mathcal{L}$ of propositional logic formed from $P := \{p_1, p_2, ..., p_n\}$ (a finite ordered set of atoms) in the usual way. And we use the standard terminology of propositional logic except for the definitions given below. A *belief base* K is a finite set of propositional formulas of $\mathcal{L}$ representing the beliefs of an agent (we identify K with the conjunction of its elements).

A *state* or *interpretation* is a function w from P to $\{1, 0\}$, these values are identified with the classical truth values *true* and *false* respectively. The set of all possible states will be denoted as $\mathcal{W}$ and its elements will be denoted by vectors of the form $(w(p_1), ..., w(p_n))$. A *model* of a propositional formula Q is a state such that $w(Q) = 1$ once w is extended in the usual way over the connectives. For convenience, if Q is a propositional formula or a set of propositional formulas then $\mathcal{P}(Q)$ denotes the set of atoms appearing in Q. $|P|$ denotes the cardinality of set P. A *literal* is an atom or its negation.

A belief *profile* E denotes the beliefs of agents $K_1, ..., K_m$ that are involved in the merging process, $E = \{\{Q_{1_1}, ..., Q_{n_1}\}, ..., \{Q_{1_m}, ..., Q_{n_m}\}\}$ where $Q_{1_i}, ..., Q_{n_i}$ denotes the beliefs in the base K_i. E is a multiset (bag) of belief bases and thus two agents are allowed to exhibit identical bases.

Two belief profiles E_1 and E_2 are said to be equivalent, denoted by $E_1 \equiv E_2$, iff there is a bijection g from E_1 to E_2 such that $K \equiv g(K)$ for every base K in E_1. With $\bigwedge E$ we denote the conjunction of the belief bases $K_i \in E$, while $\sqcup$ denotes the multiset union. For every belief profile E and positive integer n, E^n denotes the multiset union of n times E.

3   Partial Satisfiability

In order to define Partial Satisfiability without loss of generality we consider a normalized language so that each belief base is taken as the disjunctive normal form (DNF) of the conjunction of its elements. Thus if $K = \{Q_1, ..., Q_n\}$ is a belief base we will identify this base with $Q_K = DNF(Q_1 \wedge ... \wedge Q_n)$. The DNF of a formula is obtained by replacing $A \leftrightarrow B$ and $A \rightarrow B$ by $(\neg A \vee B) \wedge (\neg B \vee A)$ and $\neg A \vee B$ respectively, applying De Morgan's laws, using the distributive law, distributing $\vee$ over $\wedge$, and finally eliminating the literals repeated in each conjunct.

Example 1. Given the belief base $K = \{a \rightarrow b, \neg c\}$ it is identified with $Q_K = (\neg a \wedge \neg c) \vee (b \wedge \neg c)$.

The last part of the construction of the DNF (the minimization by eliminating literals) is important since the number of literals in each conjunct affects the satisfaction degree of the conjunct. We are not applying other logical minimization methods to reduce the size of the DNF expressions since this may affect the intuitive meaning of the formulas. A further analysis of logical equivalence and the results obtained by the Partial Satisfiability is required.

Definition 1 (Partial Satisfiability). *Let K be a belief base, w any state of $\mathcal{W}$ and $|P| = n$, we define the Partial Satisfiability of K for w, denoted as $w_{ps}(Q_K)$, as follows:*

- *if $Q_K := C_1 \wedge ... \wedge C_s$ where C_i are literals then*

$$w_{ps}(Q_K) = max\left\{\sum_{i=1}^{s}\frac{w(C_i)}{s}, \frac{n - |\mathcal{P}(Q_K)|}{2n}\right\},$$

- *if $Q_K := D_1 \vee ... \vee D_r$ where each D_i is a literal or a conjunction of literals then*

$$w_{ps}(Q_K) = max\left\{w_{ps}(D_1), ..., w_{ps}(D_r)\right\}.$$

The intuitive interpretation of Partial Satisfiability is as follows: it is natural to think that if we have the conjunction of two literals and just one is satisfied then we are satisfying 50% of the conjunction. If we generalize this idea we can measure the satisfaction of a conjunction of one or more literals as the sum of their evaluation under the interpretation divided by the number of conjuncts. However, the agent may consider only some atoms of the language, in which case the agent is not affected by the decision taken over the atoms that are not considered. In fact, we can interpret the absence of information in two different ways: the agent knows about the existence of these atoms but it is indifferent to their values since it is focused on its beliefs or the agent does not consider the atoms because it is ignoring their existence. In both cases, believing something or its opposite makes no difference, but the number of such atoms is a measure of the extent to which the agent is disinterested or wilfully ignorant of it's situation. So, in order to measure this indifference (or deliberate ignorance) we give a partial satisfaction of 50% for each atom not appearing in the agent's beliefs.

On the other hand, the agent is interested in satisfying the literals that appear in its beliefs, and we interpret this fact by assigning a satisfaction of 100% to each literal verified by the state and 0% to those that are falsified. As we can see the former intuitive idea is reflected in Definition 1 since the literals that appear in the agent's beliefs have their classical value and atoms not appearing have a value of just $\frac{1}{2}$.

Finally, if we have a disjunction of conjunctions the intuitive interpretation of the valuation is to obtain the maximum value of the considered conjunctions.

Example 2. The Partial Satisfiability of the belief base of Example 1 given $P = \{a, b, c\}$ and $w = (1, 1, 1)$ is

$$w_{ps}(Q_K) = max\left\{max\{\tfrac{w(\neg a)+w(\neg c)}{2}, \tfrac{1}{6}\}, max\{\tfrac{w(b)+w(\neg c)}{2}, \tfrac{1}{6}\}\right\} = \tfrac{1}{2}.$$

Instead of using distance measures as [6,9,7,11] the notion of Partial Satisfiability has been proposed in order to define a new merging operator. The elected states of the merge are those whose values maximize the sum of the Partial Satisfiability of the bases.

Definition 2. *Let E be a belief profile obtained from the belief bases $K_1, ..., K_m$, then the Partial Satisfiability Merge of E denoted by $PS\text{-}Merge(E)$ is a mapping from the belief profiles to belief bases such that the set of models of the resulting base is:*

$$\left\{w \in \mathcal{W} \,\middle|\, \sum_{i=1}^{m} w_{ps}(Q_{K_i}) \geq \sum_{i=1}^{m} w'_{ps}(Q_{K_i}) \; for \; all \; w' \in \mathcal{W}\right\}$$

Example 3. We now give a concrete merging example taken from [13]. The author proposes the following scenario: a teacher asks three students which among three languages, SQL, Datalog and O_2, they would like to learn. Let s, d and o be the propositional letters used to denote the desire to learn SQL, Datalog and O_2, respectively, then $P = \{s, d, o\}$. The first student only wants to learn SQL or

O_2, the second wants to learn only one of Datalog or O_2, and the third wants to learn all three languages. So, we have $E = \{K_1, K_2, K_3\}$ with $K_1 = \{(s \vee o) \wedge \neg d\}$, $K_2 = \{(\neg s \wedge d \wedge \neg o) \vee (\neg s \wedge \neg d \wedge o)\}$ and $K_3 = \{s \wedge d \wedge o\}$.

In [11] using the Hamming distance applied to the anonymous aggregation function Σ and in [6] using the operator Δ_Σ, both approaches obtain the states $(0, 0, 1)$ and $(1, 0, 1)$ as models of the merging.

We have $Q_{K_1} = (s \wedge \neg d) \vee (o \wedge \neg d)$, $Q_{K_2} = (\neg s \wedge d \wedge \neg o) \vee (\neg s \wedge \neg d \wedge o)$, and $Q_{K_3} = s \wedge d \wedge o$. As we can see in the fifth column of Table 1 the models of $PS\text{-}Merge(E)^1$ are the states $(0, 0, 1)$ and $(1, 0, 1)$.

Table 1. *PS-Merge* of Example 3 and min function

w	Q_{K_1}	Q_{K_2}	Q_{K_3}	Sum	min
$(1,1,1)$	$\frac{1}{2}$	$\frac{1}{3}$	1	$\frac{11}{6} \simeq 1.83$	$\frac{1}{3}$
$(1,1,0)$	$\frac{1}{2}$	$\frac{2}{3}$	$\frac{2}{3}$	$\frac{11}{6} \simeq 1.83$	$\frac{1}{2}$
$\mathbf{(1,0,1)}$	1	$\frac{2}{3}$	$\frac{2}{3}$	$\frac{14}{6} \simeq \mathbf{2.33}$	$\frac{2}{3}$
$(1,0,0)$	1	$\frac{1}{3}$	$\frac{1}{3}$	$\frac{10}{6} \simeq 1.67$	$\frac{1}{3}$
$(0,1,1)$	$\frac{1}{2}$	$\frac{2}{3}$	$\frac{2}{3}$	$\frac{11}{6} \simeq 1.83$	$\frac{1}{2}$
$(0,1,0)$	$\frac{1}{6}$	1	$\frac{1}{3}$	$\frac{9}{6} = 1.5$	$\frac{1}{6}$
$\mathbf{(0,0,1)}$	1	1	$\frac{1}{3}$	$\frac{14}{6} \simeq \mathbf{2.33}$	$\frac{1}{3}$
$(0,0,0)$	$\frac{1}{2}$	$\frac{2}{3}$	0	$\frac{7}{6} \simeq 1.16$	0

In [7] two classes of merging operators are defined: majority and arbitration merging. The former strives to satisfy a maximum of agents' beliefs and the latter tries to satisfy each agent beliefs to the best possible degree. The former notion is treated in the context of *PS-Merge*, and it can be refined tending to arbitration if we calculate the minimum value among the Partial Satisfiability of the bases. Then with this indicator, we have a form to choose the state that is impartial and tries to satisfy all agents as far as possible. If we again consider Example 3 in Table 1 there are two different states that maximize the sum of the Partial Satisfaction of the profile, $(1, 0, 1)$ and $(0, 0, 1)$. If we try to minimize the individual dissatisfaction these two states do not provide the same results. Using the *min* function (see 6th column of Table 1) over the partial satisfaction of the bases we get the states that minimize the individual dissatisfaction and between the states $(1, 0, 1)$ and $(0, 0, 1)$ obtained by the proposal we might prefer the state $(1, 0, 1)$ over $(0, 0, 1)$ as the Δ_{GMax} operator (an arbitration operator) does in [6].

Usually, model-based merging operators take only into account consistent belief bases but *PS-Merge* allows us to merge even inconsistent bases as we can see in the following example.

---

[1] If Δ is a merging operator, we are going to abuse the notation by referring to the models of the merging operator $mod(\Delta(E))$ and their respective belief base $\Delta(E)$ simply as $\Delta(E)$.

Example 4. Let be $K_1 = \{a \rightarrow b, \neg c\}$ and $K_2 = \{a \rightarrow \neg b, a, b\}$ two bases, the second base is inconsistent but *PS-Merge* allows us to merge these bases, in fact $PS\text{-}Merge(\{K_1, K_2\}) = \{(1,1,0),(0,1,0)\}$.

If we analyze the profile we will realize that the first base has three different models $(1,1,0)$, $(0,1,0)$ and $(0,0,0)$ which means that is "uncertain" about a and b, the second base has no model, but both together they can give us a compromise between the "uncertainly" of K_1 and the "inconsistency" of K_2 which results intuitive.

4 Properties and Comparisons

Axioms to ensure that a belief merging operator yields rational behavior have been proposed in [6,14,8,9]. In [6] Konieczny and Pino-Pérez proposed the basic properties (A1)-(A6) for merging operators.

Definition 3. *Let E, E_1, E_2 be belief profiles, and K_1 and K_2 be consistent belief bases. Let Δ be an operator which assigns to each belief profile E a belief base $\Delta(E)$. Δ is a merging operator if and only if it satisfies the following postulates:*

(A1) $\Delta(E)$ is consistent
(A2) if $\bigwedge E$ is consistent then $\Delta(E) \equiv \bigwedge E$
(A3) if $E_1 \equiv E_2$, then $\Delta(E_1) \equiv \Delta(E_2)$
(A4) $\Delta(\{K_1, K_2\}) \wedge K_1$ is consistent iff $\Delta(\{K_1, K_2\}) \wedge K_2$ is consistent
(A5) $\Delta(E_1) \wedge \Delta(E_2) \models \Delta(E_1 \sqcup E_2)$
(A6) if $\Delta(E_1) \wedge \Delta(E_2)$ is consistent then $\Delta(E_1 \sqcup E_2) \models \Delta(E_1) \wedge \Delta(E_2)$

In [12] the postulates satisfied by *PS-Merge* have been studied. Postulates A1 and A2 are satisfied by *PS-Merge*. If there is no redundant information, i.e. formulas including disjuncts of the form $a \wedge \neg a$, then the operator satisfies A3 and A4. If we extend the language of E_1 to include the atoms appearing in E_2 and vice versa, then *PS-Merge* satisfies A5.

PS-Merge is commutative and is a majority operator; if an opinion is the most popular, then it will be the opinion of the group, hence it satisfies the postulate (M7) of [6]:

(M7) $\forall K \exists n \in \mathbb{N} \quad \Delta(E \sqcup \{K\}^n) \models K$

PS-Merge yields similar results compared with existing techniques such as $CMerge$, the Δ_Σ operator[2] and MCS (Maximal Consistent Subsets) considered in [9,6,7]. A deeper comparison between these operators can be found in [12]; here we provide examples where *PS-Merge* yields more intuitive answers. Let E be in each case the belief profile consisting of the belief bases enlisted below and let P be the corresponding set of atoms ordered alphabetically.

---

[2] As stated in [6], merging operator Δ_Σ is equivalent to the merging operator proposed by Lin and Mendelzon in [9] called $CMerge$.

1. $K_1 = \{a, c\}$, $K_2 = \{a \rightarrow b, \neg c\}$ and $K_3 = \{b \rightarrow d, c\}$. In this case $MCS(E) = \{a, a \rightarrow b, b \rightarrow d\}$ that implies b and d but does not imply c. $CMerge(E) = \{a, a \rightarrow b, b \rightarrow d, c\}$ implies majority view on c, and the implicit knowledge of b and d which is equivalent to $PS\text{-}Merge(E) = \Delta_\Sigma(E) = \{(1, 1, 1, 1)\}$.

2. $K_1 = \{a, c\}$, $K_2 = \{a \rightarrow b, \neg c\}$, $K_3 = \{b \rightarrow d, c\}$ and $K_4 = \{\neg c\}$. While $MCS(E) = CMerge(E) = \{a, a \rightarrow b, b \rightarrow d\}$ which is equivalent to $\Delta_\Sigma(E) = \{(1, 1, 0, 1), (1, 1, 1, 1)\}$, $PS\text{-}Merge(E) = \{(1, 1, 0, 1)\}$. $CMerge$, MCS and the Δ_Σ operator give no information about c. Using $PS\text{-}Merge$, c is falsified and this leads us to have total satisfaction of the second and fourth bases and partial satisfaction of the first and third bases.

3. $K_1 = \{a\}$, $K_2 = \{a \rightarrow b\}$ and $K_3 = \{a, \neg b\}$. Now $MCS(E) = \{a, a \rightarrow b, b\} \vee \{a, \neg b\} \vee \{\neg b, a \rightarrow b\}$, $CMerge(E) = \{a\}$ and $\Delta_\Sigma(E) = \{(1, 1), (1, 0)\}$ but $PS\text{-}Merge(E) = \{(1, 1)\}$, this answer, however, is intuitive since the model $(1, 0)$ satisfies only two bases while the model $(1, 1)$ satisfies two bases and a "half" of the third base.

4. $K_1 = \{b\}$, $K_2 = \{a \rightarrow b\}$, $K_3 = \{a, \neg b\}$ and $K_4 = \{\neg b\}$. In this case $MCS(E) = \{a, a \rightarrow b, b\} \vee \{a, \neg b\} \vee \{\neg b, a \rightarrow b\}$ as before, $CMerge(E) = \{a \vee \neg b\}$, $\Delta_\Sigma(E) = \{(0, 0), (1, 0), (1, 1)\}$ and $PS\text{-}Merge(E) = \{(1, 1), (0, 0)\}$. The model $(1, 0)$ obtained using Δ_Σ operator satisfies only two bases, while the two options of $PS\text{-}Merge(E)$ satisfy two bases and a "half" of the third base. Then $PS\text{-}Merge$ is a refinement of the answer given by $CMerge$ and Δ_Σ.

5 Algorithm *PS-Merge*

A valuable range of model-based operators have been proposed that conform to interesting and intuitive properties. However, the implementation of such operators has been briefly addressed, partly due to the considerable computational complexity of the proposals and the considerable computational complexity of the decision problem associated with belief merging, generally situated at the second level of the polynomial hierarchy or above [5]. As far as we know there is a unique attempt in [4] where a method for implementing the operators Δ_Σ, Δ_{max}, Δ_{GMax} and Δ_{GMin} is proposed and experimentally evaluated. We propose Algorithm 1 that calculates the $PS\text{-}Merge$ of profile E which needs the representation of the profile E in the following format:

- V : Number of variables of E
- B : Number of bases of E
- D : Vector of number of disjuncts of each base in E
- L : Matrix of occurrences of literals in each disjunct of each base of E

Example 5. Consider the profile of example 3, in this example $Q_{K_1} = (s \wedge \neg d) \vee (o \wedge \neg d)$, $Q_{K_2} = (\neg s \wedge d \wedge \neg o) \vee (\neg s \wedge \neg d \wedge o)$ and $Q_{K_3} = s \wedge d \wedge o$ then we format the profile E as:

- $V = 3$
- $B = 3$
- $D = (2, 2, 1)$

$$- L = \begin{pmatrix} s & \neg s & d & \neg d & o & \neg o \\ 1 & 0 & 0 & 1 & 0 & 0 \\ 0 & 0 & 0 & 1 & 1 & 0 \\ 0 & 1 & 1 & 0 & 0 & 1 \\ 0 & 1 & 0 & 1 & 1 & 0 \\ 1 & 0 & 1 & 0 & 1 & 0 \end{pmatrix}$$

Algorithm 1. *PS-Merge*

Data:

V : Number of variables of E

B : Number of bases of E

D : Vector of number of disjuncts of each base in E

L : Matrix of occurrences of literals in each disjunct of each base of E

Result:

Solution_Set : The set of states in $PS\text{-}Merge(E)$

begin

    $Solution \leftarrow \emptyset$

    $Max\text{-}Sum \leftarrow 0$

    $W \leftarrow$ Matrix whose rows are all the possible states for V variables

    for $s = 1 \dots B$ **do**

        $ID_s \leftarrow \sum_{k=1}^{s-1} D_k + 1$

    for $i = 1 \dots 2^V$ **do**

        $Sum \leftarrow 0$

        for $s = 1 \dots B$ **do**

            $ps\text{-}disjunct \leftarrow \emptyset$

            for $d = ID_s \dots ID_s + D_s$ **do**

                $satisfied \leftarrow 0$

                $conjuncts \leftarrow 0$

                $vars\text{-}not\text{-}appearing \leftarrow 0$

                for $j = 1 \dots V$ **do**

                    if $W_{i,j} = 1$ **then**

                        $satisfied \leftarrow satisfied + L_{d,2j-1}$

                    if $W_{i,j} = 0$ **then**

                        $satisfied \leftarrow satisfied + L_{d,2j}$

                    $conjuncts \leftarrow conjuncts + L_{d,2j-1} + L_{d,2j}$

                    if $L_{d,2j} = 0$ and $L_{d,2j-1} = 0$ **then**

                        $vars\text{-}not\text{-}appearing \leftarrow vars\text{-}not\text{-}appearing + 1$

                $ps\text{-}disjunct \leftarrow ps\text{-}disjunct \cup \{max(\frac{satisfied}{conjuncts}, \frac{vars-not-appearing}{2V})\}$

            $PS \leftarrow max(ps\text{-}disjunct)$

            $Sum \leftarrow Sum + PS$

        if $Sum > MaxSum$ **then**

            $Solution \leftarrow \{i\}$

            $MaxSum \leftarrow Sum$

        else if $Sum = MaxSum$ **then**

            $Solution \leftarrow Solution \cup \{i\}$

    $Solution\_Set = \{\text{ith-row of } W | i \in Solution\}$

end

The Algorithm 1 has been implemented in Matlab[3] and tested with all the examples provided in this article and those presented in [12] with an average case performance of 0.4 sec. Other tests with a larger number of variables have been done, for example 15 variables with 35 bases and a total of 130 disjuncts with an average case performance of 1400 sec. The results indicate that problems of modest size exhibit short computation times using commodity hardware.

6 *PS-Merge* for ITC

We can see the timetabling problem designed for the First International Timetabling Competition (ITC)[4] as a particular case of multiagent problem, where multiple agents want to schedule different events under some hard and soft restrictions. In this context we propose an algorithm based on *PS*-Merge that intends to solve ITC problems. The problem consists of finding an optimal timetable within the following framework: there is a set of events $E = \{E_1, E_2, ..., E_{n_E}\}$ to be scheduled in a set of rooms $R = \{R_1, R_2, ..., R_{n_R}\}$, where each room has 45 available timeslots, nine for each day in a five day week. There is a set of students $S = \{S_1, S_2, ..., S_{n_S}\}$ who attend the events, and a set of features $F = \{F_1, F_2, ..., F_{n_F}\}$ satisfied by rooms and required by events. Each event is attended by a number of students, and each room has a given size, which is the maximum number of students the room can accommodate. A feasible timetable is one in which all events have been assigned a timeslot and a room so that the following hard constraints are satisfied:

(1) no student attends more than one event at the same time;
(2) the room is big enough for all the attending students and satisfies all the features required by the event; and
(3) only one event is scheduled in each room at any timeslot.

In contest instance files there were typically 10 rooms, hence there are 450 available places. There were typically 350-400 events, 5-10 features and 200-300 students.

The original problem penalizes a timetable for each occurrence of certain soft constraints violations, which, for the sake of brevity, are not described here. The problem may be precisely formulated as:

- let $\{F_1, F_2, ..., F_{n_F}\}$ be a set of variables representing the features;
- $R_i = \{F_1^{'}, F_2^{'}, ..., F_{n_{R_i}}^{'}\}$ where $F_j^{'} \in F$ for $j = 1, ..., n_{R_i}$ and n_{R_i} is the number of feature satisfied by room R_i;
- $\{N_1, ..., N_{n_R}\}$ be a set of integer numbers indicating the maximum of students each room can accommodate;
- $E_i = \{F_1^{''}, F_2^{''}, ..., F_{n_{E_i}}^{''}\}$ where $F_j^{''} \in F$ for $j = 1, ..., n_{E_i}$ and n_{E_i} is the number of features required by event E_i;

---

[3] The source and the examples can be downloaded from:
   http://www.utm.mx/~vero0304/PSMerge.htm
[4] http://www.idsia.ch/Files/ttcomp2002/

- $S_i = \{E'_1, E'_2, ..., E'_{n_{S_i}}\}$ where $E'_j \in E$ for $j = 1, ..., n_{S_i}$ and n_{S_i} is the number of events student S_i attends; and
- $T = \{T_1, ..., T_{45}\}$ be a set of timeslots.

Find a feasible solution, i.e. a set of pairs $\{(T'_1, R'_1), ..., (T'_{n_E}, R'_{n_E})\}$ such that:

- $T'_i \in T$ and $R'_i \in R$
- $\neg(T'_i = T'_j)$ if $E_i \in S_k$ and $E_j \in S_k$ and $\neg(i = j)$
- $E_i \subseteq R'_i$ and $|\{S_j | j = 1, ..., n_S$ and $E_i \in S_j\}| \le N_k$ where $R'_i = R_k$;
- $\neg \exists i \exists j \, (\neg(i = j) \wedge T'_i = T'_j \wedge R'_i = R'_j)$.

The rest of this section describes an algorithm based on $PS\text{-}Merge$ that is designed to solve this problem. Although the competition considers a time limit, in this first stage we omit this time constraint because our initial interest is to obtain a deterministic algorithm. It is worth noticing that the proposed solutions appearing in the competition web page were stochastic.

Before the problem is tackled by the algorithm, there is a preprocessing phase. Apart from the obvious actions such as reading the problem file, the following actions are performed:

- the number of students for each event is calculated and stored,
$n_i = |\{S_j | j = 1, ..., n_S$ and $E_i \in S_j\}|$
- a list of possible rooms is created for each event,
$e_i = \{R_j | E_i \subseteq R_j$ and $n_i \le N_j\}$

The preprocessing phase allows us to reduce the problem by eliminating the feature set, event set and capacity requirements, defining a new event set $\{e_1, ..., e_{n_E}\}$ that includes the eliminated information. The new event set will be used for satisfying hard constraint (2).

The next phase, the construction phase, creates the parameters to be used for the $PS\text{-}Merge$ operator as follows:

- create the propositional variables set $V = \{R_{i,s,r} | R_r \in e_i,\ s = 1, ..., 45$ and $i = 1, ..., n_E\}$,
- for every event e_i create the belief source $K_i = \bigvee_{R_{i,s,r} \in V} R_{i,s,r}$,
- create the source $K_{n_E+1} = \bigwedge_{R_{i_1,s,r}, R_{i_2,s,r} \in V} R_{i_1,s,r} \wedge \neg(i_1 = i_2) \to \neg R_{i_2,s,r}$. Here, we are abusing the notation for obtaining a concise representation. For example if event 5 can be allocated only in room 2 and 4, and event 8 can be allocated only in room 1 and 2, some of the conjuncts of this formula are: $\neg R_{5,1,2} \vee \neg R_{8,1,2}, \neg R_{5,2,2} \vee \neg R_{8,2,2}, ..., \neg R_{5,45,2} \vee \neg R_{8,45,2}$,
- create the source $K_{n_E+2} = \bigwedge_{R_{i_1,s,r_1}, R_{i_2,s,r_2} \in V, j \in \{1,...,n_S\}} R_{i_1,s,r_1} \wedge \neg(i_1 = i_2)$ $\wedge\ E_{i_1}, E_{i_2} \in S_j \to \neg R_{i_2,s,r_2}$. We are again abusing the notation for obtaining a concise representation. For example if room 2 can be used for allocating event 5, room 3 can be used for allocating event 3, and Student 90 attends event 5 and 3, some of the conjuncts of this formula are: $\neg R_{3,1,3} \vee \neg R_{5,1,2}, \neg R_{3,2,3} \vee \neg R_{5,2,2}, ..., \neg R_{3,45,3} \vee \neg R_{5,45,2}$.

Now, the last phase obtains the interpretations that maximize the satisfaction of the belief sources as follows: $PS\text{-}Merge(\{K_1, ..., K_{n_E+2}\})$.

It is worth noticing that the last phase can use any belief operators found in the literature. However, given the characteristics of the problem, the implementation of *PS-Merge* can use some heuristics in order to reduce the time of search. Firstly, all problem instances provided for the competition are assumed to have a feasible timetable solution, so it means that the conjunction of the sources is possible and so we reduce the search to states w such that $w_{ps}(Q_{K_i}) = 1$ for any belief source i. Hence the belief merging operator *PS-Merge* can be reduced to a classical satisfiability problem in this particular case. Then for finding the assignment of rooms and slots to the events we can use a SAT solver.

7   Conclusion

PS-Merge operator has been proposed in [3] that is not defined in terms of a distance measure on interpretations, but is Partial Satisfiability-based. It appears to resolve conflicts among the belief bases in a natural way. The idea is intended to extend the notion of satisfiability to one that includes a "measure" of satisfaction.

Unlike other approaches *PS-Merge* can consider belief bases which are inconsistent, since the source of inconsistency can refer to specific atoms and the operator takes into account the rest of the information.

The approach bears some resemblance to the belief merging framework proposed in [6,7,9,11], particularly with the Δ_Σ operator. As with those approaches the *Sum* function is used, but instead of using it to measure the distance between the states and the profile *PS-Merge* uses *Sum* to calculate the general degree of satisfiability. The result of *PS-Merge* are simply the states which maximize the *Sum* of the Partial Satisfiability of the profile and it is not necessary to define a partial pre-order. Because of this similarity between *PS-Merge* and Δ_Σ and the high complexity for implementing Δ_Σ we propose to implement the *PS-Merge* operator; given that it is not necessary to calculate a partial pre-order the computation is reduced. However, a complexity analysis of the proposed algorithm is intended to modify it and save some operations. For example, by computing the fraction $\frac{vars-not-appearing}{2V}$ one for each disjunct of each base we save an exponential number of operations.

Finally, in order to take advantage of the approach two principal problems will be considered. First ITC-problems where it is necessary to find some heuristics that allow us to reduce the memory space needed to handle these problems, and second integrate the approach in a preprocessing phase for defining consistent BDI (Belief-Desire-Intention) mental states of agents.

Concerning the ITC, we intend to use SAT(ID) solver [10] for finding the assignment of rooms and slots to the events.

References

1. Baral, C., Kraus, S., Minker, J., Subrahmanian, V.S.: Combining knowledge bases consisting of first-order theories. Computational Intelligence 1(8), 45–71 (1992)
2. Bloch, I., Hunter, A.: Fusion: General concepts and characteristics. International Journal of Intelligent Systems 10(16), 1107–1134 (2001)
3. Borja Macías, V., Pozos Parra, P.: Model-based belief merging without distance measures. In: Proceedings of the Sixth International Conference on Autonomous Agents and Multiagent Systems, Honolulu, Hawai'i, pp. 613–615. ACM, New York (2007)
4. Gorogiannis, N., Hunter, A.: Implementing semantic merging operators using binary decision diagrams. Int. J. Approx. Reasoning 49(1), 234–251 (2008)
5. Konieczny, S., Lang, J., Marquis, P.: Da2 merging operators. Artificial Intelligence Journal(AIJ) 157(1-2), 49–79 (2004)
6. Konieczny, S., Pino-Pérez, R.: On the logic of merging. In: Cohn, A.G., Schubert, L., Shapiro, S.C. (eds.) KR 1998: Principles of Knowledge Representation and Reasoning, pp. 488–498. Morgan Kaufmann, San Francisco (1998)
7. Konieczny, S., Pino Pérez, R.: Merging with integrity constraints. In: Hunter, A., Parsons, S. (eds.) ECSQARU 1999. LNCS (LNAI), vol. 1638, p. 233. Springer, Heidelberg (1999)
8. Liberatore, P., Schaerf, M.: Arbitration (or how to merge knowledge bases). IEEE Transactions on Knowledge and Data Engineering 10(1), 76–90 (1998)
9. Lin, J., Mendelzon, A.: Knowledge base merging by majority. In: Pareschi, R., Fronhoefer, B. (eds.) Dynamic Worlds: From the Frame Problem to Knowledge Management. Kluwer Academic, Dordrecht (1999)
10. Marien, M., Wittocx, J., Denecker, M., Bruynooghe, M.: Sat(id): Satisfiability of propositional logic extended with inductive definitions. In: Proceedings of the Eleventh International Conference on Theory and Applications of Satisfiability Testing, pp. 211–224 (2008)
11. Meyer, T., Parra, P.P., Perrussel, L.: Mediation using m-states. In: Godo, L. (ed.) ECSQARU 2005. LNCS (LNAI), vol. 3571, pp. 489–500. Springer, Heidelberg (2005)
12. Pozos Parra, P., Borja Macías, V.: Partial satisfiability-based merging. In: Proceedings of the Sixth Mexican International Conference on Artificial Intelligence, pp. 225–235 (2007)
13. Revesz, P.Z.: On the semantics of theory change: Arbitration between old and new information. In: Proceedings of the Twelfth ACM SIGACT-SIGMOD-SIGART Symposium on Principles of Database Systems, pp. 71–82 (1993)
14. Revesz, P.Z.: On the Semantics of Arbitration. Journal of Algebra and Computation 7(2), 133–160 (1997)

Generating Explanations Based on Markov Decision Processes

Francisco Elizalde[1], Enrique Sucar[2], Julieta Noguez[3], and Alberto Reyes[1]

[1] Instituto de Investigaciones Eléctricas, Reforma 113, Col. Palmira, Cuernavaca,
Morelos, 62490, México
[2] Instituto Nacional de Astrofísica Optica y Electrónica, Luis Enrique Erro No. 1,
Tonantzintla, Puebla, 72000, México
[3] Tec de Monterrey, Campus Cd. de México, Calle del Puente 222, Ejidos de
Huipulco, 14380, México, D.F.

Abstract. In this paper we address the problem of explaining the recommendations generated by a Markov decision process (MDP). We propose an automatic explanation generation mechanism that is composed by two main stages. In the first stage, the most *relevant variable* given the current state is obtained, based on a factored representation of the MDP. The relevant variable is defined as the factor that has the greatest impact on the utility given certain state and action, and is a key element in the explanation generation mechanism. In the second stage, based on a general template, an explanation is generated by combing the information obtained from the MDP with domain knowledge represented as a frame system. The state and action given by the MDP, as well as the relevant variable, are used as pointers to the knowledge base to extract the relevant information and fill-in the explanation template. In this way, explanations of the recommendations given by the MDP can be generated on-line and incorporated to an intelligent assistant. We have evaluated this mechanism in an intelligent assistant for power plant operator training. The experimental results show that the automatically generated explanations are similar to those given by a domain expert.

1 Introduction

An important requirement for intelligent trainers is to have an explanation generation mechanism, so that the trainee has a better understanding of the recommended actions and can generalize them to similar situations [1]. This work is motivated by an application for training power plant operators. Under emergency conditions, a power plant operator has to assimilate a great amount of information to promptly analyze the source of the problem and take the corrective actions. Novice operators might not have enough experience to take the best action; and experienced operators might forget how to deal with emergency situations, as these occur sporadically. So in both cases, an intelligent assistant can help to train the operators so they can react appropriately when an emergency situation arises; and in particular it is important that the assistant explains to the user the recommendations generated by a Markov decision process.

A. Hernández Aguirre et al. (Eds.): MICAI 2009, LNAI 5845, pp. 51–62, 2009.

Although there has been extensive work on explanation generation for rule-based systems and other representations, there is very little work on explanations using probabilistic representations, in particular for decision–theoretic models such as Markov decision processes (MDPs).

We have developed an automatic generation explanation mechanism for intelligent assistants. For this, we analyzed the explanations given by the domain expert, and designed a general template. This template is composed of three main parts: (i) the recommended action and the relevant variable in the current situation; (ii) a graphical representation of the process highlighting the relevant variable, and (iii) a verbal explanation. To generate the information required to fill the template we combine several knowledge sources. The optimal action is obtained from the MDP that guides the operator in the training session. The relevant variable, which is a key element in the explanation, is obtained from the MDP by analyzing which variable has the highest impact on the utility given the current state. The graphical part is generated from a general block diagram of the process, where the relevant variable is highlighted. The verbal explanation is obtained from a domain knowledge–base (KB) that includes the main components, actions and variables in the process, represented as a frame system. To extract the important information from the KB, we use the information from the MDP (state, action and relevant variables) as pointers to the relevant components, actions and variables; and then follow the links in a frame system to extract other relevant information. The information in the explanation template depends on the level of the operator: novice, intermediate or expert; which is obtained from a simple student model.

Although the explanations are motivated by an application in training power plant operators, the mechanism can be applied in other domains in which a person is being trained or guided by an assistant based on MDPs. For instance, training operators in other industrial environments, training pilots or drivers, etc. This work is one of the first efforts in explaining the actions recommended by an MDP, which main contributions are the selection of the relevant variable, and the combination of the MDP and domain knowledge to generate explanations. We have evaluated the explanations generated in a realistic application in the power plan domain.

2   Related Work

The work on explanations based on probabilistic models can be divided according to the classes of models considered, basically Bayesian networks (BN's) and decision-theoretic models. Two main strategies have been proposed for explanation with BN's. One strategy is based on transforming the network to a qualitative representation, and using this more abstract model to explain the relations between variables and the inference process [2,3]. The other strategy is based on the graphical representation of the model, using visual attributes (such as colors, line widths, etc.) to explain relations between nodes (variables) as well as the the inference process [4].

Influence diagrams (IDs) extend BNs by incorporating decision nodes and utility nodes. The main objective of these models is to help in the decision making process, by obtaining the decisions that maximize the expected utility. So explanation in this case has to do with understanding why some decision (or sequence of decisions) is optimal given the current evidence. There is very little work on explanations for IDs. Bielza *et al.* [5] propose an explanation method for medical expert systems based on IDs. It is based on reducing the table of optimal decisions obtained from an ID, building a list that clusters sets of variable instances with the same decision. They propose to use this compact representation of the decision table as a form of explanation, showing the variables that have the same value as a rule for certain case. It seems like a very limited form of explanation, difficult to apply to other domains. The explanation facilities for Bayesian networks proposed by Lacave et al. [4] were extended to IDs and integrated in the Elvira software [6]. The extension is based on a transformation of the ID into a BN by using a strategy for the decisions in the model. Lacave *et al.* [6] describe several facilities: incorporating evidence into the model, the conversion of the influence diagram into a decision tree, the possibility of analyzing non-optimal policies imposed by the user, and sensitivity analysis with respect to the parameters.

MDPs can be seen as an extension of decision networks, that consider a series of decisions in time (dynamic decision network). Some factored recommendation systems use algorithms to reduce the size of the state space [7] and perform symbolic manipulations required to group similarly behaving states as a pre-processing step. [8] also consider top-down approaches for choosing which states to split in order to generate improved policies [9]. Recently [10] proposed an approach for the explanation of recommendations based on MDPs. They define a set of preferred scenarios that correspond to set of states with high expected utility, and generate explanations in terms of actions that will produce a preferred scenario based on predefined templates. They demonstrate their approach in the domain of course selection for students, modeled as a finite horizon MDP with three time steps. In contrast, out approach considers an infinite-horizon and incorporates domain knowledge in the explanations generated.

In summary, there is very limited previous work on explanation generation for intelligent assistants systems based on MDPs. In particular, there is no previous work on determining the most relevant variable, which is a key element in the explanation mechanism we propose, and in integrating the information from the MDP and domain knowledge to generate explanations that are understandable for a user with no previous knowledge on probability or decision theory.

3   Factored Markov Decision Processes

A Markov decision process [11] models a sequential decision problem, in which a system evolves in time and is controlled by an agent. The system dynamics is governed by a probabilistic transition function Φ that maps states $\mathbf{S}$ and actions

A to new states **S'**. At each time, an agent receives a reward R that depends on the current state s and the applied action a. Thus, the main problem is to find a control strategy or *policy* π that maximizes the expected reward V over time. For the discounted infinite-horizon case with any given discount factor γ, there is a policy π^* that is optimal regardless of the starting state and that satisfies the *Bellman* equation [12]:

$$V^\pi(s) = max_a\{R(s,a) + \gamma \sum_{s' \in \mathbf{S}} P(s'|s,a)V^\pi(s')\} \tag{1}$$

Two methods for solving this equation and finding an optimal policy for an MDP are: (a) dynamic programming and (b) linear programming [11].

In a factored MDP [13], the set of states is described via a set of random variables $\mathbf{S} = \{X_1, ..., X_n\}$, where each X_i takes on values in some finite domain $Dom(X_i)$. A state $\mathbf{x}$ defines a value $x_i \in Dom(X_i)$ for each variable X_i. Dynamic Bayesian networks (DBN) are used to describe the transition model concisely, so the post-action nodes (at the time $t+1$) contain smaller matrices with the probabilities of their values given their parents' values under the effects of an action.

4   Automatic Explanation Generation

4.1   Overview

The explanation generation mechanism is inspired on the explanations provided by a domain expert, and it combines several knowledge sources: (i) the MDP that represents the process and defines the optimal actions, (ii) a domain knowledge base, (iii) a set of templates, and (iv) an operator model.

The explanation system consists of three main stages (see Fig. 1). The relevant variable is the key factor to build an explanation, therefore in the first stage the relevant variable is obtained and additional elements as the current state S_i and the optimal action a^*. In a second stage, according to the operator model (novice, intermediate or advanced), a template is selected. In the last stage, the template is filled–in with information from a domain knowledge–base. Each of these stages is detailed in the following sections.

4.2   Explanation Template

The explanations are based on a predefined structure or *template*. Each template has a predefined structure and is configured to receive additional information from different sources. There are three types of templates according to the user: novice, intermediate and advanced. The main difference is the amount and depth of the information provided. For novice users is more detailed while for advance operators a more concise explanation is given.

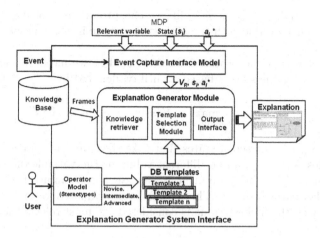

Fig. 1. Block diagram of the explanation generation system. When the trainee makes an error (event), the current state, optimal action and relevant variable are obtained from the MDP. These are used as pointers to obtain the relevant elements from the KB to fill–in an explanation template that is according to the user's model.

The explanation template has three main components (see section 5):

1. The optimal action given the current state.
2. A description in natural language of the main reasons for the previous action; which depends on the user level.
3. An schematic diagram of the process, highlighting the relevant variable.

The optimal action is obtained from the MDP given the current state of the plant, and the schematic diagram is previously defined for the process (or relevant subprocess). Thus, the two aspects that require a more complex procedure are the relevant variable and the verbal component, which are presented next.

4.3   Relevant Variable Selection

The strategy for automatic explanation generation considers as a first step, to find the most relevant variable, V_R, for certain state s and action a. All the explanations obtained from the experts are based on a variable which they considers the most important under current situation and according to the optimal policy. Examples of these explanations in the power plant domain are given later. We expect that something similar may happen in other domains, so discovering the relevant variable is an important first step for explanation generation.

Intuitively we can think that the relevant variable is the one with greatest effect on the expected utility, given the current state and the optimal policy. So as an approximation to estimating the impact of each factor X_i in the utility, we estimate how much the utility, V, will change if we vary the value for each variable, compared to the utility of the current state. This is done by maintaining all the other variables, X_j, $j \neq i$, fixed. The process is repeated for all the variables, and the variable with the highest difference in value is selected as the relevant variable.

Let us assume that the process is in state s, then we measure the relevance of a variable X_i for the state s based on utility, denoted by $rel_s^V(X_i)$, as:

$$rel_s^V(X_i) = \max_{s' \in neigh_{X_i}(s)} V(s') - \min_{s' \in neigh_{X_i}(s)} V(s') \qquad (2)$$

where $neigh_{X_i}(s)$ is the set of states that take the same the values as s for all other variables X_j, $j \neq i$; and a different value for the variable of interest, X_i. That is, the maximum change in utility when varying the value of X_i, maintaining all the other variables fixed. This expression is evaluated for all the variables, and the one with the highest value is considered the most relevant for state s:

$$X_R^V = argmax_i(rel_s^V(X_i)), \forall(i) \qquad (3)$$

4.4   Knowledge Base

To complement the information obtained from the MDP model, additional domain knowledge is required about relevant concepts, components and actions in the process. A *frame* [14] is a data structure with typical knowledge about a particular concept. Frames provide a natural way for representing the relevant elements and their relations, to fill–in the explanations templates.

In the KB, the frames store the basic knowledge about the domain components, variables, actions, and their relationships (see Fig. 2). This representation is an extension of the one proposed by [15]. The KB is conformed by three hierarchies: (i) procedures and actions, (ii) components, and (iii) variables. It also includes relationships between the frames in different hierarchies: which actions affect each component and the variables associated to each component.

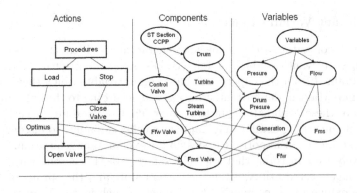

Fig. 2. The KB represented as a frame hierarchy; it is divided in 3 parts: (i) actions, (ii) components, and (iii) variables

Each frame contains general knowledge for each procedure, component or variable; which is relatively easy to obtain from written documentation or domain experts. The advantage of this representation is that the same KB can be used to generate explanations for any procedure related to the specific domain.

4.5   Filling the Template

A template is selected according to the user level. The optimal action and relevant variable are deduced from the MDP model of the process and the policy, and together with the process schematic diagram are inserted in the corresponding template (Complete algorithm is depicted in figure 3). The missing element is the textual explanation of *why* the action should be selected in the current situation. The explanation mechanism must then determine *what* should be included and *how much* detail to give to the user. *What* is determined by the template structure and *how much* by the operator level.

Deduced from the expert explanations, the following elements are defined for the textual explanation structure:

1. Optimal action, including information on what is the purpose of the action obtained from the corresponding frame.
2. Relevant variable, with information of the relevance of this variable and its relation to certain component.
3. Component, including its main characteristics and its relation to the action.

```
ExplanationGeneration (a*,V_R, s, Comp, Exp, UsrLev)
    For each event e
    For each a*(s), V_R(x_i), s ∈ S, Comp, Exp, UsrLev;
    Where: a* = optimal action; V_R = relevant variable; s = current state;
           Comp = component; Exp = explanation; UsrLev = user level;
    IF e=true THEN {
            Get from MDP: a*(s)
            Get from ProbabilisticModel: V_R
            Get from Simulator: s
            Get from OperatorModel: UsrLev
            IF UsrLev = Novice THEN {
                Select Template = novice
                Get Frame(a*, Comp, V_R)
                Fill Template = Frame(a*, Comp, V_R) & s
                Exp = Template }
            ELSE IF UsrLev = Intermediate THEN {
                Select Template = Intermediate
                Get Frame(a*, V_R)
                Fill Template = Frame(a*, V_R) & s
                Exp = Template }
            ELSE IF UsrLev = Advanced THEN {
                Select Template = Advanced
                Get Frame(a*, V_R)
                Fill Template = Frame(a*, V_R)
                Exp = Template }
        }
    e = False;
    Return Exp;
```

Fig. 3. High–level algorithm for explanation generation

The elements obtained from the MDP, state, action and relevant variables, are used as pointers to the corresponding frames in the KB from where the basic elements of the textual explanation are extracted. For intermediate and novice users the textual explanation is extended by following the links in the KB hierarchies, adding information from frames in the upper part of the hierarchy. For instance, if the important component is the *feed water valve*, it includes additional information from the *control valve* frame for intermediate users, and more general knowledge on *valves* for the novice user. In this way, the amount of detail in the explanation is adapted to the user level.

5 Experimental Results

First we describe the intelligent assistant in which we tested our method for explanation generation, and then the experiments comparing the automatic explanations against those given by a domain expert.

5.1 Intelligent Assistant for Operator Training

We have developed an intelligent assistant for operator training (IAOT) [16]. The input to the IAOT is a policy generated by an MDP, which establishes the sequence of actions that will allow to reach the optimal operation of a steam generator. Operator actions are monitored and discrepancies are detected regarding the operator's expected behavior. The process starts with an initial state of the plant, usually under an abnormal condition; so the operator should return the plant to its optimum operating condition using some controls. If the action performed by the operator deviates from the optimal plan, either in the type of action or its timing, an explanation is generated.

We considered a training scenario based on a simulator of a combined cycle power plant, centered in the drum. Under certain conditions, the drum level becomes unstable and the operator has to return it to a safe state using the control valves. The variables in this domain are: (i) drum pressure (Pd), (ii) main steam flow (Fms), (iii) feed water flow (Ffw), (iv) generation (G), and (v) disturbance (this variable is not relevant for the explanations so is not included in the experiments). There are 5 possible actions: a0–do nothing, a1–increase feed water flow, a2–decrease feed water flow, a3–increase steam flow, and a4–decrease steam flow.

We started by defining a set of explanation units with the aid of a domain expert, to test their impact on operator training. These explanation units are stored in a data base, and the assistant selects the appropriate one to show to the user, according to the current state and optimal action given by the MDP. An example of an explanation unit is given in Figure 4.

To evaluate the effect of the explanations on learning, we performed a controlled experiment with 10 potential users with different levels of experience in power plant operation. An analysis of the results [16] shows a significant

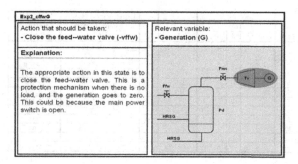

Fig. 4. An example of an explanation unit defined by a domain expert. In this example the relevant variable is *generation*, G, as the absence of generation is the main reason to close the feed–water valve.

difference in favor of the group with explanations. These results give evidence that explanations help in the learning of skills such as those required to operate an industrial plant.

Next we analyze the results of the automatic explanation generation mechanism, in terms of a comparison with the human–generated explanations.

5.2   Results

Relevant Variable Selection. In a first stage we evaluated the selection of the most relevant variable, and compared these to the ones given by the domain expert. In the power plant domain there are 5 state variables with a total of 384 states. We analyzed a random sample of 30 states, nearly 10% of the total number of states. For the 30 cases we obtained the most relevant variable(s); and compared these with the relevant variables given in the explanation units provided by the expert.

Figure 5 presents the results of 11 of the 30 cases (the rest are omitted due to space limitations). For each case we show: (i) the current state, (ii) the value, (iii) the optimal action, (iv) the variable selected according to the change in utility, including this change, and (v) the relevant variable(s) given by the expert. Note that for some cases the expert gives two relevant variables. The system selects in 100% of the 30 cases one of the most relevant variables according to the expert. In the future we plan to extend the relevant variable selection mechanism so we can obtain more than one relevant variable.

Explanation Generation. To evaluate the explanations generated by the system, we generated explanations for different situations, and compared this to the explanations given for the same state by the domain expert. Figure 6 depicts an example of an explanation template generated. It contains on the left side: (i) the optimal action, (ii) why is important to do this action, (iii) which component is related with optimal action, and (iv) a description of the current state. In the right side the relevant variable is highlighted in an schematic diagram of the process.

Test	selected S Var fms	ffw	d	pd	g	U	Π^*	Experimental results Changes in Utility $|\Delta U|$	Relevant Var Selected by Expert
1	0	0	0	0	1	2601.29	a1	g = 2132.44	g,fms
2	0	0	1	3	1	2514.00	a2	g =2235.79	g,fms
3	1	1	0	4	0	796.95	a2	fms = 2413.53	fms, g
4	2	0	0	7	1	3488.60	a4	pd =2762.60	pd, fms
5	3	0	0	0	1	3295.55	a3	g =2427.94	g, pd
6	3	1	0	2	1	3053.19	a2	g = 2109.73	g, pd
7	4	0	1	0	1	2986.18	a1	g = 2257.11	g, pd
8	4	1	1	7	0	843.27	a4	pd = 3271.43	pd, g
9	5	1	0	1	1	3632.66	a0	fms = 3720.05	fms
10	5	1	1	1	1	3287.13	a0	fms = 3642.53	fms
...	...	...	...	...	...	...	...	...	...
30	5	0	0	5	1	2761.32	a4	fms = 2116.43	fms

Fig. 5. Results for the relevant variable selection phase. Is shows for each test case: the state (random variables), the expected value (U), the optimal action (Π^*), the relevant variable with $|\Delta U|$, and the variable(s) selected by the expert.

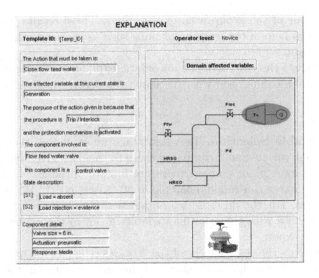

Fig. 6. An example of an explanation template generated

To evaluate explanations generated by the system, the the methodology proposed by [17] was adapted. The main idea is that an expert panel compares explanations given by the system to those given by a human, in terms of some evaluation criterium. It considers four aspects: (i) coherence, (ii) contents, (iii) organization, and (iv) precision. This not intend to generate explanations in natural language. Each criterium is evaluated in a quantitative scale: 1=bad, 2=regular, 3=good, 4=excellent. Each expert is presented with two explanations, side by side, for the same case; one given by a human expert, the other generated by the system. It is asked to grade each aspect for the generated explanations taking as reference the expert's explanations. 15 different explanations

CRITERIA	CASES	EVALUATION PANEL						SCORE	
		Eval 1	Eval 2	Eval 3	Eval 4	Eval 5	Eval 6	Average	Ranking
COHERENCE	C1 to C15	3.67	3.47	2.47	3.80	3.67	3.07	3.36	2°
CONTENT	C1 to C15	2.00	3.40	2.80	3.80	3.87	3.53	3.20	4°
ORGANIZATION	C1 to C15	2.87	2.93	2.93	4.00	3.93	3.67	3.39	1°
CORRECTNESS	C1 to C15	4.00	2.87	2.47	3.80	3.67	2.87	3.24	3°
Total		3.13	3.17	2.67	3.75	3.78	3.28	3.30	

Fig. 7. Results of the evaluation by a panel of experts. The table shows the average score per criteria–evaluator, and the totals for each dimension.

were considered in a variety of situations for the power plan domain, and asked 6 different external domain experts to evaluate them. Results are summarized in figure 7.

Results are between *good* and *excellent* in average for all the aspects and for all the experts. The criterium with highest score is *organization* and the lowest *contents*, although the gap is small among all the aspects. We consider that these are encouraging results, as we are comparing the explanations generated automatically against those given by an experienced domain expert, a very high standard. Given these results, and the previous results on the impact of the explanations on learning [16], a similar impact is expected with automatically generated explanations; however, the plan is to conduct an additional user study in the future.

6   Conclusions and Future Work

We have developed a novel explanation generation mechanism for the actions recommended by an intelligent assistant based on an MDP. For this we developed and algorithm to determine the most relevant variable for certain state–action; and combine this information with domain knowledge to generate explanations based on a general template and adapted according to the user level. We have evaluated the explanation generation system in the power plant domain with good results. We consider that this mechanism can be easily extended to other training and assistant systems based on decision-theoretic models.

Currently the explanations are centered on why the recommended action should be selected, but not on why other actions are not appropriate. Another limitation is that it considers only one optimal plan, and there could be alternative solutions. In the future we plan to extend the system to consider other types of explanations, and to apply it to other domains.

References

1. Herrmann, J., Kloth, M., Feldkamp, F.: The role of explanation in an intelligent assistant system. In: Artificial Intelligence in Engineering, vol. 12, pp. 107–126. Elsevier Science Limited, Amsterdam (1998)
2. Druzdzel, M.: Explanation in probabilistic systems: Is it feasible? Will it work? In: Intelligent information systems V, Proc. of the workshop, Poland, pp. 12–24 (1991)

3. Renooij, S., van der Gaag, L.: Decision making in qualitative influence diagrams. In: Proceedings of the Eleventh International FLAIRS Conference, pp. 410–414. AAAI Press, Menlo Park (1998)
4. Lacave, C., Atienza, R., Díez, F.J.: Graphical explanations in Bayesian networks. In: Brause, R., Hanisch, E. (eds.) ISMDA 2000. LNCS, vol. 1933, pp. 122–129. Springer, Heidelberg (2000)
5. Bielza, C., del Pozo, J.F., Lucas, P.: Optimal decision explanation by extracting regularity patterns. In: Coenen, F., Preece, A., Macintosh, L. (eds.) Research and Development in Intelligent Systems XX, pp. 283–294. Springer, Heidelberg (2003)
6. Lacave, C., Luque, M., Díez, F.J.: Explanation of Bayesian networks and influence diagrams in Elvira. IEEE Transactions on Systems, Man and Cybernetics—Part B: Cybernetics 37, 952–965 (2007)
7. Givan, R., Dean, T., Greig, M.: Equivalence notions and model minimization in markov decision processes. Artif. Intell. 147(1-2), 163–223 (2003)
8. Dean, T., Givan, R.: Model minimization in markov decision processes. In: AAAI (ed.) Proceedings AAAI 1997, pp. 106–111. MIT Press, Cambridge (1997)
9. Munos, R., Moore, A.W.: Variable resolution discretization for high-accuracy solutions of optimal control problems. In: IJCAI 1999: Proceedings of the Sixteenth International Joint Conference on Artificial Intelligence, pp. 1348–1355. Morgan Kaufmann Publishers Inc., San Francisco (1999)
10. Khan, O.Z., Poupart, P., Black, J.: Explaining recommendations generated by MDPs. In: de Roth-Berghofer, T., et al. (eds.) 3rd International Workshop on Explanation-aware Computing ExaCt 2008, Proceedings of the 3rd International ExaCt Workshop, Patras, Greece (2008)
11. Puterman, M.: Markov Decision Processes: Discrete Stochastic Dynamic Programming. Wiley, New York (1994)
12. Bellman, R.E.: Dynamic Programming. Princeton U. Press, Princeton (1957)
13. Boutilier, C., Dean, T., Hanks, S.: Decision-theoretic planning: structural assumptions and computational leverage. Journal of AI Research 11, 1–94 (1999)
14. Minsky, M.: A framework for representing knowledge. In: Winston, P. (ed.) The Psychology of Computer Vision. McGraw-Hill, New York (1975)
15. Vadillo-Zorita, J., de Ilarraza, A.D., Fernández, I., Gutirrez, J., Elorriaga, J.: Explicaciones en sistemas tutores de entrenamiento: Representacion del dominio y estrategias de explicacion, Pais Vasco, España (1994)
16. Elizalde, F., Sucar, E., deBuen, P.: A Prototype of an intelligent assistant for operator's training. In: International Colloquim for the Power Industry, Mexico, CIGRE-D2 (2005)
17. Lester, J.C., Porter, B.W.: Developing and empirically evaluating robust explanation generators: the knight experiments. Computational Linguistics 23(1), 65–101 (1997)

Compiling Multiply Sectioned Bayesian Networks: A Comparative Study

Xiangdong An and Nick Cercone

Department of Computer Science and Engineering
York University, Toronto, ON M3J 1P3, Canada
xan@cs.yorku.ca, ncercone@yorku.ca

Abstract. Inference with multiply sectioned Bayesian networks (MSBNs) can be performed on their compiled representations. The compilation involves cooperative moralization and cooperative triangulation. In earlier work, agents perform moralization and triangulation separately and the moralized subgraphs need to be made consistent to be the input of the triangulation. However, the set of moralized subnets is only an intermediate result, which is of no use except as the input to the triangulation. On the other hand, combining moralization and triangulation won't make the compilation complex but simpler and safer. In this paper, we first propose a change to the original algorithm (the revised algorithm), which is supposed to provide higher quality compilation, then we propose an algorithm that compiles MSBNs in one process (the combined compilation), which is supposed to provide lower quality compilation, however. Finally, we empirically study the performance of all these algorithms. Experiments indicate that, however, all 3 algorithms produce similar quality compilations. The underlying reasons are discussed.

1 Introduction

Multiply sectioned Bayesian networks (MSBNs) [1,2] provide a coherent framework for probabilistic inference in cooperative multiagent interpretation systems. They support object-oriented inference [3] and have been applied in many areas such as medical diagnosis [4], distributed network intrusion detection [5], and distributed collaborative design [6]. Inference in MSBNs can be performed in their compiled representations called *linked junction forests* (LJFs) [7]. A LJF is a collection of junction trees (JTs) [8] each of which corresponds to a Bayesian subnet in the MSBN. Corresponding to the set of interface nodes between each pair of adjacent agents, there exists a JT, called *linkage tree* [9]. A clique in a linkage tree is called a *linkage*. A linkage tree needs to be consistent with both JTs incident to it so that messages can be passed in both directions. This requires that multiple agents compile their Bayesian subnets cooperatively. On the other hand, to protect agents' privacy, we could not assemble these subnets in a central location and then compile the union. Similar to the compilation of BNs, moralization and triangulation are the two most important compilation steps.

In earlier work [2], a suite of algorithms is presented for each of the cooperative moralization and the cooperative triangulation. Each suite is composed of two recursive algorithms performing operations in a depth first traversal of the *hypertree*, which

A. Hernández Aguirre et al. (Eds.): MICAI 2009, LNAI 5845, pp. 63–74, 2009.

we call the *traversal moralization* or the *traversal triangulation*. Here, a hypertree is a tree structure that organizes the subnets, In this paper, we are interested in examining two issues related with the two suites of algorithms. First, in *DepthFirstEliminate* of the traversal triangulation, fill-ins received by an agent from its caller are not passed to its callees. Though such fill-ins would eventually be distributed to the relevant agents in *DistributeDlink*, it might be better to do so early since the heuristics used in triangulation could have made better decision if they were delivered early.[1] Second, the cooperative triangulation starts only after the cooperative moralization has finished. The moralization needs to be done cooperatively because the cooperative triangulation cannot work on inconsistent moral subnets and the moralized subnets need to be made consistent with each other. However, the set of consistently moralized subnets is of no use except as the input to the cooperative triangulation. By a closer look at the two suites of algorithms, they actually can be unified. After unification, the moralization process can be done locally at each subnet, and the shared moralization fill-ins can be distributed in the triangulation process. This would make the compilation process simpler. Since the moralization fill-ins would be distributed with the triangulation fill-ins, the combined compilation may require less message passings. In distributed problem domains, communication among agents are not cost, trouble or risk-free. Reduced communication would help minimize costs and risks. Nevertheless, the unification would defer the delivery of the moralization fill-ins, which could downgrade the triangulation quality as we discussed earlier. In this work, we first propose algorithms to address these issues, then we empirically study their performances. The experimental results tell us a different story, however. The underlying reasons are analyzed.

The rest of the paper is organized as follows. In Section 2, we briefly introduce the graph-theoretical terminologies used in the paper. In Section 3, we give a review of MSBNs. The requirements for MSBN compilation are then presented in Section 4. We first in Section 5 present the revised algorithm to address the first issue, and then in Section 6 present the combined compilation to address the second issue. In Section 7, we empirically study all algorithms. The conclusion is given in Section 8.

2 Graph-Theoretical Terminologies

A graph is a pair $G = (V, E)$, where V denotes a set of nodes (vertices), and E a set of edges (arcs, links), where two nodes u and v are *adjacent* if $(u, v) \in E$. A directed graph is *acyclic* if it contains no directed cycles, which is often referred to as a *DAG*. *Moralization* converts a DAG into a moral (moralized) graph by pairwise connecting all parents of a child and dropping the direction of all arcs. In a graph, a set of nodes is *complete* if they are adjacent to each other and a maximal set of nodes that are complete is called a *clique*. A node in a graph is *eliminable* if the set of its adjacent nodes is complete. An ineliminable node can be made eliminable by adding links called *fill-ins*. An eliminable node is *eliminated* if it and all the links incident to it are removed. A graph is *triangulated* if and only if all nodes can be eliminated one by one in some order without adding any fill-ins. A *chord* in a graph is a link connecting two non-adjacent nodes. G is *triangulated* if every cycle of length > 3 has a chord.

---

[1] Finding minimum fill-ins is NP-complete [10].

A *junction tree* (JT) over a set of nodes is a cluster tree where each *cluster* represents a subset of nodes and each link (called a *separator*) the intersection of its incident clusters such that the intersection of any two clusters is contained in every separator on the path between them.

3   Overview of MSBNs

A BN is a triplet (U, G, P), where U is a set of domain variables, G is a DAG whose nodes are labeled by elements of U, and P is a joint probability distribution (JPD) over U. G qualitatively while P quantitatively encodes conditional dependencies among variables in U. In an MSBN, a set of $n > 1$ agents $A_0, A_1, ..., A_{n-1}$ populates a total universe U of variables. Each A_i has knowledge over a subdomain $U_i \subset U$ encoded as a *Bayesian subnet* (U_i, G_i, P_i). The collection $\{G_0, G_1, ..., G_{n-1}\}$ of local DAGs encodes agents' qualitative knowledge about domain dependencies. Local DAGs should overlap and be organized into a tree structure, called a *hypertree*. Any two adjacent agents on the hypertree exchange their information over the overlapped variables, called their *interface*.

Definition 1. [2] *Let $G = (V, E)$ be a connected graph sectioned into connected subgraphs $\{G_i = (V_i, E_i)\}$. Let these subgraphs be organized into a connected tree Ψ where each node, called a* hypernode, *is labeled by G_i and each link between G_i and G_j, called a* hyperlink, *is labeled by the interface $V_i \cap V_j$ such that for each pair of nodes G_l and G_m, $V_l \cap V_m$ is contained in each subgraph on the path between G_l and G_m. The tree Ψ is called a* hypertree *over G.*

Definition 1 indicates that a hypertree has the running intersection property [11] of a JT. For a hyperlink to separate its two directed branches, it has to be a *d-sepset* [2].

Definition 2. [2] *Let G be a directed graph such that a hypertree over G exists. A node x contained in more than one subgraph with its parents $\pi(x)$ in G is a* d-sepnode *if there exists a subgraph that contains $\pi(x)$. An interface I is a* d-sepset *if every $x \in I$ is a d-sepnode.*

The overall structure of an MSBN is a hypertree MSDAG.

Definition 3. [2] *A hypertree MSDAG $G = \bigcup_i G_i$, where each $G_i = (V_i, E_i)$ is a DAG, is a connected DAG such that there exists a hypertree over G and each hyperlink is a d-sepset.*

In an MSBN, each agent holds its partial perspective of a large problem domain, and has access to a local evidence source (sensors). Global evidence can be obtained by communicating with other agents. Agents update their beliefs with their local evidence and global information from other agents, and then answer queries or take actions based on the updated beliefs. Figure 1 illustrates the structure of an MSBN and its hypertree organization, where in (a) each dotted box represents a Bayesian subnet and each interface node is highlighted with one circle, and in (b) each rectangular box with rounded corner denotes a hypernode. In an MSBN, the interface nodes are *public* to the corresponding agents but *private* to all other agents; all non-interface nodes are only known to their host agents, and *private* to all other agents. Agents' privacy forms the constraint of many operations in an MSBN, e.g. triangulation [9] and communication [12].

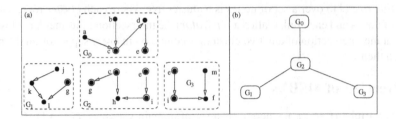

Fig. 1. (a) The structure of an MSBN; (b) The hypertree of the MSBN

4   Compilation Requirements

4.1   General Requirements

For message passing over the interfaces, Requirement 1 should be followed.

Requirement 1. *[2] For a Bayesian subnet G over a set of nodes N which has an interface I with one of its adjacent subnets, G must be eliminable in the order $(N \setminus I, I)$.*

For message passing in both directions based on the same linkage tree, Requirement 2 needs to be followed.

Requirement 2. *[2] Let H be a hypertree MSDAG $G = \bigcup_{i=0}^{n-1} G_i$. Let G'_i $(0 \leq i < n)$ be the chordal graph of G_i $(0 \leq i < n)$ obtained from the compilation of H. Then each pair of G'_i and G'_j should be graph-consistent.*

4.2   An Implicit Requirement

Fill-ins can be added in local moralization, which does not guarantee global graph consistency. For example, Figure 2 (a) shows an MSBN example where two subnets share nodes a, b, and c. During local moralization in (b), a fill-in (the dotted line) is added between nodes a and b in only one subnet.

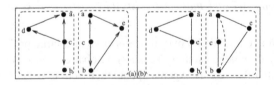

Fig. 2. (a) A hypertree MSDAG with two subnets; (b) The fill-in between nodes a and b is added in only one subnet during local moralization

However, the suite of traversal triangulation algorithms requires that the moral subgraphs of a hypertree MSDAG be graph consistent; otherwise the hypertree MSDAG may not be properly triangulated. For example, we use the subgraphs in Figure 2 (b) as the input of the traversal triangulation algorithm, and start the triangulation from the subset on the left. The elimination order d, b, c, a will not produce any fill-ins. Therefore nothing will be sent to the subnet on the right. For subnet on the right, an elimination order e, a, b, c will not produce any fill-ins and hence will not pass any fill-ins to the subnet on the left. After the triangulation finishes, the two subgraphs remain inconsistent. The similar problem would occur if all DAGs of the input MSDAG to the moralization algorithm is not graph-consistent. That is, an implicit requirement for traversal compilation is that the moralization fill-ins need to be properly distributed.

Requirement 3. *Let H be a hypertree MSDAG $G = \bigcup_{i=0}^{n-1} G_i$. Let G_i' $(0 \le i < n)$ be the moral graph of G_i $(0 \le i < n)$ obtained from the moralization of H. Then, each pair of G_i' and G_j' should be graph-consistent.*

5   The Revised Compilation

In this section, we first review the traversal compilation and the traversal triangulation [2], then propose a change to the algorithm *DepthFirstEliminate*, which is supposed to improve the compilation performance.

The traversal moralization contains 3 algorithms: *CoMoralize, CollectMlink* and *DistributeMlink*. In the beginning, the system coordinator needs to select a root agent to activate the moralization process. Then, *CollectMlink* is started at the root agent, and after it finishes, *DistributeMlink* is started at the same agent. Both *CollectMlink* and *DistributeMlink* are recursive algorithms. In *CollectMlink*, multiple agents are called one by one in a depth-first traversal order of the hypertree to perform local moralization and collect fill-ins from the callees. In *DistributeMlink*, agents are recursively called in a depth-first traversal order of the hypertree to distribute fill-ins collected in *CollectMlink*, which are maintained in variable LINK by each agent.

The traversal triangulation contains 4 algorithms: *SafeCoTriangulate, CoTriangulate, DepthFirstEliminate*, and *DistributeDlink*. From *SafeCoTriangulate, CoTriangulate* is called to perform a round of triangulation. After *CoTriangulate* finishes, whether the triangulation is finished with both Requirements 1 and 2 satisfied is examined. If not, *CoTriangulate* is restarted; otherwise the process finishes. From algorithm *CoTriangulate*, an agent needs to be selected by the system coordinator to start one round of triangulation. *DepthFirstEliminate* and *DistributeDlink* need to be activated at the same agent one after another. Both *DepthFirstEliminate* and *DistributeDlink* are recursive algorithms, where agents are called to perform triangulation or distributed fill-ins relative to each interface in an order of the depth first traversal of the hypertree.

For the readers' convenience, here we copy *DepthFirstEliminate* and *DistributeDlink* as below, where the agent that is called to execute an algorithm is denoted by A_i corresponding to the subnet G_i. A caller is either an adjacent agent or the system coordinator. If the caller is an adjacent agent, we denote it by A_i^c. Denote the additional adjacent agents of A_i by $A_i^1, ..., A_i^{m-1}$. Denote interface $V_i^c \cap V_i$ by I_i^c and interfaces $V_i \cap V_i^j$ by I_i^j $(j = 1, ..., m - 1)$. In *DepthFirstEliminate*, all fill-ins forwarded to the callees by A_i are from LINK. However, it is set to $\emptyset$ at line 4. This indicates that the set F_i^c of fill-ins A_i received from its caller A_i^c (lines 1 to 3), if there is any, would not go farther than A_i. Since at line 4 of *DistributeDlink*, LINK is set to the set of all fill-ins added to G_i so far, F_i^c will be distributed properly at that time. That is, there is a delay in the delivery of fill-ins. This delay could affect the compilation quality since the heuristics used in triangulation could have made better decision if the information has been made available at line 6 of *DepthFirstEliminate* earlier. This is illustrated in Figure 3.

Algorithm 1 (DepthFirstEliminate). *[2] When the caller calls on A_i, it does the following:*

1 if caller is an agent A_i^c, do
2   receive a set F_i^c of fill-ins over I_i^c from A_i^c;
3   add F_i^c to G_i;
4 set LINK=∅;
5 for each agent A_i^j (j=1,...,m-1), do
6   eliminate V_i in the order $(V_i \setminus I_i^j, I_i^j)$ with fill-ins F;
7   add F to G_i and LINK;
8   send A_i^j the restriction of LINK to I_i^j;
9   call A_i^j to run **DepthFirstEliminate** *and receive fill-ins F' from A_i^j when done;*
10   add F' to G_i and LINK;
11 if caller is an agent A_i^c, do
12   eliminate V_i in the order $(V_i \setminus I_i^c, I_i^c)$ with fill-ins F_c';
13   add F_c' to G_i and LINK;
14   send A_i^c the restriction of LINK to I_i^c;

Algorithm 2 (DistributeDlink). *[2] When the caller calls on A_i, it does the following:*

1 if caller is an agent A_i^c, do;
2   receive a set F_i^c of fill-ins over I_i^c from A_i^c;
3   add F_i^c to G_i;
4 set LINK to the set of all fill-ins added to G_i so far;
5 for each agent A_i^j (j=1,...,m-1), do
6   send A_i^j the restriction of LINK to I_i^j;
7   call A_i^j to run **DistributeDlink;**

In Figure 3, the 3 subnets share a set of variables $\{a, b, c, d, e\}$ as shown in (a). They are organized as a hypertree G_0–G_1–G_2. After cooperative moralization, a fill-in is added between nodes b and e as shown in (b). Assume we activate *DepthFirstEliminate* at G_0, and a fill-in between c and e (the dotted line) is added in the triangulation relative to G_1. The fill-in will be forwarded to G_1. A_1 (on G_1) takes it, adds it to G_1 (the dotted line), and performs triangulation relative to G_2. Since no fill-in is produced in the triangulation, nothing is forwarded to G_2 whereas A_2 (over G_2) could triangulation G_2 by fill-in (b, d) (the dotted line). The fill-in (b, d) would be put in the LINK of A_2 and sent to A_1 (lines 11 to 14). A_1 receives it, adds it to G_1 (not drawn), puts it in its LINK (lines 9 and 10), and sends it to A_0. Similarly, A_0 adds it to G_0 (not drawn) and its LINK. In *DistributeDlink*, (c, e) will be distributed to G_2. That is, we get a clique of 4 nodes b, c, d, e. However, if we allow (c, e) to be forwarded to G_2 in *DepthFirstEliminate*, the heuristics may not add another fill-in (b, d). In that case, we would get smaller cliques, which is preferred for computational efficiency. This change may also potentially reduces the number of times *SafeCoTriangulate* is restarted since the cooperative triangulation could converge faster. Therefore, we propose to change line 3 of *DepthFirstEliminate* to "add F_i^c to G_i and set $LINK = F_i^c$;", and line 4 to "else set LINK=$\emptyset$;".

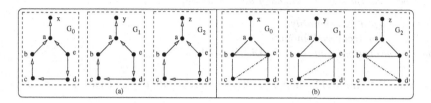

Fig. 3. (a) A hypertree MSDAG with two subnets; (b) The fill-in between nodes a and b is added in only one subnet during local moralization

6   The Unified Compilation

Since the fill-ins produced and distributed in the cooperative moralization are certain in themselves, we propose to delay the propagation of such fill-ins until the cooperative triangulation. Specifically, we combine the moralization and the triangulation in one process, and we do not propagate the fill-ins immediately after the local moralizations. Instead, we pass the local moralization fill-ins to the respective LINKs in the triangulation process so that they can be propagated with triangulation fill-ins. By the combination, we do not need to satisfy the implicit requirement for the traversal compilation. We also make the compilation process robust and fault-tolerant: any inconsistent intermediate results from faulty communications would be corrected eventually. However, since there is a delay in the distribution of the moralization fill-ins, we may expect a compilation of lower quality and slow speed.

The combined compilation is proposed as a suite of 4 algorithms: *SafeCompile*, *CoCompile*, *RecursiveEliminate*, and *DistributeDlink*.

Algorithm 3   (SafeCoCompile)

1 each agent performs local moralization and puts fill-ins in the respective LINK;
2 perform **CoCompile***;*
3 each agent performs an elimination relative to the d-sepset with each adjacent agent;
4 if no new fill-ins are produced, halt;
5 else set all LINKs to be $\emptyset$, go to 2;

In *SafeCoCompile*, local moralizations are firstly done (line 1) and the process for triangulation and fill-in propagation, called *CoCompile*, is then started (line 2). Then the compilation is examined to ensure both Requirements 1 and 2 are satisfied (lines 3-4) to finish. Otherwise, another round of compilation is started (line 5). When examining the success of compilation, any inconsistent results would also lead to another round of compilation. In *CoCompile*, both *RecursiveEliminate* and *DistributeDlink* are called one after another on the same root hypernode. In *RecursiveEliminate*, in addition to the cooperative triangulation, the local moralization fill-ins are also distributed. *DistributeDlink* is the same as Algorithm 2, by which all fill-ins are further distributed.

Algorithm 4 (CoCompile). *A hypertree MSDAG is populated by multiple agents with one at each hypernode. The system coordinator does the following:*

1 choose an agent A_* arbitrarily;
2 call A_* to run **RecursiveEliminate***;*
3 after A_* has finished, call A_* to run **DistributeDlink***;*

Algorithm 5 (RecursiveEliminate). *[2] When the caller calls on A_i, it does the following:*

1 if caller is an agent A_i^c, do
2   receive a set F_i^c of fill-ins over I_i^c from A_i^c;
3   add F_i^c to G_i and LINK;
4 for each agent A_i^j ($j=1,...,m-1$), do
5   eliminate V_i in the order $(V_i \setminus I_i^j, I_i^j)$ with fill-ins F;
6   add F to G_i and LINK;
7   send A_i^j the restriction of LINK to I_i^j;
8   call A_i^j to run **RecursiveEliminate** *and receive fill-ins F' over I_i^j from A_i^j when done;*
9   add F' to G_i and LINK;
10 if caller is an agent A_i^c, do
11   eliminate V_i in the order $(V_i \setminus I_i^c, I_i^c)$ with fill-ins F_c';
12   add F_c' to G_i and LINK;
13   send A_i^c the restriction of LINK to I_i^c;

In *CoCompile*, each of the n agents is exactly called twice. The local moralizations and particularly the local relative triangulations dominate the computation since message passing is minor compared with the moralization and particularly the elimination process. Let n be the number of hypernodes. For local moralization, let p be the cardinality of the largest possible parent set of an individual variable, and s be the cardinality of the largest subdomain. Then the moralization complexity at each agent is $O(sp^2)$, and the overall complexity of moralization is $O(nsp^2)$. For local relative triangulation, let d be the maximum degree of a node. To eliminate a node, the completeness of its adjacency is checked. The complexity of the checking is $O(d^2)$. Based on the heuristic that selects the node with the minimum number of fill-ins to eliminate, $O(s)$ nodes are checked before one is eliminated. So, the complexity of each local relative triangulation is $O(s^2d^2)$. Let m be the maximum degree of a hypernode. Then, the overall complexity of triangulation is $O(mns^2d^2)$. Therefore, the overall complexity of *CoCompile* is $O(nsp^2 + mns^2d^2)$. In the worst case, *SafeCoCompile* should be much more expensive than *CoCompile*. However, previous experiments on traversal triangulation indicated that generally one or two rounds of cooperative triangulation (by *CoTriangulate*) are needed to eventually triangulate an average case [9]. We believe this is also true for *CoCompile* to be called by *SafeCoCompile*. We would experimentally verify the belief in the next section. Note, with the combined compilation, we save a recursive process for cooperative compilation, which may help reduce the cost and the risk (e.g., faults in communication and security and privacy issues) of the communication.

For *SafeCoCompile*, we have Proposition 1.

Proposition 1. *Let* SafeCoCompile *be applied to a hypertree MSDAG. When all agents halt, Requirement 1 is satisfied.*

Proof: Line 3 of *SafeCoCompile* ensures that it would not halt until each subnet can be eliminated relative to each of its d-sepsets. It would finally halt anyway since in the worst case when all subnets are complete, the halt condition would be satisfied.   □

For *CoCompile*, we have Proposition 2.

Proposition 2. *Let* CoCompile *be applied to a hypertree MSDAG. When all agents halt, Requirement 2 is satisfied.*

Proof: For graph consistency, we need to show a fill-in (x, y) produced for a pair of disconnected d-sepnodes x and y by an agent A_i will be communicated to any other agent A_j that shares the pair. We virtually direct the hypertree from the subnet associated with $A*$ to each terminal subnet, then the path ρ between A_i and A_j can only be directed as in (a), (b), or (c) in Figure 4 since any other possible directions would result in more than one root. The direction is consistent with the forward direction of the depth-first traversal of the hypertree. Since on a hypertree MSDAG, a pair of nodes x and y shared by two subnets also appears on every subnet on the path between the two subnets, if ρ is directed as in Figure 4 (a) or (b), (x, y) will be passed to A_j in forward or backward direction of *RecursiveEliminate*. If ρ is directed as in Figure 4 (c), (x, y) would first be collected to R in *RecursiveEliminate*, and then be distributed to A_j in *DistributeDlink*. So fill-ins produced in local moralization or relative triangulation will be properly distributed to relevant agents, and each pair of adjacent subnets on the hypertree is graph-consistent.   □

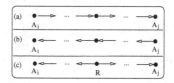

Fig. 4. Possible directions of a path on a virtually directed hypertree: (a) & (b) Consistently directed; (c) There is one and only one tail-to-tail node

7   Experiments

In this section, we experimentally investigate how the revised algorithm and the combined compilation perform compared with the original compilation. Since fill-ins assume dependencies between nodes connected, we prefer less fill-ins in triangulation (the number of fill-ins needed to moralize a hypertree MSDAG is certain) for computational efficiency. We examine the compilation quality based on the heuristic that the node with the minimum fill-ins is eliminated each time. This is also the most popular heuristic for graph triangulation. We would also examine the compilation speed.

We first test these algorithms based on a set of 8 manual MSBNs: 3dags, 5partc, big, bigb, bigcirc, circs, painulim, and tricky. Among them, 3dags, digitals, painulim, and tricky contain 3 subnets, and the rest contain 5 subnets. A maximum degree of a hypernode on a hypertree is 3. A subnet could contain as many as 51 nodes (bigcirc), and an MSBN could have as many as 91 nodes (bigcirc). The cardinality of an interface could be as large as 13 variables (bigcirc). These manual cases are small and simple. We then examine these algorithms based on a set of 10 large and complex MSBNs randomly generated [13]. The profiles of the 10 MSBNs are as shown in Figure 5.

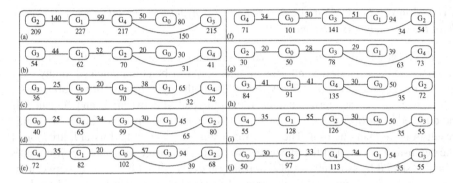

Fig. 5. Profiles of 10 randomly generated MSBNs: next to each subnet is the number of variables in the subdomain, and on each hyperlink is the size of the d-sepset

Table 1. Triangulation quality on the manual cases

MSBNs	Vars	Original	Revised	Combined
3DAGS	15	4	4	4
5PARTC	21	2	1	1
BIG	80	0	0	0
BIGB	80	11	8	8
BIGCIRC	91	3	3	3
CIRCS	45	8	8	8
PLAIULIM	83	1	1	1
TRICKY	8	1	1	1

The experimental results on the compilation quality over the 8 manual cases are summarized in Table 1 (we only consider the number of triangulation fill-ins since the number of moralization fill-ins is certain), where the second column lists the number of variables each MSBN contains (shared variables counted once), the third column lists the number of fill-ins produced by the original algorithm, the fourth column lists the number of fill-ins produced by the revised algorithm, and the last column lists the number of fill-ins produced by the combined compilation. For each case, each method just needs one round of triangulation to finish. It shows all methods produce similar amount of triangulation fill-ins on every case. Since all methods are very fast on these manual cases, we do not compare their speeds on them.

For the 10 randomly generated MSBNs, except MSBNs (a) and (b), all other cases just need one round triangulation to finish. The experimental results are as shown in Figure 6, where the compilation time includes both the moralization time and the triangulation time. It is indicated all algorithms produce similar number of fill-ins using similar time on all cases. Note since MSBN (a) is much more complex and needs much longer compilation time than the others, we do not include its compilation time in the figure. Anyway, the original, the revised, and the combined algorithms respectively use 554,256ms, 554,602ms, and 549,189ms to compile MSBN (a), which are still very similar.

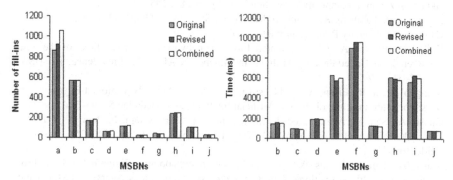

Fig. 6. Left: the number of fill-ins produced from triangulation. Right: the compilation time.

Therefore, the experimental results indicate that the triangulation heuristic is not quite affected by the delay of the delivery of fill-ins. We believe this could be because (1) the compilation is activated from the same hypernode of the same hypertree, the compilation is generally dominated by the order of the triangulation of all hypernodes, which is a depth-first traversal order of the hypertree; (2) the fill-ins that may matter are those shared, and the amount of fill-ins delayed in the revised algorithm could be too less to produce meaningful effects; (3) for the combined compilation, a triangulation is started in a subnet after a caller passes its relevant fill-ins to the subnet and the loss of shared moralization fill-ins at triangulation could be limited when a subnet has limited number of neighbors; and (4) when MSBNs are sparse (all practical MSBNs should be sparse for efficiency consideration and so are the generated MSBNs for study consideration), there exist a lot of similar triangulations. The triangulation time should be similar if the triangulation is similar.

8   Conclusion

Earlier work divides MSBN compilation into two separate processes: the moralization process and the triangulation process. The moralized subnets produced by the moralization process need to satisfy some conditions for the triangulation process to use. We first revise the original algorithm to distribute the shared fill-ins as early as possible in an attempt to improve the triangulation quality. We then propose a unified compilation, which not only provides simplified compilation with less communication, but also makes the compilation robust, though is supposed to produce lower quality triangulation. Anyway, experiments tell us a different story: all algorithms perform similarly in compilation quality and time. The underlying reasons are analyzed. This indicates that eliminating node with the minimum fill-ins is a quite stable heuristic for MSBN triangulation.

References

1. Xiang, Y.: A probabilistic framework for cooperative multi-agent distributed interpretation and optimization of computation. Artificial Intelligence 87, 295–342 (1996)
2. Xiang, Y.: Probabilistic Reasoning in Multiagent Systems: A Graphical Models Approach. Cambridge University Press, Cambridge (2002)
3. Koller, D., Pfeffer, A.: Object-oriented Bayesian networks. In: Proceedings of the 13th Conference on Uncertainty in Artificial Intelligence (UAI 1997), Providence, RI, USA, pp. 302–313 (1997)
4. Xiang, Y., Pant, B., Eisen, A., Beddoes, M.P., Poole, D.: Multiply sectioned Bayesian networks for neuromuscular diagnosis. Artificial Intelligence in Medicine 5, 293–314 (1993)
5. Ghosh, A., Sen, S.: Agent-based distributed intrusion alert system. In: Sen, A., Das, N., Das, S.K., Sinha, B.P. (eds.) IWDC 2004. LNCS, vol. 3326, pp. 240–251. Springer, Heidelberg (2004)
6. Xiang, Y., Chen, J., Havens, W.S.: Optimal design in collaborative design network. In: Proceedings of the 4th International Joint Conference on Autonomous Agents and Multiagent Systems (AAMAS 2005), Utrecht, Netherland, pp. 241–248 (2005)
7. Xiang, Y., Poole, D., Beddoes, M.P.: Multiply sectioned Bayesian networks and junction forests for large knowledge based systems. Computational Intelligence 9, 171–220 (1993)
8. Jensen, F.V., Lauritzen, S.L., Olesen, K.G.: Bayesian updating in causal probabilistic networks by local computation. Computational Statistics Quarterly 5, 269–282 (1990)
9. Xiang, Y.: Cooperative triangulation in MSBNs without revealing subnet structures. Networks 37, 53–65 (2001)
10. Yannakakis, M.: Computing the minimum fill-in is NP-complete. SIAM Journal of Algebraic and Discrete Methods 2, 77–79 (1981)
11. Koller, D., Friedman, N., Getoor, L., Taskar, B.: Graphical models in a nutshell. In: Getoor, L., Taskar, B. (eds.) Introduction to Statistical Relational Learning, pp. 13–56. MIT Press, Cambridge (2007)
12. Xiang, Y.: Belief updating in multiply sectioned Bayesian networks without repeated local propagations. International Journal of Approximate Reasoning 23, 1–21 (2000)
13. Xiang, Y., An, X., Cercone, N.: Simulation of graphical models for multiagent probabilistic inference. Simulation: Transactions of the Society for Modeling and Simulation International 79, 545–567 (2003)

Cosine Policy Iteration for Solving Infinite-Horizon Markov Decision Processes

Juan Frausto-Solis[1], Elizabeth Santiago[1], and Jaime Mora-Vargas[2]

[1] Tecnológico de Monterrey Campus Cuernavaca, Autopista del Sol Km 104+06,
Colonia Real del Puente, 62790, Xochitepec, Morelos, México
[2] Tecnológico de Monterrey Campus Estado de México
{juan.frausto,jmora}@itesm.mx, eliza.stgo@gmail.com

Abstract. Police Iteration (PI) is a widely used traditional method for solving Markov Decision Processes (MDPs). In this paper, the cosine policy iteration (CPI) method for solving complex problems formulated as infinite-horizon MDPs is proposed. CPI combines the advantages of two methods: i) Cosine Simplex Method (CSM) which is based on the Karush, Kuhn, and Tucker (KKT) optimality conditions and finds rapidly an initial policy close to the optimal solution and ii) PI which is able to achieve the global optimum. In order to apply CSM to this kind of problems, a well- known LP formulation is applied and particular features are derived in this paper. Obtained results show that the application of CPI solves MDPs in a lower number of iterations that the traditional PI.

Keywords: Markov decision processes, policy iteration, cosine simplex method, hybrid method.

1 Introduction

The traditional methods for solving infinite-horizon Markov Decision Processes (MDPs) with discount are commonly policy iteration (PI) and value iteration in the Dynamic Programming (DP) approach, some improvements have been applied in [1] and [2] and depending on the different kinds of controls and refinements of the basic MDPs (centralized, decentralized, hierarchical, factorized) new method are emerged [3] and [4]. The more general framework for MDP is applied in this paper. However, the classical PI method proposed by Howard [5] has resulted efficiently for its iterative process of rapid convergence [6]. Nonetheless, IP has disadvantages. First, in each iteration is necessary to solve an equation system that in complex problems (large number of states and actions) has an exponential growth, and second the interpretative formulation of its solution does not show a clear description of the problem. In this sense, the Linear Programming (LP) approach has an important advantage in its formulation for analyzing different solutions. This approach has used the classical Simplex Method (SM) [7].

It is known that SM and PI converge to the optimal solution and both are based on the generation of a sequence of improved solutions as in [8] and [6]. The quality of

A. Hernández Aguirre et al. (Eds.): MICAI 2009, LNAI 5845, pp. 75–86, 2009.

the solutions is measured according to the objective function for SM [7] or Bellman optimality equation for PI [9]. Though SM has an exponential behavior, it has been very successful for many practical problems [10]; however it has not been the case for MDPs, where the SM execution time is usually greater than PI [11]. An efficient strategy of LP is to begin with a good initial solution and then apply SM which is able of finding the global optimum. Cosine Simplex Method (CSM) which is based on Karush Kunh Tucker conditions [12] applies this strategy. This method proposed in [13] and [14], and redefined in [15] and [16] identifies the initial point by means of the calculation of the gradients between the objective function and each constraint. CSM has shown in [14] and [16] to be very efficient in complex LP problems, such as those presented in the Netlib [17].

In this paper, a new hybrid method named Cosine Policy Iteration (CPI) is proposed. CPI adapts as first phase CSM for finding an initial policy, afterwards the regular steps of PI are applied. Furthermore, as CSM cannot be applied directly to MDPs; hence, some important characteristics from the LP formulation for infinite-horizon MDPs [1] are analyzed in this paper. CPI can be seen as a strong modification of PI. Hence, CPI is compared in this paper with PI and MPI. According to the obtained results, CPI overcomes in runtime both of these methods.

This paper is organized as follows: in section 2, the formulation of infinite-horizon MDPs with discount is presented. In section 3, traditional policy iteration is described. Section 4 presents features derived from the LP formulation associated to MDPs. Section 5 describes the original CSM; subsequently the general scheme of CPI and its algorithm are explained. In section 6, the results among traditional PI, MPI and CPI are shown. Finally, in section 7, the conclusions and further works are presented.

2 Markov Decision Processes

The problems of probabilistic DP named Markov decision processes or MDPs can be divided into discrete or continuous time. The first ones are commonly divided into periods or stages, which can be either of finite horizon $t=\{1,...,N\}$ for some integer $N<\infty$, or of infinite horizon $t=\{1,2,...\}$ [1]. In this paper is considered the infinite horizon. MDPs are defined as a 4-tuple $(A, S, \mathbf{P}, \mathbf{R})$ where: $A=\{a_1,a_2,...,a_k\}$ is a finite set of actions, $S=\{s_1,s_2,...,s_m\}$ is a finite set of states (k and m are the number of actions and states, respectively), a matrix of transition $\mathbf{P}$, which an element $p^k_{s_i s_j}$ indicates the probability of taking a transition to state s_j when taking action k in state s_i; and the matrix of rewards $\mathbf{R}$, such that $r^k_{s_i s_j}$ represents the profit (or cost) of changing from state s_i to state s_j, when the action k is executed. For a discounted infinite-horizon MDP, the model includes the discount factor λ ($0 < \lambda < 1$). In general, the stochastic decision problems are modeled by the Bellman's optimality equation [9] as in (1) in order to find the maximum value $v(s_i)$ for each state s_i.

$$v(s_i) = \max_{k \in A_i}\left\{ r(s_i,k) + \lambda\sum_{j=1}^{m} p^k_{s_i s_j} v(s_j)\right\}, \quad i=1,2,...,m. \tag{1}$$

Where $r(s_i, k) = \sum_{j=1}^{S} p_{s_i s_j}^k r_{s_i s_j}^k$ is the expected reward for state s_i and action k. Equation (1) can be written as (2) according to [1].

$$v(s_i) \geq r(s_i, k) + \lambda \sum_{j=1}^{m} p_{s_i s_j}^k v(s_j), \quad i = 1, 2, ..., m. \tag{2}$$

In (2), $v(s_i)$ is an upper bound for the optimal value of the MDP in state s_i. Therefore, for finding the solution in (2), the minimum of $v(s_i)$ should be found. The primal problem is formulated as a minimization LP problem, and according to [1] its dual model is the maximization problem defined by the objective function (3), the technical constraints (4) and the nonnegativity constraints (5). In this model, $y(s_i, k)$ are the decision variables of the problem which are also known as technical variables. Every variable is defined by the state s_i and the action k. The coefficient of theses variables are the expected rewards. On the right hand side in (4), b is a vector of ones [1]. Actually, the number of technical constraints is equal to m and the total number of variables is $T = m*K$. In this formulation, each state corresponds to exactly one technical constraint.

$$\text{Max} \quad Z = \sum_{i=1}^{m} \sum_{k=1}^{K} r(s_i, k) y(s_i, k) \tag{3}$$

Subject to

$$\sum_{k=1}^{K} y(s_j, k) - \lambda \sum_{i=1}^{m} \sum_{k=1}^{K} p_{s_i s_j}^k y(s_i, k) \leq b_{s_j}, \quad \forall j \tag{4}$$

$$y(s_j, k) \geq 0 \quad \forall j, k. \tag{5}$$

In LP is well-known for the standardization (process that translates the constraints in equalities) in (3)-(5), it is necessary to add slack variables [8]; however, in decision problems they usually do not form part of any solution. This behavior can be seen in section 4, where only the technical variables $y(s_i, k)$ take part in the solution. Therefore, equation (6) redefines the model (3)-(4), where no slack variables are needed and only technical variables are required. In the rest of the paper, this model will be used. First, the traditional recursive method for solving (1) is described in the next section, and later the features of the LP formulation are derived using equation (6).

$$\text{Max} \quad Z = \sum_{i=1}^{m} \sum_{k=1}^{K} r(s_i, k) y(s_i, k)$$

Subject to

$$\sum_{k=1}^{K} y(s_j, k) - \lambda \sum_{i=1}^{m} \sum_{k=1}^{K} p_{s_i s_j}^k y(s_i, k) = b_{s_j}, \quad \forall j \tag{6}$$

$$y(s_i, k) \geq 0, \quad \forall i, k.$$

3 Policy Iteration

PI is a traditional method for MDPs proposed by Howard [5]. PI consists of two phases: policy evaluation and policy improvement, the algorithm is presented in Figure 1. PI initiates with an arbitrary policy π which is evaluated in the first phase in the loop; the obtained values in this phase are used to obtain the maximum value for every state (policy improvement); this process continues until two consecutive policies are equal. Equation (1) is used to derive the cost for each state s_i under the application of π. With this strategy, the optimal policy is obtained by searching iteratively in the space of policies, hence the name of this method.

```
PolicyIteration Algorithm
```
INPUT: arbitrary policy π'
OUTPUT: optimal policy π'
  Loop
    $\pi \leftarrow \pi'$
    Compute the value function of policy π'

    Solve the linear equations $v^{\pi}(s_i) = r(s_i,\pi) + \lambda \sum_{j=1}^{m} p_{s_i s_j}^{\pi} v^{\pi}(s_j)$

    For each state i

      Improve the policy $\pi'(s_i) = \arg\max_{k \in A_i} \left\{ r(s_i,k) + \lambda \sum_{j=1}^{m} p_{s_i s_j}^{k} v^{\pi}(s_j) \right\}$

  Until $\pi = \pi'$
End.

Fig. 1. Policy Iteration Algorithm

Although this algorithm converges to the optimal solution in few iterations, it is very costly in every iteration. One of its more known variants is Modified PI (MPI) [1], which is used in this paper for comparing it with CPI. Next, the LP formulation associated to MDPs is discussed.

4 Analysis of the LP Formulation of MDPs

The benefits of the LP formulation can be seen in (6). This equation is extended in (7) for analyzing the features of the MDPs. The variables in (7) are $y(s_1,a_1),...,$ $y(s_1,a_K)$, $y(s_2,a_1),..., y(s_2,a_K),.., y(s_m,a_1),..., y(s_m,a_K)$; the number of variables is $T=m*K$, and every variable represents a relationship between a state and an action. Note that among these variables, there are K variables related to state s_1, K variables related to state s_2, and so on.

$$y(s_j,a_1) + y(s_j,a_2) + ... + y(s_j,a_K) - \lambda \sum_{i=1}^{S} p_{s_i s_j}^{a_1} y(s_i,a_1) - \lambda \sum_{i=1}^{S} p_{s_i s_j}^{a_2} y(s_i,a_2) - ....$$

$$- \lambda \sum_{i=1}^{S} p_{s_i s_j}^{a_K} y(s_i,a_K) = b_{s_j}, \quad \forall j \tag{7}$$

For highlighting the usefulness of the observation made to (7), suppose that only there are two actions (i.e. $K=2$) denoted by a_1 and a_2; hence the obtained substitution in (7) is:

$$y(s_j,a_1) + y(s_j,a_2) - \lambda \sum_{i=1}^{S} p_{s_i s_j}^{a_1} y(s_i,a_1) - \lambda \sum_{i=1}^{S} p_{s_i s_j}^{a_2} y(s_i,a_2) = b_{s_j}, \quad \forall j. \tag{8}$$

On the right hand side in (8), b is a vector of ones; thus, for avoiding this constraint be violated, it is necessary to obtain a positive value on the left hand side. According to (5) all the variables must take non negative values. Therefore, it is precise to analyze the coefficients of the variables. Observe in this problem, $T=m*2$. For every constraint j, $y(s_j,a_1)$ and $y(s_j,a_2)$ are the unique variables with positive coefficients. Theses two variables are also in the summations of the third and fourth terms on the left hand side of (8); and by simplifying this equation, these two variables remain with positive coefficients. This is due to the coefficients of the summations are negative values obtained from the multiplications between the discount factor and each transition probability which is always less that one. Thus, grouping all the common terms in (8), the coefficients of $y(s_j,a_1)$ and $y(s_j,a_2)$ are always positives. In an MDP problem with only one action, every constraint only has a single positive coefficient. Hence, this feature is important at the moment of selecting the m variables in order to form a solution. The selected variables are named as basic variables (BV) and the rest as non basic [8]. Hence, every constraint has its own variable set (each variable is easily identifiable with the state number) with positive coefficients; and one of these variables belongs to the BV set. For instance, in the problem with two actions, the set $w(s_j)$ has two variables: $y(s_j,a_1)$ and $y(s_j,a_2)$ which are related to state s_j for any j, and one of them belongs to the BV set.

In general, for an MDP problem with K actions, the resulting coefficients of the variables on the left hand side in (7) are shown in Figure 2. The first block of coefficients correspond to state 1 (indicated by the first suffix of p), the second block corresponds to state 2, and so on. Observe that in every row of Figure 2, there are exactly K positive values; the rest of them are less than or equal to zero.

In summary, our observations for the LP formulation of MDP's are: 1) Every constraint has its own variable set with positive coefficients, and 2) A variable of this set is part of the BV set from which a feasible policy can be derived. Taking into account these considerations, an efficient LP algorithm for MDPs can be designed.

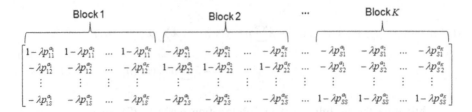

Fig. 2. Coefficients of the LP formulation for MDPs

5   Hybridization Policy Iteration-Cosine Simplex Method

The proposed hybrid method named CPI uses an adaptation of CSM in PI which is explained in this section. Firstly, the background of CSM is described, and then its integration with PI is illustrated taking into account the characteristics presented in the previous section.

5.1   Cosine Simplex Method or CSM

CSM proposed in [13] and [14] begins with a standardized LP problem which consists of two phases: cosine phase (CP) and simplex phase (SP). The goal of CP is to find an initial vertex near the optimal one and SP uses this vertex to initialize the SM. With a good initial solution, the optimum solution is reached in few iterations by the simplex method. However, CPI applies only CP of CSM in order to find an initial policy, and then PI is executed. The procedure of CP is presented in the next paragraphs.

CP is based on KKT conditions and basically consists of two main tasks [14]: 1) compute the angles formed between the gradient of the objective function and each constraint of the LP problem and 2) find the m smallest angles (number of constraints) to form the initial vertex. Basically, the latter task identifies an intersection of the constraints that have a shorter distance to the objective function measured by the angles. For applying CP to MDPs, the LP model defined in (6) is used. In this model, there are m technical constraints, T nonnegativity constraints, and m slack variables.

CP uses the inner product presented in equation (9) for determining the angle between two vectors, where $\nabla Z(y)$ is the gradient of the objective function, $\nabla g_i(y)$ is the gradient of the constraint g_i, and θ is the angle formed between both gradients. When this formula is applied to the technical and nonnegativity constraints, equations (10) and (11) are obtained. These equations are different because the gradients of the nonnegativity constraints in (5) (i.e. $y(i,k) \geq 0$) correspond to vectors formed by zeroes except in one position which takes a value equal to -1 [13]. This negative value is due to all the constraints in (5) need to be standardized to the type $\leq$. Furthermore, these constraints are related to technical variables of the LP problem. For instance, in a problem with eight variables, the gradient of its objective function and a nonnegativity constraint are respectively $\nabla Z(y)$ =[2 3 1 -2 3 -2 0 4] and $\nabla g_2(y) = $ [0 -1 0 0 0 0]. By applying (10), the numerator is equal to -3, and as $\| \nabla g_2(y) \|$=1, the denominator is equal to $\left\| \nabla Z(y) \right\| = \sqrt{47}$. Thus, for nonnegativity constraints, equation (10) is immediately reduced to (11). Equation (10) represents the technical constraints concerned to slack variables, and according to (6), the slack variables are not involved in the LP formulation for MDPs, it is obvious that only equation (11) is required to find a feasible solution.

$$\nabla Z(y) \cdot \nabla g_i(y) = \left\| \nabla Z(y) \right\| \left\| \nabla g_i(y) \right\| \cos \theta. \tag{9}$$

$$\cos \theta_i = \frac{\nabla Z(y)^T \cdot \nabla g_i(y)}{\left\| \nabla Z(y) \right\| \left\| \nabla g_i(y) \right\|} \qquad i = 1, 2, \ldots, m. \tag{10}$$

$$\cos \theta_i = \frac{-\nabla Z(y)_j}{\left\| \nabla Z(y) \right\|} \qquad i = m+1, \ldots, m+T; \quad j = 1, \ldots, T. \tag{11}$$

To illustrate the selection of an initial vertex by applying CP, a small LP problem is shown in Figure 3; it consists of four variables $Y = [h_1, h_2, y_1, y_2]$, and four constraints $U = [H_1, H_2, y_1 \geq 0, y_2 \geq 0]$. H_1 and H_2 are technical constraints and h_1 and h_2 their slack variables, respectively (as is well-known the slack variables are not drawn in the space of solutions). The last two nonegativity constraints in U do not need slack variables. The gradients of U are $\nabla g_1(y)$, $\nabla g_2(y)$, $\nabla g_3(y)$, and $\nabla g_4(y)$. Z is the objective function and $\nabla Z(y)$ is its gradient. As this problem has two technical constraints, thus two BVs are required which are found as follows. Observe that the angles formed by the gradients of H_1 and Z, and H_2 and Z (i.e. θ_1 and θ_2) are the smallest ones; therefore, the intersection of these constraints represent the optimal solution D^*. And its variables h_1 and h_2 take a value of zero. Hence, they are non-basic variables and consequently y_1 and y_2 are the initial BVs. These BVs are related to the nonnegativity constraints,

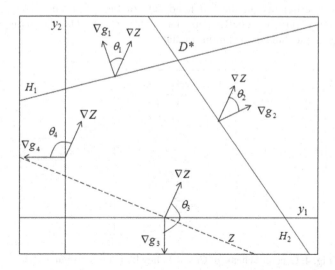

Fig. 3. Example of an LP problem with two variables and four constraints

which have the largest angles or the smallest cosines. The found initial point has the following characteristics: 1) The m smallest angles (i.e. largest cosines) identify the active constraints and the non-basic variables h_1 and h_2, and 2) the two largest angles (i.e. smallest cosines) identify the non-active constraints and the basic variables y_1 and y_2. In the next section, this procedure is applied to find the initial policy in PI.

5.2 Cosine Policy Iteration Method

CPI is a hybrid method which includes the CP described before. For this hybridization, some modifications should be applied to PI, which are related to the feasible set for every constraint introduced in section 4. As is shown in Figure 4, CPI is modeled by three steps: 1) cosine phase (CP), 2) policy evaluation and 3) policy improvement. The first step is used to find an initial policy. The last two steps are evaluated in the same recursive way than traditional PI. In contrast to the original CP, CP into CPI is slightly different as is explained in the next paragraph.

CP usually has two tasks aforementioned. However, to solve efficiently an MDP using CPI, it is only necessary to make changes in the second task to find the initial policy; this is, the way of selecting the m smallest cosines. In the original CSM, m constraints with the smallest cosines are selected of all the constraint set. Nevertheless, in CPI, every technical constraint (or every state) has its own feasible variable set. Hence to find a feasible initial solution, only an element of each set should be chosen, and it is obtained by finding the constraint with the smallest angle (i.e. the greatest cosine). Thus, the variables related with these constraints are used to form the initial policy, and then the next two steps are applied as is shown in Figure 4.

For clarifying only the first step, Figure 5 shows the values of an LP formulation related to an MDP case with $S=(s_1,s_2,s_3,s_4,s_5)$ and $A=(a_1,a_2,a_3,a_4)$. The interpretation of this figure is as follows: the LP problem consists of 5 constraints and 20 variables, which are shown in the first row. The second row presents the variables; the third row shows the coefficients of the objective function (Z) and the rows labeled with H_1 to H_5 represent the coefficients on the left hand side of the constraints. Note that every constraint has 4 positive coefficients and there are 5 sets $W=\{w_1,w_2,w_3,w_4,w_5\}$ representing each technical constraint (or state).

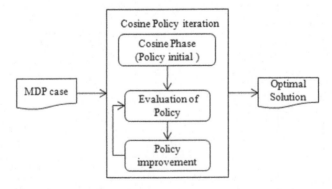

Fig. 4. General Schema of Cosine Policy Iteration for solving MDPs

Index	1	2	3	4	5	6	7	8	9	10	11	12	13	14	15	16	17	18	19	20
Var	y_{11}	y_{12}	y_{13}	y_{14}	y_{21}	y_{22}	y_{23}	y_{24}	y_{31}	y_{32}	y_{33}	y_{34}	y_{41}	y_{42}	y_{43}	y_{44}	y_{51}	y_{52}	y_{53}	y_{54}
Z	50	18.5	8.1	-1.5	9.5	12.4	-6	4.5	9.4	17	9	-6.5	2.8	2	12	-11	-8.6	-15	-10.5	-8.9
H_1	0.4	0,7	0.94	1	-0.12	0	-0.18	-0.3	0	-0.6	0	0	0	-0.18	0	0	-0.12	0	0	0
H_2	0	0	-0.24	0	1	0.52	1	1	-0.48	0	-0.6	-0.42	0	-0.24	0	0	0	0	0	-0.06
H_3	0	0	-0.06	-0.24	-0.18	0	-0.18	0	0.94	1	1	0.82	-0.42	0	-0.24	0	-0.12	0	0	0
H_4	0	-0.3	-0.3	-0.3	0	-0.12	-0.18	0	0	0	0	0	0.82	1	0.64	0.52	-0.12	0	-0.3	0
H_5	0	0	0	-0.06	-0.3	0	-0.06	-0.3	-0.06	0	0	0	0	-0.18	0	-0.12	0.76	0.4	0.7	0.46

Fig. 5. LP formulation of an MDP problem with five constraints (H_1-H_5) and twenty variables

For obtaining the initial policy, the new CP is applied. First, the cosines are calculated by (9) and presented in Table 1. Every nonnegativity constraint is shown in Table 1 with its corresponding technical variable in the first column, its cosine in the second one, and the set to which it belongs is indicated in the third column. Second, the smallest cosine value for every set w_j is selected and asterisked in Table 1. Hence, the initial solution is formed by the variables y_{11}, y_{24}, y_{32}, y_{41} and y_{51}. These variables are identified with two suffixes, where the first one represents the state and the second the action. Therefore, the initial policy is π =[1 4 2 1 1]. This result suggests taking the action a_1 for the states s_1, s_4 and s_5, the action a_4 for s_2 and finally the action a_2 for s_3. Then, the steps 2 and 3 are executed by finding the optimal solution in 2 iterations.

Hybrid algorithm is presented in Figure 6. The input data are the LP formulation and the feasible sets. These sets are not calculated due to the variables of the LP problem for MDPs have already been labeled with the state. For example, in Figure 5, the first four variables are related to state 1 (indicated by the first subindex) and belong to w_1. The CPI algorithm instead of taking an arbitrary policy as original PI, it selects the initial policy using CP. Thus, CP guides the search to a policy near the global optimum. The improvements are included in the two first lines of the algorithm. The operations in the loop are the same as that of traditional PI. The experimentation with this algorithm is shown in the next section.

Table 1. Cosine obtained between the gradient of the objetive function and each gradient of the technical constraints

Variable	Cosine	Set	Variable	Cosine	Set
y_{11}	-0.747332*	w_1	y_{33}	-0.134520	w_3
y_{12}	-0.276513	w_1	y_{34}	0.097153	w_3
y_{13}	-0.121068	w_1	y_{41}	-0.041851*	w_4
y_{14}	0.022420	w_1	y_{42}	-0.029893	w_4
y_{21}	0.141993	w_2	y_{43}	0.179360	w_4
y_{22}	-0.185338	w_2	y_{44}	0.164413	w_4
y_{23}	0.089680	w_2	y_{51}	0.128541*	w_5
y_{24}	-0.067260*	w_2	y_{52}	0.224199	w_5
y_{31}	-0.140498	w_3	y_{53}	0.156940	w_5
y_{32}	-0.254093*	w_3	y_{54}	0.133025	w_5

```
CosinePolicyIteration Algorithm
```
INPUT: LP formulation Q, technical variable sets W
OUTPUT: optimal policy π'
   Compute cosines of nonnegativity constraints in Q
   $\pi' \leftarrow$ Select the smallest cosine for each subset $w_j \in W$
   Loop
      $\pi \leftarrow \pi'$
      Compute the value function of policy π

      Solve the linear equations $v^\pi(s_i) = r(s_i, \pi) + \lambda \sum_{j=1}^{m} p^\pi_{s_i s_j} v^\pi(s_j)$

      For each state i

         Improve the policy $\pi'(s_i) = \arg\max_{k \in A_i}\left\{ r(s_i, k) + \lambda \sum_{j=1}^{m} p^k_{s_i s_j} v^\pi(s_j) \right\}$

   Until $\pi = \pi'$
End.

Fig. 6. Cosine Policy Iteration Algorithm

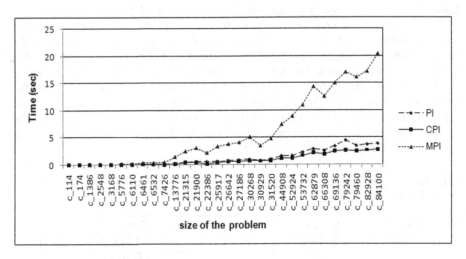

Fig. 7. Execution time of the proposed instances

6   Experimental Results

CPI is compared with PI and MPI and tested with 30 cases generated randomly which are shown in Figure 7. Every MDP case is labeled with the number of variables (multiplication of states (m) by actions (K) associated to the MDP problem) obtained from the LP formulation. For instance, the c_2548 case consists of 2548 variables, and the last instance has a total of 84100 variables. In this figure, the first ten cases are considered small, the next ten medium and the last ten large. Since all the three methods always find the global optimum, the experimentation only measured the execution time. It is important to highlight that PI executed 3 iterations in 29 cases and 4 iterations in case c_72242; meanwhile with CPI, the cases c_2548, c_6461, and c_7426 required only one iteration; 25 cases needed only 2 iterations, and the small cases, c_114 and c_174 required the same number of executions than PI. Despite PI converges in few iterations, every iteration is costly, it can be observed in the last ten cases (c_44908 to c_84100), the difference between PI and CPI in these ten large instances is increasing slightly each time. Note that MPI is the worst of the three methods. In the tested cases, on average the execution time applying PI is 1.2442 seconds, while with CPI is 0.8871 seconds. The saving time is 28.7% using CPI with respect to PI. In Figure 7, note that the distance of the time line between MPI and CPI is greater when the size of the problem increases.

7 Conclusions and Future Works

This work proposed a new hybrid method named cosine policy iteration for solving infinite-horizon discounted Markov decision processes. This hybrid method improves the classical policy iteration method with the adaptation of the cosine simplex method. In particular, the cosine phase was applied using the KKT conditions to find the initial policy. Subsequently, this information was employed by the two basic steps of policy iteration. The adaptation of cosine phase contributed to finding a policy near the optimal policy, reducing the number of iterations, and subsequently the runtime in 93% of tested cases. The remaining percentage of the cases maintained the same number of iteration than PI. The obtained results showed a significant savings of time overall in large MDPs cases in comparison to classic PI. In addition, the LP formulation allowed us to know about the particular features of the MDPs, which were considered in the design of CPI. Hereby, this hybrid method took advantage not only of the LP formulation, but also of the cosine simplex method used to solve complex problems in LP and now tested for complex MDPs cases.

In future research, we will continue using both approaches of linear and dynamic programming by exploring new schemas of solutions for infinite-horizon MDPs and to apply CPI to finite-horizon MDPs. Also, we are considering the combination of cosine simplex method with other strategies as factorized MDPs and the classical value iteration.

References

1. Putterman, L.: Markov Decision Processes Discrete Stochastic Dynamic Programming. Wiley-Interscience, Hoboken (2005)
2. Dai, P., Goldsmith, J.: Topological Value Iteration Algorithm for Markov Decision Processes. International Join Conference on Artificial Intelligence (2007)
3. Bernstein, D.S., Amato, C., Hanse, E.A., Zilberstein, S.: Policy Iteration for Decentralized Control of Markov Decision Processes. Journal of Artificial Intelligence Research, 89–132 (2009)

4. Kim, K.-E., Dean, T.L., Meuleau, N.: Approximate solutions to Factored Markov Decision Processes via Greedy Search in the Space of Finite State Controllers. American Association for Artificial Intelligence (2000)
5. Howard, R.: Dynamic Programming and Markov processes. Wiley, New York (1960)
6. Santos, M.S., Rust, J.: Convergence properties of policy iteration. SIAM Journal on Control and Optimization 42(6), 2094–2115 (2003)
7. Dantzig, G.B., Thapa, M.N.: Linear Programming. 1: Introduction. Springer Series in Operations Research and Financial Engineering, 1st edn. (1997)
8. Chvátal, V.: Linear Programming. W. H. Freeman and Company, New York (1983)
9. Bellman, R.: Dynamic Programming. Princeton University Press, Princeton (2003) Republished 2003: Dover; ISBN 0486428095
10. Bertsekas, D.P.: Network Optimization continuous and discrete models. Athena Scientific (1998)
11. Bello, D., Riano, G.: Linear Programming solvers for Markov Decision Processes. Systems and Information Engineering Design Symposium, IEEE 28(28), 90–95 (2006)
12. Nocedal, J., Wright, S.J.: Numerical Optimization, 2nd edn. Springer, Heidelberg (1999)
13. Trigos, F., Frausto-Solis, J., Rivera-Lopez, R.: A Simplex Cosine Method for Solving the Klee-Minty Cube. In: Advances in Simulation System Theory and Systems Engineering, vol. 70X, pp. 27–32. WSEAS Press (2002) ISBN 960852
14. Trigos, F., Frausto-Solis, J.: Experimental Evaluation of the Theta Heuristic for Starting the Cosine Simplex Method. In: International Conference in Computational Science and its Applications, ICCSA, Proceedings Part IV (2005)
15. Corley, H.W., Rosenberger, J., Wei-Chang, Y., Sung, T.K.: The Cosine Simplex Algorithm. The International Journal of Advanced Manufacturing Technology 27(9-10), 1047–1050 (2006)
16. Wei-Chang, Y., Corley, H.W.: A simple direct cosine simplex algorithm. Applied Mathematics and Computation 214(1), 178–186 (2009)
17. Netlib Repository at UTK and ORNL, http://www.netlib.org/lp/data/

Transformer Diagnosis Using Probabilistic Vibration Models

Pablo H. Ibargüengoytia[1], Roberto Liñan[1], and Enrique Betancourt[2]

[1] Instituto de Investigaciones Eléctricas
Av. Reforma 113, Palmira
Cuernavaca, Mor., 62490, México
{pibar,rlinan}@iie.org.mx
[2] Prolec-General Electric
Blvd. Carlos Salinas de Gortari Km. 9.25, Apodaca, N.L. 66600 México
enrique.betancourt@ge.com

Abstract. Detecting transformer failures at early basis represents enormous economical and technical advantage for a utility company. One approach reported in the literature is the vibration analysis for the detection of mechanical failures in transformers. The basic idea is the characterization of the normal vibration during the operation of the transformer, and the recognition of variations in the vibration patterns of the transformer when a failure is present.

This paper presents the development of a probabilistic vibration model used for the detection of incipient failures in transformers. Vibration measurements are taken all around of the transformer tank and a probabilistic model is constructed using automatic learning algorithms developed in the Artificial Intelligence community. The models are Bayesian networks that relate probabilistically all the variables in the experiments. Later, inference algorithms are used to estimate on-line, a probability of a failure in the transformer.

This project is in collaboration with Prolec General Electric, the largest constructor of transformers in North America. Experiments were carried out at Prolec GE laboratories on a power substation transformer (PST). A discussion of the experiments and their results are included in this paper.

Keywords: Transformers, Vibrations, Diagnosis, Uncertainty, Learning.

1 Introduction

Power transformers are some of the most important equipment for the transmission and distribution of electric power. A single failure in a transformer causes disturbances in the electric network and may cause severe conflicts in hospitals, banks, industrial installations or urban areas in general.

In Mexico, the transmission network is composed by 350 power substations and 2,580 power transformers. The capacities of these transformers are typically 375, 225 and 100 MegaVoltAmpere (MVAS), with a nominal tension of 400 kV,

A. Hernández Aguirre et al. (Eds.): MICAI 2009, LNAI 5845, pp. 87–98, 2009.
© Springer-Verlag Berlin Heidelberg 2009

230 kV and lower. Approximately 27% of these transformers have more than 30 years in operation. For this reason, it is important to observe and register the amount and type of failures that have presented the transformers in the country. Table 1 shows the type of transformer failures from 1997 to 2007 [2].

Table 1. Type of failures in transformers since 1997

Failure	1997	1998	1999	2000	2001	2002	2003	2004	2005	2006	2007	Total
winding insulation	11	6	5	10	5	2	9	4	6	5	6	69
core	0	0	0	0	0	0	0	0	0	0	2	2
bushing	3	2	5	3	1	3	1	1	5	7	5	36
on load tap changer	2	0	2	2	2	2	1	1	2	3	1	18
explosion with fire	1	3	0	0	2	0	2	0	3	0	0	11
other failures	1	0	1	0	2	4	1	2	0	0	0	11
TOTAL	18	11	13	15	12	11	14	8	16	15	14	147

Notice that the highest percentage of failures corresponds to isolation in the windings. The Mexican case is not unique. Failures in transformers in United States and Russia show similar results. Problems with insulation represent 80% of the failures for contamination, aging and core insulation. Other causes can be overvoltage or short circuits. The insulation failures can be slow degradation, while overvoltage and short circuits represent instantaneous failures.

All these common failures can be considered as mechanical failures, in contrast to other chemical based failures. Consequently, mechanical failures cause vibration.

Literature reports different methods for the diagnosis of power transformers. Some methods utilize analysis of the oil, frequency response analysis or partial discharge detection. An alternative method for detecting failures in transformers is the analysis of the vibration produced inside the transformer due to its operation. Normally, the transformer produces vibrations in the windings and the core, and these vibrations vary according to certain operative conditions. Also, in the presence of mechanical failures, the vibration pattern is different than in normal conditions. One advantage of this alternative method for detecting failures, with respect to the mentioned methods is that it is possible to design an on-line diagnosis system. This implies that the detection of the incipient failures can be achieved at all times, while the transformer is working.

This paper presents the development of a probabilistic model of the vibrations in a transformer given all the possible combinations of operating conditions. The probabilistic models are Bayesian networks (BN). The BN are directed acyclic graphs that represent the probabilistic relation between the variables in a domain. In this case, we obtain a Bayesian network that codifies the relation between the vibration signals with all the variables that conform the operative conditions. The model is obtained using automatic learning algorithms applied to historical data of the transformer working at different conditions. The BN is used to calculate the probability of a failure, given the evidence from the operative conditions.

The next section establishes the central problem in this project, namely the vibrations in a power transformer. Also, this section reviews some of the related work reported in the literature.

2 Vibration in Transformers

Transformers always vibrate while operating. Vibration can be detected at different frequencies, in different places of the transformer and caused by different sources. According to the literature, vibration below 100 Hz. is caused by cooling fans and oil pumps. Vibrations above 1000 Hz. are caused by small elements not related to the state of the core or winding [5]. Thus, the vibration frequency range of interest in transformer diagnosis is between the power frequency of 60 Hz (50 in Europe) and multiples of this up to 960 Hz.

In normal operating conditions, the main sources of vibration are the core and the winding of the power transformers. This vibration is transmitted to the transformer tank through the cooling oil and through the solid structure. Different levels of vibration can be measured at different locations of the transformer.

Vibration in the winding is caused by Lorenz forces that depend on the current density and the leakage flux density. Since the leakage flux and the current density have different directions, the winding force density has components in the radial and axial directions. Also, both components are a function of loading current, so the total Lorenz forces are quadratic functions of the current. It is worth to mention that the vibration caused by the winding is not too significant under normal operating conditions, but it is significant under several kinds of failures.

Vibrations at the core are caused by the magnetostriction process. It consists of changes of the dimension of core laminations, made by ferromagnetic materials, due to changes of orientation of the material crystals for magnetic fields. Thus, magnetostrictive effects are function of the magnitude of the applied field. Also, it is known that this magnetostrictive forces have a fundamental frequency of 120 Hz. i.e., twice the exciting frequency. Other source of core vibration is the air gap produced by the magnetic repulsion among laminations. This repulsion forces are mostly present at the corner joints of the core legs with the jokes, and it has also a fundamental frequency of 120 Hz [7].

In case of a failure like a short circuit, the mechanical integrity of the transformer can be altered. Certain changes, such as loss of winding clamping pressure, will lead to insulation deterioration, and therefore, the vibrational response will be altered. In general, most of the failures occurring in the transformer, produce mechanical deformations in the winding, and hence, a change in the vibrational signature of the transformer. Typical failures occurring at the transformer core are caused by short circuits between core laminations or between the core and the tank. Since the tank is grounded, if several contact points between core and tank exist, a current will flow and temperature will rise. Also with laminations short circuits, the temperature increases. In general, temperature changes will produce changes in the insulation system that will produce mechanical changes that provoke changes in the vibrational signature of the transformer.

Summarizing, most of the failures in the winding and core of the transformer, will produce mechanical changes and will turn out in changes in the vibration pattern of the transformers.

2.1   Related Work in Transformers Diagnosis by Vibrations

Several technical groups have found interesting and attractive the study of vibrations for identifying and diagnosing failures. The basic approach followed by all the research groups is the development of a model that represents the behavior of the transformer under different conditions. Later, when new readings are obtained, the model estimates or predicts the behavior of the transformer and a comparison with the current behavior can identify abnormal events.

At MIT, Lavalle [7] and McCarthy [8] utilized analytical models that relate the harmonic content of one input (square of the current) to one output (the vibration in the winding). The model considers the effects of initial core temperature, winding temperature, and excitation voltage. These parameters are estimated in a set of experiments made to a transformer free of failures, with some failures and at different conditions of load.

Spanish Union Fenosa proposes a model that is also based on vibrations [3,4]. They measure vibrations at the tank of the transformer, claiming that the vibrations are transmitted to the tank from two sources: the winding and the core. Thus, the vibration at the tank results from the vibration of the windings plus the vibration of the core, multiplied by a coefficient of transmission. The winding vibration is proportional to the square of the current, and the vibration at the core is proportional to the square of the voltage. The model relates oil temperature and complex parameters dependent on the geography of the transformer. Their diagnosis method consists of the estimation of the tank vibration and its comparison with the real measure. If the difference is greater that certain threshold, then a failure is detected.

The Russian experiments [5] install accelerometers in both sides of the transformer in order to acquire vibration measurements while the transformer is working properly. They executed two sets of experiments. In the first experiments, no load is included in order to detect the vibration pattern due to the core. In the second set of experiments, load is included for detecting vibration from both, core and winding. Thus, they subtract the effect of both minus the effect of the core to deduce the effects of the winding. With this information, they calculate four coefficients that reflect the clamping pressures. These coefficients describes the state of the transformer.

The approaches commented above, and our approach have similar basis. All utilize vibration measures in the tank of the transformer. All transform the vibration signals to the frequency domain in order to process the vibration components at the different frequencies. All propose a model that is utilized to estimate vibration amplitude values, and then compare with real measurements in order to detect changes in the behavior. In the related work above, models are deduced with analytical equations to define certain parameters that have to be acquired off-line over a testing transformer. Experiments are required over different operating conditions and also, in presence or absence of different failures.

The approach proposed in this paper also utilizes a model. However, this model represents the probabilistic relations between condition operational variables and vibration measurements. This implies some special advantages:

- several automatic learning algorithms are available for model construction,
- empirical human expertise can be included in the models,
- the models can be adapted constantly for each kind of transformer in its real operational condition,
- other sources of information can be included, for example, structural characteristics of a transformer.

The next section describes basis for the proposed model.

3   Probabilistic Modeling

The approach proposed in this paper consists of the use of probabilistic models that represent the probabilistic relationships between all the variables in the process. One of the formalisms that have proven to be appropriate for this is the Bayesian networks technique.

In a Bayesian network, the nodes represent the variables in the application and the arcs represent the probabilistic relation between the variables. The variable destine of the arc is probabilistically dependent of the variable at the source of the arc. This follows the Bayes theorem that allows calculating the probability of a hypothesis given certain evidence.

As an example, if we want to calculate the probability of faulty transformer hypothesis ($P(H \mid E)$) given that we observe smoke as evidence, we could easily calculate by counting the times that we observe smoke given that we knew that the transformer failed ($P(E \mid H)$).

By definition, a Bayesian network is a directed acyclic graph (DAG) representing the joint probability distribution of all variables in the domain [9]. The topology of the network gives direct information about the dependency relationship between the variables involved. In particular, it represents which variables are conditionally independent given another variable.

Given a knowledge base represented as a Bayesian network, it can be used to reason about the consequences of specific input data, by what is called probabilistic reasoning. This consists of assigning a value to the input variables, and propagating their effect through the network to update the probability of the hypothesis variables. The updating of the certainty measures is consistent with probability theory, based on the application of Bayesian calculus and the dependencies represented in the network.

One of the advantages of the use of Bayesian networks is the three ways to acquire the required knowledge. First, with the participation of human experts in the domain, who can explain the dependencies and independencies between the variables and also may calculate the conditional probabilities. Second, with a great variety of automatic learning algorithms which utilize historical data to provide the structure and the conditional probabilities corresponding to the process where data was obtained. Third, a combination of the two, i.e., using an automatic learning algorithm which allows the participation of the human expert in the definition of the structure.

This project obtains historical data from different accelerometers collocated in different parts of the prototype transformer. The transformer is operated at different conditions of load, temperature, and excitation. The data acquired is fed to an automatic learning algorithm that produces a probabilistic model of the vibrations in the transformer working under different conditions. Thus, given new readings in a testing transformer, the model calculates through probabilistic propagation, the probability of certain failure. The next section explains this process detailed.

4   Probabilistic Vibration Models

Two approaches were considered for the diagnosis of transformers based on vibration signals. The first approach consists of inserting failures in a transformer and measures the vibration pattern according to the operational conditions. The diagnosis becomes pattern recognition according to the set of failures registered. Some examples of common failures are loosening the core or loosening the windings. These failures are similar to those failures caused by strikes or short circuits. The second approach consists of the measurement of vibration signals of a correct transformer working at different operational conditions. These measures allow the creation of a vibrational pattern of the transformer working properly. Only one model is obtained in this approach. Only measures in a correct transformer are required. As a consequence, this paper reports the work carried out in the second approach, i.e., the construction of a model for the correct transformer.

Additionally, two sets of experiments were carried out. In the first, experiments considered the operational tests performed at the factory in the last steps of the construction of the transformers. This increments the number of factory acceptance tests (FAT). The second set of experiments considers the normal operational conditions of the transformer and detects abnormal behavior in site (SAT).

In this section, we include a description of the experiments executed, and the construction of the model of correct transformer. Finally, we discuss the difference between FAT and SAT models.

4.1   Experiments

The creation of a model for the correct functioning of the transformer requires correct transformers. The experiments were done at the Prolec-GE transformer factory in Monterrey, Mexico. We had access to the production line at the last step of the new transformers tests. We installed 8 sensors around the transformer: two in each side, one in the lower and the other in the upper part of every side.

This array of sensors permits us to identify the specific points of the transformer where the vibrations signals can be detected properly.

Experiments in Prolec GE factory consisted in 19 different types of operational conditions. Table 2 shows the operational conditions and the effect we wanted to study. Un represents the percent of nominal voltage applied and In represents the percentage of nominal current.

Table 2. Type of experiments in factory

Num.	Condition	Effect to study
1 - 5	Temperature = cold, excitation = core, $U_n = \{70\%, 80\%, 90\%, 100\%, 110\%\}$	Effect of voltage in core vibrations
6 - 9	Temperature = cold, excitation = winding, $I_n = \{30\%, 60\%, 100\%, 120\%\}$	Effect of current in winding packages vibration
10 - 14	Temperature = hot, excitation = winding, $I_n = \{30\%, 60\%, 100\%, 120\%\}$	Effect of current and temperature in winding packages vibration
15 - 19	Temperature = hot, excitation = core, $U_n = \{70\%, 80\%, 90\%, 100\%, 110\%\}$	Effect of voltage and temperature in core vibration

For example in experiments 1 to 5, we first excited the coil and no current, with 70% of the nominal voltage, then with 80% and so on. Experiments 10 to 14 consisted in exciting the windings to a hot transformer at 30% to 120% of nominal current without voltage. Ten samples are taken in each experimental condition.

Once that the vibrations are recorded, we need to obtain their frequency content. This is due to the characteristics of the vibration that is related to the excitation signal of 60 Hertz. We apply the discrete Fourier Transform (DFT) and obtain the graphs as shown in the following figure. Notice that the only information that we need to extract with the DFT is the frequency content of the vibration at frequencies multiple of 60Hz. In fact, we find no other components in frequencies different than these multiples.

Figure 1 shows some graphs of the experiment of cold transformer excited with current and no voltage, i.e., windings excited. The steps shown in the figure corresponds to excitations of 30% of the nominal current (lower amplitudes) and then 60%, 100% and 120%. Horizontal axis corresponds to every sample that were measured. Ten samples in each experimental condition. Horizontal axis in the figure indicates the data file used in these experiment (*no. de Archivo de prueba* in Spanish). The upper graph shows the amplitudes of the vibration detected by the all the sensors of the transformer. Each sensor was connected to a channel (canal in Spanish) of the data acquisition system. Sensors 6 and 7 detect more clearly the vibration, while sensors 3, 5 and 8 are almost constant.

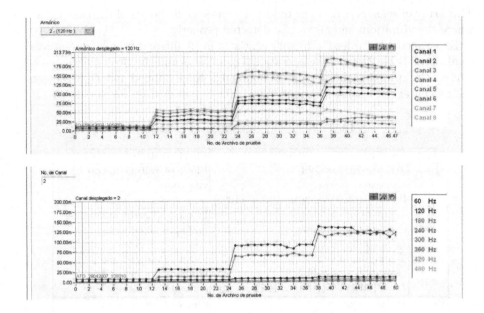

Fig. 1. Example of the recorded vibration signals. Above, amplitude of the 120 Hz. component in all sensors. Below, amplitude of sensor 2 at all the frequencies. Graphs correspond to our GUI in Spanish.

In the lower graph, the amplitude of different frequencies is shown. The only frequencies that increase their value are 120Hz. and 360 Hz. The rest signals remain constant. After these experiments, the following variables were defined for the construction of the probabilistic vibration model: current or voltage, sensors (1 to 8) and frequencies (from 60Hz to 960 Hz in multiples of 60 Hz). The next section explains the construction of the models.

4.2   Model of Correct Transformers

In the first stage of this project, the variables available for constructing the model are sensors, frequencies, temperature and excitation of the transformer (voltage or current). Following the experts' advice, we consider two possible set of models. The first is a model relating operational conditions and frequencies. One model for each sensor. The second possible set relates operational conditions and sensors. One model for each frequency. We decided to try a set of models that relates conditions and sensors, i.e, operational conditions and vibrations detected in certain parts of the transformer. Figure 2 shows the resulting model.

Actually, the complete model is formed by two BNs like the one shown in Fig. 2. One corresponding to the 120 Hz component and the second corresponding to 240 Hz. Once defined the structure, the EM (Estimation-Maximization) algorithm [6] is utilized to obtain the conditional probability tables. We used 10 experiments of each type as indicated in Table 2 and applied in 5 transformers.

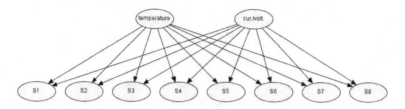

Fig. 2. Model that relates operational conditions with the amplitude measured by each sensor

In this model, the operational conditions of the transformer are character-ized by the temperature and the transformer excitation. This can be voltage or current, i.e., core and the winding. Given certain conditions, probabilities are propagated and sensor nodes provide a probability distribution of the amplitude estimated for certain frequency. This distribution is compared and decided if the behavior is abnormal. Thus, we can detect if there is a possible failure in one sensor at one frequency.

4.3 Experiments for FAT

We designed a computational program that utilize the measurements obtained in the experiments described in Table 2. We run experiments and indicate if there is a failure. Figure 3 shows the results of one experiment (in Spanish). In the upper left of the window, the operational conditions are indicated. First, load *(carga en %)* with 100% of current and cool transformer temperature *(frio)*. In the middle left of the window, there are two lights. One corresponds to a model for 120 Hz. and the other corresponds to 240 Hz. As mentioned above, we are using one model for each frequency. These lights become green if the transformer is correct, yellow if the transformer is suspicious and red if there is definite a failure. Below, in the lower left of the window, there are a little box for each sensor in the transformer. The first 8 corresponding to 120Hz and the last corresponding to 240 Hz. If the posterior probability obtained in a node (sensor) corresponds to the vibrational value currently detected, then an OK mark is described, and a NO-OK mark otherwise. Notice that the sixth sensor detected a deviation in the model of 240 Hz. In the upper right of the window, four rows of data are included. The first two correspond to the current vibration amplitude measured in the 8 sensors in the transformer. The next two rows correspond to the normalized information. This normalization is performed dividing the highest value and the measured vibration. They are actually the inputs to the BNs. The lower right part of the window displays other prototype information.

Several transformers have been tested in factory and the only transformer with factory problems was detected. The next section describes the changes made to the model in order to run SAT tests.

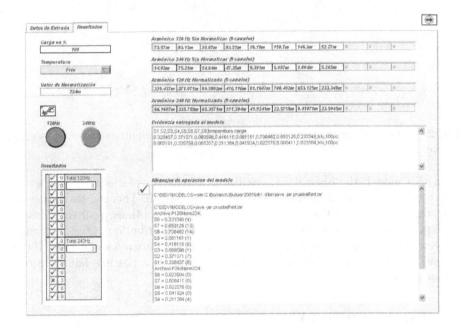

Fig. 3. User interface of the diagnosis software (in Spanish)

4.4 Preliminary Experiments for SAT

Experiments in site have certain differences with FAT experiments. The main difference is that transformers always operate at their fixed nominal voltage, e.g. 125 Kvolts. The current corresponds to the demanded power by the load.

In order to utilize the information acquired in FAT experiments, one assumption was necessary: vibration corresponds to the sum of vibration by current (produced at the winding) plus vibration by fixed voltage at 100% of nominal value (produced at the core). This assumption is valid at the transformer operational condition below the saturation condition. Voltage is always fixed at its nominal value (controlled by the grid), and the current is always tried to keep in normal conditions. In reality, we use all the information acquired for FAT experiments, modified with this assumption.

Additionally, we run experiments in power transformers working in site. Of course, we could not modify the working conditions and we took only data in certain loads.

Table 3 shows an example of the experiments carried out at the power transformer in Prolec GE substation. The transformer provides power to the entire plant. Columns indicate the measurement obtained by all every sensor. The first row indicates the real amplitude obtained by the sensor and normalized. Once normalized, the signals are discretized in 20 intervals. The second row indicates the interval number, from 0 to 19. Third row indicates the posterior probability obtained after the propagation in the probabilistic model. This number indicates

Table 3. Example of one experiment

	S1	S2	S3	S4	S5	S6	S7	S8
Real value measured	0.159	0.121	0.184	0.178	0.083	0.016	0.729	0.141
Corresponding interval	3	2	3	3	1	0	14	2
Posterior probability	0.312	0.312	0.0	0.0	0.687	0.0	0.0	0.375
Decision	0	0	1	0	0	0	1	0

the probability of being a normal measurement, so the fourth row decides if there is a failure (1 value) or there is no failure (0 value). This decision is based on the assumption that the posterior probability distribution is Gaussian given certain operational conditions. Thus, the real value measured is compared with $\pm\sigma$ from the media.

For example in Table 3 , sensor 2 measured a normalized value of 0.121 that corresponds to the interval number 2. Propagation indicates 31% of the value that corresponds to no failure. On the contrary, sensor 7 reads a normalized value of 0.729, corresponding to interval 14 and there is no probability of being correct. The decision is 1. Notice however, that sensor 6 has the same 0 probability but the standard deviation may be very wide and the decision marked 0.

The prototype was constructed using the *hugin* platform [1], so the off–line automatic learning and the on–line propagation are carried out with the Java APIs of this package.

Several tests were made in this Prolec GE substation transformer and the model resulted in a correct tool for transformer diagnosis.

5   Conclusions and Future Work

This paper has presented a probabilistic vibration model for detection of abnormal behavior in power substation transformers. Our approach utilizes Bayesian networks as the formalism for constructing and utilizing the models. We used 8 sensors situated all around the tank of the transformer. Every measure was transformed to the frequency domain and only amplitude multiples of the 60 Hz were considered. Experiments were carried out at different operational conditions to construct the models. Finally, a diagnosis program receives vibration data from a transformer, inserting it as evidence and probability propagation allows calculating the probability of proper behavior.

One of the main advantages of this approach is the facility to deal with incomplete evidence and the availability of algorithms that automatically adapt the models based on vibration in the normal life of the transformer. This means that an old transformer vibrates more than a new one even if the two have normal behavior.

Future work is needed in the determination of additional operation conditions variables, like parameters in the construction of each transformer. We can detect the vibration transmission between different parts of the transformer and identify more clearly if the behavior is normal or not.

Final results will be available after months of tests in new and old transformers, in site and at the factory.

Acknowledgments

Thanks to the anonymous referees for their comments which improved this article. This research is supported by the Prolec-IIE projet 13261-A.

References

1. Andersen, S.K., Olesen, K.G., Jensen, F.V., Jensen, F.: Hugin: a shell for building bayesian belief universes for expert systems. In: Proc. Eleventh Joint Conference on Artificial Intelligence, IJCAI, Detroit, Michigan, U.S.A., pp. 1080–1085, August 20–25 (1989)
2. Comision Federal de Electricidad, Generación termoeléctrica (2008), http://www.cfe.gob.mx/LaEmpresa/generacionelectricidad/termoelectrica
3. García, B., Burgos, J.C., Alonso, Á.M.: Transformer tank vibration modeling as a method of detecting winding deformations - Part I: Theoretical foundation. IEEE Transactions on Power Delivery 21(1), 157–163 (2006)
4. García, B., Burgos, J.C., Alonso, Á.M.: Transformer tank vibration modeling as a method of detecting winding deformations - Part II: Experimental verification. IEEE Transactions on Power Delivery 21(1), 164–169 (2006)
5. Golubev, A., Romashkov, A., Tsvetkov, V., Sokolov, V., Majakov, V., Capezio, O.H., Rojas, B., Rusov, V.: On-line vibro-acustic alternative to the frequency response analysis and on-line partial discharge measurements on large power transformers. In: Proc. TechCon Annu. Conference, New Orlans, L.A., U.S.A, February 1999, pp. 155–171. TJ/H2b, Analytical Services Inc. (1999)
6. Lauritzen, S.L.: The em algorithm for graphical association models with missing data. Computational Statistics & Data Analysis 19, 191–201 (1995)
7. Lavalle, J.C.: Failure detection in transformers using vibrational analysis. Master of science in electrical engineering, Massachusetts Institute of Technology, MIT, Boston, Mass., U.S.A, September 1986 (1986)
8. McCarthy, D.J.: An adaptive model for vibrational monitoring of power transformers. Master of science in electrical engineering, Massachusetts Institute of Technology, MIT, Boston, Mass., U.S.A (1987)
9. Pearl, J.: Probabilistic reasoning in intelligent systems: networks of plausible inference. Morgan Kaufmann, San Francisco (1988)

Intelligent Aircraft Damage Assessment, Trajectory Planning, and Decision-Making under Uncertainty

Israel Lopez and Nesrin Sarigul-Klijn

Mechanical and Aerospace Engineering Department, University of California, Davis
Davis, CA, 95616, U.S.A.

Abstract. Situational awareness and learning are necessary to identify and select the optimal set of mutually non-exclusive hypothesis in order to maximize mission performance and adapt system behavior accordingly. This paper presents a hierarchical and decentralized approach for integrated damage assessment and trajectory planning in aircraft with uncertain navigational decision-making. Aircraft navigation can be safely accomplished by properly addressing the following: decision-making, obstacle perception, aircraft state estimation, and aircraft control. When in-flight failures or damage occur, rapid and precise decision-making under imprecise information is required in order to regain and maintain control of the aircraft. To achieve planned aircraft trajectory and complete safe landing, the uncertainties in system dynamics of the damaged aircraft need to be learned and incorporated at the level of motion planning. The damaged aircraft is simulated via a simplified kinematic model. The different sources and perspectives of uncertainties in the damage assessment process and post-failure trajectory planning are presented and classified. The decision-making process for an emergency motion planning and landing is developed via the Dempster-Shafer evidence theory. The objective of the trajectory planning is to arrive at a target position while maximizing the safety of the aircraft given uncertain conditions. Simulations are presented for an emergency motion planning and landing that takes into account aircraft dynamics, path complexity, distance to landing site, runway characteristics, and subjective human decision.

Keywords: Decision-making, Dempster-Shafer theory, uncertainty, damage assessment, trajectory planning.

1 Introduction

In the past two decades, significant effort has been made towards the development of fault detection, isolation and recovery (FDIR) and prognostics health management (PHM) sub-systems in order to increase safety and performance of aircraft systems, [14]. Discrete damage events, both on ground and in-flight, represent a threat to aircraft systems creating a distress event [6]. Sources of discrete damage include hail impact, lightning strike, transport and handling, and foreign objects. In addition, the high number of flight-cycles causes permanent deteriorating damage. Under distress conditions, system dynamics may differ considerably from nominal dynamics so that the control flight performance may be significantly reduced. In such conditions, the

A. Hernández Aguirre et al. (Eds.): MICAI 2009, LNAI 5845, pp. 99–111, 2009.
© Springer-Verlag Berlin Heidelberg 2009

distressed vehicle may not be capable of performing assigned missions, including safe landing,. Under distress conditions, pilot requires integration of obstacle awareness with intelligent decision-making, path planning and trajectory generation to achieve mission completion. Decisions need be accomplished during flight under uncertain conditions, such as limited failure information and cluttered environment. In this work, the framework for in-flight re-planning of computing optimal-length trajectories to safe landing for a distressed vehicle is developed. The method presented here is designed to be used in conjunction with an integrated vehicle health monitoring (IVHM) and fault tolerance control (FTC) systems. This paper proposes a hierarchical approach for integrated decision-making, vehicle health monitoring, and motion planning under uncertain conditions. The objective is to achieve mission success, including safe landing, under the occurrence of distress events. The integrated decision, monitoring, and motion planning is demonstrated by means of 2D and 3D simulations of a simplified aircraft kinematics model.

2 Integrated Vehicle Health Monitoring

The main objective of integrated vehicle health monitoring (IVHM) is to provide up-to-date vehicle health information via sensors, software and design. The generated information is to be acted upon by decision-making mechanisms with intelligent reasoning and response to maximize a safe outcome probability. For distressed aircraft, an IVHM system would provide valuable information to an adaptive and/or reconfigurable control system for achieve effective failure and damage accommodation while updating flight performance regimes and determining the appropriate set of decisions and responses to complete mission and/or achieve safe landing [3,8]. The generation of a flyable trajectory relies upon the knowledge of the vehicle's dynamic behavior and constraints. The dynamic characteristics are always approximate due to the uncertain knowledge, or the prediction methods used especially under vehicle faults, component failures and/or structural damage. The emergence and successful applications of PHM technologies over the last decade have given rise to proactive capabilities that can perform condition monitoring, detection of anomalies (faults), overall system state, predict system impacts, contingency management, and communication of contextual situational awareness to control mechanisms and system operators of human or autopilots [14]. One of the areas of interest in implementing comprehensive IVHM functionality includes the detection and localization of impact events on key structural and flight control surfaces. In this paper we present a damage assessment performance analysis of the papers of Lopez *et al* [6,7] that can assess the location and damage level resulting from impacts that may occur during flight. Lopez *et al* [7] recently proposed a distance similarity matrix and combined dimensional reduction (D-R) technique for damage assessment. They applied the proposed technique to vibration data obtained from a base-excited cantilever beam. The beam was instrumented with 5-accelerometers at well-spaced discrete locations, and frequency response data was collected. Four different conditions were tested to simulate increasing structural damage where increasing mass quantities were placed between two of the system sensors so as to simulate increasing loss of stiffness levels. Figure 1(a) shows the damage detection and tracking results obtained from vibration data. The upper and lower 3σ standard

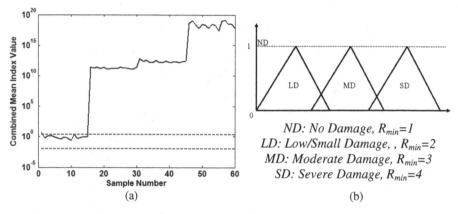

(a)   (b)

Fig. 1. (a) Ensemble of dimensional reduction methods. NoMass: sample 1-15, Mass-1: sample 16-30, Mass-2: sample 31-45, Mass-3: sample 46-60. (b) Input fuzzy damage variable.

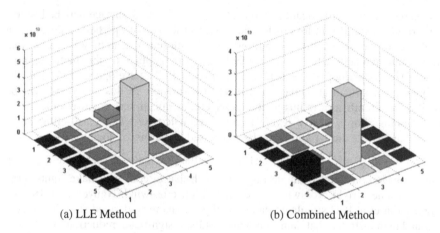

(a) LLE Method   (b) Combined Method

Fig. 2. Localization results for Dataset-2, damage induced between sensors 3 and 4. Axis X and Y represent sensor number, and Z axis represents the anomaly level.

deviation limits of nominal structure are plotted. The mean index results indicate a difference among the multiple damaged datasets. As the setup is changed from undamaged to incremental levels of damage, the anomaly indicator index jumps significantly and proportionally. The results of the distance similarity mean index in combination with the combined D-R technique demonstrate that not only can abrupt change detection can be achieved, but tracking of change progression can be performed. Figure 2 shows the damage localization obtained from the vibration experiment when the induced damage was set between sensors 3 and 4. The damage localization results of two data-driven methods, local linear embedding (LLE) and a combined method proposed in [7]. Both methods clearly indicate that large anomalies exist in relation to sensor 3 and 4. The experimental setup is representative of a damaged aircraft wing. By exploiting the vibrational characteristics of the cantilever wing-like structure, damage detection and localization was achieved by data-driven methods.

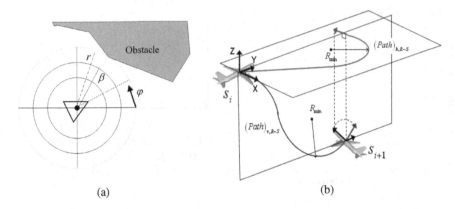

(a)   (b)

Fig. 3. (a) Vehicle guidance problem; (b) Two 2D Reeds-Shepp curves integrated to obtain 3D path (adapted from [4])

In damage events, significant portions of the aircraft's wing surface may be lost and would result in significant loss of lift and asymmetric mass properties manifesting in unstable and anomalous flight conditions. In the current research, we assume that aerodynamic damage effects are available and are applied to the path planning module. Understanding the reduction of allowable flight envelope is critical to the contingency of structural damage and mission assurance. In this study, we will be utilizing the aircraft turn radius as the damage constrained performance variable for path planning. For a nearly level steady coordinated turn and small-climb-angle approximation, the turn radius can be expressed in terms of load factor, n, and airspeed, V, as $R = V^2 / g\sqrt{n^2 - 1}$, where g is the gravitational acceleration. A turn radius damage constraint example might be such that aircraft can make rights turns only, or handle a right crosswind rather than a left crosswind. Nguyen et al [8] and Sarigul-Klijn et al [11] simulated damaged flight dynamics, derived stability derivatives, and demonstrated that damaged wing results in significant reduction of lift coefficient, which minimizes the turning radius capability, reduction in lift and pitch moment causing an inability to hold attitude and flight path angle, among other flight envelope complications. According to the results of Fig. 2, in the wing-like beam experiment, we studied and were able to detect three levels of damage. According to their degree of damage, fuzzy labels [1,13] were assigned to the undamaged state and the three damage levels as shown in Fig. 1(b). The figure shows one membership function with four linguistic variables used from which a minimum turning radius is given, $R_{\min}$. The types of failures addressed in this work are not necessarily found using derived stability derivatives or parameter state estimation. The approach here is to use an Dempster-Shafer [12] expert system to make decisions based on the sampled diagnostic monitoring readings and a rule base. Diagnostic information is inputted to the expert system as a defined "degree of damage" to which the damage detected belongs to. Knowing the limitations of turn radius will allow for shaping of a flight trajectory based on endpoint constraints on position and heading.

3 Trajectory Generation

In this section, we present the distressed vehicle guidance problem. We assume that we have perfect knowledge of the terrain. The vehicle guidance problem is shown in Figure 3(a), where the vehicle's k-th position is given by $(x_k, y_k, z_k, \theta_k, \varphi_k)$. For the initial path planning, we will assume that the vehicle can move according to a heading angle φ, flight path angle θ and a distance β. The vehicle guidance problem, where the vehicle's $(k+1)$ position is given by

$$
\begin{bmatrix} x_{k+1} \\ y_{k+1} \\ z_{k+1} \\ \varphi_{k+1} \\ \theta_{k+1} \end{bmatrix} = \begin{bmatrix} x_k \\ y_k \\ z_k \\ \varphi_k \\ \theta_k \end{bmatrix} + \begin{bmatrix} \beta \cos(\varphi)\sin(\theta) \\ \beta \sin(\varphi)\cos(\theta) \\ \beta \cos(\theta) \\ \dot{\varphi}_{k+1}\Delta t \\ \dot{\theta}_{k+1}\Delta t \end{bmatrix} + \begin{bmatrix} \Delta_\beta \cos(\Delta_\varphi)\sin(\Delta_\theta) \\ \Delta_\beta \sin(\Delta_\varphi)\cos(\Delta_\theta) \\ \Delta_\beta \cos(\Delta_\theta) \\ \Delta_\varphi \\ \Delta_\theta \end{bmatrix}
\tag{1}
$$

where z is the height variable, x and y are the ground-tracking position variables. The sum of the first two terms represent the desired position, and the third term represents path uncertainty term due to inability of vehicle in achieving desired position. The uncertainty components $\Delta_\varphi, \Delta_\theta, \Delta_\beta$ are distributed accordingly to some assumed known probability density function. The more uncertainty exists in the aircraft's performance, the larger these parameters, which would be the case given the occurrence of a distressed event. The path planning is formulated as a steepest descent problem optimization, where it's assumed that the goal location, i.e. safest landing airport, is the minimum point of the dimensional domain [9]. Goal function is given

$$
J_g(x,y) = \left[[x,y]^T - [x_g, y_g]^T \right]^T \left[[x,y]^T - [x_g, y_g]^T \right]
\tag{2}
$$

where (x_g, y_g) is the desired final goal position. To represent the obstacles, we use multiple Gaussian functions to represent a potential field or surface as given by

$$
J_o(x,y) = \max \begin{pmatrix} m_{o,1}\left((x-x_{o,1})^2 + (y-y_{o,1})^2 \right) \\ m_{o,2}\left((x-x_{o,2})^2 + (y-y_{o,2})^2 \right) \\ \vdots \\ m_{o,n}\left((x-x_{o,n})^2 + (y-y_{o,n})^2 \right) \end{pmatrix}^T
\tag{3}
$$

where $m_{o,n}$ is a shape parameter, and $(x_{o,n}, y_{o,n})$ is the center location of nth obstacle.

The objective function is chosen to be a weighted sum for a 2-Airport scenario is given by

$$J(x,y) = w_{g,1}J_{g,1}(x,y) + w_{g,2}J_{g,2}(x,y) + w_oJ_o(x,y) \tag{4}$$

where $w_{g,n}$ is a scale factor for each goal function, which determines preferred final goal, and w_o is a scale factor used to determine the risk level with respect to obstacle avoidance. The weights specify the relative importance of achieving obstacle avoidance and reaching the desired goal. The choice of these weights is rather important since it will affect the shape of trajectory and maintain avoidance of obstacles. The following section describes the analysis done via Dempster-Shafer (D-S) evidence theory to generate such weights. At each position (x_k, y_k), the objective function J is computed at N_s values (x_k^i, y_k^i), $i = 1, 2, .., N_s$, regularly spaced on a circle of radius r around the k-th vehicle position, by finding

$$J(x_k^*, y_k^*) \le J(x_k^i, y_k^i), \ i = 1, 2, ..., N_s \tag{5}$$

which provides direction θ_k for vehicle's movement. The path generated using the steepest descent method is piecewise linear and not suitable for an aircraft with kinematics and dynamic constraints. To smoothly connect this piecewise path, a cubic Bezier curve path smoothing method is utilized, for further details see [10]. After determining the control points, i.e. waypoints, Bezier curves can be fitted to path generated to obtain smooth path using Bezier interpolation. As an input, the maximum allowed square distance error between fitted path and smooth Bezier path has is given.

We study the problem of determining a length-optimal trajectory from a specified initial configuration to a specified final configuration while considering the non-holonomic constraint and a lower-bounded turning radius of an air-vehicle in a ground tracking setup. Chitsaz et al [2] studied a similar application using 3D Dubins paths. The length-optimal path, proportional to shortest time types, is based on Reeds and Shepp (R-S) vehicle that can move forward and backward at a constant velocity. Subsequently, the sufficient family of the optimal trajectories by combining the Pontryagin's Minimum Principle (PMP) with Lie algebras. The optimal trajectory presented in this paper is based on the work by Huifang [5] which used geometric local reasoning for achieving the sufficient family by PMP and a global reasoning for eliminating non-optimal trajectories within the sufficient family.

The kinematic model of the vehicle described in Eqn. (1) can be represented as

$$\dot{q} = uf(q) + vg(q) \tag{6}$$

where $q = (x, y, \theta)$, $f(q) = (\cos(\theta), \sin(\theta), 0)^T$, $g(q) = (0, 0, 1)^T$, u and v describe, respectively, the linear and angular velocities of the vehicle.

The time-optimal trajectories of the vehicle and the path curves are based on a convexified Reeds-Shepp model. The notations of the base curves for the trajectory turning directions are described by the letters L, R, and S, respectively, left ($v = 1$), right ($v = -1$) and straight ($v = 0$). The superscript + or – indicates that the motion is forward ($u = 1$) or backward ($u = -1$). We want to find an admissible control (u, v)

which, subject to Eqn. (8), minimizes the total travel time. First, the adjoint vector, β, is introduced and we define

$$\gamma(t) = \langle \beta(t), f(q) \rangle, \quad \psi(t) = \langle \beta(t), g(q) \rangle \tag{7}$$

where $\gamma(t)$ and $\psi(t)$ are called the u-switching function and v-switching function, respectively. The Hamiltonian function, H, is expressed as [2]

$$H = \langle \beta, f(q)u + g(q)v \rangle = \gamma(t)u(t) + \psi(t)v(t) \tag{8}$$

From PMP, let the optimal control obtained be $(u^*(t), v^*(t))$ and $q^*(t)$ the corresponding state trajectory. Now, let $\beta(t)$ be a nontrivial solution to the adjoint equation

$$\dot{\beta}(t) = -\nabla_q H = -\left[u \frac{\partial f}{\partial q} + v \frac{\partial g}{\partial q} \right]^T \beta(t) \tag{9}$$

Minimizing the Hamiltonian function, we get $u^*(t) = -sign(\gamma(t))$, $v^*(t) = -sign(\psi(t))$. By PMP, there is a constant β_0 such that $\beta_0 = |\gamma(t)| + |\psi(t)|$. Let $h(q) = [g(q), f(q)]$ denote the Lie bracket of the vector field g and f. Introducing $\chi(t) = \langle q(t), f(t) \rangle$, then we obtain

$$\dot{\gamma}(t) = v(t)\chi(t), \quad \dot{\psi}(t) = -u(t)\chi(t), \quad \dot{\chi}(t) = -v(t)\psi(t) \tag{10}$$

According to the above geometric reasoning, any two configurations, i.e. waypoints, of the vehicle can be linked by a minimum-length trajectory that belongs to the sufficient family of trajectories with additional global reasoning. For additional details on the optimal trajectory properties and categories please refer to [5]. The optimal trajectory work based on R-S curves is used to integrate the k waypoint configuration, (x_k, y_k, θ_k), to the next $k+1$ waypoint, $(x_{k+1}, y_{k+1}, \theta_{k+1})$, of the modeled vehicle. The waypoints are selected from the smooth Bezier curve. Reeds-Shepp curves are designed for ground tracking path computation. In order to apply Reeds-Shepp curves to 3D motion, the dimensionality needs to be adjusted. The 3D vehicle trajectory between two waypoints is decomposed onto tow orthogonal planes, i.e. xy-plane and yz-plane, see Fig. 3(b).

4 Decision-Making

Given the scenario of a damage airplane, a pilot may include various aspects of relevant information whose mapping to context information can be extremely complicated or fuzzy. Our goal is to maximize the probability or possibility of safe landing in which the vehicle's sensory data is fused with the pilot's perception and reasoning process in terms of environmental conditions and airport selection. To properly represent a situation of unknown or subjective decisions, it is necessary to allow interval-based assessment functions for which the Dempster-Shafer (D-S) theory provides a representation scheme and reasoning mechanisms for this context [1,12]. The basic entity in the D-S theory is a set of exclusive and exhaustive hypotheses about some problem domain. It is called the frame of discernment, denoted as Θ. The degree of belief in each hypothesis is represented by a real number in [0,1]. The basic belief assignment (BBA) is a function $m : \Psi \rightarrow [0,1]$, where Ψ is the set of all subsets of Θ, the power set of Θ is $\Psi = 2^\Theta$. The function m can be interpreted as distributing belief to each of element in Ψ, with the following criteria satisfied:

$$\sum_{A \in \Psi} m(A) = 1, \, m(\varnothing) = 0 \tag{11}$$

In evidence theory, we do not assign any degree of belief to the empty proposition $\varnothing$ and we ignore the possibility for an uncertain parameter to be allocated outside of the frame of discernment. Thus, the element A is assigned a basic belief number $m(A)$ describing the degree of belief that is committed exclusively to A. Note that a situation of total ignorance is characterized by $m(\Theta) = 1$. The total evidence that is attributed to A is the sum of all probability numbers assigned to A and its subsets

$$Bel(A) = \sum_{\forall E : E \subseteq A} m(E) \tag{12}$$

Given that we have n number of information sources affecting decision-making, then each information source S_i will contribute by assigning its beliefs over Θ. The assignment function of each source is denoted by m_i. The lower bound of interval is the belief function, which amounts for all evidence E_k that supports the airport selection A

$$Bel_i(A) = \sum_{E_k \subseteq A} m_i(E_k) \tag{13}$$

The upper bound of the evidence interval is the plausibility function, which accounts for all the observations that do not rule out the selection of airport A

$$Pl_i(A) = 1 - Bel_i(\overline{A}) = 1 - \sum_{E_k \cap A \neq \varnothing} m_i(E_k) \tag{14}$$

For n mass functions $m_1, m_2, ..., m_n$, the combined mass function and measure of contradiction are given by Dempster's rule

$$m(A) = (m_1 \oplus m_2 \oplus ... \oplus m_n)(A) = \frac{1}{1-C} \sum_{\cap_{i=1}^{N} E_i = A} m_1(E_1) \cdot m_2(E_2) \cdots m_n(E_n) \tag{15}$$

$$C = \sum_{\cap_{i=1}^{N} E_i = \varnothing} m_1(E_1) \cdot m_2(E_2) \cdots m_n(E_n) > 0 \tag{16}$$

5 Simulations

The following types of information sources are defined: integrated vehicle health monitoring (m_1), relative airport geography (m_2), environment (m_3), airport's operational conditions (m_4), runway conditions (m_5), path complexity (m_6), and external human decisions (m_7).For the simulations, it is assumed a low-small damage (LD) with minimum radius $R_{min} = 2$ and that information becomes incrementally available at four sections of the trajectory. Under an emergency scenario, interpretation, integration and decision-making is made incrementally since not all information is available at once. Table 1 shows the basic belief assignments (BBAs) and trajectory sections for each BBA derived for the specific airport scenario of Figure 5. For this particular airport selection, we simulate the scenario where the initial airport choice, B, is gradually changed for an abort airport, A. The abort decision is a subjective decision which is often made in real-world situations when selection advantages are not clearly quantifiable.

Figure 4 shows the aggregations obtained via a weighted Dempster's rule of the BBAs in Table 1. A weighted Dempster's rule was used due to the amount of conflict in the BBAs. Figures 4(a)-(d) represent multiple levels of information aggregation acknowledging that from the distress event occurrence, the information gathering it's done at incremental information levels. As such, all available information, m_1 to m_7, is not taken into account until the last section of the trajectory. As complete information becomes available, the decision becomes clearer.

The BBA aggregation is then used to generate the weights for the multiobjective function of Eqn. (4), which is the basis for generating the distress event-to-airport trajectory. Fig. 4(a)-(b) shows a preference for airport B, which indicates the trajectory directed towards airport B. Fig. 4(d) shows that after more information is taken into account, the airport selection changes to airport A, which becomes the landing airport for this simulation according to the BBA aggregation of all the information provided in Table 1.

Table 1. Basic belief assignment for Dempster-Shafer evidence theory analysis

Input	Trajectory Section	BBA	A	B		T =Simulation Duration
IVHM	I, II, III, IV	m1	0.4	0.6	Section I	$t : 0 \to 0.3T$
Relative airport geography	II, III, IV	m2	0.4	0.6	Section II	$t : 0.3T \to 0.5T$
Environment	III, IV	m3	0.2	0.8	Section III	$t : 0.5T \to 0.7T$
Airport operational conditions	III, IV	m4	0.6	0.4	Section IV	$t : 0.7T \to T$
Runway characteristics	IV	m5	0.6	0.4		
Path complexity	IV	m6	0.5	0.5		
Subjectivity: human decision	IV	m7	0.01	0.99		

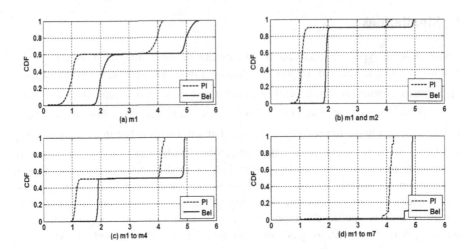

Fig. 4. Cumulative distribution function (CDF) of combined mass functions. The interval for each airport are airport-B → [1,2], and airport-A → [4,5].

For trajectory generation, we used a 2-airport selection scenario. The locations of the distress event and airports are set at $(0,0)_{distress}$, $(20,25)_A$ and $(28,10)_B$, respectively. Multiple obstacles (keepaway zones) are depicted by the high density contour lines around the distress event and near airport-A. The contour plot of the multiobjective function J shows the variances of the Gaussian obstacle functions and goal functions. The choice of the weights, obtained from the Dempster-Shafer analysis, will affect the shape of the trajectory that the vehicle will move toward its final goal position. Trajectory generation was performed at three different levels: (1) multiobjective function with uncertainty; (2) Bezier curve smoothing; and (3) optimal-length waypoint generation. The multiobjective function is used as a guide for path heading. The smooth Bezier curve serves as a basis for waypoint selection, which is then utilized by the optimal-length trajectory. To simulate constraints generated from the damage event, the performance parameters are assumed to be: turning radius $R_{min} = 2$, sensing radius $r = 3$, step size $\beta = 0.3$, location uncertainty $\Delta_\beta = N(0.1, 0.01)$, heading angle uncertainty with uniform distribution $-5^o \leq \Delta_\phi \leq 5^o$. Figure 5 shows the 2D trajectories generated using the multiobjective function, Bezier curve smoothing and the optimal-length waypoint approach. Due to the uncertainty, the path generated from the multiobjective path is irregular with sharp turns and changes in headings. By applying the Bezier curve smoothing, the sharp turns and erratic headings, such as sharp loops, are eliminated. The optimal-length trajectory by using Reeds-Shepp curves is shown with 10 waypoints, including initial and final position. Figure 5 shows that the proposed trajectory generation algorithm can avoid obstacles, and the 3-phase trajectory generation approach results in feasible trajectories under system constraints.

For the 3D trajectory generation, we used the airport selection scenario shown in Figure 6. The locations of the distress event and airports are set at $(1,1,12)_{distress}$,

$(15,25,0)_A$ and $(26,15,0)_B$, respectively. Multiple obstacles are depicted by the high density contour lines around the distress event and near airport-A. As in the 2D simulation, the weights for the multiobjective function were obtained using Dempster-Shafer analysis. To simulate constraints generated from the damage event, the performance parameters are the same as those used for the 2D case. The traveled distances obtained for each technique were 43.4 for the Path3 optimal-length trajectory, 44.15 for Path2 Bezier curve, and 45.83 for Path1 of the original potential function. It is important to note that Path3 minimum-length is optimal in between waypoints and not necessarily global minimum between initial point and landing point. Due to the uncertainty, the path generated from the multiobjective path is irregular with sharp turns and changes in headings. By applying the Bezier curve smoothing, the sharp turns and erratic headings are eliminated. Figure 6 shows that the proposed trajectory generation algorithm can avoid obstacles in a 3D scenario. The use of Dempster-Shafer theory to integrate situational awareness information and provide proportional weights provides a systematic approach for performing decision-making; thus, airport selection and trajectory generation are performed on evidential reasoning. Because trajectory of aircraft cannot be adequately represented using line segments, the proposed framework uses Reeds-Shepp curves to produce continuous

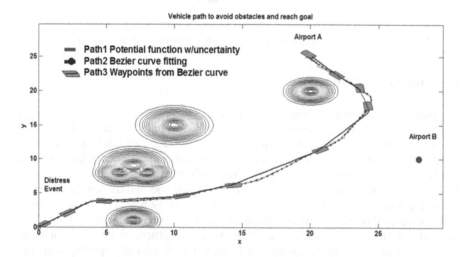

Fig. 5. 2D paths for distressed aircraft under turn constraint with decision analysis

path lines described by set of waypoints chosen based on the recommended guidance information, which approximate strategies that a pilot would use given context information. The multi-criteria decision-making landing approach produced collision-free paths and successfully integrated context information to compute path decisions for each example. In ideal situations, decision makers are assumed to maximize or choose the best available choice, but maximizing requires thorough comparison of all possible alternatives, which in practice, may not always be possible and results in significant decision uncertainty. In the case of emergency landing scenarios, consideration of what makes a decision sequence satisfactory relies primarily on the safe outcome. Therefore, optimality in emergency decisions is not required, only decisions which result in a satisfactory safe outcome. Context informational data should be integrated to support human experts and non-experts in decision-making.

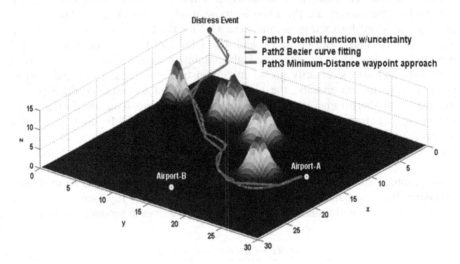

Fig. 6. 3D paths for distressed aircraft under turn constraint with decision analysis

6 Concluding Remarks

This paper presented a hierarchical and decentralized approach for integrated damage assessment and motion planning in aircraft with uncertain navigational decision-making. The proposed approach for flight trajectory architecture offers an integration of fuzzyfied damage levels with trajectory generation in particular when an abort decision is made for a secondary airport. Under a flight distress condition, distance to the landing site becomes even more important. Hence, the optimal-length trajectory generation method developed here offers improvement by reducing the path length while increasing the probability of safe landing given the occurrence of an abort situation. The system integrates the adaptability of a structural health monitoring sub-system and the inference ability of a Dempster-Shafer based expert system. The inference engine uses the inference rules to draw conclusions, which resembles more closely human reasoning.

References

1. Basir, O., Yuan, X.: Engine Fault Diagnosis Based on Multi-sensor Information Fusion using Dempster-Shafer Evidence Theory. Information Fusion 8, 379–386 (2007)
2. Chitsaz, H., LaValle, S.M.: Time-optimal Paths for a Dubins Airplane. In: 46th IEEE Conference on Decision and Control, pp. 2379–2384 (2007)

3. Fahroo, F., Doman, D.: A Direct Method for Approach and Landing Trajectory Reshaping with Failure Effect Estimation. In: AIAA Guidance, Navigation, and Control Conference and Exhibit Providence, Rhode Island (2004)
4. Hwangbo, M., Kuffner, J., Kanade, T.: Efficient Two-phase 3D Motion Planning for Small Fixed-wing UAVs. In: IEEE International Conference on Robotics and Automation (2007)
5. Huifang, W., Yangzhou, C., Soueres, P.: A Geometric Algorithm to Compute Time-Optimal Trajectories for a Bidirectional Steered Robot. IEEE Transactions on Robotics 25, 399–413 (2009)
6. Lopez, I., Sarigul-Klijn, N.: System Identification and Damage Assessment of Deteriorating Hysteretic Structures. In: 49th AIAA/ASME/ASCE/AHS/ASC Structures, Structural Dynamics, and Materials (2008)
7. Lopez, I., Sarigul-Klijn, N.: Distance Similarity Matrix Using Ensemble of Dimensional Data Reduction Techniques: Vibration and Aerocoustic Case Studies. Mechanical Systems and Signal Processing 23, 2287–2300 (2009)
8. Nguyen, N., Krishnakumar, K., Kaneshige, J., Nespeca, P.: Dynamics and Adaptive Control for Stability Recovery of Damaged Asymmetric Aircraft. In: AIAA Guidance, Navigation, and Control Conference and Exhibit, Keystone, Colorado (2006)
9. Passino, K.M.: Biomimicry for Optimization, Control, and Automation. Springer, Heidelberg (2005)
10. Sarfraz, M., Khan, M.N.: Automatic Outline Capture of Arabic Fonts. Information Sciences 140, 269–281 (2002)
11. Sarigul-Klijn, N., Nespeca, P., Marchelli, T., Sarigul-Klijn, M.: An Approach to Predict Flight Dynamics and Stability Derivatives of Distressed Aircraft. In: AIAA Atmospheric Flight Mechanics Conference and Exhibit, Honolulu, HI (2008)
12. Shafer, G.: A Mathematical Theory of Evidence. Princeton University Press, Princeton (1976)
13. Wu, P., Clothier, R., Campbell, D., Walker, R.: Fuzzy Multi-Objective Mission Flight Planning in Unmanned Aerial Systems. In: IEEE Symposium on Computational Intelligence in Multicriteria Decision Making, pp. 2–9 (2007)
14. Zhang, Y., Jiang, J.: Bibliographical Review on Reconfigurable Fault-Tolerant Control Systems. Annual Reviews in Control 32, 229–252 (2008)

A Scrabble Heuristic Based on Probability That Performs at Championship Level

Arturo Ramírez[1], Francisco González Acuña[1,2], Alejandro González Romero[3],
René Alquézar[3], Enric Hernández[3], Amador Roldán Aguilar[4],
and Ian García Olmedo[5]

[1] Centro de Investigación en Matemáticas A.C. (CIMAT)
[2] Instituto de Matemáticas, UNAM
[3] Dept. LSI, Universitat Politècnica de Catalunya (UPC)
[4] Centro de Investigación y Estudios Avanzados (CINVESTAV), IPN
[5] Universidad Nacional Autónoma de México
ramirez@cimat.mx, ficomx@yahoo.com.mx, yarnalito@gmail.com,
{alquezar,enriche}@lsi.upc.edu, amador.roldanaguilar@gmail.com,
ian.garcia@gmail.com

Abstract. The game of Scrabble, in its competitive form (one vs. one), has been tackled mostly by using Monte Carlo simulation. Recently [1], Probability Theory (Bayes' theorem) was used to gain knowledge about the opponents' tiles; this proved to be a good approach to improve even more Computer Scrabble. We used probability to evaluate Scrabble leaves (rack residues); then using this evaluation, a heuristic function that dictates a move can be constructed. To calculate these probabilities it is necessary to have a lexicon, in our case a Spanish lexicon. To make proper investigations in the domain of Scrabble it is important to have the same lexicon as the one used by humans in official tournaments. We did a huge amount of work to build this free lexicon. In this paper a heuristic function that involves leaves probabilities is given. We have now an engine, Heuri, that uses this heuristic, and we have been able to perform some experiments to test it. The tests include matches against highly expert players; the games played so far give us promising results. For instance, recently a match between the current World Scrabble Champion (in Spanish) and Heuri was played. Heuri defeated the World Champion 6-0 ! Heuri includes a move generator which, using a lot of memory, is faster than using DAWG [2] or GADDAG [3]. Another plan to build a stronger Heuri that combines heuristics using probabilities, opponent modeling and Monte Carlo simulation is also proposed.

Keywords: Scrabble, computer Scrabble, heuristics, probability, move generator, simulation.

1 Introduction

Scrabble is a popular board game played by millions of people around the world. In this article we consider its tournament variant, which is a two-player game. Competitors

A. Hernández Aguirre et al. (Eds.): MICAI 2009, LNAI 5845, pp. 112–123, 2009.
© Springer-Verlag Berlin Heidelberg 2009

make plays by forming words on a 15 x 15 grid, abiding by constraints similar to those found in crossword puzzles.

Scrabble, in contrast with other games like chess, Go, draughts, etc. is a game of imperfect information. Techniques of Artificial Intelligence have been applied to games of imperfect information, mainly card games such as Bridge and Poker [4,5]. The theoretical support for the games of strategy goes back to the work of Von Neumann and Morgenstern [6]. As to the generation of valid moves, Appel and Jacobson [2] introduced an algorithm, which proved to be the fastest and more efficient at one time. It is based on the data structure DAWG (directed acyclic word graph) derived from the entries of a reference lexicon. Later, Steve Gordon [3] introduced a variant of this data structure (GADDAG) which occupies 5 times more space than DAWG, but it duplicates the speed of move generation.

Once we have a move generator, the implementation of basic engines simply based on the move that maximizes the score in each turn, is simple. Shapiro [7] and Stuart [8] are examples of precursors in the development of engines for Scrabble. To solve the obvious deficiencies of this greedy approach, simulation techniques were applied. These techniques took into account the possible replies to a candidate move, the replies to those replies, and so on for many plies. This method, known as simulation, rests on Monte Carlo sampling for the generation of the opponent's rack, in situations with uncertainty, and it has its theoretical base in the Minimax technique of Game Theory. The program MAVEN [9] developed by Brian Sheppard is one of the references for this paradigm and its excellent results against top-level players are the best demonstration of the appropriateness of this approach. Recently Jason Katz-Brown and John O'Laughlin [10] have implemented QUACKLE, a program distributed under the GNU license, which also exploits the simulation technique.

Although simulation is an excellent mechanism to quantify how good plays are, it requires some complements to improve its efficiency. For example, the generation of the opponent's racks in MAVEN did not take into account that expert players try to keep a good *leave* (rack residue), which suggests that the Monte Carlo sampling should be biased to model this characteristic. One can also model the strategic component so as to reduce uncertainty, inferring information from the opponent's racks. For this purpose, the probabilistic models are especially adequate. As an example of this line of research, we can cite the work of Richards and Amir [1].

A difference between Heuri and other programs is that the evaluation of a leave is calculated as a sum of products of probabilities times expected rewards (see 3.2) rather than using simulation to calculate the values of individual tiles, summing them, and finally adjusting for synergy and other factors [11].

Even though Scrabble is already dominated by computer agents [11] there is still room for improvement in Computer Scrabble [1].

The present article uses probability models and techniques to select the best move. We believe that a combination of simulation with probability methods might improve even more Computer Scrabble.

The game of Scrabble is a good platform for testing Artificial Intelligence techniques. The game has an active competitive tournament scene, with national and international scrabble associations and an annual world championship for Spanish-language players.

2 Building the Scrabble Tools

2.1 Scrabble Basics and Terminology

We describe briefly the basics of the Scrabble Game for two players, which is the version used in tournaments.

The object of the game is to score more points than one's opponent. A player collects points by placing words on the 15 x 15 game board; the constraints are similar to those in crossword puzzles: not only the word placed but also all words perpendicular to it must be valid (contained in the lexicon).

Scrabble is played with 100 tiles; 98 of these tiles contain letters on them, while there are blank tiles which are denoted by #. The blanks substitute for any letter in the alphabet. Once played, a blank tile remains, for the remainder of the game, the letter for which it was substituted when first played. The values of the tiles and their distribution, in the Spanish version, are as follows:

0 Points – 2 #'s (blank tiles).
1 Point – 12 A's, 12 E's, 6 I's, 4 L's, 5 N's, 9 O's, 5 R's, 6 S's, 4 T's, 5 U's.
2 Points – 5 D's, 2 G's.
3 Points – 2 B's, 4 C's, 2 M's, 2 P's.
4 Points – 1 F, 2 H's, 1 B, 1 Y.
5 Points – 1 Ch, 1 Q.
8 Points – 1 J, 1 Ll, 1 Ñ, 1 Rr, 1 X.
10 Points – 1 Z.

Some squares on the board represent letter multipliers (*double-letter squares* and *triple-letter squares*). If a tile is placed on these squares, then the tile's value is multiplied by 2 or 3 respectively. Certain squares multiply the point value of an entire word by 2 or 3 (*double-word squares* and *triple-word squares* respectively). Extra point squares are only usable once as premium squares.

A *9-timer* (resp. *4-timer*) is a move that uses two empty triple-word squares (resp. two empty double-word squares). A *6-timer* is a move using a double-letter square and a triple-letter square or a triple-letter square and a double-word square, all empty.

Starting the game, each player begins their turn by drawing seven tiles apiece from a bag initially containing the 100 tiles. The first word placed on the board must use the central square 8H; all subsequent words placed must use at least one square adjacent to a tile already on the board.

The player can do one of three things on a turn. The player can place a word, exchange tiles or pass. Exchanging tiles allows a player to replace between one and all of the tiles on the player's rack. A player may pass at any time.

When a player places tiles on the board, that player draws new tiles from the tile bag, adding until that player's number of tiles equals 7 (if possible).

When a player is able to place all 7 tiles from the rack on the board in a single move, that player receives a 50 point bonus. This move is called a *bingo*.

The game ends when all of the tiles have been taken from the bag and one player has used all of the tiles on his rack. Once the game has ended each player subtracts the points in his rack from his final score. The player who ended the game obtains an additional bonus: the points of the tiles left on his opponent's rack.

For more details about the rules see [16].

The *endgame* is the final part of the game starting when the bag has no tiles. In this phase the game has perfect information. We call *pre-endgame* the part preceding the endgame, starting when the bag has less than 11 tiles.

Notation: As examples 3B D(A)N is a horizontal move starting at the square 3B. The vertical move J13 (A)Rr(E) starts at the square J13 . Letters inside parenthesis were already on the board.

2.2 The Lexicon

An important part to conduct research in Scrabble and to build a program that plays Scrabble, in Spanish for our case, is the construction of a lexicon. The length of it is significantly larger than the length of a corresponding lexicon in English.

A regular transitive verb in Spanish, as "amar", "temer" or "partir" has 46, 55 or 54 verbal inflections depending on whether it is an ar-verb, an er-verb or an ir-verb, whereas a verb in English, like "love", typically has only four verbal inflections (love, loves, loved, loving). Even though short words are more numerous in English than in Spanish -1350 words of length less than 4 in SOWPODS, the UK scrabble dictionary, and 609 in Spanish- the number of words of length less than 9 (the important lengths are 8 and 7) is roughly 107000 in SOWPODS and 177000 in our Spanish lexicon. Accordingly, more bingos (seven tiles plays) are expected in a Spanish game than in an English game. The total number of words (of length less than 16) is, approxi- mately, 246000 in SOWPODS and 635000 in our Spanish lexicon.

The electronic version of DRAE (Diccionario de la Real Academia Española), the official Scrabble dictionary in Spanish has approximately 83000 lemmas (entries) from which we had to construct the complete lexicon. Let us mention that petitions to RAE (La Real Academia Española) for a text file containing the collection of all conjugated verbs, or even just all verbs in infinitive form, were unsuccessful. Petitions for nouns and adjectives were also unsuccessful.

We therefore have the following two subproblems:

1. Conjugation of verbs. We constructed, in an automated way, the irregular inflec- tions of irregular verbs; similar constructions were done for regular verbs and regular inflections. Some effort was required to express all conjugations with as few collec- tions of endings as possible. For the purposes of Scrabble we group the irregular verbs in 38 families. The last family consists of the verbs "caber, caler, desosar, erguir, errar, estar, haber, ir, jugar, oler, poder and ser"; no verb in this family is conjugated like any other verb in Spanish. We gave a model for each of the remaining 37 families. Each verb in a family is conjugated like the model belonging to it. We use the following abbreviations of collections of endings:

E1={o, a, as, amos, ais, an}, E2={a, e, o, as, es, an, en},E3={e, es, en}, E4={ieron, iera, ieras, ieramos, ierais, ieran, iere, ieres, ieremos, iereis, ieren, iese, ieses, iesemos, ieseis, iesen}, E4'={E4, iendo,io,amos, ais}, E4''={E4, e, o, iste, imos, isteis}, E5={eron,era,eras,eramos,erais, eran, ere, eres, eremos, ereis, eren, ese, eses, esemos, eseis, esen}, E5'={E5, endo, o}, E5''={E5, e, o, iste, imos, isteis}, E6={a, as, an, ia, ias, iamos, iais, ian}, E6'={E6, e, emos, eis}, E6''={E6, o, amos, ais}.

Due to lack of space we only list 10 models corresponding to 10 of the 37 families. In the list that follows we give the irregular inflections of each model and also the additional inflections which come from "voseo" (Spanish of some regions including Argentina) and the number of verbs in each family.

Table 1. Models for irregular verbs

Verb Models	Irregular inflections (Irri)	"Voseo", irreg. infle. not in Irri	No. of verbs in the family
1.agradecer	agradezc(E1)		260
2. acertar	aciert(E2)	acert(a, as)	157
3. contar	cuent(E2)	cont(a,as)	126
4. construer	construy(E1,E3,E5')		55
5. sentir	sient(E2), sint(E4')		46
6. pedir	pid(E1, E3, E4')		43
7. mover	muev(E2)	mov(e, es)	33
8. entender	entiend(E2)	entend(e, es)	30
9. mullir	mull(E5')		30
10. poner	pong(E1), pus(E4''), pondr(E6'),pon,p(uesto)		27

We feel that this compact way of describing the conjugations of all Spanish verbs may have independent interest and be useful, for example, in the study of the structure and teaching of the Spanish language.

2. Plurals of nouns and adjectives. Plurals of articles are explicit in DRAE and pronouns which admit plurals are few (roughly 20). Adverbs, prepositions, conjunctions, interjections, onomatopoeias and contractions do not have plurals. If a word is not an entry but is contained in a phrase having the abreviation expr. (expression) or loc. ("locución") it cannot be pluralized. However, an adjective contained in an "envío" having a little square and the abreviation V.("véase") can be pluralized, for example "gamada".

The FISE (Federacion Internacional de Scrabble en Español) rules are clear as to which inflections of verbs are valid since old verbs cannot be conjugated (old words are those which in all its meanings have one of the abbreviations ant., desus.or germ.) and all verbs which are not old are explicitly conjugated in DRAE. Unfortunately all these conjugations cannot be copied in a text file as they are encrypted in a special format. Therefore we had to develop a way to conjugate verbs. The FISE rules indicating which nouns and adjectives admit a plural are less clear.

2.3 An Initial Spanish Scrabble Program

After a lot of work the huge task of building the lexicon was accomplished. Then we modified a free English Scrabble source program developed by Amitabh [12]. Finally after all this work we had (Scrabler II) a Spanish Scrabble program that plays with the official rules of Spanish Scrabble tournaments (the FISE rules).

In order to check the performance of the new lexicon built and the functional capabilities of the Scrabler II program a match was performed between Scrabler II and an expert human Scrabble player who is ranked No. 33 in the Spanish Scrabble League. The match followed the format of a baseball world series match, that is the player who arrives at 4 victories first wins the match. The final result was 4-2 in favour of the human expert.

As can be seen, although at first sight one could think that the computer has a huge advantage against a human because it knows all the permissible words, it turns out that this is not sufficient to beat a Scrabble expert. The reason is that Scrabbler II employs a naïve, greedy strategy. It always plays a move that maximizes the points at the current time, ignoring how this could affect the options for making moves in the future. Many times you will end up with low quality racks like {GGLNQPM} and since there is no change strategy these low quality racks will tend to prevail giving the human lots of advantage. For instance, in Scrabble one usually tries to play a *bingo*, (playing all seven tiles in one turn), since this gives you 50 bonus points. Therefore it is convenient to preserve the blank tiles which are jokers that represent any chosen permissible letter. Since Scrabbler II plays always the highest point move, it tends to get rid of the blank tiles without making any bingo.

Since all these features were observed we decided to make an improved version of Scrabbler II with an inference strategy engine. This engine will follow a heuristic function that tries to play a high scoring move as well as balancing the rack in play.

3 Research Work (Scrabble Strategy)

3.1 Heuri Move Generator

In this section Heuri's method for generating all possible moves is described. This is independent of the strategy for selecting the move to be played, which involves the evaluation function to be described in 3.2.

For each empty square s of the board we collect information on the tiles played above it (and below it, and to the right and left of it) including two collections of letters H_s y V_s. A horizontal (resp. vertical) word is valid iff each letter of it, on an empty square s, belongs to H_s (resp. V_s).

We describe now how Heuri finds all words that can be played on a row, say. Suppose a row is _ _ _RE_ _ _ _AS_ _ _S. We call RE, AS and S *molecules*. The first and second squares of the row form an *interval* and so do the 7-th and the 8-th; intervals must have length > 1 and are not next to a molecule. Besides the one-letter plays, the valid words are either contained in an interval or use one or more molecules. Clearly they must respect the H_s information.

As an example, suppose a player has the rack { D E E I M N T } and the board is shown in Fig. 1. If s is an unoccupied square of row 10 then H_s is the set of all letters { A,B,...,Z } with the following exceptions: if s = 10B (the 2nd square on row 10) then H_s = { M, N } and if s = 10G (the 7th square) then H_s = { A, B, E, I, L, N, O, S }. Some words which can be played on this row are: IRE (a one-letter play), REMENDASTEIS (a bingo containing the molecules RE, AS and S), ENTREMEDIAS (a bingo containing RE and AS), MASEEIS (containing AS and S), ENREDE (containing RE), TIMASEN (containing AS), DES (containing S). Also IN (in the interval preceding RE) and ET (in the interval between RE and AS) can be played.

```
 7    · · · · · · · · · · · · · ·
 8    · D E S A C R E D I T A D O R
 9    · O · · L · O · · R · · · · E
10    · · · R E · · · · A S · · · S
11    · · · · · · · · · · · · · · ·
```

Fig. 1. A board position

With the letters of the rack and those of a (possible empty) collection of consecutive molecules an *extended rack* is made up and Heuri finds using a lookup table, and therefore very fast, the list of anagrams of it; one keeps only the valid words.

The use of the lookup tables, molecules and intervals makes Heuri move generator very fast. These tables use 338 MB, a reasonable amount of memory for present day computers.

To illustrate Heuri's move generator speed, consider the rack { # # A S R E N }. It takes Heuri 0.4 seconds to find all the words that can be formed with it and it takes Quackle 11 minutes to do the same job.

3.2 Probability and Heuristic

An important part of a program that plays scrabble is the decision of what to leave in the rack. In a move t tiles are played or exchanged and $n - 7$ ($n = 7$, except at the end) tiles make up the leave r.

We propose to give a numerical evaluation of all potential moves as follows:

$$v = j + e - d. \tag{1}$$

where j is the number of points made by the move, in which t tiles are played[1] ($j=0$ if t tiles are changed rather than played on the board); e is the expected value of a bingo, given a leave r, if t tiles are drawn randomly from the set, that we call the *augmented bag*, that is the union of the bag and the opponent's rack; d is a nonnegative number which is zero if the move is not weak from a defensive point of view. Presently, $d=20$ if one puts a tile on the edge allowing a possible nine-timer, $d=10$ if one plays a tile next to a premium square allowing a possible six-timer, $d=5$ if one plays allowing a possible four-timer. From all potential moves one chooses one with maximal v.

To explain our estimate of e define a *septet* to be a lexicographically ordered string of seven characters of {A,B,...,Z,#} (where # is a blank) from which a 7-letter word can be constructed; for example {AAAAÑRR} (yielding ARAÑARA) and {ACEINR#} (yielding RECIBAN) are septets but {AEEQRY#} and {ADEILOS} are not.

---

[1] We assume here that the bag has at least t tiles.

There are 130065 septets. We use for the calculation of e the formula

$$e = \sum_{i=1}^{130065} p_i \left(50 + k\sigma_i\right).$$ (2)

Here p_i is the probability (which might be zero) of obtaining the i-th septet, given a leave r consisting of $7 - t$ tiles, if t tiles are drawn at random from the augmented bag; σ_i is the sum of the values of the characters of the i-th septet. The existence of premium squares, hook words, bingos of length > 7 and experimentation have led us to take presently 2.5 as the value of the constant k.

It is better to explain the calculation of p_i using an example:

Suppose that, in the beginning of the game, the first player has the rack {AAAÑHQV} and puts aside {HQV} to exchange them, keeping the leave {AAAÑ}.

What is the probability p_i of obtaining the i-th septet {AAAAÑRR} (from which one can form the 7-letter bingo ARAÑARA) if one has the leave {AAAÑ}, the augmented bag is " total bag - {AAAÑHQV}" and one draws 3 tiles from it?

Answer: If {AAAÑ} were not contained in {AAAAÑRR} p_i would be 0. However {AAAÑ} is contained in {AAAAÑRR} so we consider the difference set {AAAAÑRR}-{AAAÑ}={ARR}={AR^2} and the augmented bag:

{AAAAAAAAABBCCCC...RRRRR...XYZ##}={A^{9}B^{2}C^4 ... R^5 ...XYZ#^{2}}

and one then computes $p_i = C(9,1)*C(5,2)/C(93,3)$ (93 is the number of tiles of the augmented bag and the 3 in $C(93,3)$ is the number of tiles that are taken from the bag) where C(m, n) is the binomial coefficient:

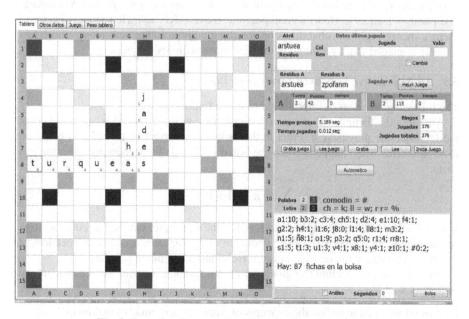

Fig. 2. Example of a game situation using Heuri's interface

$$C(m,n) = m(m-1)(m-2)\ldots(m-n+1)/n!$$

The numerator has n factors and C(m,n) is the number of n-element subsets of a set consisting of m elements. Notice that the denominator C(93,3) does not depend on the septet.

Heuri plays Scrabble in the following manner: it generates all moves that can be played and then using formulas (1) and (2) it evaluates them; finally it plays a move with highest v .

Now we give an example of a possible line of play using Heuri.

The first player has played H4 jades 42pts. Then it is Heuri's turn. Heuri has the following rack: { r h t u u q e }. Using (1) $v = j + e - d$ with $d = 0$, and using (2) to calculate e Heuri's engine gives us:

Coordinate and word or exchange and leave	j	e	v
7G he	9	34.39	43.39
7H eh	9	34.39	43.39
I4 uh	19	20.85	39.85
5G ha	5	34.39	39.39
5H ah	5	34.39	39.39
Exch. h			
Leave (e q r t u u)	0	34.39	34.39
5E huta	14	19.21	33.21

Therefore Heuri plays 7G he for 9 pts. and takes an "a" out of the bag. The first player exchanges letters and Heuri with the rack { r a t u u q e } plays 8A turqueas 106 points ! Thanks to Heuri's good previous move (7G he), Heuri nicely balanced its rack increasing the probabilities of obtaining a bingo in the next move, and Heuri actually managed to play a bingo in the next move with 8A turqueas.

3.3 An Alternative Improved Strategy or Plan

The following are ideas to improve Heuri in the future.

Although the formula (2) is a good estimate of e when the board is open, it ignores the fact that one may have a septet in the rack which cannot be placed as a bingo on the board. This often happens when the board is closed. To account for this, one can write:

$$e = \sum_{i=1}^{130065} \delta_i p_i \left(50 + k\sigma_i\right). \tag{3}$$

where δ_i, which depends on the board, is 1 if the i-th septet can be placed as a bingo on the board and 0 otherwise. The calculation of δ_i , for all i, which is independent of the rack, is a task that is feasible because Heuri has a fast move generator.

To get a better Scrabble engine, we could combine several approaches in the following way:

First of all we need to define the concepts of "open board" and "closed board".
Let us call an *open board* a board position in which there is at least one "bingo line".
A *bingo line* is a place on the board where a bingo can be inserted legally.

Let a *closed position* be a board position in which there are no bingo lines.
Then a possible strategy or plan to follow by a Scrabble engine would be:

Determine whether the board is open or closed and then:

1.   If the board is open use the heuristic that employs probabilities.
2.   If the board is closed use Monte Carlo simulation.

In both cases use opponents' modeling.

Another possible aspect in Scrabble strategy is to take into account the current score of the game. For instance, if a player is well behind in the score, a possible action would be to make *"come back moves"*, this type of moves would risk opening bingo lines, for example, to catch up. On the other hand, if a player has a good lead, he might try to hold the lead by making "blocking moves". A *blocking move* could be any move that reduces the number of bingo lines, or covers triple premium squares.

4   Results, Conclusions and Future Work

After playing 40 games against several top class players Heuri results using the heuristic described in section 3.2 are:

Heuri 27 wins and 13 loses , average per game: 511 pts.

Matches against humans are being performed by internet, this gives the humans an extra force since they can consult the words before they actually put them on the board, they also have a lot of time to play. Almost all matches follow the format of a baseball world series, the first opponent arriving at 4 victories wins the match.

In its first match against a human Heuri won 4-2 against the 2007 national champion of Colombia, Fernando Manriquez. Another important victory of Heuri was against Benjamin Olaizola from Venezuela who is the 2007 world champion. It was a very close game! (Heuri 471 Benjamin Olaizola 465). Eventually Heuri lost the match against Benjamin Olaizola 2-4.

The most important match victory of Heuri was against the current 2008 World Champion Enric Hernandez from Spain also World Champion in 2006. This was a longer match, it consisted in playing a maximum of eleven games to decide the winner. Heuri defeated the current World Champion 6-0 ! An excellent result, but since the random factor is always present in Scrabble games, we decided to extend the matches against the World Champion to achieve more accurate and significant results. These extended matches are: the best of 11, 19, 27, 35, 43 and 51 games, being the last extended match the most significant one as the luck factor decreases as more games are played. The extended match best of 19 is currently in progress.

Heuri also won 4-1 a match against Airan Perez from Venezuela one of the best players in the world (current 2008 world subchampion and 2007 world subchampion). Heuri also defeated 4-1 Aglaia Constantin the current best player in Colombia. Heuri won four matches and lost one, the rest of the games were incomplete matches against past ex world champions and national champions. Heuri also played a match against an engine who always plays the highest move, Heuri won 4-0.

The results indicate that indeed Heuri employs a good Heuristic and is probably already stronger than humans, but to confirm this we should perform many morematches. It is also convenient to hold matches against strong engines like Maven [9] and Quackle [10]. Quackle's computer agent has the same basic architecture as Maven.

Heuri has not been able to play against other engines since they still need to incorporate the Spanish Lexicon into their engines. The English lexicon is being incorporated into Heuri in order to play matches in English against other engines.

In the future the results could also help to discover new strategies that could help human players become the best Spanish scrabble players in the world, and we would also have a world competitive Spanish Scrabble program to challenge any Spanish Scrabble program in a match, perhaps this match could be held in the computer Olympiads that are held every year by the ICGA (International Computer Game Association). The technique employed can also serve to build Scrabble engines for other languages like English, French and Catalan.

Finally, in order to get a much stronger engine we could divide the game into three phases as Maven does. Maven's architecture is outlined in [11]. The program divides the game into three phases: the endgame, the pre-endgame, and the middlegame. The endgame starts when the last tile is drawn from the bag. Maven uses B*-search [13] to tackle this phase and is supposedly nearly optimal. Little information is available about the tactics used in the pre-endgame phase, but the goal of that module is to achieve a favorable endgame situation. All the work presented in this paper could be used in the middlegame and perhaps in the pre-endgame too, although the pre-endgame might require careful study to be improved [14,15]. Since the endgame phase is supposedly nearly optimal using B*-search [13], we could follow Maven steps [11] to tackle this phase of the game.

References

1. Richards, M., Amir, E.: Opponent modeling in Scrabble. In: Proceedings of the Twentieth International Joint Conference on Artificial Intelligence, Hyderabad, India, January 2007, pp. 1482–1487 (2007)
2. Appel, A.W., Jacobson, G.J.: The World's Fastest Scrabble Program. Communications of the ACM 31(5), 572–578 (1988)
3. Gordon, S.A.: A Faster Scrabble Move Generation Algorithm. Software—Practice and Experience 24(2), 219–232 (1994)
4. Ginsberg, M.L.: GIB: Steps toward an expert-level bridge-playing program. In: Proceedings of the Sixteenth International Joint Conference on Artificial Intelligence (IJCAI 1999), pp. 584–589 (1999)
5. Billings, D., Davidson, A., Schaeffer, J., Szafron, D.: The challenge of poker. Artificial Intelligence 134(1-2), 201–240 (2002)
6. von Newman, J., Morgenstern, O.: The theory of Games and Economic behavior. Princeton University Press, Princeton (1953)
7. Shapiro, S.C.: A scrabble crossword game playing program. In: Proceedings of the Sixth IJCAI, pp. 797–799 (1979)
8. Stuart, S.C.: Scrabble crossword game playing programs. SIGArt Newsletter 80, 109–110 (1982)
9. Sheppard, B.: World-championship-caliber Scrabble. Artificial Intelligence 134(1-2), 241–275 (2002)

10. Katz-Brown, J., O'Laughlin, J., Fultz, J., Liberty, M.: Quackle is an open source crossword game program released in (March 2006)
11. Sheppard, B.: Towards Perfect Play of Scrabble. PhD thesis, IKAT/Computer Science Department, Universiteit Maastricht (July 2002)
12. Amitabh, S.: Scrabbler award winning Scrabble program built as part of a programming competition at IIT Delhi (1999)
13. Berliner, H.J.: The B* tree search algorithm; A best-first proof procedure. Artificial Intelligence 12, 23–40 (1979)
14. Fisher, A., Webb, D.: How to win at Scrabble. Anova Books (2004)
15. González Romero, A., Alquézar, R., Ramírez, A., González Acuña, F., García Olmedo, I.: Human-like Heuristics in Scrabble. Accepted in CCIA 2009 Twelfth International Congress of the Catalan Association in Artificial Intelligence, to be published in the proceedings. Frontiers in Artificial Intelligence series, IOS Press, Amsterdam (October 2009)
16. FISE (Federación Internacional de Scrabble en Español). Reglamento de Juego released in (April 2009), http://www.scrabbel.org.uy/reglas/reglas.htm

From Semantic Roles to Temporal Information Representation*

Hector Llorens, Borja Navarro, and Estela Saquete

Natural Language Processing Research Group
University of Alicante, Spain
{hllorens,borja,stela}@dlsi.ua.es

Abstract. The automatic treatment of temporal elements of natural language has become a very important issue among NLP community. Recently, TimeML annotation scheme has been adopted as standard for temporal information representation by a large number of researchers. There are few TimeML resources for languages other than English whereas there exist semantic roles annotated corpora and automatic labeling tools for several languages. The objective of this paper is to study if semantic roles resources can be exploited to generate TimeML corpora. An analysis of the similarities and differences between the temporal semantic role and TimeML elements has been carried out, focusing on temporal expressions (TIMEX3). Using this analysis, an approach consisting of a set of transformation rules between semantic roles and TIMEX3 has been implemented. It has been evaluated in TIMEX3 identification for English and Spanish obtaining same quality results (76.85% $F_{\beta=1}$ AVG), which suggests that it could be also valid for other languages.

1 Introduction

Nowadays, the automatic treatment of temporal expressions (TEs), events and their relations over natural language (NL) text is receiving a great research interest [1]. This task consists of making temporal elements of NL explicit automatically through a system that identifies and annotates them following a standard scheme. Many natural language processing (NLP) areas, such as summarization or question answering, take advantage of considering temporal information [2]. Specialized conferences [3,2] and evaluation forums [4,5] reflect its relevance. There are different ways to represent time in NL. In this work we analyze TimeML and semantic roles.

TimeML annotation scheme [6] has been recently adopted as standard scheme for temporal information annotation by a large number of researchers [2].

Semantic role labeling has achieved important results in the last years [7]. The temporal semantic role represents *"when"* an event takes place in NL text.

Furthermore, in recent years, the development of language independent systems has become an important issue among NLP community. This has been

---

* This paper has been supported by the Spanish Government, project TIN-2006-15265-C06-01 where Hector Llorens is funded under a FPI grant (BES-2007-16256).

A. Hernández Aguirre et al. (Eds.): MICAI 2009, LNAI 5845, pp. 124–135, 2009.

reflected in many international conferences, such as CLEF[1]. Focusing on TE recognition field, examples of the importance of multilingual capabilities of systems can be found in [8,9].

Machine learning (ML) techniques have showed very good results in many AI areas. However, to apply this techniques we need large enough corpora. The motivation for this work lies on the lack of TimeML corpora for languages other than English against the availability of semantic roles resources in several languages. Our hypothesis is that TimeML representation can be inferred from semantic roles representation. In this manner, the objective of this paper is to provide a quantitative and a qualitative study of the advantages and the disadvantages that semantic roles show in temporal information identification. For this purpose, firstly, an analysis comparing the temporal role and TimeML elements has been carried out in order to extract the features they share and the points they differ in. Secondly, we present an approach based on semantic roles to automatically identify TimeML temporal expressions (TIMEX3). This approach is evaluated for English and Spanish to quantitatively determine the performance of semantic roles in this task. Finally, an analysis of results and errors is made to qualitatively analyze the influence of semantic roles in the different TimeML elements.

The paper is structured as follows: Section 2 focuses on the background of temporal information processing and semantic roles, Section 3 describes our proposal to obtain TIMEX3 from semantic roles, and Section 4 includes the evaluation and error analysis. Finally, conclusions and further works are presented.

2   Background

Several efforts have been done to define standard ways to represent temporal information in NL. Since temporal information extraction was included in Message Understanding Conference context, there have been three important annotation schemes for temporal information: STAG [10], TIDES [11] and TimeML [6].

TimeML is a rich specification language for events and TEs in NL that combines and extends features of STAG and TIDES schemes. It was designed to address time stamping, ordering and reasoning about TEs and events of NL. The TimeML elements mentioned in this work are:

- **TIMEX3** tag corresponds to temporal expressions (*"May, two years, etc."*).
- **EVENT** tag corresponds to events (*"went, attack, etc."*).
- **SIGNAL** tag corresponds to temporal signals (*"before, on, during, etc."*).
- **TLINK, ALINK** and **SLINK** tags correspond to different kinds of relationships between previous elements.

In the example at Fig. 1, *"came"* represents an event which is linked to the temporal expression *"Monday"* through a temporal link (TLINK), in which the temporal signal *"on"* is involved.

---

[1] http://www.clef-campaign.org/

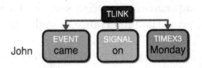

Fig. 1. TimeML example

TimeBank [12] is an English TimeML corpus which was created to illustrate an example annotation. The last version of the corpus, TimeBank 1.2, compiled following the TimeML 1.2.1 [13] specifications, is now considered a gold standard and has been published by Linguistic Data Consortium[2] (LDC). An in-depth analysis of TimeBank corpus can be found in [14,15]. Unfortunately, there are no TimeML corpora available for other languages.

There have been different works on developing systems for automatically tagging NL text following TIMEX3 specifications.

On the one hand, TTK [16] accomplishes this task using the GUTime module. This system was benchmarked on training data from TERN 2004 [4] at 85% and 78% in F for TIMEX2 identification and exact bounding respectively.

On the other hand, the work of Boguraev and Ando [15] presents an evaluation on automatic TimeML annotation over TimeBank using ML techniques. The results for TIMEX3 recognition using 5-fold cross validation were 89.6% and 81.7% in F for relaxed and strict span.

Furthermore, at the present, there are semantic role labeling tools and annotated corpora available for different languages. Many works on the application of semantic roles to other NLP fields have appeared [17,18]. Semantic roles classify semantic predicates of NL sentences as an argument (agent, patient, etc.) or adjunct role (locative, temporal, etc.). Fig. 2 illustrates how semantic roles represent temporal information through the temporal semantic role.

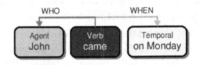

Fig. 2. Semantic roles example

Only one reference about using semantic roles for temporal information processing has been found in literature [19]. That work used them as complementary information to identify temporal relations, but not TEs.

3   Proposal

This section is divided in two parts. The first one analyzes the proposed hypothesis and points out the differences and similarities between temporal semantic role

---

[2] http://www.ldc.upenn.edu/

(TSR) and TimeML temporal expression (TIMEX3). The second one describes our proposal of transformation from TSR to TIMEX3 and its implementation.

3.1 Hypothesis Analysis

Both TSR and TIMEX3 represent temporal information in texts. However, each one of them does it in a different way.

A TSR represents a complete semantic predicate with a temporal function. That is, given an input sentence, TSR represents when the sentence event takes place. Example 1 shows TSR annotated sentences.

Example 1.
EN: I was there [yesterday TSR] **ES:** Estuve [ayer TSR]
EN: She was born [in 1999 TSR] **ES:** Nació [en 1999 TSR]

The TimeML TIMEX3 element is used to markup temporal expressions, such as times, dates, durations, etc. In this case, the full extent of a TIMEX3 tag must correspond to one of the following categories: noun phrase (*"yesterday"* NP), adjective phrase (*"3-day"* ADJP) or adverbial phrase (*"3 days ago"* ADVP). Example 2 shows TIMEX3 annotated sentences.

Example 2.
EN: I was there <TIMEX3>yesterday</TIMEX3> **ES:** Estuve <TIMEX3>ayer</TIMEX3>
EN: She was born in <TIMEX3>1999</TIMEX3> **ES:** Nació en <TIMEX3>1999</TIEX3>

As the previous descriptions show, there exists a parallelism between both kinds of temporal representation. However, there are differences between a TSR and a TIMEX3, some related to the extent and other related to the meaning. In the first sentence of the examples, the TSR corresponds exactly to a TIMEX3. In the second sentence the unique difference between the TSR and the TIMEX3 is the exclusion of the preposition *"in"* in the TIMEX3. According to TimeML specifications and TimeBank annotations, this introducing preposition corresponds to a SIGNAL element. Therefore it should be removed from the TSR to get a correct TIMEX3. This example can be extended to all introducing prepositions and adverbs, such as *"on, at, before, as of, so far, etc"*.

The explained examples only cover TSR consisting of one NP. However, there are TSR that have more than one NP (see example 3)

Example 3.
EN: He came [in 1996 and 1997 TSR]
EN: I will go [for a week from Sunday TSR]

EN: He came in <TIMEX3>1996</TIMEX3> and <TIMEX3>1997</TIMEX3>
EN: I will go for <TIMEX3>a week</TIMEX3> from <TIMEX3>Sunday</TIMEX3>

A TSR composed of more than one NP correspond to a set of related TIMEX3. There are different relation types depending on the linking element. If the link is a conjunction, like in the first sentence of example 3, it corresponds to a conjunction relation. If the link is a temporal preposition, like in the second sentence, it corresponds to an anchoring relation. These cases are generally marked with independent TIMEX3. There are two exceptions for this difference. Firstly, times *"[ten minutes to four]"*, where the preposition *"to"* is denoting a specification relation. And secondly, the preposition *"of"* (*"at the end of 1999"*), when denoting a specification as well.

In previous cases, a TSR always contain one or more TIMEX3. Nevertheless, there are cases in which TSR does not represent a TE. For example:

Example 4.
TSR: She ate [before they started the meeting TSR]
TSR: She ate [before the meeting TSR]

TIMEX3: She had ate before they started the meeting
TIMEX3: She ate before the meeting

In these cases the temporal information provided by the TSR do not correspond to TIMEX3. In the first sentence it corresponds to a subordination and can be easily detected. However, in the second sentence the TSR corresponds to an EVENT. In this case, the difficulty arises on how to automatically differentiate it from real TEs like the one in *"She ate before the night"*. If we analyze these sentences, we will note that it is not trivial using part-of-speech (PoS), syntactic and semantic roles information. Both expressions, *"the night"* and *"the meeting"*, are nouns in a NP of a TSR. Therefore, we need some extra knowledge to decide if a noun in a TSR corresponds to a TIMEX3 or an EVENT.

3.2 Proposal Implementation

Two approaches of TE identification have been implemented in order to quantitatively analyze the influence of semantic roles in this task.

Baseline. It tags each TSR as a TIMEX3 temporal expression. This implementation has been developed only as a baseline to measure how accurate are temporal roles by their own on representing TIMEX3.

TIPSem (Temporal Information Processing Based on Semantic Roles). It implements the following set of transformation rules from TSR to TIMEX3 defined according to the hypothesis analysis.

1. A subordination sentence never represents a TIMEX3. Using syntactic information, all TSRs representing a subordination are removed.
2. Semantic roles annotation can present nesting because each verb of a sentence has its own roles. If there are nested TSRs only the one representing the minimum syntactic unit (NP, ADVP or ADJP) is kept.
3. Every TSR consisting of more than one NP is split if it does not represent one of the exceptions explained in previous section.
4. All temporal prepositions and adverbs are excluded from TSRs leaving tagged only the minimum syntactic unit.
5. All resulting TSRs are tagged as TIMEX3.

In this manner, given an input document, it is processed to obtain semantic roles, PoS and syntactic information. Then, the described rules are applied. Due to the fact that semantic roles rely on verbs, nominal sentences can not be labeled. These sentences are commonly found in titles, brackets, notes, etc. To minimally cover them, TIPSem, as a post-processing step, applies a TE tagger capable of identifying explicit dates in nominal sentences. Finally, a valid TimeML document is obtained as output (see Fig. 3).

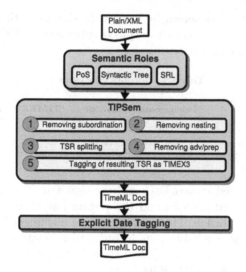

Fig. 3. TIPSem Architecture

4   Evaluation

The objective of this section is to provide a quantitative study of the presented approaches for TIMEX3, and a qualitative analysis of the presented hypothesis of temporal information representation transformation between TSR to TimeML in different languages. For that reason, an evaluation has been carried out for English and Spanish, and an analysis of results and errors has been made.

4.1   Evaluation Environment

Corpora. The presented approaches have been evaluated using TimeBank 1.2 corpus for English, and a manually annotated sample of AnCora for Spanish.

- **English (TimeBank):** TimeBank 1.2 consists of 183 news articles tagged following the TimeML 1.2.1 specification. For this evaluation, this corpus has been automatically annotated with semantic roles using the tool developed by University of Illinois CCG group[3] [20], which uses PropBank role set. This tool obtains a 77.44% $F_{\beta=1}$ in TSR labeling.

---

[3] http://l2r.cs.uiuc.edu/~cogcomp/srl-demo.php

- **Spanish (AnCora TimeML Sample):** Due to the lack of TimeML corpus
for Spanish, to carry out the evaluation, we have annotated manually a
Spanish TimeML TIMEX3 corpus sample using 30 documents of the AnCora
corpus [21]. AnCora is the largest available corpus annotated at different
linguistic levels in Spanish and Catalan. It consists of 500K words mainly
taken from newspaper texts. The corpus is annotated and manually reviewed
at: morphological level (PoS and lemmas), syntactic level (constituents and
functions), and semantic level (roles, named entities and WordNet senses).

Both corpora statistics are shown in Table 1. In the table, the *in TEXT* value
indicates the TIMEX3 tags found in corpus text (between *TEXT* tags), ignoring
explicit dates in documents headers.

Table 1. Corpora Statistics

Corpus	Documents	Words	TIMEX3 (in TEXT)
TimeBank	183	61.8K	1414 (1228)
AnCora Sample	30	7.3K	155 (125)

Criteria. The presented approaches have been tested in TE identification within
the previously described corpora and the results have been compared to the
original TIMEX3 annotation. The explicit dates of document headers have been
ignored to make a more reliable test. In order to evaluate the performance of
the implemented approaches, we applied the criteria used in TERN-2004. The
measures, inherited from TERN-2004, are:

- **POS:** Total TIMEX3 tags in the corpus
- **ACT:** TIMEX3 tags returned by the system (same as corr+inco+spur)
- Correct (**corr**): Correct instances
- Incorrect (**inco**): Identified but wrongly bounded instances
- Missing (**miss**): Not detected instances
- Spurious (**spur**): False positives
- **Precision:** corr/ACT,   **Recall:** corr/POS   and $F_{\beta=1}$: (2*P*R)/(P+R)

An adaptation to TIMEX3 of the TERN-2004 scorer[4], originally developed for
TIMEX2, has been used to calculate these measures.

4.2 Results

Table 2 shows the obtained results for English and Table 3 the ones obtained
for Spanish. For each system, span relaxed (R) and span strict (S) results are
indicated. Span strict refers to strict match of both boundaries of a TIMEX3
expression (exact extent) while span relaxed results consider as correct every tag
identifying a TIMEX3 even if it is wrongly bounded.

For English, the Baseline approach obtains 57.9% in $F_{\beta=1}$ for span relaxed
TIMEX3 identification, but it falls to 27.9% in the span strict case. Nevertheless,
TIPSem approach achieves a 73.4% and a 66.1% in $F_{\beta=1}$ for R and S.

[4] http://fofoca.mitre.org/tern.html#scorer

Table 2. TIMEX3 results for English (1)

System		POS	ACT	corr	inco	miss	spur	Precision	Recall	$F_{\beta=1}$
Baseline	R	1228	1410	764	0	464	646	54.2	62.2	**57.9**
	S	1228	1410	368	396	464	646	26.1	30.0	**27.9**
TIPSem	R	1228	1245	908	0	320	337	72.9	73.9	**73.4**
	S	1228	1245	817	91	320	337	65.6	66.5	**66.1**

Table 3. TIMEX3 results for Spanish (1)

System		POS	ACT	corr	inco	miss	spur	Precision	Recall	$F_{\beta=1}$
Baseline	R	125	147	93	0	32	54	63.3	74.4	**68.4**
	S	125	147	44	49	32	54	29.9	35.2	**32.4**
TIPSem	R	125	144	108	0	17	36	75.0	86.4	**80.3**
	S	125	144	102	6	17	36	70.8	81.6	**75.8**

For Spanish, the Baseline approach obtains 68.4% in $F_{\beta=1}$ for span relaxed TIMEX3 identification, but it falls to 32.4% in the span strict case. However, TIPSem approach achieves a 80.3% and a 75.8% in $F_{\beta=1}$ for R and S.

Results show that, Baseline, that takes all temporal roles as TIMEX3 is reasonably good in the span relaxed identification, but not in the span strict one. However, TIPSem obtains much higher results which indicates that this approach resolves many of differences between TSR and TIMEX3.

Although both corpora consist of news articles and have a similar TE distribution, English and Spanish results are not strictly comparable due to the difference in size of the corpora. Thus, prior to analyzing the results, we studied the comparability of the results. We created a TimeBank normalized corpus dividing TimeBank corpus into 10 parts whose average statics are: 18 docs, 6.1 K words, 140 TIMEX3 and 122 TIMEX3 in TEXT. We evaluated the approaches over each part and averaged the results to check to what extent the size affects to the results. Table 4 shows the averaged $F_{\beta=1}$ results.

The normalized corpus shows same quality results. As shown in Table 4, average results are comparable to the ones obtained using the complete TimeBank corpus with a 0.47% average difference. Therefore, although the English

Table 4. EN TimeBank averaged $F_{\beta=1}$ 10 fold

Evaluation	10-part AVG $F_{\beta=1}$	Full TimeBank $F_{\beta=1}$
Baseline R	58.0	57.9
Baseline S	28.5	27.9
TIPSem R	72.8	73.4
TIPSem S	65.3	65.9

and Spanish corpus have different sizes, the previous analysis showed that the normalized version of the English corpus is comparable with a bearable error percentage. Fig. 4, illustrates English and Spanish evaluation results.

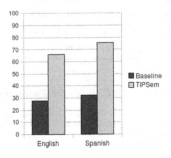

Fig. 4. $F_{\beta=1}$ results comparison

As shown in the Fig. 4 the results for both languages follow the same pattern and offer the similar quality. The Spanish evaluation achieved better results than the English evaluation. This is influenced by the fact that contrary to English, Spanish corpus role labeling has been done manually.

4.3 Error Analysis

The aim of this section is to show in which aspects is TIPSem failing and analyze if the errors are the same in English and Spanish.

- **Spurious:** (33% EN / 26% ES): This value represents the false positives. Most of spurious errors are TSRs not representing a TE, but an event. They can not be easily differentiated as described in section 3. The rest of false positives are caused mainly by role labeling errors. Example 5 shows the spurious errors for both evaluated languages.

   Example 5.
   EN: before <TIMEX3>the trial</TIMEX3>.
   ES: en <TIMEX3>el pitido final</TIMEX3>[5]

- **Missing:** (18% EN / 13% ES): This problem appears because semantic roles not always cover all possibilities of TE in NL.

   • The major problem appear in nominal sentences, NL text where verbs are not present. Semantic roles can be annotated only if there is a verb. Example 7 illustrates these errors for both languages.

   Example 6.
   EN: The 1999 results
   ES: Tres años en Francia[6]

---

[5] At final whistle (football).
[6] Three years in France.

- Also, cases in which a TE has no temporal function in the sentence (i.e., *Agent* role) but it is a TIMEX3. Example 7 illustrates these errors.

Example 7.
EN: [He A0] [spent V] [6 days A3] [on that A1]
ES: [las últimas semanas A0] [fueron V] [nefastas A1][7]

- Minor problems are caused by role labeling errors.
- **Incorrect** (7% EN / 5% ES): Span errors found are mostly produced by role labeling errors. For example, in expressions like *"[10 p.m]. [Wednesday]"* the error is produced because the last period of *"10 p.m."* has been interpreted as a sentence separation mark.

5   Conclusions and Further Work

This work presented a quantitative and qualitative study of the application of semantic roles to temporal information treatment for English and Spanish. The similarities and differences between temporal semantic role and TimeML elements have been analyzed. A Baseline and an approach (TIPSem) resolving these differences have been implemented. Both approaches have been evaluated on TIMEX3 for English and Spanish.

Baseline obtains very low $F_{\beta=1}$ results in span strict. However, the TIPSem approach much higher results, which indicates that it resolves several differences between temporal semantic roles and TIMEX3. Results for both languages follow the same pattern and offer similar quality. Besides, both face similar error types and percentages. Hence, we can confirm that the approach is valid for English and Spanish. Due to the fact that TIPSem is based on semantic roles and morphosyntactic information, it could be valid also for other languages that share several features at these levels.

Error analysis shows that some problems of the approach remain unsolved: the differentiation between TEs and events, and the identification of TEs in nominal sentences. It increased the approach errors. State of the art ML approaches obtain higher results but, contrary to our approach, they require an annotated corpus in the target language. There are not available TimeML corpora for many languages (i.e., Spanish). Therefore, the presented approach could help in the semi-automatic creation of TimeML corpus for those languages that does not have TimeML corpora but they have semantic roles resources available.

Taking into consideration these conclusions, tasks in two lines are proposed as further work:

- **Studying solutions for the encountered problems.** As error analysis showed, some TSR are not denoting TIMEX3 but events. Some extra knowledge is necessary to differentiate between them. Furthermore, the hypothesis analysis indicated that some elements of a TSR correspond to temporal signals. In that manner, next research efforts will be oriented to solve TIMEX3

---

[7] Last weeks were terrible.

problems improving current results, as well as to study the influence of semantic roles, also, in SIGNAL and EVENT elements.

- **Exploiting the benefits that this analysis showed.** Taking into account that same quality results have been obtained for English and Spanish using the same approach, this study will be extended to other languages to confirm if it can be considered multilingual. Due to the lack of TimeML corpora in languages other than English, which prevents the application of state of the art ML approaches, it will be analyzed if the presented study could be exploited as part of a semi-automatic process of TimeML corpus building for other languages.

References

1. Mani, I., James Pustejovsky, R.G. (eds.): The Language of Time: A Reader. Oxford University Press, Oxford (2005)
2. Schilder, F., Katz, G., Pustejovsky, J.: Annotating, Extracting and Reasoning about Time and Events. In: Schilder, F., Katz, G., Pustejovsky, J. (eds.) Annotating, Extracting and Reasoning about Time and Events. LNCS (LNAI), vol. 4795, pp. 1–6. Springer, Heidelberg (2007)
3. Pustejovsky, J.: TERQAS: Time and Event Recognition for Question Answering Systems. In: ARDA Workshop (2002)
4. TERN-2004: Time Expression Recognition and Normalization Evaluation Workshop (2004), http://fofoca.mitre.org/tern.html
5. Verhagen, M., Gaizauskas, R.J., Hepple, M., Schilder, F., Katz, G., Pustejovsky, J.: Semeval-2007 task 15: Tempeval temporal relation identification. In: Proceedings of the 4th International Workshop on Semantic Evaluations, pp. 75–80. ACL (2007)
6. Pustejovsky, J., Castaño, J.M., Ingria, R., Saurí, R., Gaizauskas, R.J., Setzer, A., Katz, G.: TimeML: Robust Specification of Event and Temporal Expressions in Text. In: IWCS-5, 5th Int. Workshop on Computational Semantics (2003)
7. Gildea, D., Jurafsky, D.: Automatic labeling of semantic roles. Computational Linguistics 28(3), 245–288 (2002)
8. Wilson, G., Mani, I., Sundheim, B., Ferro, L.: A multilingual approach to annotating and extracting temporal information. In: Proceedings of the workshop on Temporal and Spatial information processing, pp. 1–7. ACL, NJ (2001)
9. Moia, T.: Telling apart temporal locating adverbials and time-denoting expressions. In: Proceedings of the workshop on Temporal and Spatial information processing, pp. 1–8. ACL, NJ (2001)
10. Setzer, A., Gaizauskas, R.: Annotating Events and Temporal Information in Newswire Texts. In: LREC 2000, Athens, pp. 1287–1294 (2000)
11. Ferro, L., Gerber, L., Mani, I., Sundheim, B., Wilson, G.: TIDES 2005 Standard for the Annotation of Temporal Expressions. Technical report, MITRE (2005)
12. Pustejovsky, J., Hanks, P., Saurí, R., See, A., Gaizauskas, R.J., Setzer, A., Radev, D.R., Sundheim, B., Day, D., Ferro, L., Lazo, M.: The TIMEBANK Corpus. In: Corpus Linguistics, pp. 647–656 (2003)
13. Saurí, R., Littman, J., Knippen, R., Gaizauskas, R.J., Setzer, A., Pustejovsky, J.: TimeML Annotation Guidelines 1.2.1 (2006), http://www.timeml.org/
14. Boguraev, B., Pustejovsky, J., Ando, R.K., Verhagen, M.: TimeBank evolution as a community resource for TimeML parsing. Language Resources and Evaluation 41(1), 91–115 (2007)

15. Boguraev, B., Ando, R.K.: Effective Use of TimeBank for TimeML Analysis. In: Schilder, F., Katz, G., Pustejovsky, J. (eds.) Annotating, Extracting and Reasoning about Time and Events. LNCS (LNAI), vol. 4795, pp. 41–58. Springer, Heidelberg (2007)

16. Verhagen, M., Mani, I., Saurí, R., Knippen, R., Jang, S.B., Littman, J., Rumshisky, A., Phillips, J., Pustejovsky, J.: Automating temporal annotation with TARSQI. In: ACL, pp. 81–84. ACL, NJ (2005)

17. Moreda, P., Llorens, H., Saquete, E., Palomar, M.S.: Two Proposals of a QA Answer Extraction Module Based on Semantic Roles. In: Gelbukh, A., Morales, E.F. (eds.) AAI - MICAI 2008. LNCS (LNAI), vol. 5317, pp. 174–184. Springer, Heidelberg (2008)

18. Melli, G., Shi, Y.Z., Wang, Y.L., Popowich, F.: Description of SQUASH, the SFU Question Answering Summary Handler for the DUC 2006 Summarization Task. In: Document Understanding Conference, DUC (2006)

19. Hagège, C., Tannier, X.: XRCE-T: XIP temporal module for TempEval campaign. In: TempEval (SemEval), Prague, Czech Republic, pp. 492–495. ACL (2007)

20. Koomen, P., Punyakanok, V., Roth, D., Yih, W.-t.: Generalized inference with multiple semantic role labeling systems (shared task paper). In: CoNLL 2004, pp. 181–184 (2005)

21. Taulé, M., Martí, M.A., Recasens, M.: AnCora: Multilevel Annotated Corpora for Catalan and Spanish. In: ELRA (ed.) LREC, Marrakech, Morocco (2008)

Dependency Language Modeling Using KNN and PLSI[*]

Hiram Calvo[1,2], Kentaro Inui[2], and Yuji Matsumoto[2]

[1] Center for Computing Research, National Polytechnic Institute, DF, 07738, Mexico
[2] Nara Institute of Science and Technology, Takayama, Ikoma, Nara 630-0192, Japan
{calvo,inui,matsu}@is.naist.jp

Abstract. In this paper we present a comparison of two language models based on dependency triples. We explore using the verb only for predicting the most plausible argument as in selectional preferences, as well as using both the verb and argument for predicting another argument. This latter causes a problem of data sparseness that must be solved by different techniques for data smoothing. Based on our results on the K-Nearest Neighbor model (KNN) algorithm we conclude that adding more information is useful for attaining higher precision, while the PLSI model was inconveniently sensitive to this information, yielding better results for the simpler model (using the verb only). Our results suggest that combining the strengths of both algorithms would provide best results.

1 Introduction

The goal of Statistical Language Modeling is to build a statistical language model that can estimate the distribution of natural language as accurate as possible. By expressing various language phenomena in terms of simple parameters in a statistical model, SLMs provide an easy way to deal with complex natural language in computers; SLMs also play a vital role in various natural language applications as diverse as machine translation, part-of-speech tagging, intelligent input methods and Text To Speech systems.

We are particularly interested on sentence reconstruction tasks such as zero anaphora resolution, for which we would like to estimate the possible candidate fillers. Consider for example: *There is a ball. The grass is green. A boy is playing.* We may think of a scenario where a boy is playing with a ball on the green grass. We usually tend to connect ideas and build scenarios by looking for the most common scenarios [6, 10]. The ball is indirectly referred by *play*, as well as *garden* but no direct reference is done.

Most of the previous work in SLM has been devoted to speech recognition tasks [5, 22] using Maximum Entropy Models. Mostly because of space limitations, usually these models are limited to sequential 3-gram models. Several works have shown [8,9] that relying only on sequential n-grams is not always the best strategy. Consider the example borrowed from [8]: *[A baby] [in the next seat] cried [throughout the*

---

[*] We thank the support of Mexican Government (SNI, SIP-IPN, COFAA-IPN, and PIFI-IPN), CONACYT; and the Japanese Government; the first author is currently a JSPS fellow.

A. Hernández Aguirre et al. (Eds.): MICAI 2009, LNAI 5845, pp. 136–144, 2009.

flight]. An n-gram model would try to predict *cried* from *next seat*, whereas a dependency language model (DLM) would try to predict *cried* from *baby*.

In this work we create a DLM for obtaining feasible scenario fillers, which can be regarded as extracting selectional preferences [21] but with a broader context for each filler. We show in the next section how this additional information helps choosing the best filler candidate; then in Section 2 we present our implementation of two models for creating a DLM, one based on PLSI (Section 2.1) and one based on KNN (Section 2.2). In Section 3 we describe our experiments for comparing both algorithms in a pseudo-disambiguation task. We analyze our results in Section 4 and finally we draw our conclusions and future work in Section 5.

1.1 Feasible Scenario Fillers

Let us consider that we want to find the most feasible thing eaten given the verb *to eat*. As *eat* has several senses, this filler could be *food*, or could be a *material*, depending on who is eating. See the following tables. On Table 1, the subject is *cow*. *Count* represents the total number of counts that voted for this combination divided by the number of sources. These tables are actual output of our system (KNN DLM).

Table 1. Feasible arguments for (*eat*, subject: *cow*)

	Verb	Argument 1	Argument 2	Count	Sources
1	eat	subj: cow	obj: hay	0.89	2
2	eat	subj: cow	obj: kg/d	0.49	1
3	eat	subj: cow	obj: grass	0.42	12

On Table 2, the subject is *acid*. It is possible to see the different adequate fillers depending on the subject doing the action.

Table 2. Feasible arguments for (*eat*, subject: *acid*)

	Verb	Argument 1	Argument 2	Count	Sources
1	eat	subj:acid	obj:fiber	2	2
2	eat	subj:acid	obj:group	1	4
3	eat	subj:acid	obj:era	0.66	2
4	eat	subj:acid	away	0.25	40
5	eat	subj:acid	obj:digest	0.19	2
6	eat	subj:acid	of:film	0.18	2
7	eat	subj:acid	in:solvent	0.13	1
8	eat	subj:acid	obj:particle	0.11	4
9	eat	subj:acid	obj:layer	0.10	3

If we consider the problem of estimating $P(a_2|v,a_1)$ instead of estimating only $P(a_2|v)$—where a_1 and a_2 are arguments, and v is a verb, the data sparseness problem increases. This has been solved mainly by using external resources such as WordNet [21,16,1]; semantic-role annotated resources, *i.e.* FrameNet, PropBank [18]; a named entity recognizer, *e.g.* IdentiFinder [23]; or other manually created thesaurus [12].

A motivation of this work is to find at which extent the information from the corpus itself can be used for estimating $P(a_2|v,a_1)$ without using additional resources. For this matter, several techniques are used for dealing with the data sparseness problem. We describe both of them in the next section.

2 Models for Plausible Argument Estimation

2.1 PLSI – Probabilistic Latent Semantic Indexing

We can regard the task of finding the plausibility of a certain argument for a set of sentences as estimating a word given a specific context. Since we want to consider argument co-relation, we have

$$P(v,r_1,n_1,r_2,n_2)$$

where v is a verb, r_1 is the relationship between the verb and n_1 (noun) as subject, object, preposition or adverb. r_2 and n_2 are analogous. If we assume that n has a different function when used with another relationship, then we can consider that r and n form a new symbol, called a. So that we can simplify our 5–tuple $P(v,r_1,n_1,r_2,n_2)$ to $P(v,a_1,a_2)$.

We want to know, given a verb and an argument a_1, which a_2 is the most plausible argument, *i.e.* $P(a_2|v,a_1)$. We can write the probability of finding a particular verb and two of its syntactic relationships as:

$$P(v,a_1,a_2) = P(v,a_1) \cdot P(a_2|v,a_1),$$

which can be estimated in several ways. Particularly for this work, we use PLSI [11] because we can exploit the concept of latent variables to deal with data sparseness. For PLSI we have:

$$P(v,a_1,a_2) = \sum_{z_i} P(z) \cdot P(a_2|z) \cdot P(v,a_1|z)$$

2.2 K-Nearest Neighbors Model

This model uses the k nearest neighbors of each argument to find the plausibility of an unseen triple given its similarity to all triples present in the corpus, measuring this similarity between arguments. See Figure 1 for the pseudo-algorithm of this model.

As votes are accumulative, triples that have words with many similar words will get more votes.

for each triple <v,a₁,a₂> with observed count c,

    for each argument a_1, a_2

        Find its k most similar words $a_{1s1}...a_{1sk}$ $a_{2s1}...a_{2sk}$ with similarities $s_{1s1}, ..., s_{1sk}$ and $s_{2s1},...,s_{2sk}$.

        Add votes for each new triple: $<v,a_{1si},a_{2sj}> += c·s_{1si}·s_{2sj}$

Fig. 1. Pseudo-algorithm for the K-nearest neighbors DLM algorithm

Common similarity measures range from Euclidean distance, cosine and Jaccard's coefficient [13], to measures such as Hindle's measure and Lin's measure [14]. Weeds and Weir [24] show that the distributional measure with best performance is the Lin similarity, so we used this measure for smoothing the co-occurrence space, following the procedure as described in [14].

3 Experiments

We conducted two experiments:

1. Comparing the effect of adding context, and
2. Comparing verb performance separately.

For all the experiments the evaluation scheme consisted on comparing the two models in a pseudo-disambiguation task following [24]. First we obtained triples $\langle v,a_1,a_2 \rangle$ from the corpus. Then, we divided the corpus in training (80%) and testing (20%) parts. With the first part we trained the PLSI model and created the co-occurrence matrix. This co-occurrence matrix was also used for obtaining the similarity measure for every pair of arguments a_2, a_2'. Then we are able to calculate *Feasibility* (v,a_1,a_2). For evaluation we created artificially 4-tuples: $\langle v,a_1,a_2,a_2' \rangle$, formed by taking all the triples $\langle v,a_1,a_2 \rangle$ from the testing corpus, and generating an artificial triple $\langle v,a_1,a_2' \rangle$ choosing a random a_2' with $r_2' = r_2$, and making sure that this new random triple $\langle v,a_1,a_2' \rangle$ was not present in the training corpus. The task consisted on selecting the correct triple.

3.1 Corpora

For our experiments we used the patent corpus from the NII Test Collection for Information Retrieval System, NTCIR-5 Patent [7], we parsed 7300 million tokens with the MINIPAR parser [15], and then we extracted the chain of relationships on a directed way, that is, for the sentence: *X add Y to Z by W*, we extracted the triples:

$$\langle add,\ subj\text{-}X,\ obj\text{-}Y \rangle, \langle add,\ obj\text{-}Y,\ to\text{-}Z \rangle,\ \text{and}\ \langle add,\ to\text{-}Z,\ by\text{-}W \rangle.$$

We obtained 177M triples in the form $\langle v,a_1,a_2 \rangle$.

Table 3. Pseudo-Disambiguation Task Sample: choose the right option

verb	arg	option 1	option 2
add	subj: I	obj: gallery	obj: member
calculate	obj: flowrate	subj: worksheet	subj: income
read	obj: question	answer	stir
seem	it	just	unlikely
go	overboard	subj: we	subj: they
write	subj: he	obj: plan	obj: appreciation
see	obj: example	in: case	in: london
become	subj: they	obj: king	obj: park
eat	obj: insect	subj: it	subj: this
do	subj: When	obj: you	obj: dog
get	but	obj: them	obj: function
fix	subj: I	obj: driver	obj: goods
fix	obj: it	firmly	fresh
read	subj: he	obj: time	obj: conclusion
need	obj: help	before	climb
seem	likely	subj: it	subj: act

3.2 Experiment I—Comparing the Effect of Adding Context

For this experiment we created a joint mini-corpus consisting of 1,000 triples for each of the verbs from the patent corpus: (*add, calculate, come, do, eat, fix, go, have, inspect, learn, like, read, see, seem, write*). We want to evaluate the impact of adding more information for verb argument prediction, so that we estimate the argument's plausibility given a verb $P(a_2|v)$, then we compare with using additional information from other arguments $P(a_2|v,a_1)$ for both models.

For completely new words it is not possible to do an estimate, henceforth, we measured precision and recall. Precision measures how many attachments were correctly predicted from the covered examples, and recall measures the correctly predicted attachment from the whole test set. We are interested on measuring the precision and coverage of these methods, so that we did not implemented any back off technique.

3.3 Experiment II—Verb Performance Compared Separately

In this experiment we measured the performance separately for each verb in the full model $P(a_2|v,a_1)$ to study the algorithms' performance with different amounts of information. For this experiment, we created a sample-corpus of 1000 random triples for 32 verbs totaling 32K triples. We created separately test and train corpus for each verb covering high-frequency and low-frequency verbs. For each one we extracted all the triples $\langle v,a_1,a_2 \rangle$ present in the corpus. On average there were 11,500 triples for each verb, ranging from 30 (*eat*) to 128,000 (*have*). We experimented with the PLSI algorithm and the co-occurrence space model algorithm and with different number of

Table 4. Results for each verb for 10 neighbors (KNN) and 10 topics (PLSI)

verb	triples	EXPANSOR KNN-10		PLSI-10	
		P	R	P	R
eat	31	0.98	0.92	1.00	0.04
seem	77	0.88	0.09	0.64	0.38
learn	204	0.82	0.10	0.57	0.22
inspect	317	0.84	0.19	0.43	0.12
like	477	0.79	0.13	0.54	0.24
come	1,548	0.69	0.23	0.78	0.17
play	1,634	0.68	0.18	0.69	0.19
go	1,901	0.81	0.25	0.80	0.15
do	2,766	0.80	0.24	0.77	0.19
calculate	4,676	0.91	0.36	0.81	0.13
fix	4,772	0.90	0.41	0.80	0.13
see	4,857	0.76	0.23	0.84	0.20
write	6,574	0.89	0.31	0.82	0.15
read	8,962	0.91	0.36	0.82	0.11
add	15,636	0.94	0.36	0.81	0.10
have	127,989	0.95	0.48	0.89	0.03
average	11,401	0.85	0.30	0.75	0.16

topics for the latent variable z in PLSI, and with different number of neighbors from the Lin thesaurus for expanding the co-occurrence space. Results are shown in Figure 3. Detailed information is shown in Table 4.

4 Analysis

Operating separately on verbs (one mini-corpus per verb) yields better results for PLSI (precision above 0.8) while seems not affecting EXPANSOR KNN. For little context $P(a_2|v)$, PLSI works better than EXPANSOR KNN, for more context, $P(a_2|v,a_1)$ EXPANSOR KNN works better.

In general, PLSI prefers a small number of topics, even for a large corpus (around 20 topics for the largest corpus of experiments). EXPANSOR KNN seems to improve recall steadily when adding more neighbors, loosing a small amount of precision. Expanding with few neighbors (1~5) seems not to be very useful. Particularly it is possible to see in Figure 2 that when recall is very low, precision can go very high or very low. This is because when so few cases are solved, performance tends to be random. In general, results for recall seem low because for we did not use any back

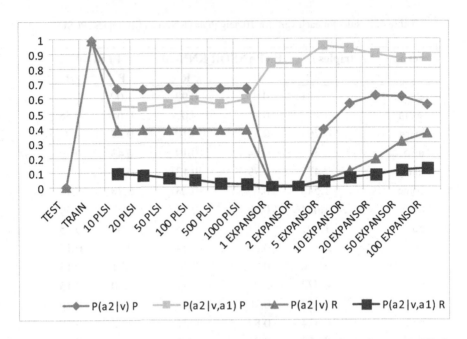

Fig. 2. Raw effect of adding more context: prediction based only on the verb versus prediction based on the verb plus one argument. EXPANSOR is the proposed KNN-based model.

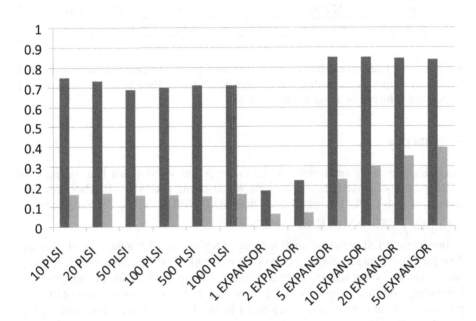

Fig. 3. Precision (dark bar) and Recall results for PLSI-model with number of topics and the KNN (Expansor) DML showing the number of neighbors

off method. If we compare the precision of the EXPANSOR KNN full model (based on more context) we can think of backing off to PLSI based on pairs $P(a_2|v)$. This would yield the best results, and it is left as future work.

5 Conclusions and Future Work

We have evaluated two different Dependency Language Models with a pseudo-disambiguation test. The KNN based model outperforms the PLSI approach when data sparseness is increased by adding more data. Effective smoothing is achieved by voting using similarity measures from the Lin distributional thesaurus.

As a future work we should explore the effect of using other similarity measures, as well as constructing a similarity table with simpler objects—a single noun instead of a composite object. We plan to explore specific applications of the obtained database, such as improving parsing [4]; Inference of meaning of unknown words—Uttering I eat *borogoves* makes us think that a *borogove* might be edible; Co-reference resolution [20]; WSD [16]; Semantic Plausibility [19], or Malapropism detection [2,3].

References

1. Agirre, E., Martinez, D.: Learning class-to-class selectional preferences. In: Workshop on Computational Natural Language Learning, ACL (2001)
2. Bolshakov, I.A., Galicia-Haro, S.N., Gelbukh, A.: Detection and Correction of Malapropisms in Spanish by means of Internet Search. In: Matoušek, V., Mautner, P., Pavelka, T. (eds.) TSD 2005. LNCS (LNAI), vol. 3658, pp. 115–122. Springer, Heidelberg (2005)
3. Budanitsky, E., Graeme, H.: Semantic distance in WorldNet: An experimental, application-oriented evaluation of five measures. In: NAACL Workshop on WordNet and other lexical resources (2001)
4. Calvo, H., Gelbukh, A., Kilgarriff, A.: Automatic Thesaurus vs. WordNet: A Comparison of Backoff Techniques for Unsupervised PP Attachment. In: Gelbukh, A. (ed.) CICLing 2005. LNCS, vol. 3406, pp. 177–188. Springer, Heidelberg (2005)
5. Clarkson, P.R., Rosenfeld, R.: Statistical Language Modeling Using the CMU-Cambridge Toolkit. In: Procs. ESCA Eurospeech (1997)
6. Foley, W.A.: Anthropological linguistics: An introduction. Blackwell Publishing, Malden (1997)
7. Fuji, A., Iwayama, M. (eds.): Patent Retrieval Task (PATENT). Fifth NTCIR Workshop Meeting on Evaluation of Information Access Technologies: Information Retrieval, Question Answering and Cross-Lingual Information Access (2005)
8. Gao, J., Suzuki, H.: Learning of dependency structure for language modeling. In: Procs. of the 41st Annual Meeting on Association for Computational Linguistics, Annual Meeting of the ACL archive, vol. 1 (2003)
9. Gao, J., Nie, J.Y., Wu, G., Cao, G.: Dependence language model for information retrieval. In: Procs. of the 27th annual international ACM SIGIR conference on Research and development in information retrieval, pp. 170–177 (2004)
10. Gelbukh, A., Sidorov, G.: On Indirect Anaphora Resolution. PACLING 1999, pp. 181–190 (1999)

11. Hoffmann, T.: Probabilistic Latent Semantic Analysis. Uncertainity in Artificial Intelligence, UAI (1999)
12. Kawahara, D., Kurohashi, S.: Japanese Case Frame Construction by Coupling the Verb and its Closest Case Component. In: 1st Intl. Conf. on Human Language Technology Research, ACL (2001)
13. Lee, L.: Measures of Distributional Similarity. In: Procs. 37th ACL (1999)
14. Lin, D.: Automatic Retrieval and Clustering of Similar Words. In: Procs. 36th Annual Meeting of the ACL and 17th International Conference on Computational Linguistics (1998)
15. Lin, D.: Dependency-based Evaluation of MINIPAR. In: Proc. Workshop on the Evaluation of Parsing Systems (1998)
16. McCarthy, D., Carroll, J.: Disambiguating Nouns, Verbs, and Adjectives Using Automatically Acquired Selectional Preferences. Computational Linguistics 29(4), 639–654 (2006)
17. McCarthy, D., Koeling, R., Weeds, J., Carroll, J.: Finding predominant senses in untagged text. In: Procs 42nd meeting of the ACL, pp. 280–287 (2004)
18. Padó, S., Lapata, M.: Dependency-Based Construction of Semantic Space Models. Computational Linguistics 33(2), 161–199 (2007)
19. Padó, U., Crocker, M., Keller, F.: Modeling Semantic Role Plausibility in Human Sentence Processing. In: Procs. EACL (2006)
20. Ponzetto, P.S., Strube, M.: Exploiting Semantic Role Labeling, WordNet and Wikipedia for Coreference Resolution. In: Procs. Human Language Technology Conference, NAACL, pp. 192–199 (2006)
21. Resnik, P.: Selectional Constraints: An Information-Theoretic Model and its Computational Realization. Cognition 61, 127–159 (1996)
22. Rosenfeld, R.: Two decades of statistical language modeling: where do we go from here? Proceedings of the IEEE 88(8), 1270–1278 (2000)
23. Salgueiro, P., Alexandre, T., Marcu, D., Volpe Nunes, M.: Unsupervised Learning of Verb Argument Structures. In: Gelbukh, A. (ed.) CICLing 2006. LNCS, vol. 3878, pp. 59–70. Springer, Heidelberg (2006)
24. Weeds, J., Weir, D.: A General Framework for Distributional Similarity. In: Procs. conf on EMNLP, vol. 10, pp. 81–88 (2003)

Supervised Recognition of Age-Related Spanish Temporal Phrases*

Sofia N. Galicia-Haro[1] and Alexander F. Gelbukh[2]

[1] Facultad de Ciencias
Universidad Nacional Autónoma de México, Mexico, D. F.
sngh@fciencias.unam.mx
[2] Centro de Investigación en Computación
Instituto Politécnico Nacional, Mexico, D. F.
gelbukh@gelbukh.com

Abstract. This paper reports research on temporal expressions shaped by a common temporal expression for a period of years modified by an adverb of time. From a Spanish corpus we found that some of those phrases are age-related expressions. To determine automatically the temporal phrases with such meaning we analyzed a bigger sample obtained from the Internet. We analyzed these examples to define the relevant features to support a learning method. We present some preliminary results when a decision tree is applied.

Keywords: temporal expressions, learning supervised method, decision tree.

1 Introduction

Some words or whole sequences of words in a text are temporal expressions: for example, *today*, *Sunday 1st*, *two weeks*, *about a year and a half* each refer to a certain period of time. Such words or sequences of words mainly share a noun or an adverb of time: *today*, *week*, *year*. This presents difficulties in automatically deciding whether a word or a sequence is a temporal expression. It is an important part of many natural language processing applications, such as question answering, machine translation, information retrieval, information extraction, text mining, etc., where robust handling of temporal expressions is necessary.

The Named Entity Recognition task of the Message Understanding Conferences considered the automatic recognition of expressions of time, whereby temporal entities were tagged as TIMEX. In [11] only "absolute" time expressions were tagged. They were categorized as three types: (1) date, complete or partial date expression, (2) time, complete or partial expression of time of day, and (3) duration, a measurement of time elapsed or a period of time during which something lasts. The absolute time expression must indicate a specific segment of time: for example, an expression of minutes must indicate a particular minute and hour, such as "20 minutes after ten",

---

* Work partially supported by the Mexican Government (CONACyT, SNI, CGPI-IPN, PIFI-IPN).

A. Hernández Aguirre et al. (Eds.): MICAI 2009, LNAI 5845, pp. 145–156, 2009.
© Springer-Verlag Berlin Heidelberg 2009

not "a few minutes after the hour," "a few minutes after ten," or "20 minutes after the hour". Determiners that introduced the expressions were not tagged. Words or phrases modifying the expressions (such as "around" or "about") were not tagged either.

Researchers have been developing temporal annotation schemes, for example [1] for English, where the authors produced a guideline intended to support a variety of applications. They were interested in the same temporal expressions: calendar dates, times of day, and durations. They broadened the annotation to introduce the specific values of the temporal expressions. For example, in the sentence: *Police said the 31-year-old Briton died Thursday*, both *31-year-old* and *Thursday* were marked as TIMEX, the first one with a value VAL="P31Y". Nevertheless they restricted the temporal expression recognition. Adverbs like meanwhile, still, before, etc. were not considered as part of the expressions, although they are temporal in their semantics, and are as a class less amenable to being pinned down to a timeline.

In this work, we analyzed other temporal expressions not considered in the previous works described. These phrases are recognized by an initial adverb: for example, *around, still*; and they end with a noun of time such as *year, month*. For example: *aún en los setentas* "still in the seventies", *alrededor de año y medio* "about a year and a half". We found that this type of phrase presents interesting cases. There is, for example, a group that describes a person's age. Automatic recognition of a person's age should be useful in question answering and machine translation tasks, inter alia. For example, consider the following sentence:

Aún a sus ocho años, Venus era muy confiada, diciendo que podía haber vencido a McEnroe. "Even at the age of eight years, Venus was very confident, saying he could have beaten McEnroe".

Spanish native speakers would understand that the phrase *aún a sus 8 años* denotes Venus's age. General machine translators, however, give the wrong phrases: 'yet his 8 years' or 'although his 8 years'. Also, this sentence would be retrieved by a question answering system to the specific question "How old was Venus?"

In this article, we present a corpus-based analysis carried out to determine the features for automatic determination. First, we present the characteristics of the temporal expressions modified by an adverb. Immediately after that, we present the automatic acquisition of the training examples. Then we present the analysis of the local context of the training examples. Finally we analyze the features required to develop a learning method and some preliminary results when a decision tree is applied.

2 Temporal Expressions Modified by an Adverb

Since the main job of adverbs is to modify verbs, adverbs include words that indicate the time in which the action of the verb takes place. These are some of the adverbs that are not based on adjectives. Adverbs of time (for example: *now, then, tomorrow*) create cohesion and coherence by forming time continuity in the events expressed in texts [9]. They are closely associated with narrative texts but they also appear in newspaper texts: for example, *The players will report today, undergo physicals Wednesday, then go on the ice for the next three days.*

Works on temporal expressions have considered adverbs of time, for example [1, 12, 13]. But they chose adverbs with precise time meaning: for example, lately, hourly, daily, monthly, etc. We chose to analyze the Spanish temporal phrases that are modified by any adverb of time (AdvT) [7] placed at the beginning of the phrase followed by a noun of time (TimeN): for example, *antes de 10 días* "before ten days", *recién 10 años* "lit. newly 10 years".

These phrases present interesting issues. For example, consider the following sentence:

A sus 80 años Juan tiene pulso de cirujano

We can modify the temporal expression *a sus 80 años* by an adverb. For example: *Aún a sus 80 años Juan tiene pulso de cirujano*; *Hoy a sus 80 años Juan tiene pulso de cirujano*; *Últimamente a sus 80 años Juan tiene pulso de cirujano*, etc. The sentences describe the same main fact: *John, who is 80 years old, has surgeon's pulse*. They tell us something else, however, when we introduce the modifier: *aún* "still" argues that in spite of his age he can have the firm pulse, *hoy* "today" concludes that today he has it, and *últimamente* "recently" suggests that in the recent past he has had a firm pulse. The adverbs make such conclusions obligatory and reinforce the meaning of time in different forms.

2.1 Examples in Newspapers

We obtained examples of these temporal expressions modified by an adverb of time from a text collection. The collection was compiled from a Mexican newspaper that is published daily on the Web almost in its entirety. The texts correspond to diverse sections, economy, politics, culture, sport, etc., from 1998 to 2002. The text collection has approximately 60 million words [2].

We wrote a program to extract the sentences matching the following pattern:

AdvT–*something*–TimeN

Adverbs of time (AdvT) are a set of 51 elements:[1] *actualmente, adelante, ahora, alrededor, anoche, antaño, anteayer, anteriormente, antes, aún, ayer, constantemente, cuando, de antemano, dentro, dentro de poco, después, en seguida, enseguida, entonces, finalmente, hasta este momento, hogaño, hoy, inmediatamente, instantáneamente, jamás, luego, mañana, más tarde, más temprano, mientras, mientras tanto, momentáneamente, nunca, ocasionalmente, por horas, posteriormente, previamente, prontamente, pronto, recién, recientemente, siempre, simultáneamente, tarde, temprano, todavía, últimamente, una vez, ya.*

Something corresponds to a sequence of up to six words[2] without punctuation marks, verbs or conjunctions.

Noun of time (TimeN) corresponds to: *año* "year", *mes* "month", *día* "day", *hora* "hour", *minuto* "minute", *segundo* "second".

The extracted sentences were analyzed in [3]. From a subset of those examples where the adverb corresponds to *actualmente* "at present', *ahora* "now', *alrededor* "around', *aún* "still', and the noun of time corresponds to *año* "year', we found some phrases expressing age of persons.

---

[1] DRAE, Real Academia Española. (1995): *Diccionario de la Real Academia Española*, 21 edición (CD-ROM), Espasa, Calpe.

[2] A larger quantity of words guarantees no relation between the AdvT and the TimeN.

2.2 Expressions Denoting Age of Persons

Usually the age of persons is described by Spanish temporal expressions including the time nouns *años* "years" and *meses* "months" (for babies). They can be recognized in the following ways:

String	Context	Example
de edad "old"	Right context	*María tiene 8 años de edad* "Mary is 8 years old"
number	delimited by commas	*María, de 8 años,*
la edad de, "the age of"	Left context	*María a la edad de 8 años*
de edad de, "the age of"	Left context	*María de edad de 8 años*

The first two cases were considered in [10], the remaining cases corresponding to syntactic variants. There are other temporal expressions, however, that describe the age of person: for example,[3] *aún a tus 15 años*, "still at yours 15 years", *ahora con casi 20 años*, "today with almost 20 years". In the sentences, these temporal phrases could denote a point in the timeline of a person, a point in the timeline of the events related in the sentence or a point in a tangential timeline.

From the previous results we manually select some examples, each one representing what we consider to be a class: a different combination of an adverb and a preposition before the quantity of years. The five resulting classes correspond to *aún a, aún con, actualmente de, alrededor de, ahora de*.

3   Automatic Acquisition of the Training Examples

Since our newspaper text collection contains a subset of all possible temporal phrases expressing the age of persons we analyzed a method to obtain a more representative group of phrases. There are different possibilities for obtaining such variants, and we consider the following ones:

- To look in other text collections for sentences that fulfill the previous pattern
- To look for collocations of the nouns of time that we choose and to generate new phrases with the previous pattern
- To look for examples on the Internet.

The first possibility is time-consuming because it would mean individual text searching since huge text collections in Spanish are not available in entirety. As regards the second option, since collocations are syntactic units we should also consider many combinations (adverb-adjective, adjective-noun, determinants, possessives, etc.). In addition, synonyms are frequently used, being determined by the context and applied with certain freedom.

---

[3] The translations of the Spanish examples are literal.

On the other hand, the third option would allow us to find in less time phrases generated by native speakers including commoner collocations. Nevertheless, we know that search the Internet has problems, as [8] has already described, principally owing to the arbitrariness of the statistics of the search engines: for example, missing pages or copies. We also know that the search engines deliver different answers to repetitions of the same question. Although these problems are not so significant in this work, it is important that the results are classified in accordance with complex and unknown algorithms, so we cannot know what predispositions have been introduced. Despite these considerations, we decided to search the Internet on the grounds that we do not know how the results are classified.

The main idea of obtaining more examples from the Internet is based on obtaining a few examples from the newspaper texts (corresponding to the five above-mentioned classes), simplifying them (eliminating determinants, adjectives, etc.) and searching for variants by including Google's asterisk facility [4]. The whole procedure consists of the following steps:

SEARCH(C)
For each phrase of type ADV-*-NounT or string-*-NounT in C
(1) Obtain 100 examples from the Internet
    (1.1) D = {examples excepting such instances where * includes verbs or punc-
          tuation}
    (1.2) Print D
(2) Classify them according to such words retrieved by *
(3) For each group of phrases sharing words retrieved by *, assign a class D_i
    (3.1) F = class Di
    (3.2) SEARCH(F)
UNTIL no new elements are obtained.

For example: for the phrase *ahora a mis 25 años*, the string when simplified becomes "*ahora a años*" and the search is "*ahora a * años*" using the Google search engine tool limited to the Spanish language where the asterisk substitutes for the eliminated words. Google returns hits where there is a string of words initiated by "*ahora a*", then a sequence of words, ending with "*años*".

The example for the whole procedure in Figure 1 is presented as follows:

SEARCH("*ahora a * años*")
Step (1) 100 examples
… Sólo mírate de *ahora a cinco años* …
Eso, viéndolo *ahora a tantos años* de distancia**...**
… *Ahora a diez años* de distancia volvemos a reflexionar**...**
ahora a mis 35 años tengo exactamente la psicología y el destino de mi heroína...
… al enfrentarnos *ahora a los años* finales de Carlos I y de Felipe II …
… Situados *ahora a 150 años* de distancia del nacimiento de Wagner …
…

Table 1. Overall results for the examples obtained from the Internet

Type of phrase	# examples	% age-related	# short snippet	# no age
aún a sus NUM años	293	96	7	5
aún hoy a sus NUM años	38	92	2	1
aún a tus NUM años	7	100	0	0
ahora a mis NUM años	182	99	1	1
aún con mis pocos NUM años	1	100	0	0
aún con mis cortos NUM años	2	100	0	0
aún con sus escasos NUM años	4	100	0	0
aún con tus casi NUM años	1	100	0	0
aún con sus casi NUM años	7	57	0	3
aún con sus NUM años	109	86	0	15
ahora de NUM años	352	86	6	45
de alrededor de NUM años	353	44	4	194
actualmente con NUM años	270	80	3	50
actualmente de NUM años	28	36	7	11
actualmente de unos NUM años	16	19	1	12
ahora con casi NUM años	118	67	0	39
ahora con más de NUM años	90	46	7	45
ahora a NUM años	112	0.9	8	103
ahora a los NUM años	132	36	2	82
alrededor de NUM años	242	16	9	194
alrededor de los NUM años	355	84	4	54

Step (2)   D= {
ahora a tantos años,
ahora a cinco años,
Ahora a diez años,
ahora a mis 35 años,
ahora a los años,
ahora a 150 años ... }
Step (3) For each one of the classes a new process is initiated
SEARCH(*"ahora a tantos * años"*)
SEARCH(*"ahora a mis * años"*)
...

The process is repeated several times until no new repeated phrases are obtained. At the end, the sequences of words are classified to obtain those appearing with greater frequency. Because of the simplicity of the method, some phrases not corresponding to the temporal phrases we are interested in are picked up: for example, *Ahora a diferencia de años anteriores* ... "Now in contrast to previous years". There is a manual identification at the end of the whole process that deletes this kind of phrase.

We manually selected 21 classes corresponding to age-related temporal phrases. They appear in the first column of Table 1, where NUM considers numbers represented by digits or letters.

Table 2. Results of context analysis for the classes maintaining the age-related meaning

	Age-related		No age	
	Right context	**Left context**	**Right context**	**Left context**
aún a sus NUM años	de vida (4) de existencia de edad (44) cumplidos (4) VERBS (148) NAMES (6) PUNCT (40)		de antigüedad de viajar de experiencia de tradición de viuda	
aún hoy a sus NUM años	de edad (4) VERBS (20) PUNCT (11)	CONJ (23) PUNCT (9)	de partida	
aún a tus NUM años	PUNCT (4)			
ahora a mis NUM años	de edad (11) PUNCT (39)	PUNCT (38) CONJ (72)	de casada	
aún con sus escasos NUM años	PUNCT (4)			
ahora de NUM años	*PUNCT (233) de edad (26)	* NAM/PN PUNCT (233) NAM/PN (18)	*ahode* list	SER (27) periodo disponer (4)

4 Analysis of the Training Examples

We manually analyzed some examples for each one of the 21 obtained classes. We found that in some cases the meaning of the age-related temporal phrase was independent of the surrounding words. To determine automatically the meaning of these temporal phrases a more thorough analysis involving a larger sample is required.

For this purpose, we obtained examples by searching again on the Internet. The quantity of pages automatically obtained for each search was limited to 50, i.e. to obtain a maximum of 500 snippets. The overall results are presented in Table 1. The first column corresponds to the 21 classes. The second column shows the quantity of examples obtained, after the elimination of phrases where there was no relation between the AdvT and the TimeN.

Since the examples were automatically obtained, some of them were not considered because of the lack of text presented in the snippet or because the phrase was missing in the snippet. Column 4 gives the number of the eliminated examples.

Columns 3 and 5 show the results after the general syntactic and semantic analysis of the context in each sentence for the recognition or otherwise of age-related meaning respectively.

We manually analyzed the local context in the examples. Table 2 summarizes the results for the cases in Table 1 with more than 85% of age-related phrases. The column "Age-related" comprises the right and left context for the phrases denoting age of person. Column "No age" comprises the right and left context for the phrases not denoting age of person. Table 3 summarizes the results for the cases in Table 1 with less than 85% of age-related phrases. The patterns appearing in the right and left context correspond to the more general and more interesting ones; for example, in the class *aún a sus* NUM *años* there are 34 phrases missing since they require many different patterns.

Table 3. Results of context analysis for the classes with age-related meaning based on context

	Age-related		No age	
	Right context	**Left context**	**Right context**	**Left context**
aún con sus casi NUM años	VERB PUNCT (2) de edad	PUNCT	en el mercado de servicio de vigencia	
aún con sus NUM años	*acsus* list	PUNCT (15)	*No_acsus* list	
de alrededor de NUM años	de edad (11)	NM/PN PT (15) NAM/PN (117)	*Nordealde* list	*Noldealde* list
actualmente con NUM años	*actcon_r* list	*actcon_l* list	*No_actcon* list	
actualmente de NUM años	de edad (5)	NAM/PN (5)		
actualmente de unos NUM años	de edad NAM/PN PUNCT			SER (12)
ahora con casi NUM años	PUNCT VERB (21) VERB (34)	¡ {CONJ, PUNCT} (30)	{¡} *Noahocca-si* list	de abandono(2)
ahora con más de NUM años	*ahconmade* list		*No_ahocmas*	hasta (4)
ahora a NUM años	de edad			
ahora a los NUM años	de edad VERB (21)	porque (4)	*No_ahoalos* list	
alrededor de NUM años	{< 100} de edad (4)	{< 100} TENER (28)		hace (47) durante (8)
alrededor de los NUM años	de edad (51)		{< 100} (22)	

In the tables: VERB means a verb related to the adverb, NAMES correspond to person's name, PUNCT comprises comma, parenthesis, semicolon, CONJ corresponds to conjunctions introducing new phrases, NAM/PN corresponds to name or personal noun, SER comprises "to be" conjugation, TENER and CONTAR the "to have" verb. An asterisk means that the left and right contexts are matched. {< 100} means NUM value lower than 100. {¡} means it excludes context for "No age" cases.

We can see that the classes *aún a sus* NUM *años* and *ahora de* NUM *años* are the best examples for context identification. The worst case is *alrededor de los* NUM *años*, where we notice that almost all phrases indicate an age but in a general form: for example, *el consumo de frutas se da sobre todo alrededor de los 60 años* "the fruit consumption is mostly seen about 60 years old".

The class *ahora de* NUM *años* shows an interesting property: many age-related examples have right and left context matching that includes punctuation, isolating the temporal phrase and giving a context independent meaning.

Contexts for phrases that are not age-related share prepositional phrases modifying the noun time: for example, *7 años de casada* "married for seven years", *7 años de cárcel* "seven years in jail".

Some of the lists indicated in both tables by italics are enumerated in the following paragraphs and the number of cases is shown in parentheses when it is greater than 1:

ahode: de antigüedad, de becas, de cárcel, de casados, de duración, más (5)

acsus: PUNCT (42), VERBO (21), CONJ (3), encima (3), a cuestas (4), a las espaldas, cumplidos, de edad (14), de vida (5)

dealde: de antiguedad (15), de duración (2, de evolución, de fallecido, de gobierno (2), de investigación, de investigar, de matrimonio, de persecución, de políticas, de prisión (2), de reformas, de trabajo, luz (9)

No_dealde: una antigüedad (7), datan (19), después (11), distancia (5), duración (2), SER (17), experiencia (3), lapso (2), luego (4), vida media (2), edad media, período (22), vida (3), vida conyugal.

The analysis sheds light on the preservation of the meaning. We wrote a program to classify the results of the left and right contexts. The results are presented in Table 4, where the examples for each class are separated into age-related examples and non-age-related examples.

Table 4. Results of context classification

	Age-related		No age	
	RC	LC	RC	LC
aún a sus NUM años	88%		100%	
aún hoy a sus NUM años	35%		100%	
aún a tus NUM años	57%			
ahora a mis NUM años		61%	100%	
aún con sus escasos NUM años	100%			
ahora de NUM años	86%			71%
aún con sus casi NUM años	100%		100%	
aún con sus NUM años	89%		100%	
de alrededor de NUM años		85%		51%
actualmente con NUM años	48%		70%	
actualmente de NUM años	50%			
actualmente de unos NUM años	100%			100%
ahora con casi NUM años	37%		74%	
ahora con más de NUM años	64%		100%	
ahora a NUM años	100%			
ahora a los NUM años	46%		18%	
alrededor de NUM años		72%		28%
alrededor de los NUM años	17%		41%	

The percentages show the number of examples which have a specific context according to the local context analysis. RC means right context and LC means left context.

We found that most of the classes have enough patterns to enable their recognition as either age-related or non-age-related. The worst cases (*aún a tus* NUM *años, actualmente de* NUM *años, ahora a los* NUM *años, alrededor de los* NUM *años*) have patterns for 40 to 60% of examples. We therefore concluded that automatic recognition can be based on the features obtained from the analysis.

5 Decision Tree as the Recognition Model

For automatic determination of the meaning of the temporal expressions that we analyzed one approach is to develop a recognition model as a binary classifier. The classifier goes through a list of possible expressions, classifying them into two classes: age meaning and no-age meaning. These two classes correspond to the concept that our system must learn.

Among the various types of learning approaches, we chose supervised inductive learning. All training examples state to which class they are assigned. After seeing all training data, the learner selects a hypothesis, one that can reliably approximate the target concept. This hypothesis is formalized as a decision tree in the decision tree learning approach that we selected [6].

In the decision tree each internal node represents a test of one of the attributes of the training data examples and the node's branches are the different values of such attributes. The decision tree contains all the necessary tests on all attributes obtained from the examples. The first attribute tested is the one with the highest information gain. The other attributes are recursively obtained according to the information gain criterion which eliminates the attributes and the examples previously classified by them. We applied DTREG Software[4] in the implementation of the decision tree. DTREG has many settings but, always trying to find the simplest tree, we chose to calculate entropy and the setting considering a mix of frequency distribution in the data set and balanced misclassification for smoothing.

Training data in this work consist of the collection of temporal expressions presented in Table 1. A number of attributes have been selected to describe the examples. These attributes are naïve approximations of linguistic properties that we described in Tables 2 and 3, and may be summarized as follows:

- The part of speech of the left and right contexts
- The prepositional phrase selected by the time noun
- The entity names in the left and right contexts
- The punctuation sign in the left and right contexts
- Specific words in context (*ser, contar, encima,* etc.)
- Number of years.

To obtain these attributes, context information was considered in this work as two words in a window surrounding the target phrase without grammatical relations of the whole sentence in an automatic form. We applied the system for automatic morphological analysis of Spanish [5] to assign parts of speech to the words of the context.We wrote a program to classify the examples according to the context, to obtain noun phrases and to eliminate temporal phrases not matching the pattern AdvT-something-TimeN.

We applied the decision tree in a sample taken from another Mexican newspaper. We search in this collection for the phrases shown in the first column of Table 4.

---

[4] http://www.dtreg.com/

Phrase	Total	Correct detected	Correct
aún a *años	100	10	11
ahora de * años	100	39	41
ahora con casi *años	6	0	0
actualmente de *años	100	20	22

Although the number of examples is small, we could observe that the analyzed context is very similar for these kinds of phrases since the precision is very near the 100 percent mark. Also, it should be noted that we used the Google search engine (www.google.com) for the Spanish language so the analyzed context in Tables 2 and 3 corresponds to examples where several dialectal variations of Spanish are considered.

In nearly every case one faulty phrase was omitted. Two errors correspond to the lack of full syntactic parsing since an adjective was not identified as part of the adverbial phrase and a verb was missing because of the introduction of a circunstancial complement. One error was owed to the entity name recognizer. It was introduced to the inability to recognize a personal noun.

These errors show that the lists of phrases corresponding to local context presented in Tables 2 and 3 should be the base for future work on lexical, morphological, syntactic, and semantic categories for these temporal expressions' structural description in order to obtain a robust system. Other work will comprise the analysis of the effect of new attributes like the relation between left and right contexts.

We have omitted in this work a learning method to recognize expressions not appearing in the search phase. A second phase will consider the annotation of plain texts with the decision tree to obtain similar patterns in context that could vary in modifiers such as determinants, adjectives, and numbers.

6 Conclusions

The variety in the structure of temporary expressions necessitates the analysis of different combinations of classes of words. We analyzed temporal expressions including the noun of time *year* that are modified by an adverb of time and the whole phrase expressing a person's age.

We first presented a method to enrich the classes of temporal phrases when only a few examples are compiled. To obtain a more representative sample we compiled examples from the Internet for each class. We manually analyzed the context surrounding them to define the specific context for such expressions in order to identify them automatically. Specific context was obtained for nine of 21 classes.

The classification was carried out by a well-known machine learning algorithm known as a decision tree and based on direct attributes. We obtained remarkable results by applying the decision tree to just a few examples from 300 sentences.

The automatic identification of these phrases and their interpretation will directly benefit natural language processing tasks including: response to questions; visualization of events in lines of time; generation of phrases; translation, etc. For example, questions on age are very common in demographic and health questionnaires and, if the interviewees feel comfortable speaking in the usual form, a system with this automatic recognition of age-related temporal expressions could translate such expressions to data repositories.

References

1. Ferro, L., Gerber, L., Mani, I., Sundheim, B., Wilson, G.: TIDES 2003 Standard for the Annotation of Temporal Expressions. MITRE Corporation (2004)
2. Galicia-Haro, S.N.: Using Electronic Texts for an Annotated Corpus Building. In: 4th Mexican International Conference on Computer Science, ENC 2003, Mexico, pp. 26–33 (2003)
3. Galicia-Haro, S.N.: Spanish temporal expressions: Some forms reinforced by an adverb. In: Gelbukh, A., Morales, E.F. (eds.) MICAI 2008. LNCS (LNAI), vol. 5317, pp. 193–203. Springer, Heidelberg (2008)
4. Gelbukh, A., Bolshakov, I.A.: Internet, a true friend of translators: the Google wildcard operator. International Journal of Translation 18(1–2), 41–48 (2006)
5. Gelbukh, A., Sidorov, G.: Approach to construction of automatic morphological analysis systems for inflective languages with little effort. In: Gelbukh, A. (ed.) CICLing 2003. LNCS, vol. 2588, pp. 215–220. Springer, Heidelberg (2003)
6. Han, J., Kamber, M.: Data Mining: Concepts and Techniques. Morgan Kaufmann Publishers, San Francisco (2006)
7. Cuadrado, H., Alberto, L.: Gramática del adverbio en español. In: Dykinson (ed.) (2006)
8. Kilgarriff, A.: Googleology is Bad Science. Computational Linguistics 33, 147–151 (2007)
9. Llido, D., Berlanga, R., Aramburu, M.J.: Extracting temporal references to assign document event-time periods. In: Mayr, H.C., Lazanský, J., Quirchmayr, G., Vogel, P. (eds.) DEXA 2001. LNCS, vol. 2113, pp. 62–71. Springer, Heidelberg (2001)
10. Mandel, M., Walter, C.: Pautas para la anotación del Tiempo para Lenguas poco enseñadas (basado en los estándares de TIMEX2) Versión 1.0 Consorcio de Datos Lingüísticos (2006)
11. Named Entity Task Definition (v2.1), Appendix C: Proceedings of the Sixth Message Understanding Conference (MUC-6). Columbia, MD, 317–332 (1995)
12. Saquete, E., Martinez-Barco, P.: Grammar specification for the recognition of temporal expressions. In: Proceedings of Machine Translation and multilingual applications in the new millennium. MT2000, Exeter, UK, pp. 21.1–21.7 (2000)
13. Saquete, E., Martinez-Barco, P., Muñoz, R.: Recognizing and tagging temporal expressions in Spanish. In: Workshop on Annotation Standards for Temporal Information in Natural Language, LREC 2002 (2002)

Using Nearest Neighbor Information to Improve Cross-Language Text Classification

Adelina Escobar-Acevedo, Manuel Montes-y-Gómez, and Luis Villaseñor-Pineda

Laboratory of Language Technologies, Department of Computational Sciences,
National Institute of Astrophysics, Optics and Electronics (INAOE), Mexico
{aescobar,mmontesg,villasen}@inaoep.mx

Abstract. Cross-language text classification (CLTC) aims to take advantage of existing training data from one language to construct a classifier for another language. In addition to the expected translation issues, CLTC is also complicated by the cultural distance between both languages, which causes that documents belonging to the same category concern very different topics. This paper proposes a re-classification method which purpose is to reduce the errors caused by this phenomenon by considering information from the own target language documents. Experimental results in a news corpus considering three pairs of languages and four categories demonstrated the appropriateness of the proposed method, which could improve the initial classification accuracy by up to 11%.

1 Introduction

Nowadays, there is a lot of digital information available from the Web. This situation has produced a growing need for tools that help people to find, filter and analyze all these resources. In particular, *text classification* (Sebastiani, 2002), the assignment of free text documents to one or more predefined categories based on their content, has emerged as a very important component in many information management tasks.

The state-of-the-art approach for text classification considers the application of a number of statistical and machine learning techniques, including Bayesian classifiers, support vector machines, nearest neighbor classifiers, and neuronal networks to mention some (Aas and Eikvil, 1999; Sebastiani, 2002). In spite of their great success, a major difficulty of this kind of supervised methods is that they require high-quality training data in order to construct an accurate classifier. Unfortunately, due to the high costs associated with data tagging, in many real world applications training data are extremely small or, what is even worst, they are not available.

In order to tackle this problem, three different classification approaches have recently proposed, each of them concerning a distinct circumstance. The first approach allow building a classifier by considering a small set of tagged documents along with a great number of unlabeled texts (Nigam et al., 2000; Krithara, et al., 2008; Guzmán-Cabrera et al., 2009). The second focuses on the construction of classifiers by reusing training sets from related domains (Aue and Gamon, 2005; Dai et al., 2007). Whereas, the third takes advantage of available training data from one language in order to construct a classifier that will be applied in a different language. In particular, this paper focuses on this last approach, commonly referred to as *cross-language text classification* (CLTC).

A. Hernández Aguirre et al. (Eds.): MICAI 2009, LNAI 5845, pp. 157–164, 2009.

As expected, one of the main problems that faces CLTC is the language barrier. In consequence, most current methods have mainly addressed different translation issues. For instance, some methods have proposed achieving the translation by means of multilingual ontologies (Olsson et al., 2005; Gliozzo and Strapparava, 2006; De Melo and Siersdorfer, 2007; Amine and Mimoun, 2007), while the majority tend to apply an automatic translation system. There are also methods that have explored the translation of complete documents as well as the translation of isolated keywords (Bel et al., 2003). In addition, there have been defined two main architectures for CLTC, based on the translation of the training and test corpus respectively (Rigutini et al., 2005; Jalam, 2003).

Although the language barrier is an important problem for CLTC, it is not the only one. It is clear that, in spite of a perfect translation, there is also a *cultural distance* between both languages, which will inevitably affect the classification performance. In other words, given that language is the way of expression of a cultural and socially homogeneous group, documents from the same category but different languages (i.e., different cultures) may concern very different topics. As an example, consider the case of news about sports from France (in French) and from US (in English); while the first will include more documents about soccer, rugby and cricket, the later will mainly consider notes about baseball, basketball and american football.

In this paper we propose a *post-processing method* for CLTC, which main purpose is to reduce the classification errors caused by the cultural distance between the source (training) and target (test) languages. This method takes advantage from the synergy between similar documents from the target corpus in order to achieve their *re-classification*. Mainly, it relies on the idea that similar documents from the target corpus are about the same topic, and, therefore, that they must belong to the same category.

The rest of the paper is organized as follows. Section 2 describes the proposed re-classification method. Section 3 details the experimental setup and shows the achieved results. Finally, Section 4 presents our conclusions and discusses some future work ideas.

2 The Re-classification Method

As we previously mentioned, our proposal for CLTC consists in applying a two stage process (refer to Figure 1). The function of the first stage is to generate an *initial classification* of the target documents by applying any traditional CLTC approach. On the other hand, the purpose of the second is to *rectify the initial classification* of each document by using information from their neighbors. Following we describe the main steps of the proposed CLTC method.

1. Build a classifier (C_l) using a specified learning method (l) and a given training set (S) in the source language.
   Depending on the used CLTC approach, the construction of this classifier may or may not consider the translation of the training corpus to the target language.

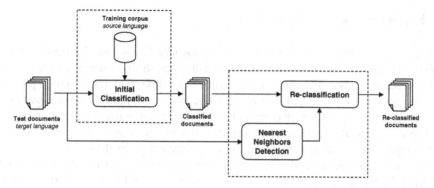

Fig. 1. Proposed two-step method for cross-language text classification

2. Classify each document (d_i) from the test set (T), in the target language, using the built classifier (C_l). The result of this step is the initial classification of the test documents. We represent the initial class of $d_i \in T$ as $c^0(d_i)$.

   Similar to the previous step, depending on the used CLTC approach, the documents from the test set may or may not be translated to the source language.

3. Determine the set of k nearest neighbors for each document $d_i \in T$, which is represented by NN_i.

   In our experiments we represented documents as set of words, and measured their similarity using the Dice coefficient. That is, the similarity between documents d_i and d_j is computed as indicated below; where $|d_x|$ indicates the number of words in document d_x, and $|d_i \cap d_j|$ their common vocabulary.

$$sim(d_i, d_j) = \frac{2 \times |d_i \cap d_j|}{|d_i| + |d_j|}$$

4. Modify the current class of each test document d_i (represented by $c^n(d_i)$), by considering information from their neighbors. We contemplate two different situations:

   a. If all neighbors of d_i belong to the same class, then:

$$c^{n+1}(d_i) = c^n(d_j): d_j \in NN_i$$

   b. In the case that the neighbors of d_i do not belong to the same class, maintain the current classification of d_i:

$$c^{n+1}(d_i) = c^n(d_i)$$

5. Iterate σ times over step 4 (being σ a user specified threshold), or repeat until no document changes their category. That is, iterate until:

$$\forall (d_i \in T): c^n(d_i) \equiv c^{n-1}(d_i)$$

3   Experiments

For the experiments we used a subset of the *Reuters Corpus RCV-1* (Lewis et al., 2004). We considered three languages: English, Spanish and French; and the news corresponding to four different classes: crime (GCRIM), disasters (GDIS), politics (GPOL) and sports (GSPO). For each language, we employed 200 news reports for training and 120 for test, which correspond to 50 and 30 news per class respectively.

The used evaluation measure was the classification *accuracy*, which indicates the percentage of test documents that were correctly categorized by the classifier.

Following we describe the performed experiments. In particular, Section 3.1 presents the results from two traditional approaches for CLTC, which correspond to our initial classification (refer to Figure 1); whereas, Section 3.2 presents the results achieved by the proposed re-classification method.

Table 1. Results from traditional approaches for cross-language text classification

Source language (training set)	Target language (test set)	Vocabulary (training set)	Vocabulary (test set)	Vocabulary intersection	Percentage intersection (w.r.t test set)	Accuracy
Translating training set to target language						
French	English	11338	7658	3700	48%	0.858
Spanish	English	9012	7658	3351	44%	0.817
French	Spanish	14684	8051	3920	49%	0.833
English	Spanish	13453	8051	3640	45%	0.717
Spanish	French	10666	9258	3793	41%	0.808
English	French	12426	9258	4131	45%	0.758
Translating test set to source language						
English	French	10892	7731	3697	48%	0.767
English	Spanish	10892	6314	3295	52%	0.750
Spanish	French	12295	9398	3925	42%	0.792
Spanish	English	12295	9190	3826	42%	0.850
French	Spanish	14071	7049	3749	53%	0.800
French	English	14071	8428	4194	50%	0.867

3.1   Traditional Cross-Language Classification

There are two traditional architectures for CLTC, one based on the translation of the training corpus to the target language, and the other based on the translation of the test corpus to the source language. Table 1 shows the results from these two approaches. In both cases, we used the Wordlingo free translator and performed the classification by means of a Naïve Bayes classifier based on word features with a Boolean weighting.

Results from Table 1 indicate that both architectures achieved similar results (around 80% of accuracy), being slightly better the one based on the translation of the test set to the source language. This table also evidences the enormous difference in the vocabularies from the training and test sets, which somehow reflects the relevance of the cultural distance problem.

Analyzing the results from cross-language text classification

Given that we had available training and test data for the three languages, we were able to perform the three monolingual classification experiments. Table 2 shows the results from these experiments. Somehow, these results represent an upper bound for CLTC methods.

Table 2. Results from the monolingual classification experiments

Source language (Training set)	Target language (test set)	Vocabulary (training set)	Vocabulary (test set)	Vocabulary intersection	Percentage intersection (w.r.t test set)	Accuracy
English	English	10892	7658	5452	71%	0.917
Spanish	Spanish	12295	8051	5182	64%	0.917
French	French	14072	9258	6000	65%	0.933

The comparison of results from Tables 1 and 2 evidences an important drop in accuracy for cross-language experiments with respect to the monolingual exercises. We presume that this effect is consequence of the small intersection between the source and target languages (30% less than for the monolingual exercises), which indicates that the training data do not contain all relevant information for the classification of the test documents. However, it is important to point out that it was not possible to establish a direct relation between the cardinality of this intersection and the classification accuracy.

With the aim of understanding the causes of the low intersection between the vocabularies of the training (source language) and test (target language) datasets, we carried out an experiment to evaluate the impact of the translation errors. In particular, we translated the training and test datasets from Spanish to English and French. Using these new datasets, we trained and evaluated two monolingual classifiers. The first classifier (with all data in English) achieved an accuracy of 91.66%, whereas, the second (with all data in French) achieved an accuracy of 90.83%. Comparing these results against the original accuracy from the Spanish monolingual exercise (91.66%), we may conclude that the lost of accuracy introduced by the translation process in practically insignificant, and, therefore, that the *cultural distance* arises as that main problem of cross-language classification; at least for these kinds of corpora. In other words, these results evidence that news from different countries (in different languages), even though belonging to the same category, tend to report very different events, which generates a great lexical discrepancy in the vocabularies, and, therefore, a noticeable decrement in the classification performance.

3.2 Results from the Re-classification Method

As we exposed in Section 2, the proposed method considers a first stage where an initial classification is performed by applying some CLTC approach, and a second stage where initial classifications are rectified by considering information from their neighbors.

In particular, the evaluation of the proposed re-classification method (second stage) considered the results from two traditional architectures for CLTC as the initial classifications (refer to Table 1), and employed information from 2 to 5 neighbors to modify or confirm these classifications. Table 3 shows the accuracy results from this experiment. In addition, it also indicates in parenthesis the number of iterations required at each case.

The results achieved by the proposed method are encouraging. In the majority of the cases they outperformed the initial accuracies, confirming the relevance of taking into account information from the own test documents (target language) for their classification. It was interesting to notice that in all cases our method obtained results better than the initial classification, and that the best results were achieved when the initial accuracy was very confident (higher than 0.80).

Results from Table 3 also indicate that the best accuracy results were achieved using only three neighbors. In this case, the average improvement was of 4.33% and the maximum was of 11.65%. For the cases where we used information from four and five neighbors the average improvement was 2.87% and the maximum improvements were 10.15% and 8.07% respectively.

Table 3. Accuracy results obtained after the re-classification process

Source language (Training set)	Target language (test set)	Initial Accuracy	Number of Neighbors		
			3	4	5
Translating training set to target language					
French	English	0.858	**0.958** (1)	0.925 (1)	0.925 (2)
Spanish	English	0.817	**0.900** (1)	**0.900** (2)	0.883 (3)
French	Spanish	0.833	**0.842** (1)	**0.842** (1)	**0.842** (1)
English	Spanish	0.717	0.725 (3)	**0.733** (4)	0.725 (1)
Spanish	French	0.808	**0.833** (1)	0.817 (1)	0.825 (1)
English	French	0.758	0.775 (1)	**0.767** (1)	**0.767** (1)
Translating test set to source language					
English	French	0.767	0.758 (2)	0.767 (1)	0.767 (1)
English	Spanish	0.750	0.750 (0)	0.750 (0)	0.750 (0)
Spanish	French	0.792	0.808 (1)	0.808 (1)	**0.817** (1)
Spanish	English	0.850	**0.908** (1)	0.892 (1)	0.892 (1)
French	Spanish	0.800	**0.817** (1)	0.808 (1)	**0.817** (1)
French	English	0.867	**0.925** (2)	0.892 (1)	0.892 (1)

Regarding the convergence of the method, the numbers in the parenthesis in Table 3 help to confirm that, due to the strong condition imposed to perform the iterations (which considers that all neighbors must belong to the same category to generate a re-classification), our method requires just a few iterations to reach the final classification. These results also show, as was expected, that augmenting the number of neighbors, the number of iterations tend to decrease.

4 Conclusions

The analysis presented in this paper showed that the problematic of *cross-language text classification* (CLTC) goes beyond the translation issues. In particular, our experiments indicated that the *cultural distance* manifested in the source and target languages greatly affects the classification performance, since documents belonging to the same category tend to concern very different topics.

In order to reduce the classification errors caused by this phenomenon, we proposed a *re-classification method* that uses information from the target-language documents for improving their classification. The experimental results demonstrated the appropriateness of the proposed method, which could improve the initial classification accuracy by up to 11%.

The results also indicated that the proposed method is independent of the approach employed for generating the initial classification, given that it achieved satisfactory results when training documents were translated to the target language as well as when test documents were translated to the source language.

Finally, it was interesting to notice that relevant improvements were only achieved when initial classification accuracies were very confident (higher than 0.80). In relation to this point, as future work we plan to apply, in conjunction with the re-classification method, a semi-supervised classification approach that allows incorporating information from the target language into the construction of the classifier.

Acknowledgements. This work was done under partial support of CONACYT (project grant 83459 and scholarship 212424).

References

1. Aas, K., Eikvil, L.: Text Categorisation: A Survey. Technical Report. Norwegian Computing Center (1999)
2. Amine, B.M., Mimoun, M.: WordNet based Multilingual Text Categorization. Journal of Computer Science 6(4) (2007)
3. Aue, A., Gamon, M.: Customizing Sentiment Classifiers to New Domains: a Case Study. In: International Conference on Recent Advances in Natural Language Processing (RANLP 2005), Borovets, Bulgaria (2005)
4. Bel, N., Koster, C., Villegas, M.: Cross-Lingual Text Categorization. In: Koch, T., Sølvberg, I.T. (eds.) ECDL 2003. LNCS, vol. 2769, pp. 126–139. Springer, Heidelberg (2003)
5. Dai, W., Xue, G., Yang, Q., Yu, Y.: Co-clustering based classification for out-of-domain documents. In: Proceedings of the 13th ACM SIGKDD international conference on Knowledge discovery and data mining (KDD 2007), San Jose, California, USA (August 2007)
6. De Melo, G., Siersdorfer, S.: Multilingual Text Classification using Ontologies. In: Amati, G., Carpineto, C., Romano, G. (eds.) ECiR 2007. LNCS, vol. 4425, pp. 541–548. Springer, Heidelberg (2007)
7. Gliozzo, A., Strapparava, C.: Exploiting Corporable Corpora and Biligual Dictionaries for Cross-Language Text Categorization. In: Proceedings of the 21st International Confer-ence on Computational Linguistics (Coling 2006), Sydney, Australia (2006)
8. Guzmán-Cabrera, R., Montes-y-Gómez, M., Rosso, P., Villaseñor-Pineda, L.: Using the Web as Corpus for Self-training Text Categorization. Journal of Information Retrieval 12(3) (June 2009)
9. Jalam, R.: Apprentissage automatique et catégorisation de textes multilingues. PhD Tesis, Université Lumiere Lyon 2, Lyon, France (2003)

10. Krithara, A., Amini, M., Renders, J.M., Goutte, C.: Semi-Supervised Document Classification with a Mislabeling Error Model. In: Macdonald, C., Ounis, I., Plachouras, V., Ruthven, I., White, R.W. (eds.) ECIR 2008. LNCS, vol. 4956, pp. 370–381. Springer, Heidelberg (2008)
11. Lewis, D., Yang, Y., Rose, T.G., Dietterich, G., Li, F.: RCV1: A New Benchmark Collection for Text Categorization Research. Journal of Machine Learning Research 5 (2004)
12. Nigam, K., Mccallum, A.K., Thrun, S., Mitchell, T.: Text classification from labeled and unlabeled documents using EM. Machine Learning 39(2/3), 103–134 (2000)
13. Olsson, J.S., Oard, D., Hajic, J.: Cross-Language Text Classification. In: Proceedings of the 28th annual international ACM SIGIR conference on research and development in information retrieval (SIGIR 2005), New York, USA (2005)
14. Rigutini, L., Maggini, M., Liu, B.: An EM based training algorithm for Cross-Language Text Categorization. In: Proceedings of the 2005 IEEE/WIC/ACM International Conference on Web Intelligence, Compiegne, France (September 2005)
15. Sebastiani, F.: Machine learning in automated text categorization. ACM Computing Surveys 34(1) (2002)

Ranking Refinement via Relevance Feedback in Geographic Information Retrieval

Esaú Villatoro-Tello, Luis Villaseñor-Pineda, and Manuel Montes-y-Gómez

Laboratory of Language Technologies, Department of Computational Sciences,
National Institute of Astrophysics, Optics and Electronics (INAOE), Mexico
{villatoroe,villasen,mmontesg}@inaoep.mx

Abstract. Recent evaluation results from Geographic Information Retrieval (GIR) indicate that current information retrieval methods are effective to retrieve relevant documents for geographic queries, but they have severe difficulties to generate a pertinent ranking of them. Motivated by these results in this paper we present a novel re-ranking method, which employs information obtained through a relevance feedback process to perform a *ranking refinement*. Performed experiments show that the proposed method allows to improve the generated ranking from a traditional IR machine, as well as results from traditional re-ranking strategies such as query expansion via relevance feedback.

1 Introduction

Information Retrieval (IR) deals with the representation, storage, organization, and access to information items[1] [1]. Given some query, formulated in natural language by some user, the IR system is suppose to retrieve and sort according to its relevance degree documents satisfying user's information needs [2].

The word *relevant* means that retrieved documents should be semantically related to the user information need. Hence, one central problem of IR is determining which documents are, and which are not relevant. In practice this problem is usually regarded as a *ranking* problem, whose goal is to define an ordered list of documents such that documents similar to the query occur at the very first positions.

Over the past years, IR models, such as: Boolean, Vectorial, Probabilistic and Language models have represented a document as a set of representative keywords (i.e., index terms) and defined a *ranking* function (or retrieval function) to associate a relevance degree for each document with its respective query [1,2]. In general, these models have shown to be quite effective over several tasks in different evaluation forums as can be seen in [3,4]. However, the ability of these models to effectively *rank* relevant documents is still limited by the ability of the user to compose an appropriate query.

---

[1] Depending on the context, items may refer to text documents, images, audio or video sequences.

A. Hernández Aguirre et al. (Eds.): MICAI 2009, LNAI 5845, pp. 165–176, 2009.
© Springer-Verlag Berlin Heidelberg 2009

In relation to this fact, IR models tend to fail when desired results have implicit information requirements that are not specified in the keywords. Such is the case of **Geographic Information Retrieval** (GIR), which is a specialized IR branch, where search of documents is based not only in conceptual keywords, but also on geographical terms (e.g., geographical references) [5]. For example, for the query: "Cities near active volcanoes", expected documents should mention explicit city and volcanoes names. Therefore, GIR systems have to interpret implicit information contained in documents and queries to provide an appropriate response to geographical queries.

Recent development on GIR systems [6] evidence that: *i)* traditional IR systems are able to retrieve the majority of the relevant documents for most queries, but that, *ii)* they have severe difficulties to generate a pertinent ranking of them. To tackle this problem, recent works have explored the use of traditional re-ranking approaches based on query expansion via either relevance feedback [7,8,9,10], or employing knowledge databases [11,12]. Although these strategies are very effective improving precision values, is known that query expansion strategies are very sensitive to the quality of the added elements, and some times may result in degradation of the retrieval performance.

In this paper we propose a novel re-ranking strategy for Geographic Information Retrieval. Since retrieving relevant documents to geographic queries is not a problem for traditional IR systems, we focus on improving the order assigned to a set of retrieved documents by employing information obtained through a relevance feedback process, i.e., *ranking refinement via relevance feedback*. Furthermore, given that geographic queries tend to show a lot of implicit information, we propose the use of complete documents instead of isolated terms in the *ranking refinement* process.

The rest of the paper is organized as follows. Section 2 discusses some related work. Section 3 shows the proposed method. Section 4 describes the experimental platform used to evaluate our ranking strategy. Section 5 presents the experimental results. Finally, section 6 depicts our conclusions and future work.

2   GIR Related Work

Formally, a geographic query (geo-query) is defined by a tuple <*what, relation, where*>[5]. The *what* part represents generic terms (non-geographical terms) employed by the user to specify its information need, it is also known as the *thematic* part. The *where* term is used to specify the geographical areas of interest. Finally, the *relation* term specifies the "spatial relation", which connects *what* and *where*.

GIR has been evaluated at the CLEF forum [3] since year 2005, under the name of the GeoCLEF task [6]. Their results evidence that traditional IR methods are able to retrieve the majority of the relevant documents for most geo-queries, but, they have severe difficulties to generate a pertinent ranking of them. Due to this situation, recent GIR methods have focused on the ranking subtask.

Common employed strategies are: *i)* query expansion through some feedback strategy, *ii)* re-ranking retrieved elements through some adapted similarity measure, and *iii)* re-ranking through some information fusion technique. These strategies have been implemented following two main approaches: first, techniques that had paid attention on constructing and including robust geographical resources in the process of retrieving and/or ranking documents. And second, techniques that ensure that geo-queries can be treated and answered employing very little geographical knowledge.

As an example of those on the first category, some works employ geographical resources in the query expansion process [11,12]. Here, they first recognize and disambiguate all geographical entities in the given geo-query by employing a GeoNER[2] system. Afterwards, they employ a geographical ontology to search for these geo-terms, and retrieve some other related geo-terms. Then, retrieved geo-terms are given as feedback elements to the GIR machine. Some others approaches that focus on the ranking refinement problem, propose algorithms that consider the existence of *Geo-tags*[3], therefore, the ranking function measures levels of *topological space proximity* among the geo-tags of retrieved documents and geo-queries [14]. In order to achieve this, geographical resources (e.g., geographical databases) are needed.

In contrast, approaches that do not depend on any robust geographical resource have proposed and applied variations of the query expansion process via relevance feedback, where no special consideration for geographic elements is made [7,8,9,10], and they have achieved good performance results. There are also works focusing on the ranking refinement problem; they consider the existence of several lists of retrieved documents (from one or many IR machines). Therefore, the ranking problem is seen as a information fusion problem, without any special processing for geo-terms contained in the retrieved documents. Some simple strategies only apply logical operators to the lists (e.g., AND) in order to generate one final re-ranked list [9], while some other works apply techniques based on information redundancy (e.g., CombMNZ or Round-Robin)[7,15,16].

Recent evaluation results indicate that there is not a notable advantage of knowledge-based strategies over methods that do not depend on any geographic resource. Motivated by these results, our proposed method does not make any special consideration for geographical terms. Our main hypothesis is that by employing information obtained through traditional relevance feedback strategies, is possible to perform an accurate ranking refinement process avoiding the drawbacks of query expansion techniques.

In addition, based on the fact that geo-queries often contain implicit information, we performed a set of experiments considering full documents (called *example documents*) in the process of re-ranking, showing that it is possible to make explicit some of the implicit information contained in the original geo-queries.

---

[2] Geographical Named Entity Recognizer.

[3] A *Geo-tag* indicates the geographical focus of certain item. As can be seen in [13], Geo-tagging and geo-disambiguating are both major problems in GIR.

3 Proposed Method

The proposed method consists of two main stages: the *retrieval stage* and the *re-ranking stage*. The goal of the first is to retrieve as many as possible relevant documents for a given query, whereas, the function of the second is to improve the final ranking of the retrieved documents by applying *ranking refinement* via *relevance feedback*. Figure 1 shows a general overview of the proposed method.

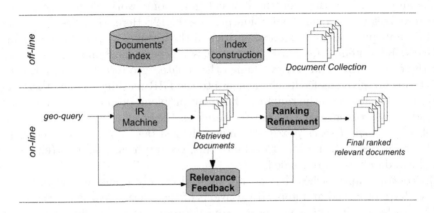

Fig. 1. Ranking refinement method

Retrieval. For this stage, we employed the vectorial space model (VSM), which is one of the most accurate and stable IR methods. In the *VSM* each document d is represented as a vector (d_i) of length equal to the vocabulary size $|V|$. Each element j from the vector d_i indicates how important is the word j inside the document d_i. The set of vectors representing all the documents contained in the collection generates a vectorial space where documents can be compared through their representations. This vectorial space is represented by a matrix (M^{TD}), usually called term-document matrix (TD), of size $N \times M$, where N is the vocabulary size in the collection, $N = |V|$, and M is the number of documents in the collection. Each entry $M_{i,j}^{TD}$ indicates the weight or contribution of term t_j in the document d_i.

We employed the *tf-idf* (*term-frequency inverse-document-frequency*) weighting scheme for each entry $M_{i,j}^{TD}$, computed as follows:

$$M_{ij}^{TD} = tf_{ij} \times log(\frac{|D|}{df_j}) \tag{1}$$

where tf_{ij} is the number of occurrences of term j inside document d_i, $|D|$ is the total number of documents in the collection and df_j is the number of documents containing the term j.

When a query arrives to the VSM, it is transformed to the same representation (i.e., a vector). Once both query and documents are in the same representation it is possible to compare the query against all the documents in the collection. For this, we employed the cosine measure, which is computed as follows:

$$sim(q, d_i) = \frac{\sum_{j=1}^{|q|} w_{qj} d_{ij}}{\sqrt{\sum_{j=1}^{|q|} (d_{ij})^2 \sum_{j=1}^{|q|} (w_{qj})^2}} \tag{2}$$

where $sim(q, d_i)$ represents the degree of similarity between the query (q) and the document d_i. w_q is the query vector while d_i is the document vector. The cosine formula measures the angle between two vectors in a space of dimension $|V|$ considering a normalization process to avoid that the vector's magnitude affects the retrieval process. Finally, the VSM method sort retrieved documents considering as its ranking score the result obtained with the cosine formula.

Re-ranking. Once a set of documents had been retrieved, we collect relevance judgments for the n top documents. Provided judgments could be either user provided (*manually*) or automatically obtained (*blindly*). Afterwards, the k most representative terms are extracted and along with the original geo-query are transformed to the VSM representation.

Our ranking refinement process considers the previous constructed vector to re-rank retrieved documents by employing the cosine formula (see formula 2), and hence, generate a final list of re-ranked documents. Notice that given this configuration, we preserve in every case the original recall levels obtained by the IR machine since we do not perform any further retrieval process.

At this point is where we also propose using complete documents (i.e., *example documents*) instead of k isolated terms in order to make explicit all implicit information contained in geo-queries. As will be seen in the experiments section, the n top documents along with the original geo-query are transformed to the VSM representation. Finally, retrieved documents are re-ranked according to its similarity degree against the *example documents* vector.

It is worth mentioning that we do not consider information contained in non-relevant documents for the relevance feedback process given that for geographical queries happens that: *i)* usually there are few relevant documents for geo-queries, i.e., there are many non-relevant documents in the retrieved set, and *ii)* non-relevant documents are not homogeneous. These reasons avoid the possibility of correctly represent non-relevant information.

4 Experimental Setup

4.1 Datasets

For our experiments we employed the GeoCLEF document collection composed from news articles from years 1994 and 1995. Articles cover as national as international events and, as a consequence, documents contain several geographic references. Table 1 shows some statistics about the collection.

Table 1. GeoCLEF Document Collection

Name	Origin	Num of Documents	Language
GH95	*The Glasgow Herald*	56,472	English
LAT94	*The Los Angeles Times*	113,005	English
		Total: 169,477	

4.2 Topics

We worked with the GeoCLEF 2008 queries. Table 2 shows the structure of each topic. The main query or title is between labels <EN-title> and </EN-title>. Also a brief description (<EN-desc>, </EN-desc>) and a narrative (<EN-narr>, </EN-narr>) are given. These last two fields usually contain more information about the requirements of the original query.

Table 2. Topic GC030: Car bombings near Madrid

```
<top>
 <num>GC030</num>
 <EN-title>Car bombings near Madrid</EN-title>
 <EN-des>Documents about car bombings occurring
 near Madrid</EN-desc>
 <EN-narr>Relevant documents treat cases of car bombings occurring
 in the capital of Spain and its outskirts</EN-narr>
</top>
```

4.3 Evaluation

The evaluation of results was carried out using two measures that have demonstrated their pertinence to compare IR systems, namely, the Mean Average Precision (MAP) and the R-prec. The MAP is defined as the norm of the average precisions ($AveP$) obtained for each query [1]. The $AveP$ for a given query q is calculated as follows:

$$AveP = \frac{\sum_{r=1}^{Ret_{docs}} P(r) \times rel(r)}{Rel_{docs}} \tag{3}$$

where $P(r)$ is the precision of the system at the r considered documents; $rel(r)$ is a binary function that indicates whether document r is relevant to the query, or not; Ret_{docs} is the number of retrieved documents, while Rel_{docs} is the number of relevant documents for the given query. Intuitively, this measure indicates how well the system puts into the first positions relevant documents. It is worth pointing out that since our IR machine was configured to retrieve 1000 documents for each query, $AveP$ values are measured at 1000 documents.

On the other hand, *R-prec* is defined as the precision reached after R documents have been retrieved, where R indicates the number of relevant documents for query q that exist in the entire document collection.

4.4 Experiments Definition

In order to achieve our goals, two experiments were performed. First, the Experiment 1 had as main goal to compare the proposed *ranking refinement* strategy against to the traditional re-ranking technique *query expansion via relevance feedback*. For this experiment, different number of documents (n) as well different number of terms (k) were considered. Additionally, we were also interested on evaluating the impact of both re-ranking strategies when selected documents were provided either by an active user (*manually selected*) or via an automatic selection process (*blindly selected*).

In the other hand, Experiment 2 had as main goal to evaluate the impact of considering full documents, which we called *example documents*, for both re-ranking strategies (*ranking refinement* and *query expansion*). Same as before, we were also interested on evaluate the performance obtained when *example documents* were manually or blindly selected.

5 Results

Experimental results are reported in Tables 3-8. Results are reported in terms of *R-prec* and *MAP*. Underlined results indicate the cases where the *baseline* was improved, while results marked in bold indicate the *best* results obtained over the different configurations.

For all our experiments, we considered as *baseline* the ranking generated by the IR machine, i.e., the rank assigned to documents employing the VSM method. For comparison purposes with the rest of the GeoCLEF 2008 participants it is worth mentioning that the best *MAP* result obtained among all the sumitted runs was of 0.3040, the median was of 0.2370 and the worst was of 0.1610.

5.1 Experiment 1

Tables 3 and 4 show results obtained when documents are blindly selected, i.e., there are no user intervention. Table 3 shows results obtained after applying a query expansion process via relevance feedback (QEviaRF). As we can observe, the *R-prec* value obtained by the baseline method is never exceeded. This means that in any configuration the query expansion process was not able to put more relevant documents among the first R retrieved documents (see section 4.3).

On the contrary, notice that when 2 documents are considered for the relevance feedback process the *MAP* values are better that the baseline, which means that the QEviaRF was able to shift some relevant documents to higher positions. However, remember that *MAP* values are measured at 1000 documents, so it is possible that the improvement implies that a document was shift from position 1000 to position 900.

Table 3. QEviaRF when documents are blindly selected

# selected terms	2 docs		5 docs		10 docs	
	R-Prec	MAP	R-Prec	MAP	R-Prec	MAP
5	0.2451	0.2435	0.2211	0.2214	0.2132	0.2064
10	**0.2545**	**0.2492**	0.2278	0.2152	0.2291	0.2025
15	0.2499	0.2405	0.2270	0.2017	0.2364	0.2209
baseline: R-Prec = 0.2610; MAP= 0.2347						

Table 4 shows results obtained after applying a *ranking refinement via relevance feedback* (RRviaRF). As we can observe, for this case it was possible to improve *R-prec* value, particularly for the case when 2 documents with 10 and 15 terms are considered for the relevance feedback process, which means that more relevant documents are been collocated among the first R retrieved documents. Notice, that results obtained with RRviaRF are slightly better that those obtained with QEviaRF (Table 3), indicating that our ranking refinement method is less sensitive to the noise contained in documents considered for the relevance feedback process.

Table 4. RRviaRF when documents are blindly selected

# selected terms	2 docs		5 docs		10 docs	
	R-Prec	MAP	R-Prec	MAP	R-Prec	MAP
5	0.2448	0.2281	0.2405	0.2313	0.2102	0.2013
10	**0.2675**	0.2402	0.2408	0.2194	0.1863	0.1914
15	0.2619	**0.2475**	0.2435	0.2257	0.2332	0.2265
baseline: R-Prec = 0.2610; MAP= 0.2347						

As general conclusions, when documents for the re-ranking process are blindly selected, better results are achieved when only two documents are considered. Also, it is important to notice that RRviaRF strategy (Table 4) is less sensitive to noise introduced by the selection of more terms from more documents than the QEviaRF (Table 3) technique. Besides, even when both strategies achieved a *MAP* close to 0.25, over a 0.23 from the *baseline*, RRviaRF is able to put more relevant documents among first positions according to the *R-prec* values.

Tables 5 and 6 show results obtained when documents are manually selected, i.e., an user intervention is considered. As expected, results get higher values under this schema. We notice in both tables that adding more terms from more documents to both QEviaRF and RRviaRF, allows to obtain better performance results than the baseline.

An interesting fact for these manual experiments, is that *R-prec* values are better for the case of QEviaRF (Table 5) than those obtained with RRviaRF (Table 6). However, observe that for QEviaRF results adding 15 or 10 terms has no noticeable impact on the *MAP*. Same phenomena occurs when applying RRviaRF, except for the case when 5 documents are considered for the relevance feedback process. Hence, one important question that emerge from these results is: *Why there is no improvement if added terms come from true relevant documents?*

Table 5. QEviaRF when documents are manually selected

# selected terms	2 docs		5 docs		10 docs	
	R-Prec	MAP	R-Prec	MAP	R-Prec	MAP
5	0.3173	0.3260	0.3257	0.3367	0.3257	0.3378
10	0.3522	0.3494	0.3500	**0.3601**	0.3491	0.3584
15	0.3490	0.3471	0.3537	0.3590	**0.3538**	0.3593
baseline: R-Prec = 0.2610; MAP= 0.2347						

As an explanation to the previous question, first we must remember that given the configuration of both strategies, as a previous step to the re-ranking phase, an expanded keyword query is constructed from selecting k terms from n documents. Second, as mentioned in Section 1, IR models are limited by the ability of the user (or an automatic process) to compose an effective keyword query. These facts, plus obtained results, make us think that isolated keyword terms are not sufficient to achieve the best performance in a GIR system since they do not effectively describe implicit information needs contained in geographical queries.

Table 6. RRviaRF when documents are manually selected

# selected terms	2 docs		5 docs		10 docs	
	R-Prec	MAP	R-Prec	MAP	R-Prec	MAP
5	0.2857	0.2760	0.2998	0.2924	0.2998	0.2932
10	0.3237	0.3178	0.3283	0.3274	0.3349	0.3313
15	0.3267	0.3236	**0.3433**	**0.3378**	0.3348	0.3348
baseline: R-Prec = 0.2610; MAP= 0.2347						

The general conclusion of these experiments is that, considering the user intervention allows to obtain better performance results; however adding more true relevant elements to the re-ranking strategies seems to have no impact at all, i.e., apparently a maximum level is achieved. Additionally, we confirmed that without any special treatment for geo-terms it is possible to achieve high recall levels ($\simeq 90\%$).

5.2   Experiment 2

The main purpose of following experiment is to employ more information than only a set of isolated keywords in the re-ranking process. For this, we based on the ideas proposed in the Image Retrieval field [17], where in order to retrieve images of some particular type, queries are usually formulated through examples (i.e., example images[4]), which is easier than formulating an effective keyword query.

Same as for image retrieval, geographic retrieval also contain many implicit information that is hard to describe with a small query. Hence, the following tables show the results obtained when *example documents* are given to both re-ranking strategies, i.e., traditional query expansion, and the proposed method based on a ranking refinement strategy.

Table 7 shows results obtained when *example documents* are blindly selected, whereas in Table 8 example documents are manually selected. Tables compare both QEviaRF considering a query-by-example approach (QEviaRF-QBE) and RRviaRF under the same circumstances (RRviaRF-QBE).

Table 7. QEviaRF-QBE VS RRviaRF-QBE (documents are blindly selected)

# selected docs	QEviaRF-QBE		RRviaRF-QBE	
	R-Prec	MAP	R-Prec	MAP
2 docs	**0.2509**	**0.2393**	**0.2678**	**0.2498**
5 docs	0.2232	0.2254	0.2656	0.2436
10 docs	0.2118	0.2057	0.2148	0.2177
baseline: R-Prec = 0.2610; MAP= 0.2347				

Table 8. QEviaRF-QBE VS RRviaRF-QBE (documents are manually selected)

# selected docs	QEviaRF-QBE		RRviaRF-QBE	
	R-Prec	MAP	R-Prec	MAP
2 docs	0.3793	0.3795	0.3549	0.3528
5 docs	0.3850	0.3983	0.3651	0.3720
10 docs	**0.3866**	**0.3995**	**0.3712**	**0.3747**
baseline: R-Prec = 0.2610; MAP= 0.2347				

Notice that when documents are blindly selected (Table 7), traditional query expansion technique is not able to improve the baseline, however, our proposed ranking refinement strategy obtains better results in two out of three cases (i.e., employing 2 and 5 example documents). Generally speaking our ranking refinement strategy allows better results than the query expansion strategy. However, when documents are selected manually (Table 8), query expansion strategy obtains better results than our ranking refinement method.

---

[4] This approach is usually known as query by example (QBE).

As our general conclusion for these experiments, we consider that employing *example documents* allows generating a more pertinent ranking, since implicit information contained in the original geo-query, is better represented by complete example documents. Obtained results also indicate that if the intervention of some user is considered, only providing 2 *example documents* is enough to reach acceptable results.

6 Conclusions

In this paper we have presented a *ranking refinement strategy via relevance feedback*. Obtained results showed that: *i)* our ranking refinement strategy considering a small set of keywords is able to improve the VSM method and also improves the traditional query expansion via relevance feedback technique, and *ii)* while more information is added to the ranking refinement strategy, a better ordering is provided.

Additionally, our experiments showed that employing full documents as relevance feedback elements (i.e., *example documents*), our ranking refinement strategy is able to obtain better performance results than those obtained when using a small set of keywords. This fact confirmed that it is possible to become explicit some of the implicit information contained in *geographical queries*.

Finally, performed experiments considering the user intervention, showed that is possible to reach a high performance results by providing only two *example documents*. This means that the user is not being overwhelmed in the process of selecting the *example documents*, since he has to mark very few documents as feedback elements from a small set of top retrieved documents.

As future work, we are planning to evaluate our method on a different data set as well as with a major number of queries, which will allow us to perform some statistical significance tests, in order to confirm the pertinence of the proposed method. Furthermore, we are interested in employing some strategy for selecting from the example documents the essential information, generating with this a more accurate ranking.

Acknowledgments. This work was done under partial support of CONACyT (scholarship 165545, and project grant 83459).

References

1. Baeza-Yates, R., Ribeiro-Neto, B.: Modern Information Retrieval. Addison Wesley, Reading (1999)
2. Grossman, D.A., Frieder, O.: Information Retrieval, Algorithms and Heuristics, 2nd edn. Springer, Heidelberg (2004)
3. Cross-lingual evaluation forum (May 2009), http://www.clef-campaign.org/
4. Text retrieval conference (trec) (May 2009), http://trec.nist.gov/
5. Henrich, A., Ldecke, V.: Characteristics of geographic information needs. In: Proceedings of Workshop on Geographic Information Retrieval, GIR 2007, Lisbon, Portugal. ACM Press, New York (2007)

6. Mandl, T., Carvalho, P., Gey, F., Larson, R., Santos, D., Womser-Hacker, C., Di Nunzio, G., Ferro, N.: Geoclef 2008: the clef 2008 cross language geographic information retrieval track overview. In: Working notes for the CLEF 2008 Workshop, Aarhus, Denmark (September 2008)
7. Larson, R.R., Gey, F., Petras, V.: Berkeley at geoclef: Logistic regression and fusion for geographic information retrieval. In: Peters, C., Gey, F.C., Gonzalo, J., Müller, H., Jones, G.J.F., Kluck, M., Magnini, B., de Rijke, M., Giampiccolo, D. (eds.) CLEF 2005. LNCS, vol. 4022, pp. 963–976. Springer, Heidelberg (2006)
8. Gillén, R.: Monolingual and bilingual experiments in geoclef 2006. In: Peters, C., Clough, P., Gey, F.C., Karlgren, J., Magnini, B., Oard, D.W., de Rijke, M., Stempfhuber, M. (eds.) CLEF 2006. LNCS, vol. 4730, pp. 893–900. Springer, Heidelberg (2007)
9. Ferrés, D., Rodríguez, H.: Talp at geoclef 2007: Results of a geographical knowledge filtering approach with terrier. In: Peters, C., Jijkoun, V., Mandl, T., Müller, H., Oard, D.W., Peñas, A., Petras, V., Santos, D. (eds.) CLEF 2007. LNCS, vol. 5152, pp. 830–833. Springer, Heidelberg (2008)
10. Larson, R.R.: Cheshire at geoCLEF 2007: Retesting text retrieval baselines. In: Peters, C., Jijkoun, V., Mandl, T., Müller, H., Oard, D.W., Peñas, A., Petras, V., Santos, D. (eds.) CLEF 2007. LNCS, vol. 5152, pp. 811–814. Springer, Heidelberg (2008)
11. Wang, R., Neumann, G.: Ontology-based query construction for geoclef. In: Working notes for the CLEF 2008 Workshop, Aarhus, Denmark (September 2008)
12. Cardoso, N., Sousa, P., Silva, M.J.: The university of lisbon at geoclef 2008. In: Peters, C., Clough, P., Gey, F.C., Karlgren, J., Magnini, B., Oard, D.W., de Rijke, M., Stempfhuber, M. (eds.) CLEF 2006. LNCS, vol. 4730, pp. 51–56. Springer, Heidelberg (2007)
13. Borges, K.A., Laender, A.H.F., Medeiros, C.B., Davis Jr, C.A.: Discovering geographic locations in web pages using urban addresses. In: Proceedings of Workshop on Geographic Information Retrieval GIR, Lisbon, Portugal. ACM Press, New York (2007)
14. Martins, B., Cardoso, N., Chaves, M.S., Andrade, L., Silva, M.J.: The university of lisbon at geoclef 2006. In: Peters, C., Clough, P., Gey, F.C., Karlgren, J., Magnini, B., Oard, D.W., de Rijke, M., Stempfhuber, M. (eds.) CLEF 2006. LNCS, vol. 4730, pp. 986–994. Springer, Heidelberg (2007)
15. Villatoro-Tello, E., Montes-y-Gómez, M., Villaseñor-Pineda, L.: Inaoe at geoclef 2008: A ranking approach based on sample documents. In: Working notes for the CLEF 2008 Workshop, Aarhus, Denmark (September 2008)
16. Ortega, J.M.P., Urea, L.A., Buscaldi, D., Rosso, P.: Textmess at geoclef 2008: Result merging with fuzzy borda ranking. In: Working notes for the CLEF 2008 Workshop, Aarhus, Denmark (September 2008)
17. Flickner, M., Sawhney, H., Niblack, W., Ashley, J., Huang, Q., Dom, B., Gorkani, M., Hafner, J., Lee, D., Petkovic, D., Steele, D., Yanker, P.: Query by image and video content: The qbic system. IEEE Computer Special Issue on Content-Based Retrieval 28(9), 23–32 (1995)

A Complex Networks Approach to Demographic Zonification

Alberto Ochoa[1], Beatriz Loranca[2], and Omar Ochoa[3]

[1] Institute of Cybernetics, Mathematics and Physics
ochoa@icmf.inf.cu
[2] BUAP School of Computer Sciences. UNAM, System Department
bety@cs.buap.mx
[3] Institute of Cybernetics, Mathematics and Physics
omar@icmf.inf.cu

Abstract. This paper presents a novel approach for the zone design problem that is based on techniques from the field of complex networks research: community detection by betweenness centrality and label propagation. A new algorithm called Spatial Graph based Clustering by Label Propagation (SGCLAP) is introduced. It can deal with very large spatial clustering problems with time complexity $O(n \log n)$. Besides, we use a parallel version of a betweenness-based community detection algorithm that outputs the graph partitioning that maximizes the so-called modularity metric. Both these methods are put at the centre of an effort to build an open source interactive high performance computing platform to assist researchers working with population data.

1 Introduction

We address the problem of demographic zonification, which is understood here as the problem of clustering geographical units (GU) according to some measure of demographic similarity. Each GU is a small geographical area for which the following is known: latitude, longitude, radius and a vector of demographic features (statistics) computed over the population living in the area. A cluster in the intended solution is expected to be compact, which means that each GU is geographically close to a large fraction of the GUs in the same cluster. Besides, the cluster should be homogeneous as to the similarity measure being used, meaning each GU has rather high similarity values with many members of its cluster. In other words, the clustering can be seen as a single objective optimization problem in the space of the non-spatial variables with the restriction of getting compact solutions with respect to the spatial variables.

In terms of computational complexity the zone design problem has been shown to be NP-Complete [8]. One of the reasons why this problem is specially difficult is the size of the solution space. The dimension of most real world problems makes finding an exact solution unfeasible. Thus, heuristic techniques seem to be the best way available to produce solutions in a reasonable computational time. Unfortunately, most of the heuristics that have been used so far have also

A. Hernández Aguirre et al. (Eds.): MICAI 2009, LNAI 5845, pp. 177–188, 2009.

a high computational cost and/or often do not provide high quality solutions. Therefore, this has been and still is an area of active research [7,4,9,1,3]. We have recently introduced the idea of using some of the techniques of complex networks research to model the zone desing problem [13].

Our work is part of an ongoing project aimed to create an open source tool for assisting *in situ* people that work interactively with population data. This means that at any moment they may change the set of features that describe the GUs, the geographical regions of interest or the used metrics. Besides they need statitistical tools for analizing their results and for preprocessing their data. We have found that very often these persons lack the necessary software and equipment, for example, parallel hardware and software. The system we are trying to build must fulfill all these requeriments.

The method we have proposed is motivated by results in complex network research [5,10,11,17,15,12]. Basically, what we do is to map the problem to an unweighted graph and then apply a community detection algorithm that produces the desired partioning of the GUs. However, the map from the distance and similarity matrices to the unweighted graph may not be so straightforward as far as a lot of weight information has to be codified in the graph topology.

The outline of the paper is as follows. Section 2 is devoted to a detailed exposition of the proposed method. The method is applied to the clustering of geographic units according to given demographic information, but with the restriction of geographic closeness. Sect. 3 shows some numerical results that illustrate the method in action. Then, the conclusions are given. Finally, the appendix of the paper presents a short overview of the first steps given in the design and building of an open source interactive high performance computing platform to assist researchers working with population data.

2   Spatially Constrained Demographic Clustering

Throughtout the paper we shall use a simplified model of the zone design problem. Our goal is just to show that a method based on network partitioning is suitable for the problem. According to the model the geographics units are assumed to be points. Figure 1 shows the distribution of a set of such units for two areas of Mexico: the metropolitan areas of the valleys of Toluca and Mexico.

The main idea of our method is to tranform the problem of clustering the GUs according to two measures: demographic similarity and euclidean distance, into the problem of community detection in an unweighted graph. As an alternative we could also work with a weighted graph, which would give a more straitghtforward representation of our heavily weighted problem. However, as far as dealing with the unweighted case is less expensive from the computational point of view we accepted the challenge of mapping the weight information into the structure of an unweighted graph. Nevertheless, the weighted case will be the topic of a forthcoming paper.

Our complex networks approach to the zone design clustering problem is encapsulated in Algorithm 1. In what follows we describe its main features.

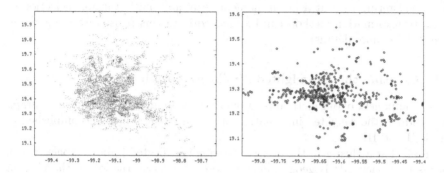

Fig. 1. Geographical distribution (latitude vs. longitude) of 4292 and 466 GUs of the metropolitan areas of the valleys of Mexico (left) and Toluca respectively. A smaller marker size has been used in the first case.

Algorithm 1. Our algorithm

Assume as given a similarity measure (cosine, correlation, ...) and an input matrix where rows codify for the geographic units (GUs) and columns for the spatial and non-spatial variables.

Phase I   Selection of variables, geographic units and the space of processing.

Phase II   Computation of the set of closest neighboors of each GU, according to the distance information.

    Step 1   Computation of the Delaunay triangulation of the set of GUs. This creates the triangulated graph, G_t, where each node represents a GU.

    Step 2   Edge and node elimination in G_t.

    Step 3   To every node, i in G_t, assign the set of its closest neighboors $Neigh_d(i)$.

Phase III   Computation of the similarity graph, G_s.

    Step 1   For every node, i, compute its pairwise similarity with every member of $Neigh_d(i)$.

    Step 2   For each node, i, construct the set, $Neigh_s(i)$ (s stand for similarity), with the k most similar neighbors taken among the members of $Neigh_d(i)$.

    Step 3   Add the edge $i \sim j$ to G_s if $j \in Neigh_s(i)$ and $i \in Neigh_s(j)$.

Phase IV   Computation of the partitioning of G_s in communities.

Let us assume that an input data matrix is given, where each row contains information about one of the GUs, whereas the columns represent the demographic and geographic variables. There are many reasons that make convenient to work with a subset of the variables and/or a subset of the available GUs. Phase I is responsible for producing an input matrix that better fulfils the requirements of the clustering task at a given moment. Any automatic variable selection or space transformation method can be plugged-in in this phase of the algorithm (see [13] for more details).

2.1   Creating the Unweighted Similarity Graph

A key component of our method is the creation of an unweighted similarity graph that codifies the weighted information of our problem. This is accomplished in Phase II and III.

Recall that a cluster in the desired solution is expected to be compact, which means that each GU is geographically close to a large fraction of the GUs in the same cluster. At the same time, the cluster should be homogeneous as to the utilized similarity measure being used, meaning each GU has rather high similarity values with many members of its cluster. In other words, the clustering can be seen as a single objective optimization problem with the restriction of getting compact solutions.

To begin with, we create a graph based on the distance information. Without loss of generality, let us assume that the second and third columns of the input matrix contain the latitute and longitude of each GU (these columns were not processed in Phase I). It is worth noting that we are not considering the radius, which means that all GUs are assumed to be geometrically equal with centers at the given coordinates. In some cases this assumption could be too strong and somehow misleading. However, it is adequate for the purposes of this paper as it is explained below.

The Delaunay triangulation (DT) is a sparse graph structure that codifies proximity relationships in a point set. Thus, the first thing we do in Phase II is the construction of the unweighted graph G_t –a DT of the set of GUs centers. The DT is connected to the Voronoi diagram of the centers, which is a partition of the space into areas, one for each GU center, so that the area for a GU center consists of that region of the space that is closer to it than to any other GU center. An edge in the DT connects two GU centers if and only if the Voronoi areas containing them share a common boundary. Hence, edges of the DT capture spatial proximity. At this point it is worth noting that for those cases where the actual area of each GU is included completely inside of the Voronoi cell of its center, the DT seems to be adequate to model the distance restriction of our problem. Regarding computational efficiency, many $O(n \log n)$ time and $O(n)$ space algorithms exist for computing the DT of a planar point set.

In the Step 2 we eliminate a certain amount of nodes and edges to get a further reduction of the graph (the first was performed with the input data in Phase I). Optionally, we eliminate GUs in regions of high density to reduce the cost of the algorithm. This is accomplished as follows. We compute a DT of the whole set and then select the GUs belonging to triangles which areas are not below the Q area quantile. A new triangulation is performed with the chosen points.

Despite the fact that edges of the DT capture spatial proximity there are some edges that might be far apart, for example the edges between the GUs in the convex hull. Therefore, we remove all edges larger than the largest one found in a set that contains for each GU the smallest among its attached edges. We term G_t the connected graph obtained after this procedure.

As the final step of Phase II we compute the set of closest neighbors of the i-th GU, $Neigh_d(i)$ (where the subscript indicates the distance measure). Considering the way G_t was constructed, it is reasonable to define as neighbors all GUs that can be reached in at most L steps from the given one. In fact, in all the experiments of this paper we use $L = 2$.

At this point the Phase III of the algorithm starts. The idea is simple, we formulate a k-nearest neighbours problem with respect to the similarity function. Fortunately, we only need to compute the similarity between each GU and its neighbors according to $Neigh_d(i)$ (Step 1). For each GU, i, construct the set, $Neigh_s(i)$ (s stands for similarity), with the k most similar neighbours taken among the members of $Neigh_d(i)$ (Step 2). At this point it is worth noting, that in the construction of $Neigh_s(i)$ one can apply other constraints, such as the fulfilment of certain aggregate properties in the neighborhood.

In the final Step 3 the unweigted graph, G_s, is constructed. The following holds: edge $i \sim j \in G_s$ if and only if $(j \in Neigh_s(i)) \wedge (i \in Neigh_s(j))$.

It is our fundamental claim that in the topology of the unweighted graph G_s there is enough information about our complex clustering problem. If we assume that, and we do, it is reasonable to accept as clusters all the connected subgraphs with high connection density. Therefore, what we need at this point is a method to detect these communities of nodes. This is a common problem in complex networks research.

2.2   The Edge betweenness Approach

In [13] we used the edge betweenness approach in the Phase IV of algorithm 1.

The shortest path edge betweenness centrality, counts the number of shortest paths that run along an edge in a graph. The highest values of betweenness are likely to be associated with bridges-edges –those that separate regions of densely connected vertexes. Thus, by removing the bridges the communities can be detected. Unfortunately, when there are several bridges between two regions, they compete for the best scores, hampering an accurate evaluation of the measure. This problem has been solved with a simple algorithm: While the network is not empty, 1) Recalculate the betweenness score of every edge. 2) Remove the edge with the highest score. Notice that the order of removal of the edges implicitly defines a hierarchical tree. Hereafter this algorithm is called GN because it was developed by Girvan and Newmann [5,11].

The fastest known version of the GN algorithm for unweighted, undirected graphs (m edges and n vertexes) has worst case time complexity $O(m^2n)$, whereas for sparse graphs $O(n^3)$. The high computational cost of the GN clustering algorithm has been an obstacle to its use on relatively large graphs. In fact, the computational cost of the algorithm is already prohibitive when the input is a graph with a few thousand edges. However, in [17] a parallel implementation of the algorithm was reported. This algorithm outputs the clustering with maximum modularity [11] and the ordered list of edge removals. A high modularity indicates that there are more intracluster and less intercluster edges than would be expected by chance. Note that this differs from partitioning a graph minimizing the number of intercluster edges.

As far as the above mentioned algorithm allows users to analyse larger networks on a distributed cluster of computers, we decide to include it in our experimental system. The ordered list of edges that it outputs can be used by the user to create a merge matrix or dendrogram from which different clusterings can be explored.

The size of the problem one can manage with this algorithm depend on the available computer power. We have achieved reasonable performance with a few desktop computers with graphs of few thousand nodes. Nevertheless, we always can restrict ourselves to the analysis of small areas as far as we are working in an interactive scenario. Besides, one should note that the very same complex networks approach can be utilized to cluster the demographic variables, as a feature selection step before running Algorithm 1 or after it to explain the obtained results. Usually, we work with less than 1000 variables.

2.3   The Label Propagation Approach

In this section we introduce a novel clustering method for mixed spatial and non-spatial data, and thus for the zone design problem. We have called it Spatial Graph based Clustering by Label Propagation (SGCLAP).

The method is intended for very large problems (graphs) and is obtained by pluggin in the label propagation method reported in [15] into the Phase IV of Algorithm 1.

As far as the label propagation algorithm has near linear time complexity the overall complexity of SGCLAP is $O(n \log n)$.

This is a fast, nearly linear time algorithm for detecting community structure in networks. It works by labeling the vertices with unique labels and then updating the labels by majority voting in the neighborhood of the vertex. As a consequence of this updating process densely connected groups of nodes form a consensus on a unique label to form communities.

The label propagation algorithm uses only the network structure to guide its progress and requires no external parameter settings. Each node makes its own decision regarding the community to which it belongs to based on the communities of its immediate neighbors. Therefore, much of the success of the SGCLAP will depend on the quality of the similarity graph. In other words, it will depend on how much information can be captured by the topology of the unweighted similarity graph.

3   Experimental Results

The aim of this section is simply to illustrate the possibilities of our method. Let us take the data for the metropolitan area of the valley of Toluca. We study 466 GUs described with a vector of 181 demographic variables according to the XII National Population Census [6].

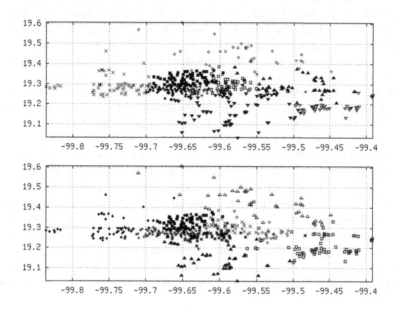

Fig. 2. Clustering results for Toluca using the whole set of 181 demographic variables. The top figure shows a clustering where only geographic distances were used, whereas for the bottom figure the euclidean distances between the vectors of demographic variables were also considered. In both cases 9 clusters were found, but there are differences.

In the first experiment we use all GUs and variables (without any prepro-cessing) and the euclidean distance as the demographic similarity measure. The distance graph was built as explained above using two levels of the triangulated graph. For the construction of the similarity graph we took the 15 most similar GUs, which produced a sparse graph with 2937 edges. The algorithm found the nine clusters shown in Fig. 2 (bottom).

For comparison purposes, in Fig. 2 (top) we show the clustering obtained using exclusively the geographic distances. To accomplish this, we applied the betweenness method to the distances' graph, which has 4289 edges. The impor-tant thing to notice is that the bottom partitioning can not be trivially obtained from the top one.

For example, the cluster represented by squares at the bottom of the figure contains GUs from three different clusters of the top clustering (three different triangles). Note that only one of these clusters is completely included in the clusters of squares. Thus, the idea of clustering with one measure and then with the other does not seem to be a good one. Our method achieves a reasonable compromise between the two measures, because the first one is utilized to constraint the solution space in such a way that the optimization of the second does not violate the requirements of the first one.

When dealing with larger problems like the one presented in Fig. 1 (left) the issue of obtaining a response in a reasonable computational time arises.

In an interactive scenario like the one we are trying to construct there are different heuristics that can be applied. Just to give an example, a simple method that consists in reducing the number of GUs for more efficient yet approximate clustering can be applied. This is accomplished by doing a two step triangulation. Only the GUs belonging to triangles which areas are not below the 0.90 area quantile in the first triangulation are considered in the second triangulation. We recall that triangulation algorithm has complexity $O(n \log n)$. In one of our experiments the number of considered GUs drops from 4292 to 835.

A second strategy, which of course can be combined with heuristics, is the use of parallel or distributed programs. We use ViPoC (see Appendix) to solve this problem in parallel. We took four clients with P-IV 2.4 GHz processors supporting hyper-threading technology in a 800 MHz bus and 512 MB of RAM. The clients were connected through a fast-ethernet network switch to a laptop where ViPoC was running. The time (in seconds) needed to complete the Phases IV of the algorithm for one, two, ..., five processors was: 854.14, 689.75, 524.19, 498.61 and 476.72, respectively. The clustered data was a rectangular area of the valley of Mexico consisting of 875 GUs and 5573 edges. Notice that an approximate time reduction from 14 to 8 minutes is achieved when 5 processors were used.

As the final experiment we take a much larger area of the valley of Mexico, exactly 1764 GUs and 11022 edges. Now we use the SGCLAP algorithm. It finds an average of 50 clusters in less than 0.25 seconds (we recall that this algorithm can output many different good solutions). For the sake of completness we report that all the solutions found by the SGCLAP for the data of the above paragraph have a similar modularity value to the one found by the betweenness method.

4   Conclusions

This paper has presented a novel approach for the zone design problem that is based on techniques from the field of complex networks research: community detection by betweenness centrality and label propagation. A new algorithm called Spatial Graph based Clustering by Label Propagation (SGCLAP) was introduced. It can deal with very large spatial clustering problems with time complexity $O(n \log n)$. The main idea behind our method is the construction of an unweighted similarity graph that codifies the weighted information of the design problem. We have shown that even a simplified model like this is able to capture enough information about the clustering structure of the data.

We should compare our approach with some related works in the area of spatial clustering [7,4,9,1,3]. Due to space constraints we leave this for a future journal paper. Nevertheless, here we present a couple of similarities with the most interesting of the reported algorithms GDBSCAN [7].

GDBSCAN can cluster point objects as well as spatially extended objects according to both, their spatial and their non-spatial attributes. Our method can do the same because it is a general graph method that is able to model any relation in its initial distance graph G_d. Both, GDBSCAN and SGCLAP have $O(n \log n)$ time complexity and create a one level clustering. We are planning to combine the SGCLAP with the betweenness method to detect simultaneously a hierarchy of clusterings. The SGCLAP algorithm will act as an initializer for the betweenness algorithm. We just mention one the advantages of SGCLAP: it has less parameters to tune.

Our work is part of an ongoing project aimed to create an open source tool for assisting *in situ* people that work interactively with population data. We are building the system on top of an interactive high performance computing platform (ViPoC) that acts as an interface between parallel hardware of different computational power and the problem software solution. This strategy should make the system useful for researchers and practitioners with low budget and at the same time should also be able to control a large cluster. What is important is that in both scenarios the user sees the same enviroment.

We believe that our approach can be used in other spatial clustering problems. For example, we are currently testing the SGCLAP algorithm with image segmentation benchmarks. The results are encouraging.

References

1. Bacao, F., Lobo, V., Painho, M.: Applying Genetic Algorithms to Zone Design. In: Soft-Computing-A Fusion of Foundations, Methodologies and Applications, vol. 2, pp. 341–348. Springer, Heidelberg (2005)
2. Csardi, G., Nepusz, T.: The igraph software package for complex network research. Inter. Journal of Complex Systems, 1695 (2006)
3. Malerba, D., Appice, A., Valaro, A., Lanza, A.: Spatial Clustering of Structured Objects. In: Kramer, S., Pfahringer, B. (eds.) ILP 2005. LNCS (LNAI), vol. 3625, pp. 227–245. Springer, Heidelberg (2005)
4. Estivill-Castro, V., Lee, I.: Fast spatial clustering with different metrics and in the presence of obstacles. In: 9th ACM International Symposium on Advances in Geographic Information Systems, pp. 142–147. ACM, New York (2001)
5. Girvan, M., Newman, M.E.J.: Community structure in social and biological networks. In: National Academy of Sciences, December 2002, vol. 99, pp. 7821–7826 (2002); arXiv:cond-mat/0112110 v1 7 December 2001
6. Instituto Nacional de Estadística, Geografía e Informática (INEGI), http://www.inegi.gob.mx
7. Sander, J., Martin, E., Kriegel, H.-P., Xu, X.: Density-based clustering in spatial databases: The algorithm GDBSCAN and its applications. Data Mining and Knowledge Discovery 2(2), 169–194 (1998)
8. Altman, M.: The Computational Complexity of Automated Redistricting: Is Automation the Answer? Rutgers Computer and Technology Law Journal 23(1), 81–142 (1997)

9. Neville, J.M., Jensen, D.: Clustering relational data using attribute and link information. In: Proceedings of the Text Mining and Link Analysis Workshop, 18th International Joint Conference on Artificial Intelligence (2003)
10. Newman, M.E.J.: Analysis of weighted networks. Physical Review E 70(056131) (November 2004); arXiv:cond-mat/0407503 v1 20 July 2004
11. Newman, M.E.J.: Fast algorithm for detecting community structure in networks. Physical Review E 69(066133) (2004); arXiv:cond-mat/0309508 v1 22 September 2003
12. Ochoa, A., Arco, L.: Differential Betweenness in Complex Networks Clustering. In: Ruiz-Shulcloper, J., Kropatsch, W.G. (eds.) CIARP 2008. LNCS, vol. 5197, pp. 227–234. Springer, Heidelberg (2008)
13. Ochoa, A., Bernábe, B., Ochoa, O.: Towards a Parallel System for Demographic Zonification Based on Complex Networks. Journal of Applied Research and Technology 7(2), 217–231 (2009)
14. Ochoa, O.: ViPoC una nueva herramienta de software libre para computación interactiva de alto rendimiento. PhD thesis, ICIMAF. Havana. Cuba, tutor: Alberto Ochoa (2009)
15. Raghavan, U.N., Albert, R., Kumara, S.: Near linear time algorithm to detect community structures in large-scale networks. Phys. Rev. E 76(036106) (2007)
16. Rodríguez, O.O., Rodríguez, A.O., Sánchez, A.: ViPoC: un clúster virtual, interactivo y portátil. Revista Ingeniería Electrónica, Automática y Comunicaciones XXVII(3) (2007)
17. Yang, Q., Lonardi, S.: A parallel edge-betweenness clustering tool for protein-protein interaction networks. Int. J. Data Mining and Bioinformatics 1(3) (2007)

Appendix

The term *interactive high performance computing* (IHPC) has to do with a system's ability to provide access to parallel hardware to run programs written in a very high language. The most common example is to acelerate Matlab by running the programs in a computer cluster. Most IHPC systems come with several mathematical libraries. We are interested in statistical and complex networks libraries. ViPoC (Virtual Interactive Portable Cluster [16,14]) is one experimental IHPC that fulfills our requirements.

We are building an experimental open source parallel tool for demographic zonification including a priori and posterior statistical analysis.

We pursue five aims: 1) allow rapid prototyping via high level languages like R and Matlab; 2) large collection of statistical and machine learning tools; 3) cheap parallel infrastructure; 4) smooth integration with existing Windows-Matlab technology; and 5) web interface with the system. ViPoC provides all these things.

A Short Overview of ViPoC

ViPoC is a flexible portable cluster because it can be used in a laptop, in a small office network or in a big organization managing a large cluster. It controls two subnetworks, one is formed by dedicated nodes behind a switch and the other may have a mixture of dedicated and no dedicated nodes –some of them may be virtual– located at the intranet of the organization. Therefore, it is easy to build a low budget cluster.

ViPoC supports the R and Octave languages, among many others. The later is a clone of Matlab, whereas the former has many statistical and machine learning functions ready to use. This is an important characteristic for the system we want to build: a large choice of statistical methods can be used in the a priori and posterior analysis of the zone design problem.

ViPoC comes with three so-called IHPC servers. They provide socket access to Octave, R and the system itself. For example, from Matlab running in the user's desktop computer it is possible to issue the following commands:

$$callViPoC\,('command',\ vipocIP) \tag{1}$$

$$callViPoCOctave\,('command',\ ListArgs,\ vipocIP) \tag{2}$$

$$callViPoCR\,('command',\ ListArgs,\ vipocIP) \tag{3}$$

where $vipocIP$ is the ip-address of the ViPoC server and $ListArgs$ is a list of arguments that depend on the concrete $command$. For example, the call

$$callViPoC\,('mpirun\ -np\ 4\ pgn\ AGEBs.pebc > getclusters.m',\ vipocIP)$$

executes in 4 processors the parallel program pgn. This program is a modification of the code reported in [17]. It takes the file $AGEBs.pebc$ (the list of edges in G_s) as input and writes to the standard output the results (in form of a Matlab program) which is redirected to the file $getclusters.m$.

In ViPoC the working directory in the user's desktop computer can be mapped to the user's working directory in the cluster (via the Samba protocol). Therefore, the Matlab function $getclusters.m$ can be called from the user's Matlab program, obtaining in this way the computed clusters.

Let us assume that the matrix $AGEBdata$ is a Matlab variable in the user's desktop computer that contains the geographic and demographic information of each GU. That matrix is usually the input to algorithm 1, but sometimes we may want to do some preprocessing. Figure 3 presents a fragment of code that

```
1   if Preprocessing
2       callViPoCR('open');
3       callViPoCR('assign', 'X', AGEBdata(:, 4:end));
4       callViPoCR('eval', ...
5                   sprintf('pc <- prcomp(X, scale=TRUE, ...
6                   center=TRUE, tol=%1.3f)', pcTol));
7       pcAGEBdata = callViPoCR('get_matrix', 'pc$x');
8       AGEBdata   = [AGEBdata(:,1:3) pcAGEBdata];
9       callViPoCR('close');
10  end
```

Fig. 3. Fragment of Matlab (or Octave) code that uses the ViPoC R server to compute the principal component of the matrix AGEBdata (each row represents one GU). Although we have removed the ip-address for the sake of space this code can be executed from the user desktop computer.

computes the principal components of the matrix, calling the R server of ViPoC (see [14] for syntactic details). This fragment of code corresponds to Phase I of algorithm 1. Notice that the same approach can be utilized after the run of our algorithm to do some statistical postprocessing using R.

At this point it should be clear for the reader that our approach to the construction of the system consists in using different languages and systems to maximize reusability. In the current version we proceeded as follows. Phase I is implemented following the ideas outlined in Figure. 3. In the step 1 of Phase II, a very efficient C++ program obtained from the Web was used. The remainder of Phase II and Phase III were programmed in Matlab (Octave). This part of the program can be run in the user's desktop computer in Matlab or Octave, or completely in the cluster in Octave. The parallel computation of Phase IV is accomplished as explained above, it returns the best clustering according to the modularity measure. Other reasonable good clusterings can be obtained, again via the ViPoC R server, using the complex network package igraph [2]. The final postprocessing is accomplished in the same way. It is worth noting that R also has the capacity to run parallel code.

Before we conclude this short presentation of the use of ViPoC as a computational platform for our system we once again stress the following fact. We are interested in providing a system that researchers and practitioners can use in their available hardware: from a single laptop to a large cluster. In the later case, it is also advisable to have a web interface for remote access. ViPoC provides this option.

Multiscale Functional Autoregressive Model for Monthly Sardines Catches Forecasting

Nibaldo Rodriguez, Orlando Duran, and Broderick Crawford

Pontificia Universidad Católica de Valparaíso, Chile
Firstname.Name@ucv.cl

Abstract. In this paper, we use a functional autoregressive (FAR) model combined with multi-scale stationary wavelet decomposition technique for one-month-ahead monthly sardine catches forecasting in northern area of Chile $(18°21'S-24°S)$. The monthly sardine catches data were collected from the database of the National Marine Fisheries Service for the period between 1 January 1973 and 30 December 2007. The proposed forecasting strategy is to decompose the raw sardine catches data set into trend component and residual component by using multi-scale stationary wavelet transform. In wavelet domain, the trend component and residual component are predicted by use a linear autoregressive model and FAR model; respectively. Hence, proposed forecaster is the co-addition of two predicted components. We find that the proposed forecasting method achieves a 99% of the explained variance with a reduced parsimonious and high accuracy. Besides, is showed that the wavelet-autoregressive forecaster is more accurate and performs better than both multilayer perceptron neural network model and FAR model.

Keywords: forecasting, wavelet decomposition, autoregression.

1 Introduction

Sardines are highly important small pelagic resources for economic development in northern Chile. The sardine fishery began to develop in Chile early 1973. However, the sardine abundance dropped drastically after 1985 and, at the same time, the anchovy abundance recovered. These historic fluctuations in sardines and anchovy stocks are associated with climate variability factors. After 1975, a warm period took place and sardine, which prey heavily on anchovy eggs, increased. The return of colder conditions after 1985 favored the recovery of anchovy and the decline of sardine, in spite of El Niño events in 1987, 1991-92, and 1997-98.

In order to develop sustainable exploitation policies, forecasting the stock and catches of pelagic species off northern Chile is one of the main goals of the fishery industry and the government. However, fluctuations in the environmental variables complicate this task. To the best of our knowledge, few publications exist on forecasting models for pelagic species. In recent years, linear regression models [1,2] and artificial neuronal networks (ANN) [3,4] have been proposed

A. Hernández Aguirre et al. (Eds.): MICAI 2009, LNAI 5845, pp. 189–200, 2009.

for forecasting models. The disadvantage of models based on linear regressions is the supposition of stationarity and linearity of the time series of pelagic species catches. Although ANN allow modeling the non-linear behavior of a time series, they also have some disadvantages such as slow convergence speed and the stagnancy of local minima due to the steepest descent learning method. To improve the convergence speed and forecasting precision of anchovy catches off northern Chile, Gutierrez [3] proposed a hybrid model based on a multilayer perceptron (MLP) combined with an autoregressive integrated moving average (ARIMA) model. The architecture of the MLP consists of an input layer with 6 nodes, two hidden layers of 15 nodes each, and an output layer with one node; the Levenberg Maquardt method was used as the learning method. This forecaster obtained a coefficient of determination R^2 of 82%, which improved slightly when combining the MLP model with the ARIMA model, reaching an R^2 of 87%. One of the disadvantages of this hybrid model is its high parsimony (230 parameters) and low forecasting precision. In this paper, the proposed forecasting model is based on multi-scale wavelet decomposition combined with autoregressive models. The multi-scale wavelet decomposition technique was selected due to its popularity in hydrological [5,6], electricity market [7], financial market [8] and smoothing methods [9,10,11]. This wavelet technique is based on the discreet wavelet transform (DWT) or the stationary wavelet transform (SWT) [12]. The advantage of these wavelet transforms in non-stationary time series analysis is their capacity to separate low frequency (LF) from high frequency (HF) components. Whereas the LF component reveals long-term trends, the HF component describes short-term fluctuations in the time series. Being able to separate these components is a key advantage in proposed forecasting strategies since the behavior of each frequency component is more regular than the raw time series.

Therefore, proposed autoregressive model for 1-month-ahead monthly sardines catches forecasting is based on an additive combination of the trend and residual components. The trend component is forecasted using a linear autoregressive (AR) model and the residual component is predicted using a functional autoregressive model and all the parameters of the autoregressive models are estimated using the least squares method.

This paper is organized as follows. In the next section, we briefly describe the multi-scale stationary wavelet transform and the proposed multi-scale functional autoregressive (MFAR) forecasting model. Section 3 presents a forecasting scheme based on MLP neural network combined with moving average smoothing technique. The simulation results and performance comparison are presented in Section 4 followed by conclusions in Section 5.

2   Proposed Forecasting Model

This section presents the proposed forecasting model for one-month-ahead sardines catches in northern Chile, which is based on a stationary wavelet transform and autoregressive models. Moreover, instead of using the original data set of past observations to predict the future value $x(n+1)$, we use its wavelet coefficients.

2.1 Stationary Wavelet Decomposition

A signal $x(n)$ can be represented at multiple resolutions by decomposing the signal on a family of wavelets and scaling functions [9,10,11]. The approximation (scaled) signals are computed by projecting the original signal on a set of orthogonal scaling functions of the form:

$$\phi_{jk}(t) = \sqrt{2^{-j}}\phi(2^{-j}t - k) \tag{1}$$

or equivalently by filtering the signal using a low pass filter of length r, $h = [h_1, h_2, ..., h_r]$, derived from the scaling functions. On the other hand, the detail signals are computed by projecting the signal on a set of wavelet basis functions of the form

$$\psi_{jk}(t) = \sqrt{2^{-j}}\psi(2^{-j}t - k) \tag{2}$$

or equivalently by filtering the signal using a high pass filter of length r, $g = [g_1, g_2, ..., g_r]$, derived from the wavelet basis functions. Finally, repeating the decomposing process on any scale J, the original signal can be represented as the sum of all detail coefficients and the last approximation coefficient.

In time series analysis, discrete wavelet transform (DWT) often suffers from a lack of translation invariance. This problem can be tackled by means of the un-decimated stationary wavelet transform (SWT). The SWT is similar to the DWT in that the high-pass and low-pass filters are applied to the input signal at each level, but the output signal is never decimated. Instead, the filters are up-sampled at each level.

Consider the following discrete signal $x(n)$ of length N where $N = 2^J$ for some integer J. At the first level of SWT, the input signal $x(n)$ is convolved with the $h_1(n)$ filter to obtain the approximation coefficients $a_1(n)$ and with the $g_1(n)$ filter to obtain the detail coefficients $d_1(n)$, so that:

$$a_1(n) = \sum_k h_1(n - k)x(k) \tag{3a}$$

$$d_1(n) = \sum_k g_1(n - k)x(k) \tag{3b}$$

because no sub-sampling is performed, $a_1(n)$ and $d_1(n)$ are of length N instead of $N/2$ as in the DWT case. At the next level of the SWT, $a_1(n)$ is split into two parts by using the same scheme, but with modified filters h_2 and g_2 obtained by dyadically up-sampling h_1 and g_1.

The general process of the SWT is continued recursively for $j = 1, ..., J$ and is given as:

$$a_{j+1}(n) = \sum_k h_{j+1}(n - k)a_j(k) \tag{4a}$$

$$d_{j+1}(n) = \sum_k g_{j+1}(n - k)a_j(k) \tag{4b}$$

where h_{j+1} and g_{j+1} are obtained by the up-sampling operator inserts a zero between every adjacent pair of elements of h_j and g_j; respectively.

Therefore, the output of the SWT is then the approximation coefficients a_J and the detail coefficients $d_1, d_2, ..., d_J$, whereas the original signal $x(n)$ is represented as a superposition of the form:

$$x(n) = a_J(n) + \sum_{j=1}^{J} d_j(n) \tag{5}$$

The wavelet decomposition method is fully defined by the choice of a pair of low and high pass filters and the number of decomposition steps J. Hence, in this study we choose a pair of haar wavelet filters [12]:

$$h = \left[\frac{1}{\sqrt{2}} \; \frac{1}{\sqrt{2}} \right] \tag{6a}$$

$$g = \left[\frac{-1}{\sqrt{2}} \; \frac{1}{\sqrt{2}} \right] \tag{6b}$$

On the other hand, a key issue for the success of any wavelet forecasting model is suitable selection of the J level decomposition. At higher J, the variability of a large number of predicted data is lower, so their prediction is easier and accurate. In our proposed model, we determine the value of J using a stopping criterion that is given as:

$$\rho = \frac{\sigma_{a_J}}{\sigma_x} < \epsilon \tag{7}$$

where σ_{a_J} and σ_x are the standard deviation of the approximation component and original data, respectively.

We stop the decomposition on the level for which the ρ ratio is substantially less than a threshold ϵ. The choice of the value of ϵ is not clear from a physical point of view and different sets of approximation coefficients will be produced by the wavelet decomposition method for different values of ϵ. In order to obtain accurate and parsimonious forecasting results, the value of ϵ was set to 0.5 in this work.

Finally, wavelet scales are such that times are separated by multiples of $2^j, j = 1, ..., J$. Our data set involves monthly observations so that the wavelet scales are such that scale 1 is associated with $1 - 2$ month dynamics, scale 2 with $2 - 4$ month dynamics, scale 3 with $4 - 8$ month dynamics, scale 4 with $8 - 16$ month dynamics, and so on.

2.2   Multi-scale Functional Autoregressive Model

In order to predict the future signal $x(n + 1)$, we can separate the original signal $x(n)$ into two components. The first component presents the trend $t(n)$ of the series and is characterized by slow dynamics, whereas the second component presents the residue $r(n)$ of the series and is characterized by fast dynamics. Therefore, our forecasting model will be the co-addition of two predicted values given as:

$$x(n + 1) = t(n + 1) + r(n + 1) \qquad (8)$$

On the one hand, the residual component is estimated by using functional autoregressive (FAR) model given as:

$$r(n + 1) = \sum_{j=1}^{J} \sum_{i=1}^{m} \alpha_{ji}[u] d_j[n - i + 1] \qquad (9a)$$

$$\alpha_{ji}[u] = \sum_{k=0}^{l} \beta_{jik} d_j^k [n - i + 1] \qquad (9b)$$

where the J value denotes the level of stationary wavelet decomposition, the m value represents the autoregressive order of the detail coefficients, the l value represents the order of the approximation polynomial and $\beta = (\beta_1, \beta_2 \ldots, \beta_K)$ is the parameter vector to be estimated, where the $K = (l + 1)mJ$ value is the total number of terms of the FAR model.

On the other hand, the trend component is estimated by using a linear autoregressive (AR) model given as:

$$t(n + 1) = \sum_{i=1}^{m} \gamma_i a_J[n - i + 1] \qquad (10)$$

We propose estimating the linear parameters $\theta = \{\beta_k, \gamma_i\}$ using the least squares method based on the Moore-Penrose pseudo-inverse. If we suppose a set of N_s training input-output samples, then we can perform N_s equations of the form of (9) and (10) as follows:

$$\Re = \beta \Phi \qquad (11a)$$

$$\Gamma = \gamma \Psi \qquad (11b)$$

where

$$\Phi = [d_1(n), \cdots, d_1(n - m + 1), d_2(n), \cdots, d_J(n), \cdots, d_J(n - m + 1)] \qquad (12a)$$

$$\Psi = [a_J(n), a_J(n - 1), \cdots, a_J(n - m + 1)] \qquad (12b)$$

The optimal values of the linear parameters β_k and γ_i and are obtained using the following residual sum of squares (RSS) function defined as:

$$RSS(\beta) = \sum_{i=1}^{N_s} \left[R_i(n + 1) - r_i(n + 1) \right]^2 \qquad (13a)$$

$$R(n + 1) = x(n + 1) - a_J(n + 1) \qquad (13b)$$

$$RSS(\gamma) = \sum_{i=1}^{N_s} \left[a_{i,J}(n + 1) - t_i(n + 1) \right]^2 \qquad (13c)$$

$$(13d)$$

The result of minimizing the RSS objective function is:

$$\beta = (\Phi^T \Phi)^\dagger \Phi^T Z \tag{14a}$$

$$\gamma = (\Psi^T \Psi)^\dagger \Psi^T Y \tag{14b}$$

where $(\cdot)^\dagger$ denotes the Moore-Penrose pseudo-inverse [13].

Once we have decided upon a forecasting structure to use, the next task is to determine the autoregressive order on the different scales. This can be done by using the Bayesian Information Criterion (BIC), which is given as [14]:

$$BIC = ln\left(\frac{RSS}{N_s - \theta}\right) + \frac{\theta log 10(N_s - \theta)}{N_s - \theta} \tag{15}$$

where the $\sharp\theta$ value represents the total number of parameters of the proposed multiscale functional autoregressive MFAR(J,m,l,m) model, which is obtained as:

$$\sharp\theta = J \times m \times (l + 1) + m \tag{16}$$

3   Forecasting without Wavelet Pre-processing

This section presents a model based on a MLP neural network for forecasting monthly catches of sardine in northern Chile. Moreover, instead of using the multi-scale stationary wavelet transform for pre-processing, we use the 3-month moving average smoothing technique [15]. The smoothed data $s(n)$ can be forecasted using a MLP neural network given as:

$$y = \sum_{j=1}^{N_h} b_j \phi_j(u_i, v_j) \tag{17}$$

where N_h is the number of hidden nodes, $u = [u_1, u_2, \ldots u_m]$ denotes the regression vector containing m lagged values, $[b_1, \ldots b_{N_h}]$ represents the linear output parameters, $[v_1, v_2, \ldots v_{N_h}]$ denotes the nonlinear parameters, and $\phi_j(\cdot)$ are hidden activation functions, which are derived as:

$$\phi_j(u_i) = \phi\left(\sum_{i=1}^{m} v_{j,i} u_i\right) \tag{18a}$$

$$\phi(u) = \frac{1}{1 + exp(-u)} \tag{18b}$$

In order to estimate both the linear and nonlinear parameters of the MLP, we use the Levenberg-Marquardt (LM) algorithm [16]. The LM algorithm adapts the $\theta = [b_1, \ldots b_{N_h}, v_{j,1}, \ldots v_{j,m}]$ parameters of the neuro-forecaster minimizing mean square error, which is defined as:

$$E(\theta) = \frac{1}{2} \sum_{i=1}^{N_s} \left(s_i(n) - y_i(u(n)) \right)^2 \qquad (19a)$$

$$(19b)$$

where $u = [s(n-1)s(n-2)\cdots s(n-m)]$ represents the regressor vector.

Finally, the LM algorithm adapts the parameter θ according to the following equations:

$$\theta = \theta + \Delta\theta \qquad (20a)$$

$$\Delta\theta = (\Upsilon\Upsilon^T + \mu I)^{-1}\Upsilon^T E \qquad (20b)$$

where Υ represents the Jacobian matrix of the error vector evaluated in θ_i and the error vector $e(\theta_i) = d_i - y_i$ is the error of the MLP neural network for i patter, I denotes the identity matrix and the parameter μ is increased or decreased at each step of the LM algorithm.

4   Experiments and Results

In this section, we apply the proposed strategy for 1-month-ahead forecasting of the monthly catches of sardines. The data set used corresponded to sardine landings off northern Chile. These samples were collected monthly from 1 January 1973 to 30 December 2007 by the National Fishery Service of Chile (www.sernapesca.cl).

The proposed multi-scale functional autoregressive (MFAR) forecasting model basically involves three stages. In the first stage, the original data set is decomposed into different wavelet scales by using stopping criterion given in (7) to separate both the trend component (approximation component) and the residual component (difference between original data and trend component). In the second stage, the trend component and residual component are forecasted independently by using an autoregressive model. In the third stage, the next sample is predicted by the co-addition of two predicted components.

The raw sardines data set have been normalized to the range from 0 to 1 by simply dividing the real value by the maximum of the appropriate set. On the other hand, the original data set was also divided into two subsets as shown in Fig.1. In the first subset, the data from 1 January 1973 to 30 December 2002 were chosen for the training phase (parameters estimation), whereas the remaining data were used for the validation phase. This normalized data set, when subjected to the stopping criterion, yielded a 5 level wavelet decomposition. The low frequency component a_5 represents the trend of the observed sardines catches data set. On the other hand, detail components $\{d_1, d_2, d_3, d_4, d_5\}$ contain high frequency components of the original data such that d_1 the highest frequency component and d_1 is considered to be more related to the noisy part of the observed data, whereas d_5 contains lower frequency information than $\{d_4, d_3, d_2\}$.

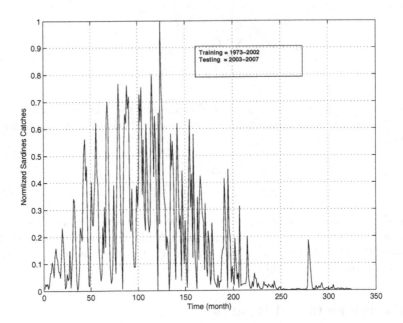

Fig. 1. Observed monthly sardine catches data

Hence, residual component forecasting is based on a functional autoregressive model combined with a third order polynomial, whereas predicting the trend component is done with a linear autoregressive model. Once we chose the multi-scale autoregressive forecasting structure to use, the next task was to determine the autoregressive order by using the method given in (14) and (15). After we applied the least squares method and the BIC, we decided to use two lagged values on each level wavelet decomposition due to the parsimony principle and precision of the proposed MFAR(J,m,l,m) forecasting model.

In this study, two criteria of forecasting accuracy were used to compare the forecasting capabilities of the MFAR, MLP and FAR forecasting models. The first measurement is the coefficient determination (R^2) given as:

$$R^2 = 1 - \frac{\sum_{i=1}^{N_s}(A_i - F_i)^2}{\sum_{i=1}^{N_s}(A_i - \bar{A})^2} \tag{21}$$

where A_i is the actual value at time i, F_i is the forecasted value at time i, $\bar{A}$ is the mean value of observed monthly catches, and N_s is the number of forecasts. If R-square is large, then the model is good. Conversely, if R-square is small, then the model is bad.

The second criterion is the mean absolute percentage error (MAPE) given as:

$$MAPE(\%) = \frac{1}{N_s} \sum_{i=1}^{N_s} \left| \frac{A_i - F_i}{F_i} \right| \times 100 \tag{22}$$

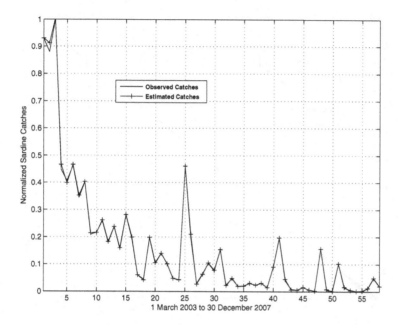

Fig. 2. Observed sardine catches vs estimated sardine catches with testing data

Figures 2 and 3 show the results obtained with the best MFAR(5,2,3,2) forecasting model during the testing phase. Fig. 2 provides data on observed monthly
sardine catches versus forecasted catches; this forecasting behavior is very accurate for testing data with a MAPE below 8.6%. Fig. 3 shows the regression
between observed and estimated monthly sardine catches. The good fit of the
data to line 1 : 1 and 99% of the explained variance can be seen in Fig.3. This
level of explained variance was achieved due to use of multi-scale stationary
wavelet decomposition.

To assess the performance of the proposed MFAR(5,2,3,2) forecaster, we compared it with two other forecasters based on MLP neural network and FAR model
combined with 3-month moving average smoothing techniques.

The MLP neural network was calibrated using 32 previous months as input
data plus one bias unit due to the 5-level wavelet decomposition used herein.
Finding the optimal number of hidden nodes is a complex problem, but in all our
experiments, the number of hidden nodes is set at half the sum of the number
of input and output nodes. In the training process, overall weights were initialized by a Gaussian random process with a normal distribution $N(0,1)$ and the
stopping criterion was a maximum number of iterations set at 200. Due to the
random initialization of the weights, we used 20 runs to find the best MLP neural
network with a low prediction error. After the training process, the best architecture was calibrated with 16 input nodes, 9 hidden nodes, and one output node;
this is denoted as MLP(16,9,1). In the testing stage, this MLP model explained
89.43% of the variance and achieved a MAPE of 137%. On the other hand, the

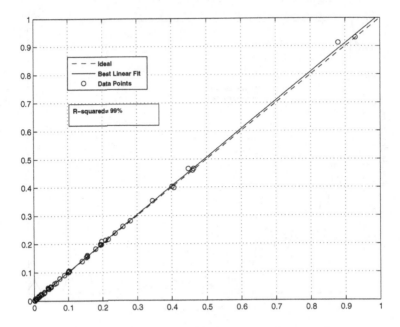

Fig. 3. Scatter for estimates monthly sardines catches

Table 1. Statistic of the forecasting models

Model	R-Squared	MAPE
MFAR(5,2,3,2)	**99%**	**6.84%**
MLP(16,9,1)	89.44%	137%
FAR(16,3)	90.94%	136%

FAR model based on a third order polynomial with 16 previous months as input had a coefficient of determination of 90.94% and a MAPE of 136% in the testing phase. These results are related to the high dispersion between the observed and estimated data sets.

Table 1 shows the best forecasting results for the MFAR forecasting model, the MLP neural network model and the FAR model. As seen from Table 1, the proposed MFAR(5,2,3,2) forecaster achieved a high R^2 and a lower MAPE for 1-month-ahead forecasting of monthly sardine catches in northern Chile. The improved accuracy of the proposed forecaster with respect to MLP(16,9,1) and FAR(16,3), in terms of R^2, is 9.7% and 8.14%, respectively. Besides, the improvement in the MAPE of the proposed method with respect to MLP(16,9,1) and FAR(16,3) is 95% and 94.97%, respectively.

Finally, it can be seen from Table 1 that the overall performance of the proposed strategy to forecast 1-month-ahead sardines catches is much better than the MLP neural network and FAR model.

5   Conclusions

In this paper was proposed a one-month-ahead monthly sardine catches forecasting strategy to improve prediction accuracy. The reason of the improvement in forecasting accuracy was due to use stationary haar wavelet decomposition to separate both the trend and residual components of the time series, since the behavior of each component is more smoothing than raw data set. The performance was compared with those using the MLP forecaster and using FAR forecaster without wavelet pre-processing. It was show that the proposed forecaster achieves a significantly less MAPE value than both MLP and FAR forecasters. Besides, proposed forecasting results showed that the 32 previous month contain valuable information to explicate a highest variance level for sardines catches forecasting. These months can be related with biological, ocean-atmospheric aspects, which have a great influence on pelagic fish fisheries in northern Chile. Finally, wavelet-autoregressive forecasting strategy can be suitable as a very promising methodology to any other pelagic specie.

References

1. Stergiou, K.I.: Prediction of the Mullidae fishery in the easterm Mediterranean 24 months in advance. Fish. Res. 9, 67–74 (1996)
2. Stergiou, K.I., Christou, E.D.: Modelling and forecasting annual fisheries catches: comparison of regression, univariate and multivariate time series methods. Fish. Res. 25, 105–138 (1996)
3. Gutierrez, J.C., Silva, C., Yaez, E., Rodriguez, N., Pulido, I.: Monthly catch forecasting of anchovy engraulis ringens in the north area of Chile: Nonlinear univariate approach. Fisheries Research 86, 188–200 (2007)
4. Garcia, S.P., DeLancey, L.B., Almeida, J.S., Chapman, R.W.: Ecoforecasting in real time for commercial fisheries:the Atlantic white shrimp as a case study. Marine Biology 152, 15–24 (2007)
5. Adamowski, J.F.: Development of a short-term river flood forecasting method for snowmelt driven floods based on wavelet and cross-wavelet analysis. Journal of Hydrology 353(3-4), 247–266 (2008)
6. Kisi, O.: Stream flow forecasting using neuro-wavelet technique. Hydrological Processes 22(20), 4142–4152 (2008)
7. Amjady, N., Keyniaa, F.: Day ahead price forecasting of electricity markets by a mixed data model and hybrid forecast method. International Journal of Electrical Power Energy Systems 30, 533–546 (2008)
8. Bai-Ling, Z., Richard, C., Marwan, A.J., Dominik, D., Barry, F.: Multiresolution Forecasting for Futures Trading Using Wavelet Decompositions. IEEE Trans. on neural networks 12(4) (2001)
9. Coifman, R.R., Donoho, D.L.: Translation-invariant denoising, Wavelets and Statistics. Springer Lecture Notes in Statistics, vol. 103, pp. 125–150. Springer, Heidelberg (1995)
10. Nason, G., Silverman, B.: The stationary wavelet transform and some statistical applications, Wavelets and Statistics. Springer Lecture Notes in Statistics, vol. 103, pp. 281–300. Springer, Heidelberg (1995)

11. Pesquet, J.-C., Krim, H., Carfantan, H.: Time-invariant orthonormal wavelet representations. IEEE Trans. on Signal Processing 44(8), 1964–1970 (1996)
12. Percival, D.B., Walden, A.T.: Wavelet Methods for Time Series Analysis. Cambridge University Press, Cambridge (2000)
13. Serre, D.: Matrices: Theory and applications. Springer, New York (2002)
14. Shumway, R.H., Stoffer, D.S.: Time series analysis and its applications. Springer, Berlin (1999)
15. Newbold, P., Carlson, W.I.: Statistics for business and economics, 5th edn. Prentice Hall, Englewood Cliffs (2003)
16. Hagan, M.T., Menhaj, M.B.: Training feedforward networks with the Marquardt algorithm. IEEE Transactions on Neural Networks 5(6), 989–993 (1996)

Discretization of Time Series Dataset with a Genetic Search

Daniel-Alejandro García-López[1] and Héctor-Gabriel Acosta-Mesa[2]

[1] Universidad del Istmo,
Ingeniería en Computación,
Ciudad Universitaria S/N, Barrio Santa Cruz, 4a. Sección
Sto. Domingo Tehuantepec, Oax., México C.P. 70760
`daniel.garcia@sandunga.unistmo.edu.mx`
[2] Universidad Veracruzana,
Departamento de Inteligencia Artificial,
Sebastián Camacho #5 Col. Centro
Xalapa, Ver., México C.P. 91000
`heacosta@uv.mx`

Abstract. In this work we propose a new approach to the discretization of time series using an approach that applies genetic algorithm operations called GENEBLA. The basic idea is to minimize the entropy of the temporal patterns over their class labels, follow a genetic search approach that allows to find good solutions more quickly to explore a wide variety of possible ways to solve the problem at the same time. The performance of GENEBLA was evaluated using twenty temporal datasets and compared to an efficient time series discretization algorithm called SAX and EBLA3 algorithm that shows similar representation.

1 Introduction

Real world measurements that are developed over time have become a common task, as a result there are applications that produce temporary information, such as medical, industrial, financial and business applications[10]. Unfortunately, this data occupies large areas of storage, which requires an efficient representation. Besides it is also required to maintain the necessary information for efficient classification, in addition to other features. Many approaches have been proposed to represent temporal data [9] [11], all of them are oriented to data compression, rather than to information maximization. That is to say, these representation transform times series of length N, into a set of n coefficients, where $n < N$. These data compression processes are only intended to reduce dimensionality on data [5]. However, they do not take into account whether the new representation preserves the relevant information in order for the observations to continue belonging to associated class labels. Discretization algorithms that maximize the data will improve efficiency when classifying.

In most of the algorithms proposed to discretize time series the user is required to specify a set of parameters to perform the transformation; for example, the

A. Hernández Aguirre et al. (Eds.): MICAI 2009, LNAI 5845, pp. 201–212, 2009.

number of segments (word size) dividing the times series length, and the number of intervals (alphabet) required to compress the time series value. Without a previous analysis of the temporal data, it is very difficult to know the proper parameter values to obtain a good discrete representation of data. However, in practice it is assumed that the parameters are known [7] [11]. In the present work, we propose a new approach to discretization of time series dataset called GENEBLA (Genetic Entropy Based Linear Approximation). Genetic algorithm has proven to be efficient to search in state space or approximate solutions to optimization. It also are very helpful when the user does not have precise domain expertise, because genetic search possess the ability to explore and learn from their domain. This proposal is based on the operations of a simple genetic algorithm, using as a fitness function the metric gain information as described in later sections. The content of the paper is presented as follows: in section 2 the related work is discussed. Section 3 describes the proposal and main concepts involved in its definition. In section 4 the algorithm GENEBLA is explained in detail. Section 5 describes the experiments carried out and the data used to evaluate the performance of the algorithm. Section 6 shows the results obtained over the experiments. Section 7 presents a discussion and finally in section 8, conclusions and future work are presented.

2   Background and Related Work

Most of the algorithms in data mining in time series assume that the time series are discrete. However, most applications generate and use floating-point data type. Consequently there are several approaches to the discretization of time series with floating-point values. They propose different approaches and different measures of utility, this is the case proposed in [12]: the purpose of the algorithm is to detect the persistent states of the time series, the algorithm does not require either parameter specification or a class label. The applicability of this algorithm is restricted to the existence of persistent states in the time series, something that is not very common in most of the world is applications. The algorithm processes a single time series at a time, so that the discretization criterion is not generalized to the complete dataset.

Dimitrova [2], represent time series as a multiconnected graph. Under this representation, similar time series are grouped into a graph model. The algorithm focuses on minimizing the number of connected nodes (graph) to represent a time series model under different criterion based on a Single-Link Clustering (SLC) algorithm plus one criterion to consider the entropy to determine which arcs on the graph are deleted. Each node in the graph represents one point in the time series. Like the previous algorithm, it only works with one time series at a time.

Another algorithm called Symbolic Aggregate Approximation (SAX) has been proposed [11]. This algorithm is based on the Piecewise Aggregate Approximation representation [6]. SAX algorithm requires the user to define an alphabet (A) with fixed interval ranges and word size (Ws) as parameters. The first parameter indicates the number of intervals to be created in the continuous domain;

the second indicates the number of segments that have the temporal time series. After a normalization procedure, times series length is partitioned into (Ws) segments. The corresponding values are mapped to one of A discrete values through the use of a normal probability density function (PDF). One advantage of this representation is that, once the data set has been transformed, a smoothed version of the original data can be recovered using the PDF. Although SAX has been created for streaming data, it has proved an efficient representation for classification and clustering. However, SAX algorithm requires the user to define parameters based in a priori information.

In [15] they discusses using the technique of genetic algorithms as a method of discretization, furthermore this algorithm is designed for datasets with attributes not related in time. This algorithm proposes the use of entropy as fitness function, using two terms: one for measuring the complexity and the other to measure its utility. This is done following the procedure of a simple genetic algorithm to find a set of cut points, although it is not clearly mentioned as the representation of individuals.

In [1] they proposed an algorithm called EBLA2 that solved the problem found automatically alphabet size and word size to maximize accuracy in classification, but only the heuristic leads to a specific solution, built from the best result in each iteration, like a greedy search. This approach requires no parameters to the search, because the outcome is deterministic.

An improved version of the algorithm EBLA2 is proposed in [4] called EBLA3. This performs a broader search than EBLA2 using a simulated annealing approach for the discretization schemes for discretized temporal datasets. This allows the result is not entirely deterministic, possible to find better discretization schemes. However, the results obtained by this approach could be improved by enabling populations to generate discretization schemes, which could quickly reach a good solution.

The aim of the present work is to find the alphabet size and word size using the genetic search approach, because there are some discretization schemes that give a good solution, but follow a single path of equivalent solutions can quickly lead to convergence of a less general solution for the representation of temporal dataset, instead of having a group equivalent to the same time, can afford to follow different paths and choose a final solution to the best of them. This new approach makes use of the operations of Genetic Algorithms: Selection, Crossover, Mutation, Replace, as detailed in section 4.

3 Proposal

3.1 Discretization

Discretization is concerned with the process of mapping variables with continuous values into discrete values. This process has been widely used to compress data to facilitate computation in terms of space and time. More formally, given the data domain $x|x \in \mathbb{R}$, where $\mathbb{R}$ is the set of reals and the discretization scheme(D) $D = \{[d_0, d1], (d_1, d_2], ..., (d_{n-1}, d_n]\}$ where d_0 and the d_n are the minimum and maximum value of x respectively. Every pair of values represents an interval, one of each maps the specific range of continuous values to one element of a discrete set $\{1..m\}$. Where m is called the discretization degree and $d_i | i = 1..n$ are the interval limits, also known as a cut points [9].

The discretization process can be split into two main jobs. The first one is to find the number of discrete groups to do the mapping from continuous to discrete. The second one is to define the range or limits of each interval in the continuous domain [9]. Both jobs are done by GENEBLA using as fitness function the principle of information gain based on entropy as obtained in [4]. The aim of the algorithm is to find the number of intervals and their limits from which the membership of the resulting discrete models are clustered with respect to the class label. As explained in the following section, the discretization scheme is computed for the whole dataset, i.e. all the time series values are considered to find the discretization scheme with minimum entropy. GENEBLA is also based on the PAA representation to reducte the dimensionality, which consists of obtaining the average values of each segment the time series is divided.

3.2   Dimensionality Reduction

A simple and efficient time series representation is Piecewise Aggregate Approximation (PAA), where each segment has equal size and the number of segments is determined for user. Often, a priori information is not available, so the goal is to find the number of intervals with different size in order to reduce of dimensionality while as maintaining the information with respect to the class label. To reduce the dimensionality must be extended PAA, so that each segment of time series has a different size, given a set of intervals. Let C be a time series with length n represented in a w-dimensional space (word) as a vector $\overline{C} = \overline{c_1}, ..., \overline{c_w}$. and $T = \{t_1, t_2, ..., t_w\}$ be the discretization scheme where t_i is the interval of time of i segment of $\overline{C}$ where the i element of $\overline{C}$, is computed as:

$$\overline{c_i} = \frac{1}{|t_i|} \sum_{j=1}^{|t_i|} C_{t_j} \qquad (1)$$

where: $|t_i|$ is the number of elements of t_i.

3.3   Fitness Function

Most discretization algorithms require heuristics to avoid the a priori definition of the alphabet size. For example, in temporal datasets, the information criterion [2], the persistence score [12], information entropy maximization (IEM), information gain, maximum entropy, Petterson-Niblett and Minimum Description Length (MDL) [9],[3] have been used. An important step in the genetic algorithms approach is defining a fitness function of individual. In this paper we propose that a fitness function to select the optimal cut points is based on the information gain measure, given a specific discretization scheme(candidate solution or individual) and its corresponding class labels.

Formally, the information gain based on entropy can be stated as:

$$Gain(S, A) = Entropy(S) - \sum_{v \in A_n} \frac{\#S_v}{\#S} Entropy(S_v) \qquad (2)$$

Where: S and A are two different set time series.
$A_n \subseteq S | a_i \in A_n \wedge a_i \notin (A_n \setminus \{a_i\}), i = 1..n$
$\#S_v$ is the number of time series with value v in S
$\#S$ is the number of time series in S
The entropy of S and S_v are calculated as detailed in [1]

Because it is possible to find different cut points with equal information gain, is proposed a heuristic to select one of the tied candidates. This metric is called Isolated_term and it is computed as detailed in [1].

Fitness of an individual is calculated as:

$$Fitness = Gain + Isolated\_term \qquad (3)$$

It is important to remark that the entire time series is considered as one attribute, and that the discretization scheme is considered for the whole dataset. It allows the algorithm to find a good overall solution that minimizes the entropy on data.

4   Algorithm

Under the fitness (3) explained above, the time series discretization process can be thought of as a search through the space of all possible discretization schemes. The discretization problem on time series can be divided into two subproblems. The first one is to find the number and range of intervals of time that maintain the relevant information (alphabet), and the second is the discretization of data on each interval of time (word size). GENEBLA has been designed to solve both problems. At each phase the objective is to find not only the number of values for the discretization, but also the ranges of each interval while maintaining a relationship with the class label. This can be thought of as a discretization in both axes, x and y respectively.

First phase. Input: $\mathbf{S_p}$, **MutationProbability**, **ReproductionRate**

```
VD = sort(Unique(Sp))
C = Percentiles(VD)
For each ci in C
      Population=Population ∪ {[MinC, ci], (ci, MaxC]}
end for
Fitness=FFitness(Population)
While iGenerational < nGenerational and Population > k
      Increase(iGenerational)
      nCross=Population * ReproductionRate
      For iCross in nCross
          Parents=FSelection(Population, Fitness)
          Children=FCrossover(Parents)
          Children=FMutation(Children, MutationProbability)
          Children=FAdd(Children, Population)
      end for
```

Fitness=FFitness(Population, S$_p$)
Population=FReplace(Population, Fitness)
end while
return BestIndividual(Population)

Where $S = \{s_i | s_i \in \mathbb{R}, i = 1..n\}$ represents the numeric values of a time series S with length n. S_p is the set of p time series. $MinC$ and $MaxC$ are the minimum and maximum value in all temporal dataset. $nGenerational$ is the maximum value of an epoch or generation of individuals. $nCross$ is the number of crossover. $Parents$ are two individuals of populations. $FSelection$ is the selected function as detailed later. $Children$ is two new individuals generated by $Parents$. $FCrossover$ is the crossover function. $FMutation$ is the Mutation function and $FAdd$ is a procedure that adds the new individuals in the population. $FFitness$ calculates the fitness of each of the individuals of population, it is computed as described in (3). $FReplace$ eliminates individuals of the population. $BestIndividual(Population)$ return the best individual of the population.

GENEBLA algorithm is based on a search through a simple genetic algorithm, in this case a gene as one of the cut points of the discretization scheme, and genes of the first phase, by the nature of data storage are represented with values of type real. Chromosome in this case is regarded as a gene, considering that each gene is values of type real. An individual is the set of genes, which is equal to the discretization scheme of dataset. The population is a collection of discretization schemes. It is important to note that the number of genes in each individual can differ.

GENEBLA requires the user to define some parameters that allow to maintain greater control over the algorithm, so that it can only be used to obtain sufficiently good results, without waiting for a refined solution. The parameters are: number of generations, minimum number of generations, reproduction rate, probability of mutation

In the first phase, the GENEBLA algorithm is used to calculate the alphabet. This phase requires finding all the different values that exist in the temporal dataset, and locating the percentiles defined from 0.0% to 100% on increments of 0.1%($Percentil$).For each of these values an individual in the form $Individual = \{[minvalue, Percentil_i], (Percentil_i, maxvalue]\}$ is created, where $minvalue$ and $maxvalue$ are the minimum and maximum values found in temporal dataset S_p. For all this initial population, calculate the fitness value of each individual as defined in equation 3. From this population iterate until it reaches the value defined in $nGenerational$. Meanwhile $nCross$ times select two individuals $Parents$ to form pairs($FSelection$). $nCross$ value is calculated from the current number of individuals in the population and reproduction rate initially defined. Crossing each of the pairs($FCrossover$) generates two individuals for each pair (children), to which it applies the mutation operator($FMutation$), given the probability of mutation set for search. Each child created add to the population($FAdd$). Then calculate the fitness of new individuals in the population. Select individuals will be removed from the population based on their fitness value($FReplace$).

The function selection of parents uses a method called roulette wheel selection which, using a single turn of the roulette wheel is proportional to the circular sectors the objective function. The probability of selection is calculated based on the population and their fitness value.

The crossover operator in the first phase: Randomly selecting a gene from one parent, which will be divided, the same procedure for is repeated the second one, resulting in two fragments obtained from each parent $[F_1, F_2]$, $[F_3, F_4]$. Now the children are formed from these four fragment. This procedure is to combine the first portion of the first parent to the second portion of the second parent, thus obtained $[F_1, F_4]$, $[F_3, F_2]$. The mutation operator in this phase first checks whether or not the individual will be mutated, a decision based on the probability of mutation set for the search and a random value. If it is decided that an individual's gene should be mutated then a gene is randomly selects and transforme into a random value in the range of temporal domain. This is calculated as follows:

$$Gen_x = (MaxC - MinC) * random(0, 1) + MinC$$
$$i = random(nGen)$$
Replace(Gen_i, Gen_x)
Where:
Gen_x is x element in one individual.
Gen_i is i element in one Individual.
Its have value between $MinC$ and $MaxC$.

It is important to verify after this procedure that the structure of the new individual is valid. Function replace: according to the median(statistics) value of the fitness of the entire population, those individuals that are above the median will be candidates to remain in the population. The second filter is based on the value of individual fitness that can be selected for removal.

Second phase. Input:
S_p, MutationProbability, ReproductionRate, MaximumFitnessFirstPhase

```
For each iS_i in nS_i
Population=Population ∪ FCreateIndividual(iS_i)
end for
Fitness = FFitness(Population, S)
While (MaximumFitnessFirstPhase <= MaximumFitness or Improve = true)
    and iGenerational < nGenerational
    Increase(iGenerational)
    nCruces=Population * ReproductionRate
    For each iCruce in nCruce
        Parents=FSelection(Population, Fitness)
        Children=FCrossover(Parents)
        Children = FMutation(Children, MutationProbability)
        Population = Add(Children, Population)
    end for
    Fitness=FFitness(Population, S_p)
    Population=FReplace(Population, Fitness)
end while
return BestIndividual(Population)
```

The representation of individuals for the second phase is different from the previous phase, because most of the applications that generate time series only performed on variables sampling point values within the temporal domain. To know this facilitates the task of representation. It is assumed that there are n samples in time. Therefore, they can be represented at each sampling point with a binary value indicating whether or not this belongs to the set of adjacent points. It is important to note that this representation is not directly related to the fitness function, so the decoding is carried out before calculating each individual's fitness value.

Initially, for convenience, the algorithm creates only n individuals. Although it can start with larger or smaller individuals. Representing the cut points in time is simpler, thanks to the binary representation used, it is only a matter of indicating which sample points will be present or absent. This can be seen as an individual's genes affecting whether or not these characteristics appear.

The selection function is similar to that described in the first phase. The crossover operator for this second stage is to randomly choose two sample points, from which both parents will be fragmented into, $[F_1, F_2, F_3]$, $[F_4, F_5, F_6]$, these six pieces will form two individuals as follows: $[F_1, F_5, F_3]$, $[F_4, F_2, F_6]$, in doing so, individuals may be more similar to their parents, so that the jumps in the solutions of the search are smoother, this would allow solutions refined.

The mutation operator first checks whether the individual will be mutated or not, based on the probability of mutation set for the search. The mutation proceeded to reverse the status of two randomly chosen sample points, transforming their values in zero or one depending on its current state. The procedure for reducing the population is similar to the first phase.

5   Experiments

The performance of the algorithm was tested using twenty datasets available on the time series classification/clustering WEB page [8]. These dataset characteristics can be consulted in table 1.

Different time series representations have been evaluated using these datasets [8]. It is important to remark that most of these representations have been developed to compress data, rather than to improve classification performance. One of the most efficient representations recently proposed is Symbolic Aggregate Approximation (SAX)[11]. Although SAX representation has been designed to discretize time series for streaming data purposes, it has shown a good performance on classification and clustering tasks. Because, as far as we know, at the time at which this work was developed there was no representation specifically designed to optimize classification performance, only EBLA3 algorithm shows a similar representation. GENEBLA has been compared with EBLA3 and SAX by having demostrated good performance in classification tasks.

One nearest neighbor (1NN) was used as a classification method. Error rate was computed using Leave-One-Out Cross Validation (LOO). The discretization scheme was obtained over the data training set. The similarity measure used in

Table 1. Datasets properties

Dataset	Number of classes	Size training	Size testing	Length TS
50Words	50	450	455	270
Adiac	37	390	391	176
Beef	5	30	30	470
CBF	3	30	900	128
Coffee	2	28	28	286
ECG200	2	100	100	96
Fish	7	175	175	463
Face(All)	14	560	1690	131
Face(Four)	4	24	88	350
Gun Point	2	50	150	150
Lighting 2	2	60	61	637
Lighting 7	7	70	73	319
Osu Leaf	6	200	242	427
Olive Oil	4	30	30	570
Swedish Leaf	15	500	625	128
Trace	4	100	100	275
Two Pattern	4	1000	4000	128
Control Chart	6	300	300	60
Wafer	2	1000	6164	152
Yoga	2	300	3000	426

1-NN, to evaluate the GENEBLA algorithm was the Euclidean distance. Classification performance was also evaluated using the raw data (continuous time series). It is important to remark to SAX uses its own similarity measure [11].

6   Results

Figure 1 shows error rates obtained on the twenty dataset, using algorithms and parameters for the classification as described above, a lower error rate indicates

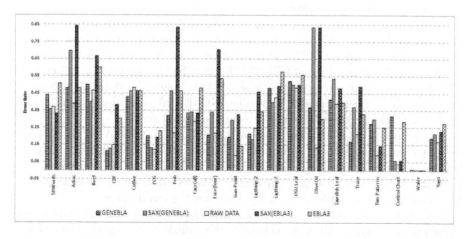

Fig. 1. Error rate obtain for GENEBLA, EBLA3 and SAX using the parameters suggested by GENEBLA and EBLA3. Error rate reached using raw data is shown as well.

better performance in classification. The graph shows that generally GENEBLA and SAX reached similar error rate using the same parameters(alphabet size and word size). With the advantage that GENEBLA does not require these parameters a priori because they are automatically calculated. In some datasets GENEBLA performance improved obtained with EBLA3, GENEBLA generally performance was better compared to SAX, when SAX uses the parameters obtained by EBLA3, GENEBLA improvement in classification performance on SAX on several datasets, even when using SAX, the parameters obtained by GENEBLA. The advantage of GENEBLA compared to SAX is that it is not required to know the alphabet and word sizes so a priori, since they are calculated automatically, using the information from class. The graph also shows the error rate of SAX classification using the parameters obtained by EBLA3 in order to compare their performance and also shows the error rate using the raw data.

SAX (GENEBLA) are the test results of the performance of SAX using the values obtained automatically by GENEBLA. SAX (EBLA3) are the test results of the performance of SAX using the values obtained automatically by EBLA3.

7 Discussion

It is noted that in some datasets, any of the three tested algorithms obtain a high error rate in classification. This may be because there are lags so any another algorithms including GENEBLA do not improve the discretization performance in the classification if they were used the raw data. In the case of GENEBLA this is because when trying to find the relevant features in the intervals on fixed sample points in time series, and looking for relationships with respect to the class, it cannot be generalized to the entire set of time series of dataset. Its choice is based on the most amount of time series discrimination. It can be seen as choosing which is more likely.

Generally, error rates reached using GENEBLA representation were similar to those obtained by SAX. Both were tested using the parameter(word size and alphabet size) found by GENEBLA. Moreover, in most datasets GENEBLA results are better than those of EBLA3. This may be because GENEBLA creates a broader search of discretization schemes that might be optimal in a local search approach with simulated annealing. GENEBLA generates populations of solutions which can take different paths at the same time.

The proposed discretization approach used by GENEBLA is an efficient technique to automatically compute the parameter for discretization (alphabet and word size), reaching competitive error rates. This characteristic makes GENEBLA very useful because sometimes it is very difficult to know a priori a good combination of alphabet and word size for a given dataset.

Note that the parameters of word size and alphabet are automatically found to be larger by GENEBLA, so generally the range of interval time series are smaller compared to those found automatically by EBLA3. However, GENEBLA gets better performance, this is because it has better opportunities to jump to detailed discretization schemes whereas EBLA3 is more general.

8   Conclusions and Future Work

In the present work a new algorithm for supervised discretization on time series data called GENEBLA has been proposed. Given a labeled dataset containting temporal data, GENEBLA algorithm automatically computes not only the alphabet size but also the limits which define the continuous intervals of each alphabet letter. GENEBLA also computes the number and size of the intervals to define the time series length.

The approach proposed in this work is to evaluate different discretization schemes following the operations of a genetic algorithm using the metric of EBLA3[4] as fitness function, which tries to minimize the entropy associated with respect to the class. The efficiency of the algorithm was evaluated using twenty datasets and compared to one of the most efficient representations of data of time series called SAX, with EBLA3 and raw data. Generally, error rates reached using GENEBLA representation were similar to those obtained by SAX tests using the same parameters (alphabet size and word size). An advantage is that GENEBLA does not requieres a priori parameters because they are automatically calculated.

GENEBLA improves the performance of EBLA3, especially in datasets where SAX improves the performance of classification using larger alphabet and word size. Hence most of the dataset improves the performance of SAX when the latter uses the parameters found by EBLA3.

Although SAX was not developed for supervised discretization, we decided to compare GENEBLA with SAX because SAX is one of the most efficient algorithms propoposed so far. GENEBLA allows greater control over the search time because the user defines a minimum value for generations so it can control the time to get a good enough solution, or perhaps get a good set of solutions as the algorithm continues its execution.

Based on fitness function used by GENEBLA it might be a measure of distance itself, which can help to achieve better performance in classification, as SAX uses its own measure of distance to reconstruct part of the original data maintaining a relationship between discrete representation and representation of continuous time series. GENEBLA can use another form of representation of the values of each interval to reduce the dimensionality. For example, instead of using the average of all values in that interval, using a wavelet function that represents these values. It is necessary to test the datasets after being discretized with GENEBLA using other classifiers that take into account the lags in time series.

Acknowledgements. The first autor wants to thank Language Center of the University of Istmo for the support translated to the paper and blind reviewers for their constructive comments.

References

1. Acosta Mesa, H.G., Nicandro, C.R., Daniel-Alejandro, G.-L.: Entropy Based Linear Approximation Algorithm for Time Series Discretization. In: Advances in Artificial Intelligence and Applications, vol. 32, pp. 214–224. Research in Computers Science
2. Dimitrova, E.S., McGee, J., Laubenbacher, E.: Discretization of Time Series Data, (2005) eprint arXiv:q-bio/0505028
3. Fayyad, U., Irani, K.: Multi-Interval Discretization of Continuous-Valued Attributes for Classification Learning. In: Proceedings of the 13th International Joint Conference on Artificial Intelligence (1993)
4. Alejandro, G.-L.D.: Algoritmo de Discretización de Series de Tiempo Basado en Entropía y su Aplicación en Datos Colposcopicos. Universidad Veracruzana (2007)
5. Han, J., Kamber, M.: Data mining, Concepts and techniques. Morgan Kaufmann, San Francisco (2001)
6. Keogh, E., Chakrabarti, K., Pazzani, M., Mehrotra, S.: Locally Adaptive Dimensionality Reduction for Indexing Large Time Series Databases. ACM Trans. Database Syst. (2002)
7. Keogh, E., Lonardi, S., Ratanamahatana, C.A.: Towards parameter-free data mining. In: Proceedings of Tehth ACM SIGKDD international Conference on Knowledge Discovery and Data Mining (2004)
8. Keogh, E., Xi, X., Wei, L., Ratanamahatana, C.A.: The UCR Time Series Classification/Clustering Homepage (2006),
   http://www.cs.ucr.edu/~eamonn/time_series_data/
9. Kurgan, L., Cios, K.: CAIM Discretization Algorithm. IEEE Transactions On Knowledge And Data Engineering (2004)
10. Last, M., Kandel, A., Bunke, H.: Data mining in time series databases. World Scientific Pub. Co. Inc., Singapore (2004)
11. Lin, J., Keogh, E., Lonardi, S., Chiu, B.: A symbolic representation of time series, with implications for streaming Algorithms. In: Proceedings of the 8th ACM SIGMOD Workshop on Research Issues in Data Mining and Knowledge Discovery (2003)
12. Mörchen, F., Ultsch, A.: Optimizing Time Series Discretization for Knowledge Discovery. In: Proceeding of the Eleventh ACM SIGKDD international Conference on Knowledge Discovery in Data Mining (2005)
13. Saito, N.:Local feature extraction and its application using a library of bases, PhD thesis, Yale University (1994)
14. Tan, P., Steinbach, M., Kumar, V.: Introduction to data mining. Addison-Wesley, Reading (2006)
15. Waldron, M., Manuel, P.: Genetic Algorithms as a Data Discretization Method. In: Proceeding of Midwest Instruction and Computing Symposium (2005)

Mining Social Networks on the Mexican Computer Science Community*

Huberto Ayanegui-Santiago[1], Orion F. Reyes-Galaviz[1], Alberto Chávez-Aragón[1], Federico Ramírez-Cruz[1,2], Alberto Portilla[1,3,4], and Luciano García-Bañuelos[1]

[1] Facultad de Ciencias Básicas, Ingeniería y Tecnología
Universidad Autónoma de Tlaxcala
Calzada Apizaquito S/N, Apizaco, Tlaxcala, México
[2] Instituto Tecnológico de Apizaco
[3] Laboratory of Informatics of Grenoble
[4] CNRS, Grenoble Institute of Technology, University Joseph Fourier
Research Center of Information and Automation Technologies
Fundación Universidad de las Américas, Puebla
{hayanegui,orionfrg,achavez,
framirez,aportilla,lgarcia}@ingenieria.uatx.mx

Abstract. Scientific communities around the world are increasingly paying more attention to collaborative networks to ensure they remain competitive, the Computer Science (CS) community is not an exception. Discovering collaboration opportunities is a challenging problem in social networks. Traditional social network analysis allows us to observe which authors are already collaborating, how often they are related to each other, and how many intermediaries exist between two authors. In order to discover the potential collaboration among Mexican CS scholars[1] we built a social network, containing data from 1960 to 2008. We propose to use a clustering algorithm and social network analysis to identify scholars that would be advisable to collaborate. The idea is to identify clusters consisting of authors who are completely disconnected but with opportunities of collaborating given their common research areas. After having clustered the initial social network we built, we analyze the collaboration networks of each cluster to discover new collaboration opportunities based on the conferences where the authors have published. Our analysis was made based on the large-scale DBLP bibliography and the census of Mexican scholars made by REMIDEC.

1 Introduction

With the fast growing Internet, social networks are emerging and more and more research efforts have been put on Social Network Analysis (SNA). SNA is an established field which proposes to analyze the relationships between social actors in a social network [1], it is also best applied in situations where the data is inherently relational.

---

* This work is partially supported by the project for consolidating the UATx Distributed and Intelligent Systems Research Group, in the context of the PROMEP-SEP program. The present analysis was performed between June and July 2009.
[1] We preferred to use the term *scholar* instead of *researcher*, because the REMIDEC census is on PhD holders even if some of them are not actively involved in research activities.

A. Hernández Aguirre et al. (Eds.): MICAI 2009, LNAI 5845, pp. 213–224, 2009.

Discovering collaboration opportunities in a social network is a challenging problem in many large social networks. Some typical problems in SNA include discovering groups of individuals sharing the same properties and evaluating the importance of individuals [2].

A social network is a structure built by nodes, representing entities from different groups that are linked with different types of relations. For this paper, we have used the data consisting on publications produced by Mexican authors, included in the DBLP bibliography database.

A community is defined as a group of entities that share similar properties or connect to each other via certain relations [3]. Identifying these properties and locating entities in different communities is an important goal of community mining and can also have various applications. Social network analysis could shed light on how a scientific community is working together. In this paper, the collaboration network among Mexican Computer Science scholars is analyzed. We are interested in finding potential collaborations by discovering communities in an author-conference social network. In this work, we use clustering techniques in order to identify hidden communities which are advisable to collaborate.

Clustering aims at finding smaller similar groups, from a larger collection of items. Computer-assisted analysis must partition the objects into groups, and it must provide an interpretation for this partitioning [4]. Many clustering methods exist to partition a data set by some natural measure of similarity. A *similarity measure* is used to place similar objects close to one another forming a group; thus, several clusters related to objects are formed. An ideal clustering algorithm is one that classifies data such that, the samples that belong to a cluster are close to each other while samples from different clusters are further away from each other. A popular clustering algorithm is Expectation Maximization (EM) [5]. This paper discusses the application of EM to the database of Mexican authors in order to discover potential collaboration among them.

This paper is organized as follows. Section 2 gives a description of the DBLP citation index used, the list from REMIDEC used to create the Mexican Authors database and PAJEK which is a software employed to build the Collaboration Network. Section 3 presents a brief overview of some recent work on social networks analysis. Section 4 presents the analysis of the Mexican community in terms of individuals for the last four decades, and Section 5 analyzes the collaboration networks among them. Finally, Section 6 presents our conclusion thoughts and outlines our future work.

2   Preliminaries

Our analysis was made based on the large-scale DBLP bibliography and the census of Mexican scholars of REMIDEC. DBLP, formerly known as "Database Systems and Logic Programming", was created by Dr. Michael Ley from the Universität Trier in Germany. This Computer Science Bibliography database, is now gradually being expanded toward other fields of CS, and its name changed to "Digital Bibliography and Library Project" or simply DBLP [6].

In order to know about the Mexican CS research community, we used a list provided by the Mexican Network on Research and Computer Development (REMIDEC) [7].

They have a record of Mexican scholars in CS and related areas that work in Universities and Institutions in Mexico. This list is being permanently under construction, and any CS Mexican scholar can join the list.

To match scholar names with published papers we had to deal with the *name authority control problem* (i.e., same authors with different spellings or different authors with the same spelling) [8]. The problem gets more complicated with Mexican Authors, who sometimes register themselves differently, by writing up to four names. For example, for an author named "Jose Juan Perez-Lopez", three or more aliases had to be made (e.g., "Jose Perez", "Jose Perez-Lopez", and "Jose J. Perez Lopez"). We found the aliases for each author, then the aliases were matched in both, the DBLP database and the REMIDEC list, extracting the paper ID tags for each author that will help us represent the *Collaboration Network* among scholars.

The social networks, were built with the aid of the program PAJEK (Program for Analysis and Visualization of Large Networks) [9], mainly used to visualize and analyze large and complex networks, which can have thousands or even millions of vertices.

3   Related Work

In this Section, we present an overview of some works based on social networks analysis. Some papers have analyzed the scientific collaboration networks in different disciplines [8,10,11]. Newman has specifically considered scientific collaboration networks based on publication information in several scientific databases [12]. Newman found that these networks display the small world characteristic, they are highly clustered, and they obey a power-law distribution with an exponential cutoff.

Wuchty et al. [13] study the evolution of the number of authors (team sizes) in publications from science and engineering, social sciences, the arts and humanities, and patents, showing that teams increasingly dominate solo authors in the production of knowledge in the four datasets. Schwartz attempted to discover shared interests using graph analysis in the early nineties [14]. His approach consist on analyzing email traffic from 15 different organizations and running various algorithms on the resulting graph to determine which individuals shared his own interests. This work did not explicitly consider which subjects the individuals shared an interest in, only that they existed. Khan et al. explicitly consider the authors and subjects of papers, but the focus was on examining existing collaborations and predicting new ones [15]. The group did not explore the utility of their system for collaborative communities.

Barabasi et al. [11] analyzes coauthorship graphs by using a database containing relevant journals in Mathematics and Neuroscience for an 8-year period (1991-1998). The authors infer the dynamic and the structural mechanisms that govern the evolution and topology of these two scientific fields. The results indicate that the network is scale-free and that the network evolution is governed by preferential attachment, affecting both internal and external links. Boner et al. [16] analyze the impact of coauthorship teams based on the number of publications and their citations on a local and global scale. The authors use a weighted graph representation that encodes coupled author-paper networks as a weighted coauthorship graph.

To the best of our knowledge, this paper is the first study that utilizes social network analysis and finds collaboration opportunities focusing on the Mexican CS community.

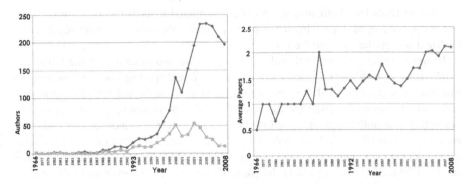

Fig. 1. a) New Scholars per year b) Average number of papers

4   Statistics about Scholars

In this Section the behavior of the Mexican CS community (i.e., Mexican scholars) is analyzed for the last four decades; considering the scholars in the REMIDEC list, which includes 462 Mexican scholars in that period of time.

The analysis performed on the data cross-referenced from DBLP and the REMIDEC list, provides a better understanding of how the scientific community is growing. Figure 1a (gray line) summarizes this analysis. To draw this graphic, the number of new scholars that joined the DBLP community, within each year, was registered. With this data we can say that, on the first twenty years, the number of new scholars remained almost the same with only a few minor changes. However, 1993 saw a significant increase in the number of new scholars. The highest number was reached in 2004, after that there is a sharp decrease, where it can be seen that there exists the same number of new scholars as there existed 15 years ago.

But not only the number of scholars that joined the DBLP community have to be considered, it is also interesting to graph the number of scholars that have remained active over the years, Figure 1a (black line) shows these numbers. This line has a similar behavior as the line shown in gray, that is, there is a little activity before 1993, after that a significant increase in the scholar's activities and finally, in the last two years, a decrease. This behavior can probably be explained in terms of the number of doctoral scholarships provided by the Mexican Science and Technology Council each year, since new and active authors most of the time are postgraduate students.

Before getting into more details, regarding the top Mexican authors, let's have a brief look at Figure 1b. This Figure shows the average number of papers that each active scholar has published in the period of time considered for this work. First, we noticed that the average number of papers per author was below one, next point shows a pronounced growth of 100%, this trend continues until the average number of papers got to a point above two in recent years. This analysis drives us to the conclusion that the scientific community is becoming more and more active in a certain sense over the years. To support the last statement, Figure 2 shows how the number of published papers has been growing year after year, there are just two very small decreases in the graphic. This Figure also shows the accumulative number of published papers.

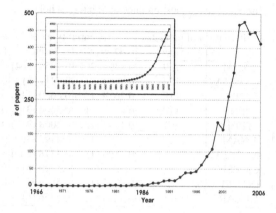

Fig. 2. Number of papers per year. Insert: Accumulative number of published papers.

To obtain these figures, the set of scholars was divided into 52 classes. This division was based on the number of papers reported per author in the DBLP citation index; where 79 authors had one published paper, 50 authors reported two papers, other 50 authors published three papers, only 32 authors had four papers, and so on, the more number of published papers the less number of authors. In the list of Mexican scholars[2] included in REMIDEC, only one person had the highest number of published papers, which was 139 papers (shown in Table 1). This analysis shows that most authors publish between one and ten papers, and a very small group of researchers have more than 40 papers. Table 1 shows the most productive scholars, which have more than 60 published papers and only six of them have more than 100 published papers.

5 Data Clustering

5.1 Preprocessing of the Matrix for the Clustering Algorithm

In this section we present the process undertaken to find new collaboration opportunities for Mexican research staff. To discover similar interests among scholars, we applied clustering algorithms to classify patterns with a similarity, in relation with the conferences where they have published.

The data which contains the information about scholars, and the conferences they usually attend to, was obtained from a series of queries from our Mexican CS scholars database. The pieces of information obtained were arranged into a numerical matrix, where each column represents a different conference listed in the DBLP database, and each row represents a different scholar cross-referenced from DBLP and from the REMIDEC list. The intersection, of an author and a conference in this matrix has a value between the interval $[0, 1]$. For any given column and row intersection, if an author didn't publish in that conference, the number should be zero, if the author has one or more published papers in any conference, the matrix's entry in that intersection

---

[2] We use directly publishing scholar names without explicit permission because they were taken from public data repositories.

Table 1. Top ten most productive Mexican scholars

Scholar	Institution	# of published papers	Centrality measures (rank)		
			Degree	Closeness	Betweenness
J. Urrutia	UNAM	139	0.1490 (207)	0.4254 (204)	0.0109 (7)
C. Coello	CINVESTAV	124	0.4254 (7)	0.5317 (8)	0.0093 (14)
O. Castillo	ITTij	118	0.2639 (139)	0.4660 (116)	0.0032 (65)
A. Guelboukh	IPN	110	0.4380 (3)	0.5404 (3)	0.0102 (11)
P. Melin	ITTij	107	0.1041 (221)	0.4068 (211)	0.0010 (169)
S. Rajsbaum	UNAM	101	0.2172 (175)	0.4492 (169)	0.0009 (171)
E. Bayro	CINVESTAV	63	0.2100 (180)	0.4378 (200)	0.0034 (59)
J. Favela	CICESE	75	0.4344 (4)	0.5340 (5)	0.0152 (3)
L. E. Sucar	INAOEP	72	0.4757 (1)	0.5579 (1)	0.0203 (2)
M. Osorio	UDLAP	69	0.4308 (5)	0.5356 (4)	0.0116 (6)

should be *the number of published papers in that conference* divided by *the overall number of published papers.*

The resulting matrix had a size of 462×844, corresponding to 462 different scholars, and 844 different conferences. Since most of the entries in the matrix have zero values, we decided to apply the Pareto's principle, which states that for a large set of events, roughly 80% of the effects come from 20% of the causes. Taking this into consideration, we analyzed the conferences where most of the scholars usually publish and realized that 20% of the conferences receive 74.84% of the work produced by them. Thus, we decided to apply a matrix pruning and as a result we obtained a smaller matrix with 462×169 entries, this matrix represents nearly 80% of the research work in the community. This matrix was the input of the clustering algorithm, in the next section this process is described in detail.

5.2 Data Clustering

In this paper, the Expectation-Maximization (EM) clustering algorithm is applied to find new collaboration opportunities for Mexican research community. For that, we use Weka system (Waikato Environment for Knowledge Analysis), a popular suite of machine learning software written in Java, developed at the University of Waikato as a free software available under the GNU General Public License. EM algorithm is part of the Weka clustering package and it is a statistical model that makes use of the finite Gaussian mixtures model. EM algorithm can be used as clustering algorithm if each instance can be seen as a random variable generated by a normal probability distribution of a mixture of k possible distributions. EM algorithm searches for the set of parameters that describe these underlying probability distributions [17].

The reason we are using EM is to fit the data better, so that clusters are compact and far from other clusters, since we initially estimate the parameters and iterate to find the maximum likelihood for those parameters. EM uses the Maximum Likelihood, in which it assumes that the parameters are fixed; the best estimate of their value is defined to be the one that maximizes the probability of obtaining the samples actually observed. EM is widely used in applications such as computer vision, speech processing and pattern recognition [18]. Unlike distance based or hard membership algorithms (such as K-Means) EM is known to be an appropriate optimization algorithm for constructing proper statistical models of the data [19].

EM algorithm in Weka assigns a probability distribution to each instance which indicates the probability of it belonging to each one of the clusters. EM can decide how many clusters to create by cross validation, or it may be specified a priori how many clusters to generate. The cross validation performed to determine the number of clusters is done in the following steps:

1. The number of clusters is set to 1.
2. The training set is split randomly into 10 folds.
3. EM is performed 10 times using the 10 folds the usual CV way.
4. The loglikelihood is averaged over all 10 results.
5. If loglikelihood has increased the number of clusters is increased by 1 and the program continues at step 2.

The number of folds is fixed to 10, as long as the number of instances in the training set is not smaller 10. If this is the case the number of folds is set equal to the number of instances.

5.3   Scholar Collaboration Network

The Table 2 shows some values of each cluster. The cluster with most scholars belonging to 36 different institutions and with most published papers is the cluster 3. The Table also shows in the fifth column the top ten scholars within each cluster according to the number of published papers (i.e., indicated between parenthesis). The last column shows the top ten institutions per cluster according to the number of scholars from each institution (i.e., shown between parenthesis).

In order to determine possible collaborations among Mexican scholars we built its collaboration networks. A collaboration network is represented by a graph. A node into the graph represents a scholar, and an edge between scholars A and B represents the fact that two scholars have published in one or more common venues. The width of an edge reflects the Jaccard distance: $| P_A \cap P_B | / | P_A \cup P_B |$, where P_A and P_B are venues sets where scholars A and B have published. We assume as a premise that scholars that have published in the same conferences or journals may collaborate. Table 3 shows the values that characterizes each collaboration network. The second column shows the density of each network, which indicates if the number of connections is much higher than the number of vertices. As you can see, the network of cluster 3 is the most dense network. Third column shows the average distance among reachable pairs of each network, the lowest value corresponds to cluster 1 and the highest value to cluster 3. The fourth column shows the distance among the most distant vertices, the lowest value comes from cluster 4 and the highest value comes from cluster 1.

Next, we analyze for each network the way that Mexican scholars may interact. Therefore, we find the scholars that play a central role in the community, acting as hubs in the network. To determine the values of centrality we considered [9]: i) the degree of edges of each node (i.e., a high value indicates that the node has many edges

Table 2. General data of each collaboration network

Cluster	# of scholars	# of institutions	# of papers	Top ten scholars (# of published papers)	Top ten institutions (# of scholars)
1	71	20	398	L. Sheremetov (46), J. A. Carrasco Ochoa (37), X. Li Zhang (29), I. Zakirzjanovich Batyrshin (23), J. C. Martínez Romo (22), M. Arias Estrada (20), L. Altamirano Robles (20), F. Ramos Corchado (20), J. A. Nolazco Flores (18), C. Torres Huitzil (14)	ITESM (15), INAOEP (7), BUAP (6), IPN (5), CICESE (4), UDG (4), UAM (4), CINVESTAV (3), UAEH (3), UNAM (3)
2	91	31	798	O. Castillo (118), P. Melin (107), J. Favela (75), E. Sucar (72), E. Bayro (63), O. Montiel Ross (37), G. Olague Caballero (37), R. Sepúlveda Cruz (35), C. A. Reyes García (32), C. A. Brizuela Rodríguez (22)	UNAM (17), ITESM (13), IPN (8), UAM (7), CINVESTAV (5), UG (5), CICESE (4), UABC (2), ITTij(2), INAOEP(2)
3	210	36	1707	J. Urrutia (139), C. Coello(124), E. Chávez (45), A. Hernández Aguirre (42), J. F. Martínez Trinidad (41), A. Sánchez (38), O. Fuentes Chávez (36), J. Pérez Ortega (33), C. Stephens (31), J. L. Marroquín (30)	ITESM (31), UNAM (23), UAM (19), IPN (18), CINVESTAV (14), CENIDET (10), CIMAT (7), ITAM(7), UDLAP (6), INAOEP (6)
4	90	27	1077	A. Guelboukh (110), S. Rajsbaum (101), M. Osorio (69), H. Sossa (58), M. Montes y Gómez (54), I. Bolshakov (40), G. Várgas Solar (33), J. A. García Macías (33), J. Frausto (28), R. Monroy (27)	ITESM (12), UNAM (11), IPN (11), UAM (10), CINVESTAV (4), CIMAT (4), ITAM (4), UDLAP (3), BUAP (3), UATx (2)

Table 3. Values of each collaboration network

Cluster	Density	Avg. distance	Max distance
All	0.1245	2.0615	6
1	0.1133	2.1460	5
2	0.1166	2.2638	5
3	0.2320	2.9288	5
4	0.3140	2.8279	4

and plays a central role), *ii)* the closeness between edges (i.e., more central nodes can quickly interact to all other), and *iii)* betweenness centrality (i.e., if a node lies on several shortest paths among other pairs of nodes).

The Table 1 shows the centrality measures into the whole network. L. Enrique Sucar with 72 published papers at the position 9 ranks as the scholar with the highest values in the centrality measures (i.e., degree-1, closeness-1, and betweeness-2).

After we continue to analyze the values of each cluster. Due to space limitations we present the analysis of most representative clusters. We considered that collaboration networks of cluster 3 and cluster 4 are the most representative because: *i)* they are the most dense networks (see Table 3), *ii)* six of the top ten Mexican scholars are included in those networks (see Table 1), and *iii)* they are the most populated networks of both, papers and scholars (see Table 2).

We analyze for cluster 3 and 4 the way that the scholars within each network interact. The Figures 3 and 4 present the collaboration network for the top ten authors of each cluster.

The Table 4 shows the data of the top scholars of the cluster 3. It can be seen that in general, top scholars maintain its centrality measures on top values. C. Coello with 124

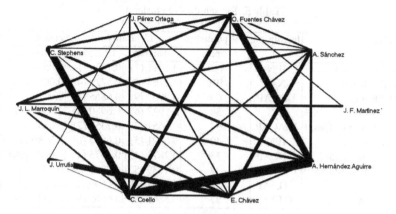

Fig. 3. Collaboration network of top ten scholars (cluster 3)

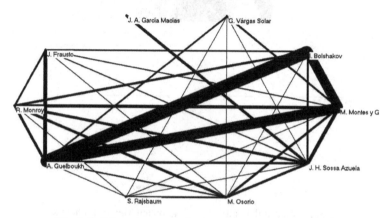

Fig. 4. Collaboration network of top ten scholars (cluster 4)

Table 4. Top scholars of cluster 3

Scholar	Institution	# of published papers	Centrality measures (rank)		
			Degree	Closeness	Betweenness
J. Urrutia	UNAM	139	0.2367 (106)	0.5123 (106)	0.0216 (9)
C. Coello	CINVESTAV	124	0.6038 (2)	0.6919 (1)	0.0351 (3)
E. Chávez	UMich	45	0.5748 (6)	0.6769 (3)	0.0281 (4)
A. Hernández Aguirre	CIMAT	42	0.5265 (22)	0.6511 (22)	0.0122 (19)
J. F. Martínez Trinidad	INAOEP	41	0.1642 (110)	0.4783 (109)	0.0042 (53)
A. Sánchez	UDLAP	38	0.5797 (4)	0.6745 (6)	0.0731 (1)
O. Fuentes Chávez	INAOEP	36	0.5748 (5)	0.6745 (5)	0.0224 (7)
J. Pérez Ortega	CENIDET	33	0.4202 (32)	0.5991 (32)	0.0269 (5)
C. Stephens	UNAM	31	0.3526 (67)	0.5581 (78)	0.0058 (44)
J. L. Marroquín	CIMAT	30	0.3913 (41)	0.5897 (35)	0.0226 (6)

Table 5. Top scholars of cluster 4

Scholar	Institution	# of published papers	Centrality measures (rank)		
			Degree	Closeness	Betweenness
A. Guelboukh	IPN	110	0.7011 (1)	0.7620 (1)	0.0561 (1)
S. Rajsbaum	UNAM	101	0.2758 (45)	0.5429 (45)	0.0057 (36)
M. Osorio	UDLAP	69	0.5862 (9)	0.6865 (10)	0.0214 (13)
H. Sossa	IPN	58	0.5977 (6)	0.6980 (4)	0.0349 (5)
M. Montes y Gómez	INAOEP	54	0.6551 (2)	0.7351 (2)	0.0348 (6)
I. Bolshakov	IPN	40	0.6091 (5)	0.6980 (6)	0.0186 (16)
G. Várgas Solar	UDLAP	33	0.2873 (43)	0.5575 (42)	0.0140 (20)
J.A. García Macías	CICESE	33	0.1839 (53)	0.4801 (57)	0.0073 (33)
J. Frausto	ITESM	28	0.5517 (19)	0.6699 (19)	0.0538 (2)
R. Monroy	ITESM	27	0.5862 (8)	0.6865 (9)	0.0340 (7)

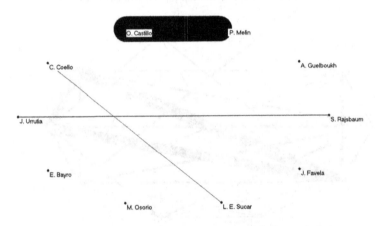

Fig. 5. Existing collaborations among top ten Mexican scholars

published papers ranks as the scholar with the highest values in the centrality measures (i.e., degree-2, closeness-1, and betweeness-9).

The Table 5 shows the data of the top scholars of the cluster 4. A. Guelboukh with 110 published papers ranks as the scholar with the highest values in the centrality measures (i.e., degree-1, closeness-1, and betweeness-1).

Finally, a way of verifying the collaboration networks we built for finding collaboration opportunities, is to compare such networks with the networks that reflect the existing collaborations (i.e., published papers written by two or more Mexican scholars). The Figure 5 and 6 shows the collaborations networks of the top ten Mexican scholars. The graph of Figure 5 represents the existing collaboration between top Mexican scholars. A node represents a scholar and an edge represents the collaboration between two scholars, where the thickness is proportional to the number of papers published between the two connected scholars. The graph of Figure 6 represents the collaboration opportunities of top tep Mexican scholars which was built using the Jaccard distance. It can be seen that, existing collaborations (see Figure 5) are augmented by the possible

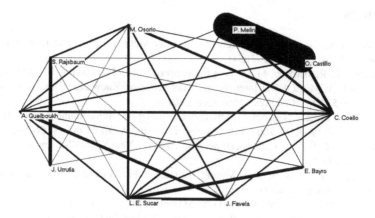

Fig. 6. Potential collaborations among top ten mexican scholars

collaboration ties (see Figure6). We argue that the collaboration opportunities network can be used for: i) to identify the CS areas being addressed by Mexican community, and ii) to determine more exactly the central scholars of the Mexican community.

6   Conclusions and Future Work

In this paper we analyzed the Mexican computer science research community. To this end, we used two publicly available data repositories, namely DBLP, and REMIDEC. Our analysis was conducted though publication data from the last four decades, as reported in the DBLP database. The obtained results by the clustering algorithm suggest a possibility of collaboration among authors who are publishing papers in related areas.

Our analysis permitted to observe the degree of collaboration that scholars have inside the community and the potential collaboration opportunities among Mexican researches. We divided or analysis into three stages. i) statics about scholars, where we reported how the community is growing, the number of papers published per year and the most productive Mexican scholars. ii) collaboration network analysis, we obtained typical measures of this kind of networks, such as: degree, closeness and betweenness. iii) hidden opportunities to collaborate, to get this results we made use of clustering algorithms obtaining four clusters of authors. This division of the community allowed us to make a more detailed analysis for the most populated cluster. We computed over the collaboration network obtained in the last stage measures like density, average distance and maximum distance. Additionally, we can know, based on our analysis, the institutions that make a important contribution to the CS community measuring this contribution on the number papers they have published.

In the near future we are going to build a website that includes more detailed information about the social networks, considering all registered names in the updated REMIDEC list, and the databases used for this paper. This information will be available at: http://ingenieria.uatx.mx/ca_sdi/social_networks/.

References

1. Wasserman, S., Faust, K.: Social Network Analysis: Methods and Applications. Cambridge University Press, Cambridge (1994)
2. Domingos, P., Richardson, M.: Mining the network value of customers. In: Proceedings of the Seventh ACM SIGKDD International Conference on Knowledge Discovery and Data Mining, pp. 57–66. ACM Press, New York (2001)
3. Zaiane, O.R., Chen, J., Goebel, R.: DBconnect: mining research community on DBLP data. In: WebKDD/SNA-KDD 2007: Proceedings of the 9th WebKDD and 1st SNA-KDD 2007 workshop on Web mining and social network analysis, New York, USA, pp. 74–81. ACM, New York (2007)
4. Berry, M.J.A., Linoff, G.S.: Data Mining Techniques. For Marketing, Sales, and Customer Support. Wiley, Chichester (1997)
5. Dempster, A.P., Laird, N.M., Rubin, D.B.: Maximum Likelihood from Incomplete Data via the EM Algorithm. Journal of the Royal Statistical Society 39(1), 1–38 (1977)
6. Ley, M.: Dblp: How the data gets in. Talk at the University of Manchester (2009)
7. REMIDEC. Investigadores y profesores con doctorado en el campo de computación en méxico. Red Mexicana de Investigación y Desarrollo en Computación (2008)
8. Newman, M.E.J.: Who is the best connected scientist? a study of scientific coauthorship networks. Phys. Rev. E64(016132) (2001)
9. Batagelj, V., Mrvar, A.: PAJEK Reference Manal Version 1.24. University of Ljubljana, Slovenia (December 2008)
10. Elmacioglu, E., Lee, D.: On six degrees of separation in dblp-db and more. SIGMOD Record 34(2), 33–40 (2005)
11. Barabasi, A.L., Jeong, H., Neda, Z., Ravasz, E., Schubert, A., Vicsek, T.: Evolution of the social network of scientific collaborations. Physica A: Statistical Mechanics and its Applications 311(3-4), 590–614 (2002)
12. Newman, M.E.J.: The structure of scientific collaboration networks. Proc. National Academy of Sciences 98(2), 404–409 (2001)
13. Wuchty, S., Jones, B.F., Uzzi, B.: The increasing dominance of teams in production of knowledge. Science 316(5827), 1036–1039 (2007)
14. Schwartz, M.F., Wood, D.C.M.: Discovering Shared Interests Using Graph Analysis. Communications of the ACM 8(36), 78–89 (1993)
15. Khan, F.M., Fisher, T.A., Shuler, L., Wu, T., Pottenger, W.M.: Mining Chat-room Conversations for Social and Semantic Interactions (2002)
16. Brner, K., Dall'Asta, L., Ke, W., Vespignani, A.: Studying the emerging global brain: Analyzing and visualizing the impact of co-authorship teams. Complexity 10(4), 57–67 (2005)
17. Mitchell, T.: Machine Learning, Paperback. McGraw-Hill Education, New York (1997)
18. Xie, L., Liu, Z.-Q.: A coupled HMM approach to video-realistic speech animation. Pattern Recogn. 40(8), 2325–2340 (2007)
19. Bradley, P.S., Fayyad, U., Reina, C.: Scaling EM (Expectation-Maximization) Clustering to Large Databases (1999)

Probabilistic Graphical Markov Model Learning: An Adaptive Strategy

Elva Diaz[1], Eunice Ponce-de-Leon[1], Pedro Larrañaga[2], and Concha Bielza[2]

[1] Computer Science Department, Autonomous University of Aguascalientes,
Ave. Universidad 940, C.P. 20100 Aguascalientes, Mexico
elva.diaz@itesm.mx,
eponce@correo.uaa.mx
[2] Department of Artificial Intelligence
Polytechnic University of Madrid 28660 Boadilla del Monte, Madrid, Spain
pedro.larranaga@fi.upm.es,
mcbielza@fi.upm.es

Abstract. In this paper an adaptive strategy to learn graphical Markov models is proposed to construct two algorithms. A statistical model complexity index ($SMCI$) is defined and used to classify models in complexity classes, sparse, medium and dense. The first step of both algorithms is to fit a tree using the Chow and Liu algorithm. The second step begins calculating $SMCI$ and using it to evaluate an index ($EMUBI$) to predict the edges to add to the model. The first algorithm adds the predicted edges and stop, and the second, decides to add an edge when the fitting improves. The two algorithms are compared by an experimental design using models of different complexity classes. The samples to test the models are generated by a random sampler (MSRS). For the sparse class both algorithms obtain always the correct model. For the other two classes, efficiency of the algorithms is sensible to complexity.

Keywords: Graphical Markov Model, Chow and Liu Algorithm, Model Learning, Entropy, Complexity.

1 Introduction

The problem of learning probabilistic graphical Markov models has two challenging tasks: learning the structure, and learning the parameters. Learning the structure involves a combinatorial search through the space of the graphical model class. Learning the parameters involves calculating marginal distributions that is a $\#P$ - complete problem [10]. For the case of the tree structures the well known Chow and Liu (CL) algorithm [4] is a polynomial-time solution. Chow and Liu constructed a polynomial-time algorithm to learn an approximate tree to a binary sample data. The CL algorithm obtains the maximum weight spanning tree using the Kruskal algorithm [11] and the mutual information values for the random variables. Chow and Liu showed that the maximum-weight spanning tree also maximizes the likelihood only in the class of tree model distributions. For

A. Hernández Aguirre et al. (Eds.): MICAI 2009, LNAI 5845, pp. 225–236, 2009.

suitably decomposable and sparse graphs, the junction tree algorithm provides a practical solution to the general problem of computing likelihoods. Unfortunately, many graphical models of practical interest are not decomposable and suitably sparse [10].

The learning problem can be defined as an optimization problem. To define an optimization problem, it is necessary to specify both a scoring function to be optimized, and a constraint set (the space of solutions) over which the optimization takes place. The space of solutions is the set of all probability distributions, characterized by a set of parameters Θ that must be estimated. The constraint set can be defined by a graph G, and in this case the graphical Markov model is represented by $M = (P(x, \Theta), G)$, where x is a vector of realization of random variables with joint probability distribution $P(x, \Theta)$, and G is an undirected graph representing the topological structure.

In this paper two algorithms are presented, that perform a graphical Markov model learning using the following adaptive strategy. In the first step, both algorithms fit a tree to the data using CL algorithm. In the second step, both algorithms calculate the edge missing upper bound index ($EMUBI$) that is used to estimate the maximum number of edges that can be added to the tree. This number of edges depends on the statistical model complexity index ($SMCI$) that adapts the algorithm performance to the information contained in the data sample. In the third step, the two algorithms differ. The first algorithm adds the edges proposed based on $EMUBI$ and stop. The second algorithm decides to add an edge when the fitting improves, and stop when has tested all the proposed edges.

The content of the paper is the following. In section 2 the complexity, the entropy and the model structure are defined and described. In section 3, the statistical model complexity index ($SMCI$) is defined and used to classified models in complexity classes. A graph similarity index (GSI) is defined too. In section 4 the two algorithms are constructed based on the indexes defined in section 3. In section 5 an experimental design is described. Samples generated by a random sampler introduced in [8] are employed. The results are presented and analyzed in the same section. In section 6 the conclusion and future work are presented. The appendix contains the algorithms performance description tables.

2   Complexity, Entropy and Model Structure

Learning from data could be seen as discovering constraints capable of leaving the noise in observed data out. The strength of the constraints can be encoded using a model structure. The model structure characterizes the constraints over the data, and the complexity of a model could be assessed thought the amount of information expressed by the model with respect to the totality of information contained in the data. Searching for a short program might be a promising way to discriminate a model from noise.

2.1   Entropy and Assessing Complexity

Most of the complexity measures are conceptually based on the Kolmogorov-Chaitin KC-complexity [14] that is defined as the length $K(m)$ of the shortest binary program p that produces a binary string m

$$K(m) = \min\{l(p) : m = C_{UT}(p)\} \tag{1}$$

where $l(p)$ is the length of the program p, $C_{UT}(p)$ is the implementation of the program p in a Universal Turing Machine. As is known the KC-complexity is not computable [14].

It is possible to approximately assess the complexity using the entropy H, as was proved in [16] cited by [1]. Let $\mathcal{M} = \{m_i\}$ be a set of N random binary strings, and let $P(m = m_i) = p(m_i)$, so, the expected value can be approximated as follows,

$$E(K(m)) \approx -H(m) = \sum_{i=1}^{N} p(m_i) \log p(m_i) \tag{2}$$

2.2   Log-Linear Models and Graphical Representation

To represent the interactions between discrete variables using log-linear models, let $V = \{v_1, v_2, ..., v_m\}$ be a finite set of classification criteria. For each $v_i \in V$ let I_{v_i} be the set of levels of the criterion v_i. The set of cells in the contingency table is the set $I = \prod_{v_i \in V} I_{v_i}$, and a particular cell is denoted by $i = (i_{v_i}, v_i \in V)$. Let O be an object characterized by the classification criteria V. If a set of n objects is classified according to the criteria, let $n(i)$ be the number of objects in the cell i. For $a \subseteq V$ let $n(i_a)$ be the number of objects in the marginal cell $i_a = (i_{v_i}, v_i \in a)$ obtained as the sum of the cell $n(i)$ for all such i that agree with i_a on the coordinates corresponding to a. Let $p(i)$ denote a parameter assigned to the cell i, let $p = (p(i), i \in I)$ be the parameter set, similarly $p_a(i_a)$ denote the parameter set for the marginal cell i_a and $p_a = (p_a(i_a), i_a \in I_a)$.

If a sample of n independent observations of individuals classified by the criteria V is taken, the classification counts of the n objects has a multinomial distribution:

$$P\{N(i) = n(i), i \in I\} = \binom{n}{n(1)\; n(2)\; ...\; n(2^n)} \Pi_{i \in I} p(i)^{n(i)} \tag{3}$$

Definition 1. *The* general log-linear interaction model M *is defined by two types of restrictions. First, the logarithm of* $p(i)$ *is expanded as,*

$$\log p(i) = \sum_{a \subseteq V} \xi_a(i_a), \tag{4}$$

where ξ_a *are functions of* i *that only depend on* i *via the coordinates in* a, *i.e., through* i_a. *This class of functions* $\xi_a(i_a)$ *is known as the class of interactions. To* *have a one-to-one correspondence between the system of functions $\{\xi_a, a \subset V\}$ and the parameter set p, let introduce the second group of constraints [9], [7]*

$$\forall b \subset a : \sum_{\{i_a : i_c = i_b\}} \xi_a(i_c) \equiv 0 \text{ for all } i_b . \tag{5}$$

These restrictions (5) mean that the sum over each interaction equals zero.

Definition 2. *A hierarchical log-linear model M_H is a log-linear interaction model where the functions ξ_a satisfies the following property:*

$$if \ \xi_a = 0 \text{ and } b \supseteq a \text{ then } \xi_b = 0 . \tag{6}$$

A hierarchical log-linear interaction model can be specified via the so-called generating class.

Definition 3. *A generating class of a hierarchical log-linear model is a set $\mathcal{E} = \{V; E_1, E_2, ..., E_k\}$ of pair-wise incomparable (w.r.t. inclusion) subsets of V to be interpreted as the maximal sets of permissible interactions, i.e., [9]*

$$\xi_a \equiv 0 \text{ iff there is no } E_i \in \mathcal{E} \text{ with } a \subseteq E_i . \tag{7}$$

Let denote a graph G by its maximal cliques $G = \{V; g_1, g_2, ..., g_k\}$ where for all $g_i \subseteq V$, they are pairwise incomparable (w.r.t. inclusion).

Definition 4. *If the subsets of the generating class $\mathcal{E}$ of a hierarchical log-linear model are in a one to one correspondence to the maximal cliques of a graph, the model receives the name of graphical Markov model and is denoted by M_G [13].*

From now on, the graphical Markov model will be denoted by M to simplify the notation.

3   Information and Model Complexity

Let $X = (X_1, X_2, ..., X_v)$ be a binary random vector with a multinomial distribution. Let $P(x)$ be a probability distributions of the random vector X. Let take a sample of size n, and let $L(\widehat{m}_n^M \mid x)$ be the likelihood statistic for the sample where $\widehat{m}_n^M = \{\widehat{m}_i^M\}$ is the *maximum likelihood estimate* assuming that the true model is M.

Definition 5. *The K-L divergence from the probability model P to the data x is given by the Kullback-Leibler information [2]:*

$$G^2(M, x) = \log(L(\widehat{m}_n^M(x))) = -2 \sum_{i=1}^{2^v} x_i \log_2 \left(\frac{\widehat{m}_i^M}{x_i}\right) . \tag{8}$$

The K-L divergence is also known as relative entropy, and can be interpreted as the amount of information in the sample x not explained by the model M, or the deviation from the model M to the data x. To obtain the $\widehat{m}_n^M$ the IPS algorithm can be employed. This algorithm is exponential, but it converges to the parameters of the model M [5][13].

3.1   Statistical Model Complexity Index (*SMCI*)

Using a short program might be a promising way to discriminate a model from noise (see 2.1). But, how much from the information contained in the sample must be saved in the model? A strategy can be to capture first the ease to catch information, and later, gain as much as possible, but not so much as to include the noise as part of the model. Given the number of nodes of a connected graph, the graph with the minimum number of edges is a tree. It is known that a graphical model that can be exactly learned in polynomial time is one whose interaction structure is a tree [4]. These two properties sustain the idea that a graphical model complexity must be a minimum for a model represented by a tree, because a tree can be learned by a polynomial time algorithm.

It is known that the uniform distribution has the maximum entropy and at the same time gives no information about the interaction structure of a model [15]. The model representing the uniform distribution is denoted by M_0, and the model represented by a tree is denoted by M_T. If a sample is generated by a model M containing more edges than a tree, the information about the model M contained in this sample and not explained, when the model structure is approximated by a tree, may be assessed by the following index.

Definition 6. *Let x be a sample generated by a model M. The* statistical model complexity index *(SMCI) of the model M is defined by the quantitative expression:*

$$SMCI(M, x \mid M_T) = \frac{G^2(M_T, x) - G^2(M, x)}{G^2(M_0, x)} . \tag{9}$$

This is the amount of information contained in the sample x and not explained by the model M with respect to the information not explained by the tree model M_T. This *SMCI* index is standardized by the information not explained by the uniform model M_0. This last information is the total amount of information contained in the sample, because the uniform model does not contain any information.

Proposition 1. *SMCI($M, x \mid M_T$) has the following properties:*

(a) $0 \leq SMCI(M, x \mid M_T) \leq 1$.

(b) If a sample is generated by a tree model M_T, the value of the complexity index $SMCI(M, x \mid M_T)$ is equal zero.

(c) If a sample x is generated by a model M with the same number of vertices but with more edges than a tree model, then the complexity index $SMCI(M, x \mid M_T)$ is greater than zero.

3.2   Sparse, Medium and Dense Model Classes

The model complexity index is used to define three types of model class complexities. The idea is that a model to describe a data sample must be as complex as the information contained in the data.

Definition 7. *Let M be a model and x be a sample generated by M, when the model complexity index $SMCI(M, x \mid M_T) \leq 0.33$, M is classified in the* sparse class type, *when* $0.33 \le SMCI(M, x \mid M_T) \le 0.66$, M *is classified in the medium class type* and *when* $SMCI(M, x \mid M_T) \ge 0.66$, M *is classified in the dense class type.*

Theorem 1. *Let* x *be a sample generated by a graphical model* M. *Let* $\{M_i\}$ *be a succession of graphical models, such that* M_{i+1} *has one edge more than* M_i, *then*

$$\lim_{M_i \longrightarrow M} SMCI(M_i, x \mid M_T) = \frac{G^2(M_T, x)}{G^2(M_0, x)} . \tag{10}$$

Proof. Substituting the definition 6 in the equation (10)

$$\lim_{M_i \longrightarrow M} SMCI(M_i, x \mid M_T) = \lim_{M_i \longrightarrow M} \frac{G^2(M_T, x) - G^2(M_i, x)}{G^2(M_0, x)} ,$$

$$= \lim_{M_i \longrightarrow M} \frac{G^2(M_T, x)}{G^2(M_0, x)} - \lim_{M_i \longrightarrow M} \frac{G^2(M_i, x)}{G^2(M_0, x)} ,$$

$$= \frac{G^2(M_T, x)}{G^2(M_0, x)} - \lim_{M_i \longrightarrow M} \frac{G^2(M_i, x)}{G^2(M_0, x)} ,$$

$$= \frac{G^2(M_T, x)}{G^2(M_0, x)} - \frac{1}{G^2(M_0, x)} \lim_{M_i \longrightarrow M} G^2(M_i, x) .$$

Let M_S be the saturated model, then $G^2(M_S, x) = 0 = L$ such that L is the under bound. Every monotone decreasing bounded sequence is convergent [12], then

$$\forall \epsilon \ge 0 \text{ and } M_j \in \{M_i\} \text{ such that } \forall i \ge j, \left| G^2(M_i, x) - L \right| < \epsilon , \tag{11}$$

then

$$\lim_{M_i \longrightarrow M} G^2(M_i, x) = 0 \text{ and the theorem is proved.} \tag{12}$$

$\square$

Let denotes

$$\frac{G^2(M_i, x)}{G^2(M_0, x)} = IP(M_i, x) . \tag{13}$$

$IP(M_i, x)$ is the information proportion contained in the data sample x and not explained by the model M_i. This will be used in the Algorithm 3 later.

3.3   Graphs Similarity Index (*GSI*)

Definition 8. *Given two graphs* G_1 *and* G_2 *a graph similarity index* $GSI(G_1, G_2)$ *between* G_1 *and* G_2 *is given by the number of common edges divided by the number of edges in the union of the two graphs [6]. Denote by* E_1 *the set of edges from* G_1 *and by* E_2 *the set of edges from* G_2, *then the similarity index is given by:*

$$GSI(G_1, G_2) = \frac{|E_1 \cap E_2|}{|E_1 \cup E_2|} , \tag{14}$$

where $|C|$ *denotes the number of elements in the set* C.

4   Graphical Markov Model Learning: An Adaptive Strategy

In this section two adaptive algorithms are introduced to learn a model from a sample data. The name adaptive for these algorithms is because their design allows that the performance depends on the complexity of the model to fit. As a first part of the algorithms a maximum expansion tree is fitted to the data using the mutual information as a measure of interaction between variables taken two by two and the Kruskal algorithm [4] is used to fit a maximum spanning tree. A prediction indicator $EMUBI$ will be defined, calculated and used, to predict the number of edges that must be added to the tree to attain an approximation to the correct model.

4.1   Edge Missing Upper Bound Index ($EMUBI$)

The most important step in both algorithms consists of a prediction of the number of edges that must be added to the maximum spanning tree to fit the model M to the data sample.

Definition 9. *Let G be the graph of the model M, and let v be the number of nodes, let $MNE(v)$ be the maximum number of edges. The edge missing upper bound index ($EMUBI$) is defined by:*

$$EMUBI(M, x \mid M_T) = (MNE(v) - v + 1)SMCI(M, x \mid M_T). \qquad (15)$$

If the data sample were generated from a tree model, the edge missing upper bound index would be almost zero. The maximum number of edges that could be added to model is proportional to the amount of information contained in the sample x and not explained by the tree model. $EMUBI$ will be used to predict the number of missing edges.

4.2   Chow and Liu Maximum Spanning Tree

The CL algorithm (See Algorithm 1) obtains the maximum weight spanning tree using the Kruskal algorithm [11] and the mutual information values for the random variables.

The mutual information measure $I_{X_i X_j}$ for all $X_i, X_j \in X$ are defined as:

$$I_{X_i X_j} = I(X_i, X_j) = \sum_{x_i, x_j} P(x_i, x_j) \log \frac{P(x_i, x_j)}{P(x_i) P(x_j)}. \qquad (16)$$

4.3   Extended Tree Adaptively Learning Algorithm (ETreeAL)

In this section the CL algorithm already described is extended. A number of edges is added based on $EMUBI$. The number of edges added, try to adapt the complexity of the learning model to the data complexity, as described in the Algorithm 2. This algorithm has a parameter τ that play a regularization role to avoid overfitting the model to the sample.

Algorithm 1. CL algorithm

Input: Distribution P over the random vector $X = (X_1, X_2, ..., X_v)$
1. Compute marginal distributions P_{X_i} , $P_{X_i X_j}$, for all $X_i, X_j \in X$
2. Compute mutual information values $I_{X_i X_j}$ for all $X_i, X_j \in X$
3. Order the values from high to low (w.r.t.) mutual information
4. Obtain the maximum weight spanning tree $M_T(CL)$ by the Kruskal algorithm [11]
Output: The maximum spanning tree $M_T(CL)$

Algorithm 2. The extended tree adaptively learning algorithm (ETreeAL)

Input: Distribution P over the random vector $X = (X_1, X_2, ..., X_v)$
1. Call Algorithm 1: CL Algorithm
2. Calculate the edge missing upper bound prediction index ($EMUBI(M, x \mid M_T(CL))$)
3. Add to $M_T(CL)$, τ per cent from missing edges in the order of the mutual information values
Output: Extended $M_{EXT}(CL)$

4.4   Global Fitting Extended Tree Adaptive Learning Algorithm (GETreeAL)

In this section the extended CL algorithm already described is enriched with a global fitting assessing, using the Kullback-Leibler divergence at each step to decide if each new edge (taken in order from the list) is added or not, after the maximum spanning tree is obtained. So the adaptation to the complexity proceed step by step, as can be seen in the Algorithm 3.

Algorithm 3. The global fitting extended tree algorithm (GETreeAL)

Input: Distribution P over the random vector $X = (X_1, X_2, ..., X_v)$
1. Call Algorithm 1: CL Algorithm
2. Calculate the $G^2(M_0, x)$ K-L divergence
3. Calculate the edge missing upper bound prediction index $EMUBI(M, x \mid M_T(CL))$
Repeat
    4. Select an edge in the order of the mutual information values
    5. $E \leftarrow E - 1$
    6. Add to the already obtained graph G_{i-1}
    7. Calculate the $G^2(M_i, x)$ K-L divergence
    8. **If** $G^2(M_i, x) < G^2(M_{i-1}, x)$ retain M_i
    Else retain M_{i-1}
Until $IP(M_i, x) < \epsilon$ for some $\epsilon > 0$
Output: Global extended $M_{GEXT}(CL)$

---

5   Experiment Design, Samples and Analysis

To perform a comparative factorial experiment of the two presented algorithms, samples are generated using the random sampler algorithm (MSRS) introduced and described in [8]. Samples of models of 16, 18 and 20 vertices and complexities sparse, medium and dense are generated (See Table 1, Appendix). A total of 9 models are used to perform the experiment. Each combination of factors is replicated 6 times for a total of 54 samples that are used with each algorithm. For ETreeAL algorithm the parameter $\tau = 0.67$. The result variables are the running time and the measure of similarity. A simple covariance analysis is performed for each result variable. The following results are obtained.

The algorithms and the experimental design were implemented in ANSI C. The result tables are in the appendix. For sparse models both algorithms obtain exactly the original models as solutions, because in this case only the CL algorithm is employed. So, they are taken out of the analysis. A covariance analysis was performed where the covariables are complexity of the model (sparse, medium, dense), and number of nodes (16, 18, 20) of the models, and the factor is the algorithm type (ETreeAL, GETreeAL). The response variables are similarity and running time. The factor "Algorithm", and the covariables "Complexity" and "Number of nodes" are statistically significant for the response variables "Run time" (in seconds) and "Similarity" (See Similarity Index, Definition 8) (See Tables 2-3, Appendix). The run time of ETreeAL is approximately 10 times faster than the run time for the GETreeAL (See Table 4, Appendix). The ETreeAL algorithm has a better performance in the case of the similarity index for dense models than medium models. For Example, for 20 nodes ETreeAL obtains complexity 3, similarity .672 and complexity 2, similarity .436 and this relationship is reversed for GETreeAL algorithm. For Example, for 20 nodes GETreeAL obtains complexity 3, similarity .992 and complexity 2, similarity 1.000 (See Table 4, Appendix).

6   Conclusions and Future Work

In this paper it is showed that it is possible to know in advance how many edges are to be added to a model to capture the information necessary to have a model whose complexity represents the data sample. Detecting the model complexity before deciding how many edges to add, permits an adaptive strategy helping the algorithms to be as efficient as possible depending on the model complexity. The learning step of the GETreeAL needs the sample evaluation at each step, to decide if an edge is added or not. It can be concluded that the ETreeAL algorithm, can be used for the sparse models class. The ETreeAL algorithm outperforms 10 times the GETreeAL in running time but GETreeAL outperforms ETreeAL 44 per cent in similarity index. The GETreeAL algorithm can be scaled for more than 20 nodes substituting the IPS algorithm by an aproximate estimate of the $\widehat{m}_n^M$.

Acknowledgments. This work is supported by Project PII 08-3, Autonomous University of Aguascalientes, Aguascalientes, Mexico. The reviewers of the paper are greatly acknowledged for their constructive comments.

234   E. Diaz et al.

References

1. Adami, C.N., Cerf, J.: Physical Complexity of Symbolic Sequences. Physica D 137, 62–69 (2000)
2. Akaike, H.: A New Look at the Statistical Model Identification. IEEE Transactions on Automatic Control 19, 716–723 (1974)
3. Chickering, M.C., Heckerman, D., Meck, C.: Large-Sample Learning of Bayesian Network is NP-Hard. Journal of Machine Learning Research 5, 1287–1330 (2004)
4. Chow, C.K., Liu, W.: Approximating Discrete Probability Distributions with Dependency Trees. IEEE Trans. Inf. Theory IT-14(3), 462–467 (1968)
5. Deming, W.E., Stephan, F.F.: On a Least Squares Adjustment of a Sampled Frequency Table When the Expected Marginal Totals are Known. Ann. of Math. Static. 11, 427–444 (1940)
6. Diaz, E.: Metaheurísticas Híbridas para el Aprendizaje de Modelos Gráficos Markovianos y Aplicaciones. Tesis para optar por el Grado de Doctor en Ciencias de la Computación. Universidad Autónoma de Aguascalientes, Ags., Mexico (2008)
7. Diaz, E., Ponce de Leon, E.: Discrete Markov model selection by a genetic algorithm. In: Sossa-Azuela, J.H., Aguilar-IbaÑÉz, C., Alvarado-Mentado, M., Gelbukh, A. (eds.) Avances en Ciencias de la Computación e Ingeniería de Cómputo, Mexico, vol. 2, pp. 315–324 (2002)
8. Diaz, E., Ponce de Leon, E.: Markov Structure Random Sampler (MSRS) algorithm from unrestricted discrete graphic Markov models. In: Gelbukh, A., Reyes, C.A. (eds.) Proceedings of the Fifth Mexican International Conference on Artificial Intelligence, pp. 199–206. IEEE Computer Society, Mexico (2006)
9. Haberman, S.J.: The Analysis of Frequency Data. The University of Chicago Press (1974)
10. Koller, D., Freedman, N., Getoor, L., Taskar, B.: Graphical models in a nutshell in Introduction to Statistical Relational Learning. In: Getoor, L., Taskar, B. (eds.) Stanford (2007)
11. Kruskal, J.B.: On the Shortest Spanning Tree of a Graph and the Traveling Salesman Problem. Proc. Amer. Math. Soc. 7, 48–50 (1956)
12. Kuratowski, K.: Introduction to Calculus. Pergamon Press, Warsaw (1961)
13. Lauritzen, S.L.: Graphical models. Oxford University Press, USA (1996)
14. Li, M., Vitanyi, P.M.B.: An Introduction to Kolmogorov Complexity and its Applications. Springer, Heidelberg (1993)
15. MacKay, J.C.: Information Theory Inference and Learning Algorithms. Cambridge Press (2003)
16. Zvonkin, A.K., Levin, L.A.: The Complexity of Finite Objects and the Development of the Concepts of Information and Randomness by Means of the Theory of Algorithms. Russ. Math. Surv. 256, 83–124 (1970)

Appendix. Algorithm Performance Description Tables

Table 1. Graphical Markov model structures

models	cliques	edges	#cliques
medium 16	ABCD,BCDE,CDEF,DEFG,EFGH, FGHI,GHIJ,HIJK, ILJK, JKLM,KLMN,LMNO,MNOP,ANOP	45	14
dense 16	ABCDEF,BCDEFG,CDEFGH, DEFGHI,EFGHIJ, FGHIJK, GHIJKL,HIJKLM,IJKLMN, JKLMNO,KLMNOP,ALMNOP	70	12
medium 18	ABCD,BCDE,CDEF,DEFG, EFGH,FGHI,GHIJ, HIJK, IJKL,JKLM,KLMN,LMNO, MNOP, NOPQ,OPQR,APQR, ABQR	53	17
dense 18	ABCDEF,CDEFGH,EFGHIJ GHIJKL, IJKLMN,KLMNOP MNOPQR,ABOPQR	77	8
medium 20	ABCD,BCDE,CDEF,DEFG,EFGH,FGHI,GHIJ, HIJK, IJKL,JKLM,KLMN,LMNO,MNOP, NOPQ,OPQR, PQRS,QRST,ARST,ABST	59	19
dense 20	ABCDEF,BCDEFG,CDEFGH,DEFGHI, EFGHIJ,FGHIJK,GHIJKL,HIJKLM, IJKLMN, JKLMNO,KLMNOP,LMNOPQ, MNOPQR,NOPQRS,OPQRST APQRST,ABQRST	94	17

Table 2. Analysis of Variance Table, Dependent Variable= RUN TIME, R Squared = .470 (Adjusted R Squared = .447)

Source	Type III Sum of Squares	df	Mean Square	F	Sig.
Corrected Model	2939636.191	3	979878.730	20.130	.000
Intercept	1651429.970	1	1651429.970	33.927	.000
NODES	1503464.143	1	1503464.143	30.887	.000
COMPLEX	439566.851	1	439566.851	9.030	.004
ALGORITHM	996605.197	1	996605.197	20.474	.000
Error	3309992.351	68	48676.358		
Total	7746170.175	72			
Corrected Total	6249628.542	71			

Table 3. Analysis of Variance Table: Dependent Variable = SIMILARITY, R Squared = .770 (Adjusted R Squared = .760)

Source	Type III Sum of Squares	df	Mean Square	F	Sig.
Corrected Model	1.910	3	.637	75.932	.000
Intercept	.690	1	.690	82.317	.000
NODES	.172	1	.172	20.545	.000
COMPLEX	.100	1	.100	11.960	.001
ALGORITHM	1.637	1	1.637	195.290	.000
Error	.570	68	8.384E-03		
Total	52.862	72			
Corrected Total	2.480	71			

Table 4. Means: Run time and similarity by algorithm, nodes number and complexity classes

ALGORITHM	NODES	COMPLEX	Mean RUN TIME	Mean SIMILARITY
ETreeAL	16	2	1.104	.730
ETreeAL	16	3	2.281	.890
ETreeAL	18	2	7.089	.656
ETreeAL	18	3	12.495	.730
ETreeAL	20	2	39.029	.436
ETreeAL	20	3	97.123	.672
GETreeAL	16	2	7.600	.979
GETreeAL	16	3	29.357	.980
GETreeAL	18	2	59.679	.994
GETreeAL	18	3	154.259	.979
GETreeAL	20	2	281.714	1.000
GETreeAL	20	3	1038.322	.992

Support Vector Optimization through Hybrids: Heuristics and Math Approach

Ariel García Gamboa, Neil Hernández Gress, Miguel González Mendoza, and Jaime Mora Vargas

Intelligent Systems Research Group
Tecnológico de Monterrey, Campus Estado de México
Km 3.5 Carretera Lago de Guadalupe
Col. Margarita Maza de Jurez
Atizapán de Zaragoza, Estado de México, Mexico, 52926
Tel.: +52 55 5864 5751
{ariel.garcia,ngress,mgonza,jmora}@itesm.mx

Abstract. This paper presents a strategy to optimize the learning phase of the Support Vector Machines algorithm (SVM). The SVM algorithm is widely used in solving different tasks like classification, regression, density estimation and clustering problems. However, the algorithm presents important disadvantages when learning large scale problems. Training a SVM involves finding the solution of a quadratic optimization problem (QP), which is very resource consuming. What is more, during the learning step, the best working set must be selected, which is a hard to perform task. In this work, we combine a heuristic approach, which selects the best working set data, with a projected conjugate gradient method, which is a fast and easy to implement algorithm that solves the quadratic programming problem involved in the SVM algorithm. We compare the performances of the optimization strategies using some well-known benchmark databases.

1 Introduction

Support Vector Machines (SVM) is a technique developed by Vapnik, [1], for solving classification, regression and density estimation problems. Training a SVM corresponds to finding the solution of a quadratic programming (QP) problem. When solving this QP problem special training patterns, called support vectors (SV), are found. Due to the QP problem nature, the SVM solution is a global optimum and not local minimums. However, an important disadvantage is that, as the number of variables in the QP problem is equal to the number of data patterns, then the problem complexity grows exponentially and uses prohibitive computational resources when is used in large scale applications. Different approaches to overcome this problem have been proposed, [1], [2], [3], [4], [5]. Most approaches are based on the fact that given a training data set, the QP optimization problem will provide the same result whether the entire data set or a reduced one (having only support vectors) is used. Different initialization strategies, that reduce considerably the training time, are showed in

A. Hernández Aguirre et al. (Eds.): MICAI 2009, LNAI 5845, pp. 237–245, 2009.

[5]. However, this approach uses a traditional QP solver, which may use much computational resources in the worst case. At this point, we propose to use an hybrid system. Hybrids have been known since the last decades, [6], [7], but recently, due to the need of well-built systems for solving hard problems, they have gained enormous attention. The goal of hybrids is to combine the strengths of the proposed methodologies and, to avoid their weakness. This way, single robust systems able to solve hard computing problems are obtained. The hybrid approach to optimize the computational resources required for training a SVM, that we propose, is composed of two main parts: heuristic and the use of a mathematical aproach. The heuristic approach consists of finding the SV's of the problem (a priori) by means of an algorithm that uses minimum computational resources and is easy to implement. The second approach is related with the QP solver. Here we propose to use a modified Projected Conjugate Gradient algorithm, [8], that is fast and easy to implement algorithm, which is faster and less resource consuming algorithm than traditional QP solvers. Both heuristics and math ensure to reduce the training time and memory resources required for the SVM algorithm. This paper is organized as follows: section 2 gives a brief description of the methods used. Section 3 shows the optimization techniques. First, the heuristic approach is described and second, the mathematical method to solve the QP problem of the SVM is presented. Section 4 shows the experiments of the hybrid methodology proposed. Finally the conclusions of this work are presented.

2   Methodological Framework

2.1   Barycentric Correction Procedure

Barycentric Correction Procedure (BCP) is an algorithm based on geometrical characteristics for training a threshold unit [9]. It is very efficient training linearly separable problems and it was proven that the algorithm rapidly converges towards a solution [10]. The algorithm defines a hyperplane $\mathbf{w}^T\mathbf{x}+\theta$ dividing the input space for each class. Thus, we can define: $I_1 = 1,\ldots,N_1$ and $I_0 = 1,\ldots,N_0$ where N_1 represents the number of patterns of target 1 and N_0 the number of patterns of target -1. Also, let $(b = b_1, b_0)$ be the barycenters of data points belongin to class $\{+1,-1\}$ respectively, and weighted by the positive coefficients $\alpha = \alpha_1,\ldots,\alpha_{N_1}$ and $\mu = \mu_1,\ldots,\mu_{N_0}$ referred as *weighting coefficients* [10]:

$$b_1 = \frac{\sum_{i \in I_1} \alpha_i x_i}{\sum_{i \in I_1} \alpha_i} \qquad\qquad b_0 = \frac{\sum_{j \in I_0} \mu_j x_j}{\sum_{j \in I_0} \mu_j} \qquad (1)$$

The weight vector $\mathbf{w}$ is defined as a vector difference $\mathbf{w} = b_1 - b_0$. At each iteration, barycenter moves towards misclassified patterns. Increasing the value of particular barycenter implies hyperplane moves on that direction. For computing the bias term θ, let's define $\vartheta : \Re^n \rightarrow \Re$ such that $\vartheta(x) = -\mathbf{w}\cdot x$ The bias term is calculated as follows: $\theta = \frac{\max \vartheta_1 + \min \vartheta_0}{2}$. Assuming the existence of $J_1 \in I_1$

and $J_0 \in I_0$ that refer to misclassified examples of target $\{+1, -1\}$, barycenter modifications are calculated by:

$$\forall_i \in J_1, \; \alpha(new)_i = \alpha(old)_i + \beta_i \quad \text{and} \quad \forall_j \in J_0, \; \mu(new)_j = \mu(old)_j + \delta_j \; (2)$$

Where $\beta = \max\{\beta_{min}, min[\beta_{min}, \frac{N_1}{N_0}]\}$ and $\delta = \max\{\delta_{min}, min[\delta_{min}, \frac{N_0}{N_1}]\}$. According to [10], β_{min} and δ_{min} can be set to 1 and β_{max} and δ_{max} set to 30.

2.2 Support Vector Machines for Classification Tasks

Support Vector Machines is a well known technique for solving classification, regression and density estimation problems, [1], [11], [4], [12]. This learning technique provides a convergence to a globally optimal solution and, for several problems, it has shown better generalization capabilities than other learning techniques.

In agreement with the inductive principle of the structural risk minimization, [1], in the statistical learning theory, a function, which describes correctly a training set X and which belongs to a set of functions with a low VC (Vapnik-Chervonenkis) dimension, will have a good generalization capacity independently of the input space dimension. Based on this principle, the Support Vector Machines [1] have a systematic approach to find a linear function, belonging to a set of functions with a low VC dimension.

The principal characteristic of SVMs for pattern recognition is to build an optimal separating hyperplane between two classes. If this is not possible, the second great property of this method resides in the projection of X space into a Hilbert space F of highest dimension, through a function $\varphi(x)$. One example is the internal product evaluated using kernel functions:

$$k(x_i, x_j) = \phi(x_i)^T \phi(x_j), \quad i, j = 1, \ldots, l. \tag{3}$$

satisfying Mercer conditions, like the Gaussian kernel:

$$k(x_i, x_j, \sigma) = \frac{1}{\sigma\sqrt{2\pi}} \exp\left[-\frac{(x_i - x_j)^2}{2\sigma^2}\right] \tag{4}$$

Thus, thanks to the freedom of using different types of kernels, the optimal separating hyperplane corresponds to different non-linear estimators in the original space.

For classification tasks, the main idea can be stated as follows: given a training data set $(\mathbf{X})$ characterized by patterns $x_i \in \Re^n$, $i = 1, \ldots, n$ belonging two possible classes $y_i \in \{1, -1\}$, there exist a solution represented by the following optimization problem:

$$\begin{aligned} Maximize \atop \alpha \quad & L_D(\alpha) = \sum_{i=1}^{l} \alpha_i \alpha_j y_i y_j k(\mathbf{x_i}, \mathbf{x_j}) & (5) \\ s.t. \quad & \sum_{i=1}^{n} y_i \alpha_i = 0, \quad 0 \le \alpha \le C & (6) \end{aligned}$$

where α_i are the Lagrange multipliers introduced to transform the original formulation of the problem with linear inequality constraints into the above representation, [1]. The parameter C controls the misclassification level on the training data and therefore the margin. The $k(\mathbf{x_i}, \mathbf{x_j})$ term represents the so called kernel trick and is used to project data into a Hilbert space F of higher dimension using simple functions for the computation of dot products of the input patterns: $k(\mathbf{x_i}, \mathbf{x_j}) = \phi(\mathbf{x_i})^T \phi(\mathbf{x_j})$, $i, j = 1, \ldots, n$. Once one has the solution, the decision function is defined as:

$$f(\mathbf{x}) = sign\left(\sum_{i=1}^{l} \alpha_i y_i k(\mathbf{x_i}, \mathbf{x}) + b \right) \qquad (7)$$

The solution to the problem formulated in (6) is a vector $\alpha_i^* \geq 0$ for which the α_i strictly greater than zero are the support vectors.

3   Optimizing the Support Vector Algorithm

Training a SVM involves the solution of a quadratic programming (QP) problem. The solution of this QP problem identifies special training patterns, contained in the original training database, called support vectors (SV). Since the number of variables in the QP problem is equal to the number of data patterns, the problem complexity grows exponentially with the number of training patterns, and thus, reaching a solution for large scale applications implies using prohibitive computational resources. To overcome this problem, different approaches have been proposed: Chunking, [1], Osuna's decomposition, [2], [13], sequential minimal optimization, [3], and different initialization strategies proposed in [5]. In this paper we propose to combine two different approaches: heuristics and math to train a SVM. The heuristic is related with an initiazation strategy to find as many support vectors as possible using algorithms that are very few resource consuming, like BCP algorithm. The second stage is related with the QP solver. Here, we propose to use the constrained conjugate gradient proposed in [14] and [8]. This algorithm takes advantage of the sparsity of the problem and solves the QP problem using less resources than traditional QP solvers, [4]. The combination of these two strategies ensures better performance of the SVM algorithm and allows us to use it in real problems whit many data patterns.

Heuristic Approach. Solving large scale applications require a decomposition technique to break up the original problem into smaller QP sub problems. The main disadvantage of this approach is that patterns for each sub problem are randomly selected (causing substantial difference in the computational resources required for the learning process). In order to obtain better sub-problems (working sets) with more support vectors, we propose to use the BCP algorithm. Using this approach, the resources for the learning process of the SVM are considerably reduced. The selection of this algorithm was based in some advantageous characteristics: the algorithm finds an optimal separating hyperplane for a given

dataset, it is easy to implement and it uses minimal processing and memory resources. Using this initialization approach ensures that the initial sub-problem is
conformed by candidate support vectors that were extracted before training the
SVM, and consequently, the computer resources for training are reduced. The
general idea is to obtain, by means of the proposed algorithms, a hyperplane that
correctly classifies the original dataset, then by simple Euclidean distance, between the hyperplane and the data patterns, we get the closest to the hyperplane
and form the active set (see figure 1)

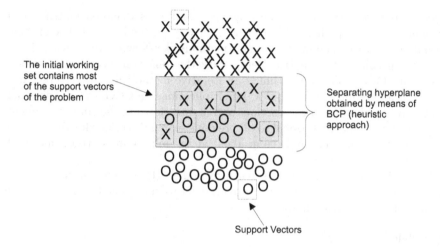

Fig. 1. A small sub-group of patterns (containing most support vectors) is selected
using BCP and euclidean distance

Mathematical Approach. Essentially, there are two types of algorithms for
the resolution of QPs: Active set and Interior point methods. Active set methods,
in one hand, are also divided in primal methods, which find dual feasibility by
keeping primal feasibility and the complementary conditions, and dual methods,
which search primal feasibility by keeping dual feasibility and the complementary
conditions. These last, are only applicable when Q, the hessian matrix of the
original QP problem, is a positive definite matrix. These methods use simple
linear programming methods to solve the QP problem. Interior point methods,
on the other hand, search feasibility of complementary conditions by keeping
primal and dual feasibility at the same time.

In this work, we propose to use a dual active set method to solve de QP
problem of the SVM algorithm, [14]. This implementation takes advantage of
the fact that the solution of the problem is sparse. The general idea is to solve
a sequence of equality constrained problems of the type:

$$\text{maximize}_\alpha = F(\alpha) = \alpha^T P^T r - \frac{1}{2}\alpha^T P^T HP\alpha \qquad (8)$$

subject to:

$$A^T \alpha = 0, \tag{9}$$

$$0 \le \alpha_0 + P\alpha \le C \tag{10}$$

Where P spans the null space of matrix A and matrix A has the form:

$$A^T = \begin{bmatrix} D & | & 0 \\ \hat{y}^T & \end{bmatrix} \tag{11}$$

where $\hat{y}$ is a permutation of the original y and D is diagonal matrix which elements are $\{1, -1\}$ when α is $\{0, C\}$ respectively (at the bounds). The problem formulated above can be solved by using methods like: Newton approach, [14] and conjugate gradient aproach, [15]. Here, we use the constrained conjugate gradient algorithm (CCG). This algorithm is an iterative method which converges in k iterations and requires at most $O(k^2)$ multiplications. In CCG a search direction is a linear combination of the previous search direction and the current gradient. The new direction is orthogonal to all previous search directions for the equality constrained sub-problem and the gradient at the minimum is orthogonal to all previous search directions.

Thus, given α, the previous serch direction $\hat{p}$, the current gradient g, the previous gradient $\hat{g}$, then the conjugate gradient to determine the next $\hat{\alpha}$, which satisfies $A^T \alpha = 0$ is as follows:

1. Compute $\sigma = \frac{\hat{g}^T P \hat{g}}{g^T P g}$
2. Set $q = -Pg + \sigma \hat{g}$
3. Compute $\tau = \frac{g^T P g}{q^T H q}$
4. Set $\hat{\alpha} = \alpha + \tau q$
5. The new gradient $\hat{g} = g + \tau H q$

The multiplication of H in step 3 is reused in step 5, then only one matrix vector multiplication involving H is necessary. Remember that there is no need to decompose the hessian matrix in $P^T H P$, since only rows or columns of H corresponding to the variables not at bounds have to be formed. Thus, the algorithm only requires vector matrix multiplication.

1. Election of an active unit A of size n_A.
2. Solve the QP (6), defined by active set A.
3. While there is any example violating $y_j g(\mathbf{x}_j) > 1$,
   (a) Shift the n_A most erroneous vectors x_j to active set A,
   (b) Shift all vectors x_i with $\alpha_i = 0$, $i \in A$, to inactive set N, return to 2.

4   Experiments and Results

We perform our test in 5 different benchmark problems commonly referred in literature,[16]: Sonar, Pima Indian Diabetes, Tic Tac Toe, Phonemes and

Table 1. The experiments were performed using a *RBF* kernel

Data Set	Size	Alg	Param	C	SV %	Ite	Acc %	Time(s)
Phonemes	1027	Random	0.5	1000	42	487	100	12.258
Phonemes	1027	T.Wen	0.5	1000	42	108	100	4.246
Phonemes	1027	BCP	0.5	1000	42	44	100	3.285
Diabetes	768	Random	0.5	1000	100	39	100	0.791
Diabetes	768	T.Wen	0.5	1000	100	39	100	0.801
Diabetes	768	BCP	0.5	1000	100	10	100	0.35
TicTacToe	958	Random	1	1000	81	43	100	2.804
TicTacToe	958	T.Wen	1	1000	81	44	100	2.193
TicTacToe	958	BCP	1	1000	81	12	100	1.832
Adult	1478	Random	5	100	44	88	94	5.608
Adult	1478	T.Wen	5	100	44	97	94	4.336
Adult	1478	BCP	5	100	45	62	94	5.327
Sonar	208	Random	5	1000	26	20	100	0.04
Sonar	208	T.Wen	5	1000	26	16	100	0.04
Sonar	208	BCP	5	1000	27	20	100	0.06

Adult. We compare the results of our proposed hybrid with two different approaches: random initialization and the Wen's initialization, [8], both using the constrained conjugate gradient (CCG) algorithm to solve the quadratic programming problem of SVM. Also, a comparison table of CCG and traditional QP solver combined with random and BCP initialization is showed. The experiments were performed using the *RBF* kernel function. The values for γ and for the regularization parameter C were in the range betwen $[0.1, 0.2, 0.5, 1, 2, 5]$ and $[10, 100, 1000]$, respectively, but the ones with the best results are shown in (see Table 1). The results showed are the standard mean of about 1000 executions. The results showed are: data set, size of the data set, algorith used, the kernel parameter (γ), the regularization parameter (C), the number of support vectors found, number of iterations to global optimum, the training accuracy and the time consumed by the algorithm.

The best results are obtained using the $BCP + CCG$ hybrid. This algorith is less time consuming than $Random + CCG$ and $Wen + CCG$ algorithms. This advantage is due to the fact that the algorithm finds an optimal solution in less iterations, because most of the support vectors are found in the heuristic step using the BCP algorithm. On the other hand, this algorithm performs better than the same heuristic (BCP initialization) and the traditional QP solver, which is much resource consuming.

In (Table 2), we show the initialization heuristic combined with the QP solver described in [4]. As can be seen, in Adult and Diabetes data sets, it is considerably reduced the training time using the $BCP + CCG$ algorithm than the $BCP + QP$. Also, in the other data sets, this algorithm performs better, but is not as visible as in the mentioned data sets.

Table 2. Performance comparison between the CCG and traditional QP algorithms (both with the heuristic initialization strategy)

Data Set	Algorithm	Time
Phonemes	BCP+CCG	3.285
Phonemes	BCP+QP	3.328
Diabetes	BCP+CCG	0.35
Diabetes	BCP+QP	5.143
TicTacToe	BCP+CCG	1.832
TicTacToe	BCP+QP	3.091
Adult	BCP+CCG	5.327
Adult	BCP+QP	24.028
Sonar	BCP+CCG	0.06
Sonar	BCP+QP	0.16

5   Conclusions

SVM is a promising methodology for solving different tasks. It has been widely and succesfully used in different research areas. However, the learning process of the SVM is a delicate problem due to time and memory requirements. This research is focused in the optimization of the SVM algorithm by means of a hybrid methodology. The hybrid is composed of an heuristic and a math approach. The heuristic cosists of an improved initialization strategy, which aims at selecting the best patterns (SV) for the initial working set, and the math is based in the constrained conjugate gradient algorithm, which is faster and less resource consuming and takes advantage of the sparsity solution of the QP problem. Then, they both work together and are able to considerably reduce the training time of the SVM algorithm. Additionally, these hybrid approache ensures better performance training large scale datasets than other methodologies like the Random-QP algorithm.

References

[1] Alur, R., Courcoubetis, C., Halbwachs, N., Henzinger, T.A., Ho, P.H., Nicollin, X., Olivero, A., Sifakis, J., Yovine, S.: The algorithmic anal- ysis of hybrid systems. Theoretical Computer Science 138(1), 3–34 (1995)
[2] Ariel, G.G., Neil, H.G.: A comparison of different initialization strategies to reduce the training time of support vector machines. In: Duch, W., Kacprzyk, J., Oja, E., Zadrożny, S. (eds.) ICANN 2005. LNCS, vol. 3697, pp. 613–618. Springer, Heidelberg (2005)
[3] Asuncion, A., Newman, D.J.: UCI machine learning repository (2007), http://www.ics.uci.edu/~mlearn/MLRepository.html
[4] Bishop, C.M.: Pattern recognition and machine learning. Springer, New York (2006)
[5] Edgar, O., Freund, R., Girosi, F.: Support vector machines: Training and applications (1997)

[6] Edgar, O., Freund, R., Girosi, F.: Training support vector machines: an application to face detection. In: 1997 IEEE Computer Society Conference on Computer Vision and Pattern Recognition, Proceedings, pp. 130–136 (1997)

[7] Goonatilake, S., Khebbal, S.: Intelligent hybrid systems. John Wiley, Inc., New York (1994)

[8] Poulard, H., Stéve, D.: Barycentric correction procedure: A fast method of learning threshold unit (1995)

[9] Poulard. H. and Stéve. D.: A convergence theorem for barycentric correction procedure. Technical Report 95180, LAAS-CNRS (1995)

[10] Kecman, V.: Learning and Soft Computing, 1st edn. The MIT Press, Cambridge (2001)

[11] Linda, K.: Solving the quadratic programming problem arising in support vector classification. In: Advances in kernel methods: support vector learning, pp. 147–167. MIT Press, Cambridge (1999)

[12] Miguel, G.M., Neil, H.G., Andr, T.: Quadratic optimization fine tuning for the learning phase of svm (2005)

[13] Platt, J.C.: Fast training of support vector machines using sequential minimal optimization. In: Bernhard SchlkopfChristopher, J.C., Smola, B.J. (eds.) Advances in Kernel Methods: Support Vector Learning. The MIT Press, Cambridge (1998)

[14] Fletcher, R.: Practical methods of optimization, 2nd edn. Wiley- Interscience, Chichester (1987)

[15] Tong, W., Alan, E., David, G.: A fast projected conjugate gradient algorithm for training support vector machines. Contemporary mathematics: theory and applications, 245–263 (2003)

[16] Vladimir, V.: The Nature of Statistical Learning Theory, 2nd edn. Springer, Heidelberg (1999)

On-Line Signature Verification Based on Genetic Optimization and Neural-Network-Driven Fuzzy Reasoning

Julio Cesar Martínez-Romo[1], Francisco Javier Luna-Rosas[1],
and Miguel Mora-González[2]

[1] Institute of Technology of Aguascalientes, Department of Electrical Engineering,
Av. A. López Mateos 1801 Ote. Col. Bona Gens,
20256 Aguascalientes, Ags. México
jucemaro@yahoo.com, fjluna@ita.mx
[2] University of Guadalajara, University Center of Los Lagos,
Av. Enrique Díaz de León 1144, Col. Paseos de la Montaña,
47460 Lagos de Moreno, Jal. México
mmora@culagos.udg.mx

Abstract. This paper presents an innovative approach to solve the *on-line signature verification* problem in the presence of skilled forgeries. Genetic algorithms (GA) and fuzzy reasoning are the core of our solution. A standard GA is used to find a near optimal representation of the features of a signature to minimize the risk of accepting skilled forgeries. Fuzzy reasoning here is carried out by Neural Networks. The method of a human expert examiner of questioned signatures is adopted here. The solution was tested in the presence of genuine, random and skilled forgeries, with high correct verification rates.

Keywords: Signature verification, fuzzy reasoning, GA, neural networks.

1 Introduction

Automatic signature verification is an active field of research with many practical applications [1]. Automatic handwritten signature verification is divided into two approaches: off-line and on-line [2]. In the off-line signature verification approach, the data of the signature is obtained from a static image utilizing a scanning device. In the on-line approach, special hardware is used to capture dynamics of the signature, therefore the on-line approach is also known as the dynamic approach and the terms on-line signature verification (OSV) and dynamic signature verification (DSV) are used interchangeably in the literature [2]. In OSV, special pens and digitizing tablets are utilized as capturing devices, and the data obtained include coordinates of position of the pen tip vs. time, pressure at the pen tip, and pen tilt [2, 3]. When full data vectors are used as features, the approach is known as the "functional approach", and the vectors are called "functions" [1], [2]. On the other hand, when parameters that describe a data vector are used as features, the approach is known as the "parameters approach"; examples are mean values, LPC coefficients, and Fourier coefficients. The latter approach has the advantage of dimensionality reduction, but it has shown to

A. Hernández Aguirre et al. (Eds.): MICAI 2009, LNAI 5845, pp. 246–257, 2009.

have worse performance than the former [1], [2]. A comprehensive survey on features, classifiers and other techniques related to on-line signature verification can be found in [2], [4], [5], [6], and [7], [8], and [9]. Given that on-line data acquisition methodology gathers information not available in the off-line mode, dynamic signature verification is more reliable [10].

Feature generation and feature selection are pattern recognition tasks related to the way an object or phenomenon is described in terms of the raw measurements available from them [11]. Feature generation refers to the transformation of the raw measurements to arrive to new descriptors of the object; feature selection refers to selecting a subset of features out of some global set of features using certain criteria. Feature generation and selection in OSV are difficult to solve in the presence of skilled forgeries because the features from genuine signatures and forgeries tend to overlap in the features space. Therefore, there exist a trend in the pattern recognition community to work on the subject of feature selection, in particular based on genetic algorithms; see [12] and [13]. A review of the use of genetic algorithms in OSV is given in the next paragraphs.

In [15], Yang used a genetic algorithm to select partial curves along a signature. The features he used were length, slope, largest angle, and curvature of partial curves along the signature. In [16], the authors reported the same technique, but the optimization was carried out over the virtual strokes of the signature (i.e., points of pen-up trajectories). In [17], Wijesoma proposed the use of genetic algorithms with emphasis on the selection of a subset of features out of a set of 24 features representing the shape and the dynamics of a signature. In [18], Galbally and others compared binary and integer encoded genetic algorithms to solve the problem of feature selection in OSV. Sub-optimal feature selection techniques have been used also, such as the floating search method; examples are [8], [9]; their objective was feature selection of global and local features of the signing process, and the cost function they used was a modified version of the Fisher ratio [11]. Please note that so far the use of genetic algorithms and other optimization techniques have been in OSV to either a) feature selection, b) compare the performance of different versions of the genetic algorithm, or c) segment a signature in partial curves whose features are used for verification (a sort of feature generation). In this paper, we introduce a different approach in which the genetic algorithm is used to create an optimal functional model of each signer enrolled in the OSV system.

2 Rationale and Global Description of the OSV System

Since the functional approach (see Sect. 1) has proven to be superior to the parameters approach for on-line signature verification, the former was adopted here as the feature representation method. In addition, our overall approach to OSV is highly based on the procedure of an expert examiner of questioned signatures. An expert examiner considers that the signing process is dominated by personalized writing habits [20]. In this context, a signature is the result of a complex combination of personalized patterns of *shape* and *rhythm*. Rhythm and dynamics will be considered synonymous in this work. it is the final balance of rhythm and form which is critical in signatures comparisons, not only the matching of individual elements [20]. Figure 1 summarizes our approach. In the *enrollment and training* phase, 16 signatures from one subject are chosen from a database; for discrimination purposes, other 80 random forgeries (4 genuine signatures from other 20 subjects) are also included, but not shown in figure 1 for simplicity. Five shape–related and five dynamics-related features are generated from every signature. Once the two sets of features are calculated, the goal is to generate a model *M* for the subject. The elements of the model *M* are: one averaged prototype function *PF*, one consistency function *CF*, and one weighting factor *W*, for each feature. The genetic algorithm is used to obtain the prototype function *PF* whose weighted distance or difference (in the rest of this paper, *error*) is minimum when compared to the respective features of genuine signatures (thus reducing the erroneously rejected genuine signatures, known as *false rejection rate, FRR*), and maximum when compared to the respective features of simple and skilled forgeries (reducing the erroneously accepted forgeries, known as *false acceptance rate, FAR*). Since it is difficult to have skilled forgeries during the training phase of any real OSV system, in this work it was decided to introduce five *synthetic skilled forgeries (SSF)* by deforming feature vectors obtained from genuine signatures: original feature vectors were phase shifted, corrupted by additive noise, and lowpass-filtered. The *grading unit* (see figure 1) grades all the signatures, genuine as well as forgeries, in the range 0-1. The distances (or errors) per feature and the grades of every signature are used to train two neural network driven fuzzy reasoning (NND-FR) systems, one for shape and another one for rhythm. Consequently, the NND-FR system *learns* to grade a questioned signature according to: a) deviations with respect to normal variations of genuine signatures, b) consistency in the signing process of the original signatures, and c) importance or contribution of every feature for the correct verification. In the *verification* phase, figure 1, a questioned signature is digitized and its respective shape and rhythm features are generated. The error between every feature and its model *M* is calculated in the respective error unit (see figure 1). The set of five shape errors are used to feed the NND-FR shape unit. The output of this unit is an indicator of the similarity of shape of the questioned signature to the genuine set of exemplars. A grade for signature rhythm is calculated in a similar manner. Finally, a 2-input fuzzy system calculates the degree of certainty (DOC) in which a signature can be considered genuine, in the range from 0 to 1. In the next sections feature generation and implementation details will be explained.

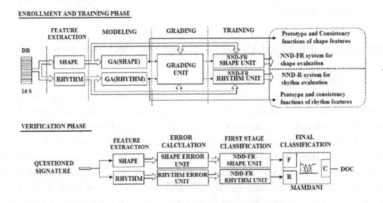

Fig. 1. Global description of our on-line signature verification system

3   Feature Generation

3.1   Shape Related Features

According to the expert signature examiner Slyter [20], shape related features should reflect the signing habits of a person in relation to gross forms, variations in the design of the signature, connective forms and micro-forms. Therefore, it should derive, from any signature, functions or parameters that reliably reflect the variations and consistency of such elements. This is still an open and difficult problem [6], [2]; however, in [7] it can be found a set of five functions that are aimed at characterizing the shape of a signature; these functions can discriminate discrepancies in slant and local shape of a signature; here we adopted those features as shape descriptors and they are shown in table 1, rows 1 through 5. For derivation and implementation details, please refer to [7]; the original name of each function was preserved.

3.2   Dynamics Related Features

Dynamics related features are listed in table 1, rows 6 through 10. The *speed*, row 6, is calculated from the coordinates of the position of the pen tip as a function of time, $x(t)$ and $y(t)$, by double differentiation and absolute value calculation. Pressure is directly obtained from the digitizer tablet, *Acecat 302 Pen & Graphics Tablet* [21], and the variations in pressure patterns are calculated by differentiation of the absolute pressure. The top and base patterns of writing are obtained by sampling the upper and lower envelopes of the signature as a digital image. Timing information is added: every point of the upper/lower envelope is associated with the time instant in which the point under consideration was drawn. This make successful forgery more difficult. The length of the feature vectors is normalized to a fixed size, enabling point-to-point averaging and comparison across different instances of a signature.

4   Prototype Functions and Pattern Recognition Setup

For mathematical purpose, consider that any feature listed in table 1 is denoted by a row vector x_j, where the subscript j stands for the j-th signature of a user in the OSV system; then a matrix $X = [\ x_1;\ x_2;\ \ldots\ x_N]$ will denote a set of several feature vectors extracted from N genuine signatures. The *prototype function* PF_x and the respective *consistency function* CF_x of the feature are calculated as:

$$PF_x = \text{mean\_col}(X) \qquad (1)$$
$$CF_x = \text{std\_col}(X) \qquad (2)$$

where   *mean_col*, is the statistical mean value of the columns of X, and
std_col, is the statistical standard deviation of the columns of X.

In this work, the OSV problem is seen as a two-class pattern recognition problem, in which for each user, one class is the "genuine" class (ω_0) and the other one is the "forgeries" class (ω_1). The training and testing sets of class ω_0 will consist of feature vectors from genuine signatures; the training and testing sets of class ω_1 will consist

Table 1. Dynamic and shape related features

Num.	Name	Meaning	Elements
1	cx(l)	x coordinate of the local center of mass	82
2	cy(l)	y coordinate of the local center of mass	82
3	T(l)	Torque exerted about origin of the seg-ment	82
4	s1(l)	Curvature ellipse measure 1	82
5	s2(l)	Curvature ellipse measure 2	82
6	Sp(t)	Speed as a function of time	256
7	P(t)	Pressure as a function of time	256
8	dP(t)	Pressure change patterns	256
9	Bow(s,t)	Base of writing with timing	84
10	Tow(s,t)	Top of writing with timing	84

of feature vectors corresponding to 4 genuine signatures of other 20 signers in the database (simulating simple and non-elaborated forgeries), and *synthetic feature vectors* of 4 virtual forgers (simulating very well executed forgeries). The quantity of signatures used in the training and testing sets of class ω_1 was established experimentally as a trade-off between processing time and accuracy. A *synthetic feature vector* -that represents a feature of a synthetic skilled forgery-, is obtained by exchanging two segments of a genuine feature vector, adding random noise, and finally filtering the resulting feature vector with a LPF IIR filter, with normalized bandpass limit of 0.8 of order 10 [22]. Figure 2 shows genuine and synthetic feature vectors of the *x* coordinate of the local center of mass cx(l) (see table 1); the solid line is a genuine feature vector; a feature vector from a real skilled forgery is drawn in dashed line . Dash-dot line is the *synthesized feature vector* that acts like a skilled forgery. Please observe the similarity between real and synthetic skilled forgeries. We propose *d* (Eq. 3, weighted l_2 metric, [11]) to determine an error measure of x with respect to PF_x:

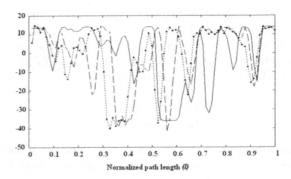

Fig. 2. Solid line: plot of a feature from a genuine signature; dashed line: from a real skilled forgery; dash-dotted line: from a synthetic skilled forgery, all in the features space. The feature displayed is cx(l).

Table 2. Parameters of the genetic algorithm

Parameter	Value
Population	First: random, constant
Selection	Elitist
Crossing method	Single, one point
Probability of mutation	MP>0.14%
Stop criterion	mean(J) = max(J) or 100 gen

$$d(PF_x, x) = \sqrt{(PF_x - x)^T B(PF_x - x)} \tag{3}$$

where B is a symmetric, positive definite matrix [11], whose diagonal elements are the inverse values of the respective consistency function CF_x.

5   Genetic Algorithms to Optimize the Prototype Function

Genetic Algorithms (GAs) are biologically inspired optimization systems, which are based upon the principles of natural selection. Fundamentals on GAs can be found in [14]. In this work, it was preferred the use of genetic algorithms because the problem to solve is combinatorial in nature. Literature review shows that standard genetic algorithm (SGA) for large scale problems and some modified versions of the SGA tend to outperform others, such as simulated annealing, floating search, and others [23], [24], [25]. The parameters of the GA used here are shown in table 2.

5.1   Chromosome Encoding and Other GA Parameters

Consider the NxL matrix $X = [\ x_1; x_2; \ ... \ x_N]$ described in Sect. 4, and suppose that the respective prototype and consistency functions PF_x and CF_x, have already been calculated; then, any of the x_j feature vectors in X is chosen and the dissimilarity measure d of Eq. 3 is calculated. Now, some subset of $M<N$ rows of X are randomly selected to recalculate a new PFx'. If $(d(PF_x',x_j)<d(PF_x,x_j))$, then it can be said that PF_x is a "better" prototype of X than PF_x is because PFx' has the ability to consider x_j more similar to itself than PF_x does. The selection of the subset of M feature vectors that will produce the best PF_x is the goal of the genetic algorithm for each one of the 10 features of table 1. To encode the chromosomes, lets consider that there are $N=16$ feature vectors in X, therefore a maximum of 4 bits are required to uniquely represent each of feature vector in X. Thus, a chromosome is composed of a binary string in which each consecutive sub-string of 4-bits represents one feature vector that will be used to compute PF_x. For example, if a part of a chromosome is '0010|0111|1010|0001|1001', then feature vectors $\{x_2, x_7, x_{10}, x_1, x_9\}$ are averaged to calculate PF_x. The objective J function to be optimized by the GA is:

$$J = (\min(ied) - \max(iad)) / \max(ied) \tag{4}$$

where: *ied* –inter-class distance- is a vector of distances of every x_j in X of the class ω_1 with respect to PF_x. and *iad* –intra-class distance- is a vector of distances of every x_j in X of the class ω_0 with respect to PFx. Fig. 3 is the plot of *ied* and *iad* of the static feature cx(l) for all the feature vector in both classes, note that the genetic algorithm increases the gap between classes ω_0 and ω_1.

5.2 Weighting the Discriminant Capability of the Features

Shape related and dynamics related features are listed in table 1. Now, lets define the matrix of *shape descriptors* $SD = [cx(l),cy(l),T(l),s1(l),s2(l)]$ of size 82x5, and *error in shape descriptors* as $ESD = [E_{cx},E_{cy},E_T,E_{s1},E_{s2}]$ (an *error in dynamics descriptors* EDD, is produced similarly), as descriptors of the behavior of a signature in the features space with respect to PF_x. The matrix SD contains the values of the five feature vectors and ESD contains the distance (or *error*, from the perspective of a human expert examiner [20]) between each feature and its corresponding PF_x optimized. As can be seen in figure 3, there exists a gap, or class separation, that is different for each feature. It is evident that such a gap is a measure of how much a feature is relevant to correctly verify a signature; to exploit this fact in benefit of the correct classification rate of the OSV system, it is assigned a weighting factor to each feature as follows:

$$wf_x = (\min(ied)\text{-}\max(iad))/\max(ied) \tag{5a}$$

$$WF = [\ wf_{cx}\ \ wf_{cy}\ \ wf_T\ \ wf_{s1}\ \ wf_{s2}\ ] \tag{5b}$$

where wf_x is the *weighting factor* of a feature *x*, and WF is a vector of personalized weighting factors. A grade is assigned to as a function of the standard deviation (σ) and the mean value (m) of the respective intra-class distances, as follows:

$$G_f = \begin{cases} 1 \Rightarrow 0 < E_f \le m - 2\sigma \\ (^{0.15}\!\!/_{(\max(iad)-m+2\sigma)})(E_f - m + 2\sigma) + 1 \Rightarrow m - 2\sigma < Ef \le \max(iad) \\ (^{0.85}\!\!/_{(\min(ied)-\max(iad))})(E_f - \max(iad)) + 0.85 \Rightarrow \max(iad) < E_f \le \min(eid) \end{cases}$$

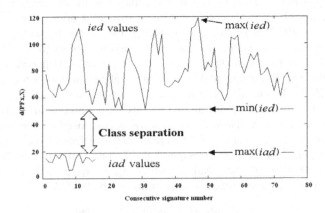

Fig. 3. Class separation as a result of the evolutionary process

putting together all the shape related features in a vector G:

$$G = [G_{cx(l)} \ G_{cy(l)} \ G_{T(l)} \ G_{s1(l)} \ G_{s2(l)}] \tag{6}$$

now, a signature receives a grade of shape GS according to:

$$GS = G * WF^T \tag{7}$$

where * denotes the product of vectors. The same process applies to generate a grade of dynamics GD for a signature using the dynamics related features. Equations 5 through 7 will be calculated by the *Grading Units*, Sect. 2, figure 1. To investigate the efficacy of the genetic algorithm and the weighting mechanism to separate classes ω_0 and ω_1, 241 signatures for one person were collected (16 genuine signatures, 220 random forgeries and 10 skilled forgeries). The process described in sections 4 through 6 was applied to the data of the signer. The **ESD** vector described at the beginning of this section was calculated for each signature and the principal component analysis PCA [11] was applied to the resulting 241 **ESD** vectors in order to determine the underlying class structure. Figure 4a shows the PCA projection; it can be seen the underlying bi-class structure, as expected. Solid diamonds correspond to the genuine signatures and hollow diamonds other 30 genuine signatures included for testing purposes. The clustering of the 16 and 30 genuine signatures reflects the fact that our procedure is able to deal with the natural intra-personal variability of the signing process of the subject. Some of the skilled forgeries are pointed. Figure 4b shows the plot of the *grade of shape* GS, Eq. 7. The high values at the left of the plot are the grades for the 16 genuine signatures; the middle zone values are the grades of the random forgeries, the grades for skilled forgeries are pointed, and *genuines 2* are the grades of the additional 30 genuine signatures.

6 NN-Driven Fuzzy Reasoning

So far, it has been established here a mathematical procedure to determine the degree of dissimilarity in the shape descriptors (**ESD**) and in dynamics descriptors (**EDD**) between a signature and its respective prototype functions; also, a numerical grade (GS and GD, respectively) has been assigned to such dissimilarities by the *grading*

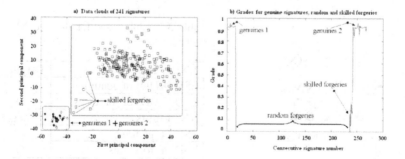

Fig. 4. a) PCA of 241 signatures processed with the proposed methodology for PF_x. b) The grades assigned to each signature with the grading units (Sect. 2).

units, thus establishing a mapping [***ESD***→GS, ***EDD***→GD], providing a perceptual measure of the similarity between a test signature and its respective patterns from the perspective of an expert examiner. Since expert examiners infer perceptually GS from ***ESD***, and GD from ***EDD***, we introduce the use of fuzzy logic to imitate them; to choose a specific fuzzy inference approach for our application, we establish the following criteria: a) to be able to learn; b) to posse proven prediction capability; both criteria are met by the Neural Network driven Fuzzy Reasoning (NND-FR), which was used to predict the chemical oxygen demand (COD) of the water in the Osaka bay [19]. NND-FR is a family of techniques in which fuzzy reasoning is implemented using neural networks [19]. Figure 5 shows a block diagram of a NND-FR system with ***ESD*** as inputs. The NND-FR use fuzzy logic of the Sugeno type. NND-FR description and implementation details can be found in [19]. The neural networks used are fully connected, trained by backpropagation. To adapt the NND-FR to our problem and infer GS (or GD) from ***ESD*** (or ***EDD***), an example of one fuzzy rule is illustrated in figure 5 for the shape-related features; the training data set of NN_{mem} of the shape-related NND-FR system is given in table 3. Again, consider a similar situation for the rhythm-related NND-FR. The error sets ***ESD*** and ***EDD*** and the grades GS and GD in table 3 are those introduced subsection 5.2. NNmem and NN_S are neural networks with three layers, in which layer 1 is comprised of 5 neurons, layer 2 of 20 and layer 3 of one single neuron. In the *verification phase* (see Fig. 1) the final decision on whether a signature is genuine or a forgery is made by a 2-inputs Mamdani system; the fuzzy logic in this stage allows to easily map the rules of combination or integration of GS and GD in a similar manner to the one of the expert examiner. The universe of discourse of the inputs goes from 0 to 1, the output range of the NND-FR systems. The universe of discourse of the output goes from 0 to 100, which will is the range of values assigned to the degree of certainty (DOC in figure 1) that a signature is genuine. The fuzzy rules base is shown in table 4. to 100, which will is the range of values assigned to the degree of certainty (DOC in figure 1) that a signature is genuine. The fuzzy rules base is shown in table 4.

7 Experimental Results and Discussion

Our database consisted of 972 genuine signatures and 1188 forgeries from 36 subjects; only 900 genuine signatures and 1000 forgeries were used in the verification

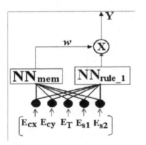

Fig. 5. NND-FR topology in the context of our application

Table 3. Training sets for the two NND-FR systems

NN_{mem} inputs	Target
ESD (EDD) of genuine	1
ESD (EDD) of forgeries	0
NN_s inputs	Target
ESD of genuine	G
ESD of forgeries	G

GENUINE SIGNATURE

FORGERIES

Fig. 6. Examples of genuine signatures and forgeries in the database

Table 4. Fuzzy rules base. Linguistic variables: VG, very good; G, good, R, regular; B, bad; VB, very bad.

		GD				
		VG	**G**	**R**	**B**	**VB**
GS	**VG**	VG	VG	G	R	B
	G	VG	G	R	R	B
	R	G	G	R	B	B
	B	R	R	R	B	VB
	VB	R	R	B	VB	VB

experiment. Figure 6 shows examples of one signature. Table 5 summarizes the parameters and results of this experiment. The genetic algorithm was run 5 times for each feature, each user, and the best performance was selected. Table 5 shows that the overall efficiency of the algorithm is very high; we will mention the main reasons we consider responsible for such a high performance, in order of relevance. The first one is that the GA gives robustness to the prototype functions against forgeries because, a) in the cost function we embedded knowledge of the class ω_1, and b) the cost function directly increases the class separability between ω_0 and ω_1, evident in figure 4a. Another reason is that, although the GA could finally arrive to a suboptimal solution, the use of NND-FR helps in avoiding classification errors due to its predictive capabilities [19]. Another important factor of success is the adaptation of the expert examiner procedure for signature verification, occurring here in two ways: number one, the overall approach depicted in figure 1, and number two, the use of fuzzy logic in both, NND-FR and the Mamdani fuzzy model, to interpret in a linguistic fashion the intermediate similarity and dissimilarity measures involved in the process. Furthermore, the expert examiner

Table 5. Settings and Results for a database of 1900 signatures. 7324 verifications.

Parameter	Value
Genuine signatures p/signer	25
Skilled Forgeries p/signer	30
Random Forgeries p/signer	220
Total amount of verification errors	10
Genetic algorithm training time p/user (average)	45
NND-FR training time p/user (average)	10
% of correct verifications	99.86%
False rejection rate (FRR)	1.15%
False acceptance rate (FAR)	0.27%
Overall Efficiency in terms of FAR and FRR	99.30%
Total training time (GA time + NND-FR time) per user: 55 seconds	
Verification time per signature < 2 seconds.	
Hardware: PC, Intel® Centrino Duo processor, 2.1GHz, 1GB RAM	
Software: Matlab, from The Mathworks®	

systematically recognizes the features of a specific signature that could be irrelevant for verification, thus making a kind of *intuitive* feature selection. In our case the *grading units* of shape and dynamics do that work, giving a personalized weighting factor to each feature for each individual. One situation in our advantage is the lack of professional forgeries in our database; however, it is worth to mention that our forgers were allowed to train during 3 weeks with the knowledge of the dynamics of their "victims". Since 2004 there exist a relatively small public database of signatures used in the *Signature Verification Contest 2004* [8], available via internet; however, because of the lack of standard methods of comparison in the research community of OSV, direct comparison of the results is unreliable. Another public database available on request [9] is MCYT proposed by Ortega-García and others, but the skilled forgeries therein are skilled only in the sense that the forgers were allowed to practice on a static image of their target, which makes it less challenging than the database we gathered in our experiment.

8 Conclusions

A high performance new approach to the problem of automatic on-line signature verification was presented. The key elements of our OSV approach are: 1) high similitude with the method used by human expert examiners, supported by the a) structure in figure 1, and b) by fuzzy logic methods for inference (NND-FR and Mamdani, Sect. 6); 2) the use of genetic algorithms with problem domain knowledge (Sect. 5) in order to make the OSV robust in the presence of random and skilled forgeries. Some contributions of this work are: 1) a method for relative feature selection (Sect. 5.2); 2) the concept of taking some already selected feature and to find the *best* representative or prototype that will minimize the misclassification error (Sect. 5, Figure 3); note that it is different of feature generation. Applications of these contributions can be found in practically any bi-class pattern recognition problem in which the functions approach is feasible. Other techniques could be used for fuzzy reasoning instead of NND-FR and, also, other optimization techniques could be utilized for the sake of time saving in the calculations, for large scale applications.

References

1. Plamondon, R., Lorette, G.: Automatic signature verification and writer identification – The state of the art. Pattern Recognition 22, 107–131 (1989)
2. Dimauro, G., Impedovo, S.: Recent Advancements in Automatic Signature Verification. In: 9th Int'l Workshop on Frontiers in Handwriting Recognition. IEEE CS, Los Alamitos (2004)
3. Yampolskiy, R.V.: Motor-Skill Based Biometrics. In: 6th Annual Security Conference 2007, pp. 33-1 – 33-12 (2007)
4. Plamondon, R.: The design of an on-line signature verification system: from theory to practice. In: Progress in Automatic Signature Verification, pp. 795–811 (1994)
5. Plamondon, R., Srihari, S.N.: On-line and off-line handwriting recognition: a comprehensive survey. IEEE Transactions on Pattern Analysis and Machine Intelligence 22, 63–84 (2000)

6. Plamondon, R., Lorette, G.: On-line signature verification: how many countries are in the race? In: International Carnahan Conference on Security Technology, pp. 183–191 (1989)
7. Nalwa, S.: Automatic online signature verification. Proceedings of the IEEE 85(2), 213–239 (1997)
8. Ketabdar, H., Richiardi, J., Drygajlo, A.: Global feature selection for on-line signature verification. In: 12th International Graphonomics Society Conference (2005)
9. Ketabdar, H., Richiardi, J., Drygajlo, A.: Global and Local Feature Selection for On-line Signature Verification. In: International Conference on Document Analysis and Recognition (2005)
10. Muralidharan, N., Wunnava, S.: Signature Verification: A Popular Biometric technology. In: Second LACCEI International Latin American and Caribbean Conference for Engineering and Technology (2004)
11. Theodoridis, S., Koutroumbas, K.: Pattern Recognition. Academic Press, California (1999)
12. Li, T.S.: Feature Selection for Classification by using a GA-Based Neural Network Approach. Journal of the Chinese Institute of Industrial Engineers 23, 55–64 (2006)
13. Zamalloa, M., Bordel, G.: Feature Selection Based on Genetic Algorithm for Speaker Recognition. In: The Speaker and Language Recognition Workshop (2006)
14. Goldberg, D.E.: Genetic Algorithms in Search, Optimization & Machine Learning. Addison-Wesley Publishing Company, Inc., Reading (1989)
15. Yang, X., Furuhashi, T.: A study on signature verification using a new approach to genetic based machine learning. In: IEEE International Conference on Systems, Man and Cybernetics 1995. Intelligent Systems for the 21st Century, vol. 5, pp. 4383–4386 (1995)
16. Yang, X., Furuhashi, T.: Constructing a High Performance Signature Verification System Using a GA Method. In: Second New Zeeland International Two-Stream Conference on Artificial Neural Networks and Expert Systems, pp. 170–173 (1995)
17. Wijesoma, W.S.: Selecting optimal personalized features for on-line signature verification using GA. In: IEEE International on Systems, Man, and Cyb., vol. 4, pp. 2740–2745 (2000)
18. Galbally, J., Fierrez, J.: Feature selection based on genetic algorithms for on-line signature verification. In: IEEE Workshop on Automatic Identification Advanced Technologies, AutoID, pp. 198-203 (2007)
19. Tsoukalas, L.H., Uhrig, R.E.: Fuzzy and Neural Approaches in Engineering. John Wiley and Sons, Inc., New York (1997)
20. Slyter, A.: Forensic Signature Examination. Thomas Publisher, U.S.A (1995)
21. ACE CAD Enterprise Co., http://www.acecad.com.tw
22. Smith, S.W.: Digital Signal Processing. In: Newness (ed.). Elsevier, Oxford (2003)
23. Oh, I.-S., Lee, J.-S.: Hybrid Genetic Algorithms for Feature Selection. IEEE Transactions on Pattern Analysis and Machine Intelligence 26(11) (2004)
24. Baluja, S.: An Empirical Comparison of Seven Iterative and Evolutionary Function Optimization Heuristics. School of Comp. Sc. Carnegie Mellon Uni. Pennsylvania (1995)
25. Pujol, O., Radeva, P., Vitria, J.: Discriminant ECOC: A Heuristic Method for Application Dependent Design of Error Correcting Output Codes. IEEE Transactions on Pattern Analysis and Machine Intelligence 28(6) (2006)

Diagnosis of Cervical Cancer Using the Median M-Type Radial Basis Function (MMRBF) Neural Network

Margarita E. Gómez-Mayorga[1], Francisco J. Gallegos-Funes[2],
José M. De-la-Rosa-Vázquez[2], Rene Cruz-Santiago[2], and Volodymyr Ponomaryov[2]

[1] National Polytechnic Institute of Mexico,
Interdisciplinary Professional Unit of Engineering and Advanced Technology
mayorgagom@hotmail.com
[2] Mechanical and Electrical Engineering Higher School
Av. IPN s/n, U.P.A.L.M. SEPI-ESIME, Edif. Z, Acceso 3, Tercer Piso,
Col. Lindavista, 07738, Mexico, D. F., Mexico
fgallegosf@ipn.mx

Abstract. The automatic analysis of Pap smear microscopic images is one of the most interesting fields in biomedical image processing. In this paper we present the capability of the Median M-Type Radial Basis Function (MMRBF) neural network in the classification of cervical cancer cells. From simulation results we observe that the MMRBF neural network has better classification capabilities in comparison with the Median RBF algorithm used as comparative.

Keywords: Median M-type Radial Basis Function neural network, Cervical cancer cell, Pap smear.

1 Introduction

Cervical cancer is the second most common cancer in women worldwide and the leading cause of cancer mortality for women in developing countries [1-3]. A Papanicolaou test, also called Pap smear or Pap test, is a medical screening method that can help prevent cervical cancer. The main purpose of the Pap smear is to detect for cell abnormalities that may occur from cervical cancer or before cancer develops. In Pap smear, sample cells are taken from the cervix and smeared onto a glass slide. These cells are stained and fixed with a preservative to keep cells from becoming airdried and distorted. The slides are then delivered to a laboratory where they are screened by a cytologist. In many developing countries where there are still inadequate numbers of cytotechnicians who can examine slides. Therefore, it is urgently necessary to develop the automated Pap smear analysis system that can help cytologists in Pap screening [4-9].

Recently, we proposed the Median M-Type Radial Basis Function (MMRBF) Neural Network for artificial data classification and mammographic image analysis [10,11]. The MMRBF neural network uses the Median M-Type (MM) estimator [12] in the scheme of radial basis function to train the neural network according with the schemes found in the references [13,14].

A. Hernández Aguirre et al. (Eds.): MICAI 2009, LNAI 5845, pp. 258–267, 2009.

In this paper, the Median M-Type Radial Basis Function (MMRBF) Neural Network is used for automatic Pap test screening process. The rest of this paper is organized as follows. Section 2 gives the information about Pap smear microscopic images. The proposed MMRBF neural network is presented in section 3. Experimental results of classification capabilities for Pap smear images by using our method and the Median RBF network used as comparative are presented in section 4. Finally, we draw our conclusions in section 5.

2 Pap Smear Microscopic Images

The Pap smear slides usually contain both of single cells and clusters of cells [1-3]. Most of cells are found with high degree of overlapping. The physical appearance of cells in an image depends how the specimen, which collected from cervix, was smeared, stained, and captured. The stained process makes cells appear in different colors. In the image acquisition process, the size of cells in an image will be large or small depends on which magnification lens used. The quality of image also depends on the resolution of a digital camera. The cervical cell microscopic images were obtained from the Pathologic Anatomy Department of 1° of October Regional Metropolitan Hospital in Mexico City by means of use of Leica DME microscopy with integrated Leica EC3 digital camera, software LAS EZ (PC), and lens Leica ∞/0.17 Hi Plan 100x/1.25 oil, the size of cell images is 2048x1536 pixels [15]. Figure 1 shows sample images which contain normal and abnormal cells.

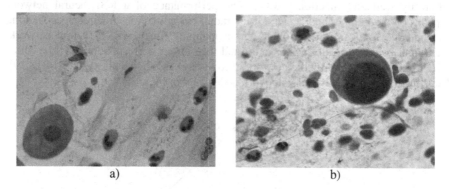

a)   b)

Fig. 1. Sample of cervical cell microscopic images, a) Normal cell, and b) Abnormal cell

The cervical cells consist of two main components. One is nucleus locates about the center of cell surrounded by the cytoplasm. Normally, nucleus shape is small and almost round. Its intensity is darker than cytoplasm. The specimens, which are taken from several areas of the cervix, most often contain cells from columnar epithelium and the squamous epithelium.

The squamous epithelium consists of 4 layers of cells: basal, parabasal, intermediate, and superficial cells. The Cervical intra-epithelial neoplasia (CIN), also known as dysplasia has the potential to become invasive cervical cancer. CIN is graded into three stages of severity, from CIN 1 (mild dysplasia) through CIN 2 (moderate dysplasia) to CIN 3 (severe dysplasia) [1-3].

In Pap smear microscopic image analysis, cervical cells can be divided into several classes categorized by cell appearance, especially in cell nucleus. Since a cell nucleus can present the significant changes when the cell is affected by a disease, the identification and quantification of these changes contribute in the discrimination of normal and abnormal cells in Pap smear images (see Figure 2).

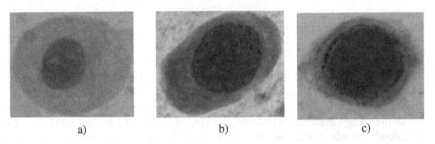

a)   b)   c)

Fig. 2. Basal cells, a) normal cell, b) moderate dysplasia (CIN 2), and c) severe dysplasia (CIN 3)

3   Median M-Type Radial Basis Function Neural Network

The Radial Basis Function (RBF) neural networks are capable of approximating arbitrarily well any function [16,17]. The performance of a RBF neural network depends on the number and centers of the radial basis functions, their shapes, and the method used for learning the input–output mapping. In the (RBF) neural network, each input is assigned to a vector entry and the outputs correspond either to a set of functions to be modeled by the network or to several associated classes [13,14,18,19]. The structure of the RBF neural network is depicted in Figure 3.

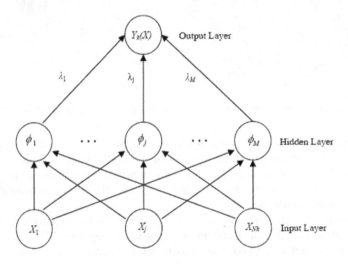

Fig. 3. Structure of Radial Basis Function Neural Network

Each of N_k components of the input vector X feeds forward to M basis functions whose outputs are linearly combined with weights $\left\{\lambda_j\right\}_{j=1}^{M}$ into the network output $Y_k(\mathbf{X})$ (see Figure 3). The output layer implements a weighted sum of hidden-unit outputs [13,14],

$$Y_k(\mathbf{X}) = \sum_{j=1}^{L} \lambda_{jk}\phi_j(\mathbf{X}) \tag{1}$$

where L is the number of hidden units, M is the number of outputs with $k=1,...,M$, the weights λ_{kj} show the distribution of the hidden unit j for modeling the output k, and $\phi_j(\mathbf{X})$ is the activation function.

We tested with different activation functions and we chose the Gaussian function [16,17] due this function provided the best results in the proposed neural network,

$$\phi_j(\mathbf{X}) = \exp\left(-\frac{\|\mathbf{X}-\mu_j\|^2}{2\sigma_j^2}\right) \tag{2}$$

where $\mathbf{X}$ is the input vector with elements x_i, μ_j is the vector determining the centre of basis function ϕ_j and σ is the standard deviation.

The input feature vector $\mathbf{X}$ is classified in k different clusters. The clustering k-means algorithm can be used to estimate the parameters of the RBF neural networks [16,17]. A new vector x is assigned to the cluster k whose centroid μ_k is the closest one to the vector. The centroid vector is updated according to,

$$\mu_k = \mu_k + \frac{1}{N_k}(\mathbf{x}-\mu_k) \tag{3}$$

where N_k is the number of vectors already assigned to the k-cluster.

The Median M-Type (MM) estimator is used as statistic estimation in the proposed Median M-Type Radial Basis Fuction (MMRBF) neural network. The non-iterative MM-estimator used as robust statistics estimate of a cluster center is given by [10-12],

$$\mu_k = \text{med}\{\mathbf{X}\tilde{\psi}(\mathbf{X}-\theta)\} \tag{4}$$

where $\mathbf{X}$ is the input data sample, $\tilde{\psi}$ is the normalized influence function ψ: $\psi(\mathbf{X}) = \mathbf{X}\tilde{\psi}(\mathbf{X})$, $\theta = \text{med}\{X_k\}$ is the initial estimate, med{·} is the median of data, $k=1, 2,...,N_k$, and $\psi(X)$ is the simple cut (skipped mean) influence function given by [12],

$$\psi_{cut(r)}(X) = \begin{cases} X, & |X| \leq r \\ 0, & \text{otherwise} \end{cases} \tag{5}$$

where X is a data sample and r is a real constant that depends of the data to process.

4   Experimental Results

The described MMRBF neural network has been evaluated, and their performance has been compared with the Median RBF neural network [13]. The Median RBF network uses the marginal median estimator based on the Marginal Median Learning Vector Quantization (MMLVQ) algorithm. It is used as statistics estimate of a cluster center given by,

$$\mu_k = \mathrm{med}\{\mathbf{X}_0, \mathbf{X}_2, ..., \mathbf{X}_{n-1}\} \tag{6}$$

where $\mathbf{X}_{n-1}$ is the last pattern assigned to k-th neuron.

To train the networks for getting the appropriate pdf's parameters was used 78 cervical cell images: 25 normal, 3 CIN 1 (mild dysplasia), 25 CIN 2 (moderate dysplasia), 20 CIN 3 (severe dysplasia), and 5 with CIS (carcinoma-in-situ) with microinvasive disease [15,16]. We do not consider the classification of squamous epithelium between the basal, parabasal, intermediate, and superficial cells. The objective of the experiment is to classify between different types of CIN.

The classification process is shown in Figure 4. Having the Pap smear microscopic image, we proceed to segment it in 2 main regions of interest [6,7]:

- Region 1 is constituted of a nucleus,
- Region 2 is constituted of a cytoplasm.

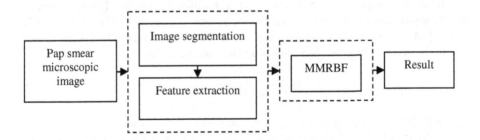

Fig. 4. Block diagram of proposed MMRBF neural network

Morphology is the study of the shape and form of objects. Morphological image analysis can be used to perform: object extraction, image filtering operations, such as removal of small objects or noise from an image, image segmentation operations, such as separating connected objects, and measurement operations, such as texture analysis and shape description [20,21].

The dilation and erosion are two fundamental morphological operations. Dilation adds pixels to the boundaries of objects in an image, while erosion removes pixels on object boundaries. The number of pixels added or removed from the objects in an image A depends on the size and shape of the structuring element B used to process the image. The dilation and erosion are defined as,

$$Dilation = A \oplus B \tag{7}$$

$$Erosion = \boxed{A \ominus B} \qquad (8)$$

The dilation and erosion are often used in combination to implement image processing operations. The morphological opening and closing of an image are defined as follows,

$$Opening = (\boxed{A \ominus B}) \oplus B \qquad (9)$$

$$Closing = (A \oplus B)\boxed{\ominus B} \qquad (10)$$

The steps used in the stage of image segmentation (see Figure 4) are the following: a) read the image; b) detect entire cell, by use a binary gradient mask to detect the edges of image; c) dilate the image, to eliminate the gaps in the lines surrounding the cell in the gradient mask image, d) fill interior gaps, there are still holes in the interior of the cell in the dilated gradient mask, these are filled; e) remove connected objects on border, it suppresses structures that are lighter than their surroundings and that are connected to the image border; and f) smoothen the cell, in order to make the segmented cell look natural, we smoothen the object by eroding the image. Figure 5 presents the final image segmentation of a cervical cell where is displayed the cytoplasm and nucleus segmented from the original cell image.

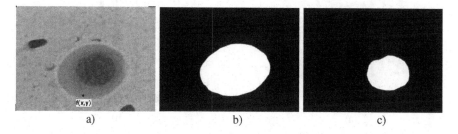

a)   b)   c)

Fig. 5. Image segmentation of a cervical cell, a) original cell, b) cytoplasm, c) nucleus

The feature extraction was applied after the segmentation stage. We obtain 9 numerical data or characteristics, which are the following [21],

- Nucleus and Cytoplasm area,

$$area = \sum_{i}^{N} \sum_{j}^{M} seg(i, j) \qquad (11)$$

where $seg(i,j)$ are pixels of segmented object.

- Nucleus and Cytoplasm perimeter. The perimeter is the sum of the pixels that form the contour of object.
- Nucleus and Cytoplasm Circularity,

$$Circularit\,y = \frac{4\pi \cdot area}{perimeter^{\,2}} \tag{12}$$

- Nucleus-Cytoplasm relation,

$$Nucleus - Cytoplasm \quad relation = \frac{Cytoplasm \quad area}{Nucleus \quad area} \tag{13}$$

- Maximum and Minimum Nucleus brightness. The image should be change to gray scale to obtain the maximum and minimum gray value by means of use of its histogram. The histogram of gray scale image is given by,

$$P(r_k) = \frac{n_k}{n} \tag{14}$$

where rk is the k-esimo gray scale level, nk is the total pixels in the image with gray scale level k, and n is the total number of pixels into the image. Figure 6 shows a gray scale cell image with its histogram. The histogram presents the distribution of data values or gray scale pixels onto image by the function P(rk).

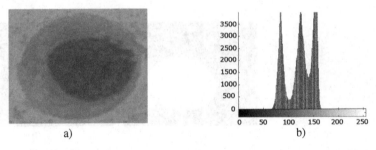

a)   b)

Fig. 6. Maximum and minimum nucleus brightness, a) gray scale cell image, b) histogram

Table 1 shows numerical data obtained in the stage of feature extraction in terms of cytoplasm area (CA), cytoplasm circularity (CC), nucleus area (NA), nucleus circularity (NC), and nucleus-cytoplasm relation (NCR), in the case of normal, CIN 2, and CIS images.

For the implementation of neural networks, in the test stage we use 20 images, these images are different form the images used in the training stage.

Table 1. Experimental results obtained in the stage of feature extraction

Cervical cell images	Numerical data				
	CA	CC	NA	NC	NCR
Normal	680029.2	0.7300	17795.2	0.852	38.9761
CIN 2	194935.1	0.7225	59347.62	0.8400	3.3854
CIS	178631.9	0.2487	40367.25	0.698	4.7378

Table 2 presents the performance results in terms of efficiency and error for Median RBF and proposed MMRBF neural networks. From this Table, we observe that the proposed MMRBF neural network outperforms the Median RBF neural network in terms of efficiency for the classification of cervical cell images.

Table 2. Results obtained with different RBF algorithms in test stage

Neural networks	Performance	Cervical cell images				
		Normal	CIN 1	CIN 2	CIN 3	CIS
Median RBF	Efficiency	100%	100%	75%	72%	100%
	Error	0%	0%	25%	28%	0%
MMRBF	Efficiency	100%	100%	80%	77%	100%
	Error	0%	0%	20%	23%	0%

To evaluate the performance of the neural networks in terms of medical purposes, we computed the sensitivity and specificity [22]. The sensitivity is the probability that a medical test delivers a positive result when a group of patients with certain illness is under study, and the specificity is the probability that a medical test delivers a negative result when a group of patients under study do not have certain illness [22],

$$Sn = TP/(TP + FN) \qquad (15)$$

$$Sp = TN/(TN + FP) \qquad (16)$$

where Sn is sensitivity, TP is the number of true positive that are correct, FN is the number of false negatives, that is, the negative results that are not correct, Sp is specificity, TN is the number of negative results that are correct and FP is the number of false positives, that is, the positive results that are not correct.

Table 3 shows the sensitivity and specificity values obtained for Median RBF and proposed MMRBF neural networks. We can observe that the sensitivity of the proposed RMRBF outperforms the Median RFB network used as comparative. In the case of specificity the proposed network has similar results in comparison with the Median RBF neural network.

From Table 2 one can see that the proposed MMRBF neural network has 20 and 23% of error in the detection of CIN 2 or 3, respectively. Table 3 presents an 83% of sensitivity and 100% of specificity for proposed network. A reason of these results is given in the failure to detect CIN 3.

A clinical definition of a false negative smear is difficult to give. The failure to detect CIN 3 is important whereas the failure to detect a low grade abnormality, many of which regress spontaneously or progress slowly is less significant provided the

Table 3. Sensitivity and specificity results for different Neural Networks

Neural Networks	Sensitivity	Specificity
Median RBF	78%	100%
Median M-type RBF	83%	100%

woman remains in the screening program. However, even with CIN 3 lesions it is the degree of under calling which is significant, i.e. if it is reported as CIN 2 or 1 [1-3,6,7].

Finally, the proposed method classifies between five classes: normal, CIN 1, CIN 2, CIN 3, and CIS cell images. The proposed method can provide high efficiency, if the classification is between 3 classes: Normal, CINs, and CIS cell images.

5 Conclusions

We present a diagnosis of cervical cancer using the Median M-Type Radial Basis Function (MMRBF) Neural Network and Pap smear (cervical cell images). The proposed neural network uses the MM-estimator in the scheme of radial basis function to train the proposed network. The results obtained with the use of the proposed MMRBF are better than the results obtained with the Median RBF algorithm used as comparative.

Acknowledgements

The authors thank the National Polytechnic Institute of Mexico and the clinical research staff (G. Vazquez-Sanchez, J. L. Bribiesca-Paramo) from the Pathologic Anatomy Department of 1° of October Regional Metropolitan Hospital for their support.

References

1. International Agency for Research on Cancer (IARC), http://www.iarc.fr/
2. Wright, T.C., Kurman, R.J., Ferenczy, A.: Cervical intraepithelial neoplasia, Pathology of the Female Genital Tract, A, Blaustein. Springer, New York (1994)
3. Kurman, R.J., Henson, D.E., Herbst, A.L., Noller, K.L., Schiffman, M.H.: Interim guidelines for management of abnormal cervical cytology. J. Am. Med. Assoc. 271, 11866–11869 (1994)
4. Mirabal, Y.N., Chang, S.K., Neely Atkinson, E., Malpica, A., Follen, M., Richards-Kortum, R.: Reflectance spectroscopy for in vivo detection of cervical precancer. Journal of Biomedical Optics. 7(4), 587–594 (2002)
5. Tumer, K., Ramanujam, N., Ghosh, J., Richards-Kortum, R.: Ensembles of radial basis function networks for spectroscopic detection of cervical precancer. IEEE Trans. Biomed. Eng. 45, 953–961 (1998)
6. Plissiti, M.E., Charchanti, A., Krikoni, O., Fotiadis, D.I.: Automated segmentation of cell nuclei in PAP smear images. In: Proc. IEEE International Special Topic Conference on Information Technology in Biomedicine, Greece, October 26-28 (2006)
7. Jantzen, J., Dounias, G.: Analysis of Pap-Smear Image Data. In: Proc. Nature-Inspired Smart Information Systems (2nd Annual Symposium), NISIS (2006)
8. Mitra, P., Mitra, S., Pal, S.K.: Staging of cervical cancer with soft computing. IEEE Trans. Biomedical Engineering 47(7), 934–940 (2000)
9. Athiar Ramli, D., Fauzan Kadmin, A., Yusoff Mashor, M., Ashidi, N., Isa, M.: Diagnosis of cervical cancer using hybrid multilayered perceptron (HMLP) network. In: Negoita, M.G., Howlett, R.J., Jain, L.C. (eds.) KES 2004. LNCS (LNAI), vol. 3213, pp. 591–598. Springer, Heidelberg (2004)

10. Moreno-Escobar, J.A., Gallegos-Funes, F.J., Ponomaryov, V., De-la-Rosa-Vazquez, J.M.: Radial basis function neural network based on order statistics. In: Gelbukh, A., Kuri Morales, Á.F. (eds.) MICAI 2007. LNCS (LNAI), vol. 4827, pp. 150–160. Springer, Heidelberg (2007)
11. Moreno-Escobar, J.A., Gallegos-Funes, F.J., Ponomaryov, V.I.: Median M-type radial basis function neural network. In: Rueda, L., Mery, D., Kittler, J. (eds.) CIARP 2007. LNCS, vol. 4756, pp. 525–533. Springer, Heidelberg (2007)
12. Gallegos, F., Ponomaryov, V.: Real-time image filtering scheme based on robust estimators in presence of impulsive noise. Real Time Imaging. 8(2), 78–90 (2004)
13. Bors, A.G., Pitas, I.: Median radial basis function neural network. IEEE Trans. Neural Networks 7(6), 1351–1364 (1996)
14. Bors, A.G., Pitas, I.: Object classification in 3-D images using alpha-trimmed mean radial basis function network. IEEE Trans. Image Process. 8(12), 1744–1756 (1999)
15. Leica Microsystems, http://www.leica-microsystems.com/
16. Haykin, S.: Neural Networks, a Comprehensive Foundation. Prentice Hall, Upper Saddle River (1994)
17. Rojas, R.: Neural Networks: A Systematic Introduction. Springer, Berlin (1996)
18. Karayiannis, N.B., Weiqun Mi, G.: Growing radial basis neural networks: merging supervised and unsupervised learning with network growth techniques. IEEE Trans. Neural Networks 8(6), 1492–1506 (1997)
19. Karayiannis, N.B., Randolph-Gips, M.M.: On the construction and training of reformulated radial basis function neural networks. IEEE Trans. Neural Networks 14(4), 835–846 (2003)
20. Ritter, G.: Handbook of Computer Vision Algorithms in Image Algebra. CRC Press, Boca Raton (2001)
21. Myler, H.R., Weeks, A.R.: The Pocket Handbook of Image Processing Algorithms in C. Prentice Hall, Englewood Cliffs (1993)
22. http://www.cmh.edu/stats/definitions/

Quasi-invariant Illumination Recognition for Appearance-Based Models, Taking Advantage of Manifold Information and Non-uniform Sampling

Diego González and Luis Carlos Altamirano

Benemérita Universidad Autónoma de Publa-FCC, Boulevard 14 sur y Avenida San Claudio, Ciudad Universitaria, Col. San Manuel, Puebla, Pue., 72570, México

Abstract. The appearance of an object can be greatly affected by the changes in illumination conditions in the scene where that object is seen. In this paper, we propose a novel method for building appearance-based models that are tolerant to lighting variations. This new approach is based on a Non-Uniform Sampling technique, so that we use a very small number of images for generating those models, and this reduces the computing time and storage space required for the modeling stage, with respect to Uniform Sampling techniques. We have tested the proposed algorithm with a comprehensive set of objects with different appearance, and the high recognition rate obtained in these experiments shows the effectiveness of the proposed method. Also, the necessary time to compute the associated manifold in eigenspace is dramatically reduced.

1 Introduction

The appearance of an object in an image can be defined as the combined effect of its geometrical and surface properties as well as its pose and the illumination conditions in the scene [1]. Since the last two properties mentioned are independent of the object characteristics, the variations due to changes in them, have been especially troubling for the appearance-based modeling and recognition systems. In this paper, we deal specifically with changes in lighting conditions.

Traditionally, the methods that take into account the variability in appearance due to lighting, are based in one of two approaches: (a) measure some property in the image of the object that is invariant or, at least, insensitive to such changes; or (b) Generate a model of the object that can predict those variations [2].

In the second of these approaches, one of the most important advances is the so called *Illumination Cone* technique. Belhumeur and Kriegman have shown that all the images of any object seen under all possible lighting conditions, form a convex cone in R^n, where n is the number of pixels in those images [2]. Besides, it was demonstrated (also in [2]), that an approximation to this cone can be obtained using few images, (where each one is produced by a single point light source), in the case of objects with *Lambertian* surfaces [3]. Georghiades et al. [4] and later Chin and Suter [5], have shown that the illumination cone method

A. Hernández Aguirre et al. (Eds.): MICAI 2009, LNAI 5845, pp. 268–279, 2009.

can be useful for face recognition, since faces have a nearly Lambertian surface. However, on objects with any other type of surface, it is not easy (and sometimes impossible), to determine the number of images that would be required to approximate the cone [2]. Also, if the images used for modeling the object have multiple unknown light sources, it is impossible to determine the illumination cone [6]. Hence, this method can not be used for any class of objects.

On the other hand, there has been many studies related to the first approach ([7,8,9] and [10], for example), and, from these works, it was shown by Chen et al., that even for objects with Lambertian surfaces it is not possible to find properties that can be considered as illumination *invariant* [11]. So, other recent works, like the one presented by Bischof et al. in [12], are focused in the creation of models based on measures such as the image gradient, which can be considered as *insensitive* to illumination changes. In this work, Bischof et al. have "expanded" a parametric eigenspace representation, like the one proposed by Nayar et al. in [1], by using a bank of filters based on the gradient of the eigenimages. The experiments performed on this approach, have reported a recognition rate of 90.6% [12]. This is a very important result, especially because these tests take into account the effects of occlusions, cluttered background and cast shadows [12]. However, this technique is compared only with standard eigenspace methods, (based in works like [1]), and not with other techniques designed specifically to handle appearance variations due to illumination changes. Besides, most of the objects used in the experiments reported in [12] seem to have similar characteristics, which might affect the generalization of the method.

Therefore, we take a different approach for obtaining a very general and effective method for generating appearance-based models tolerant to illumination variations and that can be useful even in the case of extreme lighting, and for a wide range of objects, (including some with specular surfaces, for which some other methods have reported difficulties and must deal with them like a special case [13]).

Since appearance-based methods require a set of digital images of the objects to be modeled, it is important to reduce the number of images required to generate the model of an object, specially, if we use the eigenspaces approach. This can be achieved through a *Non-Uniform Sampling (NUS)* technique [15], which is the base of the method that we will present in this paper. NUS has shown to be efficient to reduce time and space requirements to determine the images strictly necessary to build an object model [15]. Therefore, in the next section, we will summarize the basis of the NUS method. In Sect. 3, we will give the details of the proposed method, that allows generating an extended model of the object, which can be used to recognize that object under many different lighting conditions, using for this, a small number of images in the modeling phase. The high recognition rate obtained in the conducted experiments, that will be described in Sect. 4, will probe the effectiveness of the novel technique. Finally in Sect. 5, we will discuss the results of these tests and give our conclusions.

2   Non-uniform Sampling

We will summarize here the functioning of the NUS technique, which serves
as the main basis for the novel method proposed in this paper. NUS generates
an approximation, with a given precision, for an appearance-based model of an
object, using only the strictly necessary images. To do this, the real appearance
of an object is approximated through a *linear interpolation* [15], within a certain
error (or precision factor). Figure 1 shows the sequence of steps followed by the
algorithm in order to approximate some given appearance-model (shown as a
thick curve), within a given error value, which we will call ε. (The value assigned
to ε can be calculated through some of the strategies that have been previously
shown in [15]).

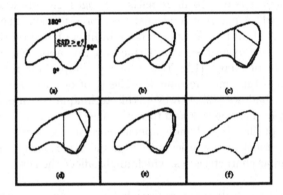

Fig. 1. Schematic view of the functioning of the NUS technique, for the interval of
images between 0° and 180° of an object model. The real model would really reside in
a multi-dimensional space, but it is shown here in 2D for illustrative purposes.

The method begins by splitting the model into two even intervals: the first one,
between 0° and 180° and the second one, between 180° and 360°. The estimation
for the first interval will be calculated first. For this, the images corresponding to
the limits of the interval, (0° and 180°), are required. These two images are used
to calculate a linear interpolation (an approximation) to the image in the middle
of the interval (90°). Then, the interpolated image is compared to the real image
by means of the *Sum of Square Differences (SSD)* [1,14,15] between them, as it
is shown in Fig 1a. If this difference is less or equal than the established error ε,
then it would be possible to estimate the real model for this first interval using only
the two initial images, (the ones for 0° and 180°). If, on the other hand, the SSD
between the interpolation and the real image is greater than the value assigned to ε,
the real image (for 90°) would be set as a necessary image to obtain an appropriate
approximation (Fig. 1b). Then, the interval is divided into two parts: the first one
for the images between 0° and 90° and the other one, between 90° and 180°, and so,
the same process can be executed again for analyzing every new range generated,
until the approximation is proper for the entire initial interval (between 0° and
180°), according to the value of precision ε (Fig. 1c - 1e).

Finally, this algorithm is repeated for the second part of the model: between 180° and 360°, so that the entire real appearance-based model is approximated within an error given by ε, (as shown in Fig. 1f), using only the essential images.

In the next section, we will show that the NUS can be useful, not just for reducing the number of images required to obtain a model, but also for generating appearance-based models that can be tolerant to lighting variations.

3   The Proposed Method

The new approach, as we have mentioned, is based on the NUS algorithm. However, as we will show next, there are several differences between the traditional NUS, (as described in Sect. 2), and our novel method, since the last one, employs information obtained from a base manifold together with object measurements in order to generate an extended model that is tolerant to lighting variations. Therefore, the new technique can be divided into two stages:

1. Using the NUS technique for building a *base model* of the object that allows us to obtain some useful information about the characteristics and properties of the object.
2. From the base model and the appearance measurements just obtained from the object, generate other models, (with different illumination conditions). This procedure will result in an *extended model*, that should enable us to perform an effective recognition, under all the lighting environments that have been modeled, using few images.

3.1   Generation of the Base Model

In this stage of the method, the NUS technique will be used to generate a base model for the object, under some given lighting condition. For this, the user should set the value of the error threshold ε. Also, it is necessary to indicate the separation angle (typically, 5° or 10°), between the images that will be used for generating the model, that is: under some given conditions, at most 72 images, (one every 5°), or 36 images, (one every 10° are captured to obtain the object model). This angle will be employed to calculate approximations, by means of a linear interpolation, for the images that are in the middle of those ones that are set by the NUS as the strictly necessary ones for generating the manifold.

The next step after generating the base manifold, will be to find certain useful information about it. Specifically, a new procedure has been designed to determine if the object is or not *symmetrical*. In this case, however, the *central-symmetry* property of an object will refer not just to the geometrical characteristics of such object, but also to the properties of it surface. With the aim of determining if an object has or not central-symmetry, all the images that would form the complete model of the object, (the ones determined by the NUS, along with the interpolated images calculated afterwards), should be divided, once again, into two even intervals. Then, it will be possible to determine whether the images that are in opposite angles of each of these two ranges or intervals have or not nearly the same "appearance". That is, (assuming a separation angle of 10°), we determine if the image for 0° is approximately the same as for 180° and if the image for 10° is approximately the same as for 190°, and so on for all the pair of images in each of the two ranges.

So, for determining if an object is symmetrical, we first seek to calculate the measure of the *total energy (TE)* for the images. This measure is defined as the sum of the intensity values of the pixels that form the image. We then compare the value of TE for each pair of images in opposite angles of the manifold. The values of the TE for all (36) images of the base model of an object with central symmetry, can be analyzed trough the plot shown in Fig. 2.

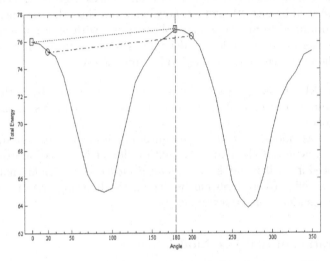

Fig. 2. Plot of the values of total energy vs. angle for the images of an object with central symmetry. (The images have been normalized to make the values smaller and easier to represent).

As can be seen, the values of TE for the images of opposite angles are an indication about whether an object is symmetrical or not, (considering a small tolerance factor in the comparison of the values, which we will call δ). However, the measure of the TE might not be sufficient in all cases, and accordingly, (for avoiding some possible errors in this analysis), we use not just the value of the TE, but also, we calculate the distance (SSD) between the compared images. In the Fig. 3 we present a plot with the values of the SSD for each pair of the described images, for an object whit central symmetry. In all cases, the difference or distance between the pair of images is very small (less than 0.05).

So, the procedure for determining if an object is or not symmetrical can be resumed in this way: let i_j, (where j is an angle between 0 and 180 degrees), be an image in the first interval of the base model of such object and i_k the image in the second interval that is in the opposite angle to j, we say that the object has central-symmetry if, for each pair of images i_j and i_k, two relations are fulfilled. The first one:

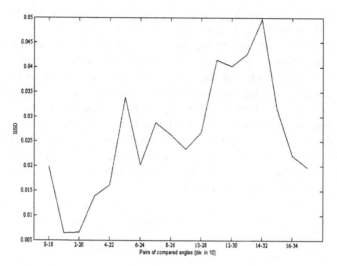

Fig. 3. Plot of the SSD values for the pair of images of opposite angles for an object with central symmetry. (Again, the images were normalized before calculating the SSD. The numbers presented in the y-axis correspond to the angles divided into 10).

$$E(i_j) - \delta \leq E(i_k) \leq E(i_j) + \delta. \tag{1}$$

Where $E(.)$ represents the value of the TE for the given image and δ is the value of tolerance for the comparison; and the second one:

$$SSD(i_j, i_k) \leq \varepsilon. \tag{2}$$

Where $SSD(.)$ is the operation that calculates the value of the SSD for the two given images and ε is a value of error or precision that can be the same as the one used by the NUS method.

If (1) and (2) are met for each one of the i_j and i_k images, (the object has central-symmetry), then we can say that the following is true for any model, (under any illumination), of the object:

$$i_k \approx i_j. \tag{3}$$

As will be seen in the next stage of the proposed technique, the information about whether or not the object has central-symmetry, will be useful for the generation of other manifolds of the object.

3.2 Generation of the Extended Model

The second step of the proposed method, allows to approximate several other manifolds for which the lighting conditions are different from the one used in the base model previously obtained, and are also different from each other. For achieving this, each new model is generated one at a time, so that the process can be repeated for building as many models as necessary.

So, for generating a new manifold, this one is divided into two even intervals, and then, the first image with the new illumination conditions, (the one for 0°) is required. Next, if (3) is met for the base model, (so that the object has central-symmetry), the image for 180° of the new lighting can be approximated by using the image for 0°, but if, otherwise, the object does no fulfill (3), the real image for 180° will also be required. Whether the object is symmetrical or not, the two images (for 0° and 180°) are used for executing the NUS technique for this new model. However, when the NUS requires a new image for the generation of this new manifold, such image could be approximated by using the information of the base model, as it is shown in Fig. 4.

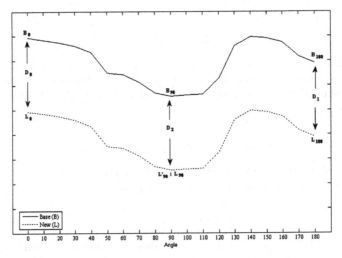

Fig. 4. Schematic view of the way in which the first interval (between 0° and 180°) of a new model (called L), is calculated from the base model (B). The models are represented here in 2D for illustrative purposes only.

As can be seen in Fig. 4, the next step of the procedure is to obtain, from the images for 0° and 180° of each model, an image that is the difference between the corresponding positions or angles. This is, let L_0, B_0 be the images for 0° and L_{180} and B_{180} the images for 180° for the new model L and the base model B, respectively, then, the described differences are calculated as follows:

$$D_0 = B_0 - L_0. \tag{4}$$

$$D_1 = B_{180} - L_{180}. \tag{5}$$

From (4) and (5) it is possible to calculate the average of the differences:

$$\bar{D} = \frac{D_0 + D_1}{2}. \tag{6}$$

This average is used to calculate an approximation for the image that is in the middle of the first interval of the manifold L, (the image for 90°), in the following way:

$$L'_{90} = B_{90} + \bar{D}. \qquad (7)$$

Next, this approximation is compared, by means of the SSD, to the linear interpolation between the images L_0 and L_{180}, (that has been previously calculated), and that we will call Li_{90}. This is:

$$SSD(Li_{90}, L'_{90}) \leq \varepsilon. \qquad (8)$$

Where ε is, once again, the error threshold previously defined. If (8) is met, then we can take the approximation L'_{90} as a valid image for generating the model, instead of requiring the real image for 90° of the L manifold, (the image L_{90}). But, if (8) is not fulfilled, the NUS will continue in the "traditional" manner, (as explained in Sect. 2), so that the real image for 90° have to be used for generating the first interval of the L model. In this case, however, the proposed method will be able to calculate a new difference: $D_3 = B_{90} - L_{90}$ and then, it will be possible to update the average of the differences $\bar{D}$, and so, we can calculate an approximation for the image between 0° and 90° in the same way as in (7), using this updated average $\bar{D}$. This same process is repeated until it is possible to generate an appropriate approximation for the first interval of the new model L.

For generating the second interval (between 180 and 360 degrees) of the manifold L, the information about whether or not the object is symmetrical becomes useful again, since, if this is true, (the object does have central-symmetry), then, according to (3), it will be possible to approximate this entire second interval using the images already obtained for the first one. However, if the object is not symmetrical, the process can be performed again in the same manner described above, for building the second part of the new model.

To probe whether the new method reduces or not the number of real images essential for the generation of the appearance-based models with respect to the original NUS technique and, more importantly, for testing the effectiveness of the defined procedures in the recognition, we have performed experiments with a comprehensive set of objects, as will be described in the next section.

4   Experiments

We have designed an experiment with 25 objects, 15 of which are used in this paper, and with 6 different illumination conditions for each one: the lighting for the base manifold and five others for extending the model. In Fig. 5 we present these 15 objects. In all cases, we have used a single point light source and some of the positions used for that source, cause extreme illumination conditions, which have allowed us to test the recognition in such environments.

Also, it is important to notice that objects shown in Fig. 5 have very different appearances. This is helpful to probe that the algorithm presented in this paper

Fig. 5. Set of 15 objects used in the experiments. For the rest of the document, objects will be numbered from left to right and top to bottom.

can be applied for objects of several different kinds. We have developed a Matlab implementation of this novel method for generating the models for the 15 objects and the 6 lighting conditions, and we established a separation of 5° in all cases and, for increasing the accuracy of the models for the later recognition process, we have obtained not only the strictly necessary real images but we have also calculated, (by means of the linear interpolation), an approximation for the images of all other positions, for completing, therefore, a total of 72 images for each single manifold. Also, we have used a value for the error threshold ε that would let us to generate a precise model, even if this implied, in some cases, a slight increase in the number of real images needed for the creation of the model. With these conditions, we performed a comparison between the new approach presented in this paper and the original NUS, with respect to the number of real images strictly required for building the models of the objects in this experiment. Table 1 summarizes the results of this comparison and presents, in the final row, the value of the error threshold ε used in both algorithms, (the range of possible values for ε is $0 < \varepsilon \leq 1$).

Table 1 shows that the proposed method allows to reduce the number of required images in more than 33% with respect to the original NUS, which demonstrates that, by means of this novel technique, it is possible to reduce the storage

Table 1. Comparison between the original *NUS* and *new* method for the number of required images for the generation of the extended model, (with 6 illumination conditions) for each one of the 15 objects presented in Fig. 5. (objs. 1 and 2 have the central-symmetry property).

Obj.	1*	2*	3	4	5	6	7	8	9	10	11	12	13	14	15	*Total*
imgs.: NUS	24	68	52	182	134	146	138	206	152	192	62	70	174	194	118	*1912*
imgs.: new	9	18	27	112	96	96	105	165	112	126	36	41	120	122	83	*1268*
ε	0.07	0.08	0.05	0.05	0.11	0.09	0.1	0.15	0.09	0.17	0.08	0.05	0.25	0.2	0.24	

space and the time required to calculate the corresponding eigenspace, (due to the complexity of the algorithms for obtaining eigenspaces [1]). In particular, for the experimental models previously described, we have created a *parametric 20-dimmension eigenspace representation* ([1]) in a 1.6 Ghz. AMD Turion 64 X2 based computer, using the strictly required images determined by the novel technique presented here and, also, the ones determined by the traditional NUS, and we have compared the computing time required in each case. Results are shown in Table 2.

Table 2. Comparison between the original *NUS* and the *new* technique for the time required to obtain a *20-D eigenspace* for the 15 objects presented in Fig. 5 and the 6 illumination conditions used

	Time (in secs).
NUS	165.95
new	45.46

For testing the recognition rate for the defined method, we have captured 36 real images (one every 10°), for each of the 6 illumination conditions and the 15 objects previously modeled. Therefore, a total set of 3240 test images was used. We assessed the recognition of the object per se and also, the lighting condition in each one of the test images. The results of the recognition test are shown in Table 3.

Table 3 shows the high efficiency of the new method for the recognition, achieving a total mean recognition rate of 99.41% and also, a high rate in the

Table 3. Recognition rates obtained in the experiment. In the 2nd. column, the percentage of success obtained in object recognition. In the 3rd. column, the rate of lighting condition identification.

Obj.	Recogn. %	Lig. cond. ident. %
1	100	94.91
2	97.69	70.37
3	100	65.28
4	100	80.56
5	93.98	64.81
6	100	97.22
7	99.54	76.85
8	100	94.44
9	100	98.15
10	100	84.72
11	100	59.26
12	100	88.43
13	100	75
14	100	70.37
15	100	81.48
Total	*99.41*	*80.12*

identification of the illumination conditions in the scene (80.12%), which confirms the validity of the proposed approach. Notice that it is possible to increase the value of the error threshold ε in the modeling stage to reduce the number of the images used and, since the recognition rate for the proposed technique is very high, the probability that the objects can still be recognized is equally very high.

5   Conclusions and Future Work

We have showed that the proposed method can be used for building appearance-based models that are tolerant to illumination changes, for any kind of objects and even for extreme lighting conditions. With this novel technique, the models can be generated using a small number of images, (even less than with pure NUS), which can be very useful for the reduction of the computational complexity when the proposed method is used together with eigenspaces techniques, (as shown in Table 2). Besides, for objects that have the central-symmetry property, (such as the objects num. 1 and num. 2 in Fig. 6), the number of images needed for building the model can be reduced even more.

The high recognition rate that we have obtained through the experiments described in Sect. 4: 99.41%, shows that the proposed algorithm is very effective and can be used for the recognition of objects that have several different kinds of appearances. We will explore in future works if, by obtaining more information about the base model generated in the first stage of the technique and also, by adding new strategies for approximating other models that have different lighting conditions, (the second stage of the method), it is possible to reduce even more the number of images strictly necessary for building the models, and, besides, if it is possible to predict the appearance of the object under some completely new illumination condition, (not-seen in the modeling process), all this, preserving the recognition rate.

We also wish to extend our tests for a larger number of objects and illumination conditions, taking into account the effects of occlusions and cluttered background. Besides, we look forward to compare the proposed method with some other techniques designed to deal with the problem of illumination changes, like the one proposed in [12], for instance. We will also explore if this novel technique, probably with some modifications, could be used to deal with pose instead of illumination or a mixture of both.

Acknowledgments. Diego González was partially supported by CONACYT (México) under grant No. 226209 and by the VIEP of the Benemérita Universidad Autónoma de Puebla.

References

1. Nayar, S.K., Murase, H., Nene, S.A.: Parametric Appearance Representation. In: Nayar, S.K., Poggio, T. (eds.) Early Visual Learning, pp. 131–160. Oxford University Press, Oxford (1996)

2. Belhumeur, P.N., Kriegman, D.J.: What is the Set of Images of an Object under All Possible Illumination Conditions? International Journal of Computer Vision 28(3), 245–260 (1998)
3. Forsyth, D., Ponce, J.: Computer Vision: a Modern Approach. Prentice Hall, Englewood Cliffs (2003)
4. Georghiades, A., Kriegman, D.J., Belhumeur, P.N.: Illumination Cones for Recognition under Variable Illumination: Faces. In: CVPR 1998, pp. 52–58. IEEE Computer Society Press, Los Alamitos (1998)
5. Chin, T.J., Suter, D.: A Study of Illumination Cones Method for Face Recognition Under Variable Illumination. Technical Report. MECSE-7-2004. Department of Electrical and Comp. Systems Engineering, Monash University (2004)
6. Jacobs, D.W., Belhumeur, P.N., Basri, R.: Comparing Images under Variable Illumination. In: CVPR 1998, pp. 610–617. IEEE Computer Society Press, Los Alamitos (1998)
7. Koenderink, J.J., van Doorn, A.J.: Photometric Invariants related to solid Shape. Optica Acta 27(7), 981–996 (1980)
8. Lamdan, Y., Schwartz, J.T., Wolfson, H.J.: Affine Invariant Model-Based Object Recognition. IEEE Trans. on Robotics and Automation 6, 578–589 (1990)
9. Wolff, L., Fan, J.: Segmentation of Surface Curvature with a Photometric Invariant. J. Opt. Soc. Am. A 11(11), 3090–3100 (1994)
10. Schmid, C., Mohr, R.: Local grayvalue invariants for image retrieval. IEEE Trans. on Pattern Analysis and Machine Intelligence 19(5), 530–535 (1997)
11. Chen, H.F., Belhumeur, P.N., Jacobs, D.W.: In Search of Illumination Invariants. In: CVPR 2000, pp. 254–261. IEEE Computer Society Press, Los Alamitos (2000)
12. Bischof, H., Wildenauer, H., Leonardis, A.: Illumination Insensitive Recognition Using Eigenspaces. Comp. Vision and Image Underst. 95, 86–104 (2004)
13. Epstein, R., Hallinan, P.W., Yuille, A.L.: 5 ± 2 Eigenimages Suffice: an Empirical Investigation of Low-Dimensional Lighting Models. Workshop on Physics-Base Modeling in Computer Vision (1995)
14. Murase, H., Nayar, S.K.: Visual Learning of Object Models from Appearance. Technical Report. CUCS-054-92. Columbia University (1992)
15. Altamirano, L.C., Altamirano, L., Alvarado, M.: Non-uniform sampling for improve appearance-based models. Pattern Recognition Letters 23(1-3), 521–535 (2003)

Automatic Camera Localization, Reconstruction and Segmentation of Multi-planar Scenes Using Two Views

Mario Santés and Javier Flavio Vigueras

Centro de Investigación en Matemáticas (CIMAT)
{santes,flavio}@cimat.mx

Abstract. This work addresses two main problems: (i) localization of two cameras observing a 3D scene composed by planar structures; (ii) recovering of the original structure of the scene, i.e. the scene reconstruction and segmentation stages. Although there exist some work intending to deal with these problems, most of them are based on: epipolar geometry, non-linear optimization, or linear systems that do not incorporate geometrical consistency.

In this paper, we propose an iterative linear algorithm exploiting geometrical and algebraic constraints induced by rigidity and planarity in the scene. Instead of solving a complex multi-linear problem, we solve iteratively several linear problems: coplanar features segmentation, planar projective transferring, epipole computation, and all plane intersections. Linear methods allow our approach to be suitable for real-time localization and 3D reconstruction. Furthermore, our approach does not compute the fundamental matrix; therefore it does not face stability problems commonly associated with explicit epipolar geometry computation.

Keywords: computer vision, unsupervised segmentation, camera localization, scene reconstruction, two-view geometry.

1 Introduction

In man-made environments, the presence of planar objects is common. These surfaces are present as walls, doors and many objects such as books, desks and other furniture. The use of planar surfaces for camera localization and structure recovery have recently received attention from the computer vision community [2,7,8,10,11]. Other tasks such as image segmentation [10] or camera calibration [11] also have been studied, furthermore, map building is an area where multi-planar scenes have been used by roboticists.

On the other hand, the camera localization problem consists of determining the camera pose where the images (scene projections) were taken. Figure 2 shows two different views (a and b) of the same multi-planar scene. A 3D global frame of reference is imposed and is used to describe the elements in the scene (structure recovery or scene reconstruction problem).

A. Hernández Aguirre et al. (Eds.): MICAI 2009, LNAI 5845, pp. 280–291, 2009.

In this paper, we propose an iterative linear algorithm exploiting geometrical and algebraic constraints induced by planarity in the scene. Our approach allows us to deal with localization and reconstruction problems using only linear systems of equations, instead of solving a multi-linear problem or a non-linear problem with the corresponding instability due to errors in the initial localization guess.

The proposed framework makes use of features extracted from two images and matched by correlation. Image features are segmented considering coplanar constraints for point transferring, furthermore, linear functions are used for epipolar geometry recovery, without explicit computation of the fundamental matrix (a 3×3 matrix which relates corresponding points in a pair of images). One geometrical fact that is exploited in our work, consists of computing the intersections between all the planar surfaces that are projected in the views. These intersections are seen as straight lines on each image, and they allow us to carry out the three problems covered by this work: camera localization, scene reconstruction and image segmentation.

This paper is organized as follows: we present related work in the area of motion and structure recovery, later we exhibit the problem definition and contributions of this work. We present later our approach to solve these problems. We describe the mathematical tools used and the developed algorithms. At the end, we show some experiments and present our conclusions and future research.

2 Related Work

Motion recovery and scene reconstruction are strongly linked tasks. One of the first works that dealt with these problems using planes was written by Faugeras and Lustman [2]; in this work, they use only one planar surface and two views. The main contribution of this work is the linear method used for motion and structure recovery and the inclusion of geometrical and algebraic constraints that are induced by rigidity and planarity in the scene. This approach solves the problem of locating the camera by using planes that was not formerly possible by means of the fundamental matrix. The authors make an analysis of the singular values of the transfer function, showing that the problem has multiple solutions. With an additional observed plane, all ambiguities are dismissed. In [8] Simon et al., present a user assisted system for tracking a plane in a marker-less scene. In this work, a user delimits the boundary of the projected planar surface. Another contribution of this work is the DLT-like algorithm for computing inter-image homographies (projective planar transformations) consistent with a rigid polyhedral structure.

For multi-planar scenes Bartoli et al., in [1], introduce the motion and structure recovery problem as a non-linear cost function with explicit epipolar geometry computation. Planar homologies were introduced by Malis and Cipolla in [6] for automatic calibration from multi-planar structures, but no segmentation stage is proposed. In [7] he shows how to recover the localization and reconstruction parameters of multi-planar scenes with minimal user assistance, which consists of manual selection of a base plane, additional planes are considered as *walls*, i. e. these additional planes are orthogonal to the base plane, helping the pose computation stability but constraining the range of applications.

Segmentation of multiplanar scenes using geometrical constraints were used in [10]. They propose a non-linear cost function that when minimized gives the maximum likelihood of the inter-image homographies and the left epipole without explicit computation of the fundamental matrix. Their approach imposes that all the transfer functions have the same localization parameters ensuring projective coherence, a similar idea as in [1] but avoiding instability due to the explicit epipolar geometry computation.

3   Problem Definition

Given two images of the same multi-planar scene, we:

- determine the position and orientation of each camera taking these images,
- compute the parameters that describe each planar surface (reconstruction),
- label the image features that belong to each plane (segmentation).

3.1   Overview

When only two views of the same multi-planar scene are considered, epipolar geometry arises as a natural mathematical tool in order to determine motion parameters of the cameras. In this work we avoid explicit epipolar geometry computation, instead, we use planar homologies which are the product of two inter-image homographies. Homographies are calculated using the RANSAC paradigm for fitting the linear model proposed in [8] on putative matches computed with the Zero-mean Normalized Cross-Correlation (ZNCC) measure.

Once the epipole and all intersections between planes are computed, the homography with the greatest support[1] is chosen as the reference homography, then an iterative linear method that computes an improved version of these vectors is triggered. All the non-reference homographies are rewritten using the reference homography, epipole and the vector that describes the intersection between the reference plane and the current plane. This stage is at the core of our method and assures projective coherence between all the homographies.

The localization of the first camera is fixed at the origin of $\mathbb{R}^3$ with null orientation. For the second one we use a Faugeras-like algorithm [2] for orientation recovery and later we use all the information computed up to now in order to recover the translation vector. Dense segmentation of the projection of each plane is carried out with the improved version of the intersections between planes and the associated homography.

3.2   Contributions

Localization, reconstruction and segmentation stages are carried out taking into account geometrical and algebraic constraints that arise from multi-planar scenes, using only linear systems of equations. Our original contributions are:

---

[1] Pairs of correspondences associated to a plane.

1. A linear system of equations for incorporating consistence in all of the computed inter-image homographies, and the algorithm to solve it. This means that all of the common parameters extracted from these homographies are the same, these parameters are related to pose estimation.
2. The first epipole and all the intersections between any two planar surfaces are used to compose the transfer functions. The camera localization and scene reconstruction are tasks solved using these functions.
3. Segmentation is achieved through the intersections of the planar surfaces in the scene.

4 Mathematical Foundations

This section presents the notation and formulations used in this paper.

- Image points are expressed in homogeneous coordinates.
- The camera model used in this work is the pinhole model [4]. The intrinsic camera matrices are K_1 and K_2 for the first and second views, respectively.
- Features considered in this paper are points and are computed via a corner detector, e. g. the Harris corner detector [3].
- Correspondences between two views are denoted by $\{x \leftrightarrow x'\}$, where x represents a detected point in the first view and x' at the second view.
- The cross product operation is often represented as a matrix operator [4], where $[a]_\times \cdot$ act as $a \times \cdot$ if:

$$[a]_\times \equiv \begin{pmatrix} 0 & -a_3 & a_2 \\ a_3 & 0 & -a_1 \\ -a_2 & a_1 & 0 \end{pmatrix}. \tag{1}$$

4.1 Inter-image Homographies

A planar homography transfers coplanar space points to the image plane of a given view. Inter-image homographies are transfer functions between two image planes.

Given two planar homographies H_π^1, H_π^2, the inter-image homography between these views can be written as: $H \equiv H_1^2 = H_\pi^2(H_\pi^1)^{-1}$. Here H maps points from the first view to the second one, and π represents the observed planar surface.

If we know the pose parameters (R, t) of the second camera, the parameters of the observed planar surface (v) and the camera matrices (K_1, K_2), it is possible to write the inter-image homography as follows [4]:

$$H \sim K_2(R - tv^T)K_1^{-1}, \tag{2}$$

H transfers the projected points from the observed planar surface on the first view to the second view, this is a 3×3 matrix. R is a 3×3 matrix, representing the rotation between the first and the second view. t is the translation vector from the first camera to the second, this vector has dimension three. $\mathbf{v}$ is the three dimensional normal vector associated to the surface, such that the plane equation is $\mathbf{v} \cdot \mathbf{X} + 1 = 0$, when $\mathbf{X}$ is a point on the planar surface.

In order to compute an homography from correspondences (at least four of them), we solve the next homogeneous linear system [8]:

$$\begin{pmatrix} x_1 & y_1 & 1 & 0 & 0 & 0 & -x_1'x_1 & -x_1'y_1 & -x_1' \\ 0 & 0 & 0 & x_1 & y_1 & 1 & -y_1'x_1 & -y_1'y_1 & -y_1' \\ & & & & \vdots & & & & \\ x_n & y_n & 1 & 0 & 0 & 0 & -x_n'x_n & -x_n'y_n & -x_n' \\ 0 & 0 & 0 & x_n & y_n & 1 & -y_n'x_n & -y_n'y_n & -y_n' \end{pmatrix} \begin{pmatrix} h_{11} \\ h_{12} \\ \vdots \\ h_{32} \\ h_{33} \end{pmatrix} = \mathbf{0} \qquad (3)$$

where h_{rs} are the entries of the homography $\mathbf{H}$, and the correspondence pairs are: $\{(x_i, y_i, 1) \leftrightarrow (x_i', y_i', 1)\}$.

In our work, scenes contain many planar surfaces. It is expected that all the computed inter-image homographies have the same pose and internal parameters [1] and only differ in the ones related to the plane equations.

We use $\mathbf{H}_j \sim \mathbf{K}_2(\mathbf{R} - \mathbf{t}\mathbf{v}_j^T)\mathbf{K}_1^{-1}$ as the inter-image homography that describes the transferring induced by the j-th planar surface and the associated views.

4.2 Epipolar Geometry

Epipolar geometry is defined between two views that project the same scene. The epipole $\mathbf{e}$ is the point of intersection of the line joining the camera centers $\mathbf{C}$ and $\mathbf{C}'$ with the image plane on the first view. For the second image the epipole is denoted by $\mathbf{e}'$.

The fundamental matrix $\mathbf{F}$ is an algebraic representation of the epipolar geometry. For any pair of correspondences $\mathbf{x} \leftrightarrow \mathbf{x}'$ in the two images, this matrix satisfies: $\mathbf{x}'^T \mathbf{F} \mathbf{x} = 0$.

When the epipolar geometry is induced by observing a planar surface, the inter-image homography $\mathbf{H}$ associated to that plane can be used in order to express the fundamental matrix: $\mathbf{F} \sim [\mathbf{e}']_\times \mathbf{H} \sim \mathbf{H}^{-T}[\mathbf{e}]_\times$. In [11] the authors report that the computed fundamental matrix is numerically more stable when using inter-image homographies than when using point correspondences from planar surfaces through classical methods [4].

4.3 Consistence between Homographies

Two homographies are used to define a planar homology [6] as follows (consider Figure 1 for a sketch of this principle): $\mathbf{M}_{ij} = \mathbf{H}_i^{-1}\mathbf{H}_j$. Therefore using relation (2) and the Sherman-Morrison formula, we obtain:

$$\mathbf{M}_{ij} \sim \mathbf{K}_1(\mathbf{R} - \mathbf{t}\mathbf{v}_i^T)^{-1}(\mathbf{R} - \mathbf{t}\mathbf{v}_j^T)\mathbf{K}_1^{-1} \sim \mathbf{I} + \mathbf{e}\mathbf{s}_{ij}^T, \qquad (4)$$

where $\mathbf{s}_{ij} \sim \mathbf{K}_1^{-T}(\mathbf{v}_i - \mathbf{v}_j)$ and $\mathbf{e} \sim \mathbf{K}_1\mathbf{R}^{-1}\mathbf{t}$ is the first epipole.

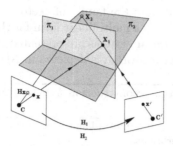

Fig. 1. The action of $\mathbf{M}_{ij} = \mathbf{H}_i^{-1}\mathbf{H}_j$ on a point $\mathbf{x}$ in the first view to transfer it to $\mathbf{x}'$ as the image of the 3D point $\mathbf{X}_2$, and then map it back as it were the image of the 3D point $\mathbf{X}_1$. Points in the first view which lie on the intersection of the planes are mapped to themselves, so are fixed points under this action.

Consequently, every homography can be written as: $\mathbf{H}_j \sim \mathbf{H}_i\mathbf{M}_{ij}$, this means that: $\mathbf{H}_j \sim \mathbf{H}_i(\mathbf{I} + \mathbf{e}\mathbf{s}_{ij}^T)$. If the second singular value of the right side is equal to one, former relation becomes an equality [2].

The reference homography is denoted by $\mathbf{H}_{ref}$, therefore all the remaining homographies can be written as:

$$\mathbf{H}_j \sim \mathbf{H}_{ref}(\mathbf{I} + \mathbf{e}\mathbf{s}_{ref,j}^T). \tag{5}$$

In this work we choose the reference plane as the one with the greatest support, i. e. , the planar surface with the biggest number of correspondences.

4.4 Intersections between Planes

The intersection between planes [7] i and j implies that a given point $\mathbf{x}$ belonging to this intersection satisfies $\mathbf{x}' \sim \mathbf{H}_i\mathbf{x}$ and $\mathbf{x}' \sim \mathbf{H}_j\mathbf{x}$. Therefore $\mathbf{H}_i\mathbf{x} \sim \mathbf{H}_j\mathbf{x}$, i. e. $\mathbf{x} \sim \mathbf{H}_i^{-1}\mathbf{H}_j\mathbf{x}$. It follows that $\mathbf{x}$ is an eigenvector of $\mathbf{M}_{ij}$.

The eigenvectors of $\mathbf{M}_{ij}$ are: (i) the epipole $\mathbf{e}$, which is common for all the elements in the scene; (ii) any vector $\mathbf{a}$ orthogonal to $\mathbf{s}_{ij}$.

The intersection between planes i and j is the set of all such vectors $\mathbf{a}$. Furthermore, by orthogonality we know that $\mathbf{s}_{ij} \cdot \mathbf{a} = 0$, then the equation of this intersection is a straight line given by the vector $\mathbf{s}_{ij}$. The intersection between the reference plane and the j-th planar surface, observed at the first image, is given by $\mathbf{s}_{ref,j} \cdot \mathbf{x} = 0$.

5 Our Approach

Our method is divided in two stages, the first computes the epipole $\mathbf{e}$ and all the required vectors $\mathbf{s}_{ref,j}$ from $\mathbf{M}_{ij}$ matrices. The second one is an iterative numerical method that incorporates the support of each homography in order to improve the previous estimation.

Our work deals with the extraction of the epipole and the vectors $\mathbf{s}_{ref,j}$ from a linear form. Our goal is to incorporate the support of the planar surfaces in order to extract the epipole and the family of vectors $\mathbf{s}_{ref,j}$ from the relation: $\mathbf{H}_j \sim \mathbf{H}_{ref}(\mathbf{I} + \mathbf{es}_{ref,j}^T)$. This is a multi-linear form for the epipole $\mathbf{e}$ and vectors $\mathbf{s}_{ref,j}$, because of this, two linear systems are presented in order to recover those vectors. An overview of our algorithm is depicted in Algorithm 1.

Algorithm 1. Localization, reconstruction and segmentation of multi-planar scenes using two views.

Require: Two views of the same multi-planar scene. The RANSAC threshold. The maximum number of iterations.
Ensure: Localization, reconstruction and dense support of the planar surfaces.
 1: Using a corner detector, image features are obtained and matched using the Zero-mean Normalized Cross-Correlation (ZNCC) measure [9].
 2: RANSAC is used to fit the linear system in (3) for each candidate plane.
 3: Selection of the reference homography (the one with the largest support).
 4: Recover the first epipole and the intersections between planes (subsection 5.1).
 5: All the non-reference homographies are rewritten (equation 5).
 6: Call Algorithm 2 for the iterative stage (subsection 5.2).
 7: The Faugeras-Lustman algorithm is used to recover the orientation of the second camera (subsection 5.3).
 8: Recovery of translation and reconstruction parameters (subsection 5.4).
 9: Dense segmentation at one view (subsection 5.5).

5.1   Epipole and Intersections between Planes from Planar Homologies

Planar homologies are calculated between any two planes, i. e. these matrices $\mathbf{M}_{ij}$ are determined through Equation (4) for all $i < j \in \{1, \cdots, n\}$, where $n > 1$ is the number of planes in the scene. We define matrices $\mathbf{G}_{ij}$ as $\mathbf{G}_{ij} = \mathbf{M}_{ij} - \mathbf{I} = \mathbf{es}_{ij}^T$. These matrices have rank equal to one.

The singular value decomposition of $\mathbf{G}_{ij}$ is:

$$\mathbf{G}_{ij} = [\mathbf{U}_1 \mathbf{U}_2 \mathbf{U}_3] \begin{pmatrix} \lambda \ 0 \ 0 \\ 0 \ 0 \ 0 \\ 0 \ 0 \ 0 \end{pmatrix} [\mathbf{V}_1^T \mathbf{V}_2^T \mathbf{V}_3^T],$$

where $\mathbf{U}_k$ and $\mathbf{V}_k$ are the k-th columns of $\mathbf{U}$ and $\mathbf{V}$ matrices, respectively.

When working with real noisy images, $\mathbf{G}_{ij}$ does not always have unitary rank. In this case we take the largest singular value and the remaining are forced to be zero. In fact, the SVD algorithm applied to $\mathbf{G}_{ij}\mathbf{G}_{ij}^T$ gives us $\mathbf{U}\mathbf{S}^2\mathbf{U}^T$, and for $\mathbf{G}_{ij}^T\mathbf{G}_{ij}$ we obtain $\mathbf{V}\mathbf{S}^2\mathbf{V}^T$ in a very stable way. Using this idea, we recover $\mathbf{e}$ and $\mathbf{s}_{ij}$ from the planar homology $\mathbf{G}_{ij}$, as follows: $\mathbf{e} = \sqrt{\lambda}\mathbf{U}_1$ and $\mathbf{s}_{ij} = \sqrt{\lambda}\mathbf{V}_1$, where λ is the largest element of $\mathbf{S}$.

For all the possible planar homologies that can be computed with the observed planes, we compose the following stack of matrices:

$$\mathbf{G} = [\mathbf{G}_{12}\mathbf{G}_{13} \cdots \mathbf{G}_{1n}\mathbf{G}_{23} \cdots \mathbf{G}_{2n} \cdots \mathbf{G}_{n-1,n}],$$

in this setting, $\mathbf{G}$ has dimension $3 \times \frac{3n(n-1)}{2}$ and rank equal to one.

Algorithm 2. Iterative stage

Require: Input parameters: correspondence pairs, vectors $\mathbf{e}$ and $\mathbf{s}'_{ref,j}$, homographies. The maximum number of iterations.
Ensure: Improved estimation of the input parameters.
1: **repeat**
2:     The correspondence pairs are refined using the segmented matches and vectors $\mathbf{s}_{ref,j}$.
3:     The vectors $\mathbf{s}_{ref,j}$ and the first epipole are re-estimated.
4:     Reference homography is re-estimated using all the matches.
5:     All the non-reference homographies are composed using the reference homography, the first epipole and the vectors $\mathbf{s}_{ref,j}$.
6: **until** maximum number of iterations reached or solution converges.

The first epipole is calculated in the same way as above, applying the SVD algorithm on $\mathbf{GG}^T$. For $\mathbf{s}_{ref,j}$ recovery, we need to extract it from $\mathbf{G}^T\mathbf{G}$. Vectors $\mathbf{s}_{ref,j}$ are stocked into the first $3n$ coefficients of $\mathbf{V}_1$.

5.2   Iterative Method Incorporating the Support of the Homographies

Computing the Intersection between Planes. From equation (5), we know that each homography can be written as $\mathbf{H}_j \sim \mathbf{H}_{ref}(\mathbf{I} + \mathbf{es}^T_{ref,j})$, where $\mathbf{H}_{ref}$ is the reference homography. Considering any planar surface and their associated support, we can write the transfer equation as:

$$\mathbf{x}' \sim \mathbf{H}_{ref}(\mathbf{I} + \mathbf{es}^T_{ref,j})\mathbf{x}, \tag{6}$$

Our goal is to transform (6) into a design matrix for a linear system. Following this idea we get: $\mathbf{H}_{ref}^{-1}\mathbf{x}' \sim \mathbf{x} + \mathbf{ex}^T\mathbf{s}_{ref,j}$.

Let $\mathbf{p} = \mathbf{H}_{ref}^{-1}\mathbf{x}'$, using the $[\cdot]_\times$ operator as defined in (1), it follows:

$$[\mathbf{p}]_\times(\mathbf{x} + \mathbf{ex}^T\mathbf{s}_{ref,j}) \sim [\mathbf{p}]_\times\left(\mathbf{H}_{ref}^{-1}\mathbf{x}'\right) = [\mathbf{p}]_\times\mathbf{p} = 0$$

Therefore: $([\mathbf{p}]_\times\mathbf{ex}^T)\mathbf{s}_{ref,j} = -[\mathbf{p}]_\times\mathbf{x}$. From this equation, we realize that we have formed a linear system $\mathbf{As}_{ref,j} = \mathbf{b}$, where $\mathbf{A} = [\mathbf{p}]_\times\mathbf{ex}^T$ is a 3×3 matrix and $\mathbf{b} = -[\mathbf{p}]_\times\mathbf{x}$ is a column vector with three elements.

Let $\mathbf{x}_i \leftrightarrow \mathbf{x}'_i$ be any pair in the support, with their associated matrices $\mathbf{A}_i$ and $\mathbf{b}_i$ defined as above, then the over determined linear system is:

$$\begin{pmatrix} \mathbf{A}_1 \\ \vdots \\ \mathbf{A}_n \end{pmatrix} \mathbf{s}_{ref,j} = \begin{pmatrix} \mathbf{b}_1 \\ \vdots \\ \mathbf{b}_n \end{pmatrix},$$

where n is the number of elements inside the support.

Computing the First Epipole. Considering that the epipole is common for all the homographies, we present a similar approach to compute it as above: $\mathbf{H}_{ref}^{-1}\mathbf{x}' \sim \mathbf{x} + \mathbf{e}(\mathbf{s}_{ref,j}^T\mathbf{x})$, Defining $\gamma \equiv \mathbf{s}_{ref,j} \cdot \mathbf{x}$, we obtain: $\mathbf{H}_{ref}^{-1}\mathbf{x}' \sim \mathbf{x} + \gamma\mathbf{e}$, following the same previous ideas, we rewrite this equivalence as: $-[\mathbf{p}]_\times\mathbf{x} = \gamma[\mathbf{p}]_\times\mathbf{e}$.

We extend our mathematical notation. Let $\mathbf{A}_i^j = \mathbf{s}_{ref,j} \cdot \mathbf{x}_{i,j}[\mathbf{H}_{ref}^{-1}\mathbf{x}_{i,j}']_\times$, be a 3×3 matrix, where $\mathbf{x}_{i,j} \leftrightarrow \mathbf{x}_{i,j}'$ is the i-th pair in the support for the j-th planar surface. And considering: $\mathbf{b}_i^j = -[\mathbf{H}_{ref}^{-1}\mathbf{x}_{i,j}']_\times\mathbf{x}_{i,j}$, as a three dimensional column vector, thus the linear system including all the support obtained is as follows:

$$\begin{pmatrix} \mathbf{A}_1^1 \\ \vdots \\ \mathbf{A}_{n_m}^m \end{pmatrix} \mathbf{e} = \begin{pmatrix} \mathbf{b}_1^1 \\ \vdots \\ \mathbf{b}_{n_m}^m \end{pmatrix},$$

where m is the total number of planar surfaces in our setting, and n_j is the total number of elements in the j-th support.

Re-estimating Homographies. The associated support that belongs to the reference plane is directly related to $\mathbf{H}_{ref}$. For all the remaining planar surfaces, their associated support holds relation (6), i. e. $\mathbf{x}_{i,j}' \sim \mathbf{H}_{ref}(\mathbf{I} + \mathbf{es}_{ref,j}^T)\mathbf{x}_{i,j}$.

This relation is used to compute $\mathbf{H}_{ref}$. For each pair $\mathbf{x}_{i,j} \leftrightarrow \mathbf{x}_{i,j}'$, the point $\mathbf{y}_{i,j}'$ is defined as: $\mathbf{y}_{i,j}' = (\mathbf{I} + \mathbf{es}_{ref,j}^T)\mathbf{x}_{i,j}$. The new pairs are $\mathbf{x}_{i,j} \leftrightarrow \mathbf{y}_{i,j}'$. The linear system (3) for computing $\mathbf{H}_{ref}$ is used with the set of correspondences $\mathbf{x}_{i,j} \leftrightarrow \mathbf{y}_{i,j}'$, in order to include all the correspondences inside and outside the reference plane.

All the remaining homographies should contain the same localization parameters, therefore these homographies are composed using relation (5).

5.3   Computing the Orientation

Given the camera matrices $\mathbf{K}_1, \mathbf{K}_2$, we obtain the associated collineation for each homography, i. e. we compute homographies expressed in canonic coordinates. Collineations are denoted by $\mathbf{C}_j$, and are defined as follows: $\mathbf{C}_j = \mathbf{K}_2^{-1}\mathbf{H}_j\mathbf{K}_1$.

The Faugeras-Lustman algorithm [2] can be applied to the reference homography only, but this requires some knowledge about the camera motion. The SVD algorithm is applied to all the collineations $\mathbf{C}_j$. Analyzing these singular values help us to reject non-feasible solutions. Last step consists of composing the rotation matrix using only singular values. In fact, the translation and reconstruction parameters can be fully obtained by means of the Faugeras-Lustman algorithm, but in our experiments with real data, these parameters were not always computed with the desired precision.

5.4   Translation and Reconstruction

Using all the computed collineations $\mathbf{C}_j$ and relation (2), we can depict that: $\mathbf{C}_j = \mathbf{R} - \mathbf{tv}_j^T$. Defining $\mathbf{D}_j = \mathbf{R} - \mathbf{C}_j$, i. e. $\mathbf{D}_j = \mathbf{tv}_j^T$, we wish to extract vectors $\mathbf{t}$ and $\mathbf{v}_j$ from $\mathbf{D}_j$, this goal is achieved by applying the ideas described in section 5.1. Extraction of $\mathbf{t}$ is the same problem as the epipole recovery and vectors $\mathbf{v}_j$ are computed in the same way as vectors $s_{ref,j}$.

5.5   Segmentation

The vectors $s_{ref,j}$ are used in order to create a partition on the first view. This partition allows us to recover the dense support for all the detected planar surfaces. This simple strategy gives a well-delimited segmentation, in the cases where only the detected planar surfaces were projected by both cameras. Thus, dense segmentation is only reliable when applied to polyhedral scenes.

6   Experiments

This experiment is carried out with the two views shown in Figure 2 (a,b). The original size of these views is 640×480. The two views shown in Figure 2 (a,b) were acquired by the same camera. The camera matrix $\mathbf{K}_1 = \mathbf{K}_2$ has the next parameters: $f = 709.676$, $f_y = 731.667$ and the principal point $(u_0, v_0)^T = (324.588, 240.653)^T$ and were obtained with the method described in [12].

Algorithm 1 is used in the following experiment. The first step consists of running a feature detector on the two views. In this work, the Harris corner detector is used with default values according to the OpenCV documentation [5].

These points are matched in our approach using a modified version of the ZNCC correlation measure [9]. The ZNCC measure takes values between $[-1, 1]$ and 1 indicates a perfect match. For real-time response, this measure has reported the best results when comparing with other similar correlation measures [9]. The ZNCC correlation measure is used to generate putative pairs of correspondences, that later are segmented by means of RANSAC. Some pairs are discarded. For the ZNCC, the window $\mathbf{W}$ is set to a 11×11 region. The RANSAC threshold is 2 pixels as defined in [4].

The vectors that are calculated in this experiment are: $\mathbf{e}$, $s_{ref,2}$, $s_{ref,3}$ and $s_{2,3}$. The $s_{2,3}$ vector is used to describe the intersection between the two non-reference planes. Figure 2 shows the resultant images of this experiment. The parameters of localization and reconstruction obtained by our algorithm are as follows:

- rotation axis between views: $(0.00810, 0.09255, 0.02354)^T$,
- angle of rotation between views: 5.49100 degrees,
- translation vector: $(0.42713, -0.02740, 0.01432)^T$,
- normal vector to the reference plane: $\mathbf{v}_{ref} = (-0.34047, -0.26459, 0.90226)^T$,
- normal vector to the 2nd plane: $\mathbf{v}_2 = (0.84644, -0.17245, 0.50379)^T$,
- normal vector to the 3rd plane: $\mathbf{v}_3 = (0.04125, 0.97386, 0.22335)^T$,
- angle between the reference plane and the 2nd plane: 102.2390 degrees,
- angle between the reference plane and the 3rd plane: 94.02559 degrees,
- angle between the 2nd and 3rd plane: 91.17444 degrees.

The last two angles should be $90°$, roughly.

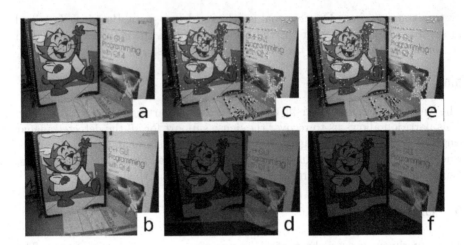

Fig. 2. (a,b) are two different views of the same multi-planar scene. (c) segmented correspondences for the detected planar surfaces. Pairs in the reference plane are red. Pairs for the second and third plane are green and blue respectively. (d) first approximation of the segmentation stage. (e) improvement of the support of each planar surface. (f) final segmentation.

Figure 2 (c) shows the support of each planar surface. This support is segmented via the RANSAC paradigm. (e) shows the improved support obtained by means of the vectors $s_{ref,j}$. The estimation of the first epipole and the family of vectors $s_{ref,j}$ is improved using this support. (d) shows the dense support of each detected planar surface. This segmentation is improved in (f) via the new estimation of the epipole e and vectors $s_{ref,j}$ after 10 iterations of Algorithm 2.

7   Conclusions and Future Research

In this paper, we have proposed a linear method to recover the localization of two cameras and the structure of the main planar surfaces projected by these cameras. The mathematical theory developed so far allows us to use minimal information (two images and their corresponding cameras) to segment coplanar features. This is an advantage over the previous work, where an expert user supplies the correspondence pairs [2,11] or others where previous knowledge of the dense support of each planar surface is required [8,10].

The linear systems presented in this work, allow us to use this framework in real-time systems for tasks such as map building, augmented reality and robot localization. As it has been used in other referred work, a *bundle adjustment* can be conducted in order to improve the precision of the localization and reconstruction parameters. The use of our framework for video processing is direct and may require non-linear filtering theory as has been shown in [2,7]. In order to obtain a more user-independent system, automatic camera calibration [11] and radial distortion correction [4] might be considered for future work.

Acknowledgments

Mario Santés was supported by a CONACyT scholarship. Flavio Vigueras is supported by CONACyT (grant 82590) and CONCyTEG (08-02-K662-119-A01).

References

1. Bartoli, A., Sturm, P., Horaud, R.: A Projective Framework for Structure and Motion Recovery from Two Views of a Piecewise Planar Scene. Research Report RR-4070, INRIA (2000)
2. Faugeras, O., Lustman, F.: Motion and Structure from Motion in a Piecewise Planar Environment. Technical Report RR-0856, INRIA (1988)
3. Harris, C., Stephens, M.: A combined corner and edge detector. In: Proc. Fourth Alvey Vision Conference, pp. 147–151 (1988)
4. Hartley, R.I., Zisserman, A.: Multiple View Geometry in Computer Vision, 2nd edn. Cambridge University Press, Cambridge (2004)
5. Intel: OpenCV: Open Source Computer Vision Library (2008), http://opencv.willowgarage.com/
6. Malis, E., Cipolla, R.: Multi-View Constraints Between Collineations: Application to Self-Calibration From Unknown Planar Structures. In: Vernon, D. (ed.) ECCV 2000. LNCS, vol. 1843, pp. 610–624. Springer, Heidelberg (2000)
7. Simon, G., Berger, M.-O.: Detection of the Intersection Lines in Multiplanar Environments: Application to Real-Time Estimation of the Camera-Scene Geometry. In: 19th International Conference on Pattern Recognition - ICPR 2008 (2008)
8. Simon, G., Fitzgibbon, A.W., Zisserman, A.: Markerless Tracking using Planar Structures in the Scene. In: Proc. International Symposium on Augmented Reality (2000)
9. Solà, J.: Towards Visual Localization, Mapping and Moving Objects Tracking by a Mobile Robot: a Geometric and Probabilistic Approach. PhD thesis, LAAS-CNRS Toulouse, France (2007)
10. Vigueras, J.F., Rivera, M.: Registration and Iteractive Planar Segmentation for Stereo Images of Polyhedral Scenes. Elsevier Pattern Recognition, Special edition on Interactive Image Processing (2009)
11. Xu, G., Ichi Terai, J., Shum, H.Y.: A linear algorithm for Camera Self-Calibration, Motion and Structure Recovery for Multi-Planar Scenes from Two Perspective Images. In: Proceedings of the Conference on Computer Vision and Pattern Recognition, pp. 474–479 (2000)
12. Zhang, Z.: A flexible new technique for camera calibration. IEEE Transactions on Pattern Analysis and Machine Intelligence 22, 1330–1334 (1998)

The Nonsubsampled Contourlet Transform for Enhancement of Microcalcifications in Digital Mammograms

Jose Manuel Mejía Muñoz[1], Humberto de J. Ochoa Domínguez[1],
Osslan Osiris Vergara Villegas[2], Vianey Guadalupe Cruz Sánchez[1],
and Leticia Ortega Maynez[1]

[1] Universidad Autónoma de Ciudad Juárez (UACJ)
Departamento de Ingeniería Eléctrica y Computación
Avenida del Charro No. 450 Norte, Ciudad Juárez, Chihuahua, México
jmejia@yahoo.com, {hochoa,vianey.cruz,lortega}@uacj.mx
[2] Universidad Autónoma de Ciudad Juárez (UACJ)
Departamento de Ingeniería Industrial y Manufactura
Avenida del Charro No. 450 Norte, Ciudad Juárez, Chihuahua, México
overgara@uacj.mx

Abstract. Microcalcifications detection plays a crucial role in the early detection of breast cancer. The enhancement of the mammographic images is one of the most important tasks during the detection process. This paper presents an algorithm for the enhancement of microcalcifications in digital mammograms. The main novelty is the application of the nonsubsampled contourlet transform and a specific edge filter to enhance the directional structures of the image in the contourlet domain. The inverse contourlet transform is applied to recover an approximation of the mammogram with the microcalcifications enhanced. Results show that the proposed method outperforms the current method based on the discrete wavelet transform.

1 Introduction

Breast cancer is the most frequent disease diagnosed in middle-aged women [1]. In addition, from 30 to 50% of the tissue surrounding malignant tumors of the breast contains groups of microcalcifications [2], [3]. Even though, the early detection decreases considerably the mortality. Mammography is a method that uses low dose of x-rays to produce a picture of the breast to help with diagnoses. This method is also known as screen-film mammography or simply mammogram and is the most common widely used technique to determine the existence of breast cancer [1].

Microcalcifications are small deposits of calcium in the breast that cannot be felt but only can be seen on a mammogram. These specks of calcium may be benign or malignant and could be a first cue of cancer. Clusters of microcalcifications have diameters from some μm up to approximately 200 μm [4]. On a

A. Hernández Aguirre et al. (Eds.): MICAI 2009, LNAI 5845, pp. 292–302, 2009.

digital mammogram, microcalfications appear as a group from one up to a few number of high intensity samples, usually considered regions of high frequency.

Among the methods used to detect microcalcifications on digital mammograms are those that use wavelet transform implemented as filter banks, [5], [6], [7], [8], [9]. Wavelets are mainly used because of their dilation and translation properties, suitable for non stationary signals [10].

In [5] a biorthogonal filter bank is used to compute four dyadic and two interpolation scales. Then, a binary threshold-operator is applied to the six scales. In [6] a hexagonal wavelet transform (HWT) is used to obtain multiscales edges at orientations of 60, 0 and -60 degrees. Afterwards, the resulting subbands are enhanced and the image reconstructed. In [7] the mammograms are decomposed into different frequency subbands and the low-frequency subband discarded. Finally, the image is reconstructed from the subbands containing only high frequencies. In [9] the integrated wavelets are derived from a model of microcalcifications for general enhance-ment of mammograms. In [17] a method aimed at minimizing image noise, while optimizing contrast of mammographic image features for more accurate detection of microcalcification clusters using contourlets is presented. The transformed image is denoised using stein's thresholding [18]. A nonlinear mapping function is applied to the set of coefficient from each level to emphasize mammographic features. However, the results presented correspond to the enhancement of regions with large masses only. In [19] an automatic mass classification system which uses the contourlet transform to detect to classify the mammograms into normal and abnormal is developed. The paper also focuses in the analysis of large masses instead of microcalcifications.

In this paper we present a novel method based on the contourlet transform (CT) [11] to decompose the digital mammogram into multidirectional and multiscale subbands, and an edge Prewitt filter to enhance the directional structures in the image. The contourlet transform allows decomposing the image in many more directions than the hexagonal wavelet transform [6]. Also, the edge set is obtained through the Prewitt method. Therefore, this allows finding a better set of edges and, consequently, recovering an enhanced mammogram with better visual characteristics. As microcalcifications have a very small size a denoising stage is not implemented in order to preserve the integrity of the injuries. This paper is organized as follows: in section 2 the method is explained. Experimental results are presented and compared with the Discrete Wavelet Transform (DWT) method in section 3. The paper concludes with section 4.

1.1   Why Contourlet?

We are interested in decomposing the mammographic image into well localized and directional components to easily capture the geometry of the image features. This is accomplished by the 2-D Contourlet Transform (2D-CT) aimed at improving the representation sparsity of images over the Discrete DWT [11], [12], [13], [14].

The 2D-CT handles singularities such as edges in a more powerful way than the DWT because the 2D-CT has basis functions at many orientations and the DWT at only three. The basis functions of the CT appear a several aspect ratios whereas the aspect ratio of WT is 1. Finally, CT similar as DWT can be implemented using iterative filter banks.

1.2   The Contourlet Transform

The CT is implemented by Laplacian pyramid followed by directional filter banks as shown in Fig. 1a); the transform decomposes the image into several directional subbands and multiple scales as in Fig. 1b). Therefore, this decomposition offers multiscale and time frequency localization and a high degree of directionality and anisotropy. Also, the cascade structure allows the multiscale and directional decomposition to be independent of each other and it is possible to decompose each scale into any arbitrary power of two's number of directions. For this end, the *maxflat* type on the filter banks of the contourlet transform is used.

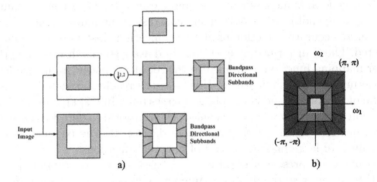

Fig. 1. The contourlet transform. a) Structure of the Laplacian pyramid together with the directional filter bank and b) frequency partitioning by the contourlet transform [11].

2   Method

The typical approach in mammography analysis is to transform the image into a more suitable representation for extracting specific features which are more difficult to extract in its original representation. For microcalcifications enhancement in digital mammograms, we use the Nonsubsampled Contourlet Transform (NSCT) [12] and the Prewitt filter. The proposed method is based on the classical approach used in transform methods for image processing. The block diagram is shown in Fig. 2.

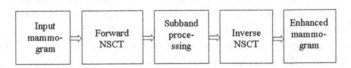

Fig. 2. Block diagram of the transform methods for images processing

The NSCT was implemented in two stages: the subband decomposition stage and the directional decomposition stages. For the subband decomposition the Laplacian pyramid is used [14], where the decomposition at each step generates a sampled low pass version of the original and the difference between the original image and the prediction. The input image is first low pass filtered using filter and then decimated to get a coarse approximation. The resulting image is interpolated and passed through a synthesis filter. The image obtained is subtracted from the original image to get a bandpass image. The process is then iterated on the coarser version of the image.

The directional filter bank (DFB) is efficiently implemented by using an *L-level* binary tree decomposition resulting in 2^L subbands with wedge-shaped frequency partitioning. The desired frequency partitioning is obtained by following a tree expanding rule for finer directional subbands [14]. An *L-level* tree-structured DFB is equivalent to a 2^L parallel channel filter bank with equivalent filters and overall sampling matrices.

2.1 Description

The input to the system is the digital mammogram; this image is transformed into the NSCT domain to obtain a multiscale and multidirectional decomposition in N levels and M_N directions per level of the image.

We denote each subband by $y_{i,j}$, where i and j are the decomposition level and the direction respectively, as depicted in Fig. 3.

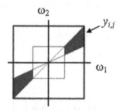

Fig. 3. Decomposition levels and directions

The processing of an image consists on applying a function to enhance the regions of interest. In multiscale analysis this could be traduced as to find a function f for each subband that emphasize the features of interest, in order to get a new set y' of enhanced subbands. Then, each of the resulting enhanced subbands can be expressed using equation 1.

$$y'_{i,j} = f(y_{i,j}) \tag{1}$$

After the enhanced subbands are obtained, the inverse transform is applied on the resulting coefficients to obtain an enhanced image.

2.2   Enhancement of the Directional Subbands

The directional subbands are enhanced using equation 2.

$$f(y_{i,j}) = \begin{cases} W_1 y_{i,j}(n1, n2) & if \quad b_{i,j}(n1, n2) = 0 \\ W_2 y_{i,j}(n1, n2) & if \quad b_{i,j}(n1, n2) = 1 \end{cases} \tag{2}$$

Where W_1, and W_2 are the weight factors for detecting the surrounding tissue and microcalcifications respectively, $b_{i,j}$ is a binary image containing the edges of the subband and (n_1, n_2) are the spatial coordinates.

Each binary edge image $b_{i,j}$ may be obtained by applying an operator to detect edges on each directional subband. In this work, a Prewitt operator is applied due to its simplicity and easy to implement, but other operators can be used without affecting the results. In order to obtain a binary image, a threshold $T_{i,j}$ for each subband is calculated. The threshold calculation is based under the premise that when mammograms are transformed into the CT domain, the microcalcifications appear on each subband over a very homogeneous background. Therefore, most of the transform coefficients, which are grouped around the mean value of the subband, correspond to the background and the coefficients corresponding to the injuries are far from this value. Therefore, a conservative threshold of $3\sigma_{i,j}$ is selected, where $\sigma_{i,j}$ is the standard deviation of the corresponding subband $y_{i,j}$. For the selection of the weights, after exhaustive tests, which consist on evaluating subjectively a set of 15 different mammograms with different combinations of values, the weights W_1, and W_2 are determined and selected as $W_1 = 3\sigma_{i,j}$ and $W_2 = 4\sigma_{i,j}$. These weights are chosen in order to keep the relationship $W_1 < W_2$, because the W factor is a gain and more gain at the edges are wanted.

After processing the subbands, the inverse CT is applied to produce an image with enhanced microcalcifications.

2.3   Metrics

In order to compare the ability of enhancement achieved by the proposed method, the Distribution Separation Measure (DSM), the Target to Background Contrast en-hancement (TBC) and the Target to Background Enhancement Measure based on Entropy (TBCE) are used [15]. These metrics are explained in following subsections.

2.3.1   Distribution Separation Measure (DSM)
The DSM represents how separated are the distributions of each mammogram and is defined by equation 3.

$$DSM = |\mu_{ucalc_E} - \mu_{tissue_E}| - |\mu_{ucalc_O} - \mu_{tissue_O}| \tag{3}$$

where:

$\mu_{ucalc_E}, \mu_{ucalc_O}$ are the mean of the microcalcification region of the enhanced and original image respectively. $\mu_{tissue_E}, \mu_{tissue_O}$ are the mean of the surrounding tissue of the enhanced and original image respectively.

2.3.2 Target to Background Contrast Enhancement Measure (TBC)

The TBC quantifies the improvement in difference between the background and the target. The TBC is defined by equation 4.

$$TBC = \frac{\frac{\mu_{ucalc_E}}{\mu_{tissue_E}} - \frac{\mu_{ucalc_O}}{\mu_{tissue_O}}}{\frac{\sigma ucalc_E}{\sigma ucalc_O}} \qquad (4)$$

where:

$\sigma ucalc_E$, $\sigma ucalc_O$ are the standard deviations of the microcalcifications region in the enhanced and original image respectively.

2.3.3 Target to Background Enhancement Measure Based on Entropy (TBCE)

This measure is an extension of the TBC metric. TBCE is based on the entropy of the regions rather than in the standard deviations and is defined by equation 5.

$$TBCE = \frac{\frac{\mu_{ucalc_E}}{\mu_{tissue_E}} - \frac{\mu_{ucalc_O}}{\mu_{tissue_O}}}{\frac{\varsigma ucalc_E}{\varepsilon ucalc_O}} \qquad (5)$$

3 Experimental Results

In this section, the experimental results of our method are presented. A set of 12 mammograms from the MIAS database was used [16]. The images were selected according to previous classification by expert radiologists who identified the microcalcifications and their positions inside the image.

The results obtained by the proposed method are compared with the well-known method based on the discrete wavelet transform, which consists on applying a DWT to decompose a digital mammogram into different subbands. Then, the lowpass wavelet band is removed (set to zero) and the remaining coefficients are enhanced. Afterwards, the inverse wavelet transform is applied to recover the enhanced mammogram containing microcalcifications [7].

Fig. 4a) shows portions of the original mammograms and, b) the enhanced images using our method and c) the enhanced images using the DWT method. For visualization purposes the ROI in the original mammogram are marked with a square. These regions contain clusters of microcalcifications (target) and surrounding tissue (background).

Table 1 shows the numerical results using DMS, TBC and TBCE metrics on the enhanced mammograms. Fig. 5a) shows a 3D plot of the numerical results. Each axis represents a metric (DSM, TBC and TBCE). Each point in the plot is normalized in the interval [0, 1]. The circles correspond to the results of the DWT method and the stars to the proposed method. In most of the mammograms tested, the proposed method gives higher results than the wavelet-based method. Therefore, the proposed method is more efficient for mammogram enhancement.

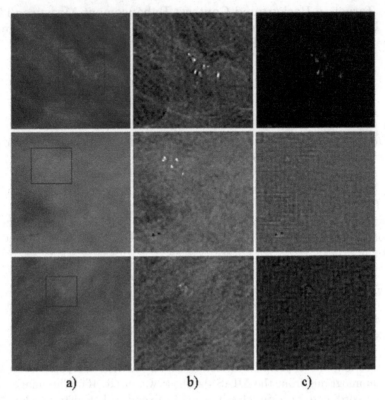

Fig. 4. ROIs of a) the original image, b) the proposed method and c) the DWT method

Table 1. Decomposition levels and directions

Proposed method			DWT method		
DSM	TBC	TBCE	DSM	TBC	TBCE
0.853	0.477	0.852	0.153	0.078	0.555
0.818	0.330	0.810	0.094	0.052	0.382
1.000	1.000	1.000	0.210	0.092	0.512
0.905	0.322	0.920	1.000	0.077	1.000
0.936	0.380	0.935	0.038	0.074	0.473
0.948	0.293	0.947	0.469	0.075	0.847
0.655	0.410	0.639	0.369	0.082	0.823
0.740	0.352	0.730	0.340	0.074	0.726
0.944	0.469	0.949	0.479	0.095	0.843
0.931	0.691	0.936	0.497	0.000	0.000
0.693	0.500	0.718	0.285	0.081	0.682
0.916	0.395	0.914	0.796	0.079	0.900

Fig. 5b), c) and d) show the plots of TBCE, TBC and DSM metrics, used to meas-ure the enhancement ability of the proposed method and the DWT method for each independent metric.

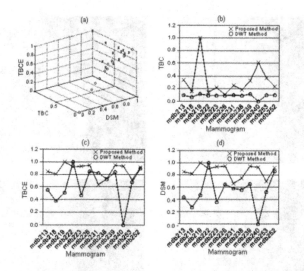

Fig. 5. Metrics plots. a) 3D plot of DSM, TBC and TBCE, b) the TBCE metric, c) the TBC metric, and d) the DSM metrics for comparison of methods

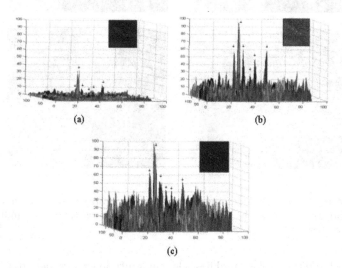

Fig. 6. Mesh plot of a ROI containing microcalcifications in a) the original mammogram, b) the enhanced mammogram using the proposed method and c) the enhanced mammogram using the DWT method

Fig. 6a) shows a mesh plot of a ROI containing a cluster of microcalcifications and inset b) the same enhanced ROI after applying our method to all the mammogram and inset c) the enhanced ROI using the DWT method. We observe that, when using our method more peaks, corresponding to microcalcifications,

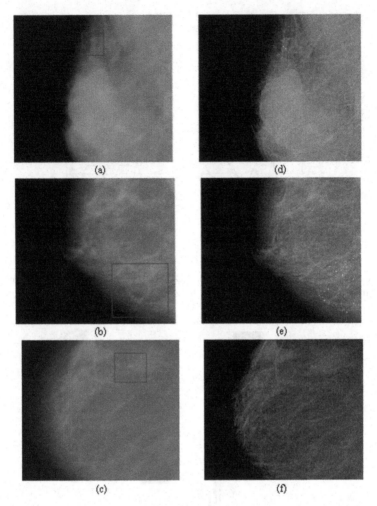

Fig. 7. Inset a), b) and c) shows the original mammograms (mdb 241, mdb245 and mdb248); and d), e) and f) the reconstructed mammograms (mdb 241, mdb245 and mdb248) after using the proposed method

are enhanced and the background has a less magnitude with respect to the peaks; therefore, the microcalcifications are more visible. This also confirms the results shown in Fig. 7.

The method also preserves the shape and other mammographic features of the image without affecting the microcalcifications as shown in Fig. 7, where the original mammograms contain clusters of microcalcifications. We see that the enhancement becomes more specific for clusters of microcalcifications than for other areas. How-ever, the disadvantage of this method is the high computational complexity because of the use of the NSCT and the Prewitt filter.

4 Conclusions

In this paper we presented a method for enhancement of microcalcifications in digital mammograms. The method proposes the combination of the nonsubsampled contourlet transform, to decompose the mammograms into several directional subbands and multiple scales, preserving the geometric characteristics of the image, and an edge filter. The method improves both, numerically and visually, the traditional method based on the discrete wavelet transform. The proposed method gives an impressive enhancement of the microcalcifications.

The exact choice of levels was three; four directions at the first level, eight directions at the third level and four directions at the finest level. The Prewitt filter with a threshold of $3\sigma_{i,j}$ was used. The surrounding tissue and microcalcifications inside a subband were enhanced by applying empiric weights of $w_1 = 3\sigma_{i,j}$ and $w_2 = 4\sigma_{i,j}$ respectively. Then, the nonsubsampled inverse contourlet transform was applied on the processed subbands to recover the enhanced mammogram. The use of contourlet transform makes the enhancement more computational extensive than previous methods. However, the presented method outperforms previous methods.

Future directions of this work include the separation of suspicious clusters containing microcalcifications, characterization of shapes of microcalcifications and the enhancement of regions containing larger abnormal masses with different shapes.

References

1. Barnes, G.T., Frey, G.D.: Screen-film Mammography Imaging: Considerations and Medical Physics Responsibilities. Medical Physics, 1–46 (1991)
2. Sickles, E.A.: Mammographic Features of Early Breast Cancer. American Journal of Roentgenology 143, 461–464 (1984)
3. Sickles, E.A.: Mammographic Features of 300 Consecutives Nonpalpable Breast Cancers. American Journal of Roentgenology 146, 661–663 (1986)
4. Alqdah, M., Rahmanramli, A., Mahmud, R.: A System of Microcalcifications Detection and Evaluation of the Radiologist: Comparative Study of the Three Main Races in Malaysia. Computers in Biology and Medicine 35, 905–914 (2005)
5. Strickland, R.N., Hahn, H.: Wavelet transforms for detecting microcalcifications in mammograms. IEEE Transactions on Medical Imaging 15, 218–229 (1996)
6. Laine, A.F., Schuler, S., Fan, J., Huda, W.: Mammographic feature enhancement by multiscale analysis. IEEE Transactions on Medical Imaging 13(4), 7250–7260 (1994)
7. Wang, T.C., Karayiannis, N.B.: Detection of Microcalcifications in Digital Mammograms Using Wavelets. IEEE Transaction on Medical Imaging 17(4), 498–509 (1989)
8. Nakayama, R., Uchiyama, Y., Watanabe, R., Katsuragawa, S., Namba, K., Doi, K.: Computer-Aided Diagnosis Scheme for Histological Classification of Clustered Microcalcifications on Magnification Mammograms. Medical Physics 31(4), 786–799 (2004)

9. Heinlein, P., Drexl, J., Wilfried, S.: Integrated Wavelets for Enhancement of Micro-calcifications in Digital Mammography. IEEE Transactions on Medical Imaging 22, 402–413 (2003)
10. Daubechies, I.: Ten Lectures on Wavelets. SIAM, Philadelphia (1992)
11. Do, M.N., Vetterli, M.: The Contourlet Transform: An efficient Directional Mul-tiresolution Image Representation. IEEE Transactions on Image Processing 14, 2091–2106 (2001)
12. Da Cunha, A.L., Zhou, J., Do, M.N.: The Nonsubsampled Contourlet Transform: Theory, Design, and Applications. IEEE Transactions on Image Processing 15, 3089–3101 (2006)
13. Burt, P.J., Adelson, E.H.: The Laplacian pyramid as a compact image code. IEEE Transactions on Communications 31(4), 532–540 (1983)
14. Park, S.-I., Smith, M.J.T., Mersereau, R.M.: A new directional filter bank for image analysis and classification. In: Proceedings of IEEE International Conference on Acoustics, Speech, and Signal Processing (ICASSP 1999), vol. 3, pp. 1417–1420 (1999)
15. Sameer, S., Keit, B.: An Evaluation on Contrast Enhancement Techniques for Mammographic Breast Masses. IEEE Transactions on Information Technology in Biomedicine 9, 109–119 (2005)
16. Suckling, J., Parker, J., Dance, D., Astley, S., Hutt, I., Boggis, C., Ricketts, I., Stamatakis, E., Cerneaz, N., Kok, S., Taylor, P., Betal, D., Savage, J.: The mammo-graphic images analysis society digital mammogram database. In: Exerpta Medica. International Congress Series, vol. 1069, pp. 375–378.
17. Lu, Z., Jiang, T., Hu, G., Wang, X.: Contourlet based mammographic image en-hancement. In: Proc. of SPIE, vol. 6534, pp. 65340M-1 – 65340M-8 (2007)
18. Yan, Y., Osadeiw, L.: Contourlet based image recovery and de-noising through wireless fading channels. In: 2005 Conference on Information Science and Systems, Maryland USA, pp. 16–18 (2005)
19. Moayedi, F., Azimifar, Z., Boostani, R., Kateb, S.: Contourlet-based mammogra-phy mass classification. In: Kamel, M.S., Campilho, A. (eds.) ICIAR 2007. LNCS, vol. 4633, pp. 923–934. Springer, Heidelberg (2007)

Denoising Intra-voxel Axon Fiber Orientations by Means of ECQMMF Method

Alonso Ramirez-Manzanares[1], Mariano Rivera[2], and James C. Gee[3]

[1] Universidad de Guanajuato, Facultad de Matematicas, Valenciana, Guanajuato,
Gto. Mexico. C.P. 36000
[2] CIMAT A.C., Callejon Jalisco S/N, Valenciana, Guanajuato,
Gto. Mexico. C.P. 36000
[3] University of Pennsylvania, PICSL, Department of Radiology
3600 Market Street, Suite 370, PA 19104, USA
{alram,mrivera}@cimat.mx, james.gee@uphs.upenn.edu

Abstract. Diffusion weighted magnetic resonance imaging is widely
used in the study of the structure of the fiber pathways in brain white
matter. In this work we present a new method for denoising intra–voxel
axon fiber tracks. In order to improve local (voxelwise) estimations, we
use the general–purpose segmentation method called Entropy–Controlled
Quadratic Markov Measure Field Models. Our proposal is capable of
spatially–regularize multiple axon fiber orientations (intra-voxel orientations). In order to provide the best as possible local axon orientations
to our spatial regularization procedure, we evaluate two optimization
methods for fitting a Diffusion Basis Function model. We present qualitative results on real human Diffusion Weighted MRI data where the
ground–truth is not available, and we quantitatively validate our results
by synthetic experiments.

Keywords: Image processing, Diffusion Weighted MRI, Axon fibers,
Image Regularization, Multi diffusion tensor.

1 Introduction

Axonal fiber pathways estimation is an important research area in neuroanatomy
due to the relationship of brain connectivity with diseases and brain development. The most widely-used approach for studying water diffusion in the human
brain is Diffusion Tensor imaging (DTI) [1,2]. Such a technique uses Diffusion
Weighted (DW) MR Images for computing the diffusion tensor's (DT), then
the main eigenvector, of the DT, corresponds to the axis of maximum diffusion. In white matter fiber tracts, the main eigenvector is aligned with the local
average orientation of the fibers, making it possible to study patterns of brain
connectivity *in-vivo* [3,4]. The DTI model have generated a flurry of research activity namely: denoising, field segmentation and field registration among others
[5,6,7,8,9,10,11,12].

The main limitation of DTI is that a DT is constrained to represent only
one maximum diffusion orientation and thus it is inadequate in voxels where

A. Hernández Aguirre et al. (Eds.): MICAI 2009, LNAI 5845, pp. 303–312, 2009.
© Springer-Verlag Berlin Heidelberg 2009

two or more fiber bundles cross, split or "kiss" [13]. This limitation represents a significant problem for diffusion tractography, where we must rely on local fiber-orientation estimates in order to reconstruct fiber pathways. This is an important limitation given that, according to [14], as many as one third of white-matter voxels contain more than one fiber bundle orientation.

A more plausible model for solving intra-voxel fiber orientations is the Gaussian Mixture Model (GMM) or multi–DT [15]:

$$S_k = S_0 \sum_{j=1}^{L} \beta_j \exp(-\mathbf{q}_k^T \mathbf{D}_j \mathbf{q}_k \tau) + \varepsilon_k; \tag{1}$$

where S_k is the attenuated signal value given the diffusion vector $\mathbf{q}_k$ for $k = 1, \ldots, M$ (i.e. for M non collinear diffusion gradients), S_0 is the signal without diffusion gradient, τ is the effective diffusion time, the anisotropic diffusion coefficients (with units equal to mm/s^2) are summarized by the positive definite symmetric 3×3 tensors $\mathbf{D}_j$, and the real–valued coefficients $\beta_j \in [0,1]$ indicate the fraction of the total diffusion associated with the tensor $\mathbf{D}_j$. Given that the MRI signal is the magnitude of a complex number with additive Gaussian noise in both real and imaginary components, the signal noise ε_k in (1) has Rician distribution. In such a case, the GMM explains quite well the diffusion phenomenon within a voxel (by assuming no exchange between fibers, i.e. the signals are independently added).

Tuch et al. [15] proposed to fix the anisotropy of the DT's $\mathbf{D}_j$ and to solve (1) by a multi–start gradient descent algorithm with high angular–resolution data. Parker and Alexander [16] used a Levenberg-Marquardt algorithm to fit the GMM. Because of the algorithmic problems related to the non-linearity of (1) both methods require a large number of diffusion images $\{S\}$ (126 are used in [15] and 54 in [16]) and it was not reported stable solution for more than 2 fiber bundles [15]. Recently, in [17] is proposed a Diffusion Basis Functions (DBF) observation model that overcome the above mentioned disadvantages. They simplified the fitting of model (1) by using a large tensor basis $\bar{\mathbf{T}}$ (isotropically distributed in the space) and by proposing the following observation model:

$$S_k = \sum_{j=1}^{N} \alpha_j \phi_{k,j} + \varepsilon_k; \tag{2}$$

where $\alpha_j \geq 0$ and the pre–computed DBF coefficients are defined as:

$$\phi_{k,j} = S_0 \exp(\mathbf{q}_k^T \bar{\mathbf{T}}_j \mathbf{q}_k \tau). \tag{3}$$

Thus, the j–th DBF $\{\phi_{k,j}, k = 1, \ldots, M\}$ is the DW–MRI signal due to a single fiber (modeled by the fixed basis tensor $\bar{\mathbf{T}}_j$) and the non-negative unknown α_j denotes its mixture contribution. The advantages of this formulation are: this model is fitted (solved for α) by solving a linear system of equations (i.e. efficient solving methods can be applied) and it is stable for recovering more than two fiber orientations per voxel. The drawback is that the recovered orientations are discrete, so that, an angular error is always present, but, by increasing the angular resolution of basis $\mathbf{T}_j$ it is possible to diminish it until it is irrelevant for practical aims.

Works in [17,18] proposed efficient methods for voxelwise fitting the DBF model (2), unfortunately, because of a low Signal to Noise Ration (SNR) and/or a limited number of DW acquisitions the local voxel computation could result in poor estimations. It is well known in the image processing area that spatial regularization (spatial average) diminishes the noxious noise effect [19,17]. In particular, we assume that the expected fiber orientations are similar for neighboring voxels, i.e. we assume smooth axon fiber pathways [20,21,22,23,24,25]. With this in mind, segmentation methods which introduce an spatial average component has been extensively used in medical image processing [26,27]. In our case, given that the α_j coefficients in (2) denote a sort of pixelwise memberships to each j-th DBF model, we can use a membership–based image segmentation approach in order to perform a computationally efficient spatial regularization. Additionally, it is desirable to introduce prior knowledge about the number of fibers that we want to recover: solutions with at most three fibers are expect for the majority of the voxels. Even though we can find in literature several membership–based segmentation methods (as for instance [26]), previous requirements can be efficiently achieved by using the Entropy–Controlled Quadratic Markov Measure Field (ECQMMF) method [27], which is a general-purpose, fast and efficient membership–based segmentation method that constrains the membership–vector's entropy (constraining, in this way, the number of α_j coefficients different from zero).

In this work we propose to perform the spatial regularization of a DBF solution by means of the general purpose ECQMMF segmentation method which can recover high quality results with only few iterations. The paper is organized as follows. In Section 2 we present the methodology for our DBF regularization proposal: we give a brief explanation of ECQMMF and show how the DBF method is integrated to our schema, then we investigate the best way for fitting the DBF model. Section 3 gives implementation details and presents quantitative and qualitative results for human DW-MR and synthetic data. Finally, we present our conclusions in Section 4.

2   Methods

The ECQMMF [27] is a method in which prior information about segmentation smoothness and low entropy of the probability distribution maps is codified in the form of a Markov Random Field with quadratic potentials. The optimal estimator is obtained by minimizing the quadratic cost function

$$U(b) = \sum_{x \in L} \sum_{j=1}^{N} b_j^2(x) \left[-\log v_j(x) - \mu \right]$$

$$+ \lambda \sum_{<x,y>} \|b(x) - b(y)\|^2, \tag{4}$$

subject to the linear constraints $b_j(x) \geq 0, \forall j, x$ and $\sum_j^N b_j(x) = 1, \forall x$. Variable x denotes the three–dimensional voxel position in the image volume L, $v_j(x)$ is the likelihood of observing model θ_j at the position x and the set of constrained unknowns b denotes a Markov Random Measure Field (MRMF) of fuzzy segmentation indicators. Moreover $< x, y >$ denotes nearest–neighbor pairs of sites. First term in (4) is composed by a data term and an entropy control term (Gini's coefficient) over the b field. Second term in (4) promotes spatial regularization. By minimizing (4) we expect to recover a spatially–regularized, low–entropy field b, so that, a predominant $b_j(x)$ value indicates that model θ_j is present at x. The amount of spatial regularization and entropy penalization is tuned by the user defined parameters λ and μ, respectively.

In case we concern here, we deal with MRI brain hydrogen diffusion models, so that, in present work we propose to use the DBF models (3) as the set of models we want to segment. For this aims we use the diffusion models as $\theta_j = \{\phi_{k,j}, \forall k \in M\}$ and we use the DBF's fitting process output as the ECQMM's input:

$$v_j(x) = \tilde{\alpha}_j(x), \forall x, j, \qquad (5)$$

where $\tilde{\alpha}(x)$ is the normalized DBF $(\alpha(x))$ vector (probability vector).

2.1   Computing Likelihoods for the ECQMMF Method

In order to achieve the highest quality likelihoods in (5) we investigate the best way to fit DBF model. Sparsity or "compact coding" scheme has been biologically motivated: in the context of biological vision models it is desirable to minimize the number of models that respond to any particular event. The prior information is that the probability of any model to be present is equal, but such probability is low for any given model [28]. In our context, sparsity in the recovered $\alpha(x)$ vectors is an important feature too: solutions that involve less basis DTs ($\tilde{\mathbf{T}}_j$) are preferable [17]. We compare the two fitting approaches proposed in [17] and [18]. In [17] it was proposed to solve DBF models by a Non–Negative Basis

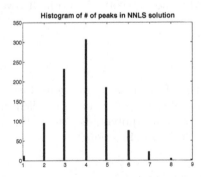 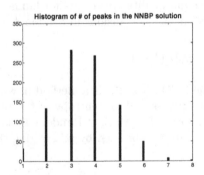

Fig. 1. Histograms about the number of significant coefficients ($\alpha_i \geq 0.05$) involved in NNLS (left) and NNBP (right) solutions. Note that NNBP method gives the sparser solution.

Table 1. Statistical values for 936 voxels that compare the development between NNLS and NNBP methods

Feature	min	max	mean	std.dev.
NNLS Fit Err.	0.929	15.820	1.616	1.058
NNBP Fit Err.	0.929	15.816	1.644	1.059
NNLS # coeffs.	1	9	3.973	1.281
NNBP # coeffs.	1	8	3.525	1.320
NNLS Time$_{sec.}$	0.002	0.190	0.068	0.037
NNBP Time$_{sec.}$	0.145	1.037	0.372	0.074

Pursuit NNBP approach, which constrains the L–1 norm of the $\alpha(x)$ vector and is minimized by the Non–Negative Merhotra's method [29]. On the other hand, in [18] it was proposed to fit it by the MATLAB-MathWorks implementation for Non–Negative Least Square (NNLS) method. In this work, our C-language implementation follows the Merhotra's method presented in [29] and also follows the strategy presented in Subsection IV-C in [17] that prevents ill–condition situations. Additionally, we force our Merhotra's algorithm to execute at least eight iterations for the sake of accuracy.

We perform, for both approaches, a statistical analysis of the following features for the computed voxelwise α vector:

1. DBF fit error

$$\sqrt{\sum_k (S_k - \sum_{j=1}^N \alpha_j \phi_{k,j})^2},$$

2. the number of $\alpha_j \geq 0.05$ coefficients involved in the solution (note that a vector solution with less involved coefficients is the sparser one), and
3. the required time.

Figure 1 and Table 1 illustrate the obtained results for 936 voxels on human brain DW-MRI data (for three axial slices in the ROI shown in Figure 3). Note that, for the same data, NNBP involves three coefficients for most voxels and NNLS involves four for most voxels, so that, based on the whole analysis we can conclude that: a) NNBP produces sparser solutions with an insignificantly bigger fitting error w.r.t. the NNLS one and b) NNBP is slower than NNLS (about 6 times slower than NNLS). For present work we prefer to use the NNBP since we consider that sparsity is more important than required time (moreover, the NNBP's fitting accuracy is not significantly worse than NNLS one).

3   Results

The DWI acquisition parameters were the following: DW–EPI on a 3.0-Tesla scanner, single-shot, spin-echo, nine images for b=$0s/mm^2$, 60 images for diffusion encoding directions isotropically distributed in space ($b=1000s/mm^2$), TR=6700ms, TE=85ms, 90° flip angle, voxel dimensions equal to $2 \times 2 \times 2 \ mm^3$.

3.1   Implementation Details

In order to fix the tensor–basis ($\overline{\mathbf{T}}$) anisotropy, we compute the mean DT on the *corpus callosum*, where it is assumed one fiber per voxel [15,30,17]; the computed mean eigen–values were $[\lambda_1, \lambda_2] = [1.4462 \times 10^{-3}, 0.4954 \times 10^{-3}]mm^2/s$. For comparison purpose, on the angular values, we use the discrete–to–continuous procedure described in [17](subsection IV-B) which collapses nearest discrete orientations (clusters) and improves the quality of the solution. Once we compute the continuous field of memberships α (described above), we normalize, per voxel, such a vector in order to compute the probability vector $\tilde{\alpha}$ in (5) which denotes the contribution of each DT in a probabilistic way.

Finally, we minimize cost function (4) by using the Gauss-Seidel method [29] (see the updating equations (16)–(18) in [27]).

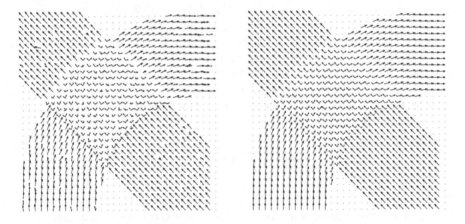

Fig. 2. Results of spatial regularization of synthetic axon fiber orientations. We show the pixelwise DBF result (left) and the regularized one by means of ECQMMF method (right).

3.2   Results on Synthetic Data

Figure 2 shows the regularization over a synthetic multi–DT field for validation purposes. We simulate a fiber crossing then we perturb the synthetic DW signals with Rician noise that results on a SNR = 7.5 (17.5dB), such a noise level is larger than the usually presents in a MRI data. Left panel shows the DBF–NNBP solution (i.e. the non–regularized one), such a multi-DT field presents the following statistical angular error (in degrees) w.r.t. the actual (ground–truth) orientations [min, max, **mean**, std.dev] = [0.00^o, 86.23^o **6.96^o** 9.54^o]. Right panel shows the regularized multi-DT field which presents the following statistical angular error [min, max, **mean**, std.dev] = [0.00^o, 22.46^o **2.90^o** 3.42^o].

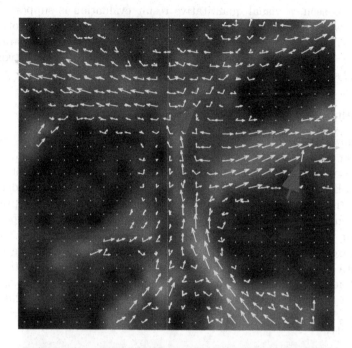

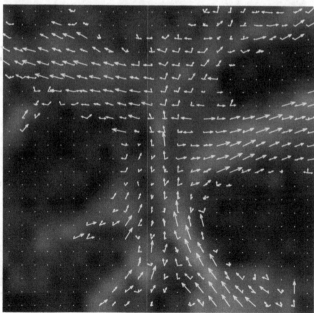

Fig. 3. Non-regularized (top) and regularized (bottom) fiber pathways for human DW-MRI data. Big red arrows indicate voxels where the regularization improves the solution in a significant way.

As we can be seen, a visual quantitative result evaluation is supported by a qualitative evaluation: the right panel presents smoother pathways in both single and multi-DT regions. Such a regularization was obtained with the parameters $[\lambda, \mu] = [4.0, 1.0]$, 6 iterations and 28.05 seconds for processing 2712 voxels in three 32×32 slices.

3.3 Results on Human Diffusion Weighted Data

Figure 3 shows non–regularized and regularized axon-fiber pathways. The ROI includes fiber crossings in the intersection of *posterior corona radiata* and *tapetum– splenium of corpus callosum* . The recover multi–orientations are plotted over color–coded DT principal diffusion direction in order to give to the reader an insight about the expected solutions. The DT orientation in a voxel is represented by red, green, and blue colors which stands for right–left, anterior–posterior, and superior–inferior orientations, respectively. Big red arrows in superior panel indicates voxels where the regularization dramatically corrects wrong estimations in both single and multiple fiber voxels. Additionally, note that the whole regularized field present smother trajectories and they agree with the expected ones. Such a regularization was computed with parameters $[\lambda, \mu] = [2.0, 5.0]$, 4 iterations and 6.03 seconds for processing 936 voxels (C implementation on a 2.80GHz CPU).

4   Discussion and Conclusions

We presented a new methodology for regularizing multiple axon fiber orientations. Our proposal does not require any prior information about the number of axon fiber within a single voxel. In order to test the proposal's performance we applied our regularization procedure to both synthetic and human DW-MR data. In the case of synthetic data, we introduce noise to the DW signals, then we show results from statistical analysis that validates the improvement of the method. In the case of real human DW-MR data, we show axon fiber orientation estimations that exhibit a qualitative improvement. We would like to emphasize that we processed high angular DW-MR data (60 orientations) with a standard b value (b=1000), and, despite the fact the high number of diffusion gradients one can realize that spatial regularization is still required in order to eliminate the noise effect.

The main contributions of this work are the following: a) We formally incorporate a voxelwise state-of-the-art method in form of likelihoods in a probabilistic spatially regularized procedure. Our approach computes high quality solutions by performing few minimization iterations that spend short time. b) We perform a statistical analysis over solutions given by two procedures for fitting the DBF model. Based on such analysis we can conclude that the highest quality solution is given by the NNBP approach and the faster one by NNLS method.

References

1. Basser, P.J., Mattiello, J., LeBihan, D.: MR diffusion tensor spectroscopy and imaging. Biophys. J. 66, 259–267 (1994)
2. Basser, P.J., Pierpaoli, C.: Microstructural and physiological features of tissues elucidated by quantitative-diffusion-tensor MRI. J. Magn. Reson. B 111, 209–219 (1996)
3. Buxton, R.: Introduction to Functional Magnetic Resonance Imaging Principles and Techniques. Cambridge University Press, Cambridge (2002)
4. Poldrack, R.A.: A structural basis for developmental dyslexia: Evidence from diffusion tensor imaging. In: Wolf, M. (ed.) Dyslexia, Fluency, and the Brain, pp. 213–233. York Press (2001)
5. Ruiz-Alzola, J., Westin, C.F., Warfield, S.K., Nabavi, A., Kikinis, R.: Nonrigid registration of 3D scalar, vector and tensor medical data. In: Proc. MICCAI, pp. 541–550 (2000)
6. Gee, J.C., Alexander, D.C., Rivera, M., Duda, J.T.: Non-rigid registration of diffusion tensor MR images. In: Press, I. (ed.) Proc. IEEE ISBI, July 2002, pp. 477–480. IEEE, Los Alamitos (2002)
7. Zhukov, L., Museth, K., Breen, D., Whitaker, R., Barr, A.: Level set modeling and segmentation of DT-MRI brain data. J. Electronic Imaging 12, 125–133 (2003)
8. Wang, Z., Vemuri, B.C.: Tensor field segmentation using region based active contour model. In: Pajdla, T., Matas, J(G.) (eds.) ECCV 2004. LNCS, vol. 3024, pp. 304–315. Springer, Heidelberg (2004)
9. Wang, Z., Vemuri, B.C.: An affine invariant tensor dissimilarity measure and its applications to tensor-valued image segmentation. In: Proc. CVPR, pp. 228–233 (2004)
10. Wang, Z., Vemuri, B.C.: DTI segmentation using an information theoretic tensor dissimilarity measure 24(10), 1267–1277 (2005)
11. Lenglet, C., Rousson, M., Deriche, R., Faugeras, O.D., Lehericy, S., Ugurbil, K.: A Riemannian approach to diffusion tensor images segmentation. In: Proc. IPMI, pp. 591–602 (2005)
12. Lenglet, C., Rousson, M., Deriche, R.: DTIsegmentation by statistical surface evolution 25(6), 685–700 (2006)
13. Alexander, D.C.: An introduction to computational diffusion MRI: the diffusion tensor and beyond. In: Weickert, J., Hagen, H. (eds.) Visualization and Image Processing of Tensor Fields. Springer, Berlin (2005)
14. Behrens, T.E.J., Berga, H.J., Jbabdi, S., Rushworth, M.F.S., Woolrich, M.W.: Probabilistic diffusion tractography with multiple fibre orientations: What can we gain? NeuroImage 34(1), 144–155 (2007)
15. Tuch, D.S., Reese, T.G., Wiegell, M.R., Makris, N., Belliveau, J.W., Wedeen, V.J.: High angular resolution diffusion imaging reveals intravoxel white matter fiber heterogeneity. Magn. Reson. Med. 48(4), 577–582 (2002)
16. Parker, J., Alexander, D.: Probabilistic Monte Carlo based mapping of cerebral connections utilising whole-brain crossing fibre information. In: Procc. IPMI., July 2003, pp. 684–695 (2003)
17. Ramirez-Manzanares, A., Rivera, M., Vemuri, B.C., Carney, P., Mareci, T.: Diffusion basis functions decomposition for estimating white matter intravoxel fiber geometry. IEEE Trans. Med. Imag. 26(8), 1091–1102 (2007)
18. Jian, B., Vemuri, B.C.: A unified computational framework for deconvolution to reconstruct multiple fibers from diffusion weighted mri. IEEE Trans. Med. Imag. 26(11), 1464–1471 (2007)

19. Chen, Y., Guo, W., Zeng, Q., He, G., Vemuri, B.C., Liu, Y.: Recovery of intra-voxel structure from HARD DWI. In: Proc. IEEE ISBI, October 2004, pp. 1028–1031 (2004)
20. Vemuri, B.C., Chen, Y., Rao, M., McGraw, T., Wang, Z., Mareci, T.: Fiber tract mapping from diffusion tensor MRI. In: Proc IEEE Workshop VLSM, pp. 81–88 (2001)
21. Weickert, J.: Diffusion and regularization methods for tensor-valued images. In: Proc. First SIAM-EMS Conf. AMCW (2001)
22. Wang, Z., Vemuri, B.C., Chen, Y., Mareci, T.: A constrained variational principle for direct estimation and smoothing of the diffusion tensor field from DWI. In: Proc. IPMI, vol. 18, pp. 660–671 (2003)
23. Wang, Z., Vemuri, B.C., Chen, Y., Mareci, T.H.: A constrained variational principle for direct estimation and smoothing of the diffusion tensor field from complex DWI 23(8), 930–939 (2004)
24. Tschumperlé, D., Deriche, R.: Vector-valued image regularization with PDE's: A common framework for different applications 27(4), 506–517 (2005)
25. Fillard, P., Arsigny, V., Pennec, X., Ayache, N.: Clinical DT-MRI estimation, smoothing and fiber tracking with log-Euclidean metrics. In: Proc. ISBI, pp. 786–789 (2006)
26. Marroquín, J.L., Santana, E.A., Botello, S.: Hidden Markov measure field models for image segmentation. IEEE Trans. Pattern Anal. Machine Intell. 25(11), 1380–1387 (2003)
27. Rivera, M., Ocegueda, O., Marroquin, J.L.: Entropy-controlled quadratic markov measure field models for efficient image segmentation. IEEE Trans. Image Processing 16(12), 3047–3057 (2007)
28. Olshausen, B.A., Field, D.J.: Sparse coding with an overcomplete basis set: A strategy employed by v1? Vision Research 37, 3311–3325 (1997)
29. Nocedal, J., Wright, S.J.: Numerical Optimization., 2nd edn. Springer Series in Operation Research (2000)
30. Tournier, J.D., Calamante, F., Gadian, D.G., Connelly, A.: Direct estimation of the fiber orientation density function from diffusion-weighted MRI data using spherical deconvolution. Neuroimage 23, 1176–1185 (2004)

A Profilometric Approach for 3D Reconstruction Using Fourier and Wavelet Transforms

Jesus Carlos Pedraza-Ortega , Efren Gorrostieta-Hurtado,
Juan Manuel Ramos-Arreguin, Sandra Luz Canchola-Magdaleno, Marco Antonio
Aceves-Fernandez, Manuel Delgado-Rosas, and Ruth Angelica Rico-Hernandez

CIDIT-Facultad de Informatica, Universidad Autonoma de Queretaro, Av. De las Ciencias
S/N, Juriquilla C.P. 76230, Queretaro, Mexico
`caryoko@yahoo.com, efren.gorrostieta@uaq.mx,`
`jramos@mecamex.net, sandracanchola@yahoo.com,`
`marco.aceves@uaq.mx, mdelgado80@hotmail.com, rico@uaq.mx`

Abstract. In this research, an improved method for three-dimensional shape measurement by fringe projection is presented. The use of Fourier and Wavelet transform based analysis to extract the 3D information from the objects is proposed. The method requires a single image which contains a sinusoidal white light fringe pattern projected on it. This fringe pattern has a known spatial frequency and this information is used to avoid the discontinuities in the fringes with high frequency. Several computer simulations and experiments have been carried out to verify the analysis. The comparison between numerical simulations and experiments has proved the validity of this proposed method.

Keywords: 3D Reconstruction, image processing, segmentation, Fourier, wavelet.

1 Introduction

In the last three decades, the idea to extract the 3D information of a scene from its 2D images has been widely investigated. Several contact and non-contact measurement techniques have been employed in many science and engineering applications to compute the 3-D surface of an object. Basically, the idea is to extract the useful depth information from an image in an efficient and automatic way. Then, the obtained information can be used to guide various processes such as robotic manipulation, automatic inspection, inverse engineering, 3D depth map for navigation and virtual reality applications [1]. Among all the diverse methodologies, one of the most widely used is the fringe projection. Fringe processing methods are widely used in non-destructive testing, optical metrology and 3D reconstruction systems. Some of the desired characteristics in these methods are high accuracy, noise-immunity and fast processing speed.

A scarcely used fringe processing methods are well-known like Fourier transform Profilometry (FTP) method [2] and phase-shifting interferometry [3]. It is also well-known that one of the main problems to overcome in these methods is the wrapped

A. Hernández Aguirre et al. (Eds.): MICAI 2009, LNAI 5845, pp. 313–323, 2009.

phase information. The first algorithm was proposed by Takeda and Mutoh in 1982[2], later Berryman [4] and Pedraza [5] proposed a modified Fourier Transform Profilometry by carrying out some global and local analysis in the wrapped phase. Then, unwrapping algorithms (temporal and spatial) were introduced and modified [6-8]. Another solution is to extract the information by the use of wavelet transform.

Due to the fact that presents a multiresolution in time and space frequency domain, wavelet transform has become a in a tool with many properties and advantages over Fourier transform [9-10]. The computation in the method could be carried out by analyzing the projected fringe patterns using wavelet transform. Mainly, this analysis consists of demodulating the deformed fringe patterns and extracting the phase information encoded into it and hence the height profile of the object can be calculated, quite similar to Fourier transform.

Different wavelet algorithms are used in the demodulation process to extract the phase of the deformed fringe patterns and can be classified into two categories: phase estimation and frequency estimation techniques.

The phase estimation algorithm employs complex mother wavelets to estimate the phase of a fringe pattern. The extracted phase suffers from 2π discontinuities and a phase unwrapping algorithm is required to remove these 2π jumps. Zhong [9] have used Gabor wavelet to extract the phase distribution where phase unwrapping algorithm is required.

The frequency estimation technique estimates the instantaneous frequencies in a fringe pattern, which are integrated to estimate the phase. The phase extracted using this technique is continuous; consequently, phase unwrapping algorithms are not required. Complex or real mother wavelets can be used to estimate the instantaneous frequencies in the fringe pattern. Dursun et al. [12] and Afifi et al. [13] have used Morlet and Paul wavelets respectively to obtain the phase distribution of projected fringes without using any unwrapping algorithms.

In the present research we present a simple profilometric approach to obtain the 3D information from an object. The main contribution of this work is to introduce a wavelet based profilometry and a comparison with the modified Fourier Transform and Profilometry is performed. Some virtual objects are created, and then computer simulations and experiments are conducted to verify the methodology.

2 Profilometric Approach

As described in the previous section, there are several fringe projection techniques which are used to extract the three-dimensional information from the objects. In this section, a Modified Fourier Transform and the Wavelet Profilometry are introduced.

2.1 Fourier Transform Profilometry

The image of a projected fringe pattern and an object with projected fringes can be represented by:

$$g(x, y) = a(x, y) + b(x, y) * \cos[2 * \pi f_0 x + \varphi(x, y)] \tag{1}$$

$$g_0(x, y) = a(x, y) + b(x, y) * \cos[2 * \pi f_0 x + \varphi_0(x, y)] \tag{2}$$

where $g(x, y)$ $g_0(x, y)$ are the intensity of the images at (x, y) point, $a(x, y)$ represents the background illumination, $b(x, y)$ is the contrast between the light and dark fringes, f_0 is the spatial-carrier frequency and $\varphi(x, y)$ and $\varphi_0(x, y)$ are the corresponding phase to the fringe and distorted fringe pattern, observed from the camera.

The phase $\varphi(x, y)$ contains the desired information, whilst $a(x, y)$ and $b(x, y)$ are unwanted irradiance variations. The angle $\varphi(x, y)$ is the phase shift caused by the object surface end the angle of projection, and it is expressed as:

$$\varphi(x, y) = \varphi_0(x, y) + \varphi_z(x, y) \tag{3}$$

Where $\varphi_0(x, y)$ is the phase caused by the angle of projection corresponding to the reference plane, and $\varphi_z(x, y)$ is the phase caused by the object's height distribution.

Considering the figure 1, we have a fringe which is projected from the projector, the fringe reaches the object at point H and will cross the reference plane at the point C. By observation, the triangles D_pHD_c and CHF are similar and

$$\frac{CD}{-h} = \frac{d_0}{l_0} \tag{4}$$

Leading us to the next equation:

$$\varphi_z(x, y) = \frac{h(x, y)2\pi f_0 d_0}{h(x, y) - l_0} \tag{5}$$

Where the value of $h(x, y)$ is measured and considered as positive to the left side of the reference plane. The previous equation can be rearranged to express the height distribution as a function of the phase distribution:

$$h(x, y) = \frac{l_0 \phi_z(x, y)}{\phi_z(x, y) - 2\pi f_0 d_0} \tag{6}$$

2.1.1 Fringe Analysis
The fringe projection equation 1 can be rewritten as:

$$g(x, y) = \sum_{n=-\infty}^{\infty} A_n r(x, y) \exp(in\varphi(x, y)) * \exp(i2\pi n f_0 x) \tag{7}$$

Where r(x,y) is the reflectivity distribution on the diffuse object [3, 4]. Then, a FFT (Fast Fourier Transform) is applied to the signal for in the x direction only. Therefore, we obtain the next equation:

$$G(f, y) = \sum_{-\infty}^{\infty} Q_n(f - nf_0, y) \tag{8}$$

Here $\varphi(x,y)$ and $r(x,y)$ vary very slowly in comparison with the fringe spacing, then the Q peaks in the spectrum are separated each other. Also it is necessary to consider that if we choose a high spatial fringe pattern, the FFT will have a wider spacing among the

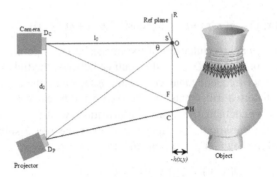

Fig. 1. Experimental setup

frequencies. The next step is to remove all the signals with exception of the positive fundamental peak f_0. The obtained filtered image is then shifted by f_0 and centered. Later, the IFFT (Inverse Fast Fourier Transform) is applied in the x direction only, same as the FFT. Separating the phase part of the result from the rest we obtain:

$$\varphi_z(x, y) = \varphi(x, y) + \varphi_0(x, y)$$
$$= \mathrm{Im}\{\log(\hat{g}(x, y)\hat{g}_0^*(x, y))\}$$

(9)

It is observed that the phase map can be obtained by applying the same process for each horizontal line. The values of the phase map are wrapped at some specific values. Those phase values range between π and $-\pi$.

To recover the true phase it is necessary to restore to the measured wrapped phase of an unknown multiple of $2\pi f_0$. The phase unwrapping process is not a trivial problem due to the presence of phase singularities (points in 2D, and lines in 3D) generated by local or global undersampling. The correct 2D branch cut lines and 3D branch cut surfaces should be placed where the gradient of the original phase distribution exceeded π *rad* value. However, this important information is lost due to undersampling and cannot be recovered from the sampled wrapped phase distribution alone. Also, is important to notice that finding a proper surface, or obtaining a minimal area or using a gradient on a wrapped phase will not work and could not find the correct branch in cut surfaces. From here, it can be observed that some additional information must be added in the branch cut placement algorithm.

2.1.2 Phase Unwrapping

As was early mentioned, the unwrapping step consists of finding discontinuities of magnitude close to 2π, and then depending on the phase change we can add or take 2π to the shape according to the sign of the phase change. There are various methods for doing the phase unwrapping, and the important thing to consider here is the abrupt phase changes in the neighbor pixels. There are a number of 2π phase jumps between 2 successive wrapped phase values, and this number must be determined. This number depends on the spatial frequency of the fringe pattern projected at the beginning of the process.

This step is the modified part in the Fourier Transform Profilometry originally proposed by Takeda [3], and represents the major contribution of this work. Another thing to consider is to carry out a smoothing before the doing the phase unwrapping, this procedure will help to reduce the error produced by the unwanted jump variations in the wrapped phase map. Some similar methods are described in [5]. Moreover, a modified Fourier Transform Profilometry method was used in [8] that include further extra analysis which considers local and global properties of the wrapped phase image.

2.2 Wavelet Transform

Wavelet transform (WT) is considered an appropriate tool to analyze non-stationary signals.

This technique has been developed as an alternative approach to the most common transforms, such as Fourier transform, to analyze fringe patterns. Furthermore, WT has a multi-resolution property in both time and frequency domains which surpasses a commonly know problem in other transforms such as resolution.

Wavelet is a small wave of limited duration (this can be real or complex). For this, two conditions must be satisfied: Firstly, it must have a finite energy. Secondly, the wavelet must have an average value of zero (admissibility condition). It is worth noting that many different types of mother wavelets are available for phase evaluation applications. The most suitable mother wavelet is probably the complex Morlet [2]. The Morlet wavelet is a plane wave modulated by a Gaussian function, and is defined as

$$\psi(x) = \pi^{1/4} \exp(icx)\exp(-x^2/2) \qquad (10)$$

where c is a fixed spatial frequency, and chosen to be about 5 or 6 to satisfy an admissibility condition [11]. Figure 3 shows the real part (dashed line) and the imaginary part (solid line) of the Morlet wavelet.

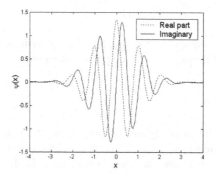

Fig. 2. Complex Morlet wavelet

The one-dimensional continuous wavelet transform (1D-CWT) of a row $f(x)$ of a fringe pattern is obtained by translation on the x axis by b (with y fixed) and dilation by s of the mother wavelet $\psi(x)$ as given by

$$W(s,b) = \frac{1}{\sqrt{s}} \int_{-\infty}^{\infty} f(x)\psi^*\left(\frac{x-b}{s}\right)dx \tag{11}$$

where * denotes complex conjugation and $W(s, b)$ is the calculated CWT coefficients which refers to the closeness of the signal to the wavelet at a particular scale.

2.2.1 Wavelet Phase Extraction Algorithms

In this contribution, phase estimation and frequency estimation methods are used to extract the phase distribution from two dimensional fringe patterns. In the phase estimation method, complex Morlet wavelet will be applied to a row of the fringe pattern. The resultant wavelet transform is a two dimensional complex array, where phase arrays can be calculated as follows:

$$abs(s,b) = |W(s,b)| \tag{12}$$

$$\varphi(s,b) = \tan^{-1}\left(\frac{\Im\{W(s,b)\}}{\Re\{W(s,b)\}}\right) \tag{13}$$

To compute the phase of the row, the maximum value of each column of the modulus array is determined and then its corresponding phase value is found from the phase array. By repeating this process to all rows of the fringe pattern, a wrapped phase map is resulted and unwrapping algorithm is needed to unwrap it.

In the frequency estimation method, complex Morlet wavelet is applied to a row of the fringe pattern. The resultant wavelet transform is a two dimensional complex array. The modulus array can be found using equation (12) and hence the maximum value for each column and its corresponding scale value can be determined. Then the instantaneous frequencies are computed using the following equation [11]

$$\hat{f}(b) = \frac{c+\sqrt{c^2+2}}{2s_{max}(b)} - 2\pi f_o \tag{14}$$

where fo is the spatial frequency. At the end, the phase distribution can be extracted by integrating the estimated frequencies and no phase unwrapping algorithm is required.

3 Proposed Method

An experimental setup shown on figure 1 is used to apply the methodology proposed on Figure 3. The first step is to acquire the image. Due to the nature of the image, sometimes a filtering to eliminate the noise is necessary. In this research, only a 9x9 Gaussian filter is used in the Fourier transform. In the wavelet transform application, none filter is applied. Next, we estimate the number of the fringes due to the fact that has a direct relationship with the fundamental frequency fo. Then, the fo is obtained by applying an algorithm. At this point, it is necessary to decide which method is going to be applied, either Fourier or else wavelet. Both methods, Fourier and wavelet analysis use the algorithms described in the previous section.

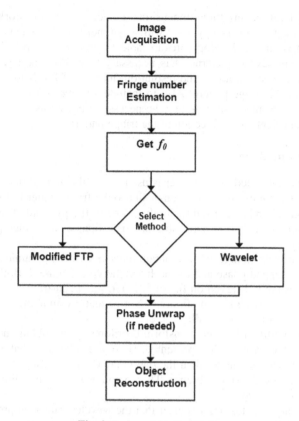

Fig. 3. Proposed Methodology

In the case of Fourier, a robust unwrapping algorithm is needed, and then an unwrapping algorithm with local and global analysis is used. The main algorithm for the local discontinuity analysis [5] is described as follows; a) first, the wrapped phase map is divided into regions and give a different weights (w_1, w_2, .., w_n) to each region, b) the modulation unit is defined and helps to detect the fringe quality and divides the fringes into regions, c) regions are grouped from the biggest to the smallest modulation value, d) next, the unwrapping process is started from the biggest to the smallest region, e) later, an evaluation of the phase changes can be carried out to avoid variations smaller than fo.

If the wavelet transform method is used then a simple unwrapping algorithm is enough to obtain the three-dimensional shape from the object. The final step is to obtain the object reconstruction and in some cases to determine the error (in case of the virtual created objects).

In the experimental setup, a high-resolution digital CCD camera can be used. The reference plane can be any flat surface like a plain wall, or a whiteboard. In the reference plane is important to consider a non reflective-surface to minimize the unwanted reflection effects that may cause some problems the image acquisition process. The object of interest can be any three-dimensional object and for this work, two objects are considered; the first one is a pyramid with a symmetrical shape and also a mask.

It is also important to develop software able to produce several different fringe patterns. To create several patterns, it is necessary to modify the spatial frequency (number of fringes per unit area), and resolution (number of levels to create the sinusoidal pattern) of the fringe pattern. Also to include into the software development a routine capable to do phase shifting may be necessary as well as to include the horizontal or vertical orientation projection of the fringe pattern.

3.1 Computer Simulation

A pyramid shape generated by computer is used to test the algorithms. The generated pyramid can be shown on figure 4. Then, a sinusoidal fringe pattern of known spatial frequency is created and later is added to the shape of the pyramid. The result image can be shown on figure 4. It is noticeable the distortions of the fringe pattern due to the pyramid shape.

The wavelet transform algorithm to obtain the shape of the pyramid is considered. As a result, the wrapped phase and its mesh are shown on figure 4. Both the wrapped phase and its mesh are observed on figure 5. Later, the reconstructed pyramid and its corresponding error can be seen on figure 6. Notice that, by applying this method, the magnitude of the error is very small.

Applying the modified Fourier Transform Profilometry, we obtain the mesh shown on figure 7; also the error mesh is presented. Here, it can be seen that even the shape of the error seems to be smaller than the one obtained by wavelet, the magnitude of the error obtained by Fourier is bigger. The computer simulation allowed us to test and validate both methods.

As a preliminary conclusion, it is clear that the wavelet transform presents a better performance than the Fourier transform for some object shapes. Therefore, the wavelet transform is selected and used. The error of the wavelet method is about 1 to 2% and using Fourier is about 4 to 6%.

Finally, the wavelet method is applied to a real object, and the results of the reconstruction can be seen on figure 8. Different views of the reconstructed object are presented.

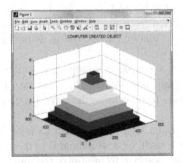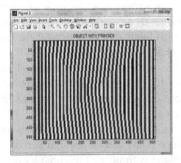

Fig. 4. Computer created pyramid and fringes projected on it

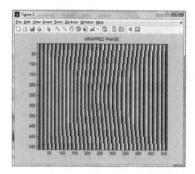
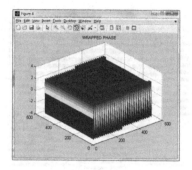

Fig. 5. Wrapped phase (image and mesh)

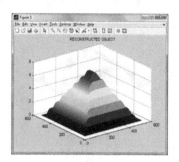
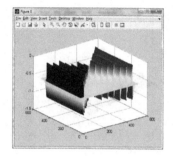

Fig. 6. Reconstructed object and error using wavelet transform

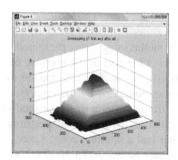
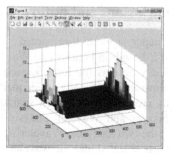

Fig. 7. Reconstructed object using Fourier transform and its error

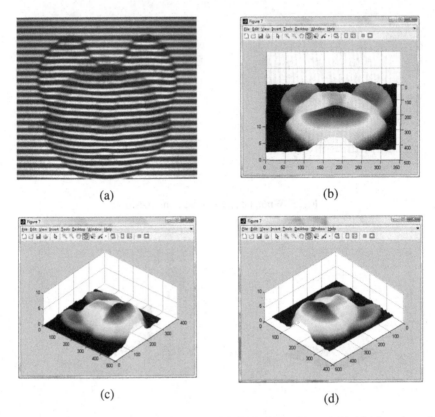

(a)

(b)

(c)

(d)

Fig. 8. Real object (a) and its reconstruction with front (b), and lateral (c) (d) views by using wavelet transform

4 Conclusions and Future Work

In this paper, a three-dimensional reconstruction methodology was presented. The method is based on Fourier and Wavelet transforms. One part of the propose method analytically obtains the high frequencies that mostly affect the performance on the phase unwrapping in the Fourier method. An object generated by the computer was virtually created and a known spatial sinusoidal fringe pattern was projected on it. Both Fourier and Wavelet analysis were conducted showing a good performance. Among the two analyses, Wavelet was the one that had a minimal error. Later, a real object was selected and the wavelet analysis was carried out and a good reconstruction of the object was achieved.

This methodology could be widely used to digitize diverse objects for reverse engineering, virtual reality, 3D navigation, and so on.

Notice that the method can reconstruct only the part of the object that can be seen by the camera, if a full 3D reconstruction (360 degrees) is needed, a rotating table is can be used and the methodology will be applied n times, where n is the rotation angle of the table.

As a future work, several tests can be carried out using objects with different shapes. Another option is to change the wavelet type and characterize the system.

One big challenge is to obtain the 3D reconstruction in real time. As a part of the solution, an optical filter to obtain the FFT directly can be used, or else, the algorithm can be implemented into a FPGA to carry out a parallel processing and minimize the processing time.

Acknowledgement

We want to acknowledge the financial support of this work through the PROMEP-SEP (PROMEP/103.5/08/3320) project.

References

1. Gokstorp, M.: Depth Computation in Robot Vision., Ph.D. Thesis, Department of Electrical Engineering, Linkoping University, S-581 83 Linkoping, Sweden (1995)
2. Takeda, M., Ina, H., Kobayashi, S.: Fourier-Transform method of fringe pattern analysis for computed-based topography and interferometry. J.Opt. Soc. Am. 72(1), 156–160 (1982)
3. Grevenkamp, J.E., Bruning, J.H.: Phase-shifting interferometry. In: Malacara, D. (ed.) Optical Shop Testing. Wiley, New York (1992)
4. Berryman, F., Pynsent, P., Cubillo, J.: A theoretical Comparison of three fringe analysis methods for determining the three-dimensional shape of an object in the presence of noise. Optics and Lasers in Engineering 39, 35–50 (2003)
5. Pedraza, J.C., Rodriguez, W., Barriga, L., et al.: Image Processing for 3D Reconstruction using a Modified Fourier Transform Profilometry Method. In: Gelbukh, A., Kuri Morales, Á.F. (eds.) MICAI 2007. LNCS (LNAI), vol. 4827, pp. 705–712. Springer, Heidelberg (2007)
6. Rastogi, P.K.: Digital Speckle Pattern Interferometry and related Techniques. Wiley, Chichester (2001)
7. Itoh, K.: Analysis of the phase unwrapping algorithm. Applied Optics 21(14), 2470–2486 (1982)
8. Lu, W.: Research and development of fringe projection-based methods in 3D shape reconstruction. Journal of Zhejiang University SCIENCE A, 1026–1036 (2006)
9. Zhong, J., Wang, J.: Spatial carrier-fringe pattern analysis by means of wavelet transform: wavelet transform profilometry. Applied Optics 43(26), 4993–4998 (2004)
10. Zang, Q., Chen, W., Tang, Y.: Method of choosing the adaptive level of discrete wavelet decomposition to eliminate zero component. Optics Communications 282, 778–785 (2009)
11. Dursun, A., Ozder, S., Ecevit, N.: Continuous wavelet transform analysis of projected fringe pattern analysis. Meas. Sci. Tech. 15, 1768–1772 (2004)
12. Afifi, M., Fassi-Fihri, M., Marjane, M., Nassim, K., Sidki, M., Rachafi, S.: Wavelet-based algorithm for optical phase distribution evaluation. Opt. Comm. 211, 47–51 (2002)
13. Gdeisat, M., Burton, D., Lalor, M.: Spatial carrier fringe pattern demodulation using a two-dimensional continuous wavelet transform. Appl. Opt. (2006)

Vector Quantization Algorithm Based on Associative Memories

Enrique Guzmán[1], Oleksiy Pogrebnyak[2], Cornelio Yáñez[2], and Pablo Manrique[2]

[1] Universidad Tecnológica de la Mixteca
eguzman@mixteco.utm.mx
[2] Centro de Investigación en Computación del Instituto Politécnico Nacional
olek@pollux.cic.ipn.mx

Abstract. This paper presents a vector quantization algorithm for image compression based on extended associative memories. The proposed algorithm is divided in two stages. First, an associative network is generated applying the learning phase of the extended associative memories between a codebook generated by the LBG algorithm and a training set. This associative network is named EAM-codebook and represents a new codebook which is used in the next stage. The EAM-codebook establishes a relation between training set and the LBG codebook. Second, the vector quantization process is performed by means of the recalling stage of EAM using as associative memory the EAM-codebook. This process generates a set of the class indices to which each input vector belongs. With respect to the LBG algorithm, the main advantages offered by the proposed algorithm is high processing speed and low demand of resources (system memory); results of image compression and quality are presented.

Keywords: vector quantization, associative memories, image coding, fast search.

1 Introduction

Vector quantization (VQ) has been used as an efficient and popular method in lossy image and speech compression [1], [2], [3]. In these areas, VQ is a technique that can produce results very close to the theoretical limits. Its main disadvantage is a high computation complexity. The most widely used and simplest technique for designing vector quantizers is the LBG algorithm by Y. Linde *et al.* [4]. It is an iterative descent algorithm which monotonically decreases the distortion function towards a local minimum. Sometimes, it is also referred as the generalized Lloyd algorithm (GLA), since it is a vector generalization of a clustering algorithm due to Lloyd [5]. New algorithms for VQ based on artificial neuronal networks (ANNs) have arisen as an alternative to traditional methods.

Sayed Bahram *et al.* in [6] proposed a method to design vector quantizers based on competitive learning neural networks. This technique is used to create a topological similarity between then input space and the index space and this similarity have been used to improve the performance of vector quantizers on the noisy channels.

A. Hernández Aguirre et al. (Eds.): MICAI 2009, LNAI 5845, pp. 324–336, 2009.

Kohonen´s algorithm, or Kohonen´s topological self-organizing map (SOM) [7], [8], [9], is a competitive-learning network that can be used for vector quantization. Two SOM features are used for vector quantization implementation; 1) SOM permits to realize the quantization of a continuous input space of stimuli into a discrete output space, the codebook; 2) The essential property of Kohonen algorithm is self-organization, since this algorithm is based on the observation of the brain that realizes self-organization of cartographies during the first stages of their development. N. Nasrabadi and Y. Feng [10] proposed a vector quantizer using Kohonen self-organizing feature maps. The obtained results were similar to the LBG algorithm when considering the quality of the compressed image. C. Amerijckx *et al.* [11], [12] proposed a new compression scheme for digital still images using the Kohonen's neural network algorithm. This new scheme is based on the organization property of Kohonen maps.

Recently, E. Guzmán *et al.* [13] proposed the design of an evolutionary codebook based on morphological associative memories to be used in a VQ process. The algorithm applied to the image for codebook generation uses the morphological autoassociative memories. An evolution process of codebook creation occurs applying the algorithm on new images, this process adds the information codified of the next image to the codebook allowing recover the images with better quality without affecting the processing speed.

In a previous work, we present the theoretical essentials to implement a fast search algorithm for vector quantization based on associative memories (FSA-EAM) [14], now this algorithm is simply named vector quantization based on associative memories (VQ-EAM). In this paper, we present the obtained results when we apply this algorithm in an images compression scheme for the particular case of the **prom** operator.

2 Theoretical Background of Extended Associative Memories

An associative memory designed for pattern classification is an element whose fundamental purpose is to establish a relation of an input pattern $\mathbf{x} = [x_i]_n$ with the index i of a class c_i (Fig. 1).

$$\mathbf{x} = \begin{pmatrix} x_1 \\ x_2 \\ \vdots \\ x_n \end{pmatrix} \longrightarrow \boxed{\begin{array}{c} \text{Associative Memory} \\ \text{for Pattern Classification} \end{array}} \longrightarrow i$$

Fig. 1. Scheme of associative memory for pattern classification

Let $\{(\mathbf{x}^1, c_1), (\mathbf{x}^2, c_2), ..., (\mathbf{x}^k, c_k)\}$ be k couples of a pattern and its corresponding class index defined as the *set of fundamental couples*. The set of fundamental couples is represented by

$$\{(\mathbf{x}^{\mu}, c_{\mu}) \mid \mu = 1, 2, \ldots, k\} \tag{1}$$

where $\mathbf{x}^{\mu} \in \mathbf{R}^{n}$ and $c_{\mu} = 1, 2, \ldots, N$. The associative memory is represented by a matrix generated from the set of fundamental couples.

In 2004, H. Sossa *et al.* proposed an associative memory model for the classification of real-valued patterns [15], [16]. This model is an extension of the Lernmatrix model proposed by K. Steinbuch [17]. The extended associative memory (EAM) is based on the general concept of learning function of associative memory and presents a high performance in pattern classification of real value data in its components and with altered pattern version. In the EAM, to the being an associative memory utilized for pattern classification, all the synaptic weights that belong to output i are accommodated at the i-th row of a matrix $\mathbf{M}$. The final value of this row is a function ϕ of all the patterns belonging to class i. The ϕ function acts as a generalizing learning mechanism that reflects the flexibility of the memory [15].

The learning phase consists on evaluating function ϕ for each class. Then, the matrix $\mathbf{M}$ can be structured as:

$$\mathbf{M} = \begin{bmatrix} \phi_1 \\ \vdots \\ \phi_N \end{bmatrix} \tag{2}$$

where ϕ_i is the evaluation of ϕ for all patterns of class i, $i = 1, 2, \ldots, N$. The function ϕ can be evaluated in various manners. The arithmetical average operator (**prom**) is used in the signals and images treatment frequently. In [15], the authors studied the performance of this operator when using EAM to evaluate the function ϕ.

2.1. Training phase of EAM

The goal of the training phase is to establish a relation between an input pattern $\mathbf{x} = [x_i]_n$, and the index i of a class c_i.

Considering that each class is composed for q patterns $\mathbf{x} = [x_i]_n$, and that $\phi_i = (\phi_{i,1}, \ldots, \phi_{i,n})$. Then, the training phase of the EAM can use the operator **prom** to evaluate the functions ϕ_i:

$$\phi_{i,j} = \frac{1}{q} \sum_{l=1}^{q} x_{j,l}, \quad j = 1, \ldots, n \tag{3}$$

The memory $\mathbf{M}$ is obtained after evaluating the functions ϕ_i. In case where N classes exist and the vectors to classify are n-dimensional, the memory $\mathbf{M} = [m_{ij}]_{N \times n}$ resultant is:

$$\mathbf{M} = \begin{bmatrix} \phi_{1,1} & \phi_{1,2} & \cdots & \phi_{1,n} \\ \phi_{2,1} & \phi_{2,2} & \cdots & \phi_{2,n} \\ \vdots & \vdots & \ddots & \vdots \\ \phi_{N,1} & \phi_{N,2} & \cdots & \phi_{N,n} \end{bmatrix} = \begin{bmatrix} m_{1,1} & m_{1,2} & \cdots & m_{1,n} \\ m_{2,1} & m_{2,2} & \cdots & m_{2,n} \\ \vdots & \vdots & \ddots & \vdots \\ m_{N,1} & m_{N,2} & \cdots & m_{N,n} \end{bmatrix} \tag{4}$$

2.2 Recalling Phase of EAM

The goal of the recalling phase is the generation of the class index to which the pattern belongs.

The pattern recall (pattern classification) by EAM is done when a pattern $\mathbf{x}^\mu \in \mathbf{R}^n$ is presented to the memory $\mathbf{M}$. It is not necessary that the pattern is one of the already used to build the memory $\mathbf{M}$.

For the case of the **prom** operator, the class to which $\mathbf{x}$ belongs is given by

$$i = \arg_l \left[\bigwedge_{l=1}^{N} \bigvee_{j=1}^{n} \left| m_{ij} - x_j \right| \right] \tag{5}$$

The operators $\vee \equiv \max$ and $\wedge \equiv \min$ execute morphological operations on the difference of the absolute values of the elements m_{ij} of $\mathbf{M}$ and the component x_j of the pattern $\mathbf{x}$ to be classified.

The **Theorem 1** and **Corollaries 1-5** from [15] govern the conditions that must be satisfied to obtain a perfect classification of a pattern.

3 Vector Quantization Algorithm Using Extended Associative Memories

For VQ applications, in this paper we propose a new algorithm, **VQ-EAM**, which uses the EAM in both codebook generation and codeword search. The proposed algorithm is divided in two stages. First, a codebook is generated using both the learning phase of the extended associative memories and the codebook generated by the LBG algorithm. This codebook is named EAM-codebook and represents a new codebook which is used in the next stage. Second, the vector quantization process is performed by means of the recalling stage of EAM using as associative memory the EAM-codebook. This process generates a set of the class indices to which each input vector belongs.

3.1. Codebook Generation using LBG Algorithm and Extended Associative Memories

In this stage, an associative network is generated applying the learning phase of the EAM between a codebook generated by the LBG algorithm and a training set. This associative network is named EAM-codebook and establishes a relation between training set and the LBG codebook.

The codebook generation based on LBG algorithm and the EAM is fundamental to obtain a fast search process for VQ. The Fig. 2 shows the scheme proposed for codebook generation.

For EAM-codebook generation, an image of $h \times w$ pixels is divided into M image blocks of $t \times l$ size, $X = \{\mathbf{x}^i : i = 1, 2, ..., M\}$, each of this image blocks represent a n-dimensional vector $\mathbf{x}=[x_j]$, $n = t \times l$.

The EAM-codebook is computed in three phases:

Phase 1. *LBG codebook generation*. This phase uses the LBG algorithm and the image blocks as the training set to generate an initial codebook: $C = \{\mathbf{y}^i : i = 1, 2,..., N\}$; this codebook is formed by a set of n-dimensional vectors $\mathbf{y} = [y_j]$, each vector is named "*codeword*".

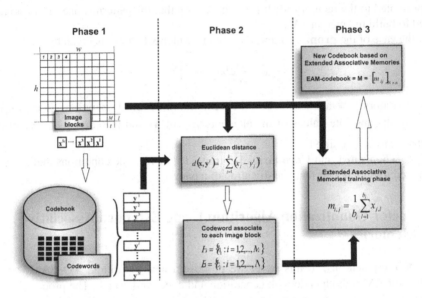

Fig. 2. Scheme of codebook generation algorithm based on LBG and EAM

Phase 2. *Determination of the codeword associated to each image block (training set) using Euclidean distance.* This phase uses the codebook generated in the previous phase and the training set.

    This stage designs a $\mathbf{Q}$ mapping and assigns an index i to each n-dimensional input pattern $\mathbf{x} = (x_1, x_2,..., x_n)$, with $\mathbf{Q}(\mathbf{x}) = \mathbf{y}^i = (y_{i1}, y_{i2},..., y_{in})$. The $\mathbf{Q}$ mapping is designed to map $\mathbf{x}$ to $\mathbf{y}^i$ with $\mathbf{y}^i$ satisfying the following condition:

$$d(\mathbf{x}, \mathbf{y}^i) = \min_j d(\mathbf{x}, \mathbf{y}^j), \text{ for } j = 1,2,3,..., N \tag{6}$$

where $d(\mathbf{x}, \mathbf{y}^j)$ is the distortion of representing the input pattern $\mathbf{x}$ by the codeword $\mathbf{y}^j$ measured by Euclidean distance.

    If one consider each codeword $\mathbf{y}^i$ as a class (existing N classes), then the result of this stage is a set of M indices that indicates to which class $\mathbf{y}^i$ each input pattern $\mathbf{x}$ belongs. Let us to denote this set of indices as $H = \{h_i : i = 1, 2,..., M\}$. This stage also generates a set of N indices that indicates the number of input patterns that integrate to each class. This set is denoted as $B = \{b_i : i = 1, 2,..., N\}$.

Phase 3. *Generation of a new codebook based on EAM.* The goal of the third stage is to generate a new codebook named EAM-codebook.

We will form a set of fundamental couples for the EAM-codebook generation, where the set of input patterns is integrated of M image blocks: $X = \{\mathbf{x}^i : i = 1, 2, ..., M\}$, $\mathbf{x}^i \in \mathbf{R}^n$, and the corresponding index of the class to which the input pattern $\mathbf{x}^i$ belongs is defined by $H = \{h_i : i = 1, 2, ..., M\}$. Finally, the number of input patterns that compose each class is contained in $B = \{b_i : i = 1, 2, ..., N\}$.

The training stage of the EAM now can be applied

$$m_{i,j} = \frac{1}{b_i} \sum_{l=1}^{b_i} x_{j,l}, \quad j = 1, ..., n$$

The result of this process is an associative network whose goal is to establish a relation of image blocks (training set) with the codebook generated by the LBG algorithm. This associative network is now a new codebook which will be used by the fast search algorithm for VQ proposed in this paper:

$$\textbf{EAM - Codebook} = \mathbf{M} = \left[m_{ij} \right]_{N \times n} \tag{7}$$

3.2 Vector Quantization Using VQ-EAM Algorithm

In a VQ process based on the similarity criterion between input patterns and codewords, each input pattern is replaced by the index of the codeword that present the nearest matching.

The vector quantization using **VQ-EAM** is a very simple process. The EAM-codebook generation is based on the EAM training phase. Next, the VQ process is performed using the EAM recalling phase. Fig. 3 shows the scheme of fast search algorithm based on EAM.

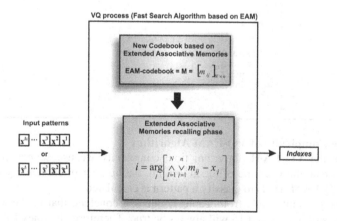

Fig. 3. Scheme of fast search algorithm based on EAM

The matrix $\mathbf{M} = \left[m_{ij}\right]_{N\times n}$ is the EAM-codebook. When an input pattern $\mathbf{x}^i \in \mathbf{R}^n$ is presented to the EAM-codebook, the index of the class to which the pattern belongs is generated. The index of the class to which $\mathbf{x}$ belongs is given by

$$i = \arg_i \left[\bigwedge_{l=1}^{N} \bigvee_{j=1}^{n} \left| m_{ij} - x_j \right| \right]$$

When a pattern not used to build the EAM-codebook is presented to **VQ-EAM** process, the EAM has the property to assign the index of the row of **M** which is the nearest to the pattern presented. This property allows to **VQ-EAM** to quantify in efficient form the patterns that do not belong to the training set.

3.3 Complexity of VQ-EAM Algorithm

The algorithm complexity allows determining the efficiency of an algorithm in the solution of a problem. It is measured by two parameters: the *time* complexity (how many steps it needs) and the *space* complexity (how much memory it requires). In this subsection, we analyze time and space complexity of the **VQ-EAM** algorithm. Table 1 shows the pseudo code of the most significant part of **VQ-EAM**.

Table 1. Pseudo code of the **VQ-EAM**

```
// FSA-EAM pattern classification subroutine
01| subroutine PCS()
02| variables
03|     x,k,j,l,aux: integer
04|     arg[N]='0': integer
05| begin
06|     for k←1 to N [operations k=k+1] do
07|         for j←1 to n [operations j=j+1]) do
08|             aux=abs(EAM_codebook[k][j]-x[j]);
09|             if (aux>arg[k]) then
10|                 arg[k]=aux;
11|             end_if
12|         end_for
13|     end_for
14|     aux=max(x);
15|     for l←1 to N [operations l=l+1] do
16|         if(arg[l]<aux)
17|             aux=arg[l];
18|             index=l;
19|         end_if
20|     end_for
21| end_subroutine
```

3.3.1 Time Complexity of VQ-EAM Algorithm
To measure the **VQ-EAM** algorithm time complexity we use several parameters. First, to estimate the run time of the algorithm, the number of elementary operations (EO) that **VQ-EAM** needs to classify a pattern is calculated.

Considering pseudo code from Table 1, one can conclude that in the *worst case*, the conditions of lines 9 and 15 will always be true. Therefore, the lines 10, 16 and 17 will be executed in all iterations, and then the internal loops realize the following number of EO:

$$\left(\sum_{j=1}^{n}(9+3) \right)+3=12\left(\sum_{j=1}^{n}1 \right)+3=12n+3$$

$$\left(\sum_{l=1}^{N}(5+3) \right)+3=8\left(\sum_{l=1}^{N}1 \right)+3=8N+3$$

The next loop will repeat $12n+3$ EO at each iteration:

$$\left(\sum_{k=1}^{N}(12n+3)+3 \right)+3=\left(\sum_{k=1}^{N}12n+6 \right)+3=N(12n+6)+3=12Nn+6N+3$$

Thus, the total number of EO that realizes the algorithm is:

$$T(n)=(12Nn+6N+3)+(8N+3)+1=12Nn+14N+7 \tag{8}$$

where N is the codebook size and n is a pattern dimension.

Second, the number and type of arithmetical operations realized for the **VQ-EAM** algorithm to quantify an image is analized. For this purpose, we know that this process is applied to image of $h \times w$ size, where h is the image height and w is the image width and considering pseudo code from Table 1. The number of operations that **VQ-EAM** needs to quantify the image depends on the image size, the codebook size N and the pattern dimension n:

$$\left(\frac{h \times w}{n} \right)[Nn\,(op\,2)+N\,(op\,1)] \tag{9}$$

where $op1=1$ comparison and $op2=1$ sum and 1 comparison; $(h \times w)/n$ is the number of patterns to quantify.

$Nn(op2)$ is the number and type of arithmetical operations used to perform a morphological dilation over the absolute difference of the EAM-codebook elements and the input pattern components; $N(op1)$ is the number and type of arithmetical operations used to perform a morphological erosion over the result of the previous process.

3.3.2 Space complexity of VQ-EAM algorithm
The **VQ-EAM** algorithm space complexity is determined by the amount of memory required for its execution.

To quantify an image of $h \times w$ size, which contains $M=(h \times w)/n$ n-dimensional patterns, a VQ process using **VQ-EAM** requires two vectors arg$[N]$, indices$[M]$ and one matrix EAM-codebook$[N][n]$. Hence, the number of memory units (mu) required for this process is:

$$\text{mu\_arg} + \text{mu\_indexes} + \text{mu\_EAM-codebook} =$$

$$\begin{aligned} N + M + (N)(n) = \\ M + N(n+1) \end{aligned} \tag{10}$$

The **VQ-EAM** algorithm uses only subtraction and comparison operations. Therefore, the result always is an integer number. For grayscale image compression, 8 bits/pixel, the **VQ-EAM** requires a variable of more than 8 bits. Compilers allow declaring variables of type *short* of 16 bit integer signed numbers. Hence, the total number of bytes required by the **VQ-EAM** algorithm is:

$$2(M + N(n+1))$$ (11)

The number of memory units depends on the image size, codebook size and the codeword size chosen for the VQ process.

4   Experimental Results

In this section, we present the experimental results obtained using the **VQ-EAM**. First, we compare the **VQ-EAM** performance with the traditional LBG algorithm. Second, we analyze the number and type of operations and the amount of memory used by the **VQ-EAM** and the traditional LBG algorithm. For this purpose, a set of test image of size 512×512 with 256 gray levels (Fig. 5) was used in simulations.

Fig. 5. Set of test images: (a) Lena, (b) Peppers, (c) Elaine, (d) Man

In order to measure the performance of both **VQ-EAM** and LBG algorithm, we used a popular objective performance criterion named peak signal-to-noise ratio (PSNR), which is defined as

$$PSNR = 10\log_{10}\left(\frac{\left(2^{n}-1\right)^{2}}{\frac{1}{M}\sum_{i=1}^{M}(p_{i}-\tilde{p}_{i})^{2}}\right)$$ (12)

where n is the number of bits per pixel, M is the number of pixels in the image, p_{i} is the i-th pixel in the original image, and $\tilde{p}_{i}$ is the i-th pixel in the reconstructed image.

The first experiment has the objective to compare the distortions that both the **VQ-EAM** and LBG algorithm add to the original image. For this purpose, the codebook was generated using the image Lena as the training set and the results were obtained using different sizes of the codebook. Then, VQ was applied to the set of the test images in order to determine the behavior of the algorithms with the patterns that do not belong to the training set. The block size used in the experiments is 4×4. To analyze the **VQ-EAM** performance Table 2 shows the results of this experiment.

From Table 2, one can make the following observations: 1) the obtained results show that the proposed method is competitive with the LBG algorithm in PSNR parameter; 2) EAM replaced an input pattern by the index of the codeword that present the nearest matching, not necessarily one of the already used to built the **M** memory. This property allows to **VQ-EAM** quantify in an efficient form the patterns that do not belong to the training set.

In the second experiment, the performance of diverse standard encoding methods applied to the quantified image with **VQ-EAM** was evaluated. These methods included statistical modeling techniques, such as arithmetical, Huffman, range, BurrowsWheeler

Table 2. Comparison of the proposed algorithm and the LBG algorithm for the set of images

	LBG Algorithm				VQ-EAM			
	Codebook size				Codebook size			
Images	64	128	256	512	64	128	256	512
	PSNR	PSNR	PSNR	PSNR	PSNR	PSNR	PSNR	PSNR
Lena	27.14	28.21	29.08	29.98	26.31	27.47	28.39	29.25
Peppers	26.38	27.18	27.64	28.21	25.12	26.06	26.54	27.04
Elaine	28.91	29.68	30.30	30.74	28.31	29.07	29.68	30.11
Man	24.20	24.93	25.49	26.02	23.28	23.96	24.55	25.09

transformation, PPM, and dictionary techniques, LZ77 and LZP. The purpose of the second experiment is to analyze **VQ-EAM** performance in image compression. To this end, a coder that implements **VQ-EAM** and diverse entropy encoding techniques was developed. The compression results of the coder on test images are expressed in Table 3.

These results show that the entropy encoding technique that offers the best results in compression are obtained on the quantified image by **VQ-EAM**, is the PPM coding. The PPM is an adaptive statistical method; its operation is based on partial equalization of chains, that is, the PPM coding predicts the value of an element basing on the sequence of previous elements.

Table 3. Performance comparison of several entropy encoding technique on the information obtained from **VQ-EAM** on test images

| Images | Entropy encoding technique | LBG Algorithm | | | VQ-EAM Algorithm (our proposal) | | |
| | | Codebook size | | | Codebook size | | |
		128	256	512	128	256	512
Elaine	PPM	0.21 bpp	0.29 bpp	0.39 bpp	0.22 bpp	0.30 bpp	0.41 bpp
	SZIP	0.22 bpp	0.30 bpp	0.40 bpp	0.23 bpp	0.31 bpp	0.41 bpp
	Burrown	0.23 bpp	0.31 bpp	0.41 bpp	0.24 bpp	0.32 bpp	0.42 bpp
	LZP	0.24 bpp	0.31 bpp	0.45 bpp	0.24 bpp	0.32 bpp	0.47 bpp
	LZ77	0.26 bpp	0.35 bpp	0.52 bpp	0.27 bpp	0.36 bpp	0.53 bpp
	Range	0.34 bpp	0.41 bpp	0.62 bpp	0.34 bpp	0.41 bpp	0.62 bpp
Lena	PPM	0.26 bpp	0.34 bpp	0.46 bpp	0.26 bpp	0.34 bpp	0.47 bpp
	SZIP	0.27 bpp	0.34 bpp	0.46 bpp	0.27 bpp	0.35 bpp	0.47 bpp
	Burrown	0.27 bpp	0.35 bpp	0.46 bpp	0.27 bpp	0.36 bpp	0.47 bpp
	LZP	0.27 bpp	0.35 bpp	0.52 bpp	0.28 bpp	0.36 bpp	0.53 bpp
	LZ77	0.30 bpp	0.38 bpp	0.56 bpp	0.30 bpp	0.39 bpp	0.57 bpp
	Range	0.39 bpp	0.46 bpp	0.66 bpp	0.38 bpp	0.45 bpp	0.65 bpp
Man	PPM	0.30 bpp	0.38 bpp	0.49 bpp	0.31 bpp	0.39 bpp	0.49 bpp
	SZIP	0.32 bpp	0.39 bpp	0.49 bpp	0.32 bpp	0.40 bpp	0.49 bpp
	Burrown	0.32 bpp	0.40 bpp	0.49 bpp	0.32 bpp	0.40 bpp	0.49 bpp
	LZP	0.33 bpp	0.41 bpp	0.57 bpp	0.34 bpp	0.41 bpp	0.57 bpp
	LZ77	0.35 bpp	0.42 bpp	0.60 bpp	0.35 bpp	0.43 bpp	0.61 bpp
	Range	0.39 bpp	0.45 bpp	0.65 bpp	0.38 bpp	0.45 bpp	0.65 bpp
Peppers	PPM	0.23 bpp	0.30 bpp	0.41 bpp	0.24 bpp	0.32 bpp	0.43 bpp
	SZIP	0.24 bpp	0.31 bpp	0.41 bpp	0.25 bpp	0.33 bpp	0.43 bpp
	Burrown	0.24 bpp	0.32 bpp	0.42 bpp	0.25 bpp	0.34 bpp	0.43 bpp
	LZP	0.25 bpp	0.32 bpp	0.47 bpp	0.26 bpp	0.34 bpp	0.49 bpp
	LZ77	0.28 bpp	0.36 bpp	0.52 bpp	0.29 bpp	0.37 bpp	0.54 bpp
	Range	0.36 bpp	0.43 bpp	0.63 bpp	0.36 bpp	0.43 bpp	0.63 bpp

Now, let us compare the complexity of the LBG and **VQ-EAM** algorithms. For this purpose, we need to determine the time complexity and space complexity of the LBG-VQ for an image $h \times w$ dimension, which contains $M = (h \times w)/n$ n-dimensional patterns. First, we analyze the number and type of arithmetical operations used by LBG-VQ. LBG-VQ is divided in two processes: distortion computation (**dc**) and comparing distortions (**cd**).

The number of operations of the **cd** process is

$$op(cd) = M[(N+1)op(dc) + N(comparison\ s)] \tag{13}$$

If the Euclidean distance is used as a distortion measure, then $op(dc)$ is defined by

$$op(dc) = n(3\ adds\ + 1\ multiplica\ tion) + 1\ sqrt \tag{14}$$

The expression (19) indicates the number of operations that LBG-VQ needs to quantify the image.

Now, we need a space complexity analysis of the LBG-VQ algorithm. The LBG-VQ uses three vectors eucli1$[n]$, eucli2$[n]$, indexes$[M]$ and one matrix LBG-codebook$[N][n]$ to quantify an image. Hence, the number of memory units (mu) required for this process is:

$$mu\_eucli1 + mu\_eucli2 + mu\_indexes + mu\_LBG\text{-}codebook =$$

$$\begin{aligned} n + n + M + (N)(n) = \\ M + n(N+2) \end{aligned} \tag{15}$$

The LBG-VQ process uses floating point operations, only the result can be expressed by integers. Then, the total number of bytes required by the LBG-VQ is

$$2M + 4n(N+2) \tag{16}$$

Table 4 summarizes computation complexities of the LBG and **VQ-EAM** algorithms in terms of the average number of operations per pixels. This experiment was

Table 4. Operations and memory required by LBG and **VQ-EAM** to quantify a gray scale image of 512×512 pixels, 8 bits/pixel

Algorithm	Pattern dimension (n)	Codebook size (N)	Required memory (bytes)	The average number of operations per pixel			
				CMP	±	×	SQRT
LBG	16	128	41088	8	387	129	8.0625
		256	49280	16	771	257	16.0625
		512	65664	32	1539	513	32.0625
	64	128	41472	2	387	129	2.015625
		256	74240	4	771	257	4.015625
		512	139776	8	1539	513	8.015625
VQ-EAM	16	128	37080	136	128	-	-
		256	41432	272	256	-	-
		512	50136	544	512	-	-
	64	128	24832	130	128	-	-
		256	41472	260	256	-	-
		512	74752	520	512	-	-

performed for $N = 128$, 256 and 512, which are the most popular codebook sizes. With regard to time complexity, this Table shows that the proposed algorithm provides considerable improvement over the LBG algorithm. Table 3 also shows that the memory required by the **VQ-EAM** is smaller that the memory needed by LBG when a VQ process is performed.

5 Conclusions

In this paper, we have proposed the use of extended associative memories in a VQ fast searching algorithm for image compression. The use of EAM at the VQ stage of an image compressor has demonstrated a high competitiveness in its efficiency in comparison to VQ traditional methods like LBG algorithm. The compression ratio and the signal to noise ratio obtained by our proposal show that the decoding quality and rate compression remains competitive with respect to the LBG algorithm. It is because the EAM is used as a pattern classifier; in this operation mode, the EAM replaces the input pattern by the index of the codeword that present the nearest matching, without taking a care that they have not been used in the construction of the EAM-codebook.

The EAM operation is based on maximums or minimums of sums, that is, it uses only operations of sums and comparisons; then, with respect to the LBG algorithm, the main advantages offered by the proposed algorithm is high processing speed and low demand of resources (system memory). For these reasons, we can conclude that the **VQ-EAM** provides a considerable improvement over the LBG algorithm.

References

1. Gray, R.M.: Vector Quantization. IEEE ASSP Magazine 1, 4–9 (1984)
2. Nasrabadi, N.M., King, R.A.: Image Coding Using Vector Quantization: A Review. IEEE Trans.on Communications 36(8), 957–971 (1988)
3. Gersho, A., Gray, R.M.: Vector Quantization and Signal Compression. Kluwer, Norwell (1992)
4. Linde, Y., Buzo, A., Gray, R.: An Algorithm for Vector Quantizer Design. IEEE Trans. on Communications 28(1), 84–95 (1980)
5. Lloyd, L.P.: Least Squares Quantization in PCM. IEEE Trans. Inform. Theory IT-28, 129–137 (1982)
6. Bahram, S., Azami, Z., Feng, G.: Robust Vector Quantizer Design Using Competitive Learning Neural Networks. In: Proc. of European Workshop on Emerging Techniques for Communications Terminals, pp. 72–75. IEEE Press, Toulouse France (1997)
7. Kohonen, T.: Automatic Formation of Topological Maps of Patterns in a Self-organizing System. In: Oja, E., Simula, O. (eds.) Proc. 2SCIA, Scand. Conf. on Image Analysis, Helsinki, Finland, pp. 214–220 (1981)
8. Kohonen, T.: The Self-Organizing Map. IEEE Proc. 78(9), 1464–1480 (1990)
9. Kohonen, T.: Self-Organizing Maps, 3rd edn. Springer, Berlin (2001)
10. Nasrabadi, N., Feng, Y.: Vector Quantization of Images based upon the Kohonen Self-Organizing Feature Maps. In: IEEE International Conference on Neural Networks, vol. 1, pp. 101–108 (1988)

11. Amerijckx, C., Verleysen, M., Thissen, P., Legat, J.-D.: Image Compression by Self-Organized Kohonen Map. IEEE Trans. on Neural Networks 9, 503–507 (1998)
12. Amerijckx, C., Legat, J.-D., Verleysen, M.: Image Compression Using Self-Organizing Maps. Systems Analysis Modelling Simulation 43(11), 1529–1543 (2003)
13. Guzmán, E., Pogrebnyak, O., Yañez, C.: Design of an Evolutionary Codebook Based on Morphological Associative Memories. In: Gelbukh, A., Kuri Morales, Á.F. (eds.) MICAI 2007. LNCS (LNAI), vol. 4827, pp. 601–611. Springer, Heidelberg (2007)
14. Guzmán, E., Pogrebnyak, O., Yañez, C.: A Fast Search Algorithm for Vector Quantization based on Associative Memories. In: Ruiz-Shulcloper, J., Kropatsch, W.G. (eds.) CIARP 2008. LNCS, vol. 5197, pp. 487–495. Springer, Heidelberg (2008)
15. Sossa, H., Barrón, R., Vázquez, A.: Real-valued Patterns Classification based on Extended Associative Memory. In: Fifth Mexican International Conference on Computer Science, ENC 2004, pp. 213–219. IEEE Computer Society, México (2004)
16. Barron, R.: Associative Memories and Morphological Neural Networks for Patterns Recall (in Spanish). PhD Thesis, Center for Computing Research (2005)
17. Steinbuch, K.: Die Lernmatrix. Kybernetik 1(1), 26–45 (1961)

A Two-Stage Relational Reinforcement Learning with Continuous Actions for Real Service Robots

Julio H. Zaragoza and Eduardo F. Morales

National Institute of Astrophysics, Optics and Electronics,
Computer Science Department,
Luis Enrique Erro 1, 72840 Tonantzintla, México
{jzaragoza,emorales}@inaoep.mx

Abstract. Reinforcement Learning is a commonly used technique in robotics, however, traditional algorithms are unable to handle large amounts of data coming from the robot's sensors, require long training times, are unable to re-use learned policies on similar domains, and use discrete actions. This work introduces $TS\text{-}RRLCA$, a two stage method to tackle these problems. In the first stage, low-level data coming from the robot's sensors is transformed into a more natural, relational representation based on rooms, walls, corners, doors and obstacles, significantly reducing the state space. We also use Behavioural Cloning, i.e., traces provided by the user to learn, in few iterations, a relational policy that can be re-used in different environments. In the second stage, we use Locally Weighted Regression to transform the initial policy into a continuous actions policy. We tested our approach with a real service robot on different environments for different navigation and following tasks. Results show how the policies can be used on different domains and perform smoother, faster and shorter paths than the original policies.

Keywords: Relational Reinforcement Learning, Continuous Actions, Robotics.

1 Introduction

Nowadays it is possible to find service robots for many different tasks. Due to the wide range of services that they provide, service robots usage has been increased, in recent years, in places like houses and offices. However, their complete use and acceptance will depend on its capability to learn new tasks.

Reinforcement Learning (RL) has been widely used and suggested as a good candidate for learning tasks in robotics. However, the use and application of traditional RL techniques has been hampered by four main aspects: (1) vast amount of data produced by the robot's sensors, (2) large search spaces, (3) the use of discrete actions, and (4) inability to re-use previously learned policies.

Robots are normally equipped with laser range sensors, rings of sonars, cameras, etc., which produce a large number of readings at high sample rates creating problems to many learning algorithms. Large search spaces produce very long

A. Hernández Aguirre et al. (Eds.): MICAI 2009, LNAI 5845, pp. 337–348, 2009.

training times which is a problem for service robotics where the state space is continuous and a description of a state may involve several variables. Researchers have proposed different strategies, normally based on a discretization of the state space with discrete actions or with function approximation techniques. However, discrete actions produce unnatural movements and slow paths and function approximation techniques tend to be computationally expensive. Also, in many approaches, once a policy has been learned to solve a particular task, it cannot be re-used on similar tasks.

In this paper, *TS-RRLCA* (*Two-Stage Relational Reinforcement Learning with Continuous Actions*) a two stage method that tackles these problems, is presented. In the first stage, low-level information from the robot's sensors is transformed into a relational representation to characterize the set of states describing the robot's environment. The policies learned with this representation framework are transferable to other similar domains. We also use Behavioural Cloning [2], i.e., traces, to induce only a subset of relevant actions per state accelerating our policy learning process. This stage produces, in few iterations, a relational control policy with discrete actions. In the second stage, the learned policy is transformed into a relational policy with continuous actions through a fast Locally Weighted Regression (*LWR*) process.

The learned policies were successfully applied to a simulated and a real service robot performing navigation and following tasks with different scenarios and goals. It is shown how the continuous actions policies are able to produce smoother and shorter paths than the original relational policies and how the performance's quality of the tasks is similar to those performed by humans.

This paper is organized as follows, Section 2 presents related work. Section 3 describes a process to reduce the data coming from the robot sensor's. Section 4 introduces our relational representation to characterize states and actions. Sections 5 and 6 describe, respectively, the first and second stages of the proposed method. Section 7 shows experiments and results and Section 8 concludes and suggests future research directions.

2   Related Work

Recently, there has been an increasing interest in addressing adaptation for control tasks in robotics. In [8], a method to build relational macros for transfer learning in robot's navigation tasks is introduced. A macro consists of a finite state machine, i.e., a set of nodes along with rulesets for transitions and action choices. In [3], a proposal to learn relational decision trees as abstract navigation strategies from example paths in presented. These approaches use relational representations to transfer learned knowledge and use training examples to speed up learning, unfortunately, they consider only discrete actions.

In [11], the authors introduced a method that temporarily drives a robot which follows certain initial policy while some user commands play the role of training input to the learning component, which optimizes the autonomous control policy for the current task. In [4], a robot is teleoperated to learn secuences of state-action pairs that show how to perform a task. These methods reduce the computational costs and times for developing its control scheme, but they use discrete actions and are unable to transfer learned knowledge. A strategy to allow the use of continuous actions is to approximate a continuous function over the state space. The work developed in [5] is a Neural Network coupled with an interpolation technique that approximates Q-$values$ to find a continuous function over all the search space. In [1], the authors use $Gaussian$ $Processes$ for learning a probabilistic distribution for a robot navigation problem. The main drawback of these methods is the computational costs and the long training times as they try to generate a continuous function over all of the search space.

Our method learns, through a relational representation, relational discrete actions policies able to transfer knowledge between similar domains. We also speed up and simplify the learning process by using traces provided by the user. Finally we use a fast LWR to transform the original discrete actions policy into a continuous actions policy. Next sections describe in detail the proposed method.

3   Natural Landmarks Representation

While performing a task, the robot senses and returns large amounts of data readings coming from its sensors. In order to produce a smaller set of meaningful information TS-$RRLCA$ uses a process based on [6,10]. In [6], the authors describe a process able to identify three kinds of natural landmarks through laser sensor readings: (1) discontinuities, defined as an abrupt variation in the measured distance of two consecutive laser readings (Figure 1(a)), (2) corners, defined as the location where two walls intersect and form and angle and (3) walls, identified using the Hough transform. We also add obstacles identified through sonars and defined as any detected object between certain range.

A natural landmark is represented by a tuple of four attributes: (DL , θL, A, T). DL and θL are, respectively, the relative distance and orientation from the landmark to the robot. T is the type of the landmark: l for left discontinuity, r for right discontinuity (Figure 1(b)), c for corner, w for wall and o for obstacle. A is a distinctive attribute and its value depends on the type of the landmark, for discontinuities A is depth and for walls A is its length.

In [10] the data from laser readings is used to feed a clustering based process which is able to identify the robot's actual location such as room, corridor and/or intersection (the location where rooms and corridors meet). Figures 1(c), 1(d) and 1(e) show examples of the resulting location clasification process.

Table 1 shows an example of the resulting data from applying this processes to the laser and sonar readings from Figure 2. The robot's actual location in this case is in-$room$.

The natural landmarks along with the robot's actual location, are used to characterize the relational states that describe the environment.

4   Relational Representations for States and Actions

In order to learn transferable policies, able to fulfill their goals under different conditions or even in different domains, a relational representation, where it is

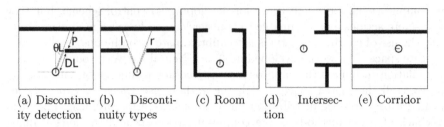

(a) Discontinu- (b)   Disconti-   (c) Room   (d)   Intersec-   (e) Corridor
ity detection   nuity types   tion

Fig. 1. Discontinuities attibutes and locations detected through a clustering processes

Table 1. Indentified natual landmarks from the sensor's readings from Figure 2

N	DL	θL	A	T
1	0.92	-17.60	4.80	r
2	1.62	-7.54	3.00	l
3	1.78	17.60	2.39	l
4	0.87	-35.70	1.51	w
5	4.62	-8.55	1.06	w
6	2.91	-6.54	1.88	w
7	1.73	23.63	0.53	w
8	2.13	53.80	2.38	w
9	5.79	-14.58	0.00	c
10	2.30	31.68	0.00	c
11	1.68	22.33	0.00	c
12	1.87	-170.00	0.00	o
13	1.63	-150.00	0.00	o
14	1.22	170.00	0.00	o
15	1.43	150.00	0.00	o

Fig. 2. Robot sensing its environment through laser and sonar sensors

easy to encode the relative position of an agent with respect to a goal or to other objects in the environment, is used. The main idea is to represent states as sets of properties that can be used to characterize a particular state and which may be common to other states. A relational state (r-$state$) is a conjuntion of first order predicates. Our states are characterized by the following predicates which receive as parameters a set of values such as those shown in Table 1.

- *place*: robot's location. Evaluates if the received location value is valid. The valid values are *in-room*, *in-door*, *in-corridor* and *in-intersection*.
- *doors_detected*: orientation and distance to doors. A door is characterized by identifying a right discontinuity (r) followed by a left discontinuity (l) from the natural landmarks. The door's orientation angle and distance values are calculated by averaging the values of the right and left discontinuities angles and distances. The values used for door orientation are: *right* (door's angle between -67.5° and -112.5°), *left* (67.5° to 112.5°), *front* (22.5° to -22.5°), *back* (157.5° to -157.5°), *right-back* (-112.5° to -157.5°), *right-front* (-22.5° to -67.5°), *left-back* (112.5° to 157.5°) and *left-front* (22.5° to 67.5°). The

values used for distance are: *hit* (door's distance between $0m.$ and $0.3m.$),
close ($0.3m.$ to $1.5m.$), *near* ($1.5m.$ to $4.0m.$) and *far* ($> 4.0m.$).
- *walls_detected*: length, orientation and distance to walls (type w landmarks).[1]
  The possible values for wall's size are: *small* (length between $0.15m.$ and
  $1.5m.$), *medium* ($1.5m.$ to $4.0m.$) and *large* ($> 4.0m.$).
- *corners_detected*: orientation and distance to corners (type c landmarks).[1]
- *obstacles_detected*: orientation and distance to obstacles (type o landmarks).[1]
- *goal_position*: relative orientation and distance between the robot and the
  current goal. Receives as parameter the robot's current position an the goal's
  current position, though a trigonometry process, the orientation and distance
  values are calculated and then discretized.[1]
- *goal_reached*: Indicates if the robot is in its goal position. Possible values are
  true or *false*.

The previous predicates tell the robot if it is in a room, a corridor or an intersection, detect walls, corners, doors, obstacles and corridors and give a rough
estimate of the direction and distance to the goal. Analogous to r-states, r-actions
are conjuntions of the following first order logic predicates. These predicates receive as parameter the odometer's speed and angle readings.

- *go*: robot's actual moving action. Its possible values are *front* (speed $>$
  $0.1m./s.$), *nil* ($-0.1m./s. <$ speed $< 0.1m./s.$) and *back* (speed $< -0.1m./s.$).
- *turn*: robot's actual turning angle. Its possible values are *right* (angular
  speed $< -0.1rad./s.$), *nil* (angular $-0.1rad./s. <$ speed $< 0.1rad./s.$) and
  left (angular speed $> 0.1rad./s.$).

Table 2 shows an r-state-r-action pair generated with the previous predicates
which corresponds to values from Table 1. As can be seen, some of the r-state
predicates (doors, walls, corners and obstacles detection) return the orientation
and distance values of every detected element. The r-action predicates return the
odometer's speed and angle values. These values are used in the second stage of
the method. The discretized values, i.e., the r-states and r-actions descriptions,
are used to develop a relational policy as described in the next section.

5   TS-RRLCA First Stage

TS-RRLCA starts with a set of human traces of the task that we want the
robot to learn. A trace ($\tau_k = \{f_{k_1}, f_{k_2}, ..., f_{k_n}\}$) is a log of all of the odometer,
laser and sonar sensor's readings of the robot while it is performing a particular
task. A trace-log is divided in frames; every frame is a register with all the low-
level values of the robot's sensors ($f_{k_j} = \{laser_1 = 2.25, laser_2 = 2.27, laser_3 = 2.29, ..., sonar_1 = 3.02, sonar_2 = 3.12, sonar_3 = 3.46, ..., speed = 0.48, angle = 0.785\}$) at a particular time.

Once the set of traces has been given ($\tau_1, \tau_2, ...\tau_m$), every frame in the traces,
is transformed into natural landmarks along with the robot's location. This

---

[1] The values used for orientation and distance are the same as with doors.

Table 2. Resulting r-state-r-action pair from the values in Table 1

r-state	r-action

place(in-room),
doors_detected([[front, close, -12.57, 1.27]]),
walls_detected([[right-front, close, medium, -35.7, 0.87],
        [front, far, small, -8.55, 4.62],
        [front, near, medium, -6.54, 2.91],
        [left-front, near, small, 23.63, 1.73],
        [left-front, near, medium, 53.80, 2.13]]),
corners_detected([[front, far, -14.58, 5.79],
        [front, near, 31.68, 2.30],
        [left-front, near, 22.33, 1.68]]),
obstacles_detected([[back, near, -170.00, 1.87],
        [right-back, near, -150.00, 1.63],
        [back, close, 170.00, 1.22],
        [left-back, close, 150.00, 1.43]]),
goal_position([right-front, far]),
goal_reached(false),
	go([nil, 0.1]),
	turn([right, 1.047]).

data is given to the first order predicates to evaluate the set of relations, i.e., generate the r-state and the r-action (as the one shown in Table 2). Every r-state-action pair is stored in a database (DB). Table 3 gives the algorithm for this Behavioural Cloning (BC) approach. At the end of this BC approach, DB contains r-state-r-action pairs corresponding to all the frames of the set of traces.

Table 3. Behavioural Cloning Approach

Given a set of trace-logs in the form of frames
For each frame
    Transform the low-level sensor data into natural landmarks and robot's location.
    Use the natural landmarks and the robot's location to evaluate the r-state and
    r-action predicates.
    $DB \leftarrow DB \cup$ new r-state-r-action pair.

As the traces correspond to different examples and they might have been generated by different persons, for the same r-state, several r-actions might have been performed. RL is used to develop a control policy that selects the best r-action in each r-state so the robot can acomplish its tasks.

5.1   Relational Reinforcement Learning

As means to select the best action for every state, an RL algorithm is applied. The RL algorithm selects, most of the time, the r-action that produces the greatest expected reward among the possible r-actions in each r-state. We used only the information from traces, so, only a smaller subset of possible actions, for

Table 4. rQ-learning algorithm

Initialize $Q(S,A)$ arbitrarily (where S is an *r-state* and A is an *r-action*)
Repeat (for each episode):
  Initialize s
  (s represents the values of the sensor's readings of the robot in the current state)
  Transform s into natural landmarks and obtain the robot's actual location.
  $S \leftarrow r\text{-}state(s)$ % Send the natural landmarks and the location to
  the first order predicates to generate the correspondig r-state.
  Repeat (for each step of episode):
    Search the r-state registers, along with its corresponding discretized
    r-actions A, in DB, which equals the S discretized values.
    Choose an action A using a *persistently exciting* policy (e.g., ϵ-greedy).
    Take action A, observe s' (the new state) and r (the new state reward).
    $S' \leftarrow r\text{-}state(s')$
    $Q(S,A) \leftarrow Q(S,A) + \alpha(r + \gamma max_A \cdot Q(S',A') - Q(S,A))$
    $S \leftarrow S'$
  until s is terminal

---

every state, will be considered which significantly reduces the search space. Table
4 gives the pseudo-code for the *rQ-learning* algorithm. This is very similar to the
Q-learning algorithm, except that the states and actions are characterized by
relations. Through this *RL* algorithm, the r-action that best serves to acomplish
the task, is selected.

Through our relational representation, policies can be transfered to different,
although similar domains or tasks. Even that learned policies can be re-used,
the actions are still discrete. In the second stage, this discrete actions policy is
transformed into a continuous actions policy.

6   TS-RRLCA Second Stage

This second stage refines the coarse actions from the discrete actions policy previously generated. This is achieved using *LWR*. The idea is to combine discrete
actions' values given by that policy with the action's values previously observed
in traces. This way the robot follows the policy, developed in the first stage,
but the actions are tuned through the *LWR* process. What we do is to detect
the robot's actual r-state, then, for this r-state the previously generated discrete
actions policy determines the action to be executed (Figure 3(a)). Before performing the action, the robot searches in *DB* for all of the registers that share
this same r-state description (Figure 3(b)). Once found, the robot gets all of the
numeric orientation and distance values from the read *DB* registers. This orientation and distance values are used to perform a triangulation process. This
process allow us to estimate the relative position of the robot from previous
traces with respect to the robot's actual position. Once this position has been
estimated, a weight is assigned to the robot from previous traces action's values.

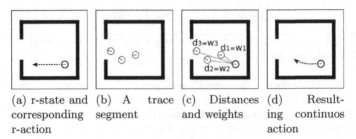

(a) r-state and (b) A trace (c) Distances (d) Result-
corresponding segment and weights ing continuos
r-action action

Fig. 3. Continuous actions developing process

This weight depends on the distance of the robot from the traces with respect
to the actual robot's position (Figure 3(c)). These weights are used to perform
the LWR that produces a continuous r-action (Figure 3(d)).

The triangulation process is performed as follows. The robot R in the actual
r-state (Figure 4(a)), senses and detecs elements E and E' (which can be a
door, a corner, a wall, etc.). Each element has a relative distance (a and b)
and a relative angle with respect to R. The angles are not directly used in this
triangulation process, what we use is the absolute difference between these angles
(α). The robot reads from DB all of the registers that share the same r-state
description, i.e., that have the same r-state discretized values. The numerical
angle and distance values from these DB registers correspond to the relative
distances (a' and b') from the robot of traces R' relative to the same elements
E and E', and the corresponding angle β (Figure 4(b)). In order to know the
distance between R and R' (d) through this triangulation process, equations 1,
2, 3 and 4 are applied.

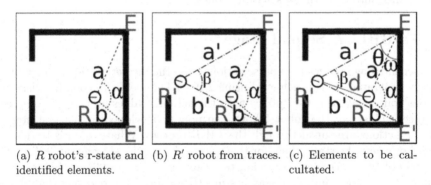

(a) R robot's r-state and (b) R' robot from traces. (c) Elements to be cal-
identified elements.   cultated.

Fig. 4. Triangulation process

$$\overline{EE'} = \sqrt{a^2 + b^2 - 2ab\ cos(\alpha)} : \ Distance\ between\ E\ and\ E'. \qquad (1)$$

$$\theta + \omega = arcsec(a'/\overline{EE'}) : \ Angle\ between\ a'\ and\ \overline{EE'}. \qquad (2)$$

$$\omega = arcsen(a/\overline{EE'}) : \ Angle\ between\ a\ and\ \overline{EE'}. \qquad (3)$$

$$d = \sqrt{a^2 + a'^2 - 2aa' \ cos(\theta)} : \ Distance \ between \ R \ and \ R'. \qquad (4)$$

These equations give the relative distance (d) between R and R'. Once this value is calculated, a Kernel is used to assign a weight (w) value. This weight is multiplied by the speed and angle values of the R' robot's r-action. The resulting weighted speed and angle values, are then added to the R robot's speed and angle values. This process is applied to every register read from DB whose r-state description is the same as R and is repeated everytime the robot reaches a new r-state. The main advantage of our approach is the simple and fast strategy to produce continuous actions policies that, as will be seen in the following section, are able to produce smooth and shorter paths in different environments.

7   Experiments

Experiments were carried out in simulation (*Player/Stage* [9]) and with a real robot[2]. Both robots (simulated and real) are equiped with a 180° front *SICK* laser sensor and an array of 4 back sonars (-170°, -150°, 150° and 170°). The laser range is 8.0m and for sonars is 6.0m. The tasks referred in these experiments were navigating through the environment and following an object.

The policy generation process was carried out in the map 1 shown in Figure 5. For each of the two tasks a set of 15 traces was generated in this map. For the navigation tasks, the robot and the goal's global position (for the goal_position predicate) were calculated by using the work developed in [6]. For the "following" tasks we used a second robot which orientation and angle were calculated through laser sensor. To every set of traces, we applied our BC approach to abstract the r-states and induce the relevant r-actions. Then, *rQ-learning* was applied to learn the policies. For generating the policies, *Q-values* were initialized to -1, ϵ = 0.1, γ = 0.9 and α = 0.1. Positive reinforcement, r, (*+100*) was given when reaching a goal (within 0.5 m.), negative reinforcement (*-20*) was given when the robot hits an element and no reward value was given otherwise (*0*). To generate the continuous actions policy, LWR was applied using a Gaussian Kernel for estimating weights. Once the policies were learned, experiments were executed in the training map with different goal positions and in two new and unknow environments for the robot (map 2 shown in Figure 5(c) and map 3 shown in Figure 6). A total of 120 experiments were performed: 10 different navigation and 10 following tasks in each map, each of this tasks executed first with the discrete actions policy from the first stage and then with the continuous actions policy from the second stage. Each task has a different distance to cover and required the robot to traverse through different places. The minimum distance was 2m. (Manhattan distance), and it was gradually increased up to 18m.

Figure 5 shows a navigation (map 1) and a following task (map 2) performed with discrete and continuous actions respectively.

Figure 6, shows a navigation and a following task performed with the real robot, with the discrete and with the continuous actions policy.

---

[2] An *ActivMedia GuiaBot*, www.activrobots.com

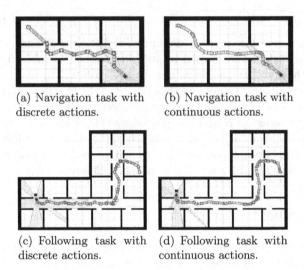

(a) Navigation task with discrete actions.

(b) Navigation task with continuous actions.

(c) Following task with discrete actions.

(d) Following task with continuous actions.

Fig. 5. Tasks examples from Maps 1 (size $15.0m. \times 8.0m.$) and 2 (size $20.0m. \times 14.0m.$)

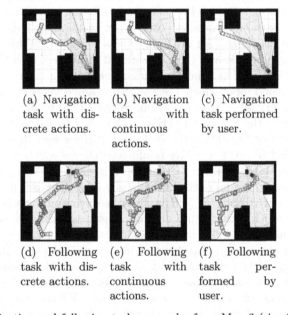

(a) Navigation task with discrete actions.

(b) Navigation task with continuous actions.

(c) Navigation task performed by user.

(d) Following task with discrete actions.

(e) Following task with continuous actions.

(f) Following task performed by user.

Fig. 6. Navigation and following tasks examples from Map 3 (size $8.0m. \times 8.0m.$)

If the robot reached an unseen r-state, it asks for guidance to the user. Through a joystick, the user indicates the robot which action to execute and the robot stores this new r-state-r-action pair. As the number of experiments increased the number of unseen r-states is reduced.

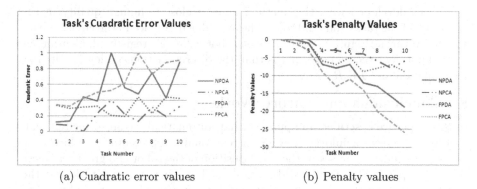

(a) Cuadratic error values   (b) Penalty values

Fig. 7. Navigation and following results of the tasks performed by the real robot

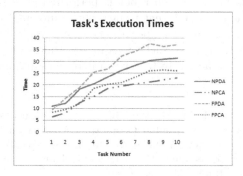

Fig. 8. Execution times results

Figure 7(a)[3] shows results in terms of quality of the performed tasks with the real robot. This comparison is made against tasks performed by humans. All of the tasks performed in experiments with the real robot, were also performed by a human using a joystick (Figures 6(c), 6(f)), and logs of the paths were saved. The graphic shows the normalized cuadratic error between these logs and the trayectories followed by the robot. Figure 7(b) shows results in terms of how much of the environment's free space, the robot uses. This comparison is made using the work developed in [7]. In that work, values were given to cells acordingly to its proximity to objects or walls. The closer the cell is to an object or wall the higher cost is given. Values were given as follows: if the cell is occupied ($0m.$ to $0.3m.$) a value of -100 was given, if the cell is close to an object ($0.3m.$ to $1.0m.$) a value of -3, if the cell is near ($1.0m.$ to $2.0m.$) a value of -1, otherwise 0. Cuadratic error and penalty values for continuous actions policies are lower than those with discrete actions.

---

[3] **NPDA:** Navigation Policy with Discrete Actions, **NPCA:** Navigation Policy with Continuous Actions
FPDA: Following Policy with Discrete Actions, **FPCA:** Following Policy with Continuous Actions.

Policies developed with this method allow a close-to-human quality in the execution of the tasks and tend to use the available free space in the environment. Execution times (Figure 8) with the real robot were also registered. Continuous actions policies execute faster paths than the discrete actions policy despite our triangulation and LWR processes.

8   Conclusions and Future Work

In this paper we described an approach that automatically transformed in real-time low-level sensor information into a relational a representation which does not require information about the exact position of the robot and greatly reduces the state space. The user provides traces of how to perform a task and the system focuses on those actions to quickly learn a control policy, applicable to different goals and environments. LWR is then used over the policy to produce a continuous actions policy as the robot is performing the task. The actions performed in the traces are sufficient to deal with different scenarios, however, rare actions may be required for special cases. We would like to include an exploration strategy to identify such cases or incorporate voice commands to indicate the robot which action to take when it reaches an unseen state.

References

1. Aznar, F., Pujol, F.A., Pujol, M., Rizo, R.: Using gaussian processes in bayesian robot programming. In: Distributed Computing, Artificial Intelligence, Bioinformatics, Soft Computing, and Ambient Assisted Living, pp. 547–553 (2009)
2. Bratko, I., Urbancic, T., Sammut, C.: Behavioural cloning of control skill. Machine Learning and Data Mining, 335–351 (1998)
3. Cocora, A., Kersting, K., Plagemanny, C., Burgardy, W., De Raedt, L.: Learning relational navigation policies. Journal of Intelligent and Robotics System, 2792–2797 (October 2006)
4. Conn, K., Peters, R.A.: Reinforcement learning with a supervisor for a mobile robot in a real-world environment. In: Proc. of the IEEE CIRA (2007)
5. Gaskett, C., Wettergreen, D., Zelinsky, A.: Q-learning in continuous state and action spaces. In: Foo, N.Y. (ed.) AI 1999. LNCS, vol. 1747, pp. 417–428. Springer, Heidelberg (1999)
6. Hernández, S.F., Morales, E.F.: Global localization of mobile robots for indoor environments using natural landmarks. In: Proc. of the IEEE 2006 ICRAM, September 2006, pp. 29–30 (2006)
7. Romero, L., Morales, E.F., Sucar, L.E.: An exploration and navigation approach for indoor mobile robots considering sensor's perceptual limitations. In: Proc. of the IEEE ICRA, pp. 3092–3097 (2001)
8. Torrey, L., Shavlik, J., Walker, T., Maclin, R.: Relational macros for transfer in reinforcement learning. In: ILP, pp. 254–268 (2008)
9. Vaughan, R., Gerkey, B., Howard, A.: On device abstractions for portable, reusable robot code. In: Proc. of the 2003 IEEE/RSJ IROS, pp. 11–15 (2003)
10. Vega, J.H.: Mobile robot localization in topological maps using visual information. Masther's thesis (to be publised, 2009)
11. Wang, Y., Huber, M., Papudesi, V.N., Cook, D.J.: User-guided reinforcement learning of robot assistive tasks for an intelligent environment. In: Proc. of the The 2003 IEEE/RSJ IROS, October 2003, pp. 27–31 (2003)

People Detection by a Mobile Robot Using Stereo Vision in Dynamic Indoor Environments

José Alberto Méndez-Polanco, Angélica Muñoz-Meléndez,
and Eduardo F. Morales

National Institue of Astrophysics, Optics and Electronics
{polanco,munoz,emorales}@inaoep.mx

Abstract. People detection and tracking is a key issue for social robot design and effective human robot interaction. This paper addresses the problem of detecting people with a mobile robot using a stereo camera. People detection using mobile robots is a difficult task because in real world scenarios it is common to find: unpredictable motion of people, dynamic environments, and different degrees of human body occlusion. Additionally, we cannot expect people to cooperate with the robot to perform its task. In our people detection method, first, an object segmentation method that uses the distance information provided by a stereo camera is used to separate people from the background. The segmentation method proposed in this work takes into account human body proportions to segment people and provides a first estimation of people location. After segmentation, an adaptive contour people model based on people distance to the robot is used to calculate a probability of detecting people. Finally, people are detected merging the probabilities of the contour people model and by evaluating evidence over time by applying a Bayesian scheme. We present experiments on detection of standing and sitting people, as well as people in frontal and side view with a mobile robot in real world scenarios.

1 Introduction

Traditionally, autonomous robots have been developed for applications requiring little interaction with humans, such as sweeping minefields, exploring and mapping unknown environments, inspecting oil wells or exploring other planets. However, in recent years significant progress has been achieved in the field of service robots, whose goal is to perform tasks in populated environments, such as hospitals, offices, department stores and museums. In these places, service robots are expected to perform some useful tasks such as helping elderly people in their home, serving as hosts and guides in museums or shopping malls, surveillance tasks, childcare, etc [1,2,3,4,5,6]. In this context, people detection by a mobile robot is important because it can help to improve the human robot interaction, perform safety path planning and navigation to avoid collisions with people, search lost people, recognize gestures and activities, follow people, and so on.

A. Hernández Aguirre et al. (Eds.): MICAI 2009, LNAI 5845, pp. 349–359, 2009.

In this work, we propose to detect people applying a mobile robot using a semi-elliptical contour model of people. The main difference between our proposed model and previously proposed models is that our model takes into account the distance between the robot and people to try to get the best fit model. Therefore, the contour model detects people without assuming that the person is facing the robot. This difference is important because we cannot expect people to cooperate with the robot to perform its task. The contour model is used to get a first estimate of the position of a person relative to the robot. However, instead of detecting people in a single frame, we apply a spatio-temporal detection scheme merging the probabilities of the contour people model over time by applying a Bayesian scheme. The idea is that people who are detected by the robot at different instants of time have a higher probability of detection. Our experimental platform is an ActivMedia PeopleBot mobile robot equipped with a stereo camera (See Figure 1).

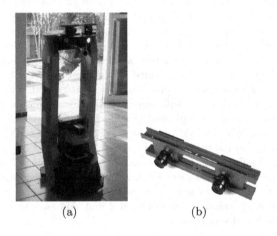

(a)   (b)

Fig. 1. (a) ActivMedia PeopleBot equipped with a stereo camera. (b) Stereo camera.

The paper is organized as follows: In section 2 we present related works. Section 3 describes our segmentation method based on the distance to detected objects. Section 4 introduces the adaptive contour model to get a first estimation of the people position relative to the robot. Section 5 presents the detection method using a spatio-temporal approach. Section 6 shows the principal experiments and results. Finally, Section 7 presents conclusions and future research directions.

2   Related Work

Depending on the specific application that integrates people detection and identification there are two princ460,13ipal approaches that can be applied over images acquired: whole human body detection [7,8,9] and part-based body detection [10,11], for instance, with single, stereo CCD or thermal cameras.

The advantages of whole human body detection approaches are the compact representation of the human body models such as human silhouettes [12,13,14]. The main drawbacks of these approaches are the sensitivity to object occlusion and cluttered environments as well as the need of robust object segmentation methods. Concerning part-based body detection approaches, the main advantage is their reliability to deal with cluttered environments and object occlusion because of their independence for whole human body parts detection. These approaches do not rely on the segmentation of the whole body silhouettes from the background. Part-based body detection approaches, in general, aim to detect certain human body parts such as face, arms, hands, legs or torso. Different cues are used to detect body parts, such as laser range- finder readings to detect legs [15,16] or skin color to detect hands, arms and faces [17,18,19].

Although people recognition using static cameras has been a major interest in computer vision [18,20,21], some of these works cannot be directly applied to a mobile robot which has to deal with moving people in cluttered environments. There are four main problems that need to be solved for people detection with mobile robots: real time response, robust object segmentation, incomplete or unreliable information cues, and the integration of spatio-temporal information. Below we briefly explain these problems.

1. **Real time people detection.** Service robots must operate robustly in real time in order to perform their tasks in the real world and achieve appropriate human robot interaction. There are some real time object detection approaches which use SIFT features [23] or the Viola and Jones algorithm for face detection [22] that have been applied for people detection purposes. However people detection is more complicated due to possible poses of the human body and the unpredictable motion of arms and legs.

2. **Object segmentation.** General object segmentation approaches assume a static background and a static camera [18,20,21]. In this way, object segmentation can be achieved subtracting the current image from the background reference resulting in a new image with the possible people detected. In the case of mobile robots, to have a background reference for each possible robot location is not feasible, so traditional segmentation is no longer applicable.

3. **Incomplete or unreliable information cues.** In the context of service robots both, people and robots move within dynamic environments. For this reason, information cues necessary for people recognition such as faces or body parts are not always available. In the case of face detection, faces are not always perceivable by the camera of a mobile robot. Concerning legs detector, their main drawbacks are the number of false positives which may occur, for instance, in situations where people's legs are indistinguishable from the legs of tables or chairs. As far as skin color detection approaches are concerned, the principal problem is that, in cluttered environments, there are typically many objects similar in color to human skin and people are not always facing the robot.

4. **Integration of spatio-temporal information.** During navigation, robots perceive the same objects from different locations at different periods of

time. A video streaming can be used to improve the recognition rate using evidence from several frames. This is conceptually different from most object recognition algorithms in computer vision where observations are considered to be independent.

3   Distance Based Image Segmentation

In this paper, we propose an image segmentation method based on distance. In order to achieve the distance information, a camera stereo is used to calculate the disparity between two images (left and right images). The idea of using a disparity image is that objects closer to the camera have a greater disparity than objects further to the camera. Therefore, the distance to the objects can be calculated with an adequate calibration. An example of a disparity image calculated from the images provided by a stereo camera is shown in Figure 2.

(a)   (b)   (c)

Fig. 2. Stereo images. (a) Left image. (b) Right image. (c) Disparity image. The color of each pixel in the disparity image indicates the disparity between the left and the right image where: clear colors indicate high disparity, dark colors indicate low disparity and black pixels indicate no disparity information.

Once the distance to the objects has been calculated, we scan one of the stereo camera images to segment objects based on the previous distance information. The idea is to form clusters of pixels with similar distances to the robot (Z coordinate) and, at the same time, near each other in the image ((X,Y) coordinate). The clusters are defined as follows:

$$C_k = \{\mu_X^k, \mu_Y^k, \mu_Z^k, \sigma_X^k, \sigma_Y^k, \sigma_Z^k, \rho_k\}, k = 1...N \tag{1}$$

where μ_X^k, μ_Y^k and μ_Z^k are the mean of the X, Y, Z coordinates of the pixels within the cluster k, σ_X^k, σ_Y^k and σ_Z^k are the variances of the X, Y, Z coordinates of the pixels within the cluster k, ρ_k is a vector containing the coordinates of the pixels in the cluster k, and N defines the maximum number of clusters.

The means and the variances of each cluster are calculated from the data provided by the disparity image. The image is then scanned and for each pixel we verify if X, Y, Z coordinates are near to one of the clusters. The segmentation consists of three main steps for each pixel in the image as described below:

1. Calculate the distances of the pixel to the clusters based on the means of the clusters:

$$d_X^k = (X - \mu_X^k)^2, d_Y^k = (Y - \mu_Y^k)^2, d_Z^k = (Z - \mu_Z^k)^2 \tag{2}$$

2. If the cluster k has the lowest distance and $d_X^k < th_X$ and $d_Y^k < th_Y$ and $d_Z^k < th_Z$ then assign the pixel to the cluster k and update the means and the variances of the cluster. In this case, th_X, th_Y and th_Z are thresholds previously defined.
3. If there are no clusters closed to the pixel create a new cluster and assign the pixel to this cluster.

We use three different thresholds to perform the segmentation to exploit the fact that, in most cases, the height of people in the images is larger than the width. Therefore, we define $th_X < th_Y$ in order to segment objects with the height greater than the width, as in the case of standing and most sitting human bodies. Further more, after segmentation, clusters with irregular proportions ($\mu_X > \mu_Y$) are eliminated to take into account only objects with human proportions. In Figure 3 two views of the coordinate system for the images shown in Figure 2 are illustrated.

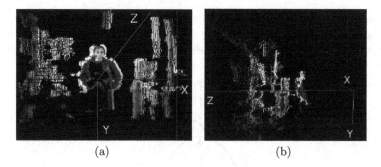

(a) (b)

Fig. 3. Coordinate system for the image shown in Figure 2. (a) Front view. (b) Top view.

Figure 4 shows examples of the object segmentation based on distance using the method described in this section. This example illustrates the reliability of our method to segment simultaneously one or more people.

4 Adaptive Semi-elliptical Contour Model

The segmentation method provides different regions where there are possible people. The next step is to determine which of those regions effectively contain people and which do not. In order to do that, we apply a semi-elliptical contour model illustrated in Figure 5, similar to the model used in [25]. The semi-elliptical contour model consists of two semi-ellipses describing the torso and the human head. The contour model is represented with an 8- dimensional state vector:

$$d_{body}^t = (x_T, y_T, w_T, h_T, x_H, y_H, w_H, h_H) \qquad (3)$$

where (x_T, y_T) is the mid-point of the ellipse describing the torso with width w_T and height h_T, and (x_H, y_H) is the mid-point of the ellipse describing the head with width w_H and height h_H.

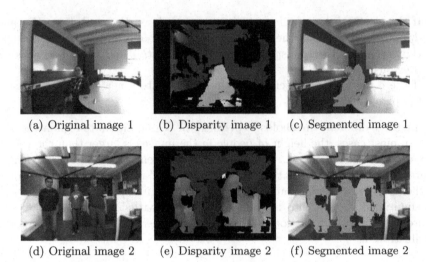

(a) Original image 1   (b) Disparity image 1   (c) Segmented image 1

(d) Original image 2   (e) Disparity image 2   (f) Segmented image 2

Fig. 4. Segmentation method. These examples illustrate how our segment method is able to segment one or multiple people. (a) and (d) are the original images, (b) and (e) are the disparity images, and (c) and (f) are the result of our segmentation process.

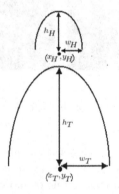

Fig. 5. Semi-elliptical contour model of people similar to the model used in [25]

For each region obtained by the segmentation method an elliptic model is fitted setting the mid-point of the torso ellipse to the center of the region. Since people width varies with distance, we determine the dimension of the elliptic model using a table containing the means of the width of torso and head for five different people at different distances to the robot. This constraint avoids considering regions whose dimensions are incompatible with the dimensions of people at specific distances. At the same time, this constraint enables the robot to achieve a better fit for the semi-ellipsis describing people.

As different people have usually different width, we adjust the contour people model by varying the mid-point of the torso as well as its width and its height. To determine which variation has the best fit, a probability of fitting for each variation is calculated. The idea is to evaluate the fitting probability as the probability of the parameters of the contour people model given a person p_i, that we denote as $P(d^v_{body}|p_i)$. The probability is calculated as follows:

$$P(d^v_{body}|p_i) = argmax \frac{(N_f|d^v_{body})}{N_T} \qquad (4)$$

where N_f is the number of points in the ellipse that fit with an edge of the image given the parameters of the model d^t_{body}, v denotes the different variations of the model and N_T is the total number of points in the contour model. The edges of the image are calculated applying the Canny edge detector [24]. The adjustment process for the contour people model is illustrated in Figure 6.

(a)   (b)   (c)

Fig. 6. Adjustment process for the contour people model. (a) Original application of the contour model. (b) adjustment process by varying the model parameters. (c) Final application of the contour model.

5   People Detection and Tracking

To detect people, we obtain first the regions of interest using the method described in Section 3. After that, we evaluate the probability of detection using the contour people model described in Section 4. At this point we can determine, with certain probability, the presence of people in the image. In this work, we improve over traditional-frame based detector incorporating cumulative evidence from several frames using a Bayesian scheme.

Once a person p_i has been detected at time t, we proceed to search if that person was previously detected at time $t-1$ calculating the Euclidean distances from the position of the current person to the positions of the previously detected people. If the distance between two people is less than a threshold, then we consider these people to be the same. Once two people have been associated we proceed to calculate the probability of a person p_i at time t denoted as $P(p^t_i)$ as follows:

$$P(p^t_i) = \frac{P(d^t_{body}|p_i)P(p^{t-1}_i)}{(P(d^t_{body}|p^t_i)P(p^{t-1}_i) + (P(d_{body}|\sim p^t_i)P(\sim p^{t-1}_i)} \qquad (5)$$

where $P(d^t_{body}|p_i)$ is the probability of fitting the contour people model at time t which is calculated applying equation 4. $P(p^{t-1}_i)$ is the probability of the person

p_i at time $t-1$. $P(d_{body}| \sim p_i^t)$ is the probability of not fitting of the contour people model at time t and $P(\sim p_i^{t-1})$ is the probability that the person p_i has not been detected at time $t-1$.

6   Experimental Results

We tested our people detection method using a mobile robot equipped with a stereo camera. The experiments were performed in a dynamic indoor environment under different illumination conditions and with sitting and standing people placed at different distances from the robot (1 to 5 m). The number of people varies from 1 to 3 people. Due to the fact that we do not use the face as a cue to detect people our method can detect people facing or not the robot. Figure 7 compares the performance of our people detection method using a single frame detection scheme against our people detection method using evidence from several frames applying a Bayesian scheme. Figure 8 shows how our people detection method is able to detect multiple people and track them over time. We consider a person as detected if $P(p_i) > 0.8$. We calculated the detection rate DR as follows:

$$DR = \frac{N_D}{N_T} \qquad (6)$$

where N_D is the number of frames where people were detected with $P(p_i) > 0.8$, and N_T is the total number of frames analysed.

Our detection method has a CR of 89.1% using $P(p_i) > 0.8$ and 96% if we use $P(p_i) > 0.6$. Table 1 presents the detection rate of our method compared with the classification rate reported in [25] which presents two different approaches to detect people using a thermal camera and a semi-elliptic contour model.

In Figure 9 one can see the results of different experiments on detection of standing and sitting people, as well as people in frontal and side view with a mobile robot in real world scenarios using our proposed people detection method are shown.

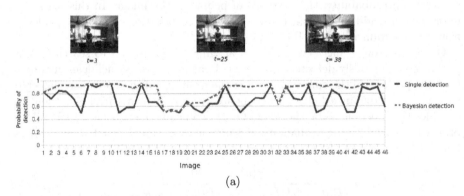

(a)

Fig. 7. People detection performance. The Bayesian people detection outperforms the single frame detection. At the top of the chart we show images at different periods of time showing the output of our people detection method.

Table 1. Comparation with other works

Method	Classification rate
Contour [25]	88.9
Combination with grey features [25]	90.0
Our Method (Spatio-temporal)	96.0

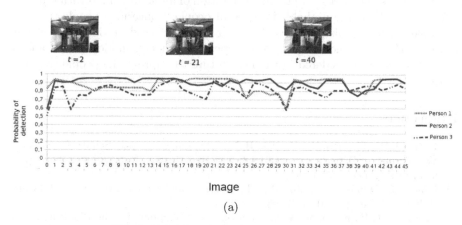

Image

(a)

Fig. 8. Multiple people detection performance

Fig. 9. Experiments performed using a mobile robot in dynamic indoor environments

7   Conclusion and Future Work

This paper addressed the problem of detecting people with a mobile robot using
a stereo camera. The proposed segmentation method takes into account human body proportions to segment people and to provide a first estimation of
people location. We presented an adaptive contour people model based on people distance to the robot. To detect people, we merged the probabilities of the

contour people model over time by applying a Bayesian scheme. According to
the experiments evidence, we show that our method is able to segment and
detect standing and sitting people in both front and lateral views. Neither previous visual model of the environment nor mandatory facing pose of people is
involved in our method. The future research directions of this work are fusing
information from diverse cues such as skin, clothes and body parts to reduce the
number of false positives, and the incorporation of a semantic map with *a priori*
information about the probable location of people. Concerning people tracking
a simple Euclidean distance based tracker has been used at this stage of our
research. However, the integration of appearance model, motion model and a
Kalman filter are considered in the near future to improve the people tracking
process.

References

1. Burgard, W., Cremers, A., Fox, D., Hähnel, D., Lakemeyer, G., Schulz, D., Steiner, W., Thrun, S.: Experiences with an Interactive Museum Tour-guide Robot. Artificial Intelligence, 3–55 (1999)
2. Osada, J., Ohnaka, S., Sato, M.: The Scenario and Design Process of Childcare Robot, PaPeRo, pp. 80–86 (2006)
3. Gharpure, C.P., Kulyukin, V.A.: Robot-assisted Shopping for the Blind: Issues in Spatial Cognition and Product Selection. Intelligent Service Robotics 1(3), 237–251 (2008)
4. Forlizzi, J., DiSalvo, C.: Service robots in the domestic environment: a study of the roomba vacuum in the home. In: Proceedings of the 1st ACM SIGCHI/SIGART Conference on Human-robot Interaction, Salt Lake City, Utah, USA, pp. 258–265 (2006)
5. Montemerlo, M., Pineau, J., Roy, N., Thrun, S., Verma, V.: Experiences with a Mobile Robotic Guide for the Elderly. In: Proceedings of the 18th national conference on Artificial intelligence, Edmonton, Alberta, Canada, pp. 587–592 (2002)
6. Gockley, R., Forlizzi, J., Simmons, R.: Interactions with a Moody Robot. In: Interactions with a Moody Robot, pp. 186–193. ACM, New York (2006)
7. Malagón-Borja, L., Fuentes, O.: Object Detection Using Image Reconstruction with PCA. Image Vision Computing 27, 2–9 (2009)
8. Cielniak, G., Duckett, T.: People Recognition by Mobile Robots. Journal of Intelligent and Fuzzy Systems: Applications in Engineering and Technology 15(1), 21–27 (2004)
9. Davis-James, W., Sharma, V.: Robust Detection of People in Thermal Imagery. In: Proceedings of 17th International Conference on the Pattern Recognition ICPR 2004, pp. 713–716 (2004)
10. Müller, S., Schaffernicht, E., Scheidig, A., Hans-Joachim, B., Gross-Horst, M.: Are You Still Following Me? In: Proceedings of the 3rd European Conference on Mobile Robots ECMR, Germany, pp. 211–216 (2007)
11. Darrell, T.J., Gordon, G., Harville, M., Iselin-Woodfill, J.: Integrated Person Tracking Using Stereo, Color, and Pattern Detection. International Journal of Computer Vision, 175–185 (2000)
12. Vinay, S., James, W.D.: Extraction of Person Silhouettes from Surveillance Imagery using MRFs. In: Proceedings of the Eighth IEEE Workshop on Applications of Computer Vision, p. 33 (2007)

13. Ahn, J.-H., Byun, H.: Human Silhouette Extraction Method Using Region Based Background Subtraction. LNCS, pp. 412–420. Springer, Berlin (2007)
14. Lee, L., Dalley, G., Tieu, K.: Learning Pedestrian Models for Silhouette Refinement. In: Proceedings of the 19th IEEE International Conference on Computer Vision, Nice, France, pp. 663–670 (2003)
15. Schaffernicht, E., Martin, C., Scheidig, A., Gross, H.-M.: A Probabilistic Multi-modal Sensor Aggregation Scheme Applied for a Mobile Robot. In: Proceedings of the 28th German Conference on Artificial Intelligence, Koblenz, Germany, pp. 320–334 (2005)
16. Bellotto, N., Hu, H.: Vision and Laser Data Fusion for Tracking People with a Mobile Robot. In: International Conference on Robotics and Biomimetics ROBIO 2006, Kunming, China, pp. 7–12 (2006)
17. Siddiqui, M., Medioni, G.: Robust real-time upper body limb detection and tracking. In: Proceedings of the 4th ACM International Workshop on Video Surveillance and Sensor Networks VSSN 2006, Santa Barbara, California, USA, pp. 53–60 (2006)
18. Wilhelm, T., Bohme, H.-J., Gross, H.-M.: Sensor Fusion for Visual and Sonar based People Tracking on a Mobile Service Robot. In: Proceedings of the International Workshop on Dynamic Perception 2002, Bochum, Germany, pp. 315–320 (2002)
19. Lastra, A., Pretto, A., Tonello, S., Menegatti, E.: Robust Color-Based Skin Detection for an Interactive Robot. In: Proceedings of the 10th Congress of the Italian Association for Artificial Intelligence (AI*IA 2007), Roma, Italy, pp. 507–518 (2007)
20. Han, J., Bhanu, B.: Fusion of Color and Infrared Video for Dynamic Indoor Environment Human Detection. Pattern Recognition 40(6), 1771–1784 (2007)
21. Muñoz-Salinas, R., Aguirre, E.: People Detection and Tracking Using Stereo Vision and Color. Image Vision Computing 25(6), 995–1007 (2007)
22. Viola, P., Jones, M.J.: Robust Real-time Object Detection, Robust Real-time Object Detection. Cambridge Research Laboratory 24 (2001)
23. Lowe, D.G.: Distinctive Image Features from Scale-Invariant Keypoints. International Journal of Computer Vision 60(2), 91–110 (2004)
24. Canny, F.: A computational approach to edge detection. IEEE Trans. Pattern Analysis and Machine Intelligence 8(6), 679–698 (1986)
25. Treptow, A., Cielniak, G., Ducket, T.: Active People Recognition using Thermal and Grey Images on a Mobile Security Robot. In: Proceedings of the IEEE/RSJ International Conference on Intelligent Robots and Systems (IROS), Edmonton, Alberta, Canada (2005)

SAT Encoding and CSP Reduction for Interconnected Alldiff Constraints

Frederic Lardeux[3], Eric Monfroy[1,2], Frederic Saubion[3],
Broderick Crawford[1,4], and Carlos Castro[1]

[1] Universidad Técnica Federico Santa María, Valparaíso, Chile
[2] LINA, Université de Nantes, France
[3] LERIA, Université d'Angers, France
[4] Pontificia Universidad Católica de Valparaíso, PUCV, Chile

Abstract. Constraint satisfaction problems (CSP) or Boolean satisfiability problem (SAT) are two well known paradigm to model and
solve combinatorial problems. Modeling and resolution of CSP is often
strengthened by global constraints (e.g., Alldiff constraint). This paper
highlights two different ways of handling specific structural information:
a uniform propagation framework to handle (interleaved) Alldiff constraints with some CSP reduction rules; and a SAT encoding of these
rules that preserves the reduction properties of CSP.

1 Introduction

During the last decades, two closely related communities have focused on the
resolution of combinatorial problems. On the one hand, the SAT community has
developed very efficient algorithms to handle the seminal Boolean satisfaction
problem which model is very simple: Boolean variables and CNF propositional
formulas. The complete resolution techniques (i.e., able to decide if an instance
is satisfiable or not) mainly rely on the DPLL procedure [6] whereas incomplete
algorithms are mostly based on local search procedures [11]. In addition, very
sophisticated techniques (e.g., symmetries detection, learning, hybrid heuristics)
were proposed to build very efficient solvers able to handle huge and very difficult benchmarks. On the other hand, the constraint programming community
has focus on the resolution of discrete constraint satisfaction problems (CSP).
This paradigm provides the user with a more general modeling framework: problems are expressed by a set of decision variables (which values belong to finite
integer domains) and constraints model relations between these variables. Concerning the resolution algorithms, complete methods aim at exploring a tree by
enumerating variables and reducing the search space using constraint propagation techniques, while incomplete methods explore the search space according
to specific or general heuristics (metaheuristics). Again, much work has been
achieved to define very efficient propagation algorithms and search heuristics.

Concerning this resolution aspect, the two paradigms share some common
principles (see [5] for a comparative survey). Here, we focus on complete methods

A. Hernández Aguirre et al. (Eds.): MICAI 2009, LNAI 5845, pp. 360–371, 2009.
© Springer-Verlag Berlin Heidelberg 2009

that aim at exploring a tree by enumerating variables (finite domain variables or Boolean ones) and reducing the search space using propagation techniques (constraint propagation or unit propagation).

The identification of typical constraints in CSP (so-called global constraints) has increased the declarativity, and the development of specialized algorithms has significantly improved the resolution performances. The first example is certainly the Alldiff constraint [15] expressing that a set of variables have all different values. This constraint is very useful since it naturally appears in the modeling of numerous problems (timetabling, planning, resource allocation, ...). Moreover, usual propagation techniques are inefficient for this constraint (due to limited domain reduction when decomposing the constraint into $n * (n-1)/2$ disequalities) and specific algorithms were proposed to boost resolution [18]. On the SAT side, no such high level modeling feature is offered. As a consequence, benchmarks (e.g., industrials ones) are often incomprehensible by users and sophisticated preprocessings should be used to simplify their structures. Hence, it could be useful to provide a more declarative approach for modeling problems in SAT by adding an intermediate layer to translate high level constraints into propositional formulas. Systematic basic transformations from CSP to SAT have been proposed [9,21,8] to ensure some consistency properties to the Boolean encodings. Even if some specific relationships between variables (e.g., equivalences) are handled specifically by some SAT solvers, global constraints must be transformed into clauses and properties can then be established according to the chosen encodings [2,3,10,12]. When modeling problems, the user has often to take into account several global constraints, which may share common variables. Therefore, while the global constraint approach was a first step from basic atomic constraints to more powerful relations and solvers a recent step consists in handling efficiently combinations of global constraints (e.g., [19,20,14]).

We focus on possibly interleaved (i.e., sharing variables) Alldiff constraints, such as it appears in numerous problems (e.g., Latin squares, Sudoku [16]). Our purpose is twofold. 1) We want to provide, on the CSP solving side, a uniform propagation framework to handle possibly interleaved Alldiff constraints. From an operational point of view, we define propagation rules to improve resolution efficiency by tacking into account specific properties. 2) We also want to generalize possible encodings of (multiple) Alldiff constraints in SAT (i.e., by a set of CNF formulas). Our purpose is to keep the reduction properties of the previous propagation rules. Therefore, our encodings are fully based on these rules. Our goal is not to compare the efficiency of CSP reductions versus their SAT encodings (nor to compete with existing solvers), but to generate CSP rules and SAT encodings that are solver independent; if one is interested in better efficiency, the solvers can then be improved (based on the CSP rules structures or on the SAT formulas structures) to take advantage of their own facilities.

2 Encoding CSP vs. SAT

CSP: Basic Notions. A CSP (X, C, D) is defined by a set of variables $X = \{x_1, \cdots, x_n\}$ taking their values in their respective domains $D = \{D_1, \cdots, D_n\}$.

A constraint $c \in C$ is a relation $c \subseteq D_1 \times \cdots \times D_n$. A tuple $d \in D_1 \times \cdots \times D_n$ is a solution if and only if $\forall c \in C, d \in c$. We consider C as a set of constraint (equivalent to a conjunction). Usual resolution processes [1,5] are based on two main components: reduction and search strategies. Search consists in enumerating the possible values of a given variable to progressively build and reach a solution. Reduction techniques are added at each node to reduce the search space (local consistency mechanisms): the idea is to remove values of variables that cannot satisfy the constraints. This approach requires an important computational effort and performances can be improved by adding more specific techniques, e.g., efficient propagation algorithms for global constraints. We recall a basic consistency notion (the seminal arc consistency is the binary subcase of this definition).

Definition 1 (Generalized Arc Consistency (GAC)). *A constraint[1] c on variables $(x_1, \cdots, x_m)$ is generalized arc-consistent iff $\forall k \in 1..m, \forall d \in D_k$, $\exists (d_1, \cdots, d_{k-1}, d_{k+1}, \cdots d_n) \in D_1 \times \cdots \times D_{k-1} \times D_{k+1} \times \cdots \times D_n$, s.t. $(d_1, \cdots, d_m) \in c$.*

Domain Reduction Rules. Inspired by [1], we use a formal system to precisely define reduction rules to reduce domains w.r.t. constraints. We abstract constraint propagation as a transition process over CSPs. A domain reduction rule is of the form:

$$\frac{(X, C, D)|\Sigma}{(X, C, D')|\Sigma'}$$

where $D' \subseteq D$ and Σ and Σ' are first order formulas (i.e., conditions of the application of the rules) such that $\Sigma \wedge \Sigma'$ is consistent. We canonically generalize $\subseteq$ to sets of domains as $D' \subseteq D$ iff $\forall x \in X \; D'_x \subseteq D_x$. Given a set of variables V, we also denote D_V the union $\bigcup_{x \in V} D_x$. $\#D$ is the set cardinality.

Given a CSP (X^k, C^k, D^k), a transition can be performed to get a reduced CSP $(X^{k+1}, C^{k+1}, D^{k+1})$ if there is an instance of a rule (i.e., a renaming without variables' conflicts):

$$\frac{(X^k, C^k, D^k)|\Sigma^k}{(X^{k+1}, C^{k+1}, D^{k+1})|\Sigma^{k+1}}$$

such that $D^k \models \bigwedge_{x \in X} x \in D_x^k \wedge \Sigma^k$, and D^{k+1} is the greatest subset of D^k such that $D^{k+1} \models \bigwedge_{x \in X} x \in D_x^{k+1} \wedge \Sigma^{k+1}$.

In the conclusion of a rule (in Σ), we use the following notations: $d \notin D_x$ means that d can be removed from the domain of the variable x (without loss of solution); similarly, $d \notin D_V$ means that d can be removed from each domain variables of V; and $d_1, d_2 \notin D_x$ (resp. D_V) is a shortcut for $d_1 \notin D_x \wedge d_2 \notin D_x$ (resp. $d_1 \notin D_V \wedge d_2 \notin D_V$).

Since we only consider here rules that does not affect constraints and variables, the sets of variables will be omitted and we highlight the constraints that are required to apply the rules by restricting our notation to $< C, D >$. We will say

---

[1] This definition is classically extended to a set of constraints.

that $< C, D >$ is GAC if C is GAC w.r.t. D. For example, a very basic rule to enforce basic node consistency [1] on equality could be:

$$\frac{< C \wedge x = d, D > | d' \in D_x, d' \neq d}{< C \wedge x = d, D' > | d' \notin D'_x}$$

This rule could be applied on $< X = 2, \{D_X \equiv \{1, 2, 3\}\} >$ with $3 \in D_X, 3 \neq 2$ to obtain $< X = 2, \{D_X \equiv \{1, 2\}\} >$; and so on.

The transition relation using a rule R is denoted $< C, D > \to_R < C, D' >$. $\to_R$ * denotes the reflexive transitive closure of $\to_R$. It is clear that $\to_R$ terminates due to the decreasing criterion on domains in the definition of the rules (see [1]). This notion can be obviously extended to sets of rules $\mathcal{R}$. Note also that we require that the result of $\to_{\mathcal{R}}$ * is independent from the order of application of the rules [1] (this is obvious with the rules that we use). From a practical point of view, it is generally faster to first sequence rules that execute faster.

SAT: Basic Notions. An instance of the SAT problem can be defined by a pair (Ξ, ϕ) where Ξ is a set of Boolean variables $\Xi = \{\xi_1, ..., \xi_n\}$ and ϕ is a Boolean formula $\phi \colon \{0, 1\}^n \to \{0, 1\}$. The formula is said to be satisfiable if there exists an assignment $\sigma \colon \Xi \to \{0, 1\}$ satisfying ϕ and unsatisfiable otherwise. The formula ϕ is in conjunctive normal form (CNF) if it is a conjunction of clauses (a clause is a disjunction of literals and a literal is a variable or its negation).

In order to transform our CSP (X, D, C) into a SAT problem, we must define how the set Ξ is constructed from X and how ϕ is obtained. Concerning the variables, we use the direct encoding [21] : $\forall x \in X, \forall d \in D_x, \exists \xi_x^d \in \Xi$ (ξ_x^d is true when x has the value d, false otherwise).

To enforce exactly one value for each variable, we use the next clauses:

$$\bigwedge_{x \in X} \bigvee_{d \in D_x} \xi_x^d \quad \text{and} \quad \bigwedge_{x \in X} \bigwedge_{\substack{d_1, d_2 \in D_x \\ d_1 \neq d_2}} (\neg \xi_x^{d_1} \vee \neg \xi_x^{d_2})$$

Given a constraint $c \in C$, one may add for all tuples $d \notin c$, a clause recording this nogood value or use other encodings based on the valid tuples of the constraint [2]. One may remark that it can be very expensive and it is strongly related to the definition of the constraint itself. Therefore, as mentioned in the introduction, several work have addressed the encodings of usual global constraints into SAT [3,10,12]. Here, our purpose is to define uniform transformation rules for handling multiple Alldiff constraints, which are often involved in many problems.

From the resolution point of view, complete SAT solvers are basically based on a branching rule that assign a truth value to a selected variable and unit propagation (UP) which allows to propagate unit clauses in the current formula [5]. This principle is very close to the propagation of constraints achieved by reduction rules to enforce consistency. Therefore, we will study the two encodings CSP and SAT from this consistency point of view. According to [21,2], we say that a SAT encoding preserves a consistency iff all variables assigned to false by unit propagation have their corresponding values eliminated by enforcing GAC. More formally, given a constraint c, UP leads to a unit clause $\neg \xi_x^d$ iff d is not GAC with c (d is removed from D_x by enforcing GAC) and if c is unsatisfiable then UP generates the empty clause (enforcing GAC leads to an empty domain).

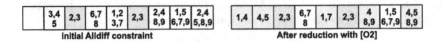

Initial Alldiff constraint   After reduction with [O2]

Fig. 1. Application of [O2]

3 Alldiff Constraints: Reduction Rules and Transformation

In the following, we classically note $Alldiff(V)$ the Alldiff constraint on a subset of variables V, which semantically corresponds to the conjunction of $n*(n-1)/2$ pairwise disequality constraints $\bigwedge_{x_i, x_j \in V, i \neq j} x_i \neq x_j$.

A Single $Alldiff$ **constraint.** We first reformulate a well known consistency property [15,18] w.r.t. the number of values remaining in the domain of the variables. This case corresponds of course to the fact that if a variable has been assigned then the corresponding value must be discarded from other domains.

$$[O1] \quad \frac{< C \wedge Alldiff(V), D > | x \in V \ \wedge \ D_x = \{d_1\}}{< C \wedge Alldiff(V), D' > | d_1 \notin D'_{V \setminus \{x\}}}$$

Property 1. If $< Alldiff(V), D > \rightarrow^*_{[O1]} < Alldiff(V), D' >$, then the corresponding conjunction $\bigwedge_{x_i, x_j \in V} x_i \neq x_j$ is GAC w.r.t. $< D' >$. Note that enforcing GAC on the disequalities with [O1] reduces less the domains than enforcing GAC on the global Alldiff constraint.

This rule can be generalized when considering a subset V' of m variables with m possible values, $1 \leq m \leq (\#V - 1)$:

$$[Om] \quad \frac{< C \wedge Alldiff(V), D > | V' \subset V \ \wedge \ D_{V'} = \{d_1, \ldots, d_m\}}{< C \wedge Alldiff(V), D' > | d_1, \ldots, d_m \notin D'_{V \setminus V'}}$$

Consider $m = 2$, and that two variables of an Alldiff only have the same two possible values. Then it is trivial to see that these two values cannot belong to the domains of the other variables (see Figure 1).

Property 2. Given $< Alldiff(V), D > \rightarrow^*_{[Om]_{1 \leq m \leq (\#V-1)}} < Alldiff(V), D' >$, then $< Alldiff(V), D' >$ has the GAC property.

The proof can be obtained from [15]. Now, the Alldiff constraints can be translated in SAT, by encoding [O1] for a variable x with a set of $\#V * (\#V - 1)$ CNF clauses:

$$[SAT - O1] \quad \bigwedge_{x \in V} \bigwedge_{y \in V \setminus \{x\}} (\neg \xi^d_x \vee (\bigvee_{f \in D_x \setminus \{d\}} \xi^f_x) \vee \neg \xi^d_y)$$

This representation preserves GAC. Indeed, if $\neg\xi_x^d$ is false (i.e. when the variable x is valued to d) and $\bigvee_{f\in D_{x_1}\backslash\{d\}} \xi_{x_1}^f$ is false (i.e., when the variable x_1 is valued to d) then $\neg\xi_{x_2}^d$ must be true to satisfy the clause (x_2 cannot be valued to d).

Generalized to a subset V' of m variables $\{x_1,...,x_m\}$ with m possible values $\{d_1,...,d_m\}$, $1 \leq m \leq (\#V-1)$, the $\#(V\backslash V')*m^{m+1}$ clauses are:

$$[SAT-Om]\ \bigwedge_{y\in V\backslash V'} \bigwedge_{k=1}^{m} \bigwedge_{p_1=1}^{m} \cdots \bigwedge_{p_m=1}^{m} [(\bigvee_{s=1}^{m} \neg\xi_{x_s}^{d_{p_s}}) \vee$$
$$(\bigvee_{i=1}^{m} \bigvee_{f\in D_{x_i}\backslash\{d_1,...,d_m\}} \xi_{x_i}^f) \vee \neg\xi_y^{d_k}]$$

Property 3. $\bigcup_{1\leq m\leq \#V-1}[SAT-Om]$ preserves the GAC property.

Proof. As mentioned above, our transformation is directly based on consistency rules and therefore Property 2 remains valid for the SAT encoding. This can be justified through the propositional rewriting of the direct encoding of $[Om]$ into $[SAT-Om]$ (not given here for lack of space). ⫞

4 Multiple Overlapping Alldiff Constraints

In presence of several overlapping Alldiff constraints, specific local consistency properties can be enforced according to the number of common variables, their possible values, and the number of overlaps. To simplify, we consider Alldiff constraints $Alldiff(V)$ such that $\#V = \#D_V$. This restriction could be weaken but it is generally needed by classical problems (e.g., Sudoku or Latin squares). We now study typical connections between multiple Alldiff. Therefore, we consider simultaneously several constraints in the design of new rules to achieve GAC.

Several Alldiff connected by one intersection. This is a simple propagation rule: if a value appears in variables of the intersection of two Alldiff, and that it does not appear in the rest of one of the Alldiff, then it can be safely removed from the other variables' domains of the second Alldiff.

$$[OI2] \quad \frac{<C\wedge Alldiff(V_1)\ \wedge\ Alldiff(V_2), D>|d\in D_{V_1\cap V_2}\ \wedge\ d\notin D_{V_2\backslash V_1}}{<C\wedge Alldiff(V_1)\ \wedge\ Alldiff(V_2), D'>|d\notin D'_{V_1\backslash V_2}}$$

$[OI2]$ is coded in SAT as $\#D_{V_1\cap V_2} * \#(V_1\cap V_2) * \#(V1\backslash V_2)$ clauses:

$$[SAT-OI2] \quad \bigwedge_{d\in D_{V_1\cap V_2}} \bigwedge_{x_1\in V_1\cap V_2} \bigwedge_{x_2\in V_1\backslash V_2} \bigvee_{x_3\in V_2\backslash V_1} (\neg\xi_{x_1}^d \vee \xi_{x_2}^d \vee \neg\xi_{x_3}^d)$$

$[OI2]$ can be extended to $[OIm]$ to handle m ($m \geq 2$) $Alldiff$ constraints connected by one intersection. Let denote by V the set of variables appearing in the common intersection: $V = \bigcap_{i=1}^{m} V_i$

$$[OIm] \quad \frac{<C\bigwedge_{i=1}^{m} Alldiff(V_i), D>|d\in D_V\ \wedge\ d\notin D_{V_1\backslash V}}{<C\bigwedge_{i=1}^{m} Alldiff(V_i), D'>|d\notin \bigcup_{i=2}^{m} D'_{V_i\backslash V}}$$

Note that this rule can be implicitly applied to the different symmetrical possible orderings of the m Alldiff.

$[OIm]$ is translated in SAT as $\#D_V * \#V * \sum_{i=2}^{m} (\#(V_i\backslash V))$ clauses:

$$[SAT-OIm] \quad \bigwedge_{d\in D_V} \bigwedge_{x_1\in V} \bigwedge_{i=2}^{m} \bigwedge_{x_3\in V_i\backslash V} \bigvee_{x_2\in V_1\backslash V} (\neg\xi_{x_1}^d \vee \xi_{x_2}^d \vee \neg\xi_{x_3}^d)$$

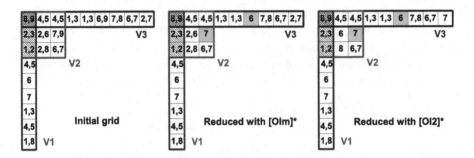

Fig. 2. $[OI2]^*$ reduces more than $[OIm]$

Property 4. Consider $m > 2$ Alldiff with a non empty intersection. Given $< C, D > \rightarrow^*_{[OIm]} < C, D' >$ and $< C, D > \rightarrow^*_{[OI2]} < C, D'' >$, then $D'' \subseteq D'$.

The proof is straightforward (see illustration on Figure 2). Consider the application of $[OIm]$: $9 \in V_1 \cap V_2 \cap V_3$, and 9 is not in the rest of V_1; thus, 9 can be removed safely from V_2 and V_3 (except from the intersection of the 3 Alldiff); no other application of $[OIm]$ is possible, leading to the second grid. Now, consider the application of $[OI2]$ on the initial grid: first between V_1 and V_2; $9 \in V_1 \cap V_2$ and 9 is not in the rest of V_1; thus, 9 can be removed from V_2; the same for 2; $[OI2]$ on V_1 and V_3 removes 9 from the rest of V_3; applying $[OI2]$ on V_3 and V_2 does not perform any effective reduction; this leads to the 3rd grid which is smaller than the second.

Although one could argue that $[OIm]$ is useless (Prop. 4) in terms of reduction, in practice $[OIm]$ can be interesting in terms of the number of rules to be applied. Moreover, $[OIm]$ can be scheduled before $[OI2]$ to reduce the CSP at low cost.

Several *Alldiff* connected by several intersections. We first consider 4 *Alldiff* having four **non-empty** intersections two by two (see Figure 3). V_{ij} (respectively $V_{i,j}$) denotes $V_i \cup V_j$ (respectively $V_i \cap V_j$). V now denotes the union of the four intersections: $V = V_{1,2} \cup V_{2,3} \cup V_{3,4} \cup V_{1,4}$.

$$[SI4.4] \quad \frac{< C \bigwedge_{i=1}^{4} Alldiff(V_i), D > | \quad V_{1,2} \neq \emptyset \wedge V_{2,3} \neq \emptyset \wedge V_{3,4} \neq \emptyset \wedge V_{1,4} \neq \emptyset \wedge d \in D_V \wedge d \notin D_{V_{13} \setminus V_{24}}}{< C \bigwedge_{i=1}^{4} Alldiff(V_i), D' > | d \notin D'_{V_{24} \setminus V_{13}}}$$

d must at least be an element of 2 opposite intersections (at least $d \in V_{1,2} \cap V_{3,4}$ or $d \in V_{2,3} \cap V_{1,4}$) otherwise, the problem has no solution. Our rule is still valid in this case, and its reduction will help showing that there is no solution.

Translated in SAT, we obtain $\#D_V * \#V * \#(V_{24} \setminus V_{13})$ clauses with $V_{1,2} \neq \emptyset \wedge V_{2,3} \neq \emptyset \wedge V_{3,4} \neq \emptyset \wedge V_{1,4} \neq \emptyset$:

$$[SAT - SI4.4] \quad \bigwedge_{d \in D_V} \bigwedge_{x_1 \in V} \bigwedge_{x_3 \in V_{24} \setminus V_{13}} \bigvee_{x_2 \in V_{13} \setminus V_{24}} (\neg \xi^d_{x_1} \vee \xi^d_{x_2} \vee \neg \xi^d_{x_3})$$

This rule can be generalized to a **ring** of $2m$ Alldiff with $2m$ **non-empty** intersections. Let V be the union of the variables of the $2m$ intersections: $V =$

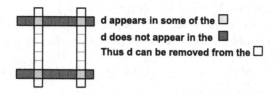

Fig. 3. Example of $[SI4.4]$

$\bigcup_{i=1}^{2m}(V_i \cap V_{(i \bmod 2m)+1})$. V_{odd} (respectively V_{even}) represents the union of the V_k such that k is odd (resp. even): $\bigcup_{i=0}^{m-1} V_{2i+1}$ (resp. $\bigcup_{i=1}^{m} V_{2i}$).

$$[SI2m.2m] \quad \frac{< C \bigwedge_{i=1}^{2.m} Alldiff(V_i), D > |}{\bigwedge_{i=1}^{2.m}(V_i \cap V_{(i \bmod 2m)+1} \neq \emptyset) \ \wedge \ d \in D_V \ \wedge \ d \notin D_{V_{odd}\setminus V}}{< C \bigwedge_{i=1}^{2.m} Alldiff(V_i), D' > |d \notin D'_{V_{even}\setminus V}}$$

These are $\#D_V * \#V * \#(V_{even}\setminus V)$ SAT clauses s.t. $\bigwedge_{i=1}^{2.m}(V_i \cap V_{(i \bmod 2m)+1} \neq \emptyset)$:

$$[SAT-SI2m.2m] \quad \bigwedge_{d\in D_V} \bigwedge_{x_1\in V} \bigwedge_{x_3\in V_{even}\setminus V} \bigvee_{x_2\in V_{odd}\setminus V}(\neg\xi_{x_1}^d \vee \xi_{x_2}^d \vee \neg\xi_{x_3}^d)$$

The reduction we obtain by applying rules for a single Alldiff and rules for several Alldiff is stronger than enforcing GAC:

Property 5. Given a conjunction of constraints $C = \bigwedge_{i=1}^{k} Alldiff(V_i)$ and a set of domains D. Given 2 sets of rules $R' = \bigcup_{l=1}^{max_i\{\#V_i\}}\{[Ol]\}$ and $R \subseteq \{[OI2],\ldots,[OIm],[SI4.4],\ldots,[SI2m.2m]\}$. Consider $< C, D >\to_{R'}^* < C, D' >$ and $< C, D >\to_{R'\cup R}^* < C, D'' >$ then $< C, D' >$ and $< C, D'' >$ are GAC and moreover $D'' \subseteq D'$.

The proof is based on the fact that $\bigcup_{l=1}^{max_i\{\#V_i\}}\{[Ol]\}$ already enforces GAC and that the $[OIm], [SI2m.2m]$ preserves GAC.

Property 6. $[SAT-OI2]$ (respectively $[SAT-OIm]$, $[SAT-SI4.4]$, and $[SAT-SI2m.2m]$) preserves the consistency property of $[OI2]$ (respectively $[OIm]$, $[SI4.4]$, and $[SI2m.2m]$).

The proof is similar to the proof of [SAT-Om] $\Longleftrightarrow$ [Om] of Property 3.

5   Evaluation

To evaluate these rules, we use them with the SAT and CSP approaches on the Sudoku problem. The Sudoku is a well-known puzzle (e.g., [17]) which can be easily encoded as a constraint satisfaction problem: it is generally played on a 9×9 partially filled grid, which must be completed using numbers from 1 to 9 such that the numbers in each row, column, and major 3×3 blocks are different.

More precisely, the $n \times n$ Sudoku puzzle (with $n = m^2$) can be modeled by $3.n$ Alldiff constraints over n^2 variables with domain $[1..n]$:

- A set of n^2 variables $\mathcal{X} = \{x_{i,j} | i \in [1..n], j \in [1..n]\}$
- A domain function D such that $\forall x \in \mathcal{X}, D(x) = [1..n]$
- A set of $3.n$ variables subsets $\mathcal{B} = \{C_1...C_n, L_1...L_n, B_{1,1}, B_{1,2}, \ldots, B_{m,m}\}$
  defined as $\forall x_{ij} \in \mathcal{X}, \ x_{ij} \in C_j, \ x_{ij} \in L_i, \ x_{ij} \in B_{((i-1)\div m)+1,((i-1)\div m)+1}$
- A set of $3.n$ Alldiff : $\forall i \in [1..n], Alldiff(C_i), \forall j \in [1..n], Alldiff(L_j),$
  $\forall k, k' \in [1..m], Alldiff(B_{kk'})$

Sudoku puzzle is a special case (i.e., more constrained) of Latin Squares, which do not require the notion of blocks nor the Alldiff constraints over the blocks.

SAT approach. We now compute the size of the SAT model for a Sudoku of size n by computing the number of generated clauses by each rule:

	Number of clauses	Complexity
Definition of the variables	$n^2 + \frac{n^3(n-1)}{2}$	$\mathcal{O}(n^4)$
[SAT-Om] $\forall m \in \{1..n-1\}$	$3n \sum_{m=1}^{n-1} \left((\frac{m}{n})^2 (n-m) m^{m+1} \right)$	$\mathcal{O}(n^{3n+3})$
[SAT-OIm] $\forall m \in \{2,3\}$	$n^3(9n - 4\sqrt{n} - 5)$	$\mathcal{O}(n^4)$
[SAT-SI2m.2m] with $m = 2$	$6n^6 - 32n^5 + \sqrt{n}\left(12n^4 - 12n^3\right) + 34n^4 - 8n^3$	$\mathcal{O}(n^6)$

In the following, we consider 9×9 grids. To encode such a Sudoku of size 9, the minimum number of clauses is 20493 (definition of the variables and [SAT-O1]) whereas encoding all the rules generates approximately 886×10^9 clauses. This increase of the number of clauses is mainly due to [SAT-Om] (see Figure 4).

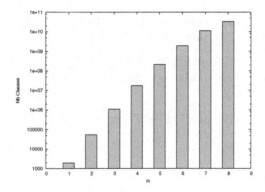

Fig. 4. Number of clauses generated by [SAT-Om] for each value of m with $n = 9$

Thus it is not practicable to generate Sudoku problems in SAT including initially all the rules. To observe the impact of these rules on the behavior of SAT solvers, we run Zchaff [13] on 9 Sudoku grids coded with:

- Definition of the variables + [SAT-O1]
- Definition of the variables + [SAT-O1] + [SAT-O2]
- Definition of the variables + [SAT-O1] + [SAT-OI2]
- Definition of the variables + [SAT-O1] + [SAT-O2] + [SAT-OI2]

Figure 5 illustrates the encoding impact on the behavior of Zchaff. For some instances, [SAT-O2] improves the results. We suppose that the performances using the other [SAT-Om] could be better. These results confirm Property 5 because [SAT-OI2] does not improve the behavior if [SAT-O2] is present. We can observe that the worst performances are obtained when [SAT-OI2] is combined to the basic definition of the problem. This rule is probably rarely used but its clauses may disrupt the heuristics. Nevertheless, the costly but powerful could be added dynamically, during the resolution process in order to boost unit propagation.

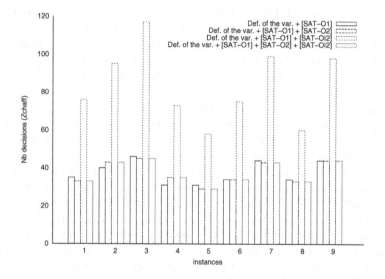

Fig. 5. Encoding impact on the behavior of Zchaff

CSP approach. From a CSP point of view, we have few rules to manage. However, the combinatoric/complexity is pushed in the rule application, and more especially in the matching: the head of the rule must be tried with all possible configurations of the constraints, and the guard must be tested. Implementing our rules in CHR (SWI-Prolog version) as propagation rules is straightforward but a generic implementation is rather inefficient: the matching of the head of the rule is too weak, and a huge number of conditions have to be tested in the guard. We thus specialized the CHR rules for arrays of variables, which is thus well suited for problems such as Latin Squares and Sudoku. The rules are also scheduled in order to first apply less complex rules, i.e., the rules that are faster to apply (strong matching condition and few conditions in the guard), and which have more chance to effectively reduce the CSP. However, we are

370 F. Lardeux et al.

still working on the implementation to improve the matching. For example, by particularizing [O7] to the Sudoku, we obtained a speed up of up to 1000 for some Sudokus. We also implemented some rules as new propagators in GeCode. The preliminary results are promising, and we plan to improve propagation time with a better scheduling of propagators, such as applying complex rules when all standard propagators have already reached a fixed point.

6   Related Work and Conclusion

Global constraints in CSP. Recent works deal with the combination of several global constraints. [19] presents some filtering algorithms for the sequence constraint and some combinations of sequence. [20] studied the conjunction of open global cardinality constraints with specific restrictions. [14] describes the cardinality matrix constraint which imposes that the same value appears p times in the variables of each row (of size n), and q times in the variables of each column (of size m). Consider some Alldiff constraints on the rows and columns of a matrix, this is a special case of the cardinality matrix constraint with $p = q = 1$. However, this constraint forces each Alldiff to be of size n or m while with our rules, they can be of different sizes.

Nevertheless, theses approaches require some specialized and complex algorithms for reducing the domains, while our approach allows us to simplify and unify the presentation of the propagation rules and attempts at addressing a wider range of possible combinations of Alldiff.

From the modeling point of view, [16] evaluate the difficulty of the Sudoku problem. To this end, various modelings using different types of constraints are proposed (e.g., the Row/Column interaction is described by the cardinality matrix global constraint; together with the row/block interaction this should compared to the application of our rule [OI2] on all intersections of a column and a row, and block and row (or column)). In our approach, we use only the classical model and do not change it: we only add more propagation rules. Moreover, our rules can be used with other problems.

Global constraints in SAT. The basic encodings of CSP into SAT have been fully studied [2,4,9,21,8,7] to preserve consistency properties and induce efficient unit propagation in SAT solvers. The specific encodings of global constraint has been also addressed, e.g., Cardinality [3,12], Among [2] or Alldiff [10]. Our transformation is based on reduction rules and extended to multiple connected Alldiff. As some of these works we proved it correctness w.r.t. GAC.

Conclusion. We have defined a set of consistency rules for general Alldiff constraints that can be easily implemented in usual constraint solvers. These rules have also been used to encode the same constraints in SAT, preserving some propagation properties through unit propagation. This work provides then an uniform framework to handle interleaved Alldiff and highlights the relationship between CSP and SAT in terms of modeling and resolution when dealing with global constraints.

We now plan to investigate other rules that could be handled in our framework, in order to solve new combinations of global constraints.

References

1. Apt, K.: Principles of Constraint Programming. Cambridge University Press, Cambridge (2003)
2. Bacchus, F.: GAC via unit propagation. In: Bessière, C. (ed.) CP 2007. LNCS, vol. 4741, pp. 133–147. Springer, Heidelberg (2007)
3. Bailleux, O., Boufkhad, Y.: Efficient cnf encoding of boolean cardinality constraints. In: Rossi, F. (ed.) CP 2003. LNCS, vol. 2833, pp. 108–122. Springer, Heidelberg (2003)
4. Bessière, C., Hebrard, E., Walsh, T.: Local consistencies in SAT. In: Giunchiglia, E., Tacchella, A. (eds.) SAT 2003. LNCS, vol. 2919, pp. 400–407. Springer, Heidelberg (2003)
5. Bordeaux, L., Hamadi, Y., Zhang, L.: Propositional satisfiability and constraint programming: A comparative survey. ACM Compututing Survey 38(4), 12 (2006)
6. Davis, M., Logemann, G., Loveland, D.: A machine program for theorem-proving. Communications of the ACM 5(7), 394–397 (1962)
7. Dimopoulos, Y., Stergiou, K.: Propagation in csp and sat. In: Benhamou, F. (ed.) CP 2006. LNCS, vol. 4204, pp. 137–151. Springer, Heidelberg (2006)
8. Gavanelli, M.: The log-support encoding of csp into sat. In: Bessière, C. (ed.) CP 2007. LNCS, vol. 4741, pp. 815–822. Springer, Heidelberg (2007)
9. Gent, I.: Arc consistency in SAT. Technical Report APES-39A-2002, University of St Andrews (2002)
10. Gent, I., Nightingale, P.: A new encoding of all different into sat. In: Proc. of 3rd Int. Work. on Modelling and Reformulating CSP, CP 2004, pp. 95–110 (2004)
11. Hoos, H.H.: Sat-encodings, search space structure, and local search performance. In: IJCAI 1999, pp. 296–303. Morgan Kaufmann, San Francisco (1999)
12. Marques-Silva, J., Lynce, I.: Towards robust cnf encodings of cardinality constraints. In: Bessière, C. (ed.) CP 2007. LNCS, vol. 4741, pp. 483–497. Springer, Heidelberg (2007)
13. Moskewicz, M., Madigan, C., Zhao, Y., Zhang, L., Malik, S.: Chaff: Engineering an efficient SAT solver. In: DAC 2001, pp. 530–535. ACM, New York (2001)
14. Régin, J.-C., Gomes, C.: The cardinality matrix constraint. In: Wallace, M. (ed.) CP 2004. LNCS, vol. 3258, pp. 572–587. Springer, Heidelberg (2004)
15. Régin, J.: A filtering algorithm for constraint of difference in csps. In: Nat. Conf. of AI, pp. 362–367 (1994)
16. Simonis, H.: Sudoku as a constraint problem. In: Proc. of the 4th CP Int. Work. on Modelling and Reformulating Constraint Satisfaction Problems, pp. 17–27 (2005)
17. Sudopedia, http://www.sudopedia.org
18. van Hoeve, W.-J., Katriel, I.: Global Constraints. In: Handbook of Constraint Programming, Elsevier, Amsterdam (2006)
19. van Hoeve, W.J., Pesant, G., Rousseau, L.M., Sabharwal, A.: New filtering algorithms for combinations of among constraints (under review, 2008)
20. van Hoeve, W.-J., Régin, J.-C.: Open constraints in a closed world. In: Beck, J.C., Smith, B.M. (eds.) CPAIOR 2006. LNCS, vol. 3990, pp. 244–257. Springer, Heidelberg (2006)
21. Walsh, T.: SAT v CSP. In: Dechter, R. (ed.) CP 2000. LNCS, vol. 1894, pp. 441–456. Springer, Heidelberg (2000)

Phase Transition in the Bandwidth Minimization Problem

Nelson Rangel-Valdez and Jose Torres-Jimenez

CINVESTAV-Tamaulipas, Information Technology Laboratory
Km. 6 Carretera Victoria-Monterrey, 87276 Victoria Tamps., Mexico
{nrangel,jtj}@tamps.cinvestav.mx

Abstract. It is known that some NP-Complete problems exhibit sharp phase transitions with respect to some order parameter. Moreover, a correlation between that critical behavior and the hardness of finding a solution exists in some of these problems. This paper shows experimental evidence about the existence of a critical behavior in the computational cost of solving the bandwidth minimization problem for graphs (BMPG). The experimental design involved the density of a graph as order parameter, 200000 random connected graphs of size 16 to 25 nodes, and a branch and bound algorithm taken from the literature. The results reveal a bimodal phase transition in the computational cost of solving the BMPG instances. This behavior was confirmed with the results obtained by metaheuristics that solve a known BMPG benchmark.

Keywords: Phase Transition, Bandwidth Minimization Problem, Connected Graphs.

1 Introduction

In general, some computational problems are easier to solve than others. A first approach to understand the hardness of such problems can be obtained using the computational complexity theory [1,2]; a computational problem can be classified as P (can be solved in polynomial time) or NP (there is no known polynomial time algorithm that solves it).

Despite the fact that a problem belongs to the class of NP problems, it is possible to find that some particular cases can be easily solved, for example the bandwidth minimization problem for graphs (BMPG) restricted to caterpillars with maximum hair length equal to 2 [3]. In the other side, there are also special cases where the BMPG problem remains NP complete such as in the case of graph caterpillar with maximum hair length equal to 3 [4]. Then, the question: Where the really hard instances are? remains open [5].

The research carried out in [5] presents the transition phenomena as a source of hard typical cases for several computational problems. This work was not pioneering on studying the phase transitions on artificial intelligence (AI). In fact, the application of statistical mechanics to NP-complete optimization problems was presented in [6], and [7] discusses the importance of such phenomena in AI.

A. Hernández Aguirre et al. (Eds.): MICAI 2009, LNAI 5845, pp. 372–383, 2009.

In the last couple of decades there has been an increasing interest in understanding the nature of the hardness of optimization problems. The phase transition phenomena has been studied in some NP-complete problems like Hamiltonian Circuit[5], Traveling Salesman Problem (TSP)[5], Chromatic Number [8], Vertex Covering [9], Independent Set[8], K-Satisfiability (K-SAT) [10] and Constraint Satisfaction Problems [11]. In [12] a work about solving instances that lies in the critical point of the phase transition of K-SAT is presented. The algorithm used was a Tabu Search. In [13] is studied the typical case of complexity of random $(2 + p)$-SAT formulas, where p varies from 0 to 1. Navarro [14] presents an approach to construct hard instances of the SAT problem; the construction of such instances is based in the location of the phase transition of that hard optimization problem.

The location of critical points of phase transitions has been mainly studied for NP-Complete problems. Some works have been performed for NP-Hard problems. Among them are TSP [5], Project Scheduling [15] Chromatic Number and MAX 2-SAT [16,17], and Number Partitioning [18,19].

In this paper we study the existence of a critical behavior in the computational cost of solving instances of BMPG. The evidence was constructed using a set of random connected graphs whose size vary from 16 to 25 nodes. The computational cost of the solution of the instances was determined through the branch and bound algorithm proposed in [20].

The rest of the paper is organized as follows. Section 2 describes the BMPG. Section 3 presents the elements used to identify the phase transition and the experiments performed so that the critical behavior can be identified. Finally, section 4 presents the conclusions of this paper.

2   Bandwidth Minimization Problem

The Bandwidth Minimization Problem for Graphs (BMPG) can be defined as follows. Let $G = (V, E)$ be an undirected graph, where V defines the set of vertices with cardinality $n = |V|$, and $E \subseteq V \times V = \{\{i, j\}|i, j \in V\}$ is the set of edges with cardinality $m = |E|$. Given a one-to-one labeling function $\tau : V \rightarrow 1, 2, ..., n$, the bandwidth β of G for τ is defined according to the Equation (1).

$$\beta_\tau = \max\{|\tau(i) - \tau(j)| : (i, j) \in E\} \qquad (1)$$

Then the BMPG consists in finding a labeling τ^* for which $\beta_{\tau^*}(G)$ is minimum, it is expressed in mathematical terms in the Equation (2).

$$\beta_{\tau^*}(G) = \min\{\beta_\tau(G) : \tau \in T\} \qquad (2)$$

where T is the set of all possible labeling functions.

The BMPG has been widely studied since 1960s [21]; and its applications range from sparse systems of equations to applications in electromagnetic industry [22], circuit design [23], hypertext reordering [24] and others.

The NP-Completeness of the BMPG has been proved in [25]. This problem remains NP-Complete even for simpler graphs such as trees [26] and carterpillars with hair length 3 [27]. In [28] have been shown that it is also NP-Hard to approximate the bandwidth of general graphs with an approximation ratio of $\frac{3}{2} - \varepsilon$ or less, where $\varepsilon > 0$.

Several exhaustive approaches have been developed to solve the BMPG problem. Del Corso and Manzini [29] developed two algorithms based on depth search to solve it, these algorithms are Minimum Bandwidth by Iterative Deepening (MB-ID) and Minimum Bandwidth by Perimeter Search (MB-PS).

Caprara and Salazar [20] show two enumeration schemes that solve the BMPG problem, *LeftToRight* and *Both*. While Del Corso and Manzini [29] present enumeration schemes based on valid labels on unlabeled nodes according with the labels of labeled neighbor nodes, Caprara and Salazar [20] generalize this definition by choosing valid labels of unlabeled nodes according with the labels of nodes at a distance d [29].

A more recent approach that solves exactly the BMPG problem is presented by Marti, et al., [30]. In this approach the solution $\overline{\phi}$ obtained by the GRASP algorithm proposed in [31] was used as an upper bound in two enumeration schemes, Branch and Bound (BB) and Branch and Bound with Perimeter search (BBP). During the enumeration schemes, when a solution $\overline{\phi}$ is found, then the value is updated by $\overline{\phi} = \overline{\phi} - 1$ but the search tree is not restarted, as done in the schemes presented in [29] and [20], because if no solution could be found with the actual upper bound, then a solution with a value smaller can not be found either.

Besides the exact approaches, several non-exhaustive approaches have been developed to solve it. Among them are Tabu Search [32], GRASP [31] and Simulated Annealing [33]. In the next section we describe the experimental design used to identify the critical behavior on BMPG.

3   Phase Transition in the Bandwidth Minimization Problem for Graphs (BMPG)

The work done on phase transition phenomena shows that hard instances of an optimization problem are found in a region near the critical point of such phenomena [5]. The critical point can be seen in a phase diagram, i.e., a curve that shows the behavior of the order parameter against a control parameter of interest for the optimization problem.

In this paper we experimentally studied the critical behavior found when solving BMPG instances using an exact approach. As it was described before, there are six known exact methods that solve the BMPG problem: MB-ID, MB-PS, *Both*, *LeftToRight*, BB and BBP. For the purpose of our research, we chose the algorithm *Both* [20] because it is a generalization of the algorithm [20], in average it performs better than the algorithms MB-ID, MB-PS, it also requires less memory than BBP and it has a simpler structure than the algorithms BB and BBP (in the sense that it does not require an heuristic algorithm to build an initial solution).

Random Connected Graph Generator. The solution of the BMPG problem for unconnected graphs is equivalent to individually solve each of its connected components. Considering this fact, we decided to work only with connected graphs. Therefore, the probabilistic model of graphs known as the Erdös-Rényi model [34] and denoted by $G(n,p)$, is adapted in this research as the source of random instances for the BMPG problem. The random graph model was implemented using two stages. The first stage generates a random tree using the Prüffer sequence[35]. The second stage applies the $G(n,p)$ model so that the graph resulting will be always connected. The pseudocode that follows this method to construct random connected graphs is shown in Algorithm 3.1.

Algorithm 3.1. RANDOMGRAPHGENERATOR(n, ρ)

$\mathcal{T} \leftarrow$ PRÜFFERSEQUENCE(n)
$\mathcal{G} \leftarrow \mathcal{T}$
$density \leftarrow \rho - (n-1)/\binom{n}{2}$
for $i \leftarrow 1$ **to** n

$\quad$**do** $\begin{cases} \textbf{if } density \leq 0 \\ \quad \textbf{then } \{\textbf{return } (\mathcal{G}) \\ \textbf{for } j \leftarrow i+1 \textbf{ to } n \\ \quad \textbf{do} \begin{cases} number \leftarrow \text{RANDOMNUMBER}() \\ \textbf{if } number \leq \rho \\ \quad \textbf{then } \begin{cases} \text{ADD}(G, \{i,j\}) \\ density \leftarrow density - 1/\binom{n}{2} \end{cases} \end{cases} \end{cases}$

$T_i \leftarrow T_i * \alpha$
return $(\mathcal{G})$

Order Parameter. An order parameter is a descriptive measure of an optimization problem instance. Examples of these metrics for graphs $G = \{V, E\}$ are the number of edges m of the graph [8], the density $\frac{m}{\binom{n}{2}}$ [15] or the coefficients $\frac{m}{n \log n}$ used by [5] to relate the number of edges m with the minimum number of edges required in a graph to be connected $n \log n$, where n is the number of nodes of the graph.

Given that the bandwidth of a graph grows with its density, the order parameter selected to control the experiment was the density of the graph $G = \{V, E\}$, i.e., the number of edges $m = |E|$ divided by the maximum number of edges that the graph can have $\binom{n}{2}$, where $n = |V|$.

Complexity Measure. The complexity of solving a BMPG instance G is defined as the number of subproblems spent by the *Both* Algorithm when solving G. A subproblem is a node that the branch and bound algorithm *Both* visits in the search tree when solving a BMPG instance. This complexity measure is the most common measure used in the literature [5,36,15,8,14].

Experimental Evidence. Following the method for constructing random graphs shown in algorithm 3.1, a set of 200000 BMPG instances was generated. The size n of the instances was in the rank from 16 to 25 nodes. The number of edges m for each graph G was set to $\Delta \times \binom{n}{2}$, where Δ is the density selected for G. The values of δ can be $\{0.05, 0.10, 0.15, ..., 1.00\}$. For each node size n and each density value Δ, a set of 1000 instances of BMPG were generated.

The computational cost of solving the set of random instances is shown in Table 1. The first column shows the density value. The rest of the columns represents the different sizes of the instances and contains the average number of subproblems required to solve each instance.

Table 1. Complexity results of solving the random set of BMPG instances. The values represent the average number of nodes in the search tree (subproblems) constructed by the algorithm *Both* before reaching the optimal solution.

Density	16	17	18	19	20	21	22	23	24	25
0.05	20.2	21.6	27.7	27.9	35.6	47.6	53.4	76.8	175.8	336.8
0.10	19.7	22.7	26.4	31.2	40.4	42.2	144.5	413.8	465.0	1175.0
0.15	27.4	38.9	49.6	91.5	98.1	178.8	218.8	452.6	879.7	1290.3
0.20	31.6	45.0	59.6	79.7	123.5	169.3	302.0	308.9	750.7	906.4
0.25	38.6	41.5	59.9	63.8	111.3	114.8	167.7	216.8	310.8	460.3
0.30	33.1	39.2	49.9	58.7	84.8	107.9	176.8	290.5	390.5	1065.9
0.35	30.7	43.2	60.1	77.3	130.1	182.5	423.0	745.5	1071.6	2791.1
0.40	41.8	62.4	86.7	175.9	220.9	399.2	724.8	1342.2	2045.6	3355.9
0.45	48.6	85.1	137.3	207.9	321.0	473.0	770.7	1231.2	1644.3	2936.1
0.50	68.1	100.3	139.8	212.4	318.6	487.4	659.5	953.5	1448.1	1972.4
0.55	63.7	92.0	135.9	197.8	279.1	406.1	521.1	703.8	1027.5	1334.0
0.60	60.7	84.6	112.6	155.2	211.4	285.2	394.2	506.9	650.4	844.2
0.65	51.8	67.6	87.4	120.4	162.6	203.3	253.0	339.5	410.4	541.9
0.70	41.8	52.8	69.4	85.9	110.5	138.3	173.5	214.2	279.7	338.7
0.75	32.1	40.4	48.4	60.4	75.4	94.1	114.9	140.9	167.1	199.3
0.80	24.0	29.0	34.7	41.4	50.1	59.1	68.8	82.1	99.2	118.4
0.85	18.6	20.0	22.0	25.2	29.5	34.8	40.5	48.5	57.1	64.2
0.90	13.6	14.3	16.5	19.0	20.2	23.0	24.4	26.4	28.4	30.8
0.95	17.2	18.1	18.5	18.8	18.6	19.0	18.2	18.0	18.2	18.5
1.00	0.0	0.0	0.0	0.0	0.0	0.0	0.0	0.0	0.0	0.0

A plot of the data presented in Table 1 (see Figure 1) reveals the existence of a density value where the average number of iterations is maximum. Moreover, the complexity curve is almost bimodal showing an easy-hard-easy-hard-easy pattern but with one of the peak larger than the other. It is precisely in that large peak where a critical behavior in the computational cost of solving the BMPG instances is claimed to exist. In the literature [5,15,8] has been established a relation between a phase diagram describing phase transitions in NP problems and the hardness of the instances in the critical point of that diagram. In our case, the pattern shown in Figure 1 allows us to identify one point where the computational cost of the instances grows significantly in comparison with other points, this can be the evidence of the existence of a more general critical behavior on BMPG (the existence of the phase transition phenomena) than can explain such behavior.

The complexity results obtained with the set of random instances were contrasted with the set of BMPG instances taken from Harwell-Boeing Sparse Matrix Collection[1] (HBSM). In order to understand the complexity nature of such benchmark we take the results given by some state-of-the-art BMPG algorithms. Tables 2 and 3 summarize the results reported by the algorithms Simulated Annealing (SA) [33], GRASP [31] and the Tabu Search [32]. The first three columns show information about the instances. The rest of the columns present the time spent by each algorithm when solving the instances. The solution achieved by each algorithm can be seen in [33].

Figure 2 shows a graph of the complexity results of the BMPG state-of-the-art algorithms solving the HBSM benchmark. This figure presents a behavior in the computational cost (measure as the time that took to each algorithm to reach its best solution) similar to the one presented in the set of random instances, in the sense that when the density increases, the computational cost of solving the instances also increases.

The experiments showed that the hardest instances were not found in the set of larger instances but in the set of medium instances of the HBSM benchmark. In addition, the average performance of the algorithms (see last row of Tables 2 and 3) over each different set of instances from the HBSM benchmark indicates that the hardest set was the set of medium size. We can note in Tables 2 and 3 that the density of the hardest instances is close to the value pointed out as the critical point in the experiment with random connected BMPG instances, this suggests that the hardest instances lie at the critical region in a phase transition phenomena for both exact and non-exact algorithms.

The analysis of the computational cost presented in this section suggests that there are two types of instances of the BMPG that requires a small computational effort to be solved. The first type corresponds to instances with a low density. The second type is formed by instances with a high density. This behavior suggests the existence of two regions, one where the BMPG instances have a low density and the other where they have a high density. The graphs with low density, as paths and cycles, can be considered under-constrained, a small number of edges are connecting the nodes resulting in a small optimal bandwidth for those graphs. These graphs are easy to solve because almost all the solutions are possible, with the possibility of finding the optimal values in a small set of integers. The graphs with high density, as the complete graph, can be consider over-constrained; the great number of edges in it allows only a few set of possible optimal solutions which also makes them easy to solve. The existence of these separated regions could explain the easy-hard-easy transition in the computational cost of solving BMPG, just as it is presented in the satisfiability problem (SAT)[10].

---

[1] http://math.nist.gov/MatrixMarket/data/

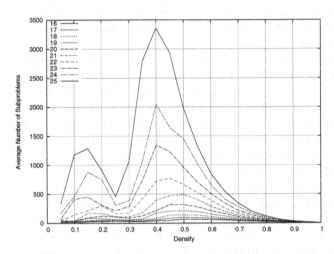

Fig. 1. Behavior of the computational cost (measured by the number of solved sub-problems) of solving random BMPG instances varying the density of the graphs. A curve for each of the 10 values for the number of nodes is shown. The x axis shows the different values of density analyzed. The y axis shows the average number of subproblems required. The forms of each curve drown follow a non-monotonic increase in the complexity revealing that the BMPG problem presents a bimodal phase transition.

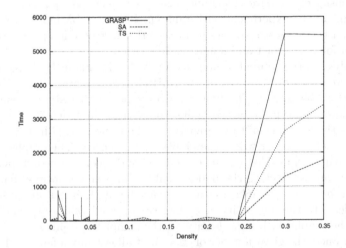

Fig. 2. Computational time to solve the HBSM benchmark using the algorithms SA, TS and GRASP. This graph relates the density of the instances (shown in x axis) with the time that each algorithm spent to solve them (shown in y axis). In this graph shows two peaks indicating a high computational cost; one of them is located between the densities 0.05 and 0.10, the other one is at the density value of 0.35.

Table 2. Computational time for solving the small size set of cases of HBSM benchmark. It shows the time (in seconds) spent by TS, GRASP and SA to solve each instance. Also, the number of nodes n, edges m and the density of each instance are presented.

Instance	n	m	Density	TS Time	GRASP Time	SA Time
dwt_234	117	162	0.02	1.2	1.9	1.1
nos1	158	312	0.03	1.3	2.6	1.1
bcspwr03	118	179	0.03	1.7	0.9	1.2
west0156	156	371	0.03	7.4	8.4	2.8
will199	199	660	0.03	12	26.9	6.1
west0167	167	489	0.04	5.8	5.6	4.8
lns_131	123	275	0.04	3.4	2.6	1.8
impcol_c	137	352	0.04	3.5	4.5	3.1
gre_185	185	650	0.04	6.2	6.1	6.8
gre_115	115	267	0.04	2.4	3.2	1.6
fs_183_1	183	701	0.04	32.4	11.8	8.7
bcsstk22	110	254	0.04	1.1	1.6	5.5
west0132	132	404	0.05	3.4	8.5	5.4
can_161	161	608	0.05	4	0.7	3.0
nos4	100	247	0.05	1.1	1.4	0.9
bcspwr02	49	59	0.05	0.2	0.6	0.2
can_144	144	576	0.06	1.7	3.1	17.6
ash85	85	219	0.06	0.7	0.4	1.1
bcspwr01	39	46	0.06	0.1	0.1	0.4
will57	57	127	0.08	0.4	0.4	1.1
arc130	130	715	0.09	4.8	1.9	23.2
curtis54	54	124	0.09	0.7	0.7	0.5
bcsstk05	153	1135	0.1	4.7	7.1	10.6
gent113	104	549	0.1	6.3	1	3.9
lund_b	147	1147	0.11	5.3	5.5	40.8
lund_a	147	1151	0.11	8.8	5.4	40.6
mcca	168	1662	0.12	23.9	10.8	81.8
steam3	80	424	0.13	0.7	1.O	8.9
bcsstk01	48	176	0.16	0.9	1	0.6
impcol_b	59	281	0.16	1.3	1.2	1.2
ibm32	32	90	0.18	0.2	0.3	0.3
bcsstk04	132	1758	0.2	16.7	5.4	79.2
pores_1	30	103	0.24	0.3	0.3	3.1
Average			0.08	4.99	4.12	11.18

Table 3. Computational time for solving the medium and large size cases of HBSM benchmark. It shows the time (in seconds) spent by TS, GRASP and SA to solve each instance. Also, the number of nodes n, edges m and the density of each instance are presented.

Medium Instance	n	m	Density	TS Time	GRASP Time	SA Time	Large Instance	n	m	Density	TS Time	GRASP Time	SA Time
494_bus	494	586	0.01	29.2	13.5	24.4	662_bus	662	906	0.01	113.8	32.9	70.4
bcsstk20	467	1295	0.01	27.8	11	44.6	nos6	675	1290	0.01	70.9	42.8	12.5
impcol_d	425	1267	0.01	21.9	30.5	77.5	lns_511	503	1425	0.01	74.2	58.3	123.8
plskz362	362	880	0.01	27.4	7.1	20.9	jpwh_991	983	2678	0.01	312.6	65.1	104.3
west0497	497	1715	0.01	78.3	88.7	137.9	nos2	638	1272	0.01	43.7	15.7	114.5
bcspwr05	443	590	0.01	24.7	16.3	26.8	nnc666	666	2148	0.01	138.5	55.8	108.5
nnc261	261	794	0.02	8.6	22.4	8.7	jagmesh1	936	2664	0.01	150.6	107.8	33.8
nos5	468	2352	0.02	60.8	103.1	121.9	fs_760_1	760	3518	0.01	101.1	97.9	95.6
hor_131	434	2138	0.02	26	27.7	154.1	fs_680_1	680	1464	0.01	43.4	33.6	39.8
gre_343	343	1092	0.02	16.6	21.2	7.4	gre_512	512	1680	0.01	77.1	92.7	14.7
saylr1	238	445	0.02	4.3	8.5	2.3	gr_30_30	900	3422	0.01	282.2	85.4	370.1
west0479	479	1889	0.02	81.2	163	40.5	west0989	989	3500	0.01	372.7	416.9	172.1
bcspwr04	274	669	0.02	9.6	5.2	28	sh1_200	663	1720	0.01	161.3	98.4	213.5
ash292	292	958	0.02	7.9	8.7	34.4	sh1_0	663	1682	0.01	153.1	110.2	211.5
pores_3	456	1769	0.02	16.8	3.2	13	west0655	655	2841	0.01	150.1	245.8	80.5
dwt_310	310	1069	0.02	15.2	8.6	13	sh1_400	663	1709	0.01	188.3	121.2	221.4
dwt_245	245	608	0.02	7.5	14.4	9.3	sherman4	546	1341	0.01	33	4.1	11.5
can_445	445	1682	0.02	64.5	47.2	114	orsirr_2	886	2542	0.01	203	42.5	164.5
dwt_419	419	1572	0.02	23.7	28	59.9	nos7	729	1944	0.01	74.5	89.9	23.9
dwt_361	361	1296	0.02	11.8	0.8	8.3	sherman1	681	1373	0.01	107.2	78	15.6
west0381	381	2150	0.03	113	185.4	38.1	bp_200	822	3788	0.01	315.2	550.3	168.4
dwt_221	221	704	0.03	5	7	23.4	bp_1600	822	4809	0.01	546.8	783.7	231.8
can_292	292	1124	0.03	19	7.9	61.5	bp_0	822	3260	0.01	386.8	483	140.7
impcol_a	206	557	0.03	5.5	3.4	5.4	bp_800	822	4518	0.01	520.9	636.9	219.9
gre_216a	216	660	0.03	7.2	9.5	3.5	bp_600	822	4157	0.01	480.9	556.8	193.3
bcsstk06	420	3720	0.04	47.6	50.6	247.9	bp_400	822	4015	0.01	355.7	560.6	180.0
bcsstm07	420	3416	0.04	40.3	113.7	215.7	bcsstk19	817	3018	0.01	174.3	86.6	222.9
str_0	363	2446	0.04	120.8	119.1	39.9	685_bus	685	1282	0.01	90.4	12.7	62.0
plat362	362	2712	0.04	24.8	3.4	189.4	bp_1400	822	4760	0.01	521	741.2	217.4
dwt_209	209	767	0.04	10.8	1.3	29.2	bp_1200	822	4698	0.01	674.8	897.1	218.2
str_200	363	3049	0.05	90.9	114.1	47.3	bp_1000	822	4635	0.01	886.7	886.2	216.2
impcol_e	225	1187	0.05	10.7	4	49.6	dwt_592	592	2256	0.01	52.9	106.8	123.7
str_600	363	3244	0.05	180.3	92	55.9	dwt_878	878	3285	0.01	195	99.6	106.7
steam1	240	1761	0.06	16.9	12.7	79.1	saylr3	681	1373	0.01	107.2	77.8	15.6
mbeause	492	36209	0.3	2637.6	5494.3	1289.5	dwt_918	918	3233	0.01	290.8	12.6	243.0
mbeaflw	487	41686	0.35	3409.4	5467.8	1744.9	can_715	715	2975	0.01	183.1	14.2	223.0
mbeacxc	487	41686	0.35	3409.1	5464.5	1774	can_838	838	4586	0.01	158.9	37.4	384.2
							fs_541_1	541	2466	0.02	54.4	26.5	82.4
							nos3	960	7442	0.02	143.8	209.1	811.6
							dwt_992	992	7876	0.02	272.5	306.4	116.4
							dwt_503	503	2762	0.02	99.3	7.9	163.5
							steam2	600	6580	0.04	242.2	182.5	687.4
							mcfe	731	15086	0.06	800.2	247.2	1868.9
Average			0.05	289.53	480.54	184.9				0.01	241.98	219.03	211.62

4   Conclusions and Future Work

In this paper we presented experimental evidence of the existence of a critical behavior in the optimization problem known as Bandwidth Minimization Problem for Graphs (BMPG).

We study the computational complexity of the BMPG instances in a set of random connected graphs. The critical behavior was observed in the computational cost of solving the BMPG instances taking as the order parameter the density of the instances and as the measure of complexity the number of subproblems required for an state-of-the-art Branch & Bound strategy to find the optimal solution.

The complexity results were contrasted with reported results of the well known BMPG benchmark of the Harwell Boeing Sparse Matrix Collection. Keeping the density as the order parameter and the time spent by an algorithm to find its best solution as the complexity measure, we presented graphs showing a similar complexity behavior that the one presented in the set of random connected graphs. Moreover, the results uncover the fact that the hardest instances do not correspond to the ones with the highest number of nodes but the ones with the density close to the point marked as the critical point in the experiment with random connected graph (there are medium sized instances). It seems for the evidence shown that the instances that lie in the critical region are hard even for non-exact algorithms.

Finally, given the results presented in this document a guideline to construct harder benchmarks for BMPG must consider the density parameter.

Currently we are working on analyzing other order parameters to better understand the hardness of the BMPG instances in the point of maximum difficulty. Also, we are studying new complexity measures for non-exact algorithms. In our future work, we also expect to increase the size of instances.

Acknowledgments. This research was partially funded by the following projects: CONACyT 58554-Cálculo de Covering Arrays, 51623-Fondo Mixto CONACyT y Gobierno del Estado de Tamaulipas.

References

1. Karp, R.: On the computational complexity of combinatorial problems. Networks 5(1), 45–68 (1975)
2. Garey, M.R., Johnson, D.S.: Computers and Intractability: A Guide to the Theory of NP Completeness. Freeman, San Francisco (1979)
3. Assman, S., Peck, G., Syslo, M., Zak, J.: The bandwidth of carterpillars with hairs of length 1 and 2. SIAM J. Algebraic and Discrete Methods 2, 387–393 (1981)
4. Monein, B.: The bandwidth minimization problem for carterpillars with hair length 3 is np-complete. SIAM J. Algebraic Discrete Methods 7(4), 505–512 (1986)
5. Taylor, W., Cheeseman, P., Kanefsky, B.: Where the really hard problems are. In: Mylopoulos, J., Reiter, R. (eds.) Proceedings of 12th International Joint Conference on AI (IJCAI 1991), vol. 1, pp. 331–337. Morgan Kauffman, San Francisco (1991)
6. Fu, Y., Anderson, P.: Application of statistical mechanics to np-complete problems in combinatorial optimisation. Journal of Physics A: Mathematical and General 19(9), 1605–1620 (1986)
7. Huberman, B., Hogg, T.: Phase transitions in artificial intelligence systems. Artif. Intell. 33(2), 155–171 (1987)
8. Barbosa, V., Ferreira, R.: On the phase transitions of graph coloring and independent sets. Physica A 343, 401–423 (2004)

9. Hartmann, A., Weigt, M.: Statistical mechanics perspective on the phase transition in vertex covering of finite-connectivity random graphs. Theoretical Computer Science 265, 199–225 (2001)
10. Kirkpatrick, S., Selman, B.: Critical behavior in the satisfiability of random Boolean expressions. Artificial Intelligence 81, 273–295 (1994)
11. Smith, B.: Constructing an asymptotic phase transition in random binary constraint satisfaction problems. Theoretical Computer Science 265, 265–283 (2001)
12. Mazure, B., Sais, L., Grégoire, E.: Tabu search for sat. In: Proceedings of the Fourteenth Natí Conf. on Artificial Intelligence (AAAI 1997), pp. 281–285. AAAI Press, Menlo Park (1997)
13. Monasson, R., Zecchina, R., Kirkpatrick, S., Selman, B., Troyansky, L.: Determining computational complexity from characteristic phase transitions. Nature 400, 133–137 (1999)
14. Navarro, J.A., Voronkov, A.: Generation of hard non-clausal random satisfiability problems. In: AAAI 2005 (2005)
15. Herroelen, W., De Reyck, B.: Phase transitions in project scheduling. Journal of the Operational Research Society 50, 148–156 (1999)
16. Slaney, J., Walsh, T.: Phase transition behavior: from decision to optimization. In: Proceedings of the 5th International Symposium on the Theory and Applications of Satisfiability Testing, SAT (2002)
17. Achlioptas, D., Naor, Y.P. A.: Rigorous location of phase transitions in hard optimization problems. Nature 435, 759–764 (2005)
18. Gent, I.P., Walsh, T.: Phase transitions and annealed theories: Number partitioning as a case study. In: ECAI, pp. 170–174 (1996)
19. Gent, I., Walsh, T.: Analysis of heuristics for number partitioning. Computational Intelligence 14(3), 430–451 (1998)
20. Caprara, A., Salazar-González, J.: Laying out sparse graphs with provably minimum bandwidth. Informs Journal on Computing 17(3), 356–373 (2005)
21. Cuthill, E.H., McKee, J.: Reducing the bandwidth of sparse symmetric matrices. In: Proc. 24th ACM National Conf. pp. 157–172 (1969)
22. Esposito, A., Catalano, M., Malucelli, F., Tarricone, L.: Sparse matrix bandwidth reduction: Algorithms, applications and real industrial cases in electromagnetics. Advances in the Theory of Computation and Computational Mathematics 2, 27–45 (1998)
23. Bhatt, S., Leighton, F.: A framework for solving VLSI graph layout problems. J. Comput. Sytem Sci. 28, 300–343 (1984)
24. Berry, M., Hendrickson, B., Raghavan, P.: Sparse matrix reordering schemes for browsing hypertext. Lectures in Applied Mathematics 32, 99–123 (1996)
25. Papadimitriou, C.: The np-completeness of the bandwidth minimization problem. Computing 16(3), 263–270 (1976)
26. Garey, M.R., Graham, R.L., Johnson, D.S., Knuth, D.E.: Complexity results for bandwidth minimization. SIAM Journal on Applied Mathematics 34(3), 477–495 (1978)
27. Monien, B.: The bandwidth minimization problem for caterpillars with hair length 3 is np-complete. SIAM J. Algebraic Discrete Methods 7(4), 505–512 (1986)
28. Blache, G., Karpinski, M., Wirtgen, J.: On approximation intractability of the bandwidth problem. Electronic Colloquium on Computational Complexity (ECCC) 5(014) (1998)
29. Del Corso, G.M., ManZini, G.: Finding exact solutions to the bandwidth minimization problem. Computing 62, 189–203 (1999)

30. Marti, R., Campos, V., Pinana, E.: A branch and bound algorithm for the matrix bandwidth minimization. EJORS 186, 513–528 (2008)
31. Pinana, E., Plana, I., Campos, V., Marti, V.: GRASP and path relinking for the matrix bandwdith minimization. European Journal of Operational Research 153, 200–210 (2004)
32. Marti, R., Laguna, M., Glover, F., Campos, V.: Reducing the bandwidth of a sparse matrix with tabu search. European Journal of Operational Research 135(2), 211–220 (2001)
33. Rodriguez-Tello, E., Hao, J., Torres-Jimenez, J.: An improved simulated annealing algorithm for bandwidth minimization. European Journal of Operational Research 185(3), 1319–1335 (2008)
34. Erdos, P., Rényi, A.: On random graphs. i. Publicationes Mathematicae 6, 290–297 (1959)
35. Prüffer, P.: Neuer beweis eines satzes über permutationen. Arch. Math. Phys. 27, 742–744 (1918)
36. Selman, B., Kirkpatrick, S.: Critical behavior in the computational cost of satisfiability testing. Artificial Intelligence 81, 273–295 (1996)

Planning for Conditional Learning Routes

Lluvia Morales, Luis Castillo, and Juan Fernández-Olivares

Department of Computer Science and Artificial Intelligence, University of Granada
Daniel Saucedo Aranda s/n 18071 Granada, Spain
{lluviamorales,L.Castillo,faro}@decsai.ugr.es

Abstract. This paper builds on a previous work in which an HTN planner is used to obtain learning routes expressed in the standard language IMS-LD and its main contribution is the extension of a knowledge engineering process that allows us to obtain conditional learning routes able to adapt to run time events, such as intermediate course evaluations, what is known as the standard IMS-LD level B.

Keywords: Planning and Scheduling, Automatic Generation of IMS-LD level B.

1 Introduction

The e-learning institutions that offer large scale courses tend to provide tens to hundreds of courses that have to be delivered to thousands of students grouped into classes [1]. For this reason, the e-learning community has created a set of standards that allows, not only the reuse of educational resources, but also the design of personalized learning sequences to introduce the concepts of the course. So that, each of them could exploit at maximum the course.

This paper is based on a standard known as Learning Design or IMS-LD [2] which describes as many custom sequences of educational resources from a single online course as the number of students profiles on it. These sequences vary depending on students' features, but also according to a set of conditional variables and its possible values, e.g. evaluations that are made to determine the level of knowledge about a given concept. Hence, the generation of a IMS-LD by hand is expected to be too expensive in both time and human effort.

One of the most effective solutions to carry out this task is by using intelligent Planning and Scheduling (P&S) techniques[3]. There are several approaches in the literature that are discussed later, but none of them takes advantage of all the IMS-LD potential, particularly the possibility of producing conditional learning sequences, what is the main concern of this paper.

Our previous work [4] describes a knowledge engineering process applied to several data encoded through e-learning standards that allows to automatically generate a planning domain and problem. These planning domain and problem describe all personalization requirements of a course and they are used by the hierarchical planner [5] to generate a IMS-LD level A. However, this process does

A. Hernández Aguirre et al. (Eds.): MICAI 2009, LNAI 5845, pp. 384–396, 2009.

not include the generation of domains and problems that allow us to generate a level B IMS-LD with conditional sequences.

Throughout this article we first give a brief introduction to the related e-learning terminology and to the planner we use to address the problem to generate a conditional learning design. Later we focus on defining a methodology that will allow us to solve that problem through modifications to the initial architecture and to improve the planning domain and problem generation algorithms. Results of experiments conducted with the planner [5] will be described in a special section. Methodology and experiments will be highlighted, discussed and compared with other current existing proposals, to finally arrive to the conclusions.

2 Preliminaries

This section shows some key concepts used in the e-learning community, and its relation with some concepts of P&S that are used in the next section. It also describes the main features of the planner which interprets the hierarchical domain and problem generated by the methodology presented in this article.

2.1 E-Learning Standards and Tools

Large scale on-line courses are usually provided through a *Learning Management System*(LMS) that includes not only the descriptive representation of the course through nested concepts by using educational resources for their full understanding and practice, but also the timing of the course, the application of evaluations, and students' features.

The *educational resources* of an online course are all those images, documents, Flash, Word, PDF, HTML, etc... with which a student interacts in some way to understand the concepts of the course, while the *concepts* are the representation of the chapters, lessons and units that can be organized in a hierarchical structure that defines educational resources at the lowest level.

The e-learning community has adopted several standards based on XML to represent features of educational resources and concepts of an online course, as well as the entities that interact with them such as questionnaires and students. The following points give a brief description of these standards that belong to the family of standards approved by IMS Global Learning Consortium[2]:

- LEARNING RESOURCE METADATA (IMS-MD) is a standard that allows us to define, by default metadata values, the features of an educational resource. It is usually edited in a tool called RELOAD[6] and it is of vital importance to define the order and hierarchy between nested concepts like chapters, subchapters, lessons, or units and educational resources in a course. Certain features and requirements of these resources as optionality, duration, resource type, language, difficulty, media resources required, etc. are also described by the standard.
- LEARNER INFORMATION PACKAGE (IMS-LIP) integrates every students' profile data in a single XML document through nested labels as identifier, name, preferences, competencies, etc.

- QUESTION AND TEST INTEROPERABILITY (IMS-QTI) standard is a XML document that represents an evaluation and its possible outcomes, besides some rules to assess the knowledge level achieved by students.
- LEARNING DESIGN(IMS-LD) standard is the main tool for designing the sequence of learning activities of a course. It helps to represent the interaction between different structures or sequences of learning activities, also called educational resources, and the different roles that bring together students from a course with similar profiles.

The Learning Design can be defined from three levels of abstraction A, B and C, where each level incorporates and extends the capabilities of the previous one. Our work aims to extend the generation of IMS-LD level A capabilities based on HTN planning techniques, already done in [4], to an IMS-LD level B which extends the ability of generate personalized learning sequences with a run-time dynamic that supports properties and conditions like intermediate evaluations.

It is also worth to mention that each course element has its equivalent within the hierarchical planning paradigm, and that it is possible to obtain relevant information to define domain and planning problems from e-learning standards.

Educational resources are represented as *durative actions* in a planning domain, while concepts and their hierarchical structure are equivalent to the *tasks*. These tasks and actions are described by the IMS-MD standard that has metadata which allows to define the duration and conditions for implementing an action, in addition to the preconditions for implementing the methods of a task and their relationships of order and hierarchy.

On the other hand, the IMS-LIP standard helps to define the initial state of a problem by representing the features of a student as predicates and/or functions that are intimately related to certain conditions and preconditions of the domain defined by IMS-MD. While the IMS-QTI standard defines an evaluation that is considered a goal within the scope of planning. These planning domain and problem structures used by the planner are described in next section.

2.2 HTN Planner

Hierarchical planners are used for most of researchers who have worked over the sequencing of educational resources. [7], [8], and [4] they agree that this type of planning allows to easily obtain a typical course concepts hierarchy without neglecting those courses that have a plain concepts structure.

The planner we use as reference for modeling the problem and domain of this work is described in [5], called SIADEX. This planner follows the hierarchical task network formalism established by SHOP2 in [9] which is based on a forward search procedure, that involves keeping the problem state during the search process. Primitive actions used by the planner are fully compatible with PDDL 2.2 level 3, and support durative actions.

Creators of this planner have defined an extension to PDDL that allows to represent hierarchical tasks and their methods within a related temporal framework that inherits their temporal restrictions to its child actions. This permits to determine deadlines for the plan execution, and other features of the planner [5]. The methodology to obtain a level B learning design by using this planner is explained in the following section.

3 Methodology

As mentioned above, in order to generate an IMS-LD level B, conditional values should be added to our previous approach, e.g. intermediate evaluations that may change during the execution of the course.

Unlike the previous proposal which generated an overall plan for the course and each student, now it will be necessary to generate as many plans as possible evaluations and outcomes that are defined for each student profile, where each profile groups the students of a course with the same features and requirements.

The mid-term evaluations of the course concepts and their possible outcomes are described by an ordered sequence of IMS-QTI documents that content the goals of a course. Moreover, each document defines particular satisfaction levels, with a range of scores related, and rules to change the requirements of a student according to his/her score in the evaluation and the level it comes in.

Furthermore, each evaluation has a main concept or task related which is fully defined by a instructor using the IMS-MD standard. This main task is composed of subtasks and actions, with relations and requirements also described by IMS-MD. These tasks have a hierarchical structure between them and must be carried out in sequence, in order to understand the main concept being taught.

Our approach takes the set of profiles and the sequence of IMS-QTI goals and follows an iterative process. For every profile, we iterate over the set of goals. Every new goal generates a planning problem, and all the resulting plans for each goal and every profile are increasingly added to the final IMS-LD.

However, to generate each of these IMS-LD "drafts" it will be necessary to design a robust hierarchical domain and planning problem able to support the planning of all the sequences of actions that could meet the goal in question. The amount of these sequences is determined, not only by the number of different profiles of students who take that course, but also by the conditions attached to the outcomes of the immediate evaluation prior to the evaluation being planned.

An IMS-LD level B includes conditions related to some properties of the course that can take different values at run time. We have used the process of applying intermediate evaluations as an example, because an evaluation can take different values at runtime according to the score obtained for each student, dynamic features of the student like the language or activity level.

In our example, possible score ranges of evaluations are used to define sequences of educational resources that are showed later under some conditions, which are defined during the planning process included in the architecture described in figure 1. This figure shows the process to generate an IMS-LD level B, summarized into the next three stages:

1. The first step is to get all the information provided by educational standards stored in a LMS called Moodle, through an XML-RPC web service.

388   L. Morales, L. Castillo, and J. Fernández-Olivares

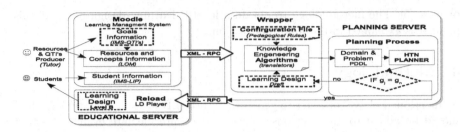

Fig. 1. IMS-LD level B generation process. The features which extend our previous work are highlighted by dotted boxes.

2. In the second stage a translation of the information obtained in first step is made through knowledge engineering algorithms and pedagogical rules described in a configuration file, in order to obtain a planning domain and problem, which satisfy the contents of the corresponding intermediate objective or goal. Sequences and conditions in IMS-LD "draft", if are not translating the first objective, are recovered and included in the planning domain.
3. Finally planning of sequences and conditions defined through the execution of a hierarchical planner is carried out to obtain a new learning design which includes these sequences and conditions, together with the previous ones.
   At the end of this stage, if the planned goal is to meet the objectives to make the last evaluation of the course, then the learning design is considered to be complete and can be shown through the RELOAD tool to the students of the course according to their profile. Otherwise, we understand that the learning design is incomplete and we will overwrite the last IMS-LD draft with this new one, and return to the second step.

In order to formalize, simplify and understand the previous process, a set of concepts have to be defined first.

Having $St = \{st_1, st_2, ..., st_l\}$ as a set of students in a course and $\mathcal{G} = \{g_1, g_2, ..., g_n\}$ as an ordered set of goals that must be met to satisfy the objectives of a course that are related to its evaluations.

Where p_i is a set that groups the students of a given course with the same features and requirements. Let us consider that for each profile p_i there are n goals $(g_1, g_2, ..., g_n)$ or intermediate evaluations to accomplish, according to the number of documents under the IMS-QTI standard that are linked to the course.

The outcome of each evaluation may be a continuous value ranging in the possible set of scores $[S_{min}, S_{max}]$. However, for practical reasons, instructors do not adapt to every possible outcome, but just for a discrete set of intervals.

Let us consider that instructors only distinguish h different levels of scores, then, they only take into account scores in the following intervals $[limit_1, limit_2)$, $[limit_2, limit_3), [limit_3, limit_4), ...[limit_{h-1}, limit_h]$ where $limit_1 = S_{min}$ and $limit_h = S_{max}$. Therefore, the set $\mathcal{L} = \{l_{j,1}, l_{j,2}, ..., l_{j,h}\}$ describes h satisfaction levels related to each goal g_j.

On the other hand, for each level $l_{j,k}$ certain changes on the features of profile p_i have to be made in order to adapt it to the new profile performance and needs.

For example, if a satisfaction level of goal g_j is in the range of scores with limits $[0,100]$ and there are 3 satisfaction levels defined within intervals $[0,50)$, $[50,80)$ and $[80,100]$, then each level must require certain changes in some features of each profile p_i, such as the following:

If g_j-level >= 80 then

  performance-level of p_i turns High.

If g_j-level < 80 and >= 50 then

  performance-level of p_i turns Medium and actions with High difficulty must be required.

If g_j-level < 50 then

  performance-level of p_i turns Low and actions with optional features must be required.

On the other hand, a course $\mathcal{C} = \{(g_1, \mathcal{SC}_1), (g_2, \mathcal{SC}_2), ..., (g_n, \mathcal{SC}_n)\}$ is defined as a set of main tasks $\mathcal{SC}_j$ that are directly related to each goal in a course.

It is important to recall that each main task $\mathcal{SC}_j$ is compound of tasks and actions with their own relations and requirements. These actions are organized in a hierarchical structure and must be realized in sequence in order to understand the concept of the main task and accomplish its related goal g_j. Hence, the definition of a set of sequences $\mathcal{S}_j = \{s_{j,1}, s_{j,2}, ..., s_{j,h}\}$ is needed to understand that each main task $\mathcal{SC}_j$ has to be sequenced on h different ways for each profile p_i in order to accomplish each goal g_j.

That h number of sequences corresponds to the number of satisfaction levels on the previous goal g_{j-1}. It means that we have to plan sequences for each possible outcome that a student could have been obtained at the previous evaluation, because features of his/her profile may change in accordance, and next sequences must be adapted to any of these changes.

In the domain generation process some conditions have to be defined in order to select the best sequence $s_{j,k}$ that can accomplish the goal g_j for each profile p_i, according to its possible outcomes in g_{j-1}.

Figure 2 shows that the first adapted sequence for every profile is unique but, since intermediate evaluations are being considered, the sequence is split into h different branches, a branch for each of the h possible levels of the previous test. So, as one can imagine, the first draft of a learning design does not include

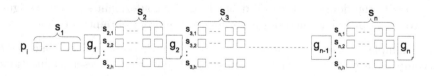

Fig. 2. Formalization of the planning process, where p_i is a profile, $\mathcal{S}_j$ is a set of adapted sequences of educational resources derived from the description of the set $\mathcal{SC}_j$, $s_{j,k}$ are the sequences of set $\mathcal{S}_j$ that are planned according to changes in the profile p_i based on the levels $l_{j,k}$ of the goal g_j that must be satisfied by any of these sequences

```
(:task Multiplication
  :parameters(?profile - profileId ?score - goalSatLevel)
  (:method Theoretical
   :precondition(learning-style ?profile ?score theoretical)
   :tasks(
     (task_addtion                        ?profile ?score)
     (action_multiplicationFormalization  ?profile ?score)
     (action_multiplicationSimulation     ?profile ?score)
     (action_multiplicationExperiment     ?profile ?score)))
  (:method Pragmatic
   :precondition(learning-style ?profile ?score pragmatic)
   :tasks(
     (task_addition                       ?profile ?score)
     (action_multiplicationExperiment     ?profile ?score)
     (action_multiplicationSimulation     ?profile ?score)
     (action_multiplicationFormalization  ?profile ?score))))
```

```
(:durative-action action_multiplicationExercise
  :parameters(?profile - profileId ?score - goalSatLevel)
  :duration(= ?duration 10)
  :condition( and
    (equipment ?profile ?score multimedia)
    (>= (language-level English ?profile ?score) 50))
  :effect(additionExercise ?profile ?score done)
)
```

Fig. 3. PDDL action and task examples

conditions for a same profile. Therefore, the way to design the hierarchical domain and planning problem, in order to generate the sequences for this design, is substantially different from what we will see in the following sections.

The algorithms to generate a hierarchical planning domain and problem are detailed in the following subsections, but it is worth noticing that only algorithms to generate the domain and problem that allow to plan sequences that meet from g_2 onwards are described because the domain and problem to meet the first goal is similar, if not equally, of that detailed described on [4].

3.1   Domain Generation

A planning domain, that describes features and relationships of the elements of a course, is generated once for every goal g_i of $\mathcal{G}$ and its generation is based on a knowledge engineering process on the features of tasks and durative actions that conforms the main task $\mathcal{SC}_i$ to be planned to meet the goal g_j.

The algorithm to implement the domain generation is the following:

1. All the features of elements that conform the main tasks $\mathcal{SC}_j$ are analyzed, dividing them into two subsets: tasks and durative actions.
2. Durative actions are generated according to its features and relations, obtaining its durations and execution conditions, in addition to a pair of auxiliar actions that helps us to retrieve the elements of IMS-LD draft.
3. Tasks and actions that are part of the same main task $\mathcal{SC}_j$ are ordered according to their hierarchy relations or pedagogical rules described in a configuration file.
4. Finally the domain tasks are defined and its conformant subtasks assigned to the appropriate methods according to the order obtained previously.

For example, elements of the main task $\mathcal{SC}_j$ which represent the concepts to accomplish, are placed in the tasks subset, and those that represents educational resources in the durative actions subset.

Durative actions definition described in step two is carried out as in figure 3 with the action_multiplicationExercise example encoded in PDDL.

Another two actions must be described to retrieve previous sequences and conditions of the IMS-LD draft. Those are:

- (initAction) that recalls from the learning design draft the profiles definition set $\mathcal{P}$, the components of past main tasks, and the sequences planned for those main tasks. Also, the property for the goal g_{j-1} is added and initialized.
- (endAction) that recovers the rest of the elements of IMS-LD draft. It is also used to define new conditions according to the values that can take the property for the goal g_{j-1} and the sequence $s_{j,k}$ that must be showed to each profile for the corresponding level $l_{j,k}$ part of that goal.

An example of a condition that can be inserted into the new IMS-LD through endAction is the following one:

```
if test-score-of-goalⱼ₋₁ > 79
    show activity-structures of every profileId with score=90 and goalⱼ₋₁
else
    hide activity-structures of every profileId with score=70 or score=30, besides goalⱼ
```

Where, when the property value test-score-of-$goal_{j-1}$ is found in execution time at level $l_{j,3}$, corresponding to the [80, 100] interval, the corresponding sequence planned for goal g_j and score 90 will be showed, and sequences generated for the same goal but another levels will be hidden.

Finally, results of the steps three and four are shown in figure 3, where the main task Multiplication is encoded in PDDL with two decomposition methods for different order of durative actions, and where the task addition, which is related with the *Required* relation to Multiplication, is the reason of its first place in the order of both methods.

Through the algorithm explained by the above examples it is possible to define a hierarchical planning domain that completely describes tasks and actions in it, to facilitate the planning process for a main task $\mathcal{SC}_j$ of the course $\mathcal{C}$.

The main difference between this generation process and previous approaches is summarized in the next two points:

- The domain is generated as many times as goals in the course, not only once for the entire course as in the previous proposal.
- Adding a pair of auxiliar actions will be possible to retrieve the information of IMS-LD draft and add information resulting of the current planning process to a new learning design.

After generating the domain will be necessary to design a planning problem. The process for creating that problem is described in following section.

3.2   Problem Generation

A hierarchical planning problem, as already mentioned above, is generated as many times as the number of elements in the set of goals $\mathcal{G}$ derived from the IMS-QTI standard.

The generation of each planning problem, to goal g_2 onwards, is divided in two steps. The first one is the initial state definition and the second of the goals that must be meet in order to satisfy the current problem.

The initial state of a planning problem for goal g_j includes a description of all the profiles in the set $\mathcal{P}$, i.e. the specification of the features of each profile by means of predicates or functions in PDDL language.

But since the features of a profile are subject to change based on the results of an intermediate evaluation, the process of generation of these initial states is a bit more complex than that described in the preceding paragraph. This process first creates as many different copies of the profiles in $\mathcal{P}$ as levels of satisfaction to the goal g_{j-1}, which will generate mxh copies that allow us to further define the same number of "initial states" each different profile-level.

```
;
; profileOne with score 90                    (:tasks-goal
; (level between 80 and 100)                     :tasks [
;                                                 (initAction)
(= (performance profileOne 90) 90)              ( and ( <= ?end 1150 )
(learning-style profileOne 90 theoretical)          (Multiplication profileOne 90) )
(= (availability profileOne 90) 1150)           ( and ( <= ?end 1750 )
(equipment profileOne 90 multimedia)                (Multiplication profileOne 70) )
;Academic trajectory                            ( and ( <= ?end 2150 )
(= (language-level en profileOne 90) 90)            (Multiplication profileOne 30) )
(= (mark profileOne 90 requiredSequence) 80)    ;
(= (mark profileOne 90 prevOptionalActions) 30)   (endAction)
(= (mark profileOne 90 prevDifficultActions) 80) ])
```

Fig. 4. PDDL initial state and goals

The definition of these copies is made because the profile of a student may change according to the results obtained in the different intermediate evaluations to be applied, and hence, personalized sequences of that profile might change too. Therefore, copies are essential to define not only the initial state, but also the goals of the problem.

Once profile copies are generated, the initial state of the problem has to be created through the assignment of a representative score value to each copy, according to its corresponding level, and the definition of its features using functions and predicates.

The PDDL profile in figure 4 describes features of the profile p_1 with score 90, meaning that its satisfaction level is in the [80,100] range which is in $l_{j,3}$ level. Therefore, its acquired features from IMS-LIP have been modified according to that level. This example shows 5 functions and 3 predicates that describe several features of the profile. Other 14 features definitions should be described in the initial state of the problem to obtain all the necessary profile copies.

The second stage consists in defining the goals of the planning problem. Thus, for each profile copy described in generation process of the initial state for planning problem, a conditional action like (and (<= ?end availability) (S_jMainTask profile_i level_k)) must be stated as a goal.

The above predicate refers to tasks and actions that belongs to the main task S_jMainTask that must be planned for p_i profile according to its features for the copy in $l_{j,k}$ level. The main task duration should not exceed that allocated for ?end variable according to the availability of this profile copy.

An example of goals definition is shown in figure 4. Where the main task SC_j has a related time that is accumulated in the variable ?end according to the duration of the actions and tasks that conform it. This time should not exceed the time established by the goals conditions for that variable.

As we can see, defined goals, to accomplish the main task for each profile, and level are between initAction and endAction. Execution of both actions can retrieve information from IMS-LD draft and add conditions for the new sequences that must be generated between them.

Table 1. Execution times of each case SC_j during the two phases of our experiment

Resources Sets for goals one to four	Case 1	Case 2	Case 3	Case 4	Average Time (sec.)
Domain & Problem Generation	1.22	1.11	1.19	1.16	1.17
Planner Execution	0.99	0.84	0.97	0.92	0.93
D&P Generation + Planner Exec Time (sec.)	2.21	1.95	2.16	2.08	**2.1**

Through this process it is possible to generate a hierarchical planning problem for each set of tasks and durative actions related to a goal. This problem must be robust enough as to require the creation of all the possible sequences for every profile in P and the different levels that can obtain the previous goal g_{j-1}.

This ends the description of the process to generate an IMS-LD level B described at the beginning of this section and will be tested through a series of experiments described in the next section.

4 Experiments

Once we have defined the methodology for the generation of IMS-LD level B, we present a set of experiments to show how this approach works in practical cases.

Description. The experiments use 4 representative examples of SC that show different scenarios for the application of our approach, e.g. Math or AI subjects. Each SC is composed of 8 basic tasks and 30 durative actions on average. These durative actions have an average duration of three hours.

We also considered students automatically classified in an average of 20 students' profiles according to 6 different features. Therefore, considering that we have divided scores into 3 different subsets, the planner has to solve an average of 60 goals for each problem.

Over every set SC we applied the algorithms for the automatic generation of domain and problem planning files, which were passed to our planner to obtain a fully standard IMS-LD level B able to be handled by tools like RELOAD.

Results. As we may see in Table 1, the average running time to generate every domain and problem file for all the sequences of the four sample sets was 1.17

seconds. Then, our planner generates all the plans for every SC in 0.93 seconds average. All these plans for every subset SC are joined into a standard IMS-LD level B document in about 2.1 seconds.

Discussion. In [1] the effort needed for an instructor to design a customized learning sequence is reported to take 20 hours of design for every hour of teaching. This effort does not consider the time needed to translate the design into standard IMS-LD.

In our case, a customized learning design, already expressed in standard IMS-LD level B, is obtained in 0.7 seconds average for every hour of teaching. Given that the labeling of learning resources with metadata and the design of intermediate tests is assumed in both cases and that there is a very small overload due to the introduction of the planner, the advantages of this approach are extremely clear.

However, there are other approaches that are intended to solve the same problem and that are commented in the next section.

5   Related Work

The standard IMS-LD first draft arose 6 years ago. During the first four years, there were no more than 40 courses encoded in IMS-LD, most of them written in the lowest level of expressiveness, the level A.

After this, several projects to ease the process of writing these learning designs at its level B[3,10] have been reported. One of them is found in [3] where an assistant is created to guide instructors in the writing of the different sections of a standard IMS-LD for given courses. However, this process is still very slow and requires the instructor to know the detailed syntax of the standard.

We also explored the possibility of using a conditional planning approach like in [11]. Our HTN planner does not support planning with conditional branches, but, since it introduces an unnecessary complexity in the planning process making the planner to branch the plan at every intermediate test, it was finally discarded since the approach presented in this paper is much easier, achieving the same results.

On the other hand, the continual planning approach of [12] can be an option for course sequencing with mid-term evaluations because sequencing could be done by generating plans for accomplish a goal at runtime according to the outcomes of the previous test. But, it is not a good approach for our problem because the IMS-LD must be completely generated before the course starts.

Finally, we also considered a plan fusion approach like in [13], but it was also discarded because the main complexity of plan fusion comes from the detection and bounding of subplans able to merge from one plan to another, and in our case they are very easily bounded by intermediate tests or well defined features of IMS-LIP.

6   Conclusions and Future Work

In summary, the approach presented in this paper is based on a former proposal [4] in which domain and problem files are automatically generated for an HTN planner to obtain customized learning designs for a given course expressed in the standard IMS-LD level A. Its main contribution is an extension of this approach able to handle with conditional sequences supported by IMS-LD level B, given the intermediate evaluations example, and the subsequent generation of conditions and branches able to adapt the learning design to a discrete set of possible outcomes of these evaluations besides the customization to the features of each student profile.

Thanks to the expressiveness and the efficiency of the planner [5] we have been able to automate the generation of the domain and problem files, needed to accomplish each goal, avoiding the need for a planning expert and increasing the independence of instructors to use the available AI planning technology.

However, there are a couple of issues that still remain open.

The first one, strictly falls within the scope of e-learning institutions and requires education institutions to increase the number of learning resources and courses described under the existing standards.

And the second one relates to the exploitation of the temporal planning capabilities of our HTN planner [5]. These capabilities would have permitted the parallelization and synchronization of sequences among different profiles, opening the door to the modeling of shared or collaborative activities, but this has been left for a future work.

References

1. Brusilovsky, P., Vassileva, J.: Course Sequencing Techinques for Large-Scale Web-Based Education. International Journal Continuing Engeenering Education and Lifelong Learning 13(1/2), 75–94 (2003)
2. OUNL: Ims Global Learning Consortium. Open University of the Netherlands (2003), http://www.imsglobal.org/
3. Heyer, S., Oberhuemer, P., Zander, S., Prenner, P.: Making Sense of IMS Learning Design Level B: From Specification to Intuitive Modeling Software. In: Duval, E., Klamma, R., Wolpers, M. (eds.) EC-TEL 2007. LNCS, vol. 4753, pp. 86–100. Springer, Heidelberg (2007)
4. Castillo, L., Morales, L., Fernández-Olivares, J., González-Ferrer, A., Palao, F.: Knowledge Engineering and Planning for the Automated Synthesis of Customized Learning Designs. In: Borrajo, D., Castillo, L., Corchado, J.M. (eds.) CAEPIA 2007. LNCS (LNAI), vol. 4788, pp. 40–49. Springer, Heidelberg (2007)
5. Castillo, L., Fernández-Olivares, J., García-Pérez, O., Palao, F.: Efficiently Handling Temporal Knowledge in an HTN Planner. In; ICAPS 2006, pp. 63–72 (2006)
6. Liber, O.: REusable eLearning Object Authoring and Delivery. University of Bolton (2007), http://www.reload.ac.uk/
7. Kontopoulos, E., Vrakas, D., Kokkoras, F., Bassiliades, N., Vlahavas, I.: An Ontology-based Planning System for e-course Generation. Expert Systems with Applications 35(1/2), 398–406 (2008)

8. Ullrich, C., Melis, E.: Pedagogically Founded Courseware Generation Based on HTN-planning. Expert Systems with Applications 36(5), 9319–9332 (2009)
9. Nau, D., Au, T.-C., Ilghami, O., Kuter, U., Murdock, J.W., Wu, D., Yaman, F.: Shop2: An HTN Planning System. Journal of Artificial Intelligence Research 20(1), 379–404 (2003)
10. Sicilia, M., Sánchez-Alonso, S., García-Barriocanal, E.: On Supporting the Process of Learning Design through Planners. CEUR Workshop Proceedings: Virtual Campus 2006 Post-Proceedings 186(1), 81–89 (2006)
11. Onder, N., Pollack, M.E.: Conditional, probabilistic planning: A unifying algorithm and effective search control mechanisms. In: Proceedings of the Sixteenth National Conference on Artificial Intelligence, pp. 577–584 (1999)
12. Myers, K.L.: Towards a framework for Continuous Planning and Execution. In: AAAI Fall Symposium on Distributed Continual Planning (1998)
13. Foulser, D., Li, M., Yang, Q.: Theory and Algorithms for Plan Merging. Artificial Intelligence 57, 143–181 (1992)

A New Backtracking Algorithm for Constructing Binary Covering Arrays of Variable Strength*

Josue Bracho-Rios, Jose Torres-Jimenez, and Eduardo Rodriguez-Tello

CINVESTAV-Tamaulipas, Information Technology Laboratory
Km. 6 Carretera Victoria-Monterrey, 87276 Victoria Tamps., Mexico
{jbracho,jtj,ertello}@tamps.cinvestav.mx

Abstract. A Covering Array denoted by $CA(N; t, k, v)$ is a matrix of size $N \times k$, in which each of the v^t combinations appears at least once in every t columns. Covering Arrays (CAs) are combinatorial objects used in software testing. There are different methods to construct CAs, but as it is a highly combinatorial problem, few complete algorithms to construct CAs have been reported. In this paper a new backtracking algorithm based on the Branch & Bound technique is presented. It searches only non-isomorphic Covering Arrays to reduce the search space of the problem of constructing them. The results obtained with this algorithm are able to match some of the best known solutions for small instances of binary CAs.

Keywords: Software testing, Covering Arrays, Branch and Bound.

1 Introduction

Within the recent years the use of software has become more and more important in our society. Most of the businesses nowadays depend on computer software, thus a failure on it can be catastrophic. As mentioned in a NIST (National Institute on Standards and Technology) report [1], the sales of software reached approximately $180 billion dollars generating a significant and high-paid workforce, composed of 697,000 software engineers and 585,000 computer programmers. It shows clearly that the software industry is a really important part of the economy, moreover if the software present errors, millionaire losses can occur affecting the economy. According to Hartman [2], the quality of the software strongly depends on the use of appropriate software testing techniques.

Software systems at the moment are numerous times more complex than before. Moreover, they usually have a lot of possible configurations that are produced by the combination of their input parameters. To totally guarantee the quality of these software products all the possible configurations must be tested, but this exhaustive approach is not a viable option. For instance, suppose that

* This research was partially funded by the following projects: CONACyT 58554-Cálculo de Covering Arrays, 51623-Fondo Mixto CONACyT y Gobierno del Estado de Tamaulipas.

A. Hernández Aguirre et al. (Eds.): MICAI 2009, LNAI 5845, pp. 397–407, 2009.

we have a simple program that takes 12 arguments and each one of the arguments take only 4 possible values. In order to verify it totally we need to make $4^{12} = 16,777,216$ tests. But, previous works have demonstrated that testing it with an interaction of level 6 (all the combinations of 6 parameters) will only require $4^6 = 14,888$ tests [3] (each test is a combination of values taken by all the parameters). This approach based on constructing economical sized test suites that provide coverage of the most prevalent configurations is known as *software interaction testing*. Covering arrays (CAs) are combinatorial structures used to represent these test suites when exhaustive testing is not feasible [4].

A Covering Array $\mathrm{CA}(N; t, k, v)$ of size N is an $N \times k$ array consisting of N vectors of length k (degree) with entries from an alphabet of size v, i.e., $\{0, 1, \ldots, v-1\}$, such that every one of the v^t possible vectors of size t (t-wise) occurs at least once in every possible selection of t elements from the vectors. The parameter t is referred to as the strength or level of interaction. The minimum N for which a $\mathrm{CA}(N; t, k, v)$ exists is known as the *covering array number* and it is defined according to (1).

$$\mathrm{CAN}(t, k, v) = min\{N : \exists\, \mathrm{CA}(N; t, k, v)\} \tag{1}$$

A $\mathrm{CA}(N; t, k, v)$ can be mapped to a software test suite as follows. In a software test we have k components, each of these has v configurations. A test suite is an $N \times k$ array where each row is a test case. Each column represents a component and the value in the column is the particular configuration.

Lei and Tai [5] demonstrated that the problem of generating the minimum pairwise test set belongs to the NP class (for non-binary alphabets). Then by reduction of the problem of vertex cover, they proved that the *pair-cover problem* is NP-complete. Colbourn [6] also showed that this problem is NP-complete by reducing it to the SAT problem.

Even though the general problem of finding a combinatorial test suite is NP [7], there are some isolated cases that can be solved in polynomial time [6]:

1. When the strength is 2 ($t = 2$) and the alphabet is 2 ($v = 2$). Sloane stated that this case was solved by Rényi with an even value of N and by Katona independently. Then, it was completely solved by Kleitman and Spencer for all N [8].
2. When the alphabet is a power of prime $v = p^\alpha$, $k \leqslant (p^\alpha + 1)$, and $p^\alpha > t$. This construction was proposed by Bush [9], he used Finite Galois Fields in order to solve this case.

Given the complexity of the optimal construction of CAs many of the algorithms developed to solve this problem are approximate methods. They try to reach solutions as close as possible to $\mathrm{CAN}(t, k, v)$, given some values for k, v, and t [10].

In this paper we introduce a new backtracking algorithm for constructing binary CAs of variable strength which is based on the Branch & Bound (B&B) technique. It incorporates some distinguished features for improving the efficiency of the search process, including: symmetry breaking techniques, partial t-wise verification and fixed blocks.

The effectiveness of the proposed backtracking algorithm is assessed using a benchmark, conformed by 14 binary covering arrays of strength $3 \leq t \leq 5$, taken from the literature. The computational results are reported and compared against those reached by some state-of-the-art methods, showing that our algorithm is able to match some of the best-known solutions for small instances of binary CAs expending in some cases less computational time.

The remainder of this work is organized as follows. In Sect. 2, a brief review is given to present some representative solution procedures for constructing binary covering arrays of variable strength. Then, the components of our backtracking algorithm are detailed in Sect. 3 and 4. Section 5 presents computational experiments and comparisons of our backtracking algorithm with respect to other previously published algorithms. Last section is dedicated to summarize the main contributions of this work.

2 Relevant Related Work

The objective of constructing a CA is to minimize the number of rows given the parameters v, k, and t. There are several reported methods for constructing these combinatorial models. Among them are: a) recursive methods [8,11], b) algebraic methods [12], c) greedy methods [13] and d) meta-heuristics such as Simulated Annealing [14] and Tabu Search [15]. These algorithms have a point in common, all of them are approximate methods.

To the best of our knowledge, there exists only one method reported in the literature which tries to make a systematic enumeration of the candidate solutions in the search space in order to guarantee to discover optimal CAs. This method, that we used in our experimental comparisons, is called EXACT (Exhaustive search of combinatorial test suites) and was proposed by Yan and Zhang [10].

EXACT is a backtracking algorithm that employs certain rules in order to eliminate isomorphic CAs. Two CAs are isomorphic if they have the same number of rows, columns and alphabet and one can be transformed into the other via permutations of rows, columns, and/or symbols [16]. Moreover, it introduces the concept of *miniblock*. A miniblock is a $mb \times t$ sub-array which values are fixed before starting the construction of a CA. It allows to reduce the size of the search space to the value given in (2). Please note that in this expression the total size of search space equals the numerator, and the denominator indicates the size of the fixed miniblock.

$$\frac{(\prod_{i=1}^{k} v_i^{N})}{(\prod_{i=1}^{t} v_i^{mb})} \tag{2}$$

A novel pruning technique, called SCEH (Sub-Combination Equalization Heuristic), is also integrated to EXACT. The authors decided to use this technique because they noted that for many CAs each symbol appears almost the same number of times in each column of the CA.

In 2008, Yan and Zhang [17] reported an improvement to EXACT. In this new version of EXACT they added a new rule: for each two rows i, j ($1 < i \leq mb$

and $j > mb$) of a CA, if these two rows have the same first t values and $R_i >_{lex} R_j$ (where $>_{lex}$ refers to a lexicographical order), then these rows are exchanged. This work has improved the best-known solution for only one instance (CA$(24; 4, 12, 2)$).

3   A New Backtracking Algorithm

We can represent a CA as a 2-dimensional $N \times k$ matrix. Each row can be regarded as a test case and each column represents some parameter of the system under test. Each entry in the matrix is called a cell, and we use M_{ij} to denote the cell at row i ($i > 0$) and column j ($j > 0$), i.e., the value of parameter j in test case i.

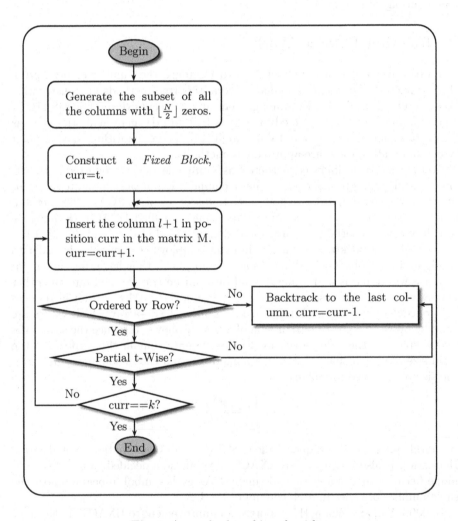

Fig. 1. A new backtracking algorithm

We apply an exhaustive search technique to this problem. Our algorithm is based on the Branch & Bound technique. The main algorithm can be described as an iterative procedure as follows.

For a given matrix $N \times k$, and a strength t, we construct the first element l belonging to the set of all possible columns with $\lfloor \frac{N}{2} \rfloor$ zeros and is inserted in the first column of the partial solution M. Then the next element $l_{i-1} + 1$ is constructed and if the row i is smaller than the row $i+1$ and it is a partial CA then the element is inserted, otherwise it tries with the next element. If no elements could be inserted in the current column of M, it backtracks to the last column inserted and tries to insert a new element. When k columns are inserted then the procedure finishes and the CA is generated. A flow chart of this algorithm is shown in Figure 1.

4   Techniques for Improving the Efficiency of the Search

The worst time complexity of the naive exhaustive search for $CA(N,k,v,t)$ is shown in (3).

$$\binom{\binom{N}{\lfloor \frac{N}{2} \rfloor}}{k} \tag{3}$$

This worst time complexity is due mainly because there exist so many isomorphic CAs. There are 3 types of symmetries in a CA: row symmetry, column symmetry and symbol symmetry. The row symmetry refers to the possibility to alter the order of the rows without affecting the CA properties. There are $N!$ possible row permutations of a CA. The column symmetry refers to permuting columns in the CA without altering it. There exist $k!$ possible column permutations of a CA. In the same way thanks to the symbol symmetry if we make one of the $(v!)^k$ possible permutations of symbols it results in an isomorphic CA. By the previous analysis we can conclude that there are a total of $N! \times k! \times (v!)^k$ number of isomorphic CAs for a particular CA. An example of two isomorphic CAs is shown in Table 1.

The covering array in Table 1(d) can be produced by the following steps over the covering array in Table 1(a): exchanging the symbols of the first column (Table 1(b)), then exchanging the first and second rows (Table 1(c)), finally exchanging the third and fourth columns.

Table 1. Isomorphic Covering Arrays

(a)

0	0	0	0
0	1	1	1
1	0	1	1
1	1	0	1
1	1	1	0

(b)

1	0	0	0
1	1	1	1
0	0	1	1
0	1	0	1
0	1	1	0

(c)

1	1	1	1
1	0	0	0
0	0	1	1
0	1	0	1
0	1	1	0

(d)

1	1	1	1
1	0	0	0
0	0	1	1
0	1	1	0
0	1	0	1

Searching within the non-isomorphic CAs can significantly reduce the search space, these symmetry breaking techniques have been previously applied in order to eliminate the row and column symmetries in [17] by Yan and Zhang. However, they only proposed an approach to eliminate row and column symmetries. In the following sections we will describe the symmetry breaking techniques that we applied within our backtracking algorithm to search only over the non-isomorphic CAs.

4.1 Symmetry Breaking Techniques

In order to eliminate the row and column symmetries in our new backtracking algorithm the restriction that within the current partial solution M the column j must be smaller than the column $j + 1$, and the row i must be smaller than the row $i + 1$.

As we have mentioned above a $CA(N; t, k, v)$ has $N! \times k!$ row and column symmetries. This generates an exponential number of isomorphic CAs. Adding the constraints mentioned above we eliminate all those symmetries and reduce considerably the search space. Another advantage of this symmetry breaking technique is that we do not need to verify that the columns are ordered, we only need to verify that the rows are still ordered as we insert new columns. This is because we are generating an ordered set of columns such that the column l is always smaller than the column $l + 1$.

Moreover, we propose a new way of breaking the symbol symmetry in CAs. We have observed, from previously experimentation, that near-optimal CAs have columns where the number of 0's and 1's are balanced or near balanced. For this reason we impose the restriction that through the whole process the symbols in the CAs columns must be balanced. In the case where N is not even, the number of 0's must be exactly $\lfloor \frac{N}{2} \rfloor$ and the number of 1's $\lfloor \frac{N}{2} \rfloor + 1$. As long as we know this is the first work in which the symbol symmetry breaking is used. In Table 2 an example of our construction is shown.

In Table 2 we can see clearly that the next column generated is automatically lexicographically greater than the previous one, so we do not need to verify for the column symmetry breaking rule. Even though that in Table 2 the rows are ordered as well, it does not happen in general so we still have to check that the rows remain ordered for each element that we try to insert in the partial solution M.

Table 2. Example of the current partial solution M after 4 column insertions

l	$l + 1$	$l + 2$	$l + 3$
0	0	0	0
0	1	1	1
1	0	1	1
1	1	0	1
1	1	1	0

4.2   Partial t-Wise Verification

Since a CA of strength $t-1$ is present within a CA of strength t we can bound the search space more quickly and efficiently if we partially verify for a CA instead of waiting until all k columns are inserted. We can test the first i columns with strength i for $2 \le i \le t-1$, where i is the current element inserted, and when i is greater or equal to t then is tested with strength t. It is important to remark that a complete CA verification is more expensive in terms of time,[1] than making the partial evaluation described above due to the intrinsic characteristics of the backtracking algorithm.

4.3   Fixed Block

The search space of this algorithm can be greatly reduced if we use a *Fixed Block* (FB). We define a FB as matrix of size N and length t in which the first $\lfloor \frac{(N-v^t)}{2} \rfloor$ rows are filled with 0's, then a CA of strength t, $k = t$ and $N = v^t$ is inserted. This CA can be easily generated (in polynomial time) by creating all the v^t binary numbers and listing them in order. An example of a $CA(v^t; t, t, v)$ is shown in Table 3. Finally, the last $\lceil \frac{(N-v^t)}{2} \rceil$ rows are filled with 1's. It can be easily verified that in a FB the rows and columns are already lexicographically ordered. This FB is constructed in this way in order to preserve the symmetry breaking rules proposed in Sect. 4.1 and is used to initialize our algorithm as shown in Figure 1.

Table 3. A $CA(v^t; t, t, v)$ example

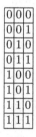

0	0	0
0	0	1
0	1	0
0	1	1
1	0	0
1	0	1
1	1	0
1	1	1

5   Computational Results

In this section, we present a set of experiments accomplished to evaluate the performance of the backtracking algorithm presented in Section 3. The algorithms were coded in C and compiled with *gcc* without optimization flags. It was run sequentially into a CPU Intel Core 2 Duo at 1.5 GHz, 2 GB of RAM with Linux operating system.

---

[1] The computational complexity of making a full verification of a $CA(N; t, k, v)$ is $N \times \binom{k}{t}$.

The test suite used for this experiment is composed of 14 well known binary covering arrays of strength $3 \le t \le 5$ taken from the literature [3,17]. The main criterion used for the comparison is the same as the one commonly used in the literature: the *best size* N found (smaller values are better) given fixed values for k, t, and v.

5.1 Comparison between Our Backtracking Algorithm and the EXACT Algorithm

The purpose of this experiment is to carry out a performance comparison of the upper bounds achieved by our backtracking algorithm (B&B) with respect to those produced by the EXACT procedure [17]. For this comparison we have obtained the EXACT algorithm from the authors. Both algorithms were run in the computational platform described in Sect. 5.

Table 4 displays the detailed computational results produced by this experiment. The first two columns in the table indicate the degree k, and strength t of the instance. Columns 3 and 5 show the best solution N found by B&B and the EXACT algorithms, while columns 4 and 6 depict the computational time T, in seconds, expended to find those solutions.

Table 4. Performance comparison between the algorithms B&B and EXACT

		B&B		EXACT	
k	t	N	T	N	T
4	3	8	**0.005**	8	0.021
5	3	10	**0.005**	10	0.021
6	3	12	**0.008**	12	0.023
7	3	12	**0.018**	12	0.024
8	3	12	0.033	12	0.023
9	3	12	0.973	12	0.022
10	3	12	0.999	12	0.041
11	3	12	0.985	12	0.280
12*	3	15	**1090.800**	15	1100.400
5	4	16	**0.020**	16	0.038
6	4	21	95.920	21	0.266
6	5	32	102.000	32	0.025

From the data presented in Table 4 we can make the following main observations. First, the solution quality attained by the proposed backtracking algorithm is very competitive with respect to that produced by the state-of-the-art procedure EXACT. In fact, it is able to consistently equal the best-known solutions attained by the EXACT method (see columns 3 and 5).

Second, regarding the computational effort, one observes that in this experiment the EXACT algorithm consumes slightly more computational time than B&B for 6 out 12 benchmark instances (shown in boldface in column 4).

We would like to point out that the instance marked with a star in Table 4 was particularly difficult to obtain using the EXACT algorithm. We have tried many different values for the parameter SCEH (Sub-Combination Equalization Heuristic), and only using a value of 1 the EXACT tool was able to find this instance consuming more CPU time than our B&B algorithm. For the rest of the experiments we have used the default parameter values recommended by the authors.

5.2   Comparison between Our Backtracking Algorithm and IPOG-F

In a second experiment we have carried out a performance comparison of the upper bounds achieved by our B&B algorithm with respect to those produced by the state-of-the-art procedure called IPOG-F [18].

Table 5 presents the computational results produced by this comparison. Columns 1 and 2 indicate the degree k, and strength t of the instance. The best solution N found by our B&B algorithm and the IPOG-F algorithm are depicted in columns 3 and 5, while columns 4 and 6 depict the computational time T, in seconds, expended to find those solutions. Finally, the difference (Δ_N) between the best result produced by our B&B algorithm compared to that achieved by IPOG-F is shown in the last column.

Table 5. Performance comparison between the algorithms B&B and IPOG-F

		B&B		IPOG-F		
k	t	N	T	N	T	Δ_N
4	3	8	**0.005**	9	0.014	-1
5	3	10	**0.005**	11	0.016	-1
6	3	12	**0.008**	14	0.016	-2
7	3	12	**0.018**	16	0.018	-4
8	3	12	0.033	17	0.019	-5
9	3	12	0.973	17	0.019	-5
10	3	12	0.999	18	0.019	-6
11	3	12	0.985	18	0.034	-6
12	3	15	1090.800	19	0.033	-4
13	3	16	1840.320	20	0.019	-4
5	4	16	0.020	22	0.016	-6
6	4	21	95.200	26	0.017	-5
7	4	24	113.400	32	0.014	-8
6	5	32	102.000	42	0.020	-10
Average		15.29		20.07		-4.79

From Table 5 we can clearly observe that in this experiment the IPOG-F procedure [18] consistently returns poorer quality solutions than our B&B algorithm. Indeed, IPOG-F produces covering arrays which are in average 31.26% worst than those constructed with B&B.

6 Conclusions

We proposed a new backtracking algorithm that implements some techniques to reduce efficiently the search space. Additionally, we have presented a new technique for breaking the symbol symmetry which allow to reduce considerably the size of the search space. Experimental comparisons were performed and show that our backtracking algorithm is able to match some of the best-known solutions for small instances of binary CAs, expending in some cases less computational time compared to another existent backtracking algorithm called EXACT. We have also carried out a comparison of the upper bounds achieved by our backtracking algorithm with respect to those produced by a state-of-the-art procedure called IPOG-F. In this comparison the results obtained by our back-tracking algorithm, in terms of solution quality, are better than those achieved by IPOG-F for all the studied instances.

Finding optimum solutions for the CA construction problem in order to construct economical sized test-suites for software interaction testing is a very challenging problem. We hope that the work reported in this paper could shed useful light on some important aspects that must be considered when solving this interesting problem. We also expect the results shown in this work incite more research on this topic. For instance, one fruitful possibility for future research is the design of new pruning heuristics in order to have the possibility to generate larger instances of CAs.

Acknowledgments. The authors would like to thank Jun Yan and Jian Zhang who have kindly provided us with an executable version of their application EXACT.

References

1. Tassey, G.: RTI: The economic impacts of inadequate infrastructure for software testing. Technical Report 02-3, National Institute of Standards and Technology, Gaithersburg, MD, USA (May 2002)
2. Hartman, A.: 10. In: Graph Theory, Combinatorics and Algorithms. Operations Research/Computer Science Interfaces Series, vol. 34, pp. 237–266. Springer, Heidelberg (2005)
3. Colbourn, C.J.: Most recent covering arrays tables, http://www.public.asu.edu/~ccolbou/src/tabby/catable.html. (Last Time Accesed November 4, 2008)
4. Kuhn, D.R., Wallance, D.R., Gallo, A.M.J.: Software fault interactions and implications for software testing. IEEE Transactions on Software Engineering 30, 418–421 (2004)
5. Lei, Y., Tai, K.: In-parameter-order: A test generation strategy for pairwise testing. In: Proceedings of the 3rd IEEE International Symposium on High-Assurance Systems Engineering, pp. 254–261. IEEE Computer Society, Washington (1998)
6. Colbourn, C.J., Cohen, M.B., Turban, R.C.: A deterministic density algorithm for pairwise interaction coverage. In: Proceedings of the IASTED International Conference on Software Engineering (2004)

7. Kuhn, D.R., Okum, V.: Pseudo-exhaustive testing for software. In: SEW 2006: Proceedings of the 30th Annual IEEE/NASA Software Engineering Workshop, pp. 153–158. IEEE Computer Society, Washington (2006)
8. Sloane, N.J.A.: Covering arrays and intersecting codes. Journal of Combinatorial Designs 1, 51–63 (1993)
9. Bush, K.A.: Orthogonal arrays of index unity. Annals of Mathematical Statistics 13, 426–434 (1952)
10. Yan, J., Zhang, J.: Backtracking algorithms and search heuristics to generate test suites for combinatorial testing. In: Computer Software and Applications Conference(COMPSAC 2006) 30th Annual International, vol. 1, pp. 385–394 (2006)
11. Hartman, A., Raskin, L.: Problems and algorithms for covering arrays. Discrete Mathematics 284, 149–156 (2004)
12. Hedayat, A.S., Sloane, N.J.A., Stufken, J.: Orthogonal Arrays: Theory and Applications. Springer, New York (1999)
13. Cohen, D.M., Fredman, M.L., Patton, G.C.: The aetg system: An approach to testing based on combinatorial design. IEEE Transactions on Software Engineering 23, 437–444 (1997)
14. Cohen, M.B., Colbourn, C.J., Ling, A.C.H.: Constructing strength three covering arrays with augmented annealing. Discrete Mathematics 308, 2709–2722 (2008)
15. Nurmela, K.J.: Upper bounds for covering arrays by tabu search. Discrete Appl. Math. 138(1-2), 143–152 (2004)
16. Hnich, B., Prestwich, S.D., Selensky, E., Smith, B.M.: Constraint models for the covering test problem. Constraints 11(2-3), 199–219 (2006)
17. Yan, J., Zhang, J.: A backtracking search tool for constructing combinatorial test suites. The journal of systems and software 81(10), 1681–1693 (2008)
18. Forbes, M., Lawrence, J., Lei, Y., Kacker, R.N., Kuhn, D.R.: Refining the in-parameter-order strategy for constructing covering arrays. Journal of Research of the National Institute of Standards and Technology 113(5), 287–297 (2008)

Pipelining Memetic Algorithms, Constraint Satisfaction, and Local Search for Course Timetabling

Santiago E. Conant-Pablos, Dulce J. Magaña-Lozano,
and Hugo Terashima-Marín

Tecnológico de Monterrey, Campus Monterrey
Centro de Computación Inteligente y Robótica
Av. Eugenio Garza Sada 2501, Monterrey, N.L. 64849 Mexico
{sconant,a01060478,terashima}@itesm.mx

Abstract. This paper introduces a hybrid algorithm that combines local search and constraint satisfaction techniques with memetic algorithms for solving Course Timetabling hard problems. These problems require assigning a set of courses to a predetermined finite number of classrooms and periods of time, complying with a complete set of hard constraints while maximizing the consistency with a set of preferences (soft constraints). The algorithm works in a three-stage sequence: first, it creates an initial population of approximations to the solution by partitioning the variables that represent the courses and solving each partition as a constraint-satisfaction problem; second, it reduces the number of remaining hard and soft constraint violations applying a memetic algorithm; and finally, it obtains a complete and fully consistent solution by locally searching around the best memetic solution. The approach produces competitive results, always getting feasible solutions with a reduced number of soft constraints inconsistencies, when compared against the methods running independently.

1 Introduction

The Course Timetabling (CTT) problem (also known as University Timetabling) appears in high education institutions and consists in choosing a sequence of reunions among instructors and students in predetermined periods of time (typically a week) that satisfies a set of restrictions [1]. Some of the main constraints that are considered when a course timetable is constructed are: the capacity and adequacy of classrooms, the availability and preparation of instructors, the course dependencies, etc. Since usually these problems involve many constraints, it is necessary to identify which are hard constraints, i.e. constraints that must be necessarily satisfied; and which are soft constraints, i.e. constraints that only express a preference of some solutions. To find general and effective solutions is really difficult, due mainly to diversity of problems, variability of constraints, and particular features of institutions [2].

A. Hernández Aguirre et al. (Eds.): MICAI 2009, LNAI 5845, pp. 408–419, 2009.

A large variety of techniques from diverse areas as operations research, human-computer interaction, and artificial intelligence (AI), have been proposed for solving CTT problems [3]. Some of these techniques include tabu search, genetic algorithms, neural networks, simulated annealing, expert systems, constraint satisfaction, integer linear programming, and decision support systems.

The solution proposed in this paper for sets of hard instances of the CTT problem combines AI techniques such as constraint-satisfaction problems (CSP), memetic algorithms and local search algorithms to obtain course timetables in a reasonable time while accomplishing all of the hard constraints and most of the soft constraints of the problems.

This paper is organized as follows: in the next section, the CTT problems are formulated and the hard and soft constraints introduced; Section 3 describes the research process and the solution algorithms; Sections 4 and 5 present the implementation details and experimental results obtained by each algorithm; and finally, conclusions and suggested future work are presented in Sect. 6.

2   Problem Definition

A set of events (courses) $E = \{e_1, e_2, ..., e_n\}$ is the basic element of a CTT problem. Also there are a set of periods of time $T = \{t_1, t_2, ..., t_s\}$, a set of places (classrooms) $P = \{p_1, p_2, ..., p_m\}$, and a set of agents (students registered in the courses) $A = \{a_1, a_2, ..., a_o\}$. Each member $e \in E$ is a unique event that requires the assignment of a period of time $t \in T$, a place $p \in P$, and a set of agents $S \subseteq A$, so that an assignment is a quadruple (e, t, p, S). A timetabling solution is a complete set of n assignments, one for each event, which satisfies the set of hard constraints defined for the CTT problem. This problem is apparently easy to solve, but possesses several features that makes it intractable, as all the NP-hard problems [4].

CTT problems have been solved using techniques as CSPs [5], local search [6] and memetic algorithms [7]. However, each of these methods have been mainly focused in obtaining solutions that satisfy certain kind of constraints (hard or soft), frequently getting bad feasible solutions or good unfeasible assignments. Since every approach has its strengths and weaknesses, this study looked for the integration of these three techniques in a hybrid algorithm that could consistently obtain very good feasible solutions in a "reasonable time". As an alternative to using the conventional algorithms alone to cope with full instances of CTT problems, this paper proposes to use a pipeline of a heuristic backtracking algorithm (HBA), a memetic algorithm (MA), and a local search algorithm (LSA), in that order.

Although different institutions confront different kinds and number of restrictions, the hard constraints included into this work are the following: courses that have students in common must not be programmed into the same period of time; the number of students assisting to an event must not exceed the capacity of the assigned classroom; the classroom assigned to an event must be equipped with all resources required; and two courses must not be programmed into the same classroom during the same period of time.

To get the best feasible solutions, the proposed algorithm would try to satisfy most of the following list of preferences: a student should not have classes during the last period of time of a day; a student should not have more than two consecutive classes; and a student should not have only one class in a given day.

3   Solution Model

In the research documented in this paper, two types of hard CTT problems were generated and solved by the proposed pipelining algorithm. To validate the proposed approach each instance was also solved by four variants of the HBA, the LSA and a conventional GA. After this, the best results obtained by each algorithm were compared. These comparisons checked the effectiveness of the methods to satisfy hard and soft constraints, and their efficiency in execution time for getting the solutions.

The variants of the HBA and the LSA formulated each CTT instance as a CSP, but while trying to get a feasible solution, their heuristics also tried to reduce the number of unsatisfied soft constraints. The genetic algorithm formulated and solved each CTT instance as a Constraint Optimization Problem (COP). Finally, three different formulations were empirically adjusted to perform the stages of the proposed pipelining algorithm.

Next, the problem formulation and the solution strategy for each of the algorithms used in this work are described.

3.1   HBA Solution

Every CTT problem instance represented as a CSP and solved using heuristic backtracking was represented as a 3-tuple (V, D, R) where:

- $V = \{v_1, v_2, ..., v_n\}$ is the finite set of variables of the CSP, one for each course (event).
- $D = \{d_1, d_2, ..., d_n\}$ represents the domain of a variable, that for this problem is the Cartesian product of $S = \{s_1, s_2, ..., s_m\}$ which is the set of classrooms and $P = \{p_1, p_2, ..., p_t\}$ which is the set of periods of time predefined for the instance. For this work, the periods of time have the same length.
- $R = \{r_1, r_2, ..., r_q\}$ is the set of constraints to be satisfied.

In this study, four variants of the algorithm that use a chronological backtracking (a kind of depth-first search) were implemented and compared. All the variants use the Fail-First heuristic to dynamically choose the next variable to be assigned. Once a variable is chosen, the order of selection for the values to be assigned is decided in agreement with one of the following heuristics:

Sequential Selection (Seq): Selects values in the order constructed by the Cartesian product that defined the variable's domains (static order).
Least Constraining Value (LCV): Selects the value that reduces the least number of values from the domains of the remaining unassigned variables.

Min-conflict in Soft Constraints (MCSC): Chooses the value that partic-
ipates in the least number of conflicts with soft constraints.
Random Selection (Rand): Chooses values in a random manner.

Each value ordering heuristic represents a variant of the algorithm. All the algorithms also implement a partial look-ahead strategy (forward-checking heuristic)
for anticipating future conflicts with hard constraints due to the assignments already done.

3.2   LSA Solution

The chosen local search technique is based on the *Min-Conflicts* heuristic. This
heuristic has been successfully applied to diverse CSPs [6], and has demonstrated
a good behavior when used for solving different kinds of scheduling problems [8].
    Since this type of algorithms uses a state formulation that represents a com-
plete solution, a list with a full set of assignments for the variables of the CTT
problem is used, one for each course. This is an iterative algorithm that in each
iteration randomly selects one variable that has conflicts with hard constraints,
and changes the value that has currently assigned by the value of its domain
that reduces the most hard constraints conflicts of the new solution. Since only
one assignment is changed in each iteration, the new solution is considered to
be a neighbor solution (close solution) to the previous, which is the key aspect
of the local search strategies.

3.3   GA Solution

The genetic algorithm (GA) used for solving the CTT problems is generational,
i.e. all individuals of a population are substituted in the next iteration of the
algorithm [9]. It was designed with the following features:

- Each individual of a population represents a timetable as a n-genes chromosome, where n is the number of courses to schedule. Each value a_i that
  gen i can take is represented by an integer number that codes a pair (s, p)
  meaning that course v_i was assigned to classroom s in the period of time p.
- The *fitness* function used to evaluate each individual is a weighted sum of
  the number of hard and soft constraints not satisfied by this solution, as
  follows:

$$F = \alpha \sum_{i=1}^{n} W(c_i) + (1 - \alpha) \sum_{i=1}^{n} Y(c_i) \tag{1}$$

  where, α represents the relative importance of unsatisfied hard constraints
  with respect to unsatisfied soft constraints; $W(c_i)$ is a function that counts
  the number of hard constraints not satisfied by course i; and $Y(c_i)$ is a
  function that counts the number of soft constraints not satisfied by course i.
- For the initial population of the GA, each individual is created through a
  random selection of values for each of the n courses of the timetable.

- The GA chooses individual for the recombination process by tournament selection. The recombination is implemented by a one-point crossover operation.
- The GA also performs a bit-by-bit mutation operation with a small probability.

This type of GA was selected because it has been frequently used to solve CTT problems, while showing good solution performance.

3.4   Proposed Algorithm Solution

As mentioned in Sect. 2, this work proposes to construct a COP algorithm that solves hard instances of the CTT problem by pipelining a heuristic backtracking algorithm, a memetic algorithm, and a local search algorithm.

Firstly, as there are many courses to schedule and the objective is to obtain good feasible solutions in a "reasonable time", the initial population for the memetic algorithm is not fully constructed at random. Instead, an empirically chosen portion of the population (i individuals) is created as follows:

1. The n courses of a CTT instance are divided in m blocks of similar size.
2. Each of the m blocks is formulated as a CSP. The bigger the m value, easier the CSPs to solve.
3. Solve $x = \left\lceil \sqrt[m]{i} \right\rceil$ times each of the m CSPs by using a HBA.
4. Combine the corresponding solved CSP solutions to get the i required individuals. If the combination produces more individuals than necessary, randomly eliminate some of them.

This procedure maintains diversity among the individuals of the initial population while getting a group of better individuals produced by the CSP solver, expecting to help the memetic algorithm to evolve better solutions in a shorter execution time.

The MA represents and evaluates the fitness of each individual in the same way as the GA of Sect. 3.3. It also uses the same selection and crossover genetic operators. The genetic operator that was changed and which transformed the GA into a MA is the mutation operator. Instead of realizing a bit-by-bit random mutation in all offspring, it randomly chooses some of them for its modification with a Min-Conflicts local search algorithm. In this way, some of the new individuals accelerate their evolution in the desired direction. Again, this is done trying to get better solutions in fewer generations.

Finally, after reaching the stop condition of the MA, the best individual obtained is post-processed by a Min-Conflicts local search algorithm similar to the algorithm explained in Sect. 3.2. This last step looks for the elimination of any remaining unsatisfied hard constraint to get a fully consistent final solution.

4   Experiments with Conventional Algorithms

This section reports the experimental results obtained when the two sets of CTT instances were solved using separately the four variants of the HBA, the LSA, and the GA. The presented results were obtained as an average by executing several times each algorithm on each of the selected problems.

4.1   CTT Problem Instances

As mentioned in Sect. 3 two types of CTT problems are used in this study for testing and validating the proposed algorithm. All problems consist in sets of events to be programmed in 45 periods of time, defined in 5 days of 9 hours each. The number of courses, classrooms, features required for the classroom of a course, and the number of registered students in each course, varies from one CTT problem instance to another. The hard and soft constraints were listed in Sect. 2.

The first type of problems, which will be referred as Type I problems, are problems proposed and used by the *International Timetabling Competition* [10]. In this competition, the goal is to obtain feasible timetables, with a minimum number of unsatisfied soft constraints, within a limited period of time. The limited time is defined for each specific computer that will run the CTT solver, this time is calculated by a benchmarking program that performs tests over this computer. The resulting limited time for the computer used for all the experiments of this work was 14 minutes. CTT instances of Type I problems were constructed with the algorithm proposed by Ben Paechter [11].

Table 1. Features of Type I and Type II experimental problems

Problem	Type I Problems				Type II Problems			
	courses	classrooms	features	students	courses	classrooms	features	students
1	400	10	10	200	200	5	5	200
2	400	10	5	300	200	5	4	1000
3	350	10	10	300	225	5	10	1000
4	350	10	10	300	200	5	3	900

The second type of problems, which will be referred as Type II problems, are problems proposed by Rhyd Lewis [12]. For these problems is particularly difficult to find a feasible solution, due to the large quantity of students considered in each instance and the purposely way in which these students were distributed among the courses. Unlike Type I problems, getting a solution for a Type II problem was not constrained to fulfill a limited time.

Table 1 presents the features of four Type I and four Type II problems that were considered the most significant instances obtained for each kind of problems, and good representatives of the full set of instances used during the experiments.

4.2   Experiments with HBAs

As mentioned in Sect. 3.1, in this work, four variants of the heuristic backtracking algorithm were implemented and tested; one for each different type of value ordering heuristic (Seq, LCV, MCSC, and Rand). Experimental results for each of the HBA variants on Type I problems are shown in Table 2.

Table 2. Experimental results for Type I and Type II problems with HBA

Type I problems												
Problem	# of Inconsistencies								Execution time (sec)			
	Hard				Soft							
	Seq	LCV	MCSC	Rand	Seq	LCV	MCSC	Rand	Seq	LCV	MCSC	Rand
1	0	0	0	0	705	671	540	682	10.86	67.24	13.72	9.27
2	59	0	0	19	971	974	848	990	840	117.76	19.67	289.86
3	0	0	6.25	0	1052	950	817	1001	11.26	55.32	205.78	10.62
4	0	0	0	0	1023	998	820	1009	5.50	26.92	9.65	5.56
Type II problems												
1	0	0	0	0	664	642	525	635	4.72	48.37	7.26	4.74
2	0	0	0	0	3332	3252	2600	3204	4.21	41.11	25.94	4.21
3	9	0	0	37	3007	2778	3047	2715	80	29.85	14.23	80
4	194	0	0	263	3355	3875	3307	3298	80	68.52	22.80	80

As it can be observed, only the *Least Constraining Value* heuristic could get a feasible solutions for all the problems, however, the *Min-Conflicts on Soft Constraints* heuristic was close with a 95% confidence of finding a feasible solution on time, resulting a better strategy for reducing soft-constraint inconsistencies, and in average running a little faster than LCV.

Experimental results with the HBA variants for the Type II problems are shown in Table 2. In these problems, again, the best heuristics were LCV and MCSC with a 100% confidence of getting a feasible solution. The Sequential and Random ordering heuristics sometimes could not find a feasible solution for a problem, even after trying for 12 hours. The average execution time does not reflect such time because for reporting purposes were bounded to 80 seconds. In general, in all experiments, if these algorithms found a solution, they did this in less than 80 sec.

4.3   Experiments with the LSA

Experimental results with the LSA for the Type I and Type II problems are shown in Table 3.

In this case, the algorithm focused only on eliminating all the hard constraint inconsistencies, which explains why it is so good to get feasible solutions, but not for fulfilling preferences. In both, Type I and Type II problems, this algorithm was capable of getting feasible solutions, but as an important difference, Type

Table 3. Experimental results for Type I and Type II problems with LSA

Problem	Type I problems				Type II problems			
	Iterations	Inconsistencies		Time(sec)	Iterations	Inconsistencies		Time(sec)
		Hard	Soft			Hard	Soft	
1	1400	0	653	155.31	370	0	622	30.13
2	1184	0	980	197.12	207	0	3314	217.55
3	891	0	1007	124.89	2500	4	2707	2060.91
4	779	0	982	103.18	1500	0	3644	1002.59

II problems take much more time in average than Type I problems, confirming they are intrinsically more difficult.

4.4   Experiments with the GA

Many experiments were done to choose good values for the GA parameters. Due to the limited execution time of Type I problems, the algorithm was set to finish with a feasible solution on time. On the contrary, for Type II problems, the GA has all the time it needed for running, then shifting the pressure from a fast solution to a good solution. After performing tests with different sizes of populations, tournament groups, crossover and mutation probabilities, the parameters presented in Table 4 were chosen. Also, the α value that define the relative importance between hard and soft inconsistencies was defined as 0.8, given much more value to restrictions than to preferences.

Experimental results for the GA on Type I and Type II problems are shown in Table 5. As it can be observed, the GA could not get feasible solutions for

Table 4. GA parameters for Type I and Type II problems

Parameter	Type I value	Type II value
Number of generations	Limited	50
Population size	10	20
Tournament group size	5	5
Crossover probability	1.0	1.0
Mutation probability	0.01	0.01

Table 5. Experimental results for Type I and Type II problems with GA

Problem	Type I problems				Type II problems			
	Generations	Inconsistencies		Time(sec)	Generations	Inconsistencies		Time(sec)
		Hard	Soft			Hard	Soft	
1	122	345	475	840	50	187	307	411.33
2	66	439	642	840	50	203	804	5512.91
3	80	354	624	840	50	164	1154	4715.05
4	81	336	536	840	50	301	800	4928.09

none of the two types of problems, with or without limited execution time. However, it was the best algorithm for getting solutions with a small number of soft constraints inconsistencies.

5   Experiments with the Proposed Approach

An algorithm with more parameters generally involves a larger tuning effort. Each of the parameters that describe and use the three pipelined methods was determined empirically. The behavior of the algorithm was tested for different percentages of individuals of the initial population to be generated with the HBA, also with different number of blocks of variables, different sizes of populations, different probabilities for the genetic operators, etc.

Table 6. Parameters for solving Type I and Type II problems with the proposed hybrid algorithm

PARAMETER	TYPE I VALUE	TYPE II VALUE
Initial population		
Population size	10	20
Individuals generated by HBA	10%	10%
HBA heuristic for value selection	MCSC	MCSC
# of blocks (m) to solve as CSPs	4	4
Execution time limit	15 sec	Unlimited
Memetic Algorithm		
# of generations	By time	20
Tournament size	5	5
Crossover probability	1	1
Mutation probability	0.1	0.5
LSA used for mutation	Min-Conflicts	Min-Conflicts
Iterations for the mutation LSA	10	5
Execution time limit	11 min 20 sec	Unlimited
LSA for final consistency		
LSA to use	Min-Conflicts	Min-Conflicts
Execution time limit	2 min 25 sec	Unlimited
Maximum # of iterations	By time	1144

The result of that empirical process for Type I and Type II problems is shown in Table 6.

The experimental results on Type I and Type II problems obtained with the two versions of the proposed hybrid pipelining algorithm (HPA) are presented in Table 7, respectively.

It can be observed that the algorithm always could get feasible solutions for both types of problems. To see the effect of each solution stage in the final results, three columns for each type of constraints were included to show the number of inconsistencies after executing the correspondent algorithm. For example, for

Table 7. Experimental results on Type I and Type II problems with the proposed algorithm

Problem	# of inconsistencies						LSA iterations	Execution time (sec)
	Hard			Soft				
	HBA	MA	LSA	HBA	MA	LSA		
Type I problems								
1	593	10	0	630	520	580	754	840
2	617	102	0	942	800	898	800	840
3	488	74	0	946	710	900	615	840
4	616	49	0	875	750	830	532	840
Type II problems								
1	223	3	0	690	480	520	200	765.41
2	200	12	0	3098	1855	2240	105	12311.56
3	168	8	0	2865	1500	1990	1090	10012
4	242	18	0	3026	2080	2689	730	9010

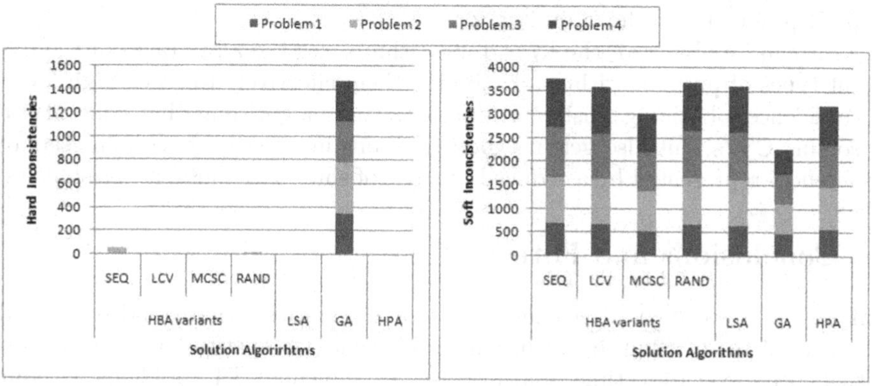

Fig. 1. Algorithms comparison in hard and soft inconsistencies for Type I problems

the Type II problem 2, after the first stage with the HBA, the number of hard constraints inconsistencies in the population was 200; after applying the MA was reduced to only 12; and these disappeared after the LSA. On the other hand, the reduction pattern was different for soft constraints inconsistencies; for example, for the same problem, after the first stage there were 3098 inconsistencies, those were reduced to 1855 by the MA, and augmented to 2240 after the LSA. As it can be noted, this pipelining strategy has a major impact in Type II problems.

A final set of four graphs are included for comparison of the efficacy and efficiency of the proposed approach against the alternative of solving CTT problems using pure strategies alone. Experimental results showing the counts of remaining hard and soft inconsistencies for all the algorithms when solving the Type I

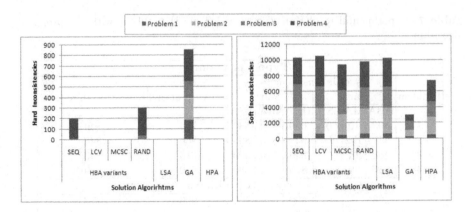

Fig. 2. Algorithms comparison in hard and soft inconsistencies for Type II problems

problems are shown in Fig. 1. And, experimental results showing the counts of remaining hard and soft inconsistencies for all the algorithms when solving the Type II problems are shown in Fig. 2.

As it can be observed, the hybrid algorithm takes the advantages of the different types of pure algorithms, combining them effectively to reach high levels of confidence, obtaining feasible solutions as those algorithms that concentrate in solving CSPs, but also getting good solutions in agreement with the sets of preferences represented by a reduced number of soft constraints inconsistencies.

6   Conclusions and Future Work

This research was focused on the creation of an algorithm that could solve hard instances of the CTT problems in reasonable time. Competition CTT problems with limited execution time, and intrinsically difficult CTT problems without time limitations were tried. The main objective in both types of problems was to find good feasible solutions, in the sense that all the hard constraints of every CTT instance and the most of its soft constraints were satisfied by the timetables obtained. The proposed approach was to use an algorithm that pipelined a heuristic backtracking algorithm, a memetic algorithm and a local search algorithm. For validating this algorithm, its effectiveness and efficiency for solving the experimental problems was compared against the results obtained when the same problems were solved completely and separately by conventional, but well-tuned, algorithms.

The comparison results showed that the proposed algorithm is capable of generating competitive solutions, feasible for all the instances of both types of CTT problems, and with a good reduction in the number of soft constraints violations. Though its results were not always the best, in average, for problems with tight time limitations, it always got good feasible solutions on time, something that is really difficult for algorithms with evolutionary processes in similar competitions.

Some ideas for extending this work include: Trying with indirect forms of representation for CTT solutions looking for reducing the complexity of their formulation as CSPs and COPs, because with the naive representations used in this research, the search spaces for the methods are very big. Tuning automatically the pipelined algorithms, maybe by using hyper-heuristics, to adapt times, parameters, and even the relative importance of the types of constraints in the objective functions, for different sizes, difficulties and types of constraints in the CTT instances to solve. And, testing a different type of memetic algorithm, such as one of stable state instead of the generational used in this study, trying to accelerate the convergence to the solution.

References

1. Lee, J., Yong-Yi, F., Lai, L.F.: A software engineering approach to university timetabling. In: Proceedings International Symposium on Multimedia Software Engineering, pp. 124–131 (2000)
2. Thanh, N.D.: Solving timetabling problem using genetic and heuristic algorithms. In: Eighth ACIS International Conference on Software Engineering, Artificial Intelligence, Networking, and Parallel/Distributed Computing, pp. 472–477 (2007)
3. Lien-Fu, L., Nien-Lin, H., Liang-Tsung, H., Tien-Chun, C.: An artificial intelligence approach to course timetabling. In: Proceedings of the 18th IEEE International Conference on Tools with Artificial Intelligence, pp. 389–396 (2006)
4. Garey, M.R., Johnson, D.S.: Computers and Intractability: A Guide to the Theory of NP-completeness. W. H. Freeman, New York (1979)
5. Zhang, L., Lau, S.: Constructing university timetable using constraint satisfaction programming approach. In: Proceedings of the International Conference on Computational Intelligence for Modelling, Control and Automation and International Conference on Intelligent Agents, Web Technologies and Internet Commerce (CIMCA-IAWTIC 2005), Washington, DC, USA, vol. 2, pp. 55–60 (2005)
6. Tam, V., Ting, D.: Combining the min-conflicts and look-forward heuristics to effectively solve a set of hard university timetabling problems. In: Proceedings of the 15th IEEE International Conference on Tools with Artificial Intelligence, ICTAI 2003 (2003)
7. Alkan, A., Ozcan, E.: Memetic algorithms for timetabling. In: Proceedings of the 2003 Congress on Evolutionary Computation, p. 1151 (2003)
8. Russell, S., Norvig, P.: Artificial Intelligence: A Modern Approach, 2nd edn. Prentice-Hall, Englewood Cliffs (2003)
9. Goldberg, D.: Genetic Algorithms in Search, Optimization, and Machine Learning. Addison Wesley, Reading (1989)
10. PATAT, WATT: International timetabling competition, http://www.cs.qub.ac.uk/itc2007/index.htm (October 2007)
11. Paechter, B.: A local search for the timetabling problem. In: Proceedings of the 4th PATAT, August 2002, pp. 21–23 (2002)
12. Lewis, R.: New harder instances for the university course timetabling problem, http://www.dcs.napier.ac.uk/~benp/centre/timetabling/harderinstances (January 2008)

Fuzzy Relational Compression Applied on Feature Vectors for Infant Cry Recognition[*]

Orion Fausto Reyes-Galaviz[1] and Carlos Alberto Reyes-García[2]

[1] Universidad Autónoma de Tlaxcala
Facultad de Ciencias Básicas, Ingeniería y Tecnología
Calzada Apizaquito, Km. 1.5 Apizaco, Tlaxcala. México
orionfrg@yahoo.com
http://ingenieria.uatx.mx/
[2] Instituto Nacional de Astrofísica, Óptica y Electrónica
Luis Enrique Erro 1, Tonantzintla, Puebla, México
kargaxxi@inaoep.mx
http://www.inaoep.mx/

Abstract. Data compression is always advisable when it comes to handling and processing information quickly and efficiently. There are two main problems that need to be solved when it comes to handling data; store information in smaller spaces and processes it in the shortest possible time. When it comes to infant cry analysis (ICA), there is always the need to construct large sound repositories from crying babies. Samples that have to be analyzed and be used to train and test pattern recognition algorithms; making this a time consuming task when working with uncompressed feature vectors. In this work, we show a simple, but efficient, method that uses Fuzzy Relational Product (FRP) to compresses the information inside a feature vector, building with this a compressed matrix that will help us recognize two kinds of pathologies in infants; Asphyxia and Deafness. We describe the sound analysis, which consists on the extraction of Mel Frequency Cepstral Coefficients that generate vectors which will later be compressed by using FRP. There is also a description of the infant cry database used in this work, along with the training and testing of a Time Delay Neural Network with the compressed features, which shows a performance of 96.44% with our proposed feature vector compression.[1]

Keywords: Fuzzy Relational Product, Infant Cry Analysis, Feature Compression, MFCC, Time Delay Neural Networks.

[*] For the present work we used the "Chillanto Database" which is property of the Instítuto Nacional de Astrofísica Óptica y Electrónica (INAOE) CONACYT, México. We like to thank Dr. Carlos A. Reyes-García, Dr. Edgar M. Garcia-Tamayo Dr. Emilio Arch-Tirado and his INR-Mexico group for their dedication of the collection of the Infant Cry data base.
[1] This work is partially supported by the project for consolidating the UATx Distributed and Intelligent Systems Research Group, in the context of the PROMEP program.

A. Hernández Aguirre et al. (Eds.): MICAI 2009, LNAI 5845, pp. 420–431, 2009.

1   Introduction

Though data we can represent information, so the same information can be represented by different amounts of data. When it comes to large data sets of information, it is always advisable to have reliable data compression algorithms. There are two main problems that need to be solved when it comes to data compression; store information in smaller spaces and processes it in lesser time. When it comes to Infant Cry Analysis (ICA), the feature vectors always tend to be large because of the sound analysis, causing with this an increase in time when training the pattern recognition algorithms, so working with them can be exhausting and may not be very practical. On pattern recognition tasks, it has been proven in several works that the algorithms work better with compressed and non-redundant data. In this paper we show and explain an algorithm designed to compress the information inside a feature vector by applying fuzzy relational operations, this algorithm uses codebooks to build a small relational matrix that represents an original vector. Since this algorithm was firstly designed to compress and decompress images, the resulting compressed matrix, along with the codebooks should hold enough information to build a lossy representation of the original image. With this taken into account, we can hypothesize that if there is enough information inside a compressed fuzzy relational matrix to restore an image by using fuzzy relations, then there should be enough information inside a matrix, created from a feature vector, to perform a good infant cry classification. We show that this algorithm is easy and fast to implement, and that the compressed vectors obtained are reliable enough to perform an infant cry classification for the identification of two kinds of pathologies: Asphyxia and Deafness.

The paper's layout is as follows; in Section 2 there is a brief overview of previous works related to this paper. Section 3 describes the automatic infant cry recognition process. In Section 4, the acoustic processing is shown, describing the Mel Frequency Cepstral Coefficient analysis used to extract the feature vectors from the cry sound waves. In Section 5 we explain the basis of the fuzzy relational compression, giving an overview of fuzzy sets and mathematical relations used to perform a fuzzy product. The system implementation is described in Section 6 followed by the experimental results which are given on Section 7. Finally, in Section 8 we give our concluding thoughts and the future trend of our work.

2   State of the Art

Recently, there has been a significant increase on the interest of a recent and flourishing area; the Infant Cry Analysis (ICA), this might be due to the research efforts that have been made; attempts that show promising results and highlight the importance of exploring this young field. In 2003 [1], Reyes & Orozco classified samples from deaf and normal babies, obtaining recognition results that go from 79.05% to 97.43%. Tako Ekkel, in 2002 experimented with cry samples from recently born babies, in these experiments he classified sounds into two categories: normal and abnormal, in order to diagnose babies with hypoxia, which occurs when there is low oxygen in the blood. Tako Ekkel reported a correct classification result of 85%, results yielded by a radial basis neural network [3]. Marco Petroni, in 1995, used neural networks to differentiate pain and no-pain cries [2]. Also, in 1999, and by using self organizing maps, Cano et al, in [4] reported experiments in which they classify infant cry units from normal and pathological babies, in these experiments they extracted 12 features directly from the sound frequencies to obtain a feature vector for each sound sample. They obtained an 85% recognition percentage on these experiments. There are other works that show a performance of 97.33%, by compressing the feature vectors with Principal Component Analysis (PCA), and classified the input vectors with a Time Delay Neural Network [5,6,7], but to the best of our knowledge, no one has applied a feature vector reduction by using fuzzy relational compression.

3   Automatic Infant Cry Recognition

The automatic infant cry recognition process (Fig. 1) is basically a pattern recognition problem, and it is similar to speech recognition. In this case, the goal is to record a baby's cry and use the sound as an input, with this at the end, obtain the type of cry or pathology detected. Generally, the Infant Cry Recognition Process is performed in two stages; the first one is the acoustic processing, or feature extraction, while the second is known as pattern processing or pattern classification. In the case of this proposed system, we have added an extra step between both of these stages called the *feature compression*. The first step in the acoustic processing is to segment the infant cry sample into one second samples. Once the sound is segmented, a feature set is extracted to build a vector which will represent that sample. Next, each vector is transformed into a matrix, which gets normalized into values between $[0, 1]$, then the matrix is compressed, and finally transformed back into a smaller vector; a vector that should hold enough information to represent the original pattern, and furthermore, the original cry sample. At the end, the compressed vectors are reinforced with an statistical analysis, as it is explained in Section 6.2. As for the pattern recognition stage, four main approaches have been traditionally used: pattern comparison,

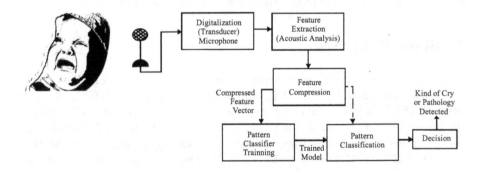

Fig. 1. Infant Cry Recognition Process

statistical models, knowledge based systems, and connectionist models. We are focusing on the use of the last one.

4   Acoustic Analysis

The acoustic analysis implies the application and selection of filter techniques, feature extraction, signal segmentation, and normalization. With the application of these techniques we try to describe the signal in terms of its fundamental components. One cry signal is complex and codifies more information than the one needed to be analyzed and processed in real time applications. For this reason, in our cry recognition system we use a feature extraction function as the first plane processor. Its input is a cry signal, and its output is a feature vector that characterizes key elements of the cry sound wave. We have been experimenting with diverse types of acoustic features, emphasizing by their utility the Mel Frequency Cepstral Coefficients (MFCC).

4.1   Mel Frequency Cepstral Coefficients

The low order cepstral coefficients are sensitive to overall spectral slope and the high order cepstral coefficients are susceptible to noise. This property of the speech spectrum is captured by the Mel spectrum. High order frequencies are weighted on a logarithmic scale whereas lower order frequencies are weighted on a linear scale. The Mel scale filter bank is a series of L triangular band pass filters that have been designed to simulate the band pass filtering believed to occur in the auditory system. This corresponds to series of band pass filters with constant bandwidth and spacing on a Mel frequency scale. On a linear frequency scale, this spacing is approximately linear up to 1 KHz and logarithmic at higher frequencies (Fig. 2) [8].

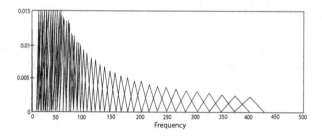

Fig. 2. Mel Filter Bank

5   Fuzzy Relational Product and Feature Compression

The fuzzy relational equations, which are the basis of the Fuzzy Relational Products (FRP), started in 1976 [9], to investigate the theoretical and applicational aspects of fuzzy set theory. The relational equations where originally intended for boolean relational equations [10].

5.1 Fuzzy Sets: A Quick Overview

In 1965 Lotfi A. Zadeh published his first article called "Fuzzy Sets", in where he establishes the fuzzy set theory basis [11]. According to Zadeh, the membership of an element to a fuzzy set is gradual. The degrees of membership values are in the interval $[0,1]$. This is why the main key of fuzzy sets are the membership values. Furthermore, the fuzzy sets include boolean or *crisp* sets [12], this because a crisp set is a fuzzy set with values 0 and 1.

The main fuzzy set operations are [13]: *Emptiness*, where a fuzzy set is empty if all candidates have 0 membership; *Complement*, the complement of a fuzzy set is the amount that the membership needs to reach 1, e.g. $\mu_{\bar{A}} = 1 - \mu_A$; *Containment*, in fuzzy sets each element must belong less to the subset than to the biggest set; *Intersection* ($\cap$), is the membership degree that two sets share: a fuzzy intersection is the lowest membership value from each element in two sets, e.g. $\mu_{A \cap B} = \min(\mu_A, \mu_B)$; and *Union* ($\cup$), where the fuzzy set union is inverse to the intersection, this is the highest value from each fuzzy element, e.g. $\mu_{A \cup B} = \max(\mu_A, \mu_B)$. With this we can formulate the following theorem: A is properly fuzzy *if and only if* $A \cap \bar{A} \neq 0$ and $A \cup \bar{A} \neq 1$. These fuzzy set operations are the basis of fuzzy relational operations, and furthermore, fuzzy relational compression.

5.2 Mathematical Relations

The intentional definition of a relation R from set x to set y is like an incomplete open phrase with two empty spaces: $R =$ "— is married to —", when the first space is filled, by inserting an element x from X, and the second space inserting an element y from Y, it results in a proposition that can be true or false:

- If its true, it is said that x is in relation R to y
- If its false, it is said that x is not in relation $\neg R$ to y

The extensional definition is given by the *satisfaction set*. The satisfaction set R_s from a R relation consists on all the pairs of elements (x, y) in which $_xR_y$, this is:

$$R_s = \{(x, y) \in X \times Y | _xR_y\} \tag{1}$$

This is clearly a subset from the Cartesian product $X \times Y$ that consists on all the possible pairs (x, y). Other properties in mathematical relations include the *reverse relation* where R^T of R is a relation in the opposite direction from X to Y: $_yR_x^T \Leftrightarrow _xR_y$. This relation is also known as the *inverse relation* and can be also represented as R^{-1}. And finally, a relation R is a sub-relation of S or is included on S ($R \subseteq S$) *if and only if* $_xR_y$ always implicates $_xS_y$: $R \rightarrow S$ [13].

5.3 Fuzzy Relational Product (FRP)

The mathematical relations are another basis of fuzzy relational operations, and
these properties can be applied to crisp or fuzzy matrices as follows. Let R
be a relation from X to Y, and S a relation from Y to Z, furthermore let be
$X = \{x_1, x_2, ..., x_n\}$, $Y = \{y_1, y_2, ..., y_n\}$, and $Z = \{z_1, z_2, ..., z_n\}$ finite sets,
there can be many binary operations applied on them, each one resulting in a
product relation from set X to set Z, operations such as [13]: *Circlet Product*
$(R \circ S)$, where x has a relation $R \circ S$ to z, *if and only if* there is at least one y
such that $_xR_y$ and $_yS_z$:

$$_x(R \circ S)_z \Leftrightarrow \exists\, y \in Y \; if \; (_xR_y \; and \; _yS_z) \tag{2}$$

Then the circlet relation $_x(R \circ S)_z$ exists *if and only if* there is a path from x
to z:

$$(R \circ S)_{x,z} = \max(\min(R_{xy}, S_{yz})) = (R \circ S)_{xz} = \bigcup(R_{xy} \cap S_{yz}), \tag{3}$$

There are other relational products that work with crisp sets, the *Subtriangle*
product exists when x is in Relation $R \triangleleft S$ to z *if and only if*, for each y, $_xR_y$
implicates that $_yS_z$:

$$(R \triangleleft S)_{xz} = \min(R_{xy} \rightarrow S_{yz}) \tag{4}$$

Where $\rightarrow$ is the boolean implication operator shown in Table 1a:

Table 1. Truth table for: *a)* the implication operator and *b)* the inverse implication
operator

	S				S		
R	0	1		R	0	1	
0	1	1	$R \rightarrow S$	0	0	0	$R \leftarrow S$
1	0	1		1	0	1	
		a)				b)	

Therefore, the *Supertriangle* is a product inverse to the previous one (Table
1b), given by:

$$(R \triangleright S)_{xz} = \min(R_{xy} \leftarrow S_{yz}) \tag{5}$$

Table 2. T-Norms and residual implications

T-Norm	R,S	Implication	Jt(R,S)
Lukasiewicz	$\max(0, R + S - 1)$	Lukasiewicz	$\min(1, 1 - R + S)$
Mamdani	$\min(R, S)$	Godel	$\begin{cases} 1, \; if \; R \leq S, \\ S, \; other \end{cases}$

The circlet product can work with both, boolean and fuzzy numbers, given the nature of the *max* and *min* operators, but since the subtriangle and supertriangle relational products can only work with boolean numbers, some authors have proposed several implication operators or "T-Norms" that work with both boolean and fuzzy numbers [10], each T-Norm has its own residual implication, these operations are shown in Table 2.

The *Supertriangle* ($\triangleright$) product can be seen as the T-Norms, and the *Subtriangle* ($\triangleleft$) product as the residual implication, these operators are the ones that will help us find the fuzzy relations between two matrices and furthermore, compress and decompress the input feature vector.

5.4 Fuzzy Relational Feature Compression

The *T-Norms* or Triangular Norms, as defined by W. Pedrycz [10], can be used to compress and decompress the information that conforms a feature vector [17,15]. This algorithm was firstly proposed for lossy compression and reconstruction of an image [14], where a still gray scale image is expressed as a fuzzy relation $R \in F(X \times Y)$ $X = \{x_1, x_2, ..., x_n\}$ and $Y = \{y_1, y_2, ..., y_n\}$ by normalizing the intensity range of each pixel from $[0, 255]$ onto $[0, 1]$. In our case, a feature vector that holds the information of an infant cry sample is also normalized onto values between $[0, 1]$, transformed into a matrix R (Fig. 3), and then compressed into $G \in F(I \times J)$ by Eq. 6.

$$G(i,j) = (R^T \triangleright A_i)^T \triangleright B_j \tag{6}$$

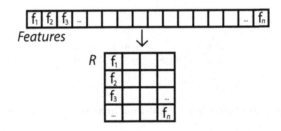

Fig. 3. Feature vector normalized and transformed into Matrix R

Where T denotes Transposition of a fuzzy set, $i = 1, 2, ..., I$, and $j = 1, 2, ..., J$ denote the final size of the matrix (Fig. 4). In [14,15,17] the authors report experiments with *square* images ($N \times N$) in grayscale, but we have performed several experiments that show a good performance on RGB images (square and rectangular), and on rectangular grayscale pictures, e.g. ($N \times M$) where $N \neq M$, and now we are applying this method to feature vectors transformed into relational matrices. The whole compression process is visually described in Fig. 4.

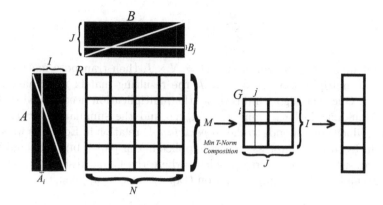

Fig. 4. Fuzzy Relational Feature compression

Where A and B are *codebooks*, a codebook is an essential process in the matrix compression; given a codebook, each block of the matrix can be represented by the binary address of its closest codebook vector. Such a strategy results in significant reduction of the information involved on the matrix transmission and storage. If the matrix is an image representation, the image is reconstructed by replacing each image block by its closest codebook vector. As a result, the quality of the reconstructed image strongly depends on the codebook design [16], but for our case, the feature reconstruction is of no interest at this point, since we are focusing on recognizing the baby's pathology, and the feature compression will help us accelerate this process, there is no need for a feature reconstruction, once the pathology has been detected. Codebooks A and B are constructed with [17]:

$$A_i(x_m) = \exp\left(- Sh\left(\tfrac{iM}{I} - m \right)^2 \right)$$
$$B_j(y_n) = \exp\left(- Sh\left(\tfrac{iN}{J} - n \right)^2 \right) \tag{7}$$

Where $A = \{A_1, A_2, ..., A_I\}$, $B = \{B_1, B_2, ..., B_J\}$, $(m = 1, 2, ..., M)$, $(n = 1, 2, ..., N)$, $(I < M)$, $(J < N)$, $A_i \in F(\subset (X))$, and $B_j \in F(\subset (Y))$. The variable $Sh(> 0)$ denotes the compression rate. In [17] a Sh value between 0.0156 and 0.0625 is recommended. For our experiments $Sh = 0.05$ and the Mamdani T-Norm were used, since these parameters yielded the best classification results.

To implement the fuzzy relational compressor, the first step is to program an algorithm that multiplies two matrices of sizes $R^T = N \times M$ and $A = M \times I$ (Fig. 4), by following the definition;

$$\sum_{m=1}^{M} R_{n,m}^{T} \times A_{m,i} \tag{8}$$

The result must be a matrix of size $Q = N \times I$, then transpose matrix $Q^T = I \times N$ and multiply it by $B = N \times J$, the resulting matrix must have a size $G = I \times J$. Once the algorithm is done, the multiplication needs to be replaced by the Mamdani T-Norm; $\min(R_{n,m}^{T}, A_{m,i})$, and the summatory by the min operator, which corresponds to the $Supertriangle$ relation in Eq. 5. As we stated before, the feature decompression is not applied in this paper, but if it should take place it consists on the inverse problem, where the Godel residual implication (Table 2), and the $Subtriangle$ relation (Eq. 4) must be applied.

6   System Implementation

The Fuzzy Relational compressor was designed and implemented in Matlab r2009a. For the infant cry classification we used a Time Delay Neural Network (TDNN) included in Matlab's Neural Networks toolbox [18]. The whole process is described in the next sections.

6.1   Infant Cry Database

The infant cries are collected by recordings obtained directly from doctors at the Mexican National Rehabilitation Institute (INR, Mexico) and at the Mexican Social Security Institute in Puebla, Mexico (IMSS, Puebla). The cry samples are labeled with the kind of cry that the collector states at the end of each cry recording. The recordings are then stored and converted into wav samples. Next, we divide each wav file into one second; these segments are then labeled with a previously established code [5].

By this way, for the present experiments, we have a repository made out of 1049 samples from normal babies, 879 from hypoacoustics (deaf), and 340 with asphyxia; all samples last 1 second. Next, the samples are processed one by one extracting their MFCC acoustic features, this process is done with Praat 4.2 [19], a windows based application for sound analysis. The acoustic features are extracted as follows: for each sample we extract 19 coefficients for every 50 milliseconds window, generating vectors with 361 features for each 1 second sample. At the end of each vector, a label L is added, where $L = 1, 2, ..., C$; C being the number of classes; in this case $C = 3$.

If n is the number of the input vector's size, whether is compressed or not, the neural network's architecture consists of n nodes in the input layer, a hidden layer with 40% less nodes than the nodes in the input layer, and an output layer of 3 nodes, corresponding to the three classes; normal, deaf, and asphyxia. The implemented system is interactively adapted; no changes have to be made to the source code to experiment with any input vector size.

The training is performed on $10,000$ epochs or until a 1×10^{-8} error is reached. Once the network is trained, we test it by using different samples separated previously from each class for this purpose; we use, 70% of the cry samples to train the neural network, and 30% for the testing phase. In these experiments, and since there are only 340 samples contained in the Asphyxia class, 340 samples are randomly selected from the Normal and Deaf classes respectively, for a total of 1020 vectors; as a result we have a Training matrix of size $(361 + 1) \times 714$ (70% from each class) and a Testing matrix of size $(361 + 1) \times 306$ (the remaining 30%).

6.2   Feature Compression

Once the MFCC features have been extracted from the segmented samples, as described in Section 6.1, each vector in both, the training and testing matrices, is unlabeled and transformed into a matrix. Since each vector holds 361 MFCCs, the resulting matrix has a size of (19×19). After applying the fuzzy relational compression, with $I = 5$, $J = 5$, and $Sh = 0.05$, the resulting matrix will have a size of $(I \times J)$. Next, the matrix its transformed back into a vector and relabeled; this vector will have a size of $(25 + 1) \times 1$. After the training and testing matrices have been compressed and rebuilt the first experiments were made, showing a low recognition performance (Table 3). To reinforce these samples, we added a vector extracted from the original uncompressed matrices; for each input vector the following statistical analysis were obtained: *maximum, minimum, standard deviation, mean, and median* values. These vectors were concatenated to the bottom of their corresponding previously compressed samples, giving us vectors of size $(30 + 1) \times 1$. With these vectors, the final compressed training and testing matrices result in a size of $(30 + 1) \times 714$ and $(30 + 1) \times 306$ respectively. In past experiments [1,5,6,7], the feature vector reduction was performed with Principal Component Analysis (PCA), reducing the size of the vectors to 50 features which is the number of principal components that had a better performance. The results obtained in the present experiments are shown in Section 4.

7   Experimental Results

In order to compare the behavior of the proposed fuzzy relational compression, a set of experiments were made:

1. Original vectors without any dimensionality reduction,
2. vectors reduced to 50 Principal Components with PCA [5,6,7],
3. vectors reduced to 25 components with FRP,
4. vectors reduced to 5 components; (*max, min, std, mean, median*), and
5. reinforced vectors with 30 components (rFRP).

Each experiment was performed five times, the average results are shown on Table 3;

Table 3. Average experimental results comparing different compression methods

Experiment No.	Compression Method	Performance
1.	No Compression	96.55%
2.	PCA	97.33%
3.	FRP	64.63%
4.	Statistics	59.39%
5.	rFRP	96.44%

8   Conclusions and Future Work

The feature compression using fuzzy relational product, and reinforcing it with a statistical analysis, becomes a simple implementation process and has satisfactory results, proving its effectiveness; given that the resulting matrices, which were compressed to almost 10% of their original size, yielded good results when classifying three different infant cry classes. This also resulted on an accelerated and efficient process for the classification algorithm, given that the training and testing phases where done in a reasonable time. The recognition performance demonstrates that the information contained in the compressed fuzzy relational matrix along with the statistical analysis is sufficient to perform the pathology recognition. The results obtained by using both, the vectors compressed with PCA and rFRP, show results almost equal to the results obtained with the uncompressed vectors, the main difference is that the neural network's training is done in less time when using compressed or non-redundant information. One of the main goals of this paper is to prove that we can rapidly build our own fast and efficient feature compressor that can achieve good results when it comes to pattern recognition and classification, and improve with this the training time without loosing recognition performance. For our future work we are planning on testing more T-Norms and residual implications, and maybe combine different T-Norms with different implications (e.g. compress with Mamdani and decompress with Lukasiewicz), for this we want to work on a Genetic Algorithm that can dynamically search for the best Sh value, T-Norm and residual implication in order to obtain a more refined and efficient fuzzy relational compressor, improving with this the recognition performance.

References

1. Orozco Garcia, J., Reyes Garcia, C.A.: Mel-Frequency Cepstrum coefficients Extraction from Infant Cry for Classification of Normal and Pathological Cry with Feedforward Neural Networks, ESANN, Bruges, Belgium (2003)
2. Petroni, M., Malowany, A.S., Celeste Johnston, C., Stevens, B.J.: Identification of pain from infant cry vocalizations using artificial neural networks (ANNs). In: The International Society for Optical Engineering, vol. 2492, Part two of two. Paper #: 2492-79 (1995)

3. Ekkel, T.: Neural Network-Based Classification of Cries from Infants Suffering from Hypoxia-Related CNS Damage, Master's Thesis. University of Twente, The Netherlands (2002)
4. Cano, S.D., Daniel, I.: Escobedo y Eddy Coello, El Uso de los Mapas Auto-Organizados de Kohonen en la Clasificación de Unidades de Llanto Infantil. In: Voice Processing Group, 1st Workshop AIRENE, Universidad Católica del Norte, Chile, pp. 24–29 (1999)
5. Galaviz, O.F.R.: Infant Cry Classification to Identify Deafness and Asphyxia with a Hybrid System (Genetic Neural). Master's Thesis on Computer Science, Apizaco Institute of Technology (ITA) (March 2005)
6. Galaviz, O.F.R., García, C.A.R.: Infant Cry Classification to Identify Hypoacoustics and Aphyxia Comparing an Evolutionary-Neural System with a Simple Neural Network System. In: Gelbukh, A., de Albornoz, Á., Terashima-Marín, H. (eds.) MICAI 2005. LNCS (LNAI), vol. 3789, pp. 949–958. Springer, Heidelberg (2005)
7. Reyes-Galaviz, O.F., Cano-Ortiz, S.D., Reyes-Garca, C.A.: Evolutionary-Neural System to Classify Infant Cry Units for Pathologies Identification in Recently Born Babies. IEEE Computer Society, Los Alamitos (2008)
8. Gold, B., Morgan, N.: Speech and Audio Signal Processing: Processing and perception of speech and music. John Wiley & Sons, Inc. Chichester (2000)
9. Sanchez, E.: Resolution of composite fuzzy relation equations. Inform. and Control (30), 38–49 (1976)
10. di Nola, A., Pedrycz, W., Sessa, S., Pei-Zhuang, W.: Fuzzy Relation Equation Under A Class Of Triangular Norms: A Survey and New Results. Stochastica 3(2) (1994)
11. Zadeh, L.A.: Fuzzy Sets. Information and Control 8(3), 338–353 (1965)
12. Mc Neill, D., Freiberger, P.: Fuzzy Logic. Simon & Shuster, New York (1993)
13. Garcia, C.A.R.: Sistemas Difusos (Fuzzy Systems): Fundamentos y Aplicaciones. In: Tutorials and Workshops of the IV Mexican International Conference on Artificial Intelligence, ITESM Monterrey, Mxico, November 14-18 (2005) ISBN 968-891-094-5
14. Hirota, K., Pedrycz, W.: Fuzzy Relational Compression. IEEE Transactions on Systems, man, and cybernetics,Part B: Cybernetics 29(3), 1–9 (1999)
15. Hirota, K., Pedrycz, W.: Data Compression with fuzzy relational equations. Fuzzy Sets and Systems 126, 325–335 (2002)
16. Karayiannis, N.B., Pai, P.I.: Fuzzy vector quantization algorithms and their application in image compression. IEEE Transactions on Image Processing 4(9), 1193–1201 (1995)
17. Nobuhara, H., Pedrycz, W., Hirota, K.: Fast Solving Method of Fuzzy Relational Equation and Its Application to Lossy Image Compression and Reconstruction. IEEE Transactions on Fuzzy Systems 8(3), 325–334 (2000)
18. Neural Network Toolbox Guide, Matlab r2009a, Developed by MathWoks, Inc. (2009)
19. Boersma, P., Weenink, D.: Praat v. 4.0.8. A system for doing phonetics by computer. Institute of Phonetic Sciences of the University of Amsterdam (February 2002)

Parametric Operations for Digital Hardware Implementation of Fuzzy Systems

Antonio Hernández Zavala[1], Ildar Z. Batyrshin[2], Imre J. Rudas[3], Luís Villa Vargas[4], and Oscar Camacho Nieto[4]

[1] Unidad Profesional Interdisciplinaria en Ingeniería y Tecnologías Avanzadas, Instituto Politécnico Nacional
antonioh@hotmail.com
[2] Instituto Mexicano del Petróleo
batyr1@gmail.com
[3] Budapest Tech
rudas@bmf.hu
[4] Centro de Investigaciones en Computación, Instituto Politécnico Nacional
{oscarc,lvilla}@cic.ipn.mx

Abstract. Operations used in fuzzy systems to realize conjunction and disjunction over fuzzy sets are commonly minimum and maximum because they are easy to implement and provide good results when fuzzy system is tuned, but tuning of the system is a hard work that can result on knowledge lost. A good choice to tune the system without loss of previously stated knowledge is by using parametric classes of conjunction and disjunction operations, but these operations must be suitable to be efficiently implemented on digital hardware. In this paper, the problem of obtain effective digital hardware for parametric operations of fuzzy systems is studied. The methods for generation of parametric classes of fuzzy conjunctions and disjunctions by means of generators and basic operations are considered. Some types of generators of parametric fuzzy operations simple for hardware implementation are proposed. New hardware implementation for parametric operations with operation selection is presented.

Keywords: Fuzzy logic, digital fuzzy hardware, conjunction, disjunction, generator, parametric conjunction.

1 Introduction

Fuzzy logic has turned to be a good mathematical tool to model knowledge based systems because of its ability for knowledge representation using a rule base where simple rules are used to realize inferences for determining satisfactory output results emulating expert knowledge. According to application demands there are different forms for final implementation of designed system, satisfying different speed requirements. Hardware implementations of fuzzy systems are the best choice to reach high speed inference rates, required on many industrial applications of fuzzy logic based intelligent systems like automated control systems, pattern recognition, expert systems, decision making among others.

A. Hernández Aguirre et al. (Eds.): MICAI 2009, LNAI 5845, pp. 432–443, 2009.

Digital fuzzy hardware has many advantages over analog counterpart [1], driving digital technology to be preferred, even more when it is reconfigurable because it can be adapted to a wide variety of applications, but digital hardware design must care about operations to use or it will become too complex that realization becomes too expensive that is useless [2].

Most difficult part on designing fuzzy systems is to tune the system to work properly and there is no rule to do this. Parameterization is used to tune fuzzy systems. Usually parameterization is realized over fuzzy sets adjusting membership functions shape but after tuning, system can result different from the one modeled by expert leading to knowledge lost and an improper system functioning. A good option is to use parameterization over fuzzy operations as parametric conjunctions and disjunctions [3-5], because they do not modify membership functions and tuning is realized at run time.

Most commonly used operations for digital hardware are minimum and maximum, but algebraic product and probabilistic sum are also used even when these operations consume many resources because of multiplication on its definition.

In [6] proposed the method for realization of parametric classes of fuzzy conjunctions and disjunctions based on operations that are suitable for an efficient hardware implementation. In this paper we present a new effective hardware solution for implementation of different parametric conjunctions based on the method proposed on [7] using t-norm and t-conorm circuits that allow selection of different operations as min, max, Lukasiewicz and Drastic. Resultant circuits are able to realize at most 27 different parametric operations with just one generator.

The paper has the following structure. Section 2 provides the basic definitions of fuzzy conjunction and disjunction operations considering the methods for its generation. In Section 3 digital representation of fuzzy operations, membership values and generators are considered. In Section 4 simple parametric families of fuzzy conjunctions defined as on [7] are considered and their digital hardware implementation is presented on section 5 along with obtained results. Finally conclusions, results and future directions of research are discussed.

2 Basic Definitions

In binary logic, the set of true values is defined as the set $L = \{0, 1\}$ with only two elements. Negation $\neg$, conjunction $\wedge$ and disjunction $\vee$ operations are defined as:

$$-0 = 1, \qquad \neg1 = 0, \tag{1}$$

$$0 \wedge 0 = 0,\ 1 \wedge 0 = 0,\ 0 \wedge 1 = 0, \qquad 1 \wedge 1 = 1, \tag{2}$$

$$0 \vee 0 = 0, \quad 1 \vee 0 = 1,\ 0 \vee 1 = 1,\ 1 \vee 1 = 1. \tag{3}$$

In fuzzy logic, the set of true values (membership values) contains continuum number of elements $L = [0, 1]$ and negation, conjunction and disjunction operations are defined as functions $N{:}L \rightarrow L$, $T{:}L \times L \rightarrow L$ and $S{:}L \times L \rightarrow L$. Let I be the maximal membership value I=1, then binary operations above turn to the following form:

$$N(0) = I, \qquad N(I) = 0 \tag{4}$$

$$T(0, 0) = 0, \quad T(I, 0) = 0, \quad T(0, I) = 0, \quad T(I, I) = I \tag{5}$$

$$S(0, 0) = 0, \quad S(I, 0) = I, \quad S(0, I) = I, \quad S(I, I) = I \tag{6}$$

Most common definitions for these operations include monotonicity properties.
Fuzzy involutive negation is defined by the following axioms:

$$N(x) \geq N(y), \quad if \ x \leq y, \quad \text{(antimonotonicity)} \tag{7}$$

$$T(x, y) \leq T(u, v), \quad S(x, y) \leq S(u, v), \quad if \ x \leq u, y \leq v. \quad \text{(monotonicity)} \tag{8}$$

Generally fuzzy negation is defined as a function $N : L \rightarrow L$ that satisfy (4) and (7).
Axioms (5), (6), (8) define a skeleton for fuzzy conjunction and disjunction operations. These operations satisfy the following conditions:

$$T(x, 0) = 0, \ T(0, y) = 0 \tag{9}$$

$$S(x, 1) = I, \ S(I, y) = I \tag{10}$$

In [4] functions T and S that satisfy (8), (9), (10) are called pseudo-conjunctions and pseudo-disjunctions respectively. In [5], fuzzy conjunction and disjunction operations are defined by (8) and following conditions:

$$T(x,1) = x, \quad T(1,y) = y, \quad \text{(boundary conditions)} \tag{11}$$

$$S(x,0) = x, \quad S(0,y) = y, \quad \text{(boundary conditions)} \tag{12}$$

Fuzzy conjunctions and disjunctions satisfy (5), (6), (9) and (10). t-norms and t-conorms are defined by (8), (11), (12) and commutativity and associativity axioms [8]. Lets consider the following pairs of simplest t-norms and t-conorms to be considered as *basic conjunction and disjunction operations*:

$$T_M(x,y) = min\{x,y\}, \quad \text{(minimum)}, \qquad S_M(x,y) = max\{x,y\}, \quad \text{(maximum)} \tag{13}$$

$$T_L(x,y)= max\{x+y -I, 0\}, \quad S_L(x,y)= min\{x+y, I\}, \qquad \text{(Lukasiewicz)} \tag{14}$$

$$T_D(x, y) = \begin{cases} 0, & if \ (x,y) \in [0,I) \times [0,I) \\ min(x, y), & otherwise \end{cases}, \quad \text{(drastic product)} \tag{15}$$

$$S_D(x, y) = \begin{cases} I, & if \ (x,y) \in (0,I] \times (0,I] \\ max(x, y), & otherwise \end{cases} \quad \text{(drastic sum)} \tag{16}$$

These operations are considered as because they have efficient hardware implementation [6] and they will be used for generation of parametric classes of conjunction and disjunction operations.

Pairs of *t*-norms and *t*-conorms considered above can be obtained one from another by means of involution $N(x) = I - x$ as follows:

$$S(x,y) = N(T(N(x),N(y))), \qquad \qquad T(x,y) = N(S(N(x),N(y))) \tag{17}$$

It is shown in [4] that fuzzy conjunctions and disjunctions satisfy the following ine-
qualities:

$$T_D(x,y) \leq T(x,y) \leq T_M(x,y) \leq S_M(x,y) \leq S(x,y) \leq S_D(x,y) \tag{18}$$

Several methods of generation of simple parametric fuzzy operations suitable for
tuning in fuzzy systems have been proposed in [3-5]. These methods are based on the
formula:

$$T(x,y) = T_2(T_1(x,y),S(x,y)), \tag{19}$$

where T_2 and T_1 are conjunctions and S is a pseudo-disjunction. Suppose h, g_1, g_2:
$L{\rightarrow}L$ are non-decreasing functions called generators. Pseudo-disjunction S can be
generated by means of generators and other pseudo-disjunctions as follows:

$$S(x,y) = S_1(g_1(x),g_2(y)), \tag{20}$$

$$S(x,y) = g_1(S_1(x,y)), \tag{21}$$

$$S(x,y) = S_2(S_1(x,y),h(y)), \tag{22}$$

where g_1 and g_2 satisfy conditions: $g_1(I) = g_2(I) = I$. As pseudo-disjunctions S in (19)
basic disjunctions or pseudo-disjunctions obtained from basic disjunctions by recursive application of (20)-(22) can be used.

Classes of parametric conjunctions suitable for tuning in fuzzy systems were proposed in [4-5] but they do not have efficient hardware implementation because of
operations used. In the present paper we solve the problem of construction of wide
class of fuzzy parametric conjunction and disjunction operations with effective hardware implementation as presented on [6-7] by the usage of dedicated circuit blocks to
realize basic t-norms and t conorms, these circuits allow us to select among basic
conjunctions and disjunctions as on (13)-(16). This way at most 27 different parametric conjunctions can be realized by final circuit implementation.

3 Digital Representation of Fuzzy Operations, Membership Values and Generators

Suppose that m bits are used to represent digital membership values. Then different
membership values are represented by 2^m numbers over the set $L_D = \{0,1,2,..., 2^m-1\}$.
Denote the maximal membership value $I = 2^m-1$ to represent full membership value
corresponding to the value 1 in traditional fuzzy sets as on section 2, where all definitions and properties of fuzzy operations presented can be transformed into the digital
case by replacing the set of membership values $L = [0,1]$ by $L_D = \{0,1,2,..., 2^m-1\}$ and
maximal membership value 1 by I. For graphical representation of digital generators
and fuzzy operations $m = 5$ bits is used, with the following set of digital membership
values $L_D = \{0, 1, 2, ..., 30, 31\}$ and maximal membership value $I = 31$. Simple t-norm
and t-conorm operations on digital representation are presented on Fig. 2.

To create parametric conjunctions as on (19)-(22) is required to have generators
that use only basic functions on their definition must be incorporate. Simplest generators depending on one parameter p that changes from 0 to I were proposed on [6].

From these generators we select one to realize tests with different t-norms and t-conorms, note that following two generators differ only on output data.

$$g(x) = \begin{cases} 0, & if \quad x < p \\ I, & otherwise \end{cases}, \qquad \qquad <0, I> \qquad (23)$$

$$g(x) = \begin{cases} 0, & if \quad x \le p \\ p, & otherwise \end{cases} \qquad \qquad <0, p> \qquad (24)$$

Fig. 1. shows these generators in digital representation with 5 bits for parameter value $p = 6$ corresponding to value 0.2 in interval [0,1] of true values.

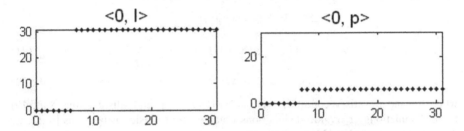

Fig. 1. Generators with parameter $p= 3$ in digital representation with 5 bits

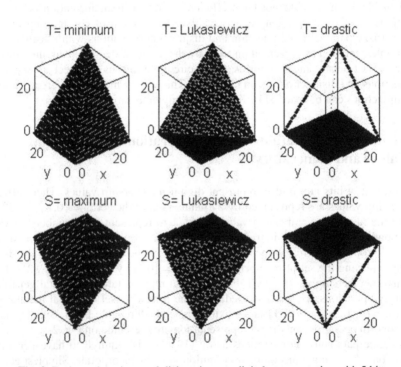

Fig. 2. Basic conjunctions and disjunctions on digital representation with 5 bits

4 Method of Generation of Fuzzy Parametric Conjunctions

Theorem 1. [7] Suppose T_1 and T_2 are conjunctions, S is a disjunction, and g is a generator then a conjunction T is defined by the formula

$$T(x,y) = min(T_1(x,y),S(T_2(x,y),g(y))),\qquad(25)$$

Proof. Monotonicity of T follows from the monotonicity property of functions in the right side of (25). From (25) we obtain: $T(I,y)= min(T_1(I,y),S(T_2(I,y),g(y))) = min(y,S(y,g(y)))$. From (12) and (8) it follows:

$$S(x,y) \geq S(x,0) = x, \qquad\qquad S(x,y) \geq S(0,y) = y. \qquad(26)$$

Hence $S(y,g(y)) \geq y$, $min(y,S(y,g(y))) = y$ and $T(I,y)= y$. $T(x,I)= min(T_1(x,I), S(T_2(x,I),g(I))) = min(x,S(x,g(I)))$. From (26) it follows $S(x,g(I)) \geq x$ and $T(x,I)= min(x,S(x,g(I))) = x$.

The method (25) generalizes the method proposed in [6], where instead of $g(y)$ a constant s was used.

We will write $T_1 \leq T_2$ if $T_1(x,y) \leq T_2(x,y)$ for all $x,y \in [0,I]$.

Proposition 2. For specific conjunctions T_1 and T_2 a conjunction T in (25) is reduced as follows:

$$\text{if } T_1 \leq T_2 \text{ then } T = T_1; \qquad(27)$$

Proof. From (26) it follows that $S(T_2(x,y),g(y)) \geq T_2(x,y)$ and if $T_2 \geq T_1$ then (25) gives $T(x,y) = T_1(x,y)$.

So, if we want to use (25) for generation of new fuzzy conjunctions we need to avoid to use conjunctions T_1 and T_2 such that $T_1 \leq T_2$. For example, from Proposition (5) and (18) it follows that for $T_1 = T_D$ and $T_2 = T_M$ a conjunction T in (25) is reduced as follows:

$$\text{if } T_1 = T_D \text{ then } T = T_D,$$

$$\text{if } T_2 = T_M \text{ then } T = T_1.$$

If we want to use in (25) basic t-norms (13)-(16) then from (27)-(29) it follows that for obtaining new conjunctions from (25) we can use the following pairs of conjunctions: $(T_1 = T_M , T_2 = T_L)$, $(T_1 = T_M , T_2 = T_D)$, $(T_1 = T_L , T_2 = T_D)$. Some examples of parametric conjunctions obtained from (25) are presented on (28) and (29).

$$T(x,y) = min(T_M(x,y),S_M(T_L(x,y),g(x))), \qquad(28)$$

$$T(x,y) = min(T_M(x,y),S_L(T_L(x,y),g(x))), \qquad(29)$$

Fig. 3 is an example of parametric fuzzy conjunction obtained by (28) using generator (24) with parameter $p=7$, and $\mathrm{T}_1 = minimum$, $\mathrm{T}_2 = Lukasiewicz$ conjunction, $S = maximum$.

5 Hardware Implementation of Fuzzy Parametric Conjunctions

To realize digital hardware implementation of fuzzy parametric conjunction (25) that allows to realize the different cases as on (28) and (29) and more, two circuits to realize basic t-norm and t-conorm are used. Each of these circuits realizes operations as on (13)-(16), what is just left to construct is generator as on (24). The circuits were realized using a Spartan 3E 3S500EFG320-5 FPGA using basic logical gates and MSI as comparator, adder and subtractor [1-2], with an 8 bit bus width for all functions.

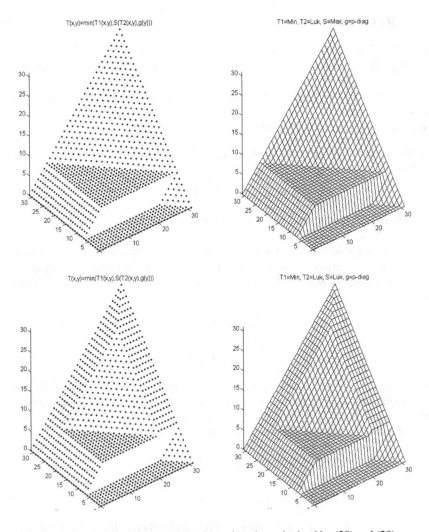

Fig. 3. Examples of fuzzy parametric conjunctions obtained by (28) and (29)

On Fig. 4. circuit realized for generator <0, p> is presented. Is constructed using one comparator and one two 8 bit channel multiplexer; x, p inputs are compared to drive zero as output if $x<p$ or another value if otherwise, this is to provide versatility to realize <0, I>, <0, p> and <p, I>.

t-norm circuit is as on Fig.5. T_SEL1 and T_SEL0 are used to select operation according to binary values 00 for minimum, 01 for Lukasiewicz and 10 for Drastic, 11 is not used. Bus width is 8 bit, but is can be set to use lower bit number by IN_I that corresponds to I value. IN_X and IN_Y corresponds to inputs x, y.

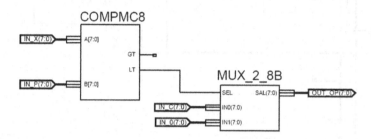

Fig. 4. Hardware implementation of generator <0, p>

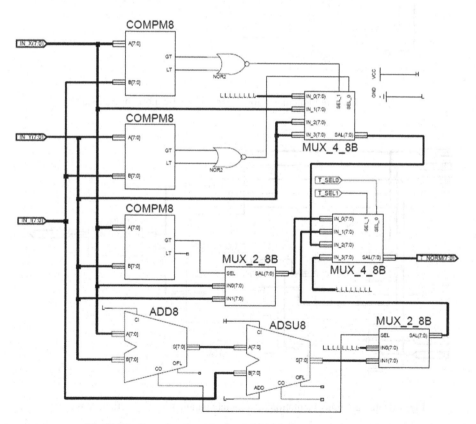

Fig. 5. Hardware implementation of circuit for t-norm operations

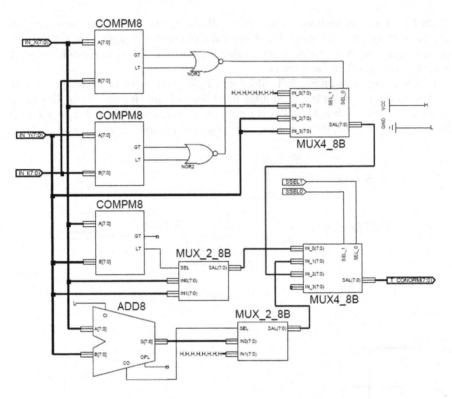

Fig. 6. Hardware implementation of circuits for t-conorm operations

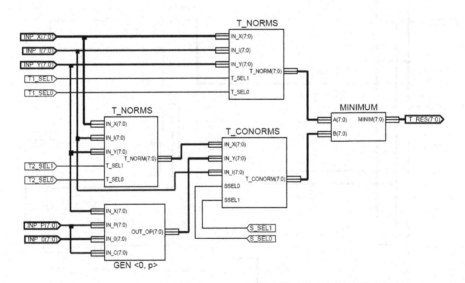

Fig. 7. Hardware implementation of complete parametric conjunction from (25)

t-conorm circuit is as on Fig. 6. S_SEL1 and S_SEL0 are used to select operation according to binary values 00 for maximum, 01 for Lukasiewicz and 10 for Drastic, 11 is not used. Bus width is 8 bit, but is can be set to use lower bit number by IN_I that corresponds to *I* value. IN_X and IN_Y corresponds to inputs *x, y*.

On Fig. 7. digital hardware realization for parametric conjunction (25) is presented using blocks mentioned before. T1 from operation is selected by T1_SEL1 and T1_SEL0, T2 is selected trough T2_SEL1 and T2_SEL0, and S is selected by S_SEL1 and S_SEL0. INP_X, INP_I, INP_Y, INP_P and INP_0 correspond to *x, I, y, p* and *zero*.

Simulations are presented on Fig. 8. to Fig. 11. where different cases are presented by selection bits for T1, T2 and S using different combinations for selection bits

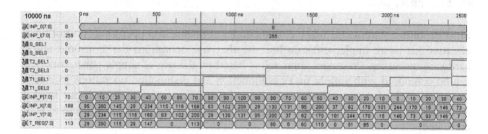

Fig. 8. Simulation for 0 to 2500 nanoseconds

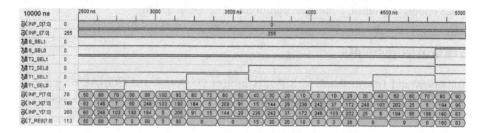

Fig. 9. Simulation for 2500 to 5000 nanoseconds

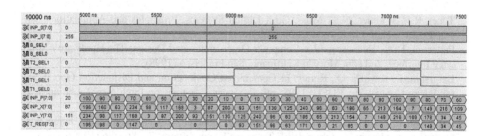

Fig. 10. Simulation for 5000 to 7500 nanoseconds

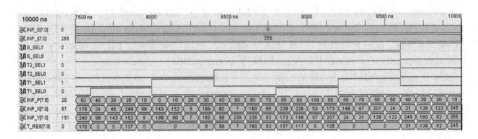

Fig. 11. Simulation for 7500 and 10000 nanoseconds

6 Conclusions

Most of known parametric fuzzy conjunction and disjunction operations do not have effective hardware implementation because they use operations that consume much hardware resources. Presented approach is based on the use of simple generators that can be composed with simple fuzzy conjunction and disjunction operations to create several classes of parametric fuzzy conjunctions. Effective digital hardware implementation of fuzzy systems with parametric conjunction and disjunction operations is presented.

Obtained results can be used in digital hardware implementation of fuzzy systems with parametric conjunctions and disjunctions extending possibilities of design of flexible embedded fuzzy systems that can be used as components of applied intelligent systems in control, pattern recognition and decision making.

Acknowledgements

Authors would like to thank to SIP-IPN project 20090668 and CONACYT project I0110/127/08 Mod. Ord. 38/08, for their financial support that made this research possible.

References

1. Tocci, R.J., Widmer, N.S., Moss, G.L.: Digital Systems: Principles and Applications, 9th edn. Prentice-Hall, Englewood Cliffs (2003)
2. Patterson, D.A., Hennessy, J.L.: Computer Organization and Design: The Hardware / Software Interface, 2nd edn. Morgan Kaufmann, San Francisco (1998)
3. Batyrshin, I., Kaynak, O., Rudas, I.: Generalized Conjunction and Disjunction Operations for Fuzzy Control. In: EUFIT 1998, 6th European Congress on Intelligent Techniques & Soft Computing, Verlag Mainz, Aachen, vol. 1, pp. 52–57 (1998)
4. Batyrshin, I., Kaynak, O.: Parametric Classes of Generalized Conjunction and Disjunction Operations for Fuzzy Modeling. IEEE Trans. on Fuzzy Systems 7, 586–596 (1999)
5. Batyrshin, I., Kaynak, O., Rudas, I.: Fuzzy Modeling Based on Generalized Conjunction Operations. IEEE Trans. on Fuzzy Systems 10, 678–683 (2002)

6. Batyrshin, I.Z., Hernández Zavala, A., Camacho Nieto, O., Villa Vargas, L.: Generalized Fuzzy Operations for Digital Hardware Implementation. In: Gelbukh, A., Kuri Morales, Á.F. (eds.) MICAI 2007. LNCS (LNAI), vol. 4827, pp. 9–18. Springer, Heidelberg (2007)
7. Rudas, I.J., Batyrshin, I., Hernández Zavala, A., Camacho Nieto, O., Villa Vargas, L., Horváth, L.: Generators of fuzzy operations for hardware implementation of fuzzy systems. In: Gelbukh, A., Morales, E.F. (eds.) MICAI 2008. LNCS (LNAI), vol. 5317, pp. 710–719. Springer, Heidelberg (2008)
8. Klement, E.P., Mesiar, R., Pap, E.: Triangular Norms. Kluwer, Dordrecht (2000)

Fuzzy Logic for Combining Particle Swarm Optimization and Genetic Algorithms: Preliminary Results

Fevrier Valdez, Patricia Melin, and Oscar Castillo

Tijuana Institute of Technology, Tijuana BC. México
epmelin@hafsamx.org, ocastillo@hafsamx.org, fvaldez@ieee.org

Abstract. We describe in this paper a new hybrid approach for mathematical function optimization combining Particle Swarm Optimization (PSO) and Genetic Algorithms (GAs) using Fuzzy Logic to integrate the results. The new evolutionary method combines the advantages of PSO and GA to give us an improved PSO+GA hybrid method. Fuzzy Logic is used to combine the results of the PSO and GA in the best way possible. The new hybrid PSO+GA approach is compared with the PSO and GA methods with a set of benchmark mathematical functions. The new hybrid PSO+GA method is shown to be superior than the individual evolutionary methods.

Keywords: PSO, GA, Fuzzy Logic.

1 Introduction

We describe in this paper a new evolutionary method combining PSO and GA, to give us an improved PSO+GA hybrid method. We apply the hybrid method to mathematical function optimization to validate the new approach. Also in this paper the application of a Genetic Algorithm (GA) [12] and Particle Swarm Optimization (PSO) [5] for the optimization of mathematical functions is considered. In this case, we are using the Rastrigin's, Rosenbrock's, Ackley's, Sphere's and Griewank's functions [4][13] to compare the optimization results between a GA, PSO and PSO+GA.

The paper is organized as follows: in section 2 a description about the Genetic Algorithms for optimization problems is given, in section 3 the Particle Swarm Optimization is presented, in section 4 we can appreciate the proposed PSO+GA method and the fuzzy system, in section 5 we can appreciate the simulation results that were obtained for this research, in section 6 we can appreciate a comparison between GA, PSO and PSO+GA, and in section 7 we can see the conclusions reached after the study of the proposed evolutionary computing methods.

2 Genetic Algorithms

John Holland, from the University of Michigan initiated his work on genetic algorithms at the beginning of the 1960s. His first achievement was the publication of *Adaptation in Natural and Artificial System* [1] in 1975.

A. Hernández Aguirre et al. (Eds.): MICAI 2009, LNAI 5845, pp. 444–453, 2009.

He had two goals in mind: to improve the understanding of natural adaptation process, and to design artificial systems having properties similar to natural systems [7].

The basic idea is as follows: the genetic pool of a given population potentially contains the solution, or a better solution, to a given adaptive problem. This solution is not "active" because the genetic combination on which it relies is split between several subjects. Only the association of different genomes can lead to the solution.

Holland's method is especially effective because it not only considers the role of mutation, but it also uses genetic recombination, (crossover) [7]. The crossover of partial solutions greatly improves the capability of the algorithm to approach, and eventually find, the optimal solution.

The essence of the GA in both theoretical and practical domains has been well demonstrated [12]. The concept of applying a GA to solve engineering problems is feasible and sound. However, despite the distinct advantages of a GA for solving complicated, constrained and multiobjective functions where other techniques may have failed, the full power of the GA in application is yet to be exploited [12] [4]. In figure 1 we show the reproduction cycle of the Genetic Algorithm.

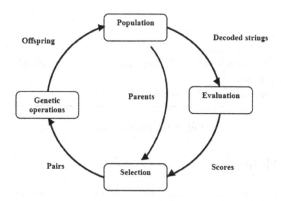

Fig. 1. The Reproduction cycle

3 Particle Swarm Optimization

Particle swarm optimization (PSO) is a population based stochastic optimization technique developed by Eberhart and Kennedy in 1995, inspired by social behavior of bird flocking or fish schooling [2]. PSO shares many similarities with evolutionary computation techniques such as Genetic Algorithms (GA) [5]. The system is initialized with a population of random solutions and searches for optima by updating generations. However, unlike GA, the PSO has no evolution operators such as crossover and mutation. In PSO, the potential solutions, called particles, fly through the problem space by following the current optimum particles [1].

Each particle keeps track of its coordinates in the problem space, which are associated with the best solution (fitness) it has achieved so far (The fitness value is also stored). This value is called *pbest*. Another "best" value that is tracked by the particle swarm optimizer is the best value, obtained so far by any particle in the neighbors of the particle. This location is called *lbest*. When a particle takes all the population as its topological neighbors, the best value is a global best and is called *gbest*.

The particle swarm optimization concept consists of, at each time step, changing the velocity of each particle toward its *pbest* and *lbest* locations (local version of PSO). Acceleration is weighted by a random term, with separate random numbers being generated for acceleration toward *pbest* and *lbest* locations. In the past several years, PSO has been successfully applied in many research and application areas. It is demonstrated that PSO gets better results in a faster, cheaper way compared with other methods [2]. Another reason that PSO is attractive is that there are few parameters to adjust. One version, with slight variations, works well in a wide variety of applications. Particle swarm optimization has been used for a wide range of applications, as well as for specific applications focused on a specific requirement. The pseudo code of PSO is illustrated in figure 2.

```
For each particle
Initialize particle
End
Do
    For each particle
        Calculate fitness value
        If the fitness value is better than the best fitness value
(pBest) in history
            set current value as the new pBest
```

Fig. 2. Pseudocode of PSO

4 PSO+GA Method

The general approach of the proposed method PSO+GA can be seen in figure 3. The method can be described as follows:

1. It receives a mathematical function to be optimized
2. It evaluates the role of both GA and PSO.
3. A main fuzzy system is responsible for receiving values resulting from step 2.
4. The main fuzzy system decides which method to take (GA or PSO)
5. After, another fuzzy system receives the Error and DError as inputs to evaluates if is necessary change the parameters in GA or PSO.
6. There are 3 fuzzy systems. One is for decision making (is called main fuzzy), the second one is for changing parameters of the GA (is called fuzzyga) in this case change the value of crossover (k1) and mutation (k2) and the third fuzzy system is used to change parameters of the PSO(is called fuzzypso) in this case change the value of social acceleration (c1) and cognitive acceleration (c2).
7. The main fuzzy system decides in the final step the optimum value for the function introduced in step 1.
8. Repeat the above steps until the termination criterion of the algorithm is met.

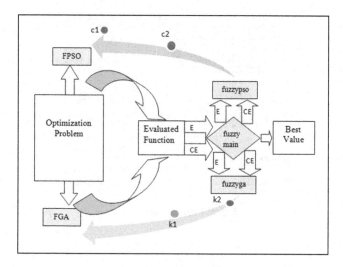

Fig. 3. The PSO+GA scheme

The basic idea of the PSO+GA scheme is to combine the advantage of the individual methods using a fuzzy system for decision making and the others two fuzzy systems to improve the parameters of the GA and PSO when is necessary.

4.1 Fuzzy Systems

As can be seen in the proposed hybrid PSO+GA method, it is the internal fuzzy system structure, which has the primary function of receiving as inputs (Error and DError) the results of the outputs GA and PSO. The fuzzy system is responsible for integrating and decides which is the best results being generated at run time of the PSO+GA. It is also responsible for selecting and sending the problem to the "fuzzypso" fuzzy system when is activated the PSO or to the "fuzzyga" fuzzy system when is activated the GA. Also activating or temporarily stopping depending on the results being generated. Figure 4 shows the membership functions of the main fuzzy

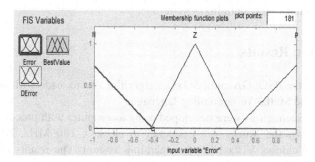

Fig. 4. Fuzzy system membership functions

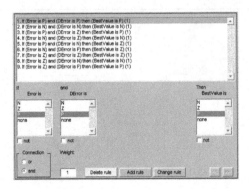

Fig. 5. Fuzzy system rules

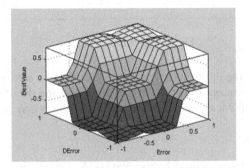

Fig. 6. Surface of the fuzzy system

system that is implemented in this method. The fuzzy system is of Mamdani type and the defuzzification method is the centroid. The membership functions are triangular in the inputs and outputs as is shown in the figure 4. The fuzzy system has 9 rules, for example one rule is if error is P and DError is P then best value is P (see figure 5). Figure 6 shows the surface corresponding for this fuzzy system. The other two fuzzy systems are similar to the main fuzzy system.

5   Simulation Results

Several tests of the PSO, GA and PSO+GA algorithms were made with an implementation done in the Matlab programming language.

All the implementations were developed using a computer with processor AMD turion X2 of 64 bits that works to a frequency of clock of 1800MHz, 2 GB of RAM Memory and Windows Vista Ultimate operating system. The results obtained after applying the GA, PSO and PSO+GA to the mathematical functions are shown in tables 1, 2, 3, 4 and 5:

The parameters used in Tables 1, 2 and 3 are:

POP= Population size
BEST= Best Fitness Value
MEAN= Mean of 50 tests
   k1= %Crossover
   k2= %Mutation
   c1= Cognitive acceleration
   c2= Social acceleration

5.1 Simulation Results with the Genetic Algorithm (GA)

From Table 1 it can be appreciated that after executing the GA 50 times, for each of the tested functions, we can see the better results and their corresponding parameters that were able to achieve the global minimum with the method. In figure 7 it can be appreciated the experimental results of table 1 are sumarized. In figure 7 it can be appreciated that the genetic algorithm was not able to find the global minimum for the Ackley's function because the closest obtained value was 2.98.

Table 1. Experimental results with GA

MATHEMATICAL FUNCTION	POP	K1	K2	BEST	MEAN
Rastrigin	100	80	2	7.36E-07	2.15E-03
Rosenbrock	150	50	1	2.33E-07	1.02E-05
Ackley	100	80	2	2.981	2.980
Sphere	20	80	1	3.49-07	1.62E-04
Griewank	80	90	6	1.84E-07	2.552E-05

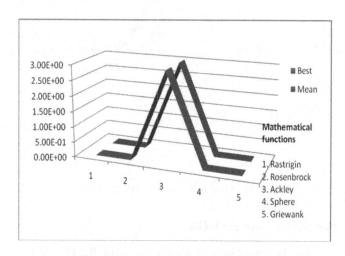

Fig. 7. Experimental results with the GA

5.2 Simulation Results with Particle Swarm Optimization (PSO)

From Table 2 it can be appreciated that after executing the PSO 50 times, for each of the tested functions, we can see the better results and their corresponding parameters that were able to achieve the global minimum with the method. In figure 8 it can be appreciated the experimental results of table 2. In figure 8 it can be appreciated that the particle swarm optimization was not able to find the global minimum for Ackley's function because the closest obtained value was 2.98.

Table 2. Experimental results with the PSO

MATHEMATICAL FUNCTION	POP	c1, c2	BEST	MEAN
Rastrigin	20	1	2.48E-05	5.47
Rosenbrock	40	0.5	2.46E-03	1.97
Ackley	30	0.8	2.98	2.98
Sphere	20	0.3	4.88E-11	8.26E-11
Griewank	40	0.7	9.77E-11	2.56E-02

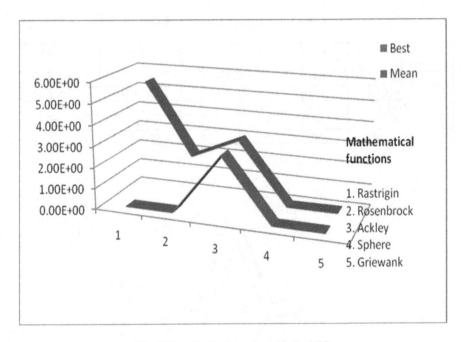

Fig. 8. Experimental results with the PSO

5.3 Simulation Results with PSO+GA

From Table 4 it can be appreciated that after executing the GA 50 times, for each of the tested functions, we can see the better results and their corresponding parameters that were able to achieve the global minimum with the method. The crossover,

mutation, cognitive and social acceleration are fuzzy parameters because are changing every time the fuzzy system decides to modify the four parameters. The population size was 100 particles for PSO and 100 individuals for GA. In figure 9 it can be
appreciated that the PSO+GA was able to find the global minimum for all test functions because the objective value was reached (in all cases was approximately 0).

Table 4. Simulation results with PSO+GA

MATHEMATICAL FUNCTION	BEST	MEAN
Rastrigin	7.03E-06	1.88E-04
Rosenbrock	3.23E-07	3.41E-04
Ackley	1.76E-04	1.84E-03
Sphere	2.80E-09	5.91E-07
Grienwak	6.24E-09	9.03E-07

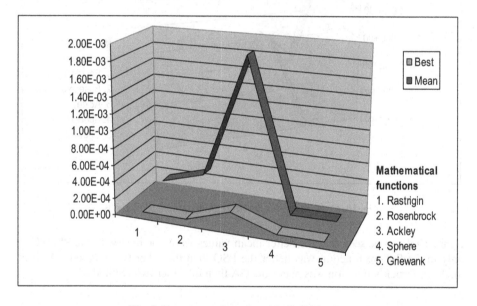

Fig. 9. Experimental results with the PSO+GA

6   Comparison Results between GA, PSO and PSO+GA

In Table 5 the comparison of the results obtained between the GA, PSO and PSO+GA
methods for the optimization of the 5 proposed mathematical functions is shown.
Table 5 shows the results of figure 10, it can be appreciated that the proposed
PSO+GA method was better than GA and PSO, because with this method all
test functions were optimized. In some cases the GA was better but in table 5 and

Table 5. Simulation results with PSO+GA

Mathematical Functions	GA	PSO	PSO+GA	Objective Value
Rastrigin	2.15E-03	5.47	1.88E-04	0
Rosenbrock	1.02E-05	1.97	3.41E-04	0
Ackley	2.980	2.98	1.84E-03	0
Sphere	1.62E-04	8.26E-11	5.91E-07	0
Griewank	2.552E-05	2.56E-02	9.03E-07	0

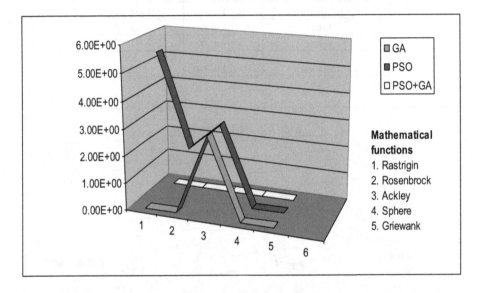

Fig. 10. Comparison results between the proposed methods

figure 10 it can be seen that the better mean values were obtained with the PSO+GA, only in the Sphere function was better the PSO than the other two methods. Also in the Rosenbrock's function was better the GA than the other two methods.

7 Conclusions

After studying the 3 methods of evolutionary computing (GA, PSO and PSO+GA), we reach the conclusion that for the optimization of these 5 mathematical functions, GA and PSO evolved in a similar form, achieving both methods the optimization of 4 of the 5 proposed functions, with values very similar and near the objectives. Also it is possible to observe that even if the GA as the PSO did not achieve the optimization of the Ackley's function, this may have happened because they were trapped in local minima. However we can appreciate that the proposed hybrid method in this paper (PSO+GA) was able of optimize all test functions. Also, in general PSO+GA had the better average optimization values. The advantage the use this method is that it incorporates a fuzzy system to improve the optimization results.

References

1. Angeline, P.J.: Evolutionary Optimization Versus Particle Swarm Optimization: Philosophy and Performance Differences. In: Porto, V.W., Waagen, D. (eds.) EP 1998. LNCS, vol. 1447, pp. 601–610. Springer, Heidelberg (1998)
2. Angeline, P.J.: Using Selection to Improve Particle Swarm Optimization. In: Proceedings 1998 IEEE World Congress on Computational Intelligence, Anchorage, Alaska, pp. 84–89. IEEE, Los Alamitos (1998)
3. Back, T., Fogel, D.B., Michalewicz, Z. (eds.): Handbook of Evolutionary Computation. Oxford University Press, Oxford (1997)
4. Castillo, O., Valdez, F., Melin, P.: Hierarchical Genetic Algorithms for topology optimization in fuzzy control systems. International Journal of General Systems 36(5), 575–591 (2007)
5. Fogel, D.B.: An introduction to simulated evolutionary optimization. IEEE transactions on neural networks 5(1) (January 1994)
6. Eberhart, R.C., Kennedy, J.: A new optimizer using particle swarm theory. In: Proceedings of the Sixth International Symposium on Micromachine and Human Science, Nagoya, Japan, pp. 39–43 (1995)
7. Emmeche, C.: Garden in the Machine. The Emerging Science of Artificial Life, p. 114. Princeton University Press, Princeton (1994)
8. Germundsson, R.: Mathematical Version 4. Mathematical J 7, 497–524 (2000)
9. Goldberg, D.: Genetic Algorithms. Addison Wesley, Reading (1988)
10. Holland, J.H.: Adaptation in natural and artificial system. The University of Michigan Press, Ann Arbor (1975)
11. Kennedy, J., Eberhart, R.C.: Particle swarm optimization. In: Proceedings of IEEE International Conference on Neural Networks, Piscataway, NJ, pp. 1942–1948 (1995)
12. Man, K.F., Tang, K.S., Kwong, S.: Genetic Algorithms: Concepts and Designs. Springer, Heidelberg (1999)
13. Montiel, O., Castillo, O., Melin, P., Rodriguez, A., Sepulveda, R.: Human evolutionary-model: A new approach to optimization. Inf. Sci. 177(10), 2075–2098 (2007)

Optimization of Type-2 Fuzzy Integration in Modular Neural Networks Using an Evolutionary Method with Applications in Multimodal Biometry

Denisse Hidalgo[1], Patricia Melin[2], Guillerrno Licea[1], and Oscar Castillo[2]

[1] School of Engineering UABC University. Tijuana, México
[2] Division of Graduate Studies Tijuana Institute of Technology. Tijuana, México
epmelin@hafsamx.org

Abstract. We describe in this paper a new evolutionary method for the optimization of a modular neural network for multimodal biometry The proposed evolutionary method produces the best architecture of the modular neural network (number of modules, layers and neurons) and fuzzy inference systems (memberships functions and rules) as fuzzy integration methods. The integration of responses in the modular neural network is performed by using type-1 and type-2 fuzzy inference systems.

Keywords: Modular Neural Network, Type-2 Fuzzy Logic, Genetic Algorithms.

1 Introduction

We describe in this paper a new evolutionary method for the optimization of modular neural networks (MNNs) for pattern recognition using fuzzy logic to integrate the responses of the modules.

The main goal of this research is to develop the evolutionary method of the complete optimization of the modular neural network applied in multimodal biometrics and show the results that we have obtained, by a previous study with type-1 and type-2 fuzzy integration.

In this paper we describe the architecture of the evolutionary method and present the simulation results. This paper is organized as follows: Section 2 shows an introduction to the theory of soft computing techniques, section 3 describes the development of the evolutionary method; in section 4 show the simulation results are presented, section 5 shows the conclusions and finally references.

2 Theory of Soft Computing Techniques

2.1 Modular Neural Networks

The neural networks inspired by biological nervous systems, many research, specially brain modelers, have been exploring artificial neural networks, a novel nonalgorithmic approach to information processing. They model the brain as a continuous-time

A. Hernández Aguirre et al. (Eds.): MICAI 2009, LNAI 5845, pp. 454–465, 2009.

non-linear dynamic system in connectionist architectures that are expected to mimic brain mechanisms and to simulate intelligent behavior. Such connectionism replaces symbolically structured representations with distributed representations in the form of weights between a massive set of interconnected neurons. It does not need critical decision flows in its algorithms [3].

In general, a computational system can be considered to have a modular architecture if it can be split into two or more subsystems in which each individual subsystem evaluates either distinct inputs or the same inputs without communicating with other subsystems. The overall output of the modular system depends on an integration unit, which accepts outputs of the individual subsystems as its inputs and combines them in a predefined fashion to produce the overall output of the system. In a broader sense modularity implies that there is a considerable and visible functional or structural division among the different modules of a computational system. The modular system design approach has some obvious advantages, like simplicity and economy of design, computational efficiency, fault tolerance and better scalability [2,7,11,13,16].

2.2 Type-2 Fuzzy Logic

The original theory of Fuzzy logic (FL) was proposed by Lotfi Zadeh [24], more than 40 years ago, and this theory cannot fully handle all the uncertainty that is present in real-world problems. Type-2 Fuzzy Logic can handle uncertainty because it can model and reduce it to the minimum their effects. Also, if all the uncertainties disappear, type-2 fuzzy logic reduces to type-1 fuzzy logic, in the same way that, if the randomness disappears, the probability is reduced to the determinism [20].

Fuzzy sets and fuzzy logic are the foundation of fuzzy systems, and have been developed looking to model the form as the brain manipulates inexact information. Type-2 fuzzy sets are used to model uncertainty and imprecision; originally they were proposed by Zadeh in 1975 and they are essentially "fuzzy-fuzzy" sets in which the membership degrees are type-1 fuzzy sets (See Figure 1) [2, 4, 19, 20, 21, 22, 23].

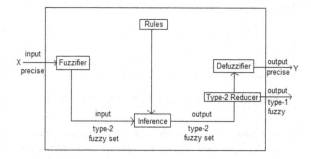

Fig. 1. Basic structure of Type-2 Fuzzy Inference System

2.3 Genetic Algorithms

To use a genetic algorithm (GA), one should represent a solution to the problem as a *genome* (or *chromosome*). The genetic algorithm then creates a population of solutions and applies genetic operators such as mutation and crossover to evolve the solutions in order to find the best one. It uses various selection criteria so that it picks the best individuals for mating (and subsequent crossover). The objective function determines the best individual where each individual must represent a complete solution to the problem you are trying to optimize. Therefore the three most important aspects of using genetic algorithms are: (1) definition of the objective function, (2) definition and implementation of the genetic representation, and (3) definition and implementation of the genetic operators [1].

3 Evolutionary Method Description

Based on the theory described above, a new method for optimization of Modular Neural Networks with Type-2 Fuzzy Integration using an Evolutionary Method with application in Multimodal Biometry is proposed. The goal of the research is the development of a general evolutionary approach to optimize modular neural networks including the module of response integration. In particular, the general method includes optimizing the type-2 fuzzy system that performs the integration of responses in the modular network, as well as the optimization of the complete architecture of the modular neural network, as number of modules, layers and neurons. The purpose of obtaining the optimal architecture is to obtain better recognition rates and improve the efficiency of the hybrid system of pattern recognition. In figure 2 we see the specific points for the general scheme of the evolutionary method.

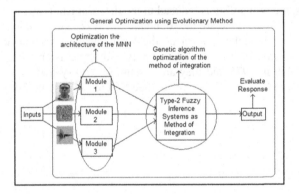

Fig. 2. General scheme of the evolutionary method

We have a modular neural network with fuzzy integration, which is optimized with the proposed evolutionary approach. First, the optimization of the complete architecture of the modular neural network, as number of modules, layers and neurons is performed. A binary hierarchical GA was used to optimize the architecture of the modular neural network with 369 gens; 9 gens are for modules, 36 gens for the layers and 324 gens for the neurons. Later, the genetic algorithm was used to optimize the method of integration, where this method of integration uses type-2 fuzzy inference systems. After obtaining the optimization of the architecture of the MNN and the we developed the new method of Optimization using General Evolutionary Method with

Application in Multimodal Biometry; the method was applied to optimize the modular neural network with type-2 fuzzy integration for multimodal biometrics and a comparison was made with results of type-1 fuzzy logic.

3.1 Optimization of the Fuzzy Systems Based on the Level of Uncertainty

For the Fuzzy Inference Systems optimization (which are the methods of response integration) based in level of uncertainty; the first step is obtain the optimal type-1 Fuzzy Inference System, which allows us to find the uncertainty of their membership functions. In this case we used ε (epsilon) for the uncertainty. For the next step we used three cases to manage the ε using genetic algorithms for all situations.

1. Equal value of uncertainty for all membership function.
2. Different value of uncertainty in each input.
3. Different value of uncertainty for each membership function.

We proposed an index to evaluate the fitness of the fuzzy systems (see equation 1).

$$Index = \frac{N}{1 + n\,\varepsilon} \qquad (1)$$

Where N = Data inside the interval output between the total output data; n = Number of fuzzy systems and ε = Value of increased uncertainty of the membership functions.

4   Simulation Results

Previously we have worked on a comparative study of type-1 and type-2 fuzzy integration for modular neural networks in multimodal biometry, optimized by genetic algorithms. We used different chromosomes structures. In figure 3 we show the General Scheme of the pattern recognition system. The input data used in the modular architecture for pattern recognition are as given in [8, 10].

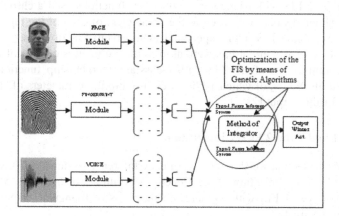

Fig. 3. General Scheme of the pattern recognition system

458   D. Hidalgo et al.

The input data used in the modular architecture for pattern recognition; _Face_:
Images of the faces of 30 different people were used to make the training of the
MNN without noise, we also use 30 images of the face of these same people but
with different gestures, to use them in the training with noise as given in [8]. Both
in face and in fingerprint, the images were obtained from a group of students of
Tijuana Institute of Technology, combined with some others of the ORL data base.
The size of these images is of 268 x 338 pixels with extension .bmp. _Fingerprint_:
Images of the fingerprints of 30 different people were used to make the training
without noise. Then it was added random noise to the fingerprint use them in the
training with noise as given in [8]. _Voice_: For the training of module of voice was
used word spoken by different persons with samples of 30 persons as with the face
and fingerprint. We applied the Mel cepstrals coefficients [10], as preprocessing for
the training in the MNN. We also have to mention that random noise was added to
the voice signals to train the MNN's with noisy signals. In the Modular Neural
Networks for the training phase we are considering three modules, one for the face,
another one for the fingerprint and finally another for the voice, each of the modules has three submodules. It is possible to mention that for each trained module
and each submodule, different architectures were used, that is to say, different number of neurons, layers, etc., and different training methods. The output of the MNN
is a vector that is formed by 30 activations (in this case because the network has
been trained with 30 different people). The fuzzy systems were of Mamdani type
with three inputs and one output, three triangular, trapezoidal or Gaussians type-1
or type-2 membership functions; therefore, we obtained various systems integrators
as fuzzy inference to test the modular neural networks. To obtain a type-1 FIS with
triangular membership functions we used a chromosome of 36-bits, of which 9 bits
were used for the parameters of each of the inputs and for the output. To obtain a
FIS with trapezoidal membership functions used a chromosome of 48-bits, of which
12 bits were used to the parameters of each of the inputs and for the output. To
obtain a FIS with Gaussian membership functions used a chromosome of 24-bits, of
which 6 bits were used to the parameters of each of the inputs and for the output.
To obtain type-2 FIS with triangular membership functions used a chromosome of
72-bits, of which 18 bits were used to the parameters of each of the inputs and for
the output. To obtain FIS with trapezoidal membership functions used a chromosome of 96-bits, of which 24 bits were used to the parameters of each of the inputs
and for the output. To obtain FIS with Gaussians membership functions used a
chromosome of 48-bits, of which 12 bits were used to the parameters of each of the
inputs and for the output [2].

4.1 Type-1 and Type-2 Fuzzy Integration

In the following section we show the best Fuzzy Inference Systems (FIS), which was
obtained with the optimization of the Genetic Algorithms. In figures 4, 5 and 6 we
show the best type-1 FIS with membership functions of triangular, trapezoidal and
Gaussians respectively.

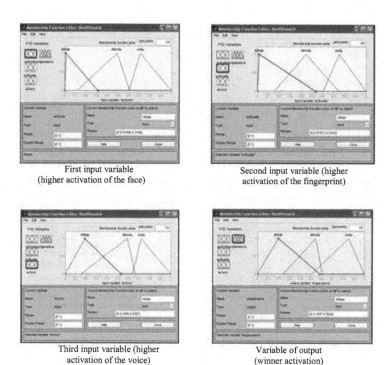

First input variable
(higher activation of the face)

Second input variable (higher
activation of the fingerprint)

Third input variable (higher
activation of the voice)

Variable of output
(winner activation)

Fig. 4. The best Type-1 FIS with triangular membership function

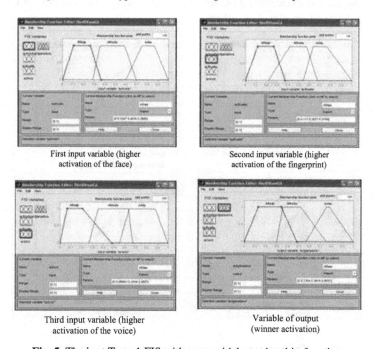

First input variable (higher
activation of the face)

Second input variable (higher
activation of the fingerprint)

Third input variable (higher
activation of the voice)

Variable of output
(winner activation)

Fig. 5. The best Type-1 FIS with trapezoidal membership function

460   D. Hidalgo et al.

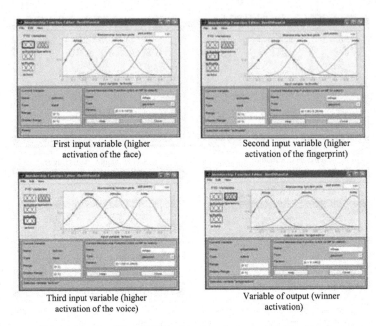

First input variable (higher
activation of the face)

Second input variable (higher
activation of the fingerprint)

Third input variable (higher
activation of the voice)

Variable of output (winner
activation)

Fig. 6. The best Type-1 FIS with Gaussian membership function

Now, in figures 7, 8 and 9 we show the best Type-2 FIS with membership functions of triangular, trapezoidal and Gaussians respectively; which were obtained whit the optimization of Genetics Algorithms.

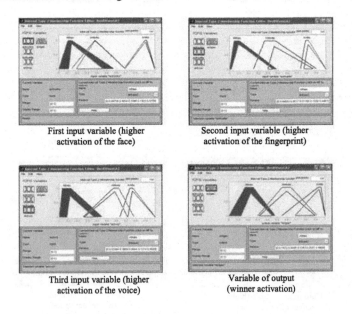

First input variable (higher
activation of the face)

Second input variable (higher
activation of the fingerprint)

Third input variable (higher
activation of the voice)

Variable of output
(winner activation)

Fig. 7. The best Type-2 FIS with triangular membership function

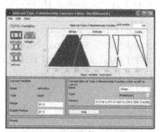

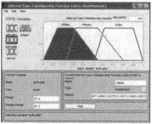

First input variable
(higher activation of the face)

Second input variable (higher
activation of the fingerprint)

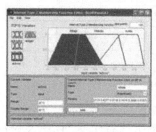

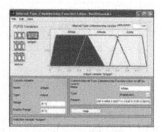

Third input variable (higher
activation of the voice)

Variable of output
(winner activation)

Fig. 8. The best Type-2 FIS with trapezoidal membership function

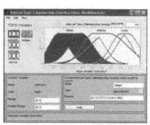

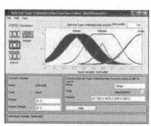

First input variable (higher
activation of the face)

Second input variable (higher
activation of the fingerprint)

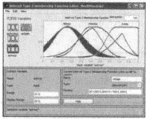

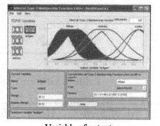

Third input variable
(higher activation of the voice)

Variable of output
(winner activation)

Fig. 9. The best Type-2 FIS with Gaussian membership function

4.2 Comparative Integration with Type-1 and Type-2 Fuzzy Inference Systems

After the modular neural network trainings were obtained, we make the integration of the modules with the type-1 and type-2 optimized fuzzy systems. Next we show the type-1 and type-2 graphics and table with the 20 modular neural network trainings and the percentage of the identification (see Figure 10). We can appreciate that in this case type-2 Fuzzy Logic is better (see table 2) [2].

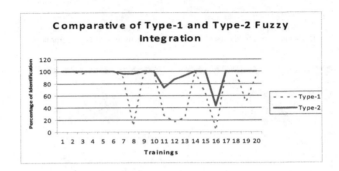

Fig. 10. Comparison of Integration with type-1 and type-2 Fuzzy Systems

Table 1 shows the average percentage of recognition of the type-1 and type-2 Fuzzy Inference Systems that we tested on the last experiment.

Table 2 shows that statistically, the difference between type-1 and type-2 was significant because when applied the t-student test, the value of T =-2.40 and P = 0.025 provide sufficient evidence of difference between the methods, where T is the value of t-student and P is the probability of error in the test. In figure 11 we can see the test t-student graphically.

Table 1. Comparative table of average percentage integration with type-1 and type-2 fuzzy inference systems

Type-1	*Type-2*
Average percentage of identification 73.50 %	Average percentage of identification 94.50 %

Table 2. Obtained values in the t-student test

	No. of Samples	*Mean*	*Stand.Dev.*	*Average Error*	
Type-1	20	0.735	0.367	0.082	**T = -2.40**
Type-2	20	0.945	0.137	0.031	**P = 0.025**

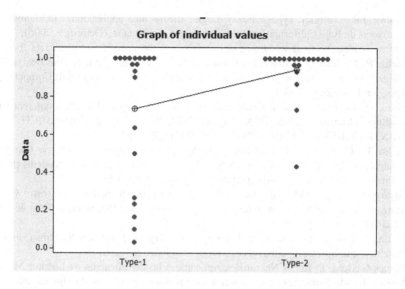

Fig. 11. Graph of individual values in the t-student test

5 Conclusions

In this paper we presented a comparison study between type-1 and type-2 fuzzy systems as integration methods of modular neural networks. The comparison was made using different simulations of modular neural networks trained with the faces, fingerprints and voices of a database of persons for recognition. Simulations results of the modular neural networks with fuzzy systems as integration modules were good. Type-2 Fuzzy Systems are shown to be a superior method for integration of responses in Modular Neural Networks application in multimodal biometrics in this case. We will continue working with results of the optimization of the complete architecture of the modular neural network.

Acknowledgements. We would like to express our gratitude to CONACYT under grant number 175883, UABC University and Tijuana Institute of Technology for the facilities and resources for the development of this research.

References

1. Man, K.F., Tang, K.S., Kwong, S.: Genetic Algorithms, Concepts and Designs. Springer, Heidelberg (1999)
2. Hidalgo, D., Melin, P., Castillo, O.: Type-1 and Type-2 Fuzzy Inference Systems as Integration Methods in Modular Neural Networks for Multimodal Biometry and its Optimization with Genetic Algorithms. Journal of Automation, Mobile Robotics & Intelligent Systems 2(1), 1897–8649 (2008)
3. Jang, J.-S.R., Sun, C.-T., Mizutani, E.: Neuro-Fuzzy and Soft Computing, A Computational Approach to Learning and Machine. Intelligence Prentice Hall, Englewood Cliffs (1997)

4. Castro, J.R.: Tutorial Type-2 Fuzzy Logic: theory and applications. In: Universidad Autónoma de Baja California-Instituto Tecnológico de Tijuana, (October 9, 2006), http://www.hafsamx.org/cis-chmexico/seminar06/tutorial.pdf

5. Melin, P., Castillo, O., Gómez, E., Kacprzyk, J., Pedrycz, W.: Analysis and Design of Intelligent Systems Using Soft Computing Techniques. In: Advances in Soft Computing 41, Springer, Heidelberg (2007)

6. The 2007 International Joint Conference on Neural Networks, IJCNN, Conference Proceedings. Orlando, Florida, USA. August 12-17, IEEE Catalog Number:07CH37922C; ISBN: 1-4244-1380-X, ISSN: 1098-7576, ©2007 IEEE (2007)

7. Melin, P., Castillo, O.: Hybrid Intelligent Systems for Pattern Recognition Using Soft Computing: An Evolutionary Approach for Neural Networks and Fuzzy Systems (Studies in Fuzziness and Soft Computing) (Hardcover - April 29) (2005)

8. Alvarado-Verdugo, J.M.: Reconocimiento de la persona por medio de su rostro y huella utilizando redes neuronales modulares y la transformada wavelet, Instituto Tecnológico de Tijuana (2006)

9. Melin, P., Castillo, O., Kacprzyk, J., Pedrycz, W.: Hybrid Intelligent Systems (Studies in Fuzziness and Soft Computing) (Hardcover - December 20) (2006)

10. Ramos-Gaxiola, J.: Redes Neuronales Aplicadas a la Identificación de Locutor Mediante Voz Utilizando Extracción de Características, Instituto Tecnológico de Tijuana (2006)

11. Mendoza, O., Melin, P., Castillo, O., Licea, P.: Type-2 Fuzzy Logic for Improving Training Data and Response Integration in Modular Neural Networks for Image Recognition. In: Melin, P., Castillo, O., Aguilar, L.T., Kacprzyk, J., Pedrycz, W. (eds.) IFSA 2007. LNCS (LNAI), vol. 4529, pp. 604–612. Springer, Heidelberg (2007)

12. Urias, J., Melin, P., Castillo, O.: A Method for Response Integration in Modular Neural Networks using Interval Type-2 Fuzzy Logic. In: FUZZ-IEEE 2007, Number 1 in FUZZ, London, UK, July 2007, pp. 247–252. IEEE, Los Alamitos (2007)

13. Urias, J., Hidalgo, D., Melin, P., Castillo, O.: A Method for Response Integration in Modular Neural Networks with Type-2 Fuzzy Logic for Biometric Systems. In: Melin, P., et al. (eds.) Analysis and Design of Intelligent Systems using Soft Computing Techniques, 1st edn., June 2007. Number 1 in Studies in Fuzziness and Soft Computing, vol. (1), pp. 5–15. Springer, Heidelberg (2007)

14. Urias, J., Hidalgo, D., Melin, P., Castillo, O.: A New Method for Response Integration in Modular Neural Networks Using Type-2 Fuzzy Logic for Biometric Systems. In: Proc.IJCNN-IEEE 2007, Orlando, USA, August 2007. IEEE, Los Alamitos (2007)

15. Melin, P., Castillo, O., Gómez, E., Kacprzyk, J.: Analysis and Design of Intelligent Systems using Soft Computing Techniques (Advances in Soft Computing) (Hardcover - July 11) (2007)

16. Mendoza, O., Melin, P., Castillo, O., Licea, P.: Modular Neural Networks and Type-2 Fuzzy Logic for Face Recognition. In: Reformat, M. (ed.) Proceedings of NAFIPS 2007, CD Rom, San Diego, June 2007, vol. (1). IEEE, Los Alamitos (2007)

17. Zadeh, L.A.: Knowledge representation in Fuzzy Logic. IEEE Transactions on knowledge data engineering 1, 89 (1989)

18. Zadeh, L.A.: Fuzzy Logic = Computing with Words. IEEE Transactions on Fuzzy Systems 4(2), 103 (1996)

19. Mendel, J.M.: UNCERTAIN Rule-Based Fuzzy Logic Systems, Introduction and New Directions. Prentice Hall, Englewood Cliffs (2001)

20. Mendel, J.M.: Why We Need Type-2 Fuzzy Logic Systems? Article is provided courtesy of Prentice Hall, by Jerry Mendel, May 11 (2001), http://www.informit.com/articles/article.asp?p=21312&rl=1

21. Mendel, J.M.: Uncertainty: General Discussions, Article is provided courtesy of Prentice Hall, by Jerry Mendel, May 11 (2001),
    http://www.informit.com/articles/article.asp?p=21313
22. Mendel, J.M., Bob-John, R.I.: Type-2 Fuzzy Sets Made Simple. IEEE Transactions on Fuzzy Systems 10(2), 117 (2002)
23. Karnik, N., Mendel, J.M.: Operations on type-2 fuzzy sets. In: Signal and Image Processing Institute, Department of Electrical Engineering-Systems. University of Southern California, Los Angeles (May 11, 2000)
24. Zadeh, L.A.: Fuzzy Logic. Computer 1(4), 83–93 (1998)
25. Hidalgo, D., Castillo, O., Melin, P.: Interval type-2 fuzzy inference systems as integration methods in modular neural networks for multimodal biometry and its optimization with genetic algorithms. International Journal of Biometrics 1(1), 114–128 (2008); Year of Publication: 2008, ISSN:1755-8301

st-Alphabets: On the Feasibility in the Explicit Use of Extended Relational Alphabets in Classifier Systems

Carlos D. Toledo-Suárez

Jabatos 150, Paseo de los Ángeles, San Nicolás
de los Garza, Nuevo León 66470, México
ctoledo@exatec.itesm.mx

Abstract. It is proposed a way of increasing the cardinality of an alphabet used to write rules in a learning classifier system that extends the idea of relational schemata. Theoretical justifications regarding the possible reduction in the amount of rules for the solution of problems such extended alphabets (*st*-alphabets) imply are shown. It is shown that when expressed as bipolar neural networks, the matching process of rules over *st*-alphabets strongly resembles a gene expression mechanism applied to a system over {0,1,#}. In spite of the apparent drawbacks the explicit use of such relational alphabets would imply, their successful implementation in an information gain based classifier system (IGCS) is presented.

1 Introduction

After the introduction of learning classifier systems (LCS's) made by Holland [8], some drawbacks in ternary $\{0, 1, \#\}$ representation of classifiers were pointed [18]:

- Their generalization capabilities are unavoidably biased.
- Disjunctions and relations among loci cannot be easily expressed.
- It is difficult for them to capture positionally independent relations between loci.

The modification of traditional ternary alphabet has been a recurrent research topic [11]. Improvements to ternary representation based on the extension of the alphabet employed to write classifiers were proposed [20, 1, 19, 3].

Exploration of new representations for classifiers has recently focused on handling real and interval based input [22], and borrowing ideas for representations from other branches of evolutionary computation [10, 12, 2] and the work on fuzzy classifier systems inspired by fuzzy logic [4, 14]. Despite these representations can allow classifiers to handle a wider range of messages than binary and avoid a fixed correspondence between the position of bits in the classifier condition, they make the theoretical study of latent relational information harder, just as genetic programming algorithms and real coded GA's operation increase the

A. Hernández Aguirre et al. (Eds.): MICAI 2009, LNAI 5845, pp. 466–477, 2009.

complexity a schema theory must have to explain it [15, 16]. In few words, the features that make these approaches different than ternary representation may not be necessary in a preliminary study of relational conditions of classifiers.

In this article an extension of the idea of relational schemata embodied by the introduction of extended alphabets named *st*-alphabets is proposed. The remainder of this article is organized as follows: Section 2 presents relational schemata as the main antecedent to the extended alphabets this article proposes. Section 3 introduces *st*-alphabets as extensions of relational schemata, together with theoretical justifications regarding the possible reduction in the amount of rules for the solution of problems and their expression and matching process as bipolar neural networks. Section 4 describes the main features of the information gain based classifier system (IGCS) designed to experiment with *st*-alphabets. Section 5 reports the results of the implementation of a IGCS explicitly using *st*-alphabets for the imitation of 11 bit multiplexer and XOR problem. Section 6 shows conclusions and future work for the research depicted in this article.

2   Relational Schemata

Collard and Escazut [3] introduced the idea of relational schemata (R-schemata) as a way to increase the expressiveness in the representation of classifiers. A relational schemata is a string over the alphabet $\{s_1, s'_1, \#\}$ where $\#$ is the don't care symbol and s_1 and s'_1 represent complementary variables: if s_1 is bounded to 0 then s'_1 is 1 and vice versa, allowing to express relations between values of different loci, whereas ternary representation expresses only values on different positions on the string, so called P-schemata.

R-schemata do not satisfy the properties [17] claimed as necessary for a useful representation in genetic search, because when they are used to explicitly express classifiers crossing two instances of a R-schema does not produce an instance of that schema, neither 1-point crossover nor uniform crossover properly assort them and they are semi closed for intersection.

These apparent limitations led Collard and Escazut to refuse making an explicit implementation of R-schemata, instead they introduced a head bit in strings over the traditional ternary alphabet to restore respect and proper assortment. Considering strings over the alphabet $\{0, 1, \#, s_1, s'_1\}$ they demonstrated that by using the head bit coding they proposed a so called RP-schemata can be implicitly implemented by a conjunction of strings over $\{0, 1, \#\}$. After implementing this head bit coding in a LCS to face the minimal parity and multiplexer problems they demonstrated the increased abstraction capabilities of their system, but without reaching optimality.

3   *st*-Alphabets

Despite the lack of practical interest in default hierarchies [24] , in this section they are used as a theoretical tool in the introduction of the extended alphabet employed to write classifiers this article focuses on. Practical justifications for it regarding an increase in the expressiveness a LCS can achieve are also considered.

It is possible to map a set of classifiers indicating non contradictory actions into a boolean function, allowing to use the imitation of boolean functions as a learning task to compare the difference between the number of classifiers needed to express the minimal boolean reduction of the function, called the *minimal normal set*, and the number of classifiers in the minimal default hierarchy that shows the same behavior as the boolean function.

The *parsimony index* of a default hierarchy Π_{DH} [23] for the learning of a boolean function is a measure of the savings of rules produced by a default hierarchy:

$$\Pi_{DH} = \frac{\min |\text{normal set}| - \min |\text{default hierarchy}|}{\min |\text{normal set}|} \qquad (1)$$

A pathological problem for which Π_{DH} is minimum consists in having a single bit action whose value corresponds to the exclusive or (XOR) of the n bits in the condition of the classifier, also known as the *parity problem*. In this case the size of the minimal normal set is 2^n, and the one of the minimal default hierarchy is $2^{n-1} + 1$, which leads to $\lim_{n \to \infty} \Pi_{DH}(n) = 0.5$.

For example, for $n = 3$ the XOR's truth table —same as the minimal normal set in this case— has eight rows and the minimal default hierarchy includes the rules

$$
\begin{array}{llllll}
A & \# & \# & \# & / & 0 \\
B & 1 & 0 & 0 & / & 1 \\
C & 1 & 1 & 1 & / & 1 \\
D & 0 & 1 & 0 & / & 1 \\
E & 0 & 0 & 1 & / & 1
\end{array}
$$

synthesized by boolean algebra equation

$$x_1 x_2' x_3' + x_1 x_2 x_3 + x_1' x_2' x_3 + x_1' x_2 x_3'. \qquad (2)$$

Is it possible to reduce the number of rules by exploiting the information present in exception rules? For example the third bit from rules B, C and D equals the AND function applied to the first two bits, but the substitution of $x_3 = x_1 x_2$ in the last equation brings $x_1 x_2 + x_1' x_2 + x_1 x_2'$, that translates into a set of classifiers which does not correspond to 3 bit XOR, because it assigns action 1 for inputs 101, 110 and 011, and does not imply a reduction of rules. This point is very important because it shows how the direct substitution of an identified relation in a rule system does not necessarily improve it. In practical terms, if x_1, x_2 and x_3 were sensors, direct substitution of $x_3 = x_1 x_2$ would imply ignoring sensor x_3. This limitation holds for every translation of equation (2) into classifiers, like messy coding or through S expressions for example [10], as long as it lacks a way to express relations other than direct substitution. If direct substitution was possible then $x_3 = x_1 x_2$ would represent what is known in inductive rule learning as a *feature* [5].

For the implementation of the idea of the AND function between first two bits we can imagine that besides characters $\{0, 1, \#\}$ the allowed alphabet for rules includes characters s_1, s_2 and t_1 [1], playing a similar role as $\#$ but with the difference that instead of meaning "don't care" they require the verification, for the values of the bits where they are located, of the relation $t_1 = s_1 s_2$. With this kind of extended alphabet, from now on to be called *st*-alphabet in this article, it is possible to express the 3 bit XOR with the rules

$$\# \; \# \; \# \; / \; 0$$
$$s_1 \; s_2 \; t_1 \; / \; 1$$
$$0 \; 0 \; 1 \; / \; 1$$

This extension implies it is possible to calculate the 3 bit XOR using two bits in seven of eight cases.

For the case $n = 5$ whose minimal default hierarchy has 17 rules the inclusion of characters s_1, s_1, t_1 and t'_1 reduces it to

$$\# \; \# \; \# \; \# \; \# \; / \; 0$$
$$0 \; 0 \; s_1 \; s_2 \; t_1 \; / \; 1$$
$$0 \; 1 \; s_1 \; s_2 \; t'_1 \; / \; 1$$
$$1 \; 0 \; s_1 \; s_2 \; t'_1 \; / \; 1$$
$$1 \; 1 \; s_1 \; s_2 \; t_1 \; / \; 1$$
$$0 \; 1 \; 0 \; 0 \; 0 \; / \; 1$$
$$0 \; 0 \; 0 \; 0 \; 1 \; / \; 1$$
$$1 \; 1 \; 0 \; 0 \; 1 \; / \; 1$$
$$1 \; 0 \; 0 \; 0 \; 0 \; / \; 1$$

In general for an n bit XOR problem the size of its default hierarchy after including s_1, s_1, t_1 and t'_1 is $2^{n-2} + 1$, and the amount of times the information of the last bit can be expressed in terms of the other two is $2^n - 2^{n-3} = (7/8)2^n$, leading to $\lim_{n \to \infty} \Pi_{DH}(n) = 0.75$.

It is possible to include more characters to a *st*-alphabet. For example the next traditional classifier system, with strings over $\{0, 1, \#\}$, defines the rules linking the characters of a *st*-alphabet

$$s_1 \; s_2 \; s_3 \; s_4$$
$$1 \; 1 \; \# \; \# \; / \; t_1$$
$$\# \; \# \; 1 \; 1 \; / \; t_2 \qquad (3)$$
$$1 \; \# \; 0 \; \# \; / \; t_3$$
$$0 \; 1 \; 0 \; \# \; / \; t_4$$

equivalent to equations $t_1 = s_1 s_2$, $t_2 = s_3 s_4$, $t_3 = s_1 s'_3$ and $t_4 = s'_1 s_2 s'_3$.

3.1   Closure, Respect and Proper Assortment

Considering the example of $s_1 s_2 t_1 \#$ and $\# t_1 s_2 s_1$, where $t_1 = s_1 s_2$, whose intersection is $\{0000, 0001, 1000, 1001, 1111\}$, we can see that for *st*-alphabets:

---

[1] s is for "source" and t for "target".

Respect does not hold. It is possible to get an instance from an schema by crossing two instances of that same schema, but not always. For example crossing 0000 and 0001 only produces offspring that still belongs to $s_1s_2t_1\#$, but 0101 and 1110 do not.

Proper assortment does not hold. For example when crossing 1110 and 0010 their offspring does not belong to the intersection.

Closure does not hold. The intersection of $s_1s_2t_1\#$ and $\#t_1s_2s_1$ cannot be expressed by the st-alphabet. This can be easily verified by the fact that the number of instances of every string can only be one or an even number and the number of instances in the intersection is five. In spite of this a st-alphabet helps a single classifier to implicitly represent disjunctions of classifiers from an alphabet with a lower cardinality, for example $s_1s_2t_1\#01$, where $t_1 = s_1s_2$, is equivalent to the disjunction of $1s_1s_1\#01$ and $0\#0\#01$.

3.2   Matching as Bipolar Neural Network

Smith and Cribbs [21] showed that the matching process of a classifier system can be accomplished by a bipolar neural network where characters $0, 1$ and $\#$ map into values $-1, 1$ and 0 respectively. Having row vector C corresponding to the bipolar representation of the condition part of classifier, and the column vector m corresponding to the bipolar representation of a message, both of length n, C is matched if the equality

$$Cm = f$$

is satisfied, where f is the number of fixed positions (0's and 1's) in C. For an extended alphabet the bipolar representation of a classifier includes the matrix M of size $S+T$ times n. 0's and 1's are represented in C as in the previous case, but if $C_j = 0$ it is necessary to check in column j in M:

$$\text{If } \sum_i |M_{i,j}| = 0, \text{ then } j \text{ represents } \#$$

$$\text{If } M_{i,j} = 1 \text{ and } 1 \le i \le S, \text{ then } j \text{ represents } s_i$$
$$\text{If } M_{i,j} = -1 \text{ and } 1 \le i \le S, \text{ then } j \text{ represents } s_i'$$
$$\text{If } M_{i,j} = 1 \text{ and } S+1 \le i \le S+T, \text{ then } j \text{ represents } t_{i-S}$$
$$\text{If } M_{i,j} = -1 \text{ and } S+1 \le i \le S+T, \text{ then } j \text{ represents } t_{i-S}'$$

with the restrictions $\sum_i |M_{i,j}| \in \{0, 1\}$, and $\sum_i |M_{i,j}| = 0$ if $|C_j| = 1$.

Table 1 shows an example of bipolar representation for the condition of a classifier written with an extended alphabet. To express the matching of this kind of classifier it is employed a Matlab like matrix notation, where logical operator functions act on numbers interpreting 0 as false and non-zero as true and together with relational operators return 0 for false and 1 for true.

Having B as a T times S matrix where ith row corresponds to the bipolar representation of the classifier defining the mapping that links t_i to its sources, a classifier defined by C and M is matched by message m if

$$(Cm == f) \,\&\, \eta \,\&\, \tau = 1 \tag{4}$$

Table 1. Example of bipolar representation for extended alphabet

	t_2	1	s_1	0	t_1'	#	s_2'	
C	0	1	0	-1	0	0	0	
M	0	0	1	0	0	0	0	s_1
	0	0	0	0	0	0	-1	s_2
	0	0	0	0	-1	0	0	t_1
	1	0	0	0	0	0	0	t_2

holds, where

$$\eta = \left(\left(\sum |Mm| \right) == \sum_{i,j} |M_{ij}| \right) \tag{5}$$

$$\alpha = (Mm > 0) - (Mm < 0)$$

$$\beta^\dagger = 2(B\alpha_S == f_B) - (|B||\alpha_S| == f_B)$$

$$\tau = \left(\beta\alpha_T == \sum \beta \& \alpha_T^\dagger \right) \tag{6}$$

α_S is the sub-vector resultant from taking the first S elements from α and α_T the one resultant from taking the last T, f_B is a column vector where its ith element tells the number of fixed (non-#) positions in the classifier that links t_i to its sources, and $\dagger$ is the transpose operator. These equations group all rules needed to interpret any classifier written with an extended alphabet as proposed.

It can be seen that after receiving a message, every classifier filters it according to M to send a new message α to system B, which has the capability to inhibit a possible matching of C interpreted as it was written with the ternary alphabet. This behavior strongly resembles a gene expression mechanism applied on C, opening the possibility to experiment with classifier systems that behave in different ways depending on a dynamic choice of B.

4   An Information Gain Based Classifier System

The imitation of boolean functions was chosen as the learning task to experiment with the extension of alphabet proposed. Main features of the information gain based classifier system (IGCS) employed in the tests (see table 2) are that every classifier lacks an action part, but has a counter for every possible output that is increased for the right one every time the classifier is matched, and that fitness is calculated as the product of the number of non-fixed positions $n - f = a$ of the classifier times the number of characters different to 0 or 1 it has times its information gain expressed by its output counters

$$1 + \sum_{o \in \mathcal{O}} Pr(o) \log_{|\mathcal{O}|} Pr(o) \tag{7}$$

where $\mathcal{O}$ is the set of outputs, $|\mathcal{O}|$ the number of outputs, o an output and $Pr(o)$ its proportion of the total sum of counters. Features of IGCS similar to XCS [25]

Table 2. Information gain based classifier system (IGCS) for the imitation of boolean functions

> (*Define alphabet*)
> Select the number of s's and t's to add to $\{0, 1, \#\}$
> Choose B defining the dependencies between s's and t's
>
> (*Initialize population*)
> Generate random population of classifiers and initialize their
>   output counters and fitness to zero
>
> (*Main cycle*)
> **repeat**
>   Generate random message m
>
>   (*Updating*)
>   **for** every classifier in match set
>   Increase the counter of the right output by 1
>   Update $Pr(o)$, $o \in \mathcal{O}$
>   Update fitness as
> $(n - f)$(no. characters different to $\{0, 1\}$)$IGain$
>   where $IGain = 1 + \sum_{o \in \mathcal{O}} Pr(o) \log_{|\mathcal{O}|} Pr(o)$
>   **end-for**
>
>   (*Covering*)
>   **if** match set is empty **then**
>   Include m in population as a new classifier
>   Change characters in m with a probability $1/4$ by $\#$'s
>   and s's chosen randomly too
>   **end-if**
>
>   (*Applying genetic algorithm*)
>   **if** age of youngest classifier in match set surpasses χ **then**
>   Select a pair of classifiers from match set proportionally
>   to their fitness
>   Apply single point crossover with probability cp
>   to generate a new classifier with fitness equal to
>   the average of its parents and output counters
>   initialized to zero
>   Mutate characters of new classifier with probability mp
>   Substitute least apt classifier in population by the new
>   **end-if**
>
>   (*Determining output*)
>   Select output as the most frequent indicated by the fittest
>   classifier in match set, or by the most frequent in the
>   whole population in the case match set is empty
>
> **until** termination criteria met

are that the genetic algorithm is applied on the set of matched classifiers and *covering*. No deletion mechanism as *subsumption* was implemented.

Some people would argue that it would have been better to directly use *st*-alphabets in the well known XCS. This is an interesting path out of the scope of this article. For the task of imitating boolean functions, information gain is a more direct way than accuracy to measure how plastic an alphabet is to allow rules group multiple sets of messages. In IGCS fittest rules are those that tend to point the same output more often, and the amount of sets of messages linked to the same output that can be grouped by classifiers is proportional to the expressiveness of their alphabet.

5 Experiments and Results

Parameters for IGCS used in all experiments reported in this section are: Size of initial population (N_0) of 10, Crossover probability (*cp*) of 1, Mutation probability (*mp*) of 0.25 and frequency of genetic algorithm (χ) of 10.

Figures 1 and 2 show the experimental results of the imitation of boolean functions performed by IGCS described in the last section. Because a subsumption mechanism [25] was not implemented, besides showing population sizes it is calculated the *effective population* as the number of classifiers in the total population whose fitness is greater or equal to the average of classifiers determining the output in the last 50 cycles.

5.1 General Remarks on Experimental Results

General remarks that can be made based on experimental results presented in this section are:

- The relation between the cardinality of an alphabet and the performance of an IGCS using it seems to be dependent on the problem, but in general performance and information gain with extended alphabets was better or at least comparable to the one of ternary, even in the multiplexer case where the optimal solution can be expressed by a purely positional alphabet. This supports the significance of theory presented in section 3, and is remarkable considering that search space for extended alphabets used is as big as 15^{11}, equivalent to the search size of problems of $\log_3(15^{11}) \approx 27$ bits faced by ternary $\{0, 1, \#\}$ alphabet.
- Population size for *st*-alphabets keeps growing in all cases, without reaching stability, and there seems to be an inverse relation between population size and alphabet's cardinality. On the other side effective population size for *st*-alphabets can reach stability but is bigger or equal to the one of ternary. A justification that can be proposed based on the theory presented in this article is that both effects can in part be expected as a side effect of the increase in alphabet's cardinality allowing the number of syntactically different rules

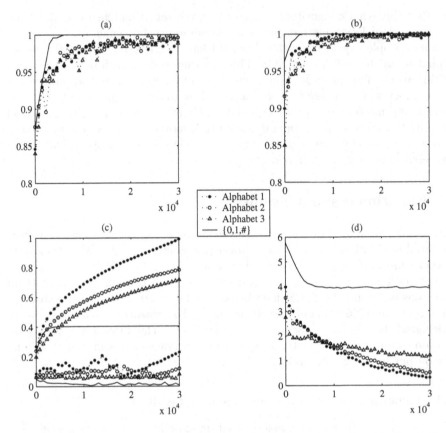

Fig. 1. 11 bit multiplexer: (a) Performance. (b) Information gain. (c) Population size/1000 (upper curves) and effective population/1000 (lower curves). (d) Average number of 0's and 1's in classifiers determining output. Statistics are calculated every 50 cycles and averaged on 10 runs. Alphabets 1, 2 and 3 are extended by $\{s_1, s_2, t_1 = s_1 s_2'\}$, $\{s_1, s_2, s_3, t_1 = s_1 s_2' s_3\}$ and $\{s_1, s_2, t_1 = s_1 s_2', s_3, s_4, t_2 = s_3 s_4'\}$ with a total of 9, 11 and 15 characters respectively.

grow: the same rule can be expressed in multiple ways, and all syntactically different versions of the same rule can survive in a population of IGCS. This side effect is similar but less dramatic to the code growth known as bloat [6] observed in genetic programming, and together with st-alphabets not satisfying respect nor proper assortment nor closure, makes the creation of a deletion mechanism such as subsumption [9] a challenge for future implementations of classifier systems using them.

- Populations with extended alphabets tend to progressively become more abstract, but the relation between the cardinality of an alphabet and the level of abstraction it can reach seems to be dependent on the problem.

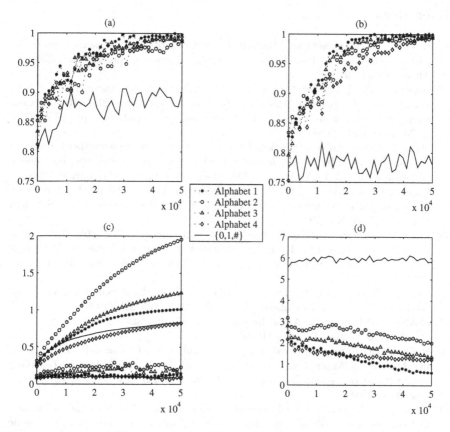

Fig. 2. 11 bit XOR: (a) Performance. (b) Information gain. (c) Population size/1000 (upper curves) and effective population/1000 (lower curves). (d) Average number of 0's and 1's in classifiers determining output. Statistics are calculated every 50 cycles and averaged on 10 runs. Alphabets 1 to 4 are extended by $\{s_1, s_2, s_3, s_4, s_5\}$, $\{s_1, s_2, t_1 = s_1 s_2'\}$, $\{s_1, s_2, s_3, t_1 = s_1 s_2' s_3\}$ and $\{s_1, s_2, t_1 = s_1 s_2', s_3, s_4, t_2 = s_3 s_4'\}$ with a total of 13, 9, 11 and 15 characters respectively.

6   Conclusions

Theoretical justifications regarding the possible reduction in the amount of rules for the solution of problems were shown. It was also shown that when expressed as bipolar neural networks, the matching process of rules over *st*-alphabets strongly resembles a gene expression mechanism applied to a LCS over $\{0,1,\#\}$.

Alphabets proposed were shown to be an extension of the idea of relational schemata introduced by Collard and Escazut [3]. The outstanding performance of their implementation in an information gain based classifier system (IGCS), in spite of the apparent drawbacks the explicit use of such relational alphabets would imply, is a computational demonstration of the advantages implied by the exploitation of latent relational information present in a problem.

References

[1] Booker, L.B.: Representing Attribute-Based Concepts in a Classifier System. In: Rawlins, G.J.E. (ed.) Foundations of Genetic Algorithms, pp. 115–127. Elsevier, Amsterdam (1991)

[2] Browne, W.N., Ioannides, C.: Investigating Scaling of an Abstracted LCS Utilising Ternary and S-Expression Alphabets. In: Proceedings of the 2007 GECCO Conference Companion on Genetic and Evolutionary Computation, pp. 2759–2764. ACM, New York (2007)

[3] Collard, P., Escazut, C.: Relational Schemata: A Way to Improve the Expressiveness of Classifiers. In: Proceedings of the 6th International Conference on Genetic Algorithms (ICGA 1995), pp. 397–404. Morgan Kaufmann, Pittsburgh (1995)

[4] Cordon, O., Gomide, F., Herrera, F., Hoffmann, F., Magdalena, L.: Ten Years of Genetic Fuzzy Systems: Current Framework and New Trends. Fuzzy Sets and Systems 141, 5–31 (2004)

[5] Flach, P.A., Lavrac, N.: The Role of Feature Construction in Inductive Rule Learning. In: Proceedings of the ICML2000 workshop on Attribute-Value and Relational Learning: crossing the boundaries, pp. 1–11. Morgan Kaufmann, San Francisco (2000)

[6] Gelly, S., Teytaud, O., Bredeche, N., Schoenauer, M.: Universal Consistency and Bloat in GP. Revue d'Intelligence Artificielle 20(6), 805–827 (2006)

[7] Goldberg, D.E.: Genetic Algorithms in Search, Optimization and Machine Learning. Addison Wesley, Reading (1989)

[8] Holland, J.H.: Escaping Brittleness: The Possibilities of General-Purpose Learning Algorithms Applied to Parallel Rule-Based Systems. In: Machine Learning: An Artificial Intelligence Approach, 2nd edn., pp. 593–623. Morgan Kaufmann, Los Altos (1986)

[9] Kovacs, T., Bull, L.: Toward a Better Understanding of Rule Initialisation and Deletion. In: Proceedings of the 2007 GECCO Conference Companion on Genetic and Evolutionary Computation, pp. 2777–2780. ACM, New York (2007)

[10] Lanzi, P.L.: Extending the Representation of Classifier Conditions Part II: From Messy Coding to S-Expressions. In: Proceedings of the Genetic and Evolutionary Computation Conference (GECCO 1999), pp. 345–352. Morgan Kaufmann, San Francisco (1999)

[11] Lanzi, P.L., Riolo, R.L.: A Roadmap to the Last Decade of Learning Classifier System Research (from 1989 to 1999). In: Lanzi, P.L., Stolzmann, W., Wilson, S.W. (eds.) IWLCS 1999. LNCS (LNAI), vol. 1813, pp. 33–62. Springer, Heidelberg (2000)

[12] Lanzi, P.L., Rocca, S., Solari, S.: An Approach to Analyze the Evolution of Symbolic Conditions in Learning Classifier Systems. In: Proceedings of the 2007 GECCO Conference Companion on Genetic and Evolutionary Computation, pp. 2795–2800. ACM, New York (2007)

[13] Lanzi, P.L., Wilson, S.W.: Using Convex Hulls to Represent Classifier Conditions. In: GECCO 2006: Proceedings of the 8th Annual Conference on Genetic and Evolutionary Computation, pp. 1481–1488. ACM, New York (2006)

[14] Orriols-Puig, A., Casillas, J., Bernadó-Mansilla, E.: Fuzzy-UCS: Preliminary Results. In: Proceedings of the 2007 GECCO Conference Companion on Genetic and Evolutionary Computation, pp. 2871–2874. ACM, New York (2007)

[15] Poli, R.: General Schema Theory for Genetic Programming with Subtree-Swapping Crossover. In: Miller, J., Tomassini, M., Lanzi, P.L., Ryan, C., Tetamanzi, A.G.B., Langdon, W.B. (eds.) EuroGP 2001. LNCS, vol. 2038, pp. 143–159. Springer, Heidelberg (2001)

[16] Poli, R., McPhee, N.F.: Exact Schema Theory for GP and Variable-length GAs with Homologous Crossover. In: Proceedings of the Genetic and Evolutionary Computation Conference (GECCO 2001), pp. 104–111. Morgan Kaufmann, San Francisco (2001)

[17] Radcliffe, N.: Forma Analysis and Random Respectful Recombination. In: Proceedings of the Fourth International Conference on Genetic Algorithms, pp. 222–229. Morgan Kaufmann, San Mateo (1991)

[18] Schuurmans, D., Schaeffer, J.: Representational Difficulties with Classifier Systems. In: Proceedings of the 3rd International Conference on Genetic Algorithms (ICGA 1989), pp. 328–333. Morgan Kaufmann, San Francisco (1989)

[19] Sen, S.: A Tale of Two Representations. In: Proceedings of the Seventh International Conference on Industrial and Engineering Applications of Articial Intelligence and Expert Systems, pp. 245–254. Gordon and Breach Science Publishers (1994)

[20] Shu, L., Schaeffer, J.: VCS: Variable Classifier System. In: Proceedings of the 3rd International Conference on Genetic Algorithms (ICGA 1989), pp. 334–339. Morgan Kaufmann, San Francisco (1989)

[21] Smith, R.E., Cribbs, H.B.: Combined Biological Paradigms: a Neural, Genetics-Based Autonomous System Strategy. Robotics and Autonomous Systems 22(1), 65–74 (1997)

[22] Stone, C., Bull, L.: For Real! XCS with Continuous-Valued Inputs. Evolutionary Computation 11, 299–336 (2003)

[23] Valenzuela-Rendón, M.: Two Analysis Tools to Describe the Operation of Classifier Systems. PhD Thesis, University of Alabama (1989)

[24] Wilson, S.W.: Bid Competition and Specificity Reconsidered. Complex Systems 2(6), 705–723 (1988)

[25] Wilson, S.W.: Generalization in the XCS Classifier System. In: Genetic Programming 1998: Proceedings of the Third Annual Conference, pp. 665–674. Morgan Kaufmann, San Francisco (1998)

Comparison of Neural Networks and Support Vector Machine Dynamic Models for State Estimation in Semiautogenous Mills

Gonzalo Acuña[1] and Millaray Curilem[2]

[1] Depto. de Ingeniería Informática, Universidad de Santiago de Chile, USACH,
Av. Ecuador 3659, Santiago, Chile
[2] Dpto. De Ingeniería Eléctrica, Universidad de La Frontera, UFRO,
Av. Francisco Salazar 01145, Temuco, Chile
gonzalo.acuna@usach.cl, millaray@ufro.cl

Abstract. Development of performant state estimators for industrial processes like copper extraction is a hard and relevant task because of the difficulties to directly measure those variables on-line. In this paper a comparison between a dynamic NARX-type neural network model and a support vector machine (SVM) model with external recurrences for estimating the filling level of the mill for a semiautogenous ore grinding process is performed. The results show the advantages of SVM modeling, especially concerning Model Predictive Output estimations of the state variable (MSE < 1.0), which would favor its application to industrial scale processes.

Keywords: Support vector regression, neural networks, dynamic systems, semiautogenous mills, NARX models.

1 Introduction

With the increasing complexities of industrial processes it becomes very difficult to build adequate first principle models in order to perform forecasting, optimization or control among other important tasks. To elaborate good dynamic models it is usually necessary a lot of work of different specialists with very deep knowledge of the process. Indeed very often the first principle models obtained are too complex and difficult to identify due to structural problems.

An alternative and fruitful approach to tackle this problem consists in designing appropriate data-driven models. In this sense in the last decade neural networks (NN) have been proven to be a powerful tool for system modeling. Many interesting applications in the field of system identification, model predictive control, observer design and forecasting can be found in the literature [1, 2].

Despite those successful results achieved with neural networks, there still remain unsolved a number of key issues such as: difficulty of choosing the number of hidden nodes, the overfitting problem, the existence of local minima solution, poor generalization capabilities and so on.

A. Hernández Aguirre et al. (Eds.): MICAI 2009, LNAI 5845, pp. 478–487, 2009.

Support Vector Machines (SVM) have shown their usefulness by improving over the performance of different supervised learning methods, either as classification models or as regression models. The SVM have many advantages such as good generalization performance, fewer free parameters to be adjusted and a convex optimization problem to be solved (non-existence of local minima solutions) [3].

Although the success of this technique a number of weak points especially concerning the use of SVM for dynamic regression tasks remain to be solved. In fact SVM has been developed mainly for solving classification and static function approximation problems. Indeed, in the case of dynamic systems almost all the work that has been done concerning Support Vector Regression (SVR) is focused in series-parallel identification methods for NARX (Nonlinear autoregressive with exogenous inputs) modeling [4].

In semiautogenous (SAG) grinding of ores the optimum operating conditions of the mills are strongly dependent on the correct determination of the relevant state variables of the process. Unfortunately, the prevailing conditions in a SAG mill make it difficult to measure these variables on line and in real time, something that is particularly critical in the case of the filling level state variable of the mill.

Software sensors have proved to be powerful tools for determining state variables that cannot be measured directly [5]. A software sensor is an algorithm for on-line estimation of relevant unmeasured process variables. This kind of on-line estimation algorithms can use any model; phenomenological, statistical, artificial intelligence based, or evens a combination of them. In general, for these kinds of sensors it is necessary to have appropriate dynamic models that account for the evolution of the relevant state variables.

In this paper, the performance of two dynamic models, one built up with the use of a NARX-type NN and the other with a NARX-type SVR are compared when acting as estimators of one of the most important state variables for SAG grinding operation.

2 Description of the Semiautogenous Grinding Process

The objective of the concentration processes in mining is to recover the particles of valuable species (copper, gold, silver, etc.) that are found in mineralized rocks. The concentration process is divided into three steps: crushing, grinding and flotation.

In the grinding process, the size of the particles from the crushing continues to be reduced to obtain a maximum granulometry of 180 micrometers (0.18 mm). The grinding process is carried out using large rotary equipment or cylindrical mills in two different ways: conventional or SAG grinding.

In SAG mills the grinding occurs by the falling action of the ore from a height close to the diameter of the mill. The SAG involves the addition of metallic grinding media to the mill, whose volumetric filling level varies from 4% to 14% of the mill's volume, as shown in Figure 1. The ore is received directly from the primary crushing in a size of about 8 inches. This material is reduced in size under the action of the same mineralized material and under the action of numerous 5-inch diameter steel balls [6].

In the operation of these mills the aim is to work under conditions that imply the maximum installed power consumption, but this means working under unstable

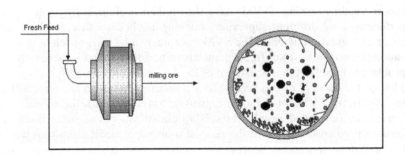

Fig. 1. SAG Mill Scheme – External and Internal look

operating conditions because an increase in the filling level of the mill beyond the point of maximum consumption leads to an overfilling condition (see Fig. 2).

Furthermore, the maximum power value that can be consumed by a SAG mill is not constant and depends mainly on the internal load density, the size distribution of the feed, and the condition of the lining. The filling level of the inner load that corresponds to the maximum power consumption is related to the filling level with grinding media and the motion of the inner load. For that reason the operators of the SAG circuit must try to conjugate these factors in order to achieve first the stabilization of the operation, and then try to improve it [6]. That is why it is important for them to have reliable and timely information on the filling level of the mill.

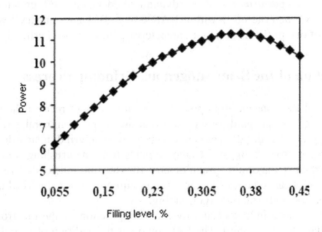

Fig. 2. Typical variation of power consumption versus filling level of SAG mill

3 Support Vector Regression

Support vector machines tackle classification and regression problems by nonlinearly mapping input data into high-dimensional feature spaces, wherein a linear decision surface is designed.

SVR algorithms are based on the results of the statistical theory of learning given by Vapnik [7], which introduces regression as the fitting of a tube of radius v to the data. The decision boundary for determining the radius of the tube is given by a small subset of training examples called Support Vectors (SV).

Vapnik's SVM Regression estimates the values of $\vec{w}$ to obtain the function

$$f(\vec{x}) = (\vec{w} \cdot \vec{x}) + b , \qquad\qquad \vec{w}, \vec{x} \in \mathbf{R}^N, b \in \mathbf{R}, \qquad\qquad (1)$$

by introducing the so called ε-insensitive loss function shown in equation 2:

$$\left| y - f(\vec{x}) \right|_\varepsilon = \max\left\{ 0, \left| y - f(\vec{x}) \right| - \varepsilon \right\} , \qquad\qquad (2)$$

which does not penalize errors smaller than ε>0 (where ε corresponds to a value chosen a priori).

The algorithm is implemented by minimizing the functional risk $\left\| \vec{w} \right\|^2$ to which is added a penalty for leaving points outside the tube (identified by slack variables ξ). In this way the risk function to be minimized is given by equation 3, where C is a constant that determines the trade-off between the complexity of the model and the points that remain outside the tube. Figure 3 shows a geometric interpretation for the case of a linear regression.

$$\theta(\vec{w}, \xi) = \frac{1}{2} \left\| \vec{w} \right\|^2 + \frac{C}{l} \sum_{i=1}^{l} \xi_i , \qquad\qquad (3)$$

minimize

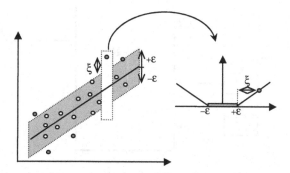

Fig. 3. Geometric interpretation of the SVR in which the regression equation is estimated by means of a tube of radius ε. The trade-off between the complexity of the model and the points left outside the regression tube is controlled by the slack variables ξ. The dark points correspond to the Support Vectors.

To solve a nonlinear regression problem it is sufficient to substitute the dot product between two independent original variables $\vec{x}_i \cdot \vec{x}_j$ by a kernel function $k(\Phi(\vec{x}_i) \cdot \Phi(\vec{x}_j))$. This function carries out the dot product in a space of higher dimension such that it ensures the linearity of the regression function in the new space via a nonlinear map Φ. Several functions may be used as kernel, such as the Gaussian $k(\vec{x}_i, \vec{x}_j) = \exp(-\|\vec{x}_i - \vec{x}_j\|^2 /(2\sigma^2))$ or the polynomial function $k(\vec{x}_i, \vec{x}_j) = (\vec{x}_i \cdot \vec{x}_j)^p$. In this way the nonlinear regression equation is given by equation 5.

$$f(\vec{x}) = k(\vec{w}, \vec{x}) + b \tag{4}$$

4 Series-Parallel Identification (NARX Models) Using SVR Methods

NARX model is the non-linear extension of the lineal ARX model [8] (eq 5)

$$y\ (k) = f(y(k-1),..., y\ (k-n),....., u(k-1),..., u(k-m),...) + w(k) \tag{5}$$

With $\{w(k)\}$ an independent, zero-mean, $\sigma 2$ variance, random variable sequence. The associated predictor is [8] (eq 6):

$$\hat{y}(k) = \psi(y(k-1),..., y\ (k-n),....., u(k-1),..., u(k-m),.....) \tag{6}$$

To identify this predictor a SVR method can be used as shown in Figure 4.

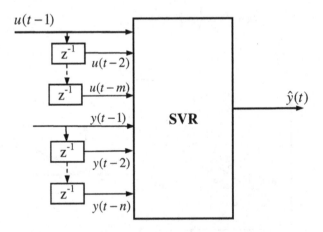

Fig. 4. SVR predictor for NARX type systems

The use of SVR methods to identify NARX type models has been a recent subject of research. San and Ge [9] discussed the use of SVM to nonlinear system modeling. They successfully use standard SVM algorithms with Gaussian radial basis function (RBF) kernel in simulation studies. The same study was performed by Rong et al. [10]. They claimed the importance of a good choice of kernel functions preferring RBF over polynomial and splines kernels. Chacon et al., [11] used SVR for modeling dynamic cerebral autoregulation with better results than neural networks.

5 Series-Parallel Identification Using Neural Networks

To implement the NN models the ideas that are introduced in the work of Nerrand *et al.* [12, 13], where it is shown that a static neural network with external recurrences is equivalent to the neural networks with internal recurrences will be used. The resultant model is similar to that shown in Figure 4 when the SVM is replaced by a static neural network.

The training algorithms for the static network correspond to variations of the back-propagation method, which is carried out by approximations of second order gradient descent methods.

Two currently used methods correspond to algorithms of the quasi-Newton type which avoid the direct calculation of the Hessian matrix. The One Step Secant method was used first; it uses an approximation for calculating the search direction of the best descending slope as the combination of the slopes calculated in previous steps. The second method corresponds to the algorithm of Levenbert-Marquardt, which uses a multiplication of Jacobian matrices to estimate the Hessian matrix.

6 Results

6.1 Data Selection, Identification and Prediction Structures

500 examples were used for training SVR and NN models. Each example has the filling level and the bearing pressure at time t as inputs and the filling level at time t+1 as the output (first order model). In this way the identified model acts as an associated predictor of a NARX type model.

Once identified, each model (SVR and NN) was used for estimating the filling level of the SAG mill one-step-ahead (OSA prediction) and multiple-step-ahead (MPO prediction) (Fig. 5) on an independent estimation set including new 1000 examples. It is usual to consider that MPO predictions are much difficult than OSA predictions and so the model which performs better as an MPO predictor will be consider to be the best [14]. The estimation error is quantified using the MSE index of Matlab.

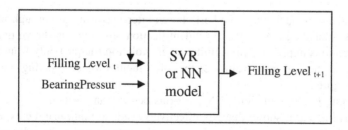

Fig. 5. SVR and NN Estimator acting as MPO predictor for Filling Level variable

6.2 SVR Training

The SVM and Kernel Methods Matlab Toolbox developed at the INSA-Rouen, France [15] was used. As mentioned before, the same training and test sets used for the NN model were use here. The SVR model was implemented with a "Gaussian" kernel. Parameter C was increased by powers of two [16]. The powers took values from -5 to 15, and a more specific research was performed between 3 to 7, where a 0.5 step was used. The best results were obtained for $C=2^4$, what means that the solution requires a low complexity for the model to achieve a good generalization when simulating the MPO prediction.

The parameter of the kernel function, called here *sigma,* also varied by powers of two, with the powers ranging from -5 to 10. The best result was obtained for *sigma=* 6.9, by a more detailed research between 5 to 8, with a 0.1 step. Parameter ε varied between 0 and 1, with 0.01 step. The best results was achieved for ε=0.15. Table 1 shows the MSE results for both the training and the test.

6.3 NN Training

The NN Toolbox of Matlab was used for training purposes with the Levenberg Marquardt optimization algorithm. A one hidden layer architecture with two neurons in the input layer and one neuron in the output layer was chosen. The number of hidden neurons was varied from 1 to 10. For each number of hidden neurons, 5 trainings were performed, beginning from random initial weights in order to avoid bad local minima. The NN with best MSE when acting as an MPO predictor over the estimating set was selected. One of the 6 hidden neurons NN models was then selected as the best model under these criteria (2-6-1 NN model).

6.4 SVR and NN Results

Figure 6 shows the results obtained when using SVR and NN models as OSA and MPO predictors respectively while Table 1 shows the result of the MSE indices for each case.

From these results it is seen that the SVR dynamic model performs better than the NN model when acting as an MPO predictor which is a rather difficult test to any dynamic model. It implies predicting the output variable (filling level) only from its initial value together with experimental input values of the other variable (bearing pressure).

Both dynamic models were the best chosen after following a thorough identification process including the NARX type identification procedure as shown in Figure 4. MPO estimations imply modeling filling level evolution (output) only from its initial condition and the real input values of the other input variable (bearing pressure) as shown in Figure 5.

Concerning OSA predictions, NN performs better than SVR in this case. This is very interesting and it shows that NN effectively tends to minimize the experimental risk while SVR in addition takes into account the structural risk thus allowing better generalization even concerning the very difficult task of performing MPO predictions from an OSA identified model in an independent set of examples.

(a) NN

(b) NN

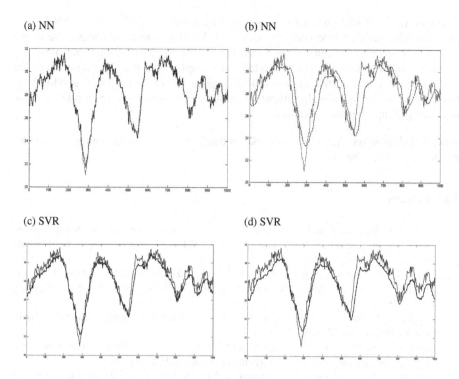

(c) SVR

(d) SVR

Fig. 6. Estimations (dashed lines) and real data (continuous lines) for One-step-ahead (OSA) Predictions of a Neural Network (a) and a SVR dynamic model (c) and multiple predictive outputs (MPO) of a Neural Network (b) and a SVR dynamic model (d)

Table 1. MSE for OSA and MPO predictions using SVR and NN dynamic models

	RNA Model		SVR Model	
	OSA prediction	MPO prediction	OSA prediction	MPO prediction
MSE	0.0505	1.5301	0.4041	0.9906

7 Conclusions

This paper presents SVR and NN dynamic models acting as state estimators of the filling level variable of a semiautogenous grinding process.

For modeling the process SVR and NN acting as NARX associated predictors, that is to say, identified with a series-parallel (OSA) method were used.

Although identified with a OSA structure and hence especially fitted for doing OSA predictions (very good MSE indices) both models were used as MPO predictors. In this case the performance revealed to be quite good, especially concerning the SVR dynamic model that clearly outperforms the NN model.

Future research will go towards implementing parallel identification of SVR models (equivalent to dynamic backpropagation or backpropagation-through-time in the case of NN) in order to improve results for the MPO estimation of state variables.

Performant blackbox dynamic models for complex industrial processes like the SAG process shown in this work could be of great relevance to design good predictive control and estimation algorithms which have proven to be of great relevance for improving complex plant operation.

Acknowledgements. Authors would like to thank partial financial support of Fondecyt under project 1090316.

References

1. Chai, M.-l., Song, S., N-n, L.: A review of some main improved models for neural network forecasting in time series. In: Proceedings of the IEEE Intelligent Vehicles Symposium, vol. (6-8), pp. 866–868 (2005)
2. Espinoza, M., Suykens, J., Belmans, R., de Moor, B.: Electric load forecasting: using kernel-based modeling for nonlinear system identification. IEEE Control Systems Magazine, 43–57 (2007)
3. Schölkopf, B., Smola, A., Williamson, R.C., Bartlett, P.L.: New support vector algorithms. Neural Computation 12, 1083–1121 (2000)
4. Suykens, J.: Nonlinear modeling and support vector machines. In: Proceedings of the IEEE Conf. on Instrum. and Measurement Technol., Budapest, Hungary (2001)
5. Alessandri, A., Cervellera, C., Sanguineti, M.: Design of asymptotic estimators: an approach based on neural networks and nonlinear programming. IEEE Trans. Neural Networks 18, 86–96 (2007)
6. Magne, L., Valderrama, W., Pontt, J.: Conceptual Vision and State of the Semiautogenous Mill Technology. In: Revista Minerales, Instituto de Ingenieros de Minas de Chile, vol. 52(218) (1997) (in Spanish)
7. Vapnik, V.: The Nature of Statistical Learning Theory. Springer, New York (1995)
8. Ljung, L.: System Identification: Theory for the user. Prentice-Hall, Englewood Cliffs (1987)
9. San, L., Ge, M.: An effective learning approach for nonlinear system modeling. In: Proceedings of the IEEE International Symposium on Intelligent Control, (2-4), pp. 73–77 (2004)
10. Rong, H., Zhang, G., Zhang, C.: Application of support vector machines to nonlinear system identification. In: Proceedings of the Autonomous Decentralized Systems, vol. (4-8), pp. 501–507 (2005)
11. Chacón., M., Díaz, D., Ríos, L., Evans, D., Panerai, R.: Support vector machine with external recurrences for modeling dynamic cerebral autoregulation. LNCS, vol. 6776(1), pp. 18–27 (2006)
12. Nerrand, O., Roussel-Ragot, P., Personnaz, L., Dreyfus, G.: Neural networks and non-linear adaptive filtering: unifying concepts and new algorithms. Neural Comput. 5, 165–199 (1993)
13. Carvajal, K., Acuña, G.: Estimation of State Variables in Semiautogenous Mills by Means of A Neural Moving Horizon State Estimator. In: Liu, D., Fei, S., Hou, Z.-G., Zhang, H., Sun, C. (eds.) ISNN 2007. LNCS, vol. 4491, pp. 1255–1264. Springer, Heidelberg (2007)

14. Leontaritis, I.J., Billings, S.A.: Input-Output Parametric Models for Non-Linear Systems; Part 1: Deterministic Non-Linear Systems; Part 2: Stochastic Non-Linear Systems. International Journal of Control 45, 303–344 (1985)
15. Canu, S., Grandvalet, Y., Guigue, V., Rakotomamonjy, A.: SVM and Kernel Methods Matlab Toolbox. Perception Systèmes et Information, INSA de Rouen, Rouen, France (2005)
16. Frohlich, H., Zell, A.: Efficient parameter selection for support vector machines in classification and regression via model-based global optimization. In: Proceedings of the IEEE International Joint Conference on Neural Networks, IJCNN 2005, vol. (3), pp. 1431–1436 (2005)

Using Wolfe's Method in Support Vector Machines Learning Stage

Juan Frausto-Solís, Miguel González-Mendoza, and Roberto López-Díaz

Instituto Tecnológico de Estudios Superiores de Monterrey
{juan.frausto,mgonza,roberto.lopez}@itesm.mx

Abstract. In this paper, the application of Wolfe's method in Support Vector Machines learning stage is presented. This stage is usually performed by solving a quadratic programming problem and a common approach for solving it, is breaking down that problem in smaller subproblems easier to solve and manage. In this manner, instead of dividing the problem, the application of Wolfe's method is proposed. The method transforms a quadratic programming problem into an Equivalent Linear Model and uses a variation of simplex method employed in linear programming. The proposed approach is compared against QuadProg Matlab function used to solve quadratic programming problems. Experimental results show that the proposed approach has better quality of classification compared with that function.

Keywords: Support Vector Machine, Classification, Simplex Method, Wolfe's Method.

1 Introduction

Support Vector Machines (SVMs) is a learning machine technique used in classification, regression and density estimation problems. Generally, SVMs obtains a global and unique solution in comparison with traditional multilayer perceptron neural network that suffer of the existence of multiple local minima solutions [1], as well as optimal topology and tuning parameters problems [2]. In addition, SVM is useful to obtain a good generalization for classification problems. Learning (training) stage of SVMs is achieved solving a Quadratic Programming Problem (QPP). Once QPP is solved, a subset of original data is obtained which defines the optimal hyperplane between two classes and it is used to classify new data. The principal drawback feature of SVM solved with a QPP algorithm is that in this kind of problems, the number of variables is equal to the number of training vectors and in this way; a dense optimization problem is obtained. Another disadvantage is that the memory requirements are the square of the number of training vectors [1][2][3]. In a previous work, an algorithm based on Wolfe's Method was proposed [4]. In this paper, we present an alternative approach for training SVMs that follows this research line; instead of solving a QPP with classical methods, Wolfe's Method is used. This method obtains an Equivalent Linear Model (ELM) of SVM and solves this linear model using a variation of classical Simplex Method (SM). This variation is referenced here as

A. Hernández Aguirre et al. (Eds.): MICAI 2009, LNAI 5845, pp. 488–499, 2009.

Modified Simplex Method for SVM (MSM). Once a solution is obtained, classification of data can be executed. This paper is organized as follows: Section two presents a description about SVMs. In Section three, the Wolfe's Method (which includes transformation of a QPP in an ELM and the Modified Simplex Method) is described. Section four presents the training approach used in this paper to train a Support Vector Machine. In sections five and six, experiments and obtained results are presented, and finally, section seven presents conclusions and future work.

2 Support Vector Machines

The learning process in SVM involves the solution of a quadratic optimization problem, which offers the architecture and parameters of a decision function representing largest possible margin [5]. These parameters are represented by the vectors in the class boundary and their associated Lagrange multipliers.

With the aim of taking into account nonlinearities, SVM can build optimal hyperplane in a Hilbert space of higher dimension than the input space. The higher dimension space is obtained transforming a data vector $x_i \in \Re^n$ through a function $\phi(x)$. In this transformation, explicit calculation of $\phi(x)$ is not necessary, instead of that; just the inner product between mapped vectors is required. For this inner product, kernel functions fulfilling the Mercer condition are usually used. An example of kernel function is show in (1).

$$k\left(\mathbf{x}_i, \mathbf{x}_j\right) = \phi\left(\mathbf{x}_i\right)^T \phi\left(\mathbf{x}_j\right) \tag{1}$$

In pattern recognition case, to build an SVM for two non-linearly separable sets (even in high dimension space), a QPP defined by (2) to (4) should be solved. This QPP is the dual representation of SVM's and it is obtained using Lagrange Dual (or Wolfe Dual) [6][7].

$$\underset{\alpha}{\text{Max}} \qquad \sum_{i=1}^{l} \alpha_i - \frac{1}{2} \sum_{i,j=1}^{l} \alpha_i \alpha_j y_i y_j k\left(\mathbf{x}_i, \mathbf{x}_j\right) \tag{2}$$

$$\text{SubjectTo} \qquad \sum_{i=1}^{l} y_i \alpha_i = 0, \tag{3}$$

$$0 \le \alpha_i \le \zeta \qquad i = 1, \dots, l \tag{4}$$

where y_i, y_j are labels for data vectors i, j respectively and α_i are the Lagrange multipliers introduced to transform the original formulation of the problem with linear inequality constraints into the above representation. The ζ parameter controls the misclassification level on the training data and therefore the margin. A large ζ corresponds to assign a higher penalty to the errors, decreasing the margin, while a small ζ tolerates more errors, growing the margin.

Once a solution is obtained, the decision rule used to classify data is defined in (5):

$$f(x) = sign\left[\sum_{i=1}^{nsv} \alpha_i y_i k(\mathbf{x}_i, \mathbf{x}) + b\right] \tag{5}$$

where b is the projection $\mathbf{x}_i$ onto the hyperplane that separates the classes [8] and only non-zero Lagrange multipliers counts for the decision rule. Consequently, just the data vectors associated to those multipliers are called support vectors (nsv defines number of support vectors). Geometrically, these vectors are at the margin defined by the separator hyperplane [1][2][3].

2.1 Solution of SVM's

The solution of SVMs is achieved using QPP solution algorithms. The most used of these algorithms are active set methods, which are divided in primal and dual methods [9]. The primal methods search dual feasibility keeping primal feasibility and the complementary conditions [2]. The main disadvantage of QPP solution algorithms is found when the analyzed problem is greater than 2000 examples (data vectors), because, depending on memory and processor capacities, can be hard to solve the problem using any of those algorithms in a standard personal computer [10][11]. Therefore, some authors propose to divide the problem in smaller subproblems, which are easier to manage and solve [2]. However, an important disadvantage of this approach is that random selection of data vectors to build subproblems [10][11], can affect the performance, giving an inferior learning rate. For this reason, different approaches that allows to increasing SVM learning rate are required. In this paper, the application of Wolfe's Method for Quadratic Programming [12] on SVM technique is described. In this approach, QPP is transformed in a Linear Model [13] and it is solved by MSM [12]. In the next section, this approach is presented.

3 Wolfe's Method

Commonly, QPP algorithms use an iterative process and some linear optimization techniques. About the latter, Simplex Method (SM) is the most used technique [14]. SM is used to solve LPPs and it has been applied to solve real problems (assignation problems, material mix problem, etc.) [15], and has presented better performance in practical problems than other techniques to solve linear programming problems [16]. The experience in practical field can be used to improve the performance in the learning of SVMs. In order to use Simplex Method with SVMs, some modifications to QPP are required. To do these changes, Wolfe's Method was used [12][17][18]. This method starts obtaining an ELM from QPP and then solves it using MSM. The process is shown in [12][17][18] and a brief description is in next section.

3.1 Equivalent Linear Model

Given a QPP as defined in (6) to (8):

$$\text{Mín} \qquad f(\alpha) = \sum_{j=1}^{l} c_j \alpha_j + \frac{1}{2} \sum_{j=1}^{l} \sum_{i=1}^{m} \alpha_j q_{ij} \alpha_i \tag{6}$$

SubjecTo $\quad g(\alpha)=\displaystyle\sum_{j=1}^{l}a_{ij}\alpha_j -b_i \le 0, \qquad i=1,\dots,m$ (7)

$$h_j(\alpha)=-\alpha_j \le 0, \qquad\qquad j=1,\dots,l$$ (8)

where α are decision variables, c is a coefficient's vector with j components, a and q are symmetric coefficient matrices and b is a vector of resources availability with i components [12].

To obtain the Equivalent Linear Model, Wolfe's Method adds λ and u artificial variables and uses Langrangian of the problem defined in (9):

$$L(\alpha,\lambda,u)=\sum_{j=1}^{l}c_j\alpha_j +\frac{1}{2}\sum_{j=1}^{l}\sum_{i=1}^{m}\alpha_j q_{ij}\alpha_i +\sum_{i=1}^{m}\lambda_i\left(\sum_{j=1}^{l}a_{ij}\alpha_j -b_i\right)+\sum_{j=1}^{l}u_j\left(-\alpha_j\right)$$ (9)

Deriving (9) with regards the variables α, λ, u, the Equivalent Linear Model is obtained and it is shown in (10) to (15):

Mín $\qquad\qquad \displaystyle\sum_{j=1}^{l}V_j$ (10)

Subject To $\quad c_j+\displaystyle\sum_{i=1}^{m}q_{ij}\alpha_i +\sum_{i=1}^{m}\lambda_i a_{ij} -u_j +V_j =0, \; j=1,\dots,l$ (11)

$$\sum_{j=1}^{l}a_{ij}\alpha_j +Y_i =b_i, \qquad i=1,\dots,m$$ (12)

$$\alpha_j \ge 0, u_j \ge 0, V_j \ge 0, \lambda_i \ge 0, \quad j=1,\dots,l, i=1,\dots,m$$ (13)

$$\lambda_i Y_i =0, \qquad i=1,\dots,m$$ (14)

$$u_j\alpha_j =0, \qquad j=1,\dots,l.$$ (15)

where V_j are artificial variables and Y_i are slack variables [12].

In this model Y_i is unrestricted in sign for $i = 1, \dots, m$, and with the addition of complementary slackness conditions defined in (14) and (15). The solution of ELM is obtained when variables V_j is equal to zero (when artificial variables V_j are not in the solution of the problem) [12]. However, the fact that there are not variables V_j in a solution not means this is the optimum, but only a feasible solution was found.

Once Equivalent Linear Model is obtained, this is solved using Modified Simplex Method, which has changes from classical Simplex Method in the way that incoming variable is selected. A variable is a candidate to be the incoming variable if this fulfills complementary slackness conditions as follows: if λ_i is an incoming variable candidate, it can be selected as the incoming one only if Y_i is not in the basis, which in Simplex Method indicates those columns (variables) that are active (usually named basic variables) and inactive in the problem solution [14]. On the other hand, if Y_i is an incoming variable candidate, it can be selected as the incoming one only if λ_i is not in the basis. Similar conditions must be fulfilled for variables α_j and u_j [12][18].

4   Proposed Approach

As was mentioned in a past section, this paper describes a different approach to use in learning stage of SVMs instead of classical methods used for this task. Unlike classical algorithms to solve QPP associated to SVM, which divide the complete problem in subproblems, selecting constrains randomly (sometimes joining one by one those constrains to the subproblem) and using some QPP method to determine the support vectors in that subproblem [10][11], the approach presented in this paper transforms QPP in a Equivalent Linear Model (without subdividing the QPP) and solves it using the Modified Simplex Method [12][18].

In Figure 1, a scheme of the approach is shown. The approach has the common stages of SVM training process: 1) Reading and Modeling data vectors using a kernel function, 2) Training, solving QPP obtained with data and 3) Classification, where a classifier is built with Support Vectors found in the training stage. This classifier will be used to classify new data. In figure 1, training stage is performed using Wolfe's Method. As was explained in section 3, the way to solve this problem is transforming it in an Equivalent Linear Model. After that transformation, MSM can be used.

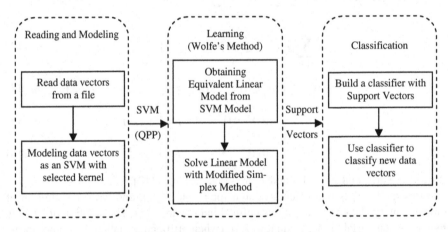

Fig. 1. Diagram of proposed approach

5   Experiments

To test the performance of the approach based on Wolfe's Method in the solution of a SVM, a program in Matlab script was developed. This program obtains the model of the data as a SVM using diverse kernel functions, then ELM is obtained from SVM and finally, the solution of the model is obtained using Modified Simplex Method.

The main objective of program was to obtain a classification equal or better than a SVM solved with a classical QPP technique. The test compares QPP solver included in Matlab (QuadProg) versus the program based on Wolfe's Method (which includes MSM). In this test, execution time of both programs was compared as well. The program based on QuadProg function is similar to figure 1, with the unique change in training stage: instead of using Wolfe's Method, QuadProg function is used; notice that QuadProg solves a Quadratic Problem of SVM.

5.1 Datasets Used for Testing

The used datasets for experimentation with Matlab script were found in UC Irvine Machine Learning Repository [19]. This benchmark is a collection of data bases, theories and data generators used by machine learning community for empirical analysis of learning algorithms. From existents datasets, a subset was selected to perform experiments with proposed method. The list of datasets is shown in Table 1.

The column Identifier indicates name identifier used in figures of section 6.

Table 1. List of datasets used for testing

Dataset Name	Identifier	# Data Vectors	# Attributes
shuttle-landing-control	Dataset 1	15	6
yellow-small+adult-stretch	Dataset 2	20	4
adult-stretch	Dataset 3	20	4
adult+stretch	Dataset 4	20	4
yellow-small	Dataset 5	20	4
echocardiogram-Attr-2	Dataset 6	131	11
hepatitis	Dataset 7	155	19
wpbc	Dataset 8	198	33
bupa	Dataset 9	345	6
ionosphere	Dataset 10	351	34
house-votes84	Dataset 11	435	16
wdbc	Dataset 12	569	31
breast-cancer-wisconsin	Dataset 13	699	9
pima-indians-diabetes	Dataset 14	768	8

5.2 Kernel Functions Used for Testing

As mentioned in section two, a kernel function is used to take data to a higher dimension that allows a linear separation of them. Table 2 presents the kernel functions used for planed experiments [1][3][4][20][21][22]. These functions were selected using a quality classification index explained in section 6. The functions included in Table 2 had obtained the higher performance of tested set. The selection of these kernel functions was made analyzing functions in the literature and using hypothesis testing for selecting those with best performance. Details about the work made for selection of the function is not reported in this paper.

494   J. Frausto-Solís, M. González-Mendoza, and R. López-Díaz

Table 2. Set of Kernel Functions selected for experiments

Kernel Function Name	Contraction	Formula
Radial Basis Function	RBF	$k(x_i, x_j) = -\dfrac{1}{\#\,\text{Data Vectors}} * \|x_i - x_j\|^2$
Radial Basis Function (Modification 1)	RBFMod	$k(x_i, x_j) = \exp^{-0.5*\|x_i - x_j\|^2}$
Radial Basis Function (Modification 2)	RBFMod2	$k(x_i, x_j) = \exp^{\frac{-(x_i - x_j)*(x_i - x_j)}{8}}$

6 Obtained Results

In this section, the experimental results obtained with the presented approach in this paper are shown. The experiments were executed in a PC with Pentium IV processor, 2 Gb of RAM and Windows XP installed. The features considered for comparison were: 1) Quality of Classification measured in percentage of correct classified data vectors, 2) Training Time used by MSM to obtain a classifier and 3) Total Time required by entire process, which includes the required time for obtaining ELM plus required training time executed by MSM and plus required time for classification. Training and total time are measured in seconds.

6.1 Comparison of Quality of Classification

Figures 2, 3 and 4 present the comparisons of Quality of Classification between the program based on QuadProg function and the program based on Wolfe's Method. The figures show obtained results with RBF, RBFMod and RBFMod2 kernel functions respectively. A Quality Classification index is measured as percentage of vectors correctly classified from total vectors of each dataset used for testing.

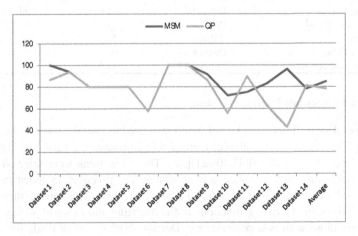

Fig. 2. Quality Classification Index (%) comparison for RBF kernel function and Datasets

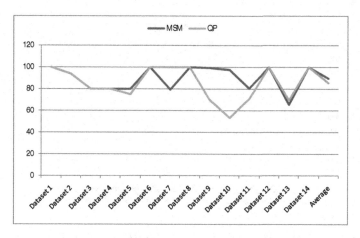

Fig. 3. Quality Classification Index (%) comparison for RBFMod kernel function and Datasets

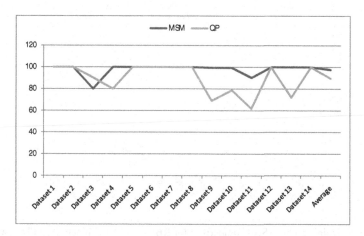

Fig. 4. Quality Classification Index (%) comparison for RBFMod2 kernel function and Datasets

In three cases (using RBF, RBFMod and RBFMod2 kernel functions), the program based on Wolfe's Method obtains a better average Quality classification index than that based on QPP solver (RBF: MSM). Results show that program based on Wolfe's Method has better performance than program based on QuadProg function.

6.2 Training Time Comparison

Figures 5, 6 and 7 present the comparisons of Training Time (measured in seconds) between the program based on QuadProg function and the program based on Wolfe's Method. Training time is defined as required time for each program to solve its model (for Quadratic Programming Model, it is QPP solver; for ELM, it is Modified Simplex Method). The Figures show obtained results with RBF, RBFMod and RBFMod2 kernel functions respectively.

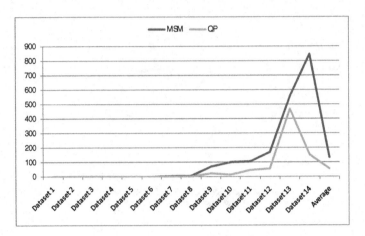

Fig. 5. Training Time (Sec) comparison for RBF kernel function and Datasets

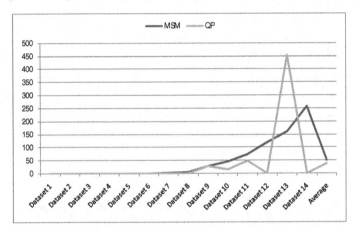

Fig. 6. Training Time (Sec) comparison for RBFMod kernel function and Datasets

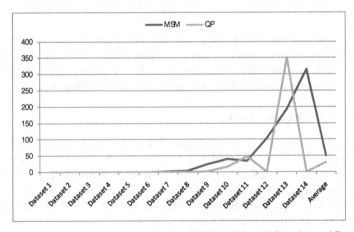

Fig. 7. Training Time (Sec) comparison for RBFMod2 kernel function and Datasets

In three cases, the program based on QPP solver obtains better average training time than the program based on Wolfe's Method. Figure 5 presents that program based on Wolfe's Method requires more time than program based on QuadProg function. Possible reasons for this situation can be: combination of RBF kernel function and pima-indian-diabetes dataset or training data vectors with large range of values. Both, kernel function and dataset will be analyzed for future work.

6.3 Total Time Comparison

Figures 8, 9 and 10 present the comparisons of Total Time between the program based on QuadProg function and the program based on Wolfe's Method. Total time is defined as required time by each program to build the SVM model with training data (in case of MSM, required time to build EML is added), plus required time for learning stage, plus required time for obtaining a classifier and plus required time to data classification with this classifier. The figures show obtained results with RBF, RBFMod and RBFMod2 kernel functions respectively.

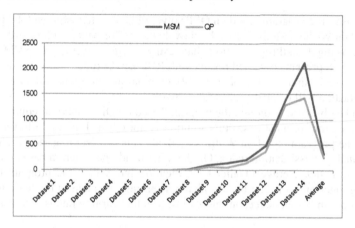

Fig. 8. Total Time (Sec) comparison for RBF kernel function and Datasets

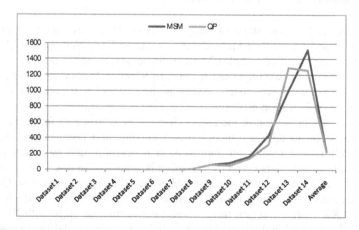

Fig. 9. Total Time (Sec) comparison for RBFMod kernel function and Datasets

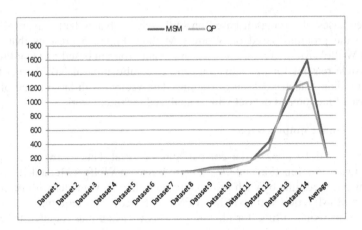

Fig. 10. Total Time (Sec) comparison for RBFMod2 kernel function and Datasets

In all cases, the program based on QPP solver obtains better average total time than that based on Wolfe's Method. Results show better performance with regards required total time using QuadProg function than using Wolfe's Method in SVM training stage. Figure 8 presents obtained results with RBF kernel functions. As is observed in Figure 5, there is an unusual time used solving pima-indian-diabetes dataset. Reasons for this situation will be analyzed as future work.

As can be observed, although the program based on QPP function require less time to find a solution, in most cases this solution is not optimal, giving a lower learning rate than the program based on Wolfe's Method. This low learning rate results in low average quality classification index. On the other hand, the program based on Wolfe's Method has better average Quality Classification Index than the program based on QuadProg function. This can be considered as an improvement in the learning model of SVMs and, in consequence, an improvement in the learning rate is obtained.

7 Conclusions and Future Work

An approach for learning stage of Support Vector Machines was presented. This approach is based on Wolfe's Method that obtains an Equivalent Linear Model and solves it by using Modified Simplex Method. Some advantages of this approach lie in the implementation of two important features of Simplex Method in a SVM solution algorithm: the practical experience as well as the possibility of obtaining the global optimum of an optimization problem.

As is observed in section 6, the approach presents, a better performance than the program based on QuadProg function when Quality of Classification Average is compared. With these results, an improvement in learning rate was obtained and a greater amount of new data correctly classified is achieved. In this way, an improvement of 6.4% as average can be observed. However, the program based on QuadProg function has better execution time in less than 38 seconds as average, which scarcely represents 1.8% more execution time. Large time in training stage with RBF kernel function and pima-indian-diabetes dataset is under analysis to find the reason of this situation. Because the amount of new data correctly classified is an important requirement for SVMs, obtained solution by the approach based on Wolfe's Method becomes attractive. In addition, a research opportunity area is open, mainly to develop new faster linear programming methods for SVM. Therefore, presented results can be used as an initial point of future research where modifications to the approach can be applied. The modifications will have as objective to require less required time in learning stage and improving obtained Quality Classification Index.

References

1. Suykens, J., Van Gestel, T., De Brabanter, J., De Moor, B., Vandewalle, J.: Least Square Support Vector Machines. World Scientific Publishing Co., Singapore (2002)
2. García-Gamboa, A., Hernández-Gress, N., González-Mendoza, M., Ibarra-Orozco, R., Mora-Vargas, J.: Heuristics to reduce the training time of SVM algorithm. In: First International Conference in Neural Network and Associative Memories (2006)
3. Burges, C.: A Tutorial on Support Vector Machines for Pattern Recognition. In: Data Mining and Knowledge Discovery. Kluwer Academic Publishers, Netherlands (1998)
4. Sideris, A., Estévez, S.: A Proximity Algorithm for Support Vector Machine Classification. In: Proceedings of the 44th IEEE Conference on Decision and Control (2005)
5. Vapnik, V.: Computational Learning Theory. John Wiley & Sons, Chichester (1998)
6. Joachims, T.: Training Linear SVMs in Linear Time. In: Proceedings of the ACM Conference on Knowledge Discovery and Data Mining, KDD (2006)
7. Keerthi, S.S., Shevade, S.K.: SMO Algorithm for Least Squares SVM Formulations. Neural Computation 15(2) (2003)
8. Welling, M.: Support Vector Machines, Class notes. University of Toronto, Department of Computer Science
9. Fletcher, R.: Practical Methods of Optimization. John Wiley & Sons, Chichester (2000)
10. Platt, J.: Fast Training of Support Vector Machines using Sequential Minimal Optimization. Microsoft Research (1998)
11. Scheinberg, K.: An Efficient Implementation of an Active Set Method for SVMs. IBM T. J. Watson Research Center, Mathematical Science Department (2006)
12. Prawda, J.: Methods and Models in Operations Research (Spanish Edition), vol. 1. Limusa (2005)
13. Mangasarian, O.: Generalized Support Vector Machines. Advances in Large Margin Classifiers. MIT Press, Cambridge (2000)
14. Chvátal, V.: Linear Programming. W. H. Freeman and Company, New York (1983)
15. McMillan Jr, C.: Mathematical Programming. John Wiley and Sons, Inc., Chichester (1970)
16. Bertsekas, D.: Network Optimization: Continuous and Discrete Models. Athena Sc. (1998)
17. Winston, W.: Operations Research Applications and Algorithms. Thomson (2005)
18. Wolfe, P.: The Simplex Method for Quadratic Programming. Econometrica 27 (1959)
19. UC Irvine Machine Learning Repository, http://archive.ics.uci.edu/ml/
20. Amari, S., Wu, S.: Improving Support Vector Machine Classifiers by Modifying Kernel Functions. Neural Networks 12, 783–789 (2000)
21. Cristianini, N., Shawe-Taylor, J.: An Introduction to Support Vector Machines and other kernel-based learning methods. Cambridge University Press, Cambridge (2000)
22. Cristianini, N.: Support Vector and Kernel Machines, Tutorial. In: International Conference in Machine Learning (2001)

Direct Adaptive Soft Computing Neural Control of a Continuous Bioprocess via Second Order Learning

Ieroham Baruch[1], Carlos-Roman Mariaca-Gaspar[1], and Josefina Barrera-Cortes[2]

[1] Department of Automatic Control, CINVESTAV-IPN, Av. IPN No 2508,
07360 Mexico City, Mexico
{baruch,cmariaca}@ctrl.cinvestav.mx
[2] Department of Biotechnology and Bioengineering, CINVESTAV-IPN, Av. IPN No 2508,
07360 Mexico City, Mexico
jbarrera@cinvestav.mx

Abstract. This paper proposes a new Kalman Filter Recurrent Neural Network (KFRNN) topology and a recursive Levenberg-Marquardt (L-M) second order learning algorithm capable to estimate parameters and states of highly nonlinear bioprocess in a noisy environment. The proposed KFRNN identifier, learned by the Backpropagation and L-M learning algorithm, was incorporated in a direct adaptive neural control scheme. The proposed control scheme was applied for real-time soft computing identification and control of a continuous stirred tank bioreactor model, where fast convergence, noise filtering and low mean squared error of reference tracking were achieved.

1 Introduction

The recent advances in understanding the working principles of artificial neural networks has given a tremendous boost to the derivation of nonlinear system identification, prediction and intelligent soft computing control tools [1], [2]. The main network property, namely the ability to approximate complex non-linear relationships without prior knowledge of the model structure, makes them a very attractive alternative to the classical modeling and control techniques [3]. This property is particularly useful in applications where the complexity of the data or tasks makes the design of such functions by hand impractical. Among several possible network architectures, the ones most widely used are the Feedforward NN (FFNN) and Recurrent NN (RNN). In a FFNN, the signals are transmitted only in one direction, which requires applying tap delayed global feedbacks and tap delayed inputs to achieve a Nonlinear Autoregressive Moving Average neural dynamic plant model [3]. An RNN has local feedback connections to some of the previous layers. Such a structure is a suitable alternative to the FFNN when the task is to model dynamic systems; its main advantage is the reduced complexity of the network structure. However, the analysis of the state of the art in the area of classical RNN-based modeling and control has also shown some of their inherent limitations: 1. The RNN input vector consists of a number of past system inputs and outputs and there is not a systematic way to define the optimal number of past values, and usually, the method of trials and errors is performed [2]; 2. The RNN

A. Hernández Aguirre et al. (Eds.): MICAI 2009, LNAI 5845, pp. 500–511, 2009.

model is naturally formulated as a discrete model with a fixed sampling period; there-fore, if the sampling period is changed, the neural network has to be trained again; 3. It is assumed that the plant order is known [1]; 4. The managing of noisy input/output plant data is required to augment the filtering capabilities of the identification RNNs, [4]. Driven by these limitations, a new Kalman Filter Recurrent Neural Network (KFRNN) topology and the recursive Backpropagation (BP) learning algorithm in vector-matrix form has been derived and its convergence has been studied [5]. But the recursive BP algorithm, applied for KFRNN learning, is a gradient descent first order learning algorithm which does not allow to augment the precision or accelerate the learning [4]. Therefore, the aim of this paper was to use a second order learning algo-rithm for the KFRNN, as the Levenberg-Marquardt (L-M) algorithm [6]. The KFRNN with L-M learning was applied for Continuous Stirred Tank Reactor (CSTR) model identification [7], [8]. The application of KFRNNs [5] together with the recursive L-M [7] could avoid all the problems caused by the use of the FFNN; thus improving the learning and the precision of the plant state and parameter estimation in presence of noise. Here, the data obtained system from the KFRNN identifier states will be used in order to design a Direct Adaptive Neural Control (DANC) of CSTR bioprocess plant. Other related examples can be found in [4], [5].

2 Kalman Filter RNN

This section is dedicated to the KFRNN topology, the recursive Backpropagation and the recursive Levenberg-Marquardt algorithms for the KFRNN learning. The KFRNN is applied as a state and parameter estimator of nonlinear plants.

2.1 Topology of the KFRNN

The KFRNN topology (see Fig. 1) is given by the following vector-matrix equations:

$$X(k+1) = A_1 X(k) + BU(k) - DY(k) \tag{1}$$

$$Z(k) = G[X(k)] \tag{2}$$

$$V_1(k) = CZ(k) \tag{3}$$

$$V(k+1) = A_2 V(k) + V_1(k) \tag{4}$$

$$Y(k) = F[V(k)] \tag{5}$$

$$A_1 = block\text{-}diag\,(A_{1,i})\,;\left|A_{1,i}\right| < 1 \tag{6}$$

$$A_2 = block\text{-}diag\,(A_{2,i})\,;\left|A_{2,i}\right| < 1 \tag{7}$$

Where: Y, X, and U are output, state and input l n, m vectors; A_1 and A_2 are (nxn) and (lxl) block-diagonal local feedback weight matrices; $A_{1,i}$ and $A_{2,i}$ are i-th diagonal block of A_1 and A_2. Equations (6) and (7) represent the local stability conditions imposed on all blocks of A_1 and A_2; B and C are (nxm) and (lxn) input and output weight matrices; D is a (nxl) global output feedback weight matrix; G[.] and F[.] are vector-valued sigmoid or hyperbolic tangent-activation functions; Z, V_1, V are vector variables with the corresponding dimensions; the integer k is a discrete-time variable.

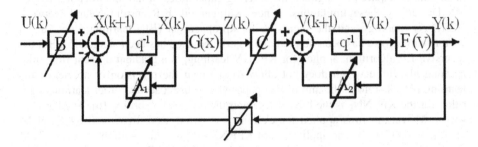

Fig. 1. Block-diagram of the KFRNN topology

2.2 BP Learning of the KFRNN

The general BP learning rule is given by the following equation:

$$W(k+1) = W(k) + \eta \Delta W(k) + \alpha \Delta W(k-1) \tag{8}$$

Where W is the weight matrix modified (A_1, A_2, B, C, D); ΔW is the weight matrix correction (ΔA_1, ΔA_2, ΔB, ΔC, ΔD); η and α are learning rate parameters. Applying the diagrammatic method [9] and using the block-diagram of the KFRNN topology (Fig. 1), it was possible to design an error predictive adjoint KFRNN (Fig. 2). Following this adjoint KFRNN block diagram, the following matrix weight updates were obtained:

$$\Delta C(k) = E_1(k)Z^T(k) \tag{9}$$

$$\Delta A_2(k) = E_1(k)V^T(k) \tag{10}$$

$$E_1(k) = F'[Y(k)]E(k); E(k) = Y_p(k) - Y(k) \tag{11}$$

$$\Delta B(k) = E_3(k)U^T(k) \tag{12}$$

$$\Delta A_1(k) = E_3(k)X^T(k) \tag{13}$$

$$\Delta D(k) = E_3(k)Y^T(k) \tag{14}$$

$$E_3(k) = G'[Z(k)]E_2(k); E_2(k) = C^T E_1(k) \tag{15}$$

$$\Delta v A_1(k) = E_3(k) \oplus X(k); \Delta v A_2(k) = E_1(k) \oplus V(k) \tag{16}$$

Where ΔA_1, ΔA_2, ΔB, ΔC, ΔD are weight corrections of the learned matrices A_1, A_2, B, C, D, respectively; E is an error vector of the output KFRNN layer, Yp is a desired target vector and Y is a KFRNN output vector, both with dimensions l; X is a state vector, and E1, E2, and E3 are error vectors, shown in Fig. 2; F'(.) and G'(.) are diagonal Jacobean matrices with appropriate dimensions, whose elements are derivatives of the activation functions. Equations (10) and (12) represent the learning of the feedback weight matrices, which are supposed to be full (lxl) and (nxn) matrices. Equation (16) gives the learning solution when these matrices are diagonal vA_1, vA_2.

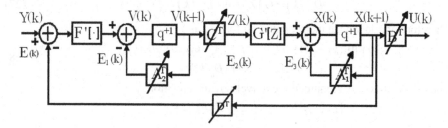

Fig. 2. Block-diagram of the adjoint KFRNN topology

2.3 Recursive Levenberg-Marquardt Learning of the KFRNN

The general recursive L-M algorithm of learning is derived from an optimization procedure (a deeper analysis can be found in [4], [6]), and is given by the following equations:

$$W(k+1) = W(k) + P(k)\nabla Y[W(k)]E[W(k)] \tag{17}$$

$$Y[W(k)] = g[W(k), U(k)] \tag{18}$$

$$E^2[W(k)] = \{Y_p(k) - g[W(k), U(k)]\}^2 \tag{19}$$

$$DY[W(k)] = \frac{\partial}{\partial W} g[W, U(k)]\Big|_{W=W(k)} \tag{20}$$

Where W is a general weight matrix (A_1, A_2, B, C, and D) under modification; P is the covariance matrix of the estimated weights updated; DY[.] is an nw-dimensional gradient vector; Y is the KFRNN output vector which depends of the updated weights and the input; E is an error vector; Yp is the plant output vector, which is in fact the target vector. Using the same KFRNN adjoint block diagram (see Fig.2), it was possible to obtain the values of the gradients DY[.] for each updated weight, propagating the value D(k) = I through it. Following the block diagram of Fig. 2, equation (20) was applied for each element of the weight matrices (A_1, A_2, B, C, D) in order to be updated. The corresponding gradient components are as follows:

$$DY[C_{ij}(k)] = D_{1,i}(k)Z_j(k) \tag{21}$$

$$DY[A_{2ij}(k)] = D_{1,i}(k)V_j(k) \tag{22}$$

$$D_{1,i}(k) = F_i'[Y_i(k)] \tag{23}$$

$$DY[A_{1ij}(k)] = D_{2,i}(k)X_j(k) \tag{24}$$

$$DY[B_{ij}(k)] = D_{2,i}(k)U_j(k) \tag{25}$$

$$DY[D_{ij}(k)] = D_{2,i}(k)Y_j(k) \tag{26}$$

$$D_{2,i}(k) = G_i'[Z_i(k)]C_iD_{1,i}(k) \tag{27}$$

Therefore, the Jacobean matrix could be formed as:

$$\begin{aligned} DY[W(k)] = [&DY(C_{ij}(k)), DY(A_{2ij}(k)), DY(B_{ij}(k)) \\ &DY(A_{1ij}(k)), DY(D_{ij}(k))] \end{aligned} \tag{28}$$

The P(k) matrix was computed recursively by the equation:

$$P(k) = \alpha^{-1}(k)\{P(k-1) - P(k-1).$$

$$\Omega[W(k)]S^{-1}[W(k)]\Omega^T[W(k)]P(k-1)\} \tag{29}$$

Where the S(.), and Ω(.) matrices were given as follows:

$$S[W(k)] = \alpha(k)\Lambda(k) + \Omega^T[W(k)]P(k-1)\Omega[W(k)] \tag{30}$$

$$\Omega^T[W(k)] = \begin{bmatrix} & \nabla Y^T[W(k)] & \\ 0 & \cdots & 1 & \cdots & 0 \end{bmatrix};$$

$$\Lambda(k)^{-1} = \begin{bmatrix} 1 & 0 \\ 0 & \rho \end{bmatrix}; 10^{-4} \leq \rho \leq 10^{-6}; \tag{31}$$

$$0.97 \leq \alpha(k) \leq 1; 10^3 \leq P(0) \leq 10^6$$

The matrix Ω(.) had a dimension (nwx2), whereas the second row had only one unity element (the others were zero). The position of that element was computed by:

$$i = k \bmod(nw) + 1; k > nw \tag{32}$$

After this, the given up topology and learning were applied for the CSTR system identification.

3   Recurrent Trainable NN

This section is dedicated to the topology, the BP and the L-M algorithms of RTNN learning. The RTNN was used as a feedback/feedforward controller.

3.1   Topology of the RTNN

The RTNN model and its learning algorithm of dynamic BP-type, together with the explanatory figures and stability proofs, are described in [5], so only a short description will be given here. The RTNN topology, derived in vector-matrix form, was given by the following equations:

$$X(k+1) = A_1 X(k) + BU(k) \tag{33}$$

$$Z(k) = G[X(k)] \tag{34}$$

$$Y(k) = F[CZ(k)] \tag{35}$$

$$A = block\text{-}diag\,(a_{ii})\,;\,|a_{ii}| < 1 \tag{36}$$

Where Y, X, and U are, respectively, output, state and input vectors with dimensions l, n, m; A = block-diag (a_{ii}) is a (nxn)- state block-diagonal weight matrix; a_{ii} is an i-th diagonal block of A with (1x1) or (2x2) dimensions. Equation (36) represents the local stability condition imposed on all blocks of A; B and C are (nxm) and (lxn) input and output weight matrices; G[.] and F[.] are vector-valued sigmoid or hyperbolic tangent-activation functions; the integer k is a discrete-time variable.

3.2 BP Learning of the RTNN

The same general BP learning rule (8) was used here. Following the same procedure as for the KFRNN, it was possible to derive the following updates for the RTNN weight matrices:

$$\Delta C(k) = E_1(k)Z^T(k) \tag{37}$$

$$E_1(k) = F'[Y(k)]E(k); E(k) = Y_p(k) - Y(k) \tag{38}$$

$$\Delta B(k) = E_3(k)U^T(k) \tag{39}$$

$$\Delta A(k) = E_3(k)X^T(k) \tag{40}$$

$$E_3(k) = G'[Z(k)]E_2(k); E_2(k) = C^T E_1(k) \tag{41}$$

$$\Delta vA(k) = E_3(k) \oplus X(k) \tag{42}$$

Where ΔA, ΔB, ΔC are weight corrections of the learned matrices A, B, and C, respectively; E, E1, E2, and E3 are error vectors; X is a state vector; F'(.) and G'(.) are diagonal Jacobean matrices, whose elements are derivatives of the activation functions. Equation (40) represents the learning of the full feedback weight matrix of the hidden layer. Equation (42) gives the learning solution when this matrix is diagonal vA, which is the present case.

3.3 Recursive Levenberg-Marquardt Learning of the RTNN

The general recursive L-M algorithm of learning [4], [6] is given by equations (17)-(20), where W is the general weight matrix (A, B, C) under modification; Y is the RTNN output vector which depends on the updated weights and the input; E is an error vector; Yp is the plant output vector, which is in fact the target vector. Using the RTNN adjoint block diagram [4], [5], it was possible to obtain the values of DY[.] for each updated weight propagating D=I. Applying equation (20) for each element of the weight matrices (A, B, C) in order to be updated, the corresponding gradient components are as follows:

$$DY[C_{ij}(k)] = D_{1,i}(k)Z_j(k) \tag{43}$$

$$D_{1,i}(k) = F_i'[Y_i(k)]$$ (44)

$$DY[A_{ij}(k)] = D_{2,i}(k) X_j(k)$$ (45)

$$DY[B_{ij}(k)] = D_{2,i}(k) U_j(k)$$ (46)

$$D_{2,i}(k) = G_i'[Z_i(k)] C_i D_{1,i}(k)$$ (47)

Therefore the Jacobean matrix could be formed as

$$DY[W(k)] = [DY(C_{ij}(k)), DY(A_{ij}(k)), DY(B_{ij}(k))]$$ (48)

The P(k) matrix was computed recursively by equations (29)-(32). Next, the given up RTNN topology and learning were applied for CSTR system control.

4 Direct Adaptive Neural Control of CSTR System

This section describes the direct adaptive CSTR control using KFRNN as plant identifier and RTNN as a plant controller (feedback and feedforward). The block-diagram of the control system is given in Fig. 3. The following study describes the linearized model of that closed-loop control system.

Let us present the following z-transfer function representations of the plant, the state estimation part of the KFRNN, and the feedback and feedforward parts of the RTNN controller:

$$W_p(z) = C_p (zI - A_p)^{-1} B_p$$ (49)

$$P_i(z) = (zI - A_i)^{-1} B_i$$ (50)

$$Q_1(z) = C_c (zI - A_c)^{-1} B_{1c}$$ (51)

$$Q_2(z) = C_c (zI - A_c)^{-1} B_{2c}$$ (52)

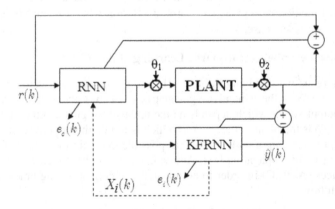

Fig. 3. Block-diagram of the closed-loop neural control system

The control systems z-transfer functions (49)-(52) are connected by the following equation, which is derived from Fig. 3, and is given in z-operational form:

$$Y_p(z) = W_p(z)\left[I + Q_1(z)P_i(z)\right]^{-1}Q_2(z)R(z) + \theta(z) \tag{53}$$

$$\theta(z) = W_p(z)\theta_1(z) + \theta_2(z) \tag{54}$$

Where $\theta(z)$ represents a generalized noise term. The RTNN and the KFRNN topologies were controllable and observable, and the BP algorithm of learning was convergent, [5], so the identification and control errors tended to zero:

$$E_i(k) = Y_p(k) - Y(k) \to 0; \ k \to \infty \tag{55}$$

$$E_c(k) = R(k) - Y_p(k) \to 0; \ k \to \infty \tag{56}$$

This means that each transfer function given by equations (49)-(52) was stable with minimum phase. The closed-loop system was stable and the feedback dynamical part of the RTNN controller compensated the plant dynamics. The feedforward dynamical part of the RTNN controller was an inverse dynamics of the closed-loop system one, which assured a precise reference tracking in spite of the presence of process and measurement noises.

5 Description of the CSTR Bioprocess Plant

The CSTR model given in [7], [8] was chosen as a realistic example of KFRNN and RTNN applications for system identification and control of biotechnological plants. Numerical values for the parameters and nominal operating conditions of this model are given in Table 1. The CSTR is described by the following continuous time nonlinear system of ordinary differential equations:

$$\frac{dC_A(t)}{dt} = \frac{Q}{V}\left(C_{Af} - C_A(t)\right) -$$
$$-k_0 C_A(t)\exp\left(-\frac{E}{RT(t)}\right) \tag{57}$$

$$\frac{dT(t)}{dt} = \frac{Q}{V}(T_f - T(t)) + \frac{(-\Delta H)C_A(t)}{\rho C_p}\exp\left(\frac{E}{RT(t)}\right)$$
$$+ \frac{\rho_c C_{pc}}{\rho C_p V}Q_c(t)\left[1 - \exp\left(\frac{-hA}{Q_c(t)\rho_c C_{pc}}\right)\right](t_{ef} - T(t)) \tag{58}$$

In this model is enough to know that within the CSTR, two chemicals are mixed and that they react in order o produce a product compound A at a concentration $C_A(t)$, and the temperature of the mixture is $T(t)$. The reaction is exothermic and produces heat which slows down the reaction. By introducing a coolant flow-rate $Q_c(t)$, the temperature can be varied and hence the product concentration can be controlled. Here C_{Af} is the inlet feed concentration; Q is the process flow-rate; T_f and T_{ef} are the inlet feed and coolant temperatures, respectively; all of which are assumed constant at nominal

Table 1. Parameters and operating conditions of the CSTR

Parameters	Parameters	Parameters
$Q = 100 \quad (L/\min)$	$E/R = 9.95 \times 10^3 \quad (K)$	$Q_{c0} = 103.41 \quad (L/\min)$
$C_{Af} = 1.0 \quad (mol/L)$	$-\Delta H = 2 \times 10^5 (cal/mol)$	$hA = 7 \times 10^5 \quad (cal/\min K)$
$T_f = T_{fC} = 350 \quad (K)$	$\rho \cdot \rho_c = 1000 \quad (g/L)$	$T_0 = 440.2 \quad (K)$
$V = 100 \quad (L)$	$C_p C_{pc} = 1 \quad (cal/gK)$	$k_0 = 7.2 \times 10^{10} \quad (1/\min)$

values. Likewise, k_0, E/R, V, ΔH, ρ, C_{pc}, C_p, and ρ_c are thermodynamic and chemical constants related to this particular problem. The quantities Q_{c0}, T_0, and C_{A0}, shown in Table 1, are steady values for a steady operating point in the CSTR. The objective was to control the product compound A by manipulating $Q_c(t)$. The operating values were taken from [7] and [8], where the performance of a NN control system is reported.

6 Simulation Results

Results of detailed comparative graphical simulation of CSTR KFRNN plant identification by means of the BP and the L-M learning are given in Fig.4 and Fig.5. A 10% white noise was added to the plant inputs and outputs and the behavior of the plant identification was studied accumulating some statistics of the final MSE% (ξ_{av}) for KFRNN BP and L-M learning. The results for 20 runs are given in Tables 3 and 4.

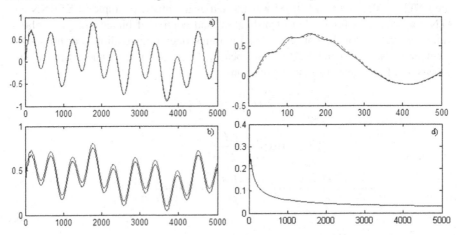

Fig. 4. Graphical results of identification using BP KFRNN learning. a) Comparison of the plant output (continuous line) and KFRNN output (pointed line); b) state variables; c) comparison of the plant output (continuous line) and KFRNN output (pointed line) in the first instants; d) MSE% of identification.

The mean average cost for all runs (ε) of KFRNN plant identification, the standard deviation (σ) with respect to the mean value, and the deviation (Δ) are presented in Table 2 for the BP and L-M algorithms. They were computed by following these formulas:

$$\varepsilon = \frac{1}{n}\sum_{k=1}^{n}\xi_{av_k} \, , \quad \sigma = \sqrt{\frac{1}{n}\sum_{i=1}^{n}\Delta_{i}^{2}} \, , \Delta = \xi_{av} - \varepsilon \tag{59}$$

The numerical results given in Tables 2, 3, and 4 are illustrated by the bar-graphics in Figures 6 a, and b. The comparative results showed inferior MSE%, ε, and σ for the L-M algorithm with respect to the BP one. The results of the DANC using the L-M algorithm of learning are presented in Figure 7 where the final MSE% was 0.873 for L-M.

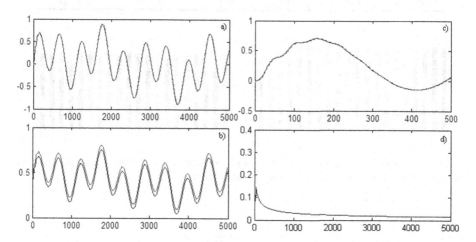

Fig. 5. Graphical results of identification using L-M KFRNN learning. a) Comparison of the plant output (continuous line) and KFRNN output (pointed line); b) state variables; c) comparison of the plant output and KFRNN output in the first instants; d) MSE% of identification.

Table 2. Standard deviations and mean average values of identification validation using the BP and L-M algorithms of KF RNN learning

BP algorithm	L-M algorithm
$\varepsilon = 0.9457$	$\varepsilon = 0.8264$
$\sigma = 0.0416$	$\sigma = 0.0188$

Table 3. MSE% of 20 runs of the identification program using the KFRNN BP algorithm

No	1	2	3	4	5
MSE%	0.9559	0.9654	0.8821	0.9614	0.8798
No	6	7	8	9	10
MSE%	0.9444	0.9591	0.9700	0.9685	1.0034
No	11	12	13	14	15
MSE%	0.8523	0.8105	0.9863	0.9038	1.0122
No	16	17	18	19	20
MSE%	0.9688	0.8630	0.8624	0.8521	0.8898

Table 4. MSE% of 20 runs of the identification program using the KFRNN L-M algorithm

No	1	2	3	4	5
MSE%	0.8123	0.8001	0.8553	0.8360	0.8149
No	6	7	8	9	10
MSE%	0.8072	0.8072	0.8285	0.8236	0.8037
No	11	12	13	14	15
MSE%	0.8659	0.8105	0.8269	0.8218	0.8118
No	16	17	18	19	20
MSE%	0.8628	0.8226	0.8514	0.8288	0.8280

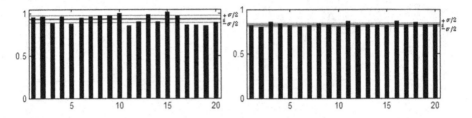

Fig. 6. Comparison between the final MSE% for 20 runs of the identification program: a) using BP algorithm of learning, b) using L-M algorithm of learning

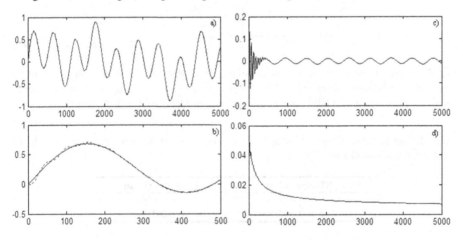

Fig. 7. Detailed graphical simulation results of CSTR plant DANC using L-M learning a) comparison between the plant output and the reference signal; b) comparison between the plant output and the reference signal in the first instants; c) control signal; d) MSE% of control

7 Conclusions

This paper proposes a new KFRNN model for system identification and state estimation of nonlinear plants. The KFRNN is learned by the first order BP and by the second order L-M recursive learning algorithms. The validating results of system identification reported here gave priority of the L-M algorithm of learning over the

BP one which is paid by augmented complexity. The estimated states of this recurrent neural network model are used for direct adaptive trajectory tracking control system design. The applicability of the proposed neural control system, learned by the BP and L-M algorithms, was confirmed by simulation results with a CSTR plant. The results showed good convergence of the two algorithms applied. The identification and DANC results showed that the L-M algorithm of learning is more precise but more complex then the BP one.

Acknowledgements

The Ph.D. student Carlos-Roman Mariaca-Gaspar is thankful to CONACYT for the scholarship received during his studies at the Department of Automatic Control, CINVESTAV-IPN, MEXICO.

References

1. Narendra, K.S., Parthasarathy, K.: Identification and Control of Dynamical Systems Using Neural Networks. IEEE Transactions on Neural Networks 1(1), 4–27 (1990)
2. Hunt, K.J., Sbarbaro, D., Zbikowski, R., Gawthrop, P.J.: Neural Network for Control Systems (A survey). Automatica 28, 1083–1112 (1992)
3. Haykin, S.: Neural Networks, a Comprehensive Foundation, 2nd edn., vol. 7458, Section 2.13, 84–89; Section 4.13, 208–213. Prentice-Hall, Upper Saddle River (1999)
4. Baruch, I.S., Escalante, S., Mariaca-Gaspar, C.R.: Identification, Filtering and Control of Nonlinear Plants by Recurrent Neural Networks Using First and Second Order Algorithms of Learning. Dynamics of Continuous, Discrete and Impulsive Systems, International Journal, Series A: Mathematical Analysis, Special Issue on Advances in Neural Networks-Theory and Applications, (S1, Part 2), 512-521. Watam Press, Waterloo (2007)
5. Baruch, I.S., Mariaca-Gaspar, C.R., Barrera-Cortes, J.: Recurrent Neural Network Identification and Adaptive Neural Control of Hydrocarbon Biodegradation Processes. In: Hu, Xiaolin, Balasubramaniam, P. (eds.) Recurrent Neural Networks, I-Tech Education and Publishing KG, Vienna, Austria, ch. 4, pp. 61–88 (2008) ISBN 978-953-7619-08-4
6. Ngia, L.S., Sjöberg, J.: Efficient Training of Neural Nets for Nonlinear Adaptive Filtering Using a Recursive Levenberg Marquardt Algorithm. IEEE Trans. on Signal Processing 48, 1915–1927 (2000)
7. Zhang, T., Guay, M.: Adaptive Nonlinear Control of Continuously Stirred Tank Reactor Systems. In: Proceedings of the American Control Conference, Arlington, June 25-27, 2001, pp. 1274–1279 (2001)
8. Lightbody, G., Irwin, G.W.: Nonlinear Control Structures Based on Embedded Neural System Models. IEEE Trans. on Neural Networks 8, 553–557 (1997)
9. Wan, E., Beaufays, F.: Diagrammatic Method for Deriving and Relating Temporal Neural Network Algorithms. Neural Computations 8, 182–201 (1996)

A Kohonen Network for Modeling Students' Learning Styles in Web 2.0 Collaborative Learning Systems

Ramón Zatarain-Cabada, M. Lucia Barrón-Estrada, Leopoldo Zepeda-Sánchez,
Guillermo Sandoval, J. Moises Osorio-Velazquez, and J.E. Urias-Barrientos

Instituto Tecnológico de Culiacán
Juan de Dios Bátiz s/n, col. Guadalupe, Culiacán, Sinaloa, 80220, México
rzatarain@itculiacan.edu.mx

Abstract. The identification of the best learning style in an Intelligent Tutoring System must be considered essential as part of the success in the teaching process. In many implementations of automatic classifiers finding the right student learning style represents the hardest assignment. The reason is that most of the techniques work using expert groups or a set of questionnaires which define how the learning styles are assigned to students. This paper presents a novel approach for automatic learning styles classification using a Kohonen network. The approach is used by an author tool for building Intelligent Tutoring Systems running under a Web 2.0 collaborative learning platform. The tutoring systems together with the neural network can also be exported to mobile devices. We present different results to the approach working under the author tool.

Keywords: Intelligent Tutoring System, Web 2.0, Authoring Tool, M-Learning.

1 Introduction

Learner or Student Models are the core of the personalization of Intelligent Tutoring Systems. They make available tutoring tailored or adapted to the needs of the individual students [1]. Many approaches and implementations have been developed in recent years in order to model students' learning styles [2, 3]. Most of those implementations use Bayesian Networks [2], Linear Temporal Logic [4], or neuro-fuzzy networks [5]. In the case of using a learning model like Felder-Silverman [6], we also use the Index of Learning Style Questionnaire (ILSQ) [7].

In this work we propose a different approach for selecting a student´s learning style using self-organising feature maps (Kohonen neural networks) [8]. Advantages of Kohonen networks include implementation simplicity, execution speed and a shorter training process; however maybe the most important advantage of these unsupervised neural networks is that they do not require an external teacher for presenting a training set. During a training session, our Kohonen network receives a number of different input patterns (the student learning style obtained from the ILSQ, the course learning style, and the student's grade in the course), discovers significant features in these patterns (felder-Silverman learning styles) and learns how to classify input.

A. Hernández Aguirre et al. (Eds.): MICAI 2009, LNAI 5845, pp. 512–520, 2009.

In the last years many research groups in the field of education are using Web 2.0 technologies, such as wikis, blogs, recommendation systems and social networking [9]. The e-learning tutor centered is shifting to become more learner centered, where learners are also part of a community of authors and users of the learning resources [10].

In this context, we have designed and implemented a software tool (EDUCA) to create adaptive learning material in a Web 2.0 collaborative learning environment. The material is initially created by a tutor/instructor and later maintained and updated by the user/learner community to each individual course. The courses can dynamically recognize user learning characteristics and be displayed on mobile computers (cell phones, PDA, etc.). EDUCA makes use of Web 2.0 technologies as a recommendation system for filtering future Web learning resources, and Web mining for discovering such resources.

The arrangement of the paper is as follows: Section 2 describes the general structure of the tool. Section 3 presents the neural network used in the tool. Tests and results are shown in Section 4. Comparison to related work is given in section 5. Conclusions and future work are discussed in Section 6.

2 Educa General Structure

Figure 1 presents the general structure of Educa. As shown in Figure 1, there are two main authors: the instructor and the learners. An instructor creates an intelligent tutoring (adaptive) system by first building a knowledge base (a learning object container) using a visual editor. A course is created by importing already prepared learning material in different standard formats like html, pdf, doc or SCORM learning objects from any type of source. The author can also insert learning material by using a simple editor included in the tool. The learning material is classified into four different types of learning objects: text, image, audio, and video. In each section the author inserts learning objects and defines a Felder-Silverman Result Score (FSRS) for every dimension: Input (visual/verbal), Perception (sensitive/intuitive), Understanding (sequential/global) and Processing (active/reflective). The scores defined by the authors will decide what learning objects are presented to the students depending of their learning style. Another way to understand this approach is that the authors create a Knowledge set which contains all the possible learning objects in different learning styles; the learners only "see" part of the set (a knowledge subset) depending of their own learning styles.

Another important learning object the authors insert into the knowledge repository is a quiz. These can be in every part of each section. The quiz is essential for the dynamic courseware generation because from the test results, the neural networks classify learning styles.

After the author create the knowledge base of the ITS, she/he can save it and export it to a Mobile Learning format used to display tutoring systems on mobile devices. The saved/exported file will enclose three elements: a XML file corresponding to the learning style model, a predictive engine for navigation purposes, and the Kohonen Neural Network for learning style classification. Another option to the output of the visual editor is to export the learning material to SCORM format. The benefit of this format is the availability of the material in any distance learning environment.

The other author in Educa is the Learners. When a student or learner reads a course or ITS for the first time, a user profile is created in the system. This profile contains student data like student's learning style, academic records and GPA, past uploaded resources (recommendation), and general academic information.

Once an ITS has been created, the module Published Course stores it in a Course Repository. We know that learners usually read tutoring systems stored in some kind of repository, but they also consult other learning resources in different web sites. For example a student who needs to implement a LR parser in a compiler course, could use the ITS stored in the course repository, but she/he could also consult extra resources in the Web. This material found by the student would be rated and recommended to be added to the regular compiler course. Educa uses a Hybrid Recommendation System [11] which stores new resources into a Resource Repository. Last, Educa uses a Data or Text Mining Subsystem used to search resources in the Web.

All of the modules or subsystems in the bottom middle part of Educa (User Profile, Recommendation System, Data mining, and Repositories) stand for the Web 2.0 Collaborative Learning System of Educa.

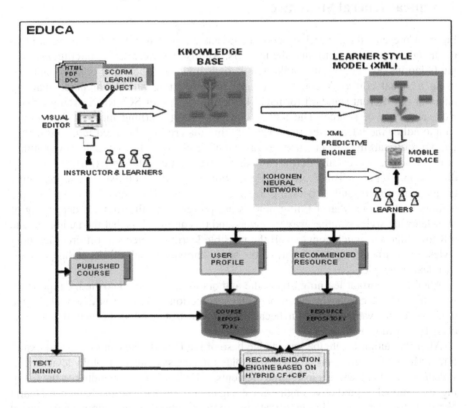

Fig. 1. General Structure of Educa

3 Educa General Structure

A self-organizing feature map or Kohonen neural network is trained using unsupervised or competitive learning [8]. In competitive learning, neurons compete among themselves to be activated. The Kohonen model provides topological mapping, placing a fixed number of input patterns from the input layer into a higher-dimension output or Kohonen layer (figure 2). The Kohonen layer consists of a single layer of computation neurons, with two types of different connections: forward connections, from input layer to the Kohonen layer, and lateral connections between neurons in the Kohonen layer. During training, the neural network receives new input patterns presented as vectors. Every neuron in the Kohonen layer receives a modified copy of the input pattern. The lateral connections produce excitatory or inhibitory effects, depending on the distance from the winning neuron.

3.1 Training the Kohonen Network

An input vector in our neural network is described as:

$$D = [d_{fs}\ d_a\ p]$$

where D is the input vector, and it is formed by three elements: d_{fs} which is the student learning style identified by applying the Felder-Silverman ILSQ questionnaire; d_a which is the learning style used in the learning material read by the student; and p is the student grade obtained in the test when the student has learning style *dfs* and the course was offered using learning style d_a.
Vectors d_{fs} and da are also composed as:

$$d_a = d_{fs} = [c_1\ c_2\ c_3]$$

where $c_1\ c_2\ c_3$ represents three scores for Perception, Input, and Understanding Dimension in the ILSQ questionnaire.

Figure 2 shows part of the Kohonen network architecture. We can see that the input layer has seven neurons, corresponding to three-dimension vectors d_{fs} and d_a, and the student grade p. Vector's input data vary between -11 to +11, which is massaged to the range -1 to +1 with the equation:

$$\text{Massaged value}_{11} = vi\ /\ |vimax|$$

where *vi* is a ILSQ score, and *vimax* is the maximum ILSQ score value (+11). In figure 2, as we observe, -3, an ILSQ score for visual/verbal learning style, is mapped to -0.272. On the other hand, the student grade, which varies between 0 and 100, is also massaged to the range -1 to +1, with equation:

$$vnj = vj \times 2\ /\ vjmax - 1$$

where vj is the student grade or score, and *vjmax* is the maximum student grade (+100). For example, 70 is mapped to 0.4 (figure 2).

The Kohonen layer has 1600 neurons arranged in a structure of a 40x40 hexagonal grid. The output of the network consists of three signals or values ranging between -1 to +1. These data are then decoded to the ILSQ range (-11 to +11); they represent the learning style of the student.

516 R. Zatarain-Cabada et al.

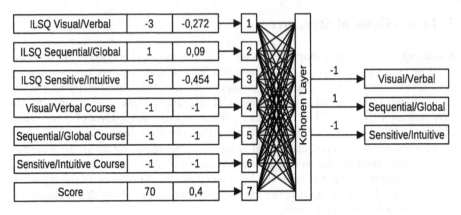

Fig. 2. The Kohonen Network for Learning Styles

The procedure to train the neural network consisted in three steps. The sample for this experiment was 140 high-school students. During the first step we applied the ILSQ questionnaire in order to identify the learning style of each student. In the second step we created three different introductory courses: Computer Science, Photography, and Eolithic Energy. For each course we produced eight versions, one for each learning style (visual/verbal, sequential/global, sensing/intuitive, and active/reflective). Randomly, the students received each of the three courses, in only one learning style. We recorded the assigned learning styles of each student in the three courses. In the third step and after the students took the courses, we applied tests in order to evaluate the performance of the students in each course.

With these three elements (student learning styles, course learning styles, and student evaluation or results) we created 140 input vectors for training the network. We trained the network with 6000 iterations, an initial neighbourhood ratio of 50% of the lattice size and a learning rate of 0.1.

The network has 1600 neurons arranged in the form of a two-dimensional lattice with 40 rows and 40 columns, and every input vector has seven elements, as we established previously.

3.2 Using the Kohonen Network

As we explained before the network is applied to identify learning styles of students. The network is then exported together with the ITS and the predictive engine to the mobile (see figure 1). When the student is displaying an ITS in a time t, the learning material shown to the students is using a learning style g. Whenever the student answers a quiz as part of the learning contents, he receives a grade k.

Therefore, the network r takes as input the learning style of the displayed contents, and performs a search to find the winner-takes-all (best-matching) neuron j_x at iteration p, using the minimum-distance Euclidean criterion:

$$j_x(p) = \min \|X - W_j(p)\| = [\Sigma(x_i - w_{ij})^2]^{1/2}$$

where n is the number of neurons in the input layer, m is the number of neurons in the Kohonen layer, X is the input pattern vector and W is the weight vector.

After, finding neuron j_x, the output of the network is the student learning style which is defined as

$$g(t + 1) = r(g(t), k)$$

This process of identifying the student learning style is performed every time the student answers a quiz on the ITS.

4 Preliminary Experiments

We conducted a set of preliminary experiments with Educa. The tool was used by a group of different users (authors), which consisted mainly of university professors/students. They produced different intelligent tutoring systems like Compiler Constructions, Introduction to Computer Science, teaching the Maya Language, and an introduction to Elementary mathematics. The intelligent systems were initially created by a professor and later the students were using and adding new learning resources with the recommendation system. Figure 3 shows the main interface of Educa for lecture 3 "Parsing" in a Compiler course. As we can observe, there are

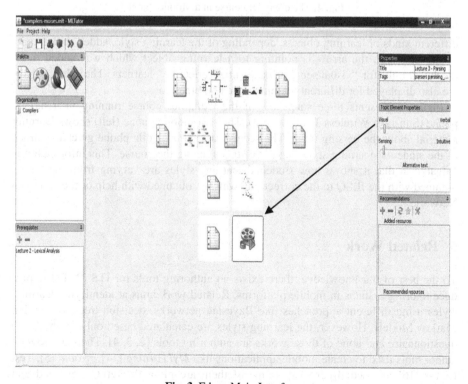

Fig. 3. Educa Main Interface

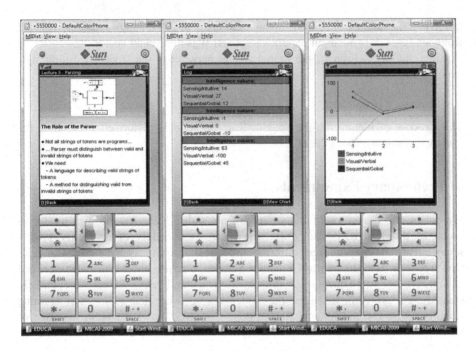

Fig. 4. The compiler course in a Mobile Phone

different kinds of learning objects, depending of the learning style, added to the topic. In our example, the arrow is pointing to a learning object which was added to be displayed only for "visual/sensitive/sequential/reflective" learners. The other objects are also displayed for different learning style combinations.

Figure 4 presents three snapshots of the Compiler course running in a mobile phone (Sun Java Wireless Toolkit 2.5.2). The first mobile phone (left) shows learning material about the parsing topic. The second and third mobile phone gives us a trace of the students' learning styles (in three stages) along the course. This information is valuable in that it shows how students' learning styles are varying from the one is obtained with the ILSQ to the correct one which is obtained with help of the Kohonen Network.

5 Related Work

To the best of our knowledge, there exists no authoring tools for ITS [2, 13] to produce tutoring systems in mobile platforms. Related work aims at identifying learning styles using different approaches like Bayesian networks, decision trees, or Hidden Markov Models. However, the learning styles are calculated based only on the ILSQ questionnaire and none of those works are authoring tools [2, 3, 4]. There are several author tools used to create mobile applications like *MyLearning* [14], *zirada*[15], Test Editor [16], or mediaBoard[17]. Some of them are more PocketPC's oriented and some are focused to quiz editing or game-based learning. None of those author tools however, have the ability of adapting to the user's learning style, nor have portability across different computer and operating system platforms.

6 Conclusions and Future Work

This paper presents an approach for modeling Students' Learning Styles in Web 2.0 Collaborative Learning Systems using a Kohonen Network. The approach was implemented under an author tool that allows the production of personalized learning material to be used under collaborative and mobile learning environments. The software tool was implemented with Java version 1.6 and XML. The learning material produced with the tool is platform independent and it has been tested with different software emulators (NHAL Win32 Emulator, J2ME Wireless Toolkit and Sony Ericsson SDK), cell and smart phones (Sony Ericsson and Nokia). We present a test with one of such mobile devices. Currently empirical studies are taking place to examine the students' reaction to the learning material produced using **Educa**. Future work should focus on continuing testing with official and complete material for course programs in public and private schools in Mexico.

Acknowledgments. The work described in this paper is fully supported by a grant from the DGEST (Dirección General de Educación Superior Tecnológica) in México under the program "support and development of academic bodies" [Academic Body: Research in Software Engineering].

References

1. Kerly, A., Bull, S.: Children's Interactions with Inspectable and Negotiated Learner Models. In: Woolf, B.P., Aïmeur, E., Nkambou, R., Lajoie, S. (eds.) ITS 2008. LNCS, vol. 5091, pp. 132–141. Springer, Heidelberg (2008)
2. Carmona, C., Castillo, G., Millán, E.: Designing a Dynamic Bayesian Network for Modeling Student's Learning Styles. In: Díaz, P., Kinshuk, A.I., Mora, E. (eds.) ICALT 2008, pp. 346–350. IEEE Computer Society, Los Alamitos (2008)
3. Graf, S., Kinshuk, Liu, T.: Identifying Learning Styles in Learning Management Systems by Using Indications from Students' behavior. In: Díaz, P., Kinshuk, A.I., Mora, E. (eds.) ICALT 2008. IEEE Computer Society, Los Alamitos (2008)
4. Limongelli, C., Sciarrone, F., Vaste, J.: LS-PLAN: An Effective Combination of Dynamic Courseware Generation and Learning Styles in Web-based Education. In: Nejdl, W., Kay, J., Pu, P., Herder, E. (eds.) AH 2008. LNCS, vol. 5149, pp. 133–142. Springer, Heidelberg (2008)
5. Zatarain-Cabada, R., Barrón-Estrada, M.L., Sandoval, G., Osorio, M., Urías, E., Reyes-García, C.A.: Authoring Neuro-fuzzy Tutoring Systems for M and E-Learning. In: Gelbukh, A., Morales, E.F. (eds.) MICAI 2008. LNCS (LNAI), vol. 5317, pp. 789–796. Springer, Heidelberg (2008)
6. Felder, R.M., Silverman, L.K.: Learning and Teaching Styles in Engineering Education. Engineering Education 78, 674–681 (1988)

7. Felder, R.M., Solomon, B.A.: Index of Learning Styles Questionnaire (2004),
   http://www.engr.ncsu.edu/learningstyles/ilsweb.html
8. Negnevitsky, M.: Artificial Intelligence: A guide to Intelligent Systems, 2nd edn. Addison-Wesley, Reading (2005)
9. O'Reilly, T.: What is Web 2.0,
   http://oreilly.com/pub/a/oreilly/tim/news/
   2005/09/30/what-is-web-20.html
10. Hage, H., Aimeur, E.: Harnessing Learner's Collective Intelligence: A Web 2.0 Approach to E-Learning. In: Woolf, B.P., Aïmeur, E., Nkambou, R., Lajoie, S. (eds.) ITS 2008. LNCS, vol. 5091, pp. 438–447. Springer, Heidelberg (2008)
11. Burke, R.: Hybrid Recommender Systems: Survey and Experiments. In: User Modeling and User-Adapted Interactions, vol. 12(4), pp. 331–370 (2002)
12. Murray, T.: Authoring Intelligent Tutoring Systems: An analysis of the state of the art. International Journal of Artificial Intelligence in Education 10, 98–129 (1999)
13. Murray, T., Blessing, S., Ainsworth, S.: Authoring Tools for Advanced Technology Learning Environments. Kluwer Academic Publishers, Dordrecht (2003)
14. Attewell, J.: Mobile technologies and learning: A technology update and mlearning project summary, http://www.m-learning.org/reports.shtml
15. Zirada Mobile Publisher. The Premier Mobile Content Creation Tool,
    http://www.Zirada.com
16. Romero, C., Ventura, S., Hervás, C., De Bra, P.: An Authoring Tool for Building both Mobile Adaptable Tests and Web-based Adaptive or Classic Tests. In: Wade, V.P., Ashman, H., Smyth, B. (eds.) AH 2006. LNCS, vol. 4018, pp. 203–212. Springer, Heidelberg (2006)
17. Attewell, J.: From Research and Development to Mobile Learning: Tools for Education and Training Providers and their Learners,
    http://www.mlearn.org.za/papers-full.html

Teaching-Learning by Means of a
Fuzzy-Causal User Model

Alejandro Peña Ayala[1,2]

[1] WOLNM, [2] ESIME-Z and [2] CIC – [2] Instituto Politécnico Nacional
[1] 31 julio 1859 #1099B Leyes Reforma, Iztapalapa, D. F., 09310, Mexico
apenaa@ipn.mx

Abstract. In this research the teaching-learning phenomenon that occurs during an E-learning experience is tackled from a fuzzy-causal perspective. The approach is suitable for dealing with intangible objects of a domain, such as personality, that are stated as linguistic variables. In addition, the bias that teaching content exerts on the user's mind is sketched through causal relationships. Moreover, by means of fuzzy-causal inference, the user's apprenticeship is estimated prior to delivering a lecture. This supposition is taken into account to adapt the behavior of a Web-based education system (WBES). As a result of an experimental trial, volunteers that took options of lectures chosen by this user model (UM) achieved higher learning than participants who received lectures' options that were randomly selected. Such empirical evidence contributes to encourage researchers of the added value that a UM offers to adapt a WBES.

1 Introduction

Usually, common knowledge pursues to describe objects and phenomena that occur around the world. Scientific knowledge also claims to explain: why do such events happen? Moreover, one of science's challenges is: to predict the behavior of a given phenomenon with the aim to manage and control it.

According to those scientific grounds, the field of this quantitative research is: adaptive and intelligent Web based education systems (AIWBES). The object of study is: the teaching-learning phenomenon. The research pursues enhancing the learning acquired by users of an AIWBES by means of a predictive and static UM. Hence, the research problem is stated as: how to depict, explain, and predict causal effects that lectures delivered by an AIWBES produce on the users' apprenticeship?

Given those guidelines, a causal and multi-varied hypothesis is set: the user's learning is enhancing when a specific option for every lecture is delivered by an AIWBES as a result of the selection achieved by a UM according to: user's attributes, lecture's attributes, causal relations among attributes, and predictions about causal outcomes.

Thereby, in this work a fuzzy-causal UM is proposed to describe, explain and predict the causal bias that a lecture delivered by an AIWBES exerts upon the user. As a result of the application of this approach the user's learning is enhanced.

Hence, the research method is based on the design and the exploration of models. The testing of the approach is achieved by a field experiment in an AIWBES

A. Hernández Aguirre et al. (Eds.): MICAI 2009, LNAI 5845, pp. 521–532, 2009.

environment. As a result of the statistical analysis, empirical evidence is found out to demonstrate: how this kind of UM is able to enhance the apprenticeship of users.

This research contributes to the UM field, because it introduces cognitive maps (CM) as a tool for depicting the teaching-learning phenomenon like a cause-effect model. As regards the innovative aspects of the work, they are: a proposal of a UM based on fuzzy-causal knowledge and reasoning; a predictive UM that outcomes a qualitative decision model; empirical information to demonstrate the value of the UM; a UM approach that shows how to represent intelligent behavior and adapt a WBES to users. Also, the research is original because it proposes CM as a model to choose suitable options of lectures to teach users and demonstrates their effectiveness in the AIWBES field. This work is relevant because: it constitutes an approach to enhance the student's learning. The significance of this research results from: it describes, explains, and predicts the teaching-learning from a UM view.

Thus, in order to describe the nature and the outcomes of this research, the following sections of the paper are devoted to: resume the related work, point out the conceptual model, represent the UM, explain how the UM works, show how the predictions are made, outline the trial's account, discuss the results, and set the future work.

2 Related Work

AIWBES pursue to adapt themselves to student's needs. They also attempt to behave in an intelligent way by carrying out some tasks traditionally executed by human tutors [1]. Thereby, AIWBES require a UM in order to meet particular needs of any user. A UM is part of an AIWBES, so that it bears the responsibility for modeling relevant aspects of the user [2]. Among the kinds of UM, a predictive UM is able to anticipate specific aims and attributes of the student, such as: goals, actions, and preferences [3].

Some predictive UM require a phase of supervised training, where cases are tagged with labels that represent the status of unobservable variables [4]. Other predictive UM set probabilistic relational models through Bayesian networks [5]. Also, there are predictive UM that apply random and experience sampling to collect assessments of subjects' intentions, and affective states [6]. In addition, some predictive UM develop cluster models that are described in terms of student-system interaction attributes to estimate possible behaviors of students [7]. Moreover, several predictive UM apply data mining to anticipate some issues of credit assignment according to the user profile [8].

Regarding causality, there is work that depicts cognitive-affective factors of academic achievement [9]. Another approach is a causal model oriented to predict the overall satisfaction of senior students with their baccalaureate nursing programs [10].

As regards fuzzy logic, Kavcic developed a UM for dealing with uncertainty in the assessment of students [11]. Also, Nykänen built a predictive fuzzy model for student classification [12]. Furthermore, fuzzy epistemic logic is used to represent student's knowledge state, while the course content is modeled by the context [13].

3 Conceptual Model

The conceptual model of this proactive UM embraces theoretical foundations from three fields: modeling, causality, and fuzzy logic. Hence, in order to provide an overall baseline, a resume of the three underlying elements is presented as follows:

3.1 Modeling

In logic, the analogy is a kind of reasoning by which: the similarity of unknown properties is induced from two similar entities (objects or phenomena) that hold key common properties. Thereby, modeling takes into account analogy with the aim to describe, explain, and predict the behavior of the entity of analysis. As a result, a model is outcome to characterize, study, and control such entity.

User modeling is a manual and computational process devoted to describe a system's user. The result, a UM, is an analogical representation of the user. It is tailored from a particular viewpoint to answer: who is a specific user? A UM, contains a knowledge base and an engine. Knowledge items are facts or beliefs according to the level of certainty. Both items concern concepts and judgments about the user. The engine is a module that achieves queries and inferences from the knowledge base.

3.2 Causality

Causality is a post-hoc explanation of events. The baseline of causality rests on a monotonic philosophical principle which states that: any fact has a cause, and given the same conditions, the same causes produce the same consequences. Causality deals with events that happen in everyday life. Causality aims to set relations between one or more events that are able to trigger the occurrence of another event. Events involved in a causal relation mean transformations that occur on the state of physical or abstract entities. Such entities are stated by concepts. So a causal relation is: a judgment about how some concepts exert a causal bias on the state of another concept.

As regards the concept's state, it is an attribute that is qualitative measured. Such quantity reveals a kind of *level* or a *variation*. A level gives away the difference between a state's value and its normal value in a given point of time. Variation reveals a tendency of change on the concept's state during a period. Thus, concepts' state is transformed during the time according to the causal relationships between concepts.

3.3 Fuzzy Logic

Uncertainty issue is tackled as a fuzzy matter by Zade [14]. Fuzzy knowledge depicts the membership degree (MD) that a value reveals in regards to a criterion. Such a value is chosen to qualify an entity's property. The MD is represented by a real number that ranges from 0 to 1. Hence, an entity's property is a linguistic variable; whereas, the value attached to the property is instantiated by linguistic terms.

Usually, a linguistic term is shaped by a membership function that outcomes a closed polygon. The set of all linguistic terms that are candidates for being attached to a linguistic variable according to a criterion is called: a universe of discourse (UD).

As regards fuzzy inference, it pursues to identify the linguistic term(s) and the respective(s) MD that is (are) attached to a linguistic variable as a result of a fuzzy operation. Hence, any linguistic term, that is a member of the linguistic variable's UD and whose MD is greater than 0, is a value attached to the linguistic variable.

An approach used to deal with fuzzy knowledge and reasoning is the fuzzy rules system. A fuzzy rule is a relationship composed by an antecedent and a consequent. Antecedent depicts a situation to be satisfied. Consequent reveals a conclusion that happens when a condition succeeds. Usually, antecedent holds just one condition, but it might have several conditions that are conjunctively or disjunctively joined. Consequent normally embraces just one conclusion, however when it contains more conclusions, they are conjunctively linked. A condition claims a situation where a linguistic variable holds a particular linguistic term. A conclusion assigns a specific linguistic term to a linguistic variable and outcomes a MD. The estimation of the conclusion's MD takes into account the MD attached to the condition(s).

4. User Model Representation

In this section the representation of the UM is given by means of a formal description of the UM, a profile about CM and the statement of concepts as follows:

4.1 User Model Formal Representation

A UM is aware of the user according to the knowledge that it holds about her/him. Facts and beliefs are outlined by propositions of the propositional calculus. They can be assessed as true or false. Therefore, the set of prepositions (p) that a UM (m) asserts (A) about a user (U) is outlined as: $A_m(U) = \{p \mid A_m p(U)\}$ [15].

However, UM knowledge is organized into several domains. A domain gives away a specific viewpoint of study. A domain contains a set of concepts that focuses on particular targets of analysis. Such target is a concept's property that is called *state*. The state is qualitatively measured in terms of levels and/or variations. Hence, any domain, as the personality domain (P), holds a set of propositions $(A_m p)$ which asserts what is true about the user, such as: $P_m(U) = \{p \mid A_m p(U) \cap p \in P\}$.

Thus, given such underlying elements, the UM introduced in this work embraces five domains: content (T), learning preferences (L), cognitive skills (C), personality (P), and knowledge acquired (K). Thereby the knowledge of the UM contains the union of sets of propositions asserted by the UM as in (1).

$$UM \ (U) = T_m(U) \cup L_m(U) \cup C_m(U) \cup P_m(U) \cup K_m(U). \tag{1}$$

4.2 Cognitive Maps Profile

A CM is a mental model devoted to depict a causal phenomenon [16]. A CM is drawn as a digraph, whose nodes represent entities, and its arcs reveal cause-effect relations. The arrowhead shows how an entity exerts a causal bias over another entity. A bias means that according to the state of a *cause* entity a shift occurs over the state of the *effect* entity. Entities are described as concepts, and its state as a property. A concept's state is a linguistic variable, which is instantiated by linguistic terms that reveal levels, variations or both. A relation is set by a fuzzy-rules base (FRB), where there is just one rule for each linguistic term of the UD held by the cause concept's state.

A CM owns an engine to carry out causal inferences. The engine estimates how a concept's state is altered by causal relations that point to it. The process begins with the assignment of initial level or/and variation value(s) to concepts' state. Afterwards, a cycle starts to compute causal influences that bias concepts' state. At the end of a cycle, concepts' state is updated. When current states' values are the same as those of prior iterations the process ends, otherwise a new cycle is done. Causal behavior is revealed by the evolution of concepts' state at each point of time, whereas causal outcome corresponds to the set of final values attached to concepts' state.

4.3 Concept Description

Concepts of the domains identified in (1) are semantically defined in an ontology [16]. Also, they are characterized by a state property. A state property is a linguistic variable. The linguistic variable owns a term to label the state, and a value to measure it. The value is expressed by one or more linguistic terms of one UD. A *level* UD shows degrees of levels, such as: {quite low, low, medium, high, quite high}. A *variation* UD depicts several intensities of variations, such as: {decreases: high, moderate, low, null, increases: low, moderate, high}. But, some linguistic variables are *bipolar* because they hold level and variation linguistic terms to depict the concept's state.

As regards the teaching domain, it corresponds to content delivered by WEBS. A level UD is attached to its concepts to measure learning theories (e.g. objectivism and constructivism), media (e.g. sound, text, video, and image), complexity, and interactivity (e.g. dynamic, static). Content is manually measured by content authors.

Learning preferences domain contains eight styles stated in the Gardner's Theory of Multiple Intelligence, such as: logical, spatial, and linguistic. They are bipolar linguistic variables, whose initial level values come from a test of 80 questions.

Concepts of the cognitive skills domain correspond to some intellectual attributes that are measured by the Wechsler Adult Intelligent Scale version III, such as: short memory, reasoning, and intellectual quotation (IQ). They are also bipolar linguistic variables. Their initial level values are outcome from a test composed by 11 trials that evaluate verbal and performance scales. Results are shifted to values of a level UD.

Personality domain holds 45 concepts that reveal behavior patterns and mental problems. They are bipolar linguistic variables. Initial level values are measured by the Minnesota Multiphase Personality Inventory version 2. Thus, a test of 567 questions is applied to the user. Its diagnostic reveals the level of presence of an issue.

Knowledge acquired domain is measured by Bloom's Taxonomy of Educational Objectives (TEO). Any concept of the teaching domain is evaluated by a scale that holds six tiers of ascending mastering as follows: knowledge, comprehension, application, analysis, synthesis, and evaluation. The assignment of a tier depends on the evidences given by the user, who must be able to accomplish specific cognitive tasks.

5. User Model Explanation

This section is devoted to explain how concepts, UD, and inferences over causal relations are achieved. Hence, their mathematical process is stated as follows:

5.1 Concepts and Universe of Discourse Description

A linguistic term is shaped as a fuzzy set (A) according to a membership function (μ), where $\mu_A(x): X \rightarrow [0, 1]$, $x \in X$, and X is the UD attached to a linguistic variable. As a result, the A is drawn as the trapezoid shown in Fig. 1^a. It holds the next attributes: $area_A$: is the surface of the fuzzy set; xC_A: is the axis of the centre of mass; $Supp_A$: is the support set outcome from $\{x \in X \mid \mu_A(x) > 0\}$; $xSize_A$: is the size of the support set; $xCore_A$: is the core set of A that reveals the set of x points in the UD of A, whose MD is 1, such as: $\{x \in X \mid \mu_A(x) = 1\}$; $xTopof_A$: is the size of the core; iB_A: is the internal base that reveals the difference between the absolute minimal x point of the core, and the absolute minimal x point of the support; eB_A: is the external base that represents the difference between the absolute maximal x point of the core, and the absolute maximal x point of the support; iS_A: is the internal slope of the absolute quotient between $1/iB_A$; eS_A: is the external slope of the absolute quotient between $1/eB_A$ [17].

Fuzzy sets that reveal the linguistic terms of a UD meet the next constraints: 1) the point of intersection x for two neighboring fuzzy sets owns a MD: $y = 0.5$ for both; 2) for any point x, a maximum of two fuzzy sets are attached to it; 3) for any point x, the MD of the sum of its corresponding fuzzy sets is: $y = 1$; 4) as much as the intensity revealed by the fuzzy set is, the length of its support set and its area are larger. Such conditions are drawn in Fig. 1^b and 1^c for variation and level UD respectively.

5.2 Fuzzy-Causal Inference Estimation

Based on the inference mechanisms set in sections 3.3, 4.2 and in [16, 17] a fuzzy-causal relationship between a pair of concepts is achieved as follows: For an iteration t, the FRB that defines a relation is tested. The process identifies which rules *fire*. This means that the rule' antecedent is met. Due to the constraints earlier stated for the fuzzy terms of a UD, at least one rule fires, or a maximum of two rules fire. As a result, one of the following situations happens: 1) when two rules fire and the linguistic term of the consequents is the same; 2) two rules fire but the consequent linguistic terms are different; 3) when just one rule fires. The result is outcome by transformation, aggregation, and normalization tasks as follows.

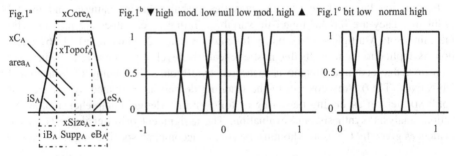

Fig. 1. Shows the graphical description of: 1^a) a fuzzy set; 1^b) a variation UD; 1^c) a level UD

In the first case, the transformation uses the MaxDot method to produce two *equivalent* sets (E^1, E^2). The attributes mean the proportion revealed by the corresponding MD. Both equivalent sets are aggregated to outcome a *result* set (R) with a MD: $y = 1$. The result set represents a *causal output* set (COS) as it is illustrated in Fig. 2.

In the second case, transformation achieves two equivalent sets (E^1, E^2). They hold proportional attributes to the former ones according to their corresponding MD. The second task generates an extra-area as a result of the aggregation effect that is part of the result set (R) with a MD: $y \leq 1$. Finally, the normalization produces a COS that takes into account the attributes of the consequent fuzzy sets, as it is shown in Fig. 3.

Due to just one fuzzy rule fires in the third situation, the consequent fuzzy set corresponds to the equivalent set, the result set, and the COS, with a MD: $y = 1$.

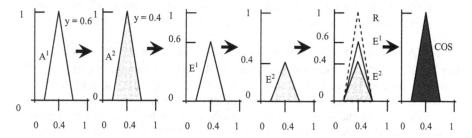

Fig. 2. Draws the result of a relation where two rules fire with the same consequent fuzzy set

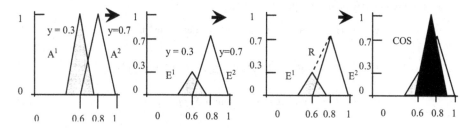

Fig. 3. Draws the result of a relation where two rules fire with different consequent fuzzy sets

6   User Model Prediction

Prediction is made by the estimation of the bias that concepts' state exerts each other during the time. So in this section the approach devoted to compute causal influences on a concept is shown. Also, the algorithm to achieve the simulation over a CM is set.

6.1   Fuzzy-Causal Inference over a Concept

A set of $COS_{1\text{-}e...c\text{-}e}$ is achieved as a result of the relationships $1\text{-}e...c\text{-}e$ that exert a concept's state e in a given point of time t. This set is used to outcome a variable output set (VOS) through a fuzzy carry accumulation (FCA). Such VOS_e means the accumulated fuzzy-causal influence that e meets in t. In consequence, a new value updates e, and such value is used in the next increment of time $t+1$.

A VOS_{1-e} is instantiated with the first COS_{1-e}. Next, a process begins during $n-1$ cycles, where n is the number of cause concepts c. At each cycle i, where $1 < i <= n$, a comparison between COS_{i-e} and VOS_{i-1-e} is made to identify the highest degree of bias, as in Fig. 4[a]. Thus, the fuzzy set with lower variation degree is shifted towards the one with the highest degree to outcome a VOS_{i-e}. Due to the accumulative effect, an extra area is estimated. This extra area is sketched by dotted lines in Fig. 4[b]. Later on, the saturation of the VOS_{i-e} is examined. This issue occurs when the support set of the VOS_{i-e} is beyond the limit of the x-axis, but its axis of the centre of mass is lesser than the limit, as in Fig. 4[c]. Thus, the area that exceeds the x-axis's limit is estimated, and the magnitude of such area is cut off from both sides of the VOS_{i-e}, as appears in Fig. 4[d]. At the end of the n iteration, the VOS_{n-e} corresponds to the linguistic term(s) that update(s) e by a numeric-linguistic conversion as in (2), where w^+ and w^- are the total area of positive and negative $COS_{1...c-e}$ that outcome VOS^+ and VOS^- respectively

A FCA is achieved as follows: given two fuzzy sets $(A, B \in F(X))$, which are positive VOS^+, and the variation intensity stated by A is greater than the one of B, and X is a discrete interval between $[0, 1]$ that is extended beyond the limit, a FCA between $A@B$ is set by (3). Where, x_i is a discrete point that acquires values from 0 until the maximum value of the support set of the VOS, as a result of the saturation process achievement. Also, the *shift* function estimates in (4) the displacement of B towards A, as the distance resulted from the difference between the minimal value of A' core set, and the minimal value of B' support set. Also, a *carry* function in (5) outcomes the *accumulation* effect in each point x_i. [17].

$$x_i = (xC_{vos^-} * w^- + xC_{vos^+} * w^+) / (w^- + w^+). \tag{2}$$

$$\mu_{A@B}(x_i) = \min\{ 1, \mu_A(x_i) + \mu_B(x_{i-shiftB}) + carry\ (x_{i-1})\}. \tag{3}$$

$$shiftB = \min(xCore_A) - \min(Supp_B). \tag{4}$$

$$carry\ (x_i) = \max\{ 0, \mu_A(x_i) + \mu_B(x_{i-shiftB}) + carry\ (x_{i-1}) - 1\};\ carry\ (x_{-1}) = 0. \tag{5}$$

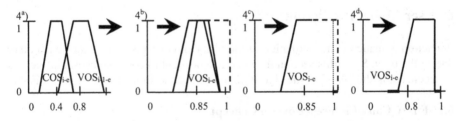

Fig. 4. Shows the graphical generation of a VOS; 4[a]) compares COS_{i-e} and VOS_{i-1-e}; 4[b]) shifts COS_{i-e}; 4[c]) identifies the extra-area in VOS_{i-e}; 4[a]) cuts of an extra-area in both sides of VOS_{i-e}

6.2 Fuzzy-Causal Inference over a Cognitive Map

In this section the algorithm devoted to select the best option of lecture is introduced. However, it is necessary the previous satisfaction of the following constraints:

- the user profile is already outcome with the knowledge acquired (background), the cognitive skills, the personality, and the learning preferences of the student,
- every lecture is authored through four options that meet content domain criteria,
- each lecture's option is already measured based on the content domain concepts.

Algorithm devoted to choose the best lecture's option from Peña, (2008) PhD Thesis... [16]

```
Retrieve the user profile that corresponds to user u
Retrieve the content profile of the current lecture l
For each set of criteria that tailor a lecture's option
  Outcome an instance c of CM called: CM_c
  Carry out the simulation of CM_c:
    Set time = 1, iterations = 100, stability = true
    While (time < iterations and stability) do:
      For each concept e of the CM_c:
        For each causal relation that points to e:
          Estimate the effect on e to outcome a COS_e
        Estimate the new state's value of e by a VOS_e
        Track behavior and new state's value of e
      If(the new state of each CM_c concept already exists)
        Set stability = false
      Set time++
    Track the causal behavior and outcomes of CM_c
  Set b = 1
  Choose CM_b as the current best option (the first CM_i)
  For every instance i (from the 2^{nd} to the last one):
    Compare the simulation outcomes of CM_b against CM_i
    If (CM_i reveals better learning achievement than CM_b)
      Set b as i
  Deliver the option b of the lecture l to user u
  Evaluate the apprenticeship of user u
  Track the real learning achievement of user u
```

7 Study Case

The approach was tested during a trial [16]. The experiment embraced seven stages: implementation, acquisition, pre-stimulus, pre-measure, stimulus, post-measure, and analysis. The results achieved during each stage are stated next.

In the *implementation*, several deliverables were fulfilled, such as: a Web based marketing campaign; the recruiting of 200 volunteers (undergraduate and postgraduate students and professors from several disciplines and states of Mexico); four tools for measuring learning preferences, personality, cognitive, and knowledge domains; teaching content about "The Scientific Research Method". Hence, ten *key concepts* were taught, e.g. law, theory, and hypothesis. So for each lecture about a key concept, four options were authored according to a learning theory and a type of content.

In the *acquisition* stage four tests were consecutively applied to participants through the Web. However, during the time many volunteers progressively deserted without accomplishing the four tests. The assessment reveals that: 113 subjects answered the learning preferences test; but, only 102 achieved the personality exam. Afterwards, just 71 of the remaining volunteers fulfilled the cognitive quiz. At the end, only 50 subjects applied the former knowledge evaluation. Therefore, 50 people were considered as the universe of the experiment.

The first action of the *pre-stimulus* was devoted to train the population by introductory lectures delivered by the AIWBES. The training subject concerned philosophy and sciences history. Later on, a sample of eighteen volunteers was randomly chosen. The sample's size reveals a standard error of 0.05 for a population of 50. Afterwards, the sample was randomly split into two teams of nine people each: *experimental* (*E*) and *control* (*C*). Both teams were balanced according to two main factors: academic level distribution and number of duties (e.g. job, school, sport).

In the *pre-measure*, the background knowledge that subjects held for the ten key topics was estimated. The test had six questions for each key concept. The level of the TEO that a participant got for a key term is the number of questions that were consecutively answered correctly. Once the volunteer answered the test, a total score was estimated for the key concepts. Thus, the score ranged from 0 to 60. Also, a team's total score was outcome from the accumulation of the score achieved by its members.

During the *stimulus* stage, volunteers were encouraged to learn, or reinforce, ten key concepts. Members of team *C* took lectures whose options were randomly chosen by the AIWBES. Participants of team *E* were taught according to the selection of the best option advised by the UM, whose algorithm is introduced in section 6.2.

A *post-measure* was made to the sample for estimating its apprenticeship. So the same questions and criteria were applied to subjects. The learning for a key topic grows from the difference between its post and pre measures.

A statistical process was fulfilled in the *analysis*. A sample of measures appear in Table 1, where *E* team got lower pre-measure for the ten key concepts than *C* team; but at the end, it achieved a higher learning level than *C*. Although, the IQ of *E* team's members is lower than the IQ of their peers in team *C*. Also, a linear regression was done to identify the causal bias between former and acquired knowledge with a significance level of 0.05. So team *C* got a *P* value of 0.126 against 0.006 for *E* team. Thus, the source of variation was only significant for team *E*, as appears in Fig. 5.

Table 1. A sample of measures outcome by experimental and control teams during the trial

Criterion	Experimental team	Control team
Intelligence quotient	11% high, 22% medium, 66% low	44% high, 22% medium, 33% low
Logical learning style	44% quite high, 44% high, 11 medium	44.4% quite high, 55% high
Maturity personality	11% high, 22% medium, 66% low	22% high, 22% medium, 55% low
Pre-measure 10 topics	Total: 38; Mean 4.22	Total: 42; Mean 4.67
Post-measure10 topics	Total: 198; Mean 22	Total: 174; Mean 19.33
Learning achievement	Total: 160; Mean 17.78	Total: 132; Mean 14.67

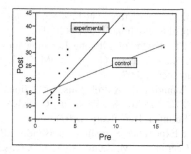

Fig. 5. Linear regression diagram of the tendency for control and experimental teams

8 Discussion and Future Work

User modeling is a complex task that deals with incertitude, imprecision, and incompleteness. The study of any domain considered in a UM is still target of research for several disciplines. However, the UM is a key module in AIWBES in order to adapt education to users in an intelligent way. Thereby, it is necessary to use paradigms, findings, and tools outcome by sciences such as psychology and pedagogy. However, most of the deliverables achieved from such fields are designed for manual application. Therefore, research is desirable to shift the on site application of psychological and pedagogical models to virtual environments in order to spread their use.

In conclusion, this approach asserts that the apprenticeship of people is enhanced by means of the fuzzy-causal support of a UM. The findings accomplished by this kind of user modeling are quite useful to encourage the use of psychological tools and adapt them to the Web-based education field.

As a future work, new challenges must be met, such as: the innovation of virtual psychology models, the development of a virtual supervisor, the use of friendly user-system interfaces, and the design of psychological tests oriented to the Web. Moreover, the conception of new paradigms to adapt the behavior of a WBES is required. Moreover, research is needed to aggregate empirical evidence about advantages and limitations of a UM. What is more, the proposal of new paradigms for predicting UM is welcome in order to provide a proactive attitude in AIWBES. Finally, a dynamic version for this kind of UM is considered.

Acknowledgments

The author gives testimony of the strength given by his Father, Brother Jesus and Helper, as part of the research projects of World Outreach Light to the Nations Ministries (WOLNM). Also, the work owns a support from grants given by: CONACYT-SNI-36453, IPN-SIP-EDI: DOPI/3189/08, IPN-SIP-20090068, IPN-COFAA-SIBE.

References

1. Brusilovsky, P., Peylo, Ch.: Adaptive and Intelligent Web-based Educational Systems. Int. J. Artificial Intelligence in Education 13, 156–169 (2003)
2. Kay, J.: Life-long Learning, Learner Models and Augmented Cognition. In: Woolf, B.P., Aïmeur, E., Nkambou, R., Lajoie, S. (eds.) ITS 2008. LNCS, vol. 5091, pp. 3–5. Springer, Heidelberg (2008)

3. Zukerman, I., Albrecht, D.W.: Predictive Statistical Models for User Modeling. Int. J. User Modeling and User-Adapted Interaction 11, 5–18 (2001)
4. Kapoor, A., Horvitz, E.: Principles of Lifelong Learning for Predictive User Modeling. In: Conati, C., McCoy, K., Paliouras, G. (eds.) UM 2007. LNCS (LNAI), vol. 4511, pp. 37–46. Springer, Heidelberg (2007)
5. Noguez, J., Sucar, L.E., Espinoza, E.: A Probabilistic Relational Student Model for Virtual Laboratories. In: Conati, C., McCoy, K., Paliouras, G. (eds.) UM 2007. LNCS (LNAI), vol. 4511, pp. 303–308. Springer, Heidelberg (2007)
6. Kapoor, A., Horvitz, E.: Experience Sampling for Building Predictive User Models: A Comparative Study. In: Czerwinski, M., Lund, A., Tan, D. (eds.) Proc. 26th SIGCHI C. on Human Factors in Computing Systems, Florence, Italy, April 05–10, pp. 657–666. ACM, New York (2008)
7. Legaspi, R., Sison, R., Fuki, K., Numao, M.: Cluster-based Predictive Modeling to Improve Pedagogic Reasoning. Int. J. Computers in Human Behavior 24(2), 153–172 (2008)
8. Chen-S., L.: Diagnostic, Predictive and Compositional Modeling with Data Mining in Integrated Learning Environments. Int. J. Computers and Education 49(3), 562–580 (2005)
9. Valle, A., Gonzalez, R., Nuñez, J.C., Rodriguez, S., Pineiro, I.: A Causal Model of the Cogntive-Affective Factors of the Academic, Achievement. J. of General and Applicated Psychology 52(4), 499–519 (1999) (in Spanish)
10. Liegler, R.M.: Predicting Student Satisfaction in Baccalaureate Nursing Programs: Testing a Causal Model. J. of Nursing Education 36(8), 357–364 (1997)
11. Kavcic, A.: Fuzzy Student Model in InterMediActor Platform. In: Proc. 26th Int. C. on Information Technology Interfaces, Cavtat, Croatia, June 7-10, vol. 10, pp. 297–302 (2004)
12. Nykänen, O.: Inducing Fuzzy Models for Student Classification. J. Educational Technology and Sociaty 9(2), 223–234 (2006)
13. Xu, D., Wang, H., Su, K.: Intelligent Student Profiling with Fuzzy Models. In: Proc. 35th Int. C. on System Sciences, HICSS 2002, Hawaii, USA, January 07-10, p. 81 (2002)
14. Zadeh, L.: Towards a Theory of Fuzzy Information Granulation and its Centrality in Human Reasoning and Fuzzy Logic. Int. J. Fuzzy Sets and Systems 19 (1997)
15. Self, J.: Formal Approaches to Student Modeling. Tech-Report AI-59, Lancaster (1991)
16. Peña, A.: Student Model based on Cognitive Maps. PhD Thesis, National Polytechnic Institute, Mexico (January 2008)
17. Carvalho, J.P.: Rule Base-based Cognitive Maps: Qualitative Dynamic Systems Modeling and Simulation. PhD Thesis, Lisboa Technical University, Portugal (October 2001)

Inferring Knowledge from Active Learning Simulators for Physics

Julieta Noguez[1], Luis Neri[1], Víctor Robledo-Rella[1], and Karla Muñoz[2]

[1] Tecnológico de Monterrey, Campus Ciudad de México,
Calle del Puente 222, Col. Ejidos de Huipulco, Tlalpan 14380 México, D.F., México
{jnoguez,neri,vrobledo}@itesm.mx
[2] University of Ulster, Magee Campus
munoz_esquivel-k@ulster.ac.uk

Abstract. Active Learning Simulators (ALS) allow students to practice and carry out experiments in a safe environment - at any time, and in any place. Furthermore, well-designed simulations may enhance learning, and provide the bridge from conceptual to practical understanding. By adding an Intelligent Tutoring System (ITS), it is possible to provide personal guidance to students. The main objective of this work is to present an ALS suited for a Physics scenario in which we incorporate elements from ITS, and where a Probabilistic Relational Model (PRM) based on a Bayesian Network is used to infer student knowledge, taking advantage of relational models. A discussion of the methodology is addressed and preliminary results are presented. Ours first results go in the right direction as proved by a relative learning gain.

Keywords: intelligent tutoring systems, probabilistic relational models, relational student model, uncertainty, active learning simulators.

1 Introduction

It is a common fact that students enrolled in non-major Physic bachelors usually tend to learn Physics as a group of disjoint concepts. Very often students do not understand neither they are encouraged to understand the coherent structure underlying Physics. Based on our teaching experience, we believe that one of the main factors that makes Physics traditionally "difficult" is the lack of this understanding, which also decreases the level of confidence of the students on this subject.

There are several Active Learning Simulators (ALS) aimed to enhance the learning of Physics [1], [2]. Simulators respond to students' actions to foster their understanding based on a context of specific situations. There are several reasons to incorporate ALS in a learning environment, besides increasing practicing and carrying experiments in a safe environment, regardless of time and place, among them:

- Students construct their own understanding about phenomena and physics laws.
- The simulator can separate and handle different parameters allowing students to explore variables and their behaviors, visualizing their effects, so to improve students' comprehension of fundamental concepts.

A. Hernández Aguirre et al. (Eds.): MICAI 2009, LNAI 5845, pp. 533–544, 2009.
© Springer-Verlag Berlin Heidelberg 2009

- The use of different multimedia tools (images, animations, graphs, and sounds) which are useful to understand concepts, relations and processes.
- To investigate phenomena which are not feasible to carry out in the classroom or in the laboratory.

Most simulators allow students to explore the effect and behavior that different variables may have on a specific scenario. These kinds of systems are called Open Learning Environments (OLE) [3]. Sometimes, students repeat many times a given experiment without a clear comprehension of the underlying phenomena. In an OLE it is hard to infer the acquisition of students' knowledge, because specific objectives are needed in order to attain and enable an effective assessment of the learning goals [4]. Even in this case, it is hard to detect how much the student does actually know, which knowledge is not known, and what skills have or have not been acquired based only on the student's interaction with the system. Given the uncertainty inherent in these processes, a model that includes the handling of uncertainty is required.

To tackle the Physics learning problems mentioned above, we propose in this work to incorporate elements from intelligent tutoring systems (ITS), inside ALS, capable of giving students benefits of an adaptive learning environment, and taking care of the balance between free exploration capabilities versus the tutoring labor. This tutor will provide the best pedagogical action to be taken such as the decision to interrupt a given experiment, as well as to keep track of the student performance.

To handle uncertainty, student models based on Bayesian Networks (BN) [5] have been developed for ITS (see below). These Bayesian Networks are useful for diagnosing – or inferring – the current cognitive state of the student, using observable data [6], [7]. However, the effort required to define the network structure, to obtain and define the parameters involved, and to manage the computational complexity of the inference algorithms, makes the application of these types of models very difficult. This is true, particularly in real time situations, such as simulators [8]. Finding a general model for varied types of experiments and domains represents an additional challenge. In terms of development, efforts to create tutors for different simulator types are very elaborated and time consuming.

The main objective of this work is to present an ALS suited for a Physics scenario in which we incorporate elements from ITS, and where a Probabilistic Relational Model (PRM) based on a Bayesian Network is used to infer student knowledge, taking advantage of relational models [9].

2 Intelligent Tutoring System

In order to connect the ITS with our ALS, we faced several challenges:

1. To determine a student's level of comprehension based on student's interactions with the ALS.
2. To enrich the simulated virtual environment by allowing students to explore several variables, to visualize displayed graphic results, and to improve their knowledge.
3. To define a generic model, in order to adapt it to diverse student profiles, including different levels of knowledge and skills.

4. To follow student progress during experiment repetition, in order to decide the most appropriate pedagogical action.
5. To develop new experiments that can easily be added to the ALS site, without changing the student model.
6. To reduce the time needed to develop new ALSs for other domains.

The ITS keeps track of student's actions in the ALS while they conduct the experiment. It monitors student performance, updates the student model (see below), gives appropriate feedback if required, and determines future experiments. When a student carries out an experiment in this learning environment, the student model propagates the evidence to the knowledge-objects in the knowledge-base. Based on this, and other accumulated information obtained from evidence of previous experiments, a behavior module updates the student model as needed. After each experiment is carried out, the results are used then by the tutor module to decide the best feedback to be displayed to the student.

The ITS is based on probabilistic relational models [10]. A general structure for each module of the system was designed. The first step was to identify the classes in the model. The next step was to define the dependency model at class level, allowing it to be used for any knowledge-object in the class. This helped us to understand and build the student model. The classes and their relationships provide a general schema. Based on this schema, a skeleton is derived which integrates the relevant variables and their dependencies into a Bayesian network. Finally, a particular Bayesian network is obtained from the skeleton for each specific experiment in certain domains. In the following sections, the methodology that we follow to incorporate the main elements in the ITS is illustrated.

3 Probabilistic Relational Model

The method we use to infer student's knowledge is based on probabilistic relational models (PRM) that provide a new approach to modeling and integrating the predicting power of Bayesian networks and the advantages of relational models. PRM allow representing a given domain in terms of their entities, properties, and relationships [10]. Our ITS has four modules that use PRM as described below.

a. Student model

In order to apply PRM to student modeling [9], the main knowledge-objects involved in the domain were identified. Next, the dependency model at the class level was defined. The general schema for the student model and for the present case of study is depicted in figure 1.

For each class, a number of attributes (information variables and random variables) is defined, as shown in figure 2.

Once the model is specified at the class level (including the attributes and their dependencies), it is possible to extract a skeleton; that is, a general Bayesian network model as a fragment of the main model. A skeleton obtained from the model in figure 2 is depicted in figure 3. This network represents the dependencies among random variables. Knowledge-objects are represented by different levels of granularity, and by the results of the experiments – in terms of performance and experiment results. Note that the general schema allows the creation of a skeleton representing a subset of random variables that may be of interest at a given time. It also supplies information variables used to build the database entities, in order to develop the PRM. From each skeleton, it is possible to define different instances according to the values of specific variables in the probabilistic model

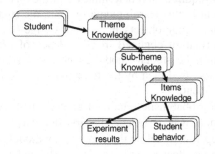

Fig. 1. A PRM schema for the student model

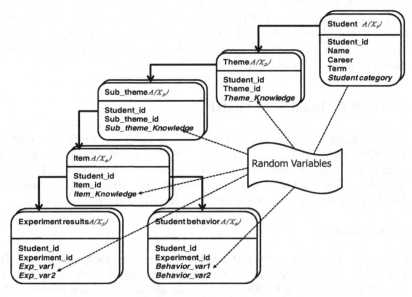

Fig. 2. A general probabilistic relational student model. The model specifies the main classes of objects, their attributes (information and random variables) and their dependencies.

The parameters of the PRM consist of a Conditional Probability Table (CPT) for each variable (attribute) given its defined parents. These parameters are defined based on the skeleton of the relational schema.

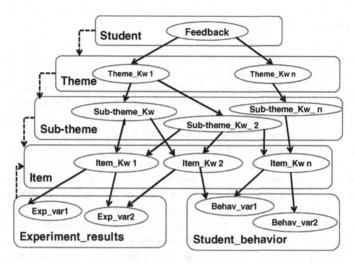

Fig. 3. A general skeleton obtained from the schema in figure 2. This skeleton specifies a general model for any experiment, which is later instantiated to a particular experiment.

b. The knowledge-base

The student's knowledge-base has different levels of granularity. It is organized in a hierarchical structure, from themes to sub-themes, and from sub-themes to knowledge items, and knowledge sub-items. When a student performs an experiment in the learning environment, the student model propagates the evidence from the experiment evaluation to the knowledge-objects in the knowledge-base.

c. Interaction and behavior analysis

To evaluate the student knowledge in each experiment, a set of exploration variables (student behavior), performance variables (experiment results), and knowledge items is defined. This conform a general schema for interaction analysis. When the student finishes one experiment the variable values are converted to probabilistic values in order to propagate the evidence and determine the student's inferred knowledge about each knowledge object.

d. The tutor model

After the student carries out an experiment, the inferred results by the student model are used by the tutor module to decide the most appropriate feedback for the student at any given moment. For example, for a good variable selection, the probability that a student knows a particular item, sub-theme, or theme increases. Then the tutor sends an appropriate congratulation message. The decision rules for the feedback are based on the abduction method of maximum likelihood [5] that the student actually knows a given knowledge object.

4 Experiment Design

In order to design a learning simulator that fosters comprehension and learning, the instructor should consider an appropriate scenario with a challenging goal and with specific initial conditions. The goal should be motivating enough to enable student engagement through the interaction with the simulator. The solution to the problem we designed involves the selection and manipulation of several physical quantities by the student. The physical quantities can take any value within specific ranges previously defined by the instructor. To attain the learning goal, students have to explore the consequences of choosing different values for each physical quantity. In each trial the student will learn more about the role of the chosen physical quantity. In each attempt the system will give appropriate feedback to the student, so that the student will learn more each time about the role of that chosen physical quantity. As a result, we expect that the student will make a better selection of parameters on successive attempts, until he or she finally attains the desired goal. It is necessary to be careful when designing a scenario. First, it must be non trivial, i.e., it should not be easily solved from the beginning. Also, to prevent simple memorization, the initial conditions must change randomly within specific ranges at each attempt. In addition, the initial conditions influence to some extent the level of challenge presented by the problem, as we mention later. Nevertheless, the values of physical quantities themselves are not so important: the main goal is the overall student understanding of the physical phenomena modeled by the simulator.

4.1 Case of Study

Conservation of Linear Momentum is one of the key topics of the *Physics I* course at university level. Therefore, a scenario where students can recognize and apply this conservation law was chosen. The concept of linear momentum (or momentum) of an object is defined as the product of its mass and its velocity, the latter being expressed as a vector [12]. Also, the student must realize that the total momentum of a given system is conserved, only if the resultant external force on the system is zero. Hence, a scenario was envisaged describing the story of an astronaut in trouble [13], [14]. The astronaut is located in interplanetary space far from his spaceship with a limited oxygen supply in his tank. The astronaut attempts to save himself by returning to the spaceship before the oxygen supply is over. Initially, he is at rest relative to the spaceship and carries with him some hand tools. The hand tools should be thrown in such a way that the astronaut moves towards the spaceship. The problem statement is the following: *"The astronaut Neri Vela is at mission in an asteroid near Europa, one of the four Galilean moons of Jupiter. He is outside of his spaceship repairing a sensor device set in the surface of the asteroid, when his primary source of oxygen supply, a cable connected to the spaceship unfortunately brakes up; so now he just has his secondary source of oxygen supply, which is a tank with an original capacity of 1.46 lt at a pressure of 41000 kPa that only contains at that time only a fraction of the total oxygen. If the astronaut only possesses a screwdriver, a pipe wrench and an adjustable wrench, how could he use these tools so to arrive to his spaceship before his secondary source of oxygen is completely exhausted? You should help the astronaut get back to his space ship on time by selecting the astronaut's mass, the tools that must be thrown, along with their mass and their velocity and direction"*. In this scenario, it is assumed that in interplanetary space, far away from massive bodies, the force of gravity on the system composed by the astronaut and his tools is small enough to be neglected in a first approximation. Hence, the total linear momentum of the system is conserved. In addition, it is implicitly assumed that the astronaut will be saved if he is able to arrive to the spaceship *before* his oxygen supply is exhausted. A value range of some the variables was selected so to offer a challenge problem to the student. The selected restriction variables were the distance between the astronaut and the spaceship, and the amount of oxygen available in the tank. We made several previous test calculations to determine the values of maximum distance to the spaceship and the minimum oxygen exhaustion time in order to provide a non trivial solution for this scenario. The instance of the general schema obtained is shown in figure 4.

Using this class diagram an experiment instance was defined for this scenario as shown in figure 5.

The system assigns then, within previously defined ranges, random values to the initial distance from the astronaut to the spaceship, D, and the remaining time for the oxygen to be exhausted, T_{Oxygen}. The corresponding ranges are: 20 m $\leq D \leq$ 25 m and 40 s $\leq T_{Oxygen} \leq$ 60 s. These values change every time the student initializes or resets the simulation. In this sense, the ALS is "dynamic", hindering the student to just memorize values. The corresponding simulator interface is shown in Figure 6.

In order to save the astronaut, the student is asked to select: i) the tools to be thrown, ii) the tools´ throwing direction, and iii) suitable values for the astronaut's mass, the mass of each tool and the tools´ throwing speed. These variables allow having a semi-open learning environment [4] and also they are used to infer knowledge as explained above. The allowed mass ranges for the tools are: 0.80 kg $\leq M_{Pipe\ Wrench} \leq$ 1.2 kg; 0.30 kg $\leq M_{Adjustable\ Wrench} \leq$ 0.70 kg and 0.10 kg $\leq M_{Scredriver} \leq$ 0.30 kg. The directions available to throw the tools can be set *towards the spaceship*, "→", or *away from the spaceship*, "←". The astronaut's mass range is 60.0 kg $\leq M_{Astronaut} \leq$ 100 kg, and the speed range to throw the tools is 0 m/s $\leq V_{Tools} \leq$ 20m/s.

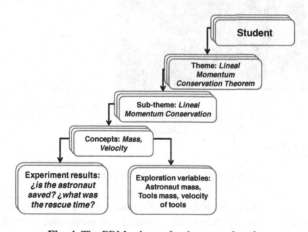

Fig. 4. The PRM schema for the case of study

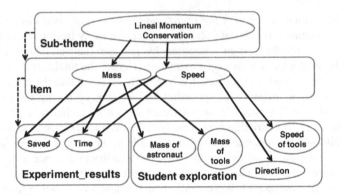

Fig. 5. Conservation of Linear Momentum Experiment Instance obtained from the general skeleton showed in Fig 3

If students fully understand the physical meaning of the law of conservation of linear momentum as well as the properties of linear momentum as a vector, they should recognize from the beginning that the tools must be thrown *away* from the spaceship, to enable the astronaut to move *towards* the spaceship. Also they should realize that the best choice corresponds to simultaneously throw all the tools with their maximum velocities and to select the minimum astronaut's mass. The chosen set of values for the exploration parameters by the student, allow us to infer the student comprehension of knowledge-items, sub-theme and theme.

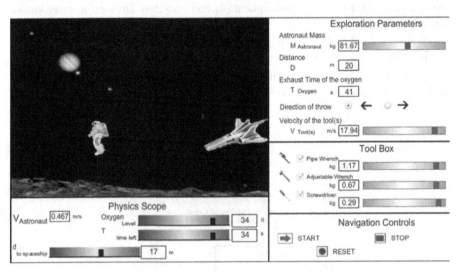

Fig. 6. Active Learning Simulator interface

To assess the cognitive state of the student, based on the results of the experiments as well as the student's interaction with the simulator, each value is evaluated according with a grade carefully defined by certain values ranges. Both the experiment results as the student interaction with the simulator are used to propagate evidence in the relational model. Each CPT was defined by a group of expert professors teaching Physics for the whole Conservation of linear momentum experiment instance model. Table 1 shows the CPT used to assess the $M_{Astronaut}$ knowledge-item.

Table 1. The CPT to propagate knowledge evidence based on student selection of the Astronaut Mass: P(Selected Astronaut Mass I Mass Concept)

Selected Astronaut Mass (kg)	P(Mass Concept=Known)	P(Mass Concept=Unknown)
$60 \leq M_A < 70$	**0.80**	0.1
$70 \leq M_A < 80$	**0.15**	0.2
$80 \leq M_A < 100$	**0.05**	0.7

In a similar way, Table 2 shows the CPT used to infer the *Speed* knowledge-item.

Table 2. The CPT to propagate knowledge evidence based on student selection of the Speed of the tools: P(Selected Speed of tools I Speed Concept)

Speed of tools (m/s)	P(Speed Concept=Known)	P(Speed Concept=Unknown)
$0 \leq V_T < 8$	**0.1**	0.75
$8 \leq V_T < 15$	**0.2**	0.15
$15 \leq V_T < 20$	**0.7**	0.10

After the evidence is propagated, the tutor module gives proper feedback using specific messages according to the student's performance. For example, if the tools' throwing direction is *towards* the spaceship "→", the ALS, will give the following feedback: *"YOU FAILED! The astronaut died.....You should repeat the experiment. In order to save the astronaut, in what direction must the tools be thrown?* However, if students choose the correct tools' throwing direction, "←" (i.e., away from the spaceship) but they do not select suitable values for the remaining exploration parameters, then the velocity acquired by the astronaut towards the spaceship will not be fast enough to arrive to the spaceship before the oxygen is exhausted. If they select unsuitable values for these variables and therefore the astronaut dies, the system gives the following feedback: *"YOU FAILED! the astronaut died....You should repeat the experiment. Please review the impact that the mass of the tools and the mass of the astronaut has on the Conservation of Momentum of the astronaut – tools- spaceship system. Are lighter tools or heavier tools better? Is a slim astronaut better than a fat one? Please review the impact that the tools throwing velocity has on the Conservation of Momentum of this system. Is high velocity or low velocity better?"*

It is important to note that the probability of success depends to some extent on the random initialization of the distance to the spaceship, D, and the remaining time to finish the oxygen supply, T_{Oxygen}, both set by the ALS. For the most restrictive case, that is, maximum D and minimum T_{Oxygen}, only the best selection of parameters will save the astronaut. In contrast, for the least restrictive case, minimum D and maximum T_{Oxygen}, there is a wider range of parameter combinations that will enable the astronaut to be saved. However, even in the latter case, the parameter ranges were selected in such a way that the astronaut will be saved by only 30% of the student's possible selections of parameter values, leaving the problem non trivial.

5 Evaluation Process

In our study we have analyzed learners' parameter explorations while interacting with the virtual learning environment; insight has been obtained about the general effectiveness of the experiment performance. We applied a study to a group of undergraduate students from different engineering majors taking the Physics I course at the Tecnológico de Monterrey, Campus Ciudad de México during the January – May term of 2009. The process of evaluation was as follow:

- **The pre-phase.** A written *Pre-test* including four questions related to the law of Conservation of Linear Momentum was applied to the student sample. To avoid the lecturer's influence over the results, the professors did not give lectures for this theme during the period when students worked with our ALS.
- **Participants.** A total of N_{tot} = 45 students participated in the study. They were divided randomly in two groups: one group interacted with the ALS (hereafter, the focus group, N_{focus} = 20) while the other group (hereafter, the control group, $N_{control}$ = 25) did not.
- **Experiment design.** The ALS was available for the focus group during about one week. This system also stored activity log files of the students' sessions and results.
- **The post-phase.** As with the pre-test, this consisted of a written test of four questions, or *Post-Test*, very similar to those applied in the *Pre-Test*. The post-tests were applied to the whole student sample once the focus group ended working with the ALS.
- **Analysis of data and results.** Learning gains were calculated using the grades derived from the Pre-test and Post-test. In order to compare our results we computed a relative learning gain for each student as defined by [15] and given in Equation (1):

$$G_{rel} = \frac{(PostTest - PreTest)}{(100 - PreTest)} \quad (1)$$

Also, for comparison purposes, we have calculated a "simple" gain, as follows:

$$G = (PostTest - PreTest) \quad (2)$$

Additionally, in order to assess how many attempts the students made before they arrived to save the astronaut, we have calculated an "efficiency" of use defined as the number of successful attempts relative to the total attempts made by the students.

Table 3. Summary of average learning gains and efficiency results for the focus group (N = 20)

Pre-Test	Post-Test	G_{rel}	G	Attempts	Efficiency
56 ± 21	76 ± 22	0.44 ± 0.32	20 ± 5	11	0.34

Table 4. Summary of average learning gains for the control group (N = 25)

Pre-Test	Post-Test	G_{rel}	G
60 ± 18	69 ± 13	0.13 ± 0.20	9 ± 5

We summarize our results in Tables 3 and 4, where we show the average values for the Pre-Test, the Post-Test, the Relative gain and the simple gain, for the focus (N_{focus} = 20) and control ($N_{control}$ = 25) groups, respectively, with their corresponding standard deviations. We also present in table 3 the average values of the total number of attempts made by the students and the corresponding efficiency, calculated as explained above. As can be seen, students using the ALS had higher relative and simple learning gains as compared to those students that did not used the ALS. Although N_{total} is still small, these results are encouraging and go in the right direction. We will increase the student sample size with forthcoming experiments to ensure confidence about these results.

It is interesting to note also that students in the average made a little more than 10 attempts and their average efficiency was about of one third. This shows that the proposed scenario was not trivial from the very start and presented some challenge to most students. This observation together with the larger gains obtained by students of the focus group as compared with those of the control group reinforces the assertion that the ALS promoted a better comprehension of the linear conservation law.

6   Conclusions and Future Work

Active Learning Simulators allow students to practice and carry out experiments in a safe environment - at any time, and in any place. ALS enhance learning, and provide the bridge from concept to practical understanding. We have incorporated relational probabilistic models in Bayesian networks in order to infer the most probable understanding level of physics concepts of students interacting with an ALS. By adding an Intelligent Tutoring System, it is possible to provide personal ad hoc feedback to each student based upon a probabilistic model as well as the student interaction and behavior within the ALS. We have calculated students learning gains from the comparison of Post-test and Pre-test grades for a sample of 45 students. Our derived results go in the right direction in the sense that the learning gains for students using our ALS are larger than those obtained bay students who did not, although our sample number size is still rather small. We plan to increase our student sample size so to give statistical significance to our derived learning gains. Our preliminary results give us confidence to elaborate other physics scenarios under an appropriate ALS to enhance student learning of Physics concepts.

References

1. Esquembre, F.: Easy Java Simulations: a software tool to create scientific simulations in Java. Computer Physics Communications 156(2) (2004)
2. Koo-Chul, L., Julian, L.: Programming physics software in Flash. Computer Physics Communications 177 (2007)

3. Bunt, A., Conati, C.: Probabilistic Student Modelling to Improve Exploratory Behaviour. Journal of User Modeling and User-Adapted Interaction 13(3), 269–309 (2003)
4. Noguez, J., Sucar, E.: A semi-open learning environment for virtual laboratories. In: Gelbukh, A., de Albornoz, Á., Terashima-Marín, H. (eds.) MICAI 2005. LNCS (LNAI), vol. 3789, pp. 1185–1194. Springer, Heidelberg (2005)
5. Pearl, J.: Probabilistic Reasoning in Intelligent Systems. Morgan Kauffman, San Mateo (1988)
6. VanLehn, K.: Bayesian student modeling, user interfaces and feedback: a sensitivy analysis. International Journal of Artificial Intelligence in Education, 154–184 (2001)
7. Mayo, M., Mitrovic, A.: Optimising ITS behaviour with Bayesian networks and decision theory. International Journal of Artificial Intelligence in Education 12(2), 124–153 (2001)
8. Adams, W.K., Reid, S., LeMaster, R., McKagan, S.B., Perkins, K.K., Wieman, C.E.: A study of educational simulations - interface design. Journal of Interactive Learning and Research 19 (2007)
9. Sucar, L.E., Noguez, J.: "Student Modeling" in Bayesian Networks: A Practical Guide to Applications. In: Pourret, Naim, Marcot (eds.), March 2008, pp. 173–185. J. Wiley & Sons, Chichester (2008)
10. Koller, D., Friedman, N.: Probabilistic Graphical Models: Principles and Techniques. MIT Press, Cambridge (2009)
11. Getoor, L., Friedman, N., Koller, D., Pfeffer, A., Taskar, B.: Probabilistic Relational Models. In: Getoor, L., Taskar, B. (eds.) Introduction to Statistical Relational Learning (2007)
12. Young, H.D., Freedman, R.A.: University Physics, 12th edn., vol. I. Pearson, Addison Wesley (2008)
13. Mc Kevitt, P., Muñoz, K., Noguez, J., Lunney, T.: Combining educational games and virtual learning environments for teaching Physics with the Olympia architecture. In: Proceedings of the 15th International Symposium on Electronic Art (ISEA 2009), August 23-September 1. University of Ulster, North Ireland (in press, 2009)
14. Muñoz, K., Noguez, J., Mc Kevitt, P., Neri, L., Robledo-Rella, V., Lunney, T.: Adding Features of Educational Games for Teaching Physics. In: Proceedings of the 39th IEEE International Conference on Frontiers in Education (FIE 2009), San Antonio, Texas. USA, Octubre 18-21 (in press, 2009)
15. Hake, R.R.: Interactive-engagement versus traditional methods: A six-thousand-student survey of mechanic test data for introductory physics courses. Am. J. Phys. 66(1), 64–74 (1998)

Leukocyte Recognition Using EM-Algorithm

Mario Chirinos Colunga[1], Oscar Sánchez Siordia[2], and Stephen J. Maybank[1]

[1] Department of Computer Science and Information Systems, Birkbeck,
University of London, Malet Street, London WC1E 7HX
[2] Centro Nacional de Investigación y Desarrollo Tecnológico (CENIDET)
Cuernavaca, Morelos, México

Abstract. This document describes a method for classifying images of
blood cells. Three different classes of cells are used: Band Neutrophils,
Eosinophils and Lymphocytes. The image pattern is projected down to
a lower dimensional sub space using PCA; the probability density function for each class is modeled with a Gaussian mixture using the EM-
Algorithm. A new cell image is classified using the maximum a posteriori
decision rule.

Key words: Leukocyte recognition, EM-Algorithm, PCA.

1 Introduction

Blood cell analysis is an important diagnostic tool because it can help to detect a
width range of diseases. Two types of blood cell analysis are performed: complete
blood count and differential blood count. In complete blood count the numbers
of erythrocytes (red cells), leukocytes (white cells) and platelets in the peripheral
blood (the blood in the circulatory system) are counted to obtain a concentration
of cells per unit volume. In a differential blood count the different classes of
leukocytes in the peripheral blood or bone marrow are counted to provide a
more detailed diagnosis.

The cell counts for peripheral blood are obtained manually or by an automatic flow cytometer, whereas the counts for bone marrow are always obtained
manually. The manual counting is done via visual inspection by technicians under a microscope. It has the advantage of being accurate, in that the counts are
repeatedly close to the correct values, with little bias; On the other hand it has
the disadvantage of being a highly time consuming process. Only a few hundred
cells are counted. This reduces the precision of the counts of cells that have a low
probability of occurring. The precision of the low cell counts can also be affected
by subjectivity, different results can be obtained by different technicians or by
the same technician in a second count. In the automatic flow cytometer the cells
pass one by one through a channel where sensors measure volume, conductivity
and scattered light. This method is accurate and precise for normal blood samples because it can count up to 80 times more cells than for manual counting.
However it cannot be used for pathological blood samples. In such cases it is
necessary to revert to manual counting

A. Hernández Aguirre et al. (Eds.): MICAI 2009, LNAI 5845, pp. 545–555, 2009.
© Springer-Verlag Berlin Heidelberg 2009

For some time researchers have been developing automatic visual inspection systems using computer vision and pattern recognition methods. This make it possible to obtain cell count automatically from images, reducing the cost of analysis [5]. The process of blood cell recognition described in this document involves three steps: segmentation, feature selection and classification. A classifier is implemented by modelling the probability distribution of each class of cells with a mixture of Gaussian functions. The parameters of the Gaussian functions are obtained using the EM-Algorithm on the data from each class. The EM-Algorithm is initialized using K-Means clustering and the data is projected down into a lower dimensional space using PCA. The class conditional densities are found using Bayes' theorem. Classification is done by choosing the class with the highest probability given the data.

The document is structured as follows: section 2 contains the background theory: K-Means clustering, EM-Algorithm and principal component analysis. Section 3 describes the method for cell classification and its implementation. Section 4 presents the experimental results and section 5 offer some conclusions about the experiment.

2   Background Theory

2.1   EM-Algorithm for Gaussian Mixtures

The expectation maximization algorithm or EM-algorithm is a method to find the maximum likelihood solution for models with latent variables. The name was given in [4] which points out that the method had been proposed many times by other authors. The model is a Gaussian mixture distribution (1) and the latent variables z_k are labels that indicate from which component k of the mixture the measurements came from [6].

The probability density function for a mixture model with K Gaussian distributions $N(x|\mu_k, \Sigma_k)$ with means μ_k and covariance matrices Σ_k is defined by:

$$p(x) = \sum_{k=1}^{K} \pi_k N(x|\mu_k, \Sigma_k) \tag{1}$$

The parameters $\pi_k \in [0,1]$ are the mixing coefficients for the Gaussian densities, they are subject to the restriction:

$$\sum_{k=1}^{K} \pi_k = 1 \tag{2}$$

The conditional probability of z_k given x can be seen as the responsibility that the component k takes for explaining the observation x [2]. This conditional probability is given by:.

$$p(z_k|x) = \frac{\pi_k N(x|\mu_k, \Sigma_k)}{\sum_{j=1}^{K} \pi_j N(x|\mu_j, \Sigma_j)} \tag{3}$$

The log of the likelihood function for a data set $X = (x_1, .., x_N)^T$ with N data points under the Gaussian mixture model is given by:

$$\ln p(X|\pi, \mu, \Sigma) = \sum_{n=1}^{N} \ln \left(\sum_{k=1}^{K} \pi_k N(x_n|\mu_k, \Sigma_k) \right) \tag{4}$$

There is no closed form solution for the values of π, μ and Σ which maximize the log likelihood in (4).

We can find an estimates of π, μ and Σ by first setting to zero the derivative of $\ln p(X|\pi, \mu, \Sigma)$ with respect to the mean μ_k of the Gaussian component.

$$\frac{\partial}{\partial \mu_k} \ln p(X|\pi, \mu, \Sigma) = 0 \tag{5}$$

$$\sum_{n=1}^{N} \frac{\partial}{\partial \mu_k} \ln \left(\sum_{j=1}^{K} \pi_j N(x_n|\mu_j, \Sigma_j) \right) = 0 \tag{6}$$

Using the chain rule and equations (1), (3) and (4), we have:

$$\sum_{n=1}^{N} \frac{(\pi_k N(x_n|\mu_k, \Sigma_k))}{p(x_n)} \left(-\Sigma_k^{-1}(x_n - \mu_k) \right) = 0 \tag{7}$$

$$-\Sigma_k^{-1} \sum_{n=1}^{N} p(z_k|x_n)(x_n - \mu_k) = 0 \tag{8}$$

therefore:

$$\sum_{n=1}^{N} p(z_k|x_n)(x_n - \mu_k) = 0 \tag{9}$$

which yields:

$$\mu_k \sum_{n=1}^{N} p(z_k|x_n) = \sum_{n=1}^{N} p(z_k|x_n)x_n \tag{10}$$

If we define the effective number of points N_k assigned to cluster k as:

$$N_k = \sum_{n=1}^{N} p(z_k|x_n) \tag{11}$$

we find that μ_k is given by:

$$\mu_k = \frac{1}{N_k} \sum_{n=1}^{N} p(z_k|x_n)x_n \tag{12}$$

Now we set the derivative of $\ln p(X|\pi, \mu, \Sigma)$ with respect to the covariance matrix Σ_k to zero,

$$\frac{\partial}{\partial \Sigma_k} \ln p(X|\pi, \mu, \Sigma) = 0 \tag{13}$$

548   M.C. Colunga, O.S. Siordia, and S.J. Maybank

thus:

$$\sum_{n=1}^{N} \frac{\partial}{\partial \Sigma_k} \ln \left(\sum_{j=1}^{K} \pi_j N(x_n|\mu_j, \Sigma_j) \right) = 0 \qquad (14)$$

which using (1) yields:

$$\sum_{n=1}^{N} \frac{\pi_k}{p(x_n)} \frac{\partial N(x_n|\mu_k, \Sigma_k)}{\partial \Sigma_k} = 0 \qquad (15)$$

defining $M(x)^2 \equiv (x - \mu_k)^T \Sigma^{-1}(x - \mu_k)$ and $\alpha \equiv \frac{1}{(2\pi)^{D/2}}$, where D is the dimension of the vector x, we have:

$$\sum_{n=1}^{N} \frac{\pi_k \alpha}{p(x_n)} \frac{\partial}{\partial \Sigma_k} \left(\frac{e^{(-\frac{1}{2}M(x_n)^2)}}{|\Sigma_k|^{1/2}} \right) = 0 \qquad (16)$$

Where $|\Sigma_k|$ is the determinant of the matrix Σ_k. Equation (16) yields:

$$\sum_{n=1}^{N} \frac{\pi_k \alpha}{p(x_n)} \left(\left(\frac{\partial}{\partial \Sigma_k} \frac{1}{|\Sigma_k|^{1/2}} \right) e^{(-\frac{1}{2}M(x_n)^2)} + \frac{1}{|\Sigma_k|^{1/2}} \frac{\partial}{\partial \Sigma_k} e^{(-\frac{1}{2}M(x_n)^2)} \right) = 0 \qquad (17)$$

On using $\frac{\partial}{\partial X}|X| = |X|X^{-T}$, where X is a square matrix with a non-zero determinant. Equation (17) yields:

$$\sum_{n=1}^{N} \frac{\pi_k \alpha}{p(x_n)} \left(\frac{-|\Sigma_k|\Sigma_k^{-T} e^{(-\frac{1}{2}M(x_n)^2)}}{2|\Sigma_k|^{3/2}} - \frac{e^{(-\frac{1}{2}M(x_n)^2)}}{|\Sigma_k|^{1/2}} \frac{\partial}{\partial \Sigma_k} \frac{1}{2} M(x_n)^2 \right) = 0 \quad (18)$$

On using $\frac{\partial}{\partial X} a^T X^{-1} b = -X^{-T} a b^T X^{-T}$, where a and b are vectors and X is any invertible matrix, (18) yields:

$$-\sum_{n=1}^{N} \frac{\pi_k \alpha}{p(x_n)} \left(\frac{\Sigma_k^{-1} e^{(-\frac{1}{2}M(x_n)^2)}}{2|\Sigma_k|^{1/2}} - \frac{\Sigma_k^{-1} e^{(-\frac{1}{2}M(x_n)^2)}}{2|\Sigma_k|^{1/2}} (x_n - \mu_k)(x_n - \mu_k)^T \Sigma_k^{-1} \right) = 0 \qquad (19)$$

Factorizing and multiplying both sides of the equation by Σ_k:

$$\sum_{n=1}^{N} \frac{\pi_k \alpha |\Sigma_k|^{-1/2} e^{(-\frac{1}{2}M(x_n)^2)}}{p(x_n)} (I - (x_n - \mu_k)(x_n - \mu_k)^T \Sigma_k^{-1}) = 0 \qquad (20)$$

It follows from (3) and (20) that:

$$\sum_{n=1}^{N} p(z_k|x_n)(I - (x_n - \mu_k)(x_n - \mu_k)^T \Sigma_k^{-1}) = 0 \qquad (21)$$

thus:

$$N_k I = \sum_{n=1}^{N} p(z_k|x_n)(x_n - \mu_k)(x_n - \mu_k)^T \Sigma_k^{-1} \qquad (22)$$

from which follows that Σ_k is given by:

$$\Sigma_k = \frac{1}{N_k} \sum_{n=1}^{N} p(z_k|x_n)(x_n - \mu_k)(x_n - \mu_k)^T \tag{23}$$

Let λ be the Lagrage multiplier associated with the constraint (2) on the π_k and define V by:

$$V = \ln p(X|\pi, \mu, \Sigma) + \lambda \left(1 - \sum_{j=1}^{K} \pi_j\right) \tag{24}$$

it follows that:

$$\frac{\partial V}{\partial \pi_k} = -\lambda + \sum_{n=1}^{N} \frac{\partial}{\partial \pi_k} \ln \left(\sum_{j=1}^{K} \pi_j N(x_n|\mu_j, \Sigma_j)\right) \tag{25}$$

On setting $\frac{\partial V}{\partial \pi_k} = 0$ it follows that:

$$\sum_{n=1}^{N} \frac{1}{p(x_n)} \frac{\partial \pi_k N(x_n|\mu_k, \Sigma_k)}{\partial \pi_k} = \lambda \tag{26}$$

thus:

$$\sum_{n=1}^{N} \frac{1}{p(x_n)} N(x_n|\mu_k, \Sigma_k) = \lambda \tag{27}$$

which yields:

$$\pi_k = \frac{1}{\lambda} \sum_{n=1}^{N} \frac{\pi_k N(x_n|\mu_k, \Sigma_k)}{p(x_n)} = \frac{1}{\lambda} \sum_{n=1}^{N} p(z_k|x_n) = \frac{N_k}{\lambda} \tag{28}$$

Substituting into the constraint (2):

$$\frac{1}{\lambda} \sum_{k=1}^{K} N_k = 1 \tag{29}$$

It follows from (28) and (29) that:

$$\pi_k = \frac{N_k}{N} \tag{30}$$

This results do not constitute a closed form solution for $\ln p(X|\pi, \mu, \Sigma)$, since these parameters appear in $p(z_k|x)$, however a solution for the maximum likelihood problem can be obtained by iteratively updating the parameters.

The EM – Algorithm begins by first choosing initial values for π_k, μ_k and Σ_k and then iterates two steps to update their values. It is common to run a

K-Means algorithm in order to find suitable initial values. In the first step called E-step (expectation) the current values of the parameters are used to evaluate the a posteriori probabilities (3). Then these probabilities are used in the M-step (maximization) to re-estimate the parameters π_k, μ_k and Σ_k. In each iteration the log likelihood is guaranteed to increase [2]. The iterative process stops when the change in the log likelihood or in the parameters falls below a threshold.

There are some difficulties with the EM-Algorithm. It is inclined to get stuck in a local maximum, and there is no guarantee that it converges to a the global maximum [6]. Another problem happens if one of the Gaussians collapses to a single point [2]. This severe overfitting can be avoided using a Bayesian approach in which a prior density $p(\pi, \mu, \Sigma)$ is defined. A MAP (maximum a posteriori) solution is found [2].

2.2   K-Means

K-Means algorithm is one of the most popular clustering methods. It is also called c-means, iterative reallocation or basic iso data [16]. The K clusters are formed such that the distances between the points in a cluster are relatively small compared to the distances to data outside the cluster. The aim is to find an assignment of the data to clusters that minimizes the sum of squared distances of each data point to its closer prototype or cluster mean μ_k.

Data assignment is indicated via a matrix of flags (1-of-K coding scheme) with elements r_{nk} that describe to which of the K clusters the element $\mathbf{x_n}$ is assigned. If $\mathbf{x_n}$ is assigned to cluster k then $r_{nk} = 1$ and $r_{nj} = 0$ $\forall j \neq k$. The goal is to find the values for r_{nk} and μ_k that minimize an objective function or distortion J [2].

This is done by a two steps that are iterated. The steps correspond to successive optimizations of r_{nk} and μ_k. In the first step each data point is assigned to the cluster which has the closest prototype. The first set μ of cluster centres is chosen randomly before the iteration begins. In the second step new μ_k are calculated using the assignation of data points to clusters obtained in the previous step. The process stops when no movement of data from a cluster to another reduces the objective function J or a maximum number of iterations is reached. The objective function is the sum of squared distances for each data point $\mathbf{x_n}$ to its corresponding cluster prototype μ_k.

$$ J = \sum_{n=1}^{N} \sum_{k=1}^{K} r_{nk} \|\mathbf{x_n} - \mu_k\|^2 \tag{31} $$

Since in each step the objective function is reduced the convergence is assured, nevertheless it can converge into a local minimum instead of a global minimum, and it may even produce empty clusters. This means that the result is a deterministic function of its initial parameters. One way to deal with this problem is to run K-means many different times and chose the best result [6].

Variants of K-Means algorithm are obtained by replacing the Euclidean distance with another measure of distance. In [13] a distance based on symmetry is used successfully to produce spherical, elliptical and ring shaped clusters. The detection of ring shaped clusters in digital images is important in industrial applications.

2.3 Lower Dimensional Space Mapping Using PCA

Working in high dimensional space leads to many problems. As the dimension of the feature space grows the number of sample points needed to reliably estimate the parameters of a classifier grows exponentially. This is known as the curse of dimensionality. It causes a poor generalization ability and thereby reduces the accuracy of a classifier. However, real data will often be confined to a region of the feature space having a lower effective dimension (intrinsic dimensionality) [2]. It is desirable to project the data to a feature space with a dimension equal to the effective dimension of the data. Feature extraction methods determine an appropriate subspace of dimension M in the original space of dimension D where $M \leq D$.

PCA is perhaps the most used technique for feature extraction. It involves finding an orthogonal projection P of the data X onto a lower dimensional linear space, where the variance of the projected data is maximized [2]. This means that the projection $Y = P^T X$ diagonalize the covariance matrix [3]:

$$\Sigma_Y = P^T \Sigma_X P \qquad (32)$$

The covariance matrix Σ_X is a symmetric matrix. Symmetry guarantees that all eigenvalues of Σ_X are real and that there is an orthonormal basis of eigenvectors [7]. Then the selected projection is a real orthogonal matrix of eigen vectors Φ that diagonalize the covariance matrix:

$$\Phi^T \Sigma_X \Phi = \Lambda = \mathrm{diag}(\lambda_1, ..., \lambda_D) \qquad (33)$$

By ranking the eigenvectors by their corespondent eigenvalues and selecting the first M principal components, is generated an orthogonal decomposition of the vector space $\Re^D$ into two complementary and orthogonal subspaces, the principal subspace $F = \{\Phi_i\}_{i=1}^M$ and its orthogonal complement $\overline{F} = \{\Phi_i\}_{i=M+1}^D$. The components in $\overline{F}$ can be interpreted as noise present in the signal [9].

PCA is a partial Karhunen-Loève transformation, a term that include all the transformations based on the covariance matrix [16], which extract a subspace of lower dimension, $\mathbf{y} = \Phi_M^T \mathbf{x} : \Re^D \rightarrow \Re^M$, corresponding to the maximum eigenvalues [8].

Two of the most wide spread applications of PCA in computer vision are face detection and face recognition. A. Turk and P. Pentland [14] introduced for the first time the term "eigenfaces", by projecting the space of the sample faces in a subspace whose axes or egienfaces, are the principal components of the training data set.

3 Leukocyte Recognition Implementation

The leukocyte images used for this experiment were acquired from "Hemosurf - an Interactive Hematology Atlas" [15] with permission of the authors to use them for research proposes. The images were taken with a Sony DXC-3000A 3CCD video camera and a Zeiss Axioskop microscope using a 100X objective and a 10x eyepiece. Judging by the size of the erythrocytes, which normally exhibit high uniformity in diameter [1], the resolution of the images is approximately 160nm/pixel.

A "ground truth" for each one of the images was created manually by labeling the regions of interest: nucleus, cytoplasm and background, figure 1, a similar manual segmentation has been created in [10] using 431 images to evaluate image segmentation accuracy; in [11] a manual selection of regions of interest is used to learn the parameters of a image segmentation. Leukocytes are extracted from the images by centring a square window at the center of mass of the cytoplasm. The window size is set at the diameter of the largest cell, in this case the neutrophil with a maximum diameter of 104 pixels. The numbers of leukocytes in each image and the corresponding pixels for each leukocyte are found using connected components, a description of some connected components algorithms can be found in [3]. After the extraction each window containing a leukocyte is dawn sampled and saved as a independent image.

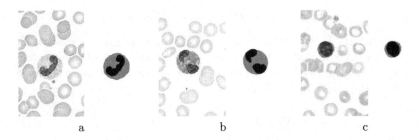

a   b   c

Fig. 1. "Ground truth" images manually created to label the regions of interest: nucleus, cytoplasm and background. a) The original image of a Neutrophil and its correspondence "ground truth" image. b) Eosinophil c) Lymphocyte.

The class conditional density function $p(x|\omega_c)$ for each leukocyte class ω_c is modelled using a mixture of K Gaussian densities. To find the parameters of each mixture, the means μ_{ck} and the covariances Σ_{ck}, the EM-Algorithm is applied independently to each class. The parameters for the EM-algorithm are initialized using the best result from a set of K-Means clustering trials.

The training set to learn the class conditional densities is created as follows: each cell example is rotated at a fixed number of different angles to generate more images, figure 2; in this way the classifier is made invariant to rotation i.e. the classifier learn each example at different rotations. In other works the cells are rotated to have a uniformly aligned training and test set. In [12] the cells are rotated based on the vector created from the center of mass of the cytoplasm and the center of mass of the nucleus and in [17] nucleus centroids are aligned with a reference image and the image is rotated until the absolute difference between the images is a minimum.

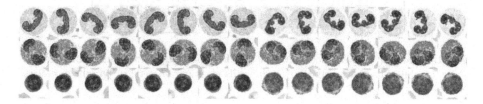

Fig. 2. Each image is rotated several times to make the classifier invariant to rotation

After training, a new leukocyte image is classified using the maximum posteriory rule, i.e. the new image is assigned to class ω_i if $p(\omega_i|x) > p(\omega_j|x)\,\forall\,i \neq j$. The posterior probabilities are found using the Bayes' theorem. Both the training and test images have been contrast stretched.

4   Experimental Results

Three classes of leukocytes (band neutrophils, eosinophils and lymphocyte) were used in the experiment. They were chosen because are the three most dissimilar classes. At this point the data set consists only of 15 examples per class. To create the training set each image is rotated [0, 45, 90, 135, 180, 225, 270, 315] degrees and then down sampled at scale $\sigma = 0.25$. The data set is projected down into a lower dimensional space using PCA and the first 7 principal components, figure 3; the number of principal component used was found empirically which is the maximum number of principal components that allow to have nonsingular covariance matrices for all the mixture components. The number of Gaussian functions that form the mixture of each class was set arbitrarily to $K = 4$. The parameters of the EM-Algorithm were initialized using the best result from 10 trials of the K-Means clustering. The parameters obtained for each one of the Gaussian mixtures can be seen in figure 4; it can be seen that the means correspond to different representation of a cell class as variation in rotation, shape or size. The classifier is tested using cross validation leave one out method.

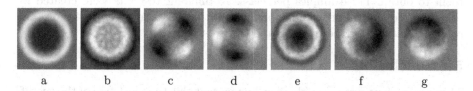

a   b   c   d   e   f   g

Fig. 3. First seven principal components for the leukocyte images data

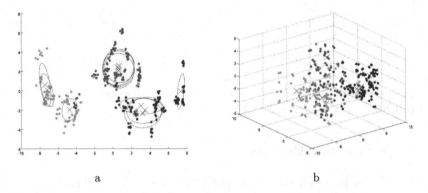

ω_1				ω_2				ω_3			
μ_1	μ_2	μ_3	μ_4	μ_1	μ_2	μ_3	μ_4	μ_1	μ_2	μ_3	μ_4
Σ_1	Σ_2	Σ_3	Σ_4	Σ_1	Σ_2	Σ_3	Σ_4	Σ_1	Σ_2	Σ_3	Σ_4

Fig. 4. Means and covariance matrices for the four components Gaussian mixture. a) neutrophils class parameters. b) eosinophils class parameters. c) lymphocyte class parameters.

a   b

Fig. 5. a) Cell data plotted using the first two principal components, red dots indicate neutrophils samples, green correspond to Eosinophils and blue to lymphocytes; ellipses correspond to the influence of each Gaussian. b) Same data plotted using the three principal components.

Confusion matrix for 15 training sets			15 test sets using leave one out cross validation		
	ω_1	ω_2	ω_3		ω_1	ω_2	ω_3

	ω_1	ω_2	ω_3		ω_1	ω_2	ω_3
ω_1	0.82	0.06	0.01	ω_1	0.80	0.013	0.06
ω_2	0.05	0.89	0.05	ω_2	0	0.86	0.13
ω_1	0	0	1	ω_1	0.06	0.06	0.86

Fig. 6. Confusion matrices for the test and training sets

Fifteen different training sets were created by using 14 cells per class and rotated them to obtain 112 examples per class. 15 different test sets were created by using the remaining one example per class without rotation. Figure 4 shows the confusion matrix for the training and test sets.

5   Conclusions

This was my first test using blood cell images; at this time just tree of the six different types of leukocyte in peripheral blood were used, neutrophils, eosinophils and lymphocytes. Just 15 samples per class were available in this experiment. The lack of samples meant that only a few principal component could be used to represent the data and that the number of Gaussian per mixture was small. Currently 787 images of blood slides containing 957 leukocytes have been acquired and have to be manually labeled and segmented to evaluate classification and segmentation performance.

A extension to this experiment could be to use a diferent feature extraction method to improve class separability, use a method to find the optimal numbers of components for each Gaussian mixture and use a different image descriptor.

References

1. Hoffbrand, A.V., Pettit, J.E.: Essential Haematology., 3rd edn. Blackwell Science, Malden (1993)
2. Bishop, C.M.: Pattern Recognition and Machine Learning (Information Science and Statistics). Springer, Heidelberg (2006)
3. Davies, E.R.: Machine Vision: Theory, Algorithms, Practicalities. Morgan Kaufmann Publishers Inc., San Francisco (2004)
4. Dempster, A.P., Laird, N.M., Rubin, D.B.: Maximum likelihood from incomplete data via the em algorithm. Journal of the Royal Statistical Society. Series B (Methodological) 39(1), 1–38 (1977)
5. Fluxion Biosciences: Microfluidic flow cytometer with stop-flow capability provides enhanced cellular imaging. Technical report, Fluxion Biosciences, QB3- Mission Bay; 1700 4th St., Suite 219; San Francisco, CA 94143 (2005)
6. Forsyth, D.A., Ponce, J.: Computer Vision: A Modern Approach. Prentice Hall Professional Technical Reference, Englewood Cliffs (2002)
7. Golub, G.H., Van Loan, C.F.: Matrix Computations, 3rd edn. Johns Hopkins University Press (October 11, 1996)
8. Moghaddam, B., Pentland, A.: Probabilistic visual learning for object detection. In: ICCV 1995, vol. 1, pp. 786–793 (1995)
9. Moghaddam, B., Pentland, A.: Probabilistic visual learning for object representation. In: PAMI, vol. 19, pp. 696–710 (July 1997)
10. Theera-Umpon, N.: Patch-Based White Blood Cell Nucleus Segmentation Using Fuzzy Clustering. Transactions on Electrical Eng., Electronics, and Communications, ECTI-EEC 3, 5–19 (2005)
11. Sabino, D.M.U., da Fontoura Costa, L., Rizzatti, E.G., Zago, M.A.: A texture approach to leukocyte recognition. Real-Time Imaging 10, 205–216 (2004)
12. Sanei, S., Lee, T.M.K.: Cell Recognition Based on PCA and Bayesian Classification. In: Fourth International Symposium on Independent Component Analysis and Blind Signal Separation. ICA, April 1 (2003)
13. Su, M.C., Chou, C.H.: A modified version of the k-means algorithm with a distance based on cluster symmetry. IEEE Trans. Pattern Anal. Mach. Intell. 23(6), 674–680 (2001)
14. Turk, M.A., Pentland, A.P.: Face recognition using eigenfaces, pp. 586–591 (1991)
15. Woermann, U., Montandon, M., Tobler, A., Solemthaler, M.: HemoSurf - An Interactive Hematology Atlas. Division of Instructional Media, Bern (2004)
16. Webb, A.R.: Statistical pattern recognition. John Wiley & Sons, Chichester (1999)
17. Yampri, P., Pintavirooj, C., Daochai, S., Teartulakarn, S.: White Blood Cell Classification based on the Combination of Eigen Cell and Parametric Feature Detection. In: IEEE (ed.) Industrial Electronics and Applications (May 24, 2006)

An Automaton for Motifs Recognition in DNA Sequences

Gerardo Perez[1], Yuridia P. Mejia[1], Ivan Olmos[1], Jesus A. Gonzalez[2],
Patricia Sánchez[3], and Candelario Vázquez[3]

[1] Facultad de Ciencias de la Computación,
Benemérita Universidad Autónoma de Puebla,
14 sur y Av. San Claudio, Ciudad Universitaria,
Puebla, México
{onofre.gerardo,yuripmt,ivanoprkl}@gmail.com
[2] Instituto Nacional de Astrofísica, Óptica y Electrónica,
Luis Enrique Erro No. 1, Sta. María Tonantzintla, Puebla, México
jagonzalez@inaoep.mx
[3] Departamento de Ciencias Microbiológicas
Benemérita Universidad Autónoma de Puebla,
Ciudad Universitaria,
Puebla, México
mpgalon@siu.buap.mx, rnahelix@yahoo.com.mx

Abstract. In this paper we present a new algorithm to find inexact motifs (which are transformed into a set of exact subsequences) from a DNA sequence. Our algorithm builds an automaton that searches for the set of exact subsequences in the DNA database (that can be very long). It starts with a preprocessing phase in which it builds the finite automaton, in this phase it also considers the case in which two different subsequences share a substring (in other words, the subsequences might overlap), this is implemented in a similar way as the KMP algorithm. During the searching phase, the algorithm recognizes all instances in the set of input subsequences that appear in the DNA sequence. The automaton is able to perform the search phase in linear time with respect to the dimension of the input sequence. Experimental results show that the proposed algorithm performs better than the Aho-Corasick algorithm, which has been proved to perform better than the naive approach, even more; it is considered to run in linear time.

1 Introduction

Finding interesting subsequences from DNA nucleotide or protein databases is a complex biological problem. The problem's complexity arises because the subsequence to be found in a much longer sequence may have variations, and may also be found repeated times (they may even overlap) to be considered interesting [2,4]. The importance of finding this type of subsequences is related to biological problems such as finding genes' functionality or genes' regulatory subsequences [3]. Such interesting subsequences that deserve further analysis are known as motifs, this is the DNA motif search problem [1,7].

A. Hernández Aguirre et al. (Eds.): MICAI 2009, LNAI 5845, pp. 556–565, 2009.

The DNA motif search problem consists of the search of a pattern P (the motif to search for) in a text T (the DNA database created from the nucleotides alphabet: A for adenine, C for cytosine, G for guanine, and T for thymine) where we output the positions in T where P appears, the instances of P in T. However, P is formed from an extended alphabet defined under the IUPAC (International Union of Pure and Applied Chemistry) rule where a character in this alphabet may represent more than one nucleotide, this means that P may represent a set of patterns to search for in T.

In order to solve this problem, the variations in a subsequence P to search for can be transformed to a set of subsequences to search for in T which is known as the **exact set matching problem** [1,6] with two restrictions: every subsequence derived from P has the same length and the alphabet of such subsequences is limited to A, C, G, and T. One of the most representative approach to solve the exact set matching problem is with the Aho-Corasick (AC) algorithm [1] that we use in this work to compare with our new approach, the **Motif Finding Automaton**, MFA for short.

This paper is organized as follows. In section 2 we present the notation and formal definition of the problem. In section 3 we present and describe the MFA algorithm, in section 4 we show our empirical results with 4 real world databases, including a comparison of the MFA algorithm with the AC algorithm. Finally in section 5 we present our conclusions and future work.

2 Notation and Problem Description

In this section, we introduce the notation on strings, sequences, and automata, that we use along this work.

An *alphabet*, denoted by Σ is a finite nonempty set of letters. A *string* on alphabet Σ is a finite subset of elements on Σ, one letter after another. As an example, if $\Sigma = \{A, C, G, T\}$ then, examples of strings on this alphabet are AAC, ACC, and AAG. We denote by ε the zero letter sequence, called the *empty string*. The *length* of a string x is defined as the number of letters on x, denoted by $|x|$. With $x[i]$, $i = 1, \ldots, |x|$, we denote the letter at position i on x. A *substring* γ of a string α is a string where: $|\gamma| \leq |\alpha|$, and $\gamma[i] = \alpha[i + k]$, $i = 1, \ldots, |\gamma|$ and $0 \leq k \leq |\alpha| - |\gamma|$.

The *concatenation* of two strings x and y is the string composed of the letters of x followed by the letters of y, denoted by xy. A string α is a *prefix* of the string x, denoted by $\alpha \sqsubseteq x$, if $x = \alpha y$. On the other hand, β is a *suffix* of x, denoted by $\beta \sqsupseteq x$, if $x = y\beta$. The empty string ε is both a prefix and a suffix of every string (not empty).

A *finite sequence* of strings $\{x_1, x_2, \ldots, x_n\}$ is a function f where strings are placed in order, denoted by $\langle f(x_1), f(x_2), \ldots, f(x_n) \rangle$. If Seq is a finite sequence of strings with length n, then Seq_i is the ith string of the sequence Seq, $1 \leq i \leq n$.

In this work we denote a *finite automaton* M as a 5-tuple $(Q, q_0, A, \Sigma, \delta)$ where: Q is a finite set of *states*, $q_0 \in Q$ is the *start state*, $A \subseteq Q$ is a distinguished set of *accepting states*, Σ is a finite *input alphabet* and $\delta : Q \times \Sigma \rightarrow Q$ is the *transition function* of M.

Let P be a pattern (string) to search for. In biochemistry, P is a string called a *motif*, wherein all of its letters are defined by the union of two alphabets: a main alphabet $\Sigma^B = \{A, C, G, T\}$ (every letter represents a nuceotide as described before), and the IUPAC alphabet, (denoted as the extended alphabet in this work) used to represent ambiguities in the pattern. The IUPAC established an ambiguity alphabet, represented by $\Sigma^E = \{R, Y, K, M, S, W, B, D, H, V, N\}$, where $R = \{G, A\}$, $Y = \{T, C\}$, $K = \{G, T\}$, $M = \{A, C\}$, $S = \{G, C\}$, $W = \{A, T\}$, $B = \{G, T, C\}$, $D = \{G, A, T\}$, $H = \{A, C, T\}$, $V = \{G, C, A\}$, and $N = \{A, G, C, T\}$.

Based on the above mentioned, the **Motif Searching Problem** (MSP) is formally defined as follows. Let P be a motif to search for based on alphabet $\Sigma = \Sigma^B \cup \Sigma^E$, and let T be a string to search in, that represents a DNA database on Σ^B. The **MSP** consists of finding all substrings γ in T, $|\gamma| = |P|$, where $\forall \gamma[i], i = 1, \ldots, |\gamma| : \gamma[i] = P[i]$ if $P[i] \in \Sigma^B$, or $\gamma[i] \in P[i]$ if $P[i] \in \Sigma^E$. As an example, if $P = AMS$ and $T = TAACGA$, then all possible patterns generated from AMS are $\{AAG, AAC, ACG, ACC\}$. Thus, two strings are found, $\gamma = AAC$ (at position 2) and $\gamma = ACG$ (at position 3).

Due to the fact that P represents a set of patterns to search for, the **MSP** is equivalent to the **exact set matching problem** [1] with two restrictions: every string in P have the same length, and the alphabet is limited to $\Sigma = \Sigma^B \cup \Sigma^E$.

In the next section we present our approach called the Motif Finding Automaton, which is capable to solve the **MSP**.

3   The MFA Algorithm

In this section we describe how the MFA algorithm works, which was designed under the idea of the KMP algorithm to retrieve previous suffixes that can be used as prefixes of other patterns. Our method, as well as the AC algorithm, creates an automaton named M_{MFA} that recognizes each element belonging to seqP. However, in M_{MFA}, all transitions are labeled in order to avoid unnecessary transition steps during the search of elements of P on T.

The MFA algorithm consists of two phases: the preprocessing and the searching phases. The preprocessing phase is performed in three stages: it first expands pattern P in all its possible combinations (it builds seqP); after that, it builds a matrix $matQ$ to store the states of the new automaton, indicating which of those states are final states. Finally, it generates the automaton transition matriz δ. After the preprocessing phase, the searching phase starts, which processes each element on T using transition function δ and identifying all the existing patterns. These steps are shown in algorithm 1.

Now we describe in detail each of the steps of the MFA algorithm. In *line 1* of the algorithm 1, the seq$P$ strings sequence is built from P. *seqP* is built through a expansion process in which strings with letters in Σ^E are replaced by strings defined with letters on Σ^B, this replacement is done from right to left for each string. For example, if we consider $P = AMS$, where $M = \{A, C\}$ and $S = \{C, G\}$. We initialize seqP = $\{\langle AMS \rangle\}$. seqP is treated as a queue in which we delete the first string α of seqP if α contains letters on Σ^E. In other case, it means that all the elements on seqP are strings defined on Σ^B, and we do not modify seqP.

If α is a string that was eliminated from seqP, then, it is expanded by replacing the last letter in Σ^E by its corresponding letters in Σ^B and the resulting strings are queued in order in seqP. Returning to our example, given that seqP = $\{\langle AMS \rangle\}$, then AMS is eliminated from seqP replacing it by the strings AMC and AMG, which are queued in seqP, where seqP = $\{\langle AMC \rangle, \langle AMG \rangle\}$. If we iteratively follow this process at the end we obtain: seqP = $\{\langle AAC \rangle, \langle ACC \rangle, \langle AAG \rangle, \langle ACG \rangle\}$

Algorithm 1. The MFA Algorithm

Input: P (pattern to search for), T (text to search in)
Output: positions where instances of $seqP$ are found
1: $seqP \leftarrow expandP(P)$
2: $matQ \leftarrow buildQ(P)$
3: $nStates \leftarrow$ get #states from $matQ$
4: $\Delta'[1] = \Delta''[1] = |P[1]|$
5: **for** $i = 2$ to $|P|$ **do**
6:      $\Delta'[i] = |P[i]|$
7:      $\Delta''[i] = \Delta'[i-1] * |P[i]|$
8: **end for**
9: $\delta \leftarrow build\delta(nStates, seqP, matQ, \Delta', \Delta'')$
10: $Accepted \leftarrow$ get the id of the minimium state accepted by the automaton δ (from $matQ$)
11: $q \leftarrow 0$
12: **for** $i = 1$ to $|T|$ **do**
13:      $q \leftarrow \delta(q, T[i])$
14:      **if** $Accepted \leqslant q$ **then**
15:          print "Pattern occurs at position $i - |P| + 1$"
16:      **end if**
17: **end for**

In *line 2* of algorithm 1 we build *matQ*, which stores all states of M_{MFA}. The number of columns and rows of *matQ* will be $|P|$ and $|seqP|$ respectively. Besides, the final states of the automaton are stored in the last column of this matrix. The construction of *matQ* is performed secuentialy from top to bottom with respect to rows, and from left to right with respecto to columns.

Let Δ'' be a vector of dimension $|P|$, where $\Delta''[i] = \prod_{k=1}^{i} |P[k]|$, $1 \leq i \leq |P|$. Based on Δ'', we give values to the cells of *matQ* from 1 to $nStates = \sum_{i=1}^{|P|} \Delta''[i]$, in descending way by columns, where for the $i - th$ column we only fill up to the $\Delta''[i]$ row. For example, for $P = AMS$, the result of *matQ* is shown in figure 1.

matQ:

	A	M	S
1	1	2	4
2	0	3	5
3	0	0	6
4	0	0	7

Fig. 1. *matQ* Matrix for $P = AMS$

In *line 9*, of algorithm 1 we call function buildδ (algorithm 2) to build the transitions matrix δ. This construction is performed row by row (*line 3*, algorithm 2). Each row is built by function *getStates*, which is shown in algorithm 3. This function receives as input the string associated to the expansion vertex (the string associated to the expansion vertex is created by the concatenation of the edge labels, each label corresponding to

---

Algorithm 2. Function buildδ

Input: $nStates$ (#states for δ), seq$\mathbb{P}$ (sequence of patterns), $matQ$ (matrix with the states of the M_{MFA}), Δ', Δ''

Output: δ (The Automaton transition matrix)

1: Initialize $\delta_{nStatesx4}$ with $0's$
2: **for** $i = 0$ to $nStates$ **do**
3:     $\delta_i \leftarrow$ getStates(i,seq$\mathbb{P}$,$matQ$, Δ', Δ'')
4: **end for**

Algorithm 3. Function getStates

Input: i (state), seq$\mathbb{P}$ (sequence of patterns), $matQ$ (matrix with the states of M_{MFA}), Δ', Δ''

Output: $vector_{1x4}$ (vector of states)

1: Initialize $vector_{1x4}$ with $0's$
2: $x \leftarrow [a, c, g, t]$
3: L_i = concatenation of characters (edge's labels) from $root$ to i
4: **for** $j = 1$ to 4 **do**
5:     $c \leftarrow L_i x[j]$
6:     $vector[j] \leftarrow prefix(c, seq\mathbb{P}, matQ, \Delta', \Delta'')$
7: **end for**

Algorithm 4. Function prefix

Input: c (string), seq$\mathbb{P}$ (sequence of patterns), $matQ$ (matrix with the states of M_{MFA}), Δ', Δ''

Output: $state$ (state where the automaton make a transition)

1: $i \leftarrow 1, k \leftarrow 1, j \leftarrow 1, r \leftarrow 1, found \leftarrow$ false, $accept \leftarrow$ false, $cn = 1$
2: **while** $found$ = false and $r \leq |c|$ **do**
3:     forward ($i, j, k, r, accept, \Delta', \Delta''$, seq$\mathbb{P}$, cn, c)
4:     **if** $accept$ = true and k > $|c|$ **then**
5:         $found$ = true
6:     **end if**
7: **end while**
8: **if** $accept$ = true **then**
9:     return $matQ[j][i-1]$
10: **else**
11:     return 0
12: **end if**

---

a letter in the Σ^B alphabet, from the initial state of the automaton to the expansion vertex), and concatenates it to each letter on Σ^B (generating four combinations), *line 5*. This string is processed by the prefix function (algorithm 4).

Function *prefix* detects a string s, where s is the largest suffix of the input string c, such that s is a preffix of a string associated to a vertex in M_{MFA}. This is better described with the following example: consider the automaton of figure 3 a), in which function *getStates* is called in *line 3* of function buildδ with node 4. *getStates* then generates the values of vector δ_4, which are associated to different transitions created from vertex 4 with letters A, C, G, T (Σ^B).

$$forward\ (i, j, k, r, accept, \Delta', \Delta'', seqP, cn, c) = \begin{cases} accept \leftarrow true, k \leftarrow k + 1, \\ i \leftarrow i + 1, cn \leftarrow 1 & If\ (c[k] = seqP_j[i]) \\ \\ j \leftarrow j + \Delta''[i\text{-}1], cn \leftarrow cn + 1, \\ accept \leftarrow false & If\ (c[k] \neq seqP_j[i])\ and\ (cn < \Delta'[i]) \\ \\ r \leftarrow r + 1, i \leftarrow 1, cn \leftarrow 1, \\ j \leftarrow 1, k \leftarrow r, accept \leftarrow false & If\ (c[k] \neq seqP_j[i])\ and\ (cn = \Delta'[i]) \end{cases}$$

Fig. 2. Function to compute the position of the strings in *seqP*

For each of the possible transitions that can be generated from vertex 4, *getStates* calls function *prefix* with string $L_4\alpha$, where $\alpha \in \Sigma^B$ (we concatenate to L_4 each of the letters on the base alphabet, *line 5, algorithm 3.*

Function *prefix* identifies a node *x*, such that the largest suffix of label $L_4\alpha$ is a prefix of *x*. Given that seqP was ordered in such a way that it is possible to predict the positions of the strings that share common preffixes from the cardinalities of the letters in *P*, *prefix* speeds up the identification process, verifying only those strings where a prefix of $L_4\alpha$ may exist, avoiding the strings in which it is not possible to find such a preffix. The prediction of the position of the strings is computed with the function called *forward* shown in figure 2, in *line 3* of algorithm 4.

If we continue with our example, for node 4 of figure 3 a), we first call function *prefix* with $L_4\alpha$, where $L_4 = AAC$ and $\alpha = A$, to compute the node which preffix is equal to the longest suffix and at the same time is the closest to the initial state of the automaton. For our example, the node that satisfies these restrictions is 1, given that $L_1 = A$ and *A* is the longest suffix of $L_4\alpha$. In figure 3 a) we show with doted lines the jumps that are generated from node 4 and in figure 3 b) we show the restrictions that are used to determine the values of vector δ_4.

If we apply the MFA algorithm to pattern *P*, we generate the automaton shown in figure 4.

Once we built the automaton with the MFA algorithm, the patterns search process in an input text *T* is performed in the same way as with a traditional automaton. For example, considering text $T = AACGAACC$ and pattern $P = AMS$ to search for, from the transitions function generated by MFA (figure 4), it processes letter by letter of such a text, where for each letter the transition function returns a state. In figure 5 we can see all the states returned by the transition function from text *T* as input. Note

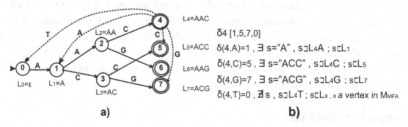

Fig. 3. Example to build the transitions of vertex 4: a) graphical represesentation of the expansion process for vertex 4; b) restrictions to determine the values of vector δ_4

	A	C	G	T
0	1	0	0	0
1	2	3	0	0
2	2	4	6	0
3	1	5	7	0
4	1	5	7	0
5	1	0	0	0
6	1	0	0	0
7	1	0	0	0

$$Q = \{0, 1, \ldots, n \mid n = \sum_{i=1}^{|P|} \prod_{j=1}^{i} |P_j|\}$$
$$q_0 = 0$$
$$A = \{i \mid i = matQ[j][|P|], j = 1, 2, \ldots, |seq\mathbb{P}|\}$$
$$\Sigma = \{A, C, G, T\}$$

$\delta:$

Fig. 4. Automata created by the MFA algorithm from pattern AMS

that identifying a valid pattern inside T is translated to get the states that the transition function returns that belong to the automaton set of final states, line 14 of algorithm 1. In this example, the states accepted by the automaton are enclosed with a circle, then the patterns found are: AAC (at position 1), ACG (at position 2), AAC (at position 5), and ACC (at position 8).

	A	A	C	G	A	A	C	C
0	1	2	④	⑦	1	2	④	⑤

Fig. 5. Search process according to the transition function

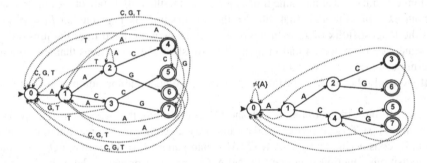

Fig. 6. Automaton generated by the MFA and AC algorithms, respectively

In figure 6 we show the automata generated by the MFA and AC algorithms, note the MFA generates an automaton where all the edges are labeled with letters from the base alphabet while the automaton generated by the AC algorithm contains unlabeled edges (all the failure edges). This is the way in which the MFA algorithm improves the performance of the AC algorithm because for each visited transition it always processes a new letter from T. The impact of this difference is reflected in the performance of both algorithms as we show in the results section.

4   Empirical Results

In this section, we test the performance of the MFA algorithm with four real DNA databases: candida albicans (with 14'361,129 letters), ustilago maydis (19'683,350 letters), aspergillus nidulans (30'068,514 letters), and neurospora crassa (39'225,835 letters). These databases represent the text to search in denoted by T. These databases are freely available through the Center for Genome Research (fungi section at http://www.broad.mit.edu/).

In our experiments we compare the MFA algorithm with the AC algorithm, because it is capable to solve the exact set matching problem in $O(n + m + k)$, where $|P| = n$, $|T| = m$, and k is the number of occurrences in T of the patterns from $seqP$ [1]. It is clear that the exact set matching problem can be solved with exact matching approaches [5,1], such as the KMP algorithm. However, the exact matching approach must be executed z times, where $|seqP| = z$, hence the exact set matching problem is solved in time $O(n + zm)$. Therefore, we do not perform tests with exact matching approaches.

In our experiments, we were interested to test our algorithm in different scenarios, where it is possible to find from a few to large number of patterns. If we change the number of patterns to search for ($|seqP|$), then it is possible to increase or decrease the number of patterns to find. Moreover, there is a relation between $|seqP|$ and $|P|$, because $|seqP|$ is gotten from all valid substitutions of letters from Σ^E in P by letters from Σ^B. Thus, it is possible to have more patterns in $|seqP|$ if we gradually increase $|P|$, or we fix $|P|$ with a large value. For the sake of simplicity and without losing of generality, in our experiments we fix $|P| = 32$, because in biochemistry, motifs typically have a dimension less than 32, and this dimension produces up to 4^{32} different combinations [1].

We performed our experiments in an Intel Core Duo at 1.8 Ghz PC, with 4 GBytes of RAM with the Windows Vista Operating System. Both, the MFA and the AC algorithms were implemented in C++. We downloaded a free available implementation of the AC algorithm written by Barak Weichselbaum from http://www.komodia.com/.

Fig. 7 shows the runtime of the MFA algorithm (left column) and the AC algorithm (right column), divided in total runtime (first row), preprocessing runtime (second row), and searching runtime (third row). Both of these algorithms run in two phases: a first phase, where the automaton is built (preprocessing phase), and a second phase, where the patterns are searched for. Hence, the total runtime corresponds to the preprocessing runtime plus the searching runtime.

We start by analyzing the preprocessing runtime. Both algorithms share the construction of an automaton, which have similarities between them. First, they have the same number of nodes but not the same number of edges (AC has less edged than MFA). Second, each vertex of each automaton represents a string that is a proper prefix of some element in $seqP$. However, the automaton edges are defined in a different way: some AC edges are labeled with an alphabet letter (keyword tree edges), and remaining edges are called "failure links", where each failure link connects a vertex v with a vertex v' in the automaton, and v' is a vertex labeled with the longest suffix of label v. Besides, no label is associated to the failure links. On the other hand, the MFA builds an automaton where each edge has a label (a letter on the alphabet), and it includes, as a subgraph, the keyword tree of the AC automaton.

---

[1] This information was supplied by Patricia Sánchez and Candelario Vázquez, biologists at the Microbiology Science Department of the Benemérita Universidad Autónoma de Puebla, México.

564   G. Perez et al.

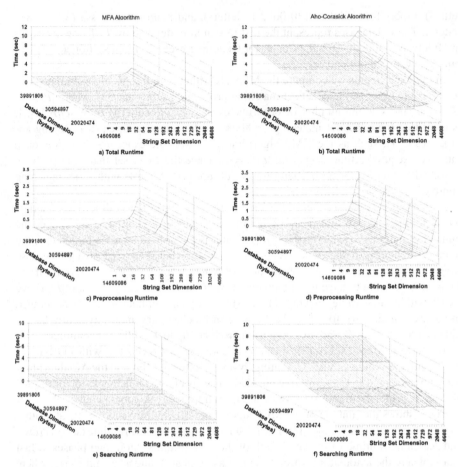

Fig. 7. Emphirical results of the MFA algorithm (left column) and the AC Algorithm (right column) in four real world databases. First row: total runtime (preprocesing runtime + searching runtime); second row: preprocesing runtime; third row: searching runtime.

Because no edges in the automaton built with the MFA are unlabeled, its construction is a little more complex. However, as we can see in Fig. 7 c) and d), the MFA preprocessing runtime is faster than the AC preprocesing runtime in each of our tests with different databases. However, based on the shape of the graph we can see that both approaches have a preprocessing phase with a similar temporal complexity, but the constant associated to the MFA is lower than the corresponding constant of the AC.

Nevertheless, the MFA algorithm shows an advantage over the AC algorithm in the searching phase. Remember that the AC automaton includes "failure links" without labels. This means that no new letter is processed from the input string T when a "failure link" is used as transition during the searching phase. Therefore, if "k" represents the total number of "failure links" visited during the searching phase, then the total number of visited transitions is $k + m$, where $|T| = m$. On the other hand, since all transitions in M_{MFA} are labeled (with an alphabet letter), this means that each letter in T is processed by each visited transition in M_{MFA}. Consequently, the search time of M_{MFA} only depends of the number of letters in the input string. This analysis is reflected in our experiments, as shown in Fig. 7 e) and f). It is clear that the AC algorithm search run-time increases faster than that of the MFA algorithm because the total number of visited failure transitions are probably increasing. As consequence of this, the MFA algorithm has a better performance than the AC algorithm with larger input strings.

5   Conclusions and Further Work

In this paper we introduced a new algorithm called MFA, to solve the exact set matching problem applied to the motif finding problem. Our experimental results show that the performance of the MFA improve over the performance of the AC algorithm that solves the same problem. This performance improvement is due to the fact that for each visited transition of the MFA a letter from the input string is processed, which does not happen in the AC automaton. Because of this improvement the MFA algorithm obtains better performance with larger databases than the AC algorithm. This is a very important property because the size of DNA databases can be extremely large. Our biology domain experts are currently validating the motifs found in our experiments.

In our future work we will apply our algorithm to different larger DNA databases to find motifs to be tested in laboratory by our biology domain experts. Finding interesting motifs saves time and economical resources.

References

1. Gusfiled, D.: Algorithms on Strings, Trees, and Sequences. Computer Science and Computational Biology. Cambridge University Press, Cambridge (1994)
2. Navarro, G.: A Guide Tour to Approximate String Matching. ACM Computing Surveys 33(1), 31–88 (2001)
3. Schmollinger, M., et al.: ParSeq: Searching motifs with Structural and Biochemical Properties. Bioinformatics Applications Note 20(9), 1459–1461 (2004)
4. Baeza-Yates, R., Gonnet, G.H.: A New Approach to Text Searching. Comunications of the ACM 35(10) (1994)
5. Boyer, R.S., et al.: A Fast String Searching Algorithm. Communications of the ACM 20(10), 726–772 (1977)
6. Crochemore, M., et al.: Algorithms on Strings. Cambridge University Press, Cambridge (2001)
7. Aluru, S.: Handbook of Computational Molecular Biology. Champan & All/Crc Computer and Information Science Series (2005) ISBN 1584884061

A New Method for Optimal Cropping Pattern

Juan Frausto-Solis, Alberto Gonzalez-Sanchez, and Monica Larre

Tecnologico de Monterrey Campus Cuernavaca, Autopista del Sol km 104, Colonia Real del
Puente, Xochitepec, Morelos, Mexico
{juan.frausto,albertogonzalez,monica.larre}@itesm.mx

Abstract. This work proposes the GenSRT method for the Cropping Pattern
Optimization (CPO) problem. GenSRT applies Genetic Algorithms, Simplex
Method, and Regression Trees. The purpose is to maximize the net income of
every cropping season. Simplex Method (SM) is the traditional approach for
solving the problem; however, CPO is complicated, because the crop yield has
a non-linear behavior which SM cannot consider directly. In GenSRT, regression trees are applied to non-linear regression models construction. The models
are provided to GenSRT to evolve efficient cropping patterns and to maximize
the benefits for sowing area distribution through a genetic search. Results show
that GenSRT overcomes Simplex maximization by obtaining better resource
distribution and a higher net profit.

Keywords: cropping pattern optimization, regression trees, genetic algorithms,
simplex method.

1 Introduction

Cropping pattern optimization (CPO) is a common issue for irrigation areas. Agricultural managers have to assign a certain amount of area to each crop for cultivation [1],
fixing the quantity of other resources required for the selected crops. These crops and
its sowing areas conforms a cropping pattern, which is considered as optimal if it
maximizes the difference between the gross income and the production costs to specific constraints imposed for the agricultural production system [2]. CPO is very
important, because economic income of every cropping season depends on a correct
distribution of sowing areas and resources, and the agricultural managers want to earn
as much as possible by using the farmland effectively [1]. Thus, CPO is searching for
economic efficiency, where a high economic return can be obtained with limited
water resources, if the proper crops are chosen [3].

The most common technique used for CPO is Linear Programming (LP) [4]. LP
([2][5]) and other traditional methods ([6][7]) have had many limitations to obtain
good solutions [7]. Other works combine LP with climate-soil-plant systems simulation ([8][9]), but they require of a great quantity of technical data that are not always
available. Therefore, new non-linear methods for CPO have been developed. Genetic
Algorithms (GAs) [10] have extensively been applied to CPO ([11][7]); Simulated
Annealing (SA) [12] also has some CPO applications ([13][7]) obtaining similar
results [7].

A. Hernández Aguirre et al. (Eds.): MICAI 2009, LNAI 5845, pp. 566–577, 2009.
© Springer-Verlag Berlin Heidelberg 2009

There is a critical factor involved in CPO. Any CPO model requires an estimation of each crop yield (the product ratio obtained per unit of surface). Regrettably, the yield is a non-linear factor depending on many variables [14]; every crop yield is affected differently by each possible resource distribution, and the number of possible scenarios is highly increased. Therefore, to build more efficient and realistic optimization models, resource limitation affecting crop yield and other external factors should be considered.

In this paper, a method named GenSRT (Genetic Simplex Regression Trees) to solve CPO that considers crop yield nonlinear behavior is proposed. The method applies regression trees (RTs) to model crop yield behavior, using a GA to perform a global search for the optimum cropping pattern. An RT is a nonparametric model which looks for the best local prediction of a continuous response through the recursive partitioning of the space of the predictor variables [15]. On the other hand, GAs are numerical algorithms inspired by the natural evolution process and random search techniques [16], which have been used successfully in many optimization problems [7]. GA and others non-lineal optimization methods have been applied previously to the CPO problem [11][13][7], but these have use an approach where non lineal relationships of crops' physiologic processes are used to obtain the crops' yield data. GenSRT uses a different approach; it deals with non linear crops´ yield by applying RT instead of physical models. Simplex Method (SM) is the most common method to solve the problem in a simplified way. Because GenSRT is a hybrid method which uses GA and SM, it is expected that it overcomes SM. Thus, GenSRT was developed looking for a more suitable method able to handle the crop yield non linearity. In the paper, GenSRT is tested with data from an irrigation district and its quality solution is compared versus SM.

2 A Mathematical Model for the Problem

A general mathematical model for a CPO problem can be seen in equation (1) [5][9]:

$$NI = \sum_{i=1}^{n} (P_i \cdot Y_i - C_i)A_i \tag{1}$$

where n is the quantity of crops in the pattern, NI is the total net income (\$), P_i is the sale price of crop i (\$/ton), Y_i is the crop's yield i (ton/ha), C_i is the production cost (\$/ha) and A_i is the sowing area of crop i (ha). Traditionally, P_i, C_i, and Y_i are treated as constant values, while A_i always remains as a decision variable for the model.

Many constraints are involved in CPO [5]; some of them which are used in the present work are (2), (3), and (4). Equation (2) shows the total area available for sowing constraint (TASC), where the sum of sowing areas assigned to each crop cannot be more than the total area available for sowing (TAS):

$$A_1 + A_2 + \cdots + A_n \le TAS \tag{2}$$

Equation (3) shows the crops sowing area constraints (CSACs). For some reasons (special preparation of land, irrigation systems availability, etc.), the sowing area for each crop i is not freely available, but rather it moves within a range defined by a lower limit (LLA) and an upper limit (ULA):

$$LLA_i \leq A_i \leq ULA_i \quad \text{for } i = 1, 2, 3, \ldots, n \tag{3}$$

Equation (4) shows the crops irrigation water depth constraints (CIWDCs). This kind of constraint indicates the minimum water requirement for the crop and the maximum that can be applied, which is expressed as a range defined by a lower limit $LLIWD$ and an upper limit ($ULIWD$) for the irrigation water depth ratio (IWD):

$$LLIWD_i \leq IWD_i \leq ULIWD_i \quad \text{for } i = 1, 2, 3, \ldots, n \tag{4}$$

At first sight, the problem and its constraints could be seen as a candidate to solve using LP techniques, but in a realistic (and not linear) world, it must be considered that yield behavior of each crop is not linear ([14][17]). There are several conditions affecting the crop yield for example, the quantity of sowing area, the quantity of irrigation water depth applied, and the weather behavior during the crop growth season. For this reason, yield must be considered as another function within the objective function, just as it can be seen in equation (5):

$$NI = \sum_{i=1}^{n} (P_i \cdot YF_i(A_i, IWD_i, W) - C_i)A_i \tag{5}$$

where YF_i represents a non-linear yield prediction function for crop i (ton/ha), and W represents weather conditions (a meta-variable that integrates monthly average measures of temperature, rain and evaporation). This consideration turns the model into a non-linear model [2], and solving it with only LP techniques becomes impractical. The work explained in this paper deals with non-linearity of YF function with RT models, providing a method to use them in a GA. The purpose of the method is finding an optimal combination of yields to maximize the net income while the constraints are respected.

3   Materials and Methods

Two stages compose the GenSRT solution method. The first consists of an RT model construction for crop yielding prediction. This technique and its approach to GenSRT method is described in Section 3.1.The second stage consists of the use of the RT models in an adapted GA. GAs and the encoding procedure for GenSRT are described in section 3.2. Section 3.3 provides a complete view of the GenSRT method.

3.1   Regression Trees

A regression tree (Breinman, 1984 [18]) is a model structure with a decision tree (DT) form capable of predicting continuous values in response to several input values. A DT is formed by nodes. One node is named the root node, and it has no incoming edges. All the other nodes have exactly one incoming edge. A node with outgoing edges is called a test node, and a node without outgoing edges is called a leaf or terminal node. Each internal node in the tree splits the instance space into two or more subspaces based on a condition of the input attribute values [19]. Each leaf is assigned to one class representing the most appropriate target value. Instances are classified by navigating them from the root of the tree down to a leaf, according to the outcome of the tests along the path [19]. In RTs case, class at the leaf nodes assigns a real value to the tested instance. A sample RT is shown in figure 1.

There are several algorithms to built RTs [18][20]; a brief CART[18] description is summarized in the next three steps [21]:

Step 1. The decision tree begins with a root node t derived from whichever variable in the feature space minimizes a measure of the impurity of the two sibling nodes. The measure of the impurity or entropy at node t, denoted by $i(t)$, is computed by

$i(t) = -\sum_{j=1}^{k} p(w_j|t) \log p(w_j|t)$, where $p(w_j \mid t)$ is the proportion of patterns $\mathbf{x}_i$ allocated

to class w_j at node t.

Step 2. Each non-terminal node is then divided into two further nodes, t_L and t_R, such that p_L, p_R are the proportions of entities passed to the new nodes, t_L and t_R respectively. The best division which maximizes the difference is calculated as $\Delta i(s,t) = i(t) - p_L i(t_L) - p_R i(t_R)$.

Step 3. The decision tree grows by means of the successive sub-divisions until there is no significant decrease in the measure of impurity when a further additional division s is implemented. Then, the node t is not sub-divided further, and automatically becomes a terminal node. The class w_j associated with the terminal node t is that which maximizes the conditional probability $p(w_j \mid t)$.

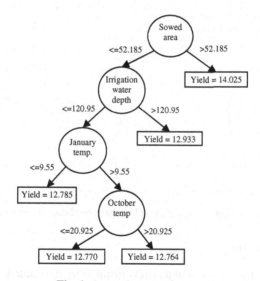

Fig. 1. A sample regression tree

In the GenSRT method, an RT per each crop is used to supply YF_i functions in (5) and make it a linear objective function. There is an analogy between the $YF_i(A_i, IWD_i, W)$ function and a RT, such as it appears in figure 1. It depends on the values of the sowed area (A_i), irrigation water depth applied (IWD_i) and some climatic variables (the June and October average temperature) (W). However, it must be considered that there are not single values for the A_i and IWD_i variables, but rather that there are a range of values (which is the CSACs and CIWDCs constraints case). This allows for obtaining more than one yield value as output. To illustrate this case with the RT in figure 1, lets assume there is a sowed area constraint of $20 \leq A_i \leq 60$ ha, using an irrigation water depth of 100 mm and a January temperature of 8 °C. Following the tree, it is estimated there will be a yield of 12.77 ton/ha from a sowed area between 20.925 and 52.185 ha, and a yield of 14.025 ton/ha for the range between 52.185 and 60 ha. It results in two different sub-ranges with two different yield values starting from an input range of values for the sowed area. The number of sub-ranges (and yields) that are obtained depends on tree structure and the given range of values for the other involved variables. To obtain the yields and its associated sub-ranges (also called sub-constraints), in GenSRT a recursive procedure is implemented (figure 2). This procedure performs a depth-first-search (DFS) through the tree, obtaining a list of yields that can be achieved with the specified constraints.

```
Input: croptree (a regression tree), constList (a list of constraints)
Output: yieldList (a list of yields and its associated sub-constraints)

function GetReachableYields(croptree,  constList):yieldList {
    List condList,yieldList
    node = GetRootNode(croptree)
    GetNodeYields(node, constList, condList, yieldList)
    return yieldList
}
procedure GetNodeYields(node, constList, condList, yieldList) {
    if (IsNotALeafNode(node)) {
        constraint=GetConstraint(constList, node.attribute)
        if (constraint.lowerLimit <= node.splitValue) {
            AddCondition(condList,node.attribute,constraint.lowerLimit,
Min(node.splitValue, constraint.upperLimit))
            GetYieldNode(node.leftChild, constList, condList, yieldList)
            RemoveLastCondition(condList)
        } else {
            AddCondition(condList,node.attribute,Max(constraint.lowerLimit,
                node.splitValue), constraint.upperLimit))
            GetNodeYields (node.rightChild, constList, condList, yieldList)
            RemoveLastCondition(condList)
        }
    } else {
        AddYieldAndConditionsToList(yieldList, condList, node.yield)
    }
}
```

Fig. 2. A pseudo-code for obtaining yields and its associated conditions from a regression tree

3.2 Genetic Algorithms

GAs are numerical optimization algorithms inspired by the natural evolution process and random search techniques [16]. GAs are able to "evolve" solutions to real world problems, if they have been suitably encoded [22].

Although implementation details vary among different GAs, they all share the following structure: GAs work with a population of *individuals*, each representing a possible solution to a given problem. Each individual is assigned a *fitness score* according to how good a solution to the problem it is. The algorithm operates cyclically, upgrading a group of *individuals* called a population. In each iteration, members of a population are evaluated according to a *fitness function*. A new population is generated by probabilistic selection of the most adaptable *individuals*; some of them pass intact to the next generation, while others are selected by applying genetic operators as *crossover* and *mutation*. The process finishes when there are not significant changes in the *generations* (it is said that it *converges*) or a maximum of *generations* is reached.

In the case of GenSRT method, a *chromosome* is encoded as an array of integer values. Figure 3 shows a chromosome representation for *n* crops′ problem. Each I_i value represents an index in a list of crop yields LY_i , where each node *j* of the list *i* store a yield Y_{jci}, a sowing area sub-constraint AR_{jci} and an irrigation water depth constraint $IWDR_{jci}$. The yields, sowing area and irrigation water depth ranges of each crop are extracted from each crop RT just as it was described in point 3.1.

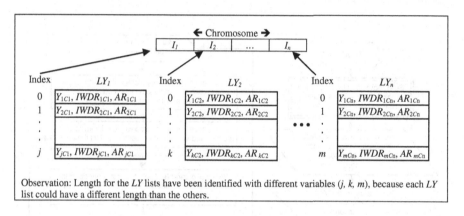

Fig. 3. A chromosome representation for GenSRT

For GenSRT, a *fitness function* corresponds to the SM maximization results of objective function indicated in (5). To do that, yield functions are replaced by constant yield values, which are indicated by the gens in the chromosome to evaluate. Sub-constraints associated to each crop yield are also included in the maximization. Yields and sub-constraints are taken from each crop yield list. Price and production cost variables are specified as input parameters. The objective function in (5) remains

as $NI = \sum_{i=1}^{n} Coef_i A_i$, where $Coef_i$ results from the $(Y_i \times P_i - C_i)$ operation. Thus, NI computation is trivial if SM is used.

3.3 The GenSRT Method

A complete view of GenSRT method operation is provided in figure 4. As input data, the method requires the RT models of each crop involved in the pattern, the CASCs, CIWDCs and TASC constraints (see equations 2,3,4), the crops sale price, the crops production cost and the weather information for the irrigation district. The size of the problem is then from *n* decision variables and 2*n*+1 constraints. After that input data is introduced, GenSRT crosses the CASCs and CIWDCs constraints with RT information. The yields, areas and irrigation water depths satisfying constraints are stored in a sorted list for each crop. Next, the method verifies if a Simplex maximization using yields at the beginning of each list (highest yields) produces a satisfactory output. A satisfactory output is the use of almost all sowing area available, which is specified through an input parameter initialized at 95%. If the sum of the sowing areas estimated by SM do not reach the minimum area required, the method starts a genetic cycle. For its operation, the GA in GenSRT method requires the following parameters: size of population, crossing probability, mutation probability and maximum number of generations. The search in the GA is guided by the maximization result with yields and constraints indicated by chromosome in turn. At the end of the genetic cycle, GA throws the best combination of yields (it produces a higher income that any other proven combination).

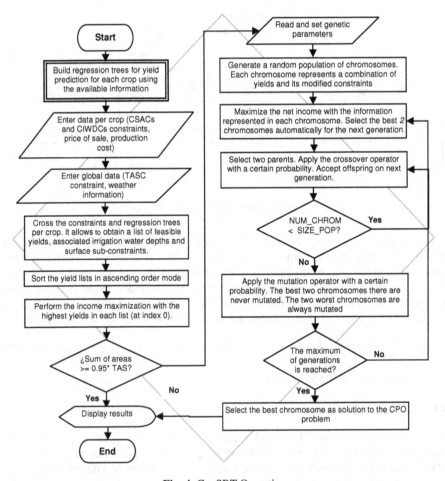

Fig. 4. GenSRT Operation

As output, the GenSRT method provides per crop: the sowing area, the estimated yield, the irrigated water depth and the optimum net income. The yields are obtained from each crop RT. The other outputs are obtained through maximization result with selected yields and its associated sub-constraints.

4 Experimental Results

The method was applied in the 038 irrigation district in Sonora, México. To test the method, an 11 year (1995-2006) historical data bank with information about 15 crops was conformed. Also, a database including weather information was structured to be integrated as part of the explanatory variables in RT models, providing the average monthly values of precipitation, evaporation and maximum and minimum temperature. The Weka [23] Data Mining tool was used to build the RTs; the method was encoded in the Delphi programming language.

4.1 Constraints and Input Parameters

To establish the CASCs and CIWDCs constraints, ranges 20% below and 20% above each variable's average value were selected. The average value was also used to establish the sale price and production cost. Table 1 shows the input data per crop used.

Table 1. Constraints per crop used to test the algorithm

Crop	Sowing area (ha)		Irrigation water depth (mm)		Production cost (M$/ha)	Sale price (M$/ton)
	Lower limit	Upper limit	Lower limit	Upper limit		
Alfalfa	84.09	126.14	82.64	123.96	5.00	1.53
Cotton	227.06	340.60	68.11	102.17	10.00	4.08
Knapweed	513.58	770.36	29.19	43.79	5.00	2.25
Pea	67.78	101.66	61.99	92.98	9.00	2.53
Forage	109.70	164.55	76.83	115.25	4.00	1.56
Bean	146.12	219.18	61.69	92.53	7.00	5.55
Fruit-bearing	17.26	25.89	66.46	99.69	10.00	1.21
Chickpea	218.87	328.31	32.84	49.26	7.00	5.65
Vegetables	240.94	361.41	66.55	99.83	16.00	2.07
Corn	638.04	957.06	65.77	98.66	7.00	1.37
Potato	348.37	522.56	63.32	94.98	37.00	3.24
Sorghum	44.32	66.47	63.48	95.22	6.00	1.18
Tomato	81.43	122.15	82.48	123.72	20.00	4.36
Wheat	2,740.06	4,110.10	61.88	92.83	7.00	1.50
Hay	32.19	48.28	86.54	129.81	8.00	2.05

For the weather information, a data bank with monthly information about precipitation, evaporation and the minimum, maximum and average temperature for the selected period of time was integrated. TASC constraint was used to model two scenarios. The first one corresponds to an abundant state, where there is more sowing area available than sowing area allowed by constraints. The second scenario simulates a scarce state, where there is not enough area for sowing and its correct distribution has a larger relevance. Due to CASCs constraints, the maximum sowing area possible is about of 8,264.72 ha. Taking this as a base, the method intends in the first scenario for a total area available of 10,000 ha, while for the second, a total area of 6,000 ha is used. The values determined by experimentation for the GA parameters were the following: population's size =10, crossing probability=0.7, maximum of generations= 50, mutation probability= 0.003. SM was compared with GenSRT. SM maximizes the function indicated in (5), but using average yields for each crop instead of yield functions. The original TASC, CASCs and IWDCs constraints were used, ignoring any information provided by RT models.

4.2 Results

Scenario 1: Maximum Sowing Area Availability of 10,000 ha.

Results of GenSRT execution for a TASC constraint of 10,000 ha are shown in table 7. After 50 generations, GenSRT found that best chromosome was [2,0,3,0,0,1,1,1,0,0,0,2,4,1,0]. As can be seen, not only the highest yields (0 values) integrate the chromosome, there are many cases where a non-optimum yield contributed to achieve the highest global benefit. Table 2 shows the yields, sowing areas and irrigation water depth that was estimated resulting from the optimization, together with the production totals, gross income, total cost and estimated net income.

Table 2. GenSRT results for an available area of 10,000 ha

Crop	Sowing area (ha)	Irrigation water depth (mm)	Estimated yield (ton/ha)	Estimated production (ton)	Gross income (M$)	Total cost (M$)	Net income (M$)
Alfalfa	126.14	103.30	13.96	1,760.65	2,693.80	630.71	2,063.09
Cotton	262.50	79.66	3.05	801.47	3,270.78	2,625.00	645.78
Knapweed	770.36	33.01	2.07	1,598.35	3,601.09	2,311.09	1,289.99
Pea	101.66	77.48	5.40	548.55	1,389.47	914.98	474.50
Forage	164.55	82.25	10.55	1,736.44	2,715.80	658.19	2,057.60
Bean	156.10	91.99	2.10	327.13	1,815.56	1,092.67	722.90
Fruit-bearing	25.89	83.07	18.69	483.73	585.31	258.85	326.46
Chickpea	328.31	41.05	1.91	628.68	3,554.56	2,298.17	1,256.39
Vegetables	361.41	80.15	15.99	5,778.17	11,966.59	5,782.54	6,184.05
Corn	957.06	85.29	6.32	6,045.87	8,282.84	6,699.45	1,583.39
Potato	522.56	71.72	32.91	17,196.69	55,665.69	19,334.57	36,331.12
Sorghum	66.09	85.02	4.64	306.43	362.82	264.35	98.47
Tomato	122.15	87.54	18.20	2,222.86	9,693.90	2,442.92	7,250.98
Wheat	4,110.10	77.35	5.50	22,625.67	33,983.76	28,770.68	5,213.08
Hay	48.28	103.75	12.76	616.19	1,261.35	386.22	875.13
Total	8,123.16	78.84			140,843.32	74,470.39	66,372.93

On the other hand, table 3 shows the results of the SM and the crops average yields. The CASCs constraints are the same used with the GenSRT method, previously shown in table 1. Comparing table 2 and 3, it is observed that surface distribution and yields estimated by the GenSRT method produces a net income of 66,372.93 M$, while the use of SM with average yields obtains an income of 64,447.38 M$. The difference is about 1,925.55 M$. Also, the total area calculated by GenSRT is 146.56 ha smaller, which shows the effectiveness of the GenSRT method.

Table 3. SM results using average yields and a total area available of 10,000 ha

Crop	Sowing area (ha)	Irrigation water depth (mm)	Estimated yield (ton/ha)	Estimated production (ton)	Gross income (M$)	Total cost (M$)	Net income (M$)
Alfalfa	126.14	103.3	14.65	1,847.73	2,827.02	630.71	2,196.31
Cotton	340.6	85.14	2.86	974.44	3,976.70	3,405.95	570.75
Knapweed	770.36	36.49	2.13	1,637.02	3,688.21	2,311.09	1,377.12
Pea	101.66	77.48	5.72	581.01	1,471.70	914.98	556.72
Forage	164.55	96.04	10.22	1,681.85	2,630.41	658.19	1,972.21
Bean	219.18	77.11	2.11	462.91	2,569.15	1,534.27	1,034.89
Fruit-bearing	25.89	83.07	20.04	518.71	627.64	258.85	368.79
Chickpea	328.31	41.05	1.93	633.31	3,580.73	2,298.17	1,282.56
Vegetables	361.41	83.19	17.44	6,303.70	13,054.95	5,782.54	7,272.41
Corn	957.06	82.22	6.76	6,471.67	8,866.18	6,699.45	2,166.74
Potato	522.56	79.15	29.74	15,542.91	50,312.39	19,334.57	30,977.81
Sorghum	66.47	79.35	4.89	325.33	385.19	265.9	119.29
Tomato	122.15	103.1	19.81	2,419.22	10,550.23	2,442.92	8,107.31
Wheat	4,110.10	77.35	5.55	22,823.37	34,280.70	28,770.68	5,510.02
Hay	48.28	108.17	13.36	645.17	1,320.67	386.22	934.45
Total	8,264.72	80.81			140,141.87	75,694.49	64,447.38

Scenario 2: Maximum Sowing Area Availability of 6,000 ha.

Table 4 shows results of the GenSRT method execution when TASC constraint is limited to 6,000 ha. The best chromosome corresponds to the [2,0,3,0,0,0,1,1,0,1,0,1,0,1,0] combination. Table 5 shows SM results using average yields and the CASCs constraints shown in table 1. It can be observed that the difference in the net income between GenSRT and SM results is equal to 2,512.89 M$, which indicates GenSRT's superiority.

Table 4. Results of the GenSRT method for an area availability of 6,000 ha

Crop	Sowing area (ha)	Irrigation water depth (mm)	Estimated yield (ton/ha)	Estimated production (ton)	Gross income (M$)	Total cost (M$)	Net income (M$)
Alfalfa	126.14	103.3	13.96	1,760.65	2,693.80	630.71	2,063.09
Cotton	262.5	79.66	3.05	801.47	3,270.78	2,625.00	645.78
Knapweed	770.36	33.01	2.07	1,598.35	3,601.09	2,311.09	1,289.99
Pea	101.66	77.48	5.4	548.55	1,389.47	914.98	474.5
Forage	164.55	82.25	10.55	1,736.44	2,715.80	658.19	2,057.60
Bean	219.18	91.99	2.1	459.53	2,550.42	1,534.27	1,016.15
Fruit-bearing	25.89	83.07	18.69	483.73	585.31	258.85	326.46
Chickpea	328.31	41.05	1.91	628.68	3,554.56	2,298.17	1,256.39
Vegetables	361.41	80.15	15.99	5,778.17	11,966.59	5,782.54	6,184.05
Corn	957.06	68.84	6.31	6,041.56	8,276.94	6,699.45	1,577.49
Potato	522.56	71.72	32.91	17,196.69	55,665.69	19,334.57	36,331.12
Sorghum	66.47	72.03	4.73	314.62	372.51	265.9	106.61
Tomato	122.15	87.54	18.2	2,222.86	9,693.90	2,442.92	7,250.98
Wheat	1,923.48	67.74	5.52	10,607.99	15,933.20	13,464.35	2,468.84
Hay	48.28	103.75	12.76	616.19	1,261.35	386.22	875.13
Total	6,000.00	76.24			123,531.41	59,607.21	63,924.18

Table 5. Results of the SM with average yields and a total area of 6,000 ha

Crop	Sowing area (ha)	Irrigation water depth (mm)	Estimated yield (ton/ha)	Estimated production (ton)	Gross income (M$)	Total cost (M$)	Net income (M$)
Alfalfa	126.14	103.3	14.65	1,847.73	2,827.02	630.71	2,196.31
Cotton	340.6	85.14	2.86	974.44	3,976.70	3,405.95	570.75
Knapweed	770.36	36.49	2.13	1,637.02	3,688.21	2,311.09	1,377.12
Pea	101.66	77.48	5.72	581.01	1,471.70	914.98	556.72
Forage	164.55	96.04	10.22	1,681.85	2,630.41	658.19	1,972.21
Bean	219.18	77.11	2.11	462.91	2,569.15	1,534.27	1,034.89
Fruit-bearing	25.89	83.07	20.04	518.71	627.64	258.85	368.79
Chickpea	328.31	41.05	1.93	633.31	3,580.73	2,298.17	1,282.56
Vegetables	361.41	83.19	17.44	6,303.70	13,054.95	5,782.54	7,272.41
Corn	957.06	82.22	6.76	6,471.67	8,866.18	6,699.45	2,166.74
Potato	522.56	79.15	29.74	15,542.91	50,312.39	19,334.57	30,977.81
Sorghum	66.47	79.35	4.89	325.33	385.19	265.9	119.29
Tomato	122.15	103.1	19.81	2,419.22	10,550.23	2,442.92	8,107.31
Wheat	1,845.38	77.35	5.55	10,247.42	15,391.62	12,917.69	2,473.93
Hay	48.28	108.17	13.36	645.17	1,320.67	386.22	934.45
Total	6,000.00	80.81			121,252.79	59,841.50	61,411.29

5 Conclusions

This paper presents a new method to solve the CPO. The method is based on the use of RTs to model the crop yield and the use of GAs to find optimal yield/constraints relationships to insure a higher net income. The method was applied in an irrigation district and compared with SM. The following conclusions were obtained: 1) GAs and RTs can be effectively mixed to produce a new solution to solve the CPO problem, with the additional benefit that this new approach allows to take into account the crop yield's non-linear behavior. 2) Resource optimization carried out by the GenSRT method is more realistic than an optimization carried out by SM using average yields. This is because GenSRT incorporates non-linear regression models which are better predictors for the yield than the average. 3) GenSRT method offers moderate advantages over the optimization with the SM. However, the method has several points of improvement (parameter tuning, individuals' selection mechanism, fitness function evaluation, penalization operator, etc.). Further research can enhance the algorithm to obtain the highest benefits. 4) Although for now it has only been applied to the CPO problem, the method can be used in any other kind of optimization problem that includes a non-linear function depending on other variables. Further research includes a comparison of GenSRT with GA, GA and non lineal relationships of crops' physiologic processes and GA plus RT without Simplex usage.

References

1. Toyonaga, T., Itoh, T., Ishii, H.: A Crop Planning Problem with Fuzzy Random Profit Coefficients. In: Fuzzy Optimization and Decision Making, vol. 4, pp. 51–69. Springer Science+Business Media, The Netherlands (2005)
2. Frizzone, J.A., Coelho, R.D., Dourado-Neto, D., Soliant, R.: Linear Programming Model to Optimize the Water Resource Use in Irrigation Projects: An Application to the Senator Nilo Coelho Project. Sci. Agric. Piracicaba, 54 (Special Number), Junio, 136-148 (1997)

3. Playán, E., Mateos, L.: Modernization and optimization of irrigation systems to increase water productivity. In: IV International Crop Science Congress, Brisbane, Australia (2004)
4. Dantzig, G.: Linear Programming and Extensions. Princeton University Press, NJ (1967)
5. Palacios, V.E.: Strategies to Improve Water Management in Mexican Irrigation Districts: A Case Study in Sonora. PhD. Thesis, The University of Arizona, Tucson, Az (1976)
6. Paudyal, G.N., Gupta, A.D.: Irrigation planning by multilevel optimization. Journal of Irrigation and Drainage Engineering, ASCE 116, 273–291 (1990)
7. Kuo, S.-F., Liu, C.-W.: Comparative study of optimization techniques for irrigation project planning. JAWRA 39, 1; ProQuest Agriculture Journals (February 2003)
8. Sonmez, F.K., Altin, M.: Irrigation Scheduling and Optimum Cropping Pattern with Adequate and Deficit Water Supply for Mid-sized Farms of Harran Plain. Pakistan Journal of Biological Sciences 7(8), 1414–1418 (2004)
9. Borges Jr, J.C.F., et al.: Computational Modeling For Irrigated Agriculture Planning. Part I: General Description and Linear Programming. Eng. Agríc., Jaboticabal 28(3), 471–482 (2008)
10. Goldberg, D.E.: Genetic Algorithms. In: Search, Optimization and Machine Learning, Addison-Wesley, New York (1989)
11. Raju, K.S., Kumar, D.N.: Irrigation Planning using Genetic Algorithms. In: Water Resources Management, vol. 18, pp. 163–176. Kluwer Academic Publishers, Dordrecht (2004)
12. Kirkpatrick, S., Gelatt, C.D., Vecchi, M.P.: Optimization by Simulated Annealing. Science 220(4598), 671–680 (1983)
13. Georgiou, P.E., Papamichail, D.M.: Optimization Model of an Irrigation Reservoir for Water Allocation and Crop Planning. Irrigation Science 26(6), 487–504 (2008)
14. Liu, J., Goering, C.E., Tian, L.: A Neural Network for Setting Target Corn Yields. In: 1999 ASAE Annual Meeting, Paper No. 99–3040 (1999)
15. da Rosa, J.M.C., Lima Filho, A.V., Medeiros, M.C.: Tree-Structured Smooth Transition Regression Models Based on Cart Algorithm. Encontro Brasileiro de Econometria (2004)
16. Sen, Z., Öztopal, A.: Genetic Algorithms for the Classification and Prediction of Precipitation Occurrence. Hydrologicat Sciences-Journal-des Sciences Hydrologiques 46(2) (2001)
17. Drummond, S.T., Sudduth, K.A., Joshi, A., Birrel, S.J., Kitchen, N.R.: Statistical and Neural Methods For Site-Specfied Yield Prediction. Transactions of the ASAE 46(1) (2003)
18. Breiman, L., Friedman, J.H., Olshen, R.A., Stone, C.J.: Classification and Regression Trees. Wadsworth, Belmont-CA (1984)
19. Rokach, L., Maimon, O.: Top-Down Induction of Decision Trees Classifiers – A Survey. IEEE Transactions on Systems, Man and Cybernetics: Part C 1(11) (2002)
20. Quinlan, J.R.: Learning with continuous classes. In: Adams, Sterling (eds.) Proc. AI 1992, 5th Australian Joint Conference on Artificial Intelligence, Singapore, pp. 343–348 (1992)
21. Bittencourt, H.R., Clarke, R.T.: Feature Selection by Using Classification and Regression Trees (CART). In: The International Archives of the Photogrammetry, Remote Sensing and Spatial Information Sciences, Istanbul (2004)
22. Beasley, D., Bull, D.R., Martin, R.R.: An Overview of Genetic Algorithms: Part I, Fundamentals. University Computing 15, 58–69 (1993)
23. Witten, I.H., Frank, E., Trigg, L., Hall, M., Holmes, G., Cunningham, S.J.: Weka: Practical Machine Learning Tools and Techniques with Java Implementations. In: Kasabov, H., Ko, K. (eds.) ICONIP/ANZIIS/ANNES 1999 (1999)

MultiQuenching Annealing Algorithm for Protein Folding Problem

Juan Frausto-Solis[1], Xavier Soberon-Mainero[2], and Ernesto Liñán-García[3]

[1] Tecnológico de Monterrey, Campus Cuernavaca, Autopista del Sol Km 104+06,
Colonia Real del Puente, 62790, Xochitepec, Morelos. México
juan.frausto@itesm.mx
[2] IBT UNAM
soberon@ibt.unam.mx
[3] Universidad Autónoma de Coahuila
ernesto_linan_garcia@mail.uadec.mx

Abstract. This paper presents a new approach named MultiQuenching Annealing (MQA) for the Protein Folding Problem (PFP). MQA has two phases: Quenching Phase (QP) and Annealing Phase (AP). QP is applied at extremely high temperatures when the higher energy variations can occur. AP searches for the optimal solution at high and low temperatures when the energy variations are not very high. The temperature during the QP is decreased by an exponential function. Both QP and AP are divided in several sub-phases to decrease the temperature parameter until a dynamic equilibrium is detected by measuring its quality solution. In addition, an efficient analytical method to tune the algorithm parameters is used. Experimentation presented in the paper shows that MQA can obtain high quality of solution for PFP.

Keywords: Peptide, Protein Folding, Simulated Annealing.

1 Introduction

Protein Folding Problem (PFP) is one of the most difficult problems in the computational biology domain. The proteins are generated after a transcription-translation process of DNA, which genes are expressed as an amino acids sequence, which in turn forms proteins. The protein folding process starts with a certain initial conformation of the amino acids' atoms, followed by intermediate states, and ends in a final state, which is known as Native Structure (NS). The natural protein folding process is not yet completely understood, however, it is known that the NS is characterized by the minimal Gibbs free energy of conformation. The protein follows an unknown path from the initial state until its native structure [1]. It seems that in natural folding, the protein does not explore all its possible states [2].

When PFP is solved by computational methods, the objective is finding the NS of a protein, which has the minimal free energy. In addition, these methods avoid generating all its possible states. Ab Inition methods are used to predict the native conformation of a protein from its sequence. The energy of a conformation depends on the interaction

A. Hernández Aguirre et al. (Eds.): MICAI 2009, LNAI 5845, pp. 578–589, 2009.

among the atoms and their relative positions, and this can be calculated by the torsion angles and the distance among atoms. The optimum of PFP can be determined by several heuristic methods. Simulated Annealing (SA) [3],[4] is one of the most efficient algorithms for PFP problem [5-8].

Recently, a new efficient Simulated Annealing algorithm named Quenching Annealing algorithm (QA) [9] for PFP was developed using two phases (a quenching and an annealing phase). In this paper, an new approach named MultiQuenching Annealing (MQA) for PFP is presented. MQA is an efficient implementation of QA, in which the two phases (QP and AP) are divided into sub-phases. As is shown by experimentation results, MQA can obtain high quality of solution for PFP.

This paper is organized as follows: In section 2, formulation for analytical tuning methods is described. In section 3, the quenching phases in QA and MQA algorithms are formulated. In section 4, the MQA approach is explained. In section 5, experimental results are presented. Finally, in section 6, the conclusions are presented.

2 Formulations for Analytical Tuning of MQA Parameters

2.1 Setting Temperatures

Similar to QA, MQA is tuned using several formulas for setting temperature values and Metropolis Cycles. As in any SA algorithm, MQA allows deterioration of the cost function during the entire execution time. The probability of accepting any new solution should be closer to 1 at the highest temperature, because at this temperature any solution is accepted. Therefore, the deterioration of the cost function must be maximum at the highest temperature. MQA starts with an initial temperature named $C(1)$, which is associated with the maximum deterioration admitted and the defined acceptance probability (closer to one).

Let S_i be the current solution, S_j a new proposed one, $Z(S_i)$ and $Z(S_j)$ the associated costs to S_i and S_j respectively; and finally ΔZ_{max} and ΔZ_{min} represent the maximum and minimum deteriorations from a random solution to another one. We also use ΔZ to represent any deterioration. Following this notation we use $P(\Delta Z_{max})$ and $P(\Delta Z_{min})$ to represent the probabilities of accepting a new solution with the maximum or minimum deterioration respectively. As it is known, these probabilities follow the Boltzmann distribution $P(\Delta Z)=\exp(-\Delta Z/C)$. Therefore, C_i can be calculated from $C_i= \Delta Z/\ln(P(\Delta Z_i))$.

The initial temperature is determined as the maximum deterioration case, i.e. C_i and ΔZ_i should be replaced by $C(1)$ and ΔZ_{max} respectively. Due to the fact that the acceptance probability is very close to one and ΔZ_{max} is very high, (10) is used to calculate the initial temperature, which is an extremely high value. Similarly, the final temperature $C(f)$ is determined with the formula (11). Now, $C(f)$ is established according to the probability of accepting a new solution with the minimum deterioration or $P(\Delta Z_{min})$; this probability is very near to zero. On (11), the final temperature has a very small value.

2.2 The Markov Chain Length

In MQA, the length of the Metropolis Chain (MC) is defined as the number of iterations into the Metropolis Loop (ML), and can be modeled by constant or variable MCs. In a constant MC, any ML has the same length each time MQA is executed; on other hand, when a variable MC is used, the ML has different lengths during the MQA execution.

Let $L(k)$ be the number of iterations at the temperature number k in ML; it can be set as a multiple of variables of the instance of PFP. In MQA with constant MC, $L(k)$ is set as a constant for all the temperatures. In other implementations, ML is stopped by a certain number of accepted solutions. On other hand, analytical methods determine L(k) with a simple Markov model [10]; at high temperatures, only a few iterations are required because the stochastic equilibrium is quickly reached. Nevertheless, at low temperatures, a more exhaustive exploration is needed, and therefore, a larger L(k) is used. Let L(1) be L(k) at C(1) and L_{max} be the maximum MC length; C(k) is decreased by the cooling function (9), where the α parameter is between 0.7 and 0.99 [3],[4]. Similarly, L(k) is calculated from (1) where β is the increment coefficient of MC (>1). Therefore, L(k+1) > L(k), L(1)=1, and L(f) is equal to L_{max}. The calculate of C(k+1) and L(k+1) are applied successively in MQA from C(1) to C(f). Consequently, C(f) and L_{max} can be obtained by(2), and (3), where n is the step number, or the number of MC's, from C(1) to C(f). Therefore, n is obtained by (4), and β is calculated by (5). This tuning approach prevents the following problems: a) MQA spends a large amount of time making computations even though the stochastic equilibrium is indeed reached or b) MQA stops long before the equilibrium state. Therefore, MQA becomes faster than other implementations. Metropolis parameters depend only on the definition of the C(1) and C(f), as we have shown in section 2.1. L_{max} must be set to a value that allows for a good exploration (between 1 to 4 times the neighborhood size or 63% to 99%) [10].

3   Quenching Phases in QA and MQA Algorithms

In QA, a general cooling function is used for decreasing the temperature number k. This cooling function is formulated as (6), where γ_k is calculated by (7), and τ_k is a quadratic recursive equation defined as (8), with $0< \tau_k <1$. τ_k is close to one (for example 0.999), so γ_k converges to 1. Therefore, $C(k+1)$ converges to a final temperature defined as (9).

$$L(k+1) = \beta L(k). \tag{1}$$
$$C(f) = \alpha^n C(1). \tag{2}$$

$$L_{max} = \beta^n L(1). \tag{3}$$
$$n = \frac{(Ln(C(1) - Ln(C(1)))}{Ln(\alpha)}. \tag{4}$$

$$\beta = \exp^{(Ln(L max) - Ln(L(1))}/_n. \tag{5}$$
$$C(k+1) = \alpha \gamma_k C(k). \tag{6}$$

$$\gamma_k = (1 - \tau_k). \tag{7}$$
$$\tau_k = \tau^2_{k-1}. \tag{8}$$

$$C(k+1) = \alpha C(k). \qquad (9) \qquad C(1) = -\Delta Z_{max} \big/ Ln(P(\Delta Z_{max})). \qquad (10)$$

$$C(f) = -\Delta Z_{min} \big/ Ln(P(\Delta Z_{min})). \qquad (11)$$

These equations (6)-(8) model the temperature cycle. The initial temperature is determined as was explained in section 2. Notice that for a certain r value, the cooling function converges to the classical cooling function, which is commonly used in SA algorithms [3],[5],[6],[13] where the latter α parameter is $0.7 < \alpha < 1$. Once the temperature $C(k)$ converges to a certain $C(k_2)$ value, this one can be set as an initial temperature with another r value, and another $C(k3)$ is reached. Model (6)-(8) is somewhat more complex than the classical (9). However, it is able to explore the space solution in several temperature slices.

A quenching phase is a sub-process of a MQA algorithm, which is applied to extremely and very high temperatures. In this phase, the temperature is suddenly decreased. In a similar way, a multi-annealing phase is a sub-process of MQA Algorithm, which is applied to high and low temperatures; in this phase, the temperature is gradually decreased. The equation (6) can be used for establishing a multi-quenching phase and a multi-annealing phase; the γ_k variable indicates the current phase, if γ_k is less than one then MQA is in the quenching phase. If γ_k converges to one, then MQA is in the annealing phase and equation (6) becomes equation (9).

When MQA is in the quenching phase (i.e. $\gamma_k < 1$), there are several sub-phases; these sub-phases are indicated with the α values defined in the quenching phase. MQA can change from one sub-phase to other one by changing its α values. This change is made by each quasi-stable state. Each time α changes a new sub-phase is executed.

At very high temperatures, the energies computed from one solution to another can be extremely high with enormous variations; therefore MQA is in a chaos phase. While the temperature is decreasing, the energy variations are less chaotic. Each time the temperature is decreased, a MC is repeated. Every MC is finished when a quasi-stable is detected by measuring the energy variations. Immediately after, the temperature reaches a quasi-stable state, a new sub-phase is started. During the quenching phase, the α values are set in a range of (0.7, 0.8) and the number of sub-phases is controlled by the number of αs. Once the energy variations are less unstable, MQA is ready to start a classical simulated annealing process; this is done by a stabilization phase. In this phase of stabilization, the energy variation is less abrupt than during the normal quenching phase.

4 The MQA Approach

MQA is an efficient implementation of QA that also has the Quenching Phase (QP) and the Annealing Phase (AP). QP searches solutions at extreme and very high temperatures when the higher energy variations can occur in a chaotic way; so QP is also called the chaos phase. The temperature during the QP is decreased by an exponential function, which was modeled by (6)-(8). The chaos phase reaches the annealing phase when the energy variations from one proposed solution to another are less unstable.

AP looks for the optimal solution at high and low temperatures until the energy variations reach a dynamic equilibrium. In MQA, both QP and AP are divided into several
sub-phases by decreasing the temperature parameter until a quasi-stable state is
reached in the first phase or a dynamic equilibrium in the second one (also known as a
freeze state in SA literature). When MQA reaches the freeze state, the algorithm is
stopped.

In MQA, an efficient analytical method to tune the algorithm parameters is used by
a Markov approach. First, the initial and final temperatures are determined as in Satisfiability problems [10]; the number of iterations in every sub-phase is also tuned with
an analytical model. During the quenching phase, the equations (6)-(8) are applied
with different α values according to the accepted solutions versus proposed solutions
ratio. These values are in the range (0.7, 0.8).

In order to obtain high quality solutions, it is important to apply a second phase
after the quenching phase. This phase can be useful to find better PFP solutions closer
to the NS. During the multi-annealing phase, the temperature is decreased gradually
by applying equation (9), but with α values different at the quenching phase, these
values are in the range (0.9, 0.98). In MQA, it is observed that τ determines the
current phase. As we mentioned before, the initial τ value in every sub-phase of QA is
very close to one (such as 0.999). Because after every Metropolis Cycle τ is updated
using equation (8) and eventually converges to zero, this value is a target to change
the α value from its lower value (such as 0.7) to the greatest value (such as 0.98).
Therefore, once α reaches it greatest value, at a very low τ value. Conventionally, we
can define QP and AP with a simple rule: If τ is greater than zero, the quenching
phase is applied, but if τ is very close to zero, the annealing phase is applied. Therefore, this algorithm is more general than QA and the classical SA algorithm. In this
way, the MQA complexity (and its execution time) is not greater than QA because
only some simple instructions are added to QA in order to obtain MQA. Therefore, it
would not be necessary to make any efficiency tests with MQA; nevertheless, some
experimentation presented in the next section confirms this issue. MQA is implemented using the following algorithm:

MultiQuenching-Simulated Annealing Algorithm

```
C1 = apply equation (2)
Cf = apply equation (3)
Initial Energy = function_energy()
While (Temperature_k > criteria_stop) do
    Metropolis_cycle()
    Temperature_{k+1} = α (1- τ )temperature_k
    τ = τ²
    If τ > 0 then
        Apply_quenching phase()
    Else
        Apply_annealing_phase()
    End if
End do
```

5 Experimental Results

MQA was tested with four proteins (Met[5]–enkephalin, C-Peptide, Proinsulin and BPTI). The Met[5]-enkephalin has 5 amino acids. The C-peptide has a sequence with 13 amino acids, the Proinsulin has 31 amino acids, and BPTI has 58 amino acids (see table 1) [17].

Table 1. Amino acid sequences of small proteins

Small Protein	Amino acid sequences
Met[5]enkephalin	TGGFM
C-peptide	KETAAAKFERQHM
Proinsulin	EAEDLQVGQVELGGGPGAGSLQPLALEGSLQ
BPTI	RPDFCLEPPYTGPCKARIIRYFANAKAGLCQTFVYGGCRAKRNNFKSAEDCMR TCGGA

ECEPP/2 was used for evaluating the conformation energy function. Neighbor solutions were selected randomly (angles in [-180°, 180°]), and $C(1)$ and $C(f)$ were calculated using $P(\Delta Z_{max})=0.7$ and $P(\Delta Z_{min})=0.3$. The initial temperature ($C(1) = 1.76x10^{25}$) is extremely high because the high values of the energies; therefore, ΔZ_{max} has extremely high or low values, and $C(f)$ is set as 0.001. Simulations were made with several combinations of α values for the QP and AP phases. The simulation results are shown in table 2-10. MQA was executed thirty times, where AP was implemented using a variable Markov Chain. The average (best, and worst) results were registered and are shown in tables 3, 7, and 9. These results are average energies and average processing times, best and worst solutions (energy computed) of processed proteins. The α values are shown in the columns QP and AP. The twelve average results of met[5]-enkephalin are shown in table 2. The best average energy (-8.44 kcal/mol in 8.50 minutes) of met[5]-enkephalin is obtained by two sub-phases applied in quenching, and three sub-phases in annealing phases. These phases are defined with 0.7 and 0.8 into quenching phase, and 0.98, 0.95, and 0.90 in the annealing phase.

Table 2. MultiQuenching and MultiAnnealing results of met[5]-enkephalin

QP α values			AP α values			Average Energy (kcal/mol)	Average Time
0.7	0.75		0.98			-7.67	5.87
0.7	0.75		0.98		0.95	-8.07	5.85
0.7	0.75		0.98	0.95	0.90	-6.95	5.87
0.7	0.8		0.98			-7.21	8.98
0.7	0.8		0.98		0.95	-7.80	9.06
0.7	0.8		0.98	0.95	0.90	-8.44	8.50
0.75	0.8		0.98			-7.58	5.33
0.75	0.8		0.98		0.95	-7.79	5.32
0.75	0.8		0.98	0.95	0.90	-7.89	5.29
0.7	0.75	0.8	0.98			-7.73	5.81
0.7	0.75	0.8	0.98		0.95	-7.38	9.00
0.7	0.75	0.8	0.98	0.95	0.90	-7.44	5.38

Table 3 shows the best and worst solutions obtained for met[5]-enkephalin: the best solution is -10.72 kcal/mol, with a processing time equal to 9.45 minutes; the worst solution is -3.21 kcal/mol with 5.39 minutes. It is important to indicate that the average energy and the best energy were obtained using two sub-phases in the quenching phase, and three in the annealing phase.

Table 3. Best and Worst Results of met[5]-enkephalin

QP α values			AP α values			Best Solutions		Worst Solutions	
						Energies (kcal/mol)	Time (minutes)	Energies (kcal/mol)	Time (minutes)
0.7	0.75		0.98			-9.51	5.63	-5.40	5.63
0.7	0.75		0.98		0.95	-10.71	6.28	-5.26	5.62
0.7	0.75		0.98	0.95	0.9	-10.28	6.28	-5.50	5.62
0.7	0.8		0.98			-10.64	9.76	-5.11	9.67
0.7	0.8		0.98		0.95	-10.71	8.52	-5.66	9.53
0.7	0.8		0.98	0.95	0.9	-10.72	9.45	-5.00	8.55
0.75	0.8		0.98			-10.58	5.30	-4.51	5.16
0.75	0.8		0.98		0.95	-10.70	5.18	-5.07	5.34
0.75	0.8		0.98	0.95	0.9	-10.69	5.55	-3.21	5.39
0.7	0.75	0.8	0.98			-10.70	6.28	-4.07	5.63
0.7	0.75	0.8	0.98		0.95	-9.52	8.10	-6.22	8.29
0.7	0.75	0.8	0.98	0.95	0.9	-10.69	5.42	-6.04	5.44

The average results of C-peptide are shown in table 6. The best and worst solutions of C-peptide are shown in table 7. The best average energy (-84.44 kcal/mol in 72.24 minutes) of C-peptide is obtained by two sub-phases applied in the quenching and the annealing phases. These phases are defined with 0.7 and 0.8 into quenching phase, and 0.98 and 0.95 in the annealing phase. The best solution obtained of C-peptide is -101.49 kcal/mol, with a processing time equal to 73.82 minutes; the worst solution is -62.79 kcal/mol with 43.18 minutes. In figure 1, the Native Structures of met[5]-enkephalin, and Proinsulin are shown; this figure includes the comparison with the best reported solutions. The Native structure is colored, and the best solution is gray. In the tables 4 and 5, the angles are shown.

 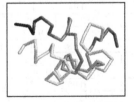

Fig. 1. Comparison of NS of met[5]-enkephalin, and Proinsulin

Table 4. Angles of NS of Proinsulin, and Best solution

Residue	Best Solution		Native Structure	
	Phi	Psi	Phi	Psi
Glu		159		172.7
Ala	-67	-43.3	-110.5	-5.8
Glu	-93.8	59.3	-78.8	-11.4
Asp	-169.1	156.8	-102.2	0.6
Leu	-61.9	-36.2	-87.7	-23.9
Gln	-79.5	130.6	-139.2	-44
Val	-126	98.3	53.3	0.5
Gly	92	-63.4	67.6	34.7
Gln	-80.2	74.4	-166.1	33.2
Val	-79.9	97.2	-159.9	-26.1
Glu	-72.9	-45.1	-158.1	-41.8
Leu	-140.7	168.4	31.5	-71.2
Gly	-135.2	33.1	68.2	52
Gly	86.1	108.1	-151.9	-150.1
Gly	125.2	-72.9	-153.5	-77.6
Pro	-75.2	83.5	-71.1	-60.8
Gly	125.5	136.1	-90.5	38.4
Ala	-54.6	-58.3	-110.9	-108.5
Gly	154.2	55.1	-114.5	46.9
Ser	-67.1	-34	72.7	-165.5
Leu	-132.4	29.7	-66.8	152.4
Gln	-169.4	88.4	-55.8	-54.3
Pro	-75	-21.8	-64.9	-35
Leu	-81.7	66.8	-90.1	18.5
Ala	-161.8	-53.6	-109.2	-7.6
Leu	-142.5	-62.2	50.6	29.3
Glu	-79.2	72.1	-162.9	-138.8
Gly	154.8	-46.7	81.1	28
Ser	-78.6	-61.5	75.8	10.6
Leu	-77.4	-48.1	83.1	42
Gln	-75.3		51.4	

Table 5. Angles of NS of met[5]-enkephalin and Best solution

Residue	Best Solution		Native Structure	
	Phi	Psi	Phi	Psi
Tyr		153.8		173.7
Gly	-161.4	70.5	158.0	-20.7
Gly	64.4	-93.4	115.6	50.1
Phe	-81.8	-29.0	-110.0	-36.8
Met	-81.0		-73.6	

586   J. Frausto-Solis, X. Soberon-Mainero, and E. Liñán-García

Table 6. MultiQuenching and MultiAnnealing results of C-Peptide

QP α values			AP α values			Average Energy (kcal/mol)	Average Time (minutes)
0.7	0.75		0.98			-79.22	52.03
0.7	0.75		0.98		0.95	-82.02	72.96
0.7	0.75		0.98	0.95	0.90	-79.17	38.98
0.7	0.80		0.98			-80.83	50.75
0.7	0.80		0.98		0.95	-84.44	72.24
0.7	0.80		0.98	0.95	0.90	-80.32	35.95
0.75	0.80		0.98			-79.23	48.49
0.75	0.80		0.98		0.95	-84.11	69.18
0.75	0.80		0.98	0.95	0.90	-78.79	35.91
0.7	0.75	0.8	0.98			-81.57	54.12
0.7	0.75	0.8	0.98		0.95	-83.96	65.18
0.7	0.75	0.8	0.98	0.95	0.90	-78.09	31.26

Table 7. Best and Worst Results of C-Peptide

AP α values			AP α values			Best Solutions		Worst Solutions	
						Energies (kcal/mol)	Time (minutes)	Energies (kcal/mol)	Time (minutes)
0.7	0.75		0.98			-95.24	44.42	-69.62	45.83
0.7	0.75		0.98		0.95	-92.42	79.69	-72.33	79.44
0.7	0.75		0.98	0.95	0.9	-89.64	37.77	-67.17	38.13
0.7	0.80		0.98			-96.77	62.06	-72.37	56.78
0.7	0.80		0.98		0.95	-101.49	73.82	-73.62	67.71
0.7	0.80		0.98	0.95	0.9	-94.38	36.05	-68.13	35.42
0.75	0.80		0.98			-90.43	48.85	-62.79	43.18
0.75	0.80		0.98		0.95	-98.40	62.38	-73.61	56.78
0.75	0.80		0.98	0.95	0.9	-90.90	34.84	-68.17	36.01
0.7	0.75	0.8	0.98			-95.83	47.58	-71.75	52.88
0.7	0.75	0.8	0.98		0.95	-96.07	58.70	-69.79	80.54
0.7	0.75	0.8	0.98	0.95	0.9	-94.79	29.17	-65.80	29.70

Table 8. MultiQuenching and MultiAnnealing results of Proinsulin

QP α values			AP α values			Average Energy (kcal/mol)	Average Time (minutes)
0.7	0.75		0.98			-135.61	265.67
0.7	0.75		0.98		0.95	-139.55	281.71
0.7	0.75		0.98	0.95	0.90	-135.09	278.62
0.7	0.80		0.98			-144.78	382.80
0.7	0.80		0.98		0.95	-140.56	390.96
0.7	0.80		0.98	0.95	0.90	-141.89	377.73
0.75	0.80		0.98			-132.18	206.62
0.75	0.80		0.98		0.95	-134.53	204.18
0.75	0.80		0.98	0.95	0.90	-135.61	211.01
0.7	0.75	0.8	0.98			-135.61	280.36
0.7	0.75	0.8	0.98		0.95	-142.76	384.59
0.7	0.75	0.8	0.98	0.95	0.90	-136.00	206.30

The average results of Proinsulin are shown in table 8. The best and worst solutions of Proinsulin are shown in table 9. The best average energy (-144.78 kcal/mol in 382.80 minutes) of Proinsulin is obtained by using two sub-phases applied in the quenching and one annealing phase. These phases are defined with 0.7 and 0.80 in the quenching phase and 0.98 in the annealing phase. Table 9 shows the best quality solution obtained for the Proinsulin (-173.77 kcal/mol) with 246.83 minutes of processing time; the worst solution is –105.80 kcal/mol with 318.64 minutes.

Table 9. Best and Worst Results of Proinsulin

QP α values			AP α values			Best Solutions		Worst Solutions	
						Energies (kcal/mol)	Time (minutes)	Energies (kcal/mol)	Time (minutes)
0.7	0.75		0.98			-173.77	246.83	-116.17	263.55
0.7	0.75		0.98		0.95	-161.41	300.55	-121.47	314.01
0.7	0.75		0.98	0.95	0.9	-155.85	282.65	-105.80	318.64
0.7	0.8		0.98			-167.33	362.78	-122.36	391.93
0.7	0.8		0.98		0.95	-160.65	411.46	-112.48	424.88
0.7	0.8		0.98	0.95	0.9	-170.62	377.08	-118.13	386.40
0.75	0.8		0.98			-157.79	204.66	-114.34	206.15
0.75	0.8		0.98		0.95	-157.83	200.55	-118.93	213.02
0.75	0.8		0.98	0.95	0.9	-146.57	203.16	-120.56	214.09
0.7	0.75	0.8	0.98			-159.55	290.67	-118.49	240.88
0.7	0.75	0.8	0.98		0.95	-155.05	435.42	-127.36	360.76
0.7	0.75	0.8	0.98	0.95	0.9	-148.88	207.79	-119.74	210.89

In table 10, MQA is compared with QA using four instances of PFP. The average results of met[5]-enkephalin, C-Peptide, and Proinsulin have high quality of solution when MQA is applied. The quality of solutions between MQA and QA are similar when they are applied to BPTI Protein. In this table, the processing time of QA is lower than that of MQA. Specifically for BPTI, the QA processing time is greater than MQA.

Table 10. MQA versus QA

	Average Energy (kcal/mol)		Average Time (minutes)	
	QA	MQA	QA	MQA
Met[5]-enkephalin	-5.50	-7.66	1.5	6.68
C-Peptide	-68.31	-80.98	11.14	52.25
Pronsulin	-110.95	-137.84	41.53	289.21
BPTI	-266.58	-262.59	191.52	81.12

6 Conclusions

In this work a new approach named MultiQuenching Annealing (MQA) for Protein Folding Problem was presented. MQA is an unconventional because a chaotic phase is applied by using a simple algebraic model and an old algorithm like simulated annealing. MQA has several sub-phases for both a Quenching phase (QP) and an Annealing

Phase (or AP) of a Simulated Annealing (SA) algorithm. By experimentation with some proteins, in general, the quality of the solution of MQA is better than using only one quenching with one annealing phase. After experimentation with some proteins, it is observed that MQA obtained better results than QA. It would not have been necessary to compare the execution time of MQA vs QA, since they have the same complexity order. However, we present some experimental results.

The paper also shows how to adapt an analytical tuning method that determines the algorithm parameters by using a simple Markov model joined with a dynamic cooling scheme that determines the final temperature of every sub-phase during the MQA execution. In order to change from one sub-phase to another, a simple algebraic model is used. Furthermore, if the user wants to use only a simple quenching and/or only one annealing phase, it is enough to set a simple coefficient. Therefore, it is possible to obtain a Quenching Annealing algorithm and the classical Simulated Annealing algorithm only by handling the MQA parameters.

References

1. Anfinsen, C.: Principles that govern the folding of protein chains. Science 181, 223–230 (1973)
2. Levinthal, C.: Are there pathways for protein folding? J. Chem. Phys. 65, 44–45 (1968)
3. Kirkpatrick, S., Gelatt, C., Vecchi, M.: Optimization by simulated annealing. Science 4598, 220, 4598, 671–680 (1983)
4. Cerny, V.: Thermodynamical approach to the traveling salesman problem: An efficient simulation algorithm. Journal of Optimization Theory and Applications 45(1), 41–51 (1985)
5. Morales, L., Garduño, R., Romero, D.: Application for simulated annealing to the multiple – minima problem in small peptides. J. Biomol. Str. And Dyn. 8, 1721–1735 (1991)
6. Morales, L., Garduño, R., Romero, D.: The multiple – minima problem in small peptide revisited. The threshold accepting approach. J. Biomol. Str. And Dyn. 9 (1992)
7. Garduño, R., Romero, D.: Heuristic Methods in conformational space search of peptides. J. Mol. Str. 308, 115–123 (1994)
8. Chen, W., Li, K., Liu, J.: The simulated annealing method applied to protein structure prediction. In: Third international conference on machine learning and cybernetics, Shanghai (2004)
9. Frausto-Solis, J., Román, E.F., Romero, D., Soberon, X., Liñan-García, E.: Analytically Tuned Simulated Annealing applied to the Protein Folding Problem. In: Shi, Y., van Albada, G.D., Dongarra, J., Sloot, P.M.A. (eds.) ICCS 2007. LNCS, vol. 4488, pp. 370–377. Springer, Heidelberg (2007)
10. Sanvicente, H., Frausto-Solís, J.: A Method to Establish the Cooling Scheme in Simulated Annealing Like Algorithms. In: Laganá, A., Gavrilova, M.L., Kumar, V., Mun, Y., Tan, C.J.K., Gervasi, O. (eds.) ICCSA 2004. LNCS, vol. 3045, pp. 755–763. Springer, Heidelberg (2004)
11. Eisenmenger, F., Hansmann, U., Hayryan, S., Hu, C.: An Enhanced Version of SMMP – Open source software package for simulation of proteins. Comp. Phys. Comm., 174–422 (2006)
12. Sanvicente-Sanchez, H.: Metodología de paralelización del ciclo de temperatura en algoritmos tipo recocido simulado. Tesis doctoral, ITESM Campus Cuernavaca, México (2003)

13. Sanvicente-Sanchez, H., Frausto-Solis, J.: Optimización de los diámetros de las tuberías de una red de distribución de agua mediante algoritmos de recocido simulado. Ingeniería hidráulica en México XVIII(1), 105–118 (2003)
14. Sanvicente-Sanchez, H., Frausto-Solis, J., Imperial, F.: Solving SAT Problems with TA Algorithms Using Constant and Dynamic Markov Chains Length. In: Megiddo, N., Xu, Y., Zhu, B. (eds.) AAIM 2005. LNCS, vol. 3521, pp. 281–290. Springer, Heidelberg (2005)
15. Frausto-Solis, J., Sanvicente-Sanchez, H., Imperial, F.: ANDYMARK: An analytical method to establish dynamically the length of the Markov chain in simulated annealing for the satisfability problem. LNCS. Springer, Heidelberg (2006)
16. Eisenmenger, F., Hansmann, U., Hayryan, S., Hu, C.: SMMP: A modern Package for Protein Simulation. Comp. Phys. Comm. 138, 192 (2001)
17. Protein Data Bank web page, http://www.rcsb.org

Outlier Detection with a
Hybrid Artificial Intelligence Method

Manuel Mejía-Lavalle[1], Ricardo Gómez Obregón[2], and Atlántida Sánchez Vivar[1]

[1] Instituto de Investigaciones Eléctricas, Reforma 113, 62490 Cuernavaca, Morelos, México
mlavalle@iie.org.mx, atlantidasv@gmail.com
[2] Comisión Federal de Electricidad, Reforma 164, México DF, México
ricardo.gomez@cfe.gob.mx

Abstract. We propose a simple and efficient hybrid artificial intelligence method to detect exceptional data. The proposed method includes a novel end-user explanation feature. After various attempts, the best design was based on an unsupervised learning schema, which uses an hybrid adaptation of the Artificial Neural Network paradigms, the Case Based Reasoning methodology, the Data Mining area, and the Expert System shells. In our method, the cluster that contains the smaller number of instances is considered as outlier data. The method provides an explanation to the end user about why this cluster is exceptional regarding to the data universe. The proposed method has been tested and compared successfully not only with well-known academic data, but also with a real and very large financial database that contains attributes with numerical and categorical values.

1 Introduction

Outlier detection has become a fast growing topic in Data Mining [1]. This is because there are important diverse of applications where is required to know if exceptional data is immersed in a very large database.

Typical applications are related with financial frauds, but also we can cite other applications of great interest for industry, medicine, astronomy, communications, and many more [1].

Although currently multiple techniques have been proposed and published to attack the problem [2], in most cases the algorithmic complexity is high, and consequently processing times is high too: this is a serious drawback if we consider that typical applications are given in databases with thousands or millions of records. Additionally, in many of the proposed methods in specialized literature, the end user receives the outlier information in a "black box" fashion. Indeed, we observed in the state-of-the-art published works for outlier detection, fraud analysis and cluster methods, that authors do not include an automated explanation facility. At most, they include information about the centroids that were formed and a post-run human result interpretation [3-11].

To solve this problem, we propose a simple and efficient hybrid artificial intelligence method, with low algorithmic complexity, which not only detects outlier data, but also provides to the end user a reasoned explanation of why this data is anomalous with respect to the data universe.

A. Hernández Aguirre et al. (Eds.): MICAI 2009, LNAI 5845, pp. 590–599, 2009.
© Springer-Verlag Berlin Heidelberg 2009

Before obtaining the method proposed here, we tried with various designs, some of them were taken directly from specialized literature and some others were realized with adaptations. At the end, the most efficient, simple and adequate method for our necessities resulted to be one based on an unsupervised learning schema, that takes ideas from: a) the Case Based Reasoning (CBR) methodology, b) the Artificial Neural Networks (ANN) paradigms, c) the Data Mining area, and d) the Expert Systems explanation facilities. Specifically, we made diverse novel hybrid adaptations to the Adaptive Resonance Theory (ART) artificial neural network paradigm [12] applying ideas from CBR and Data Mining to obtain clusters and prototypes; to obtain the explanation facility, we adapted ideas emerged from the Expert Systems area [13]. In our case, the proposed explanation facility allows the end user to reach a better problem understanding and consequently, it helps to take better, well-supported and informed actions.

We tested and compared our method with well-known academic data and we obtained very competitive results. In addition, we tested the method with a real and very large Mexican financial database, with more than 30 million transactions with numerical and categorical attributes. With this real database, we observe the method scalability and good performance, not only to detect the outlier data, but also to provide an articulated explanation related with the applied reasons by our method in considering those transactions as exceptional.

We think that the hybrid method that we propose can be of interest and of immediate application in a great variety of domains where it is needed to know, in an informed way, if exceptional transactions exist.

To develop these ideas, our paper is organized in the following way. In Section 2, the proposed method is described and illustrated. In Section 3, experiments and obtained results are shown, and finally in Section 4 conclusions and future works are addressed.

2 Proposed Hybrid Method

To present the proposed hybrid artificial intelligence method we will explain firstly the basic algorithm that we choose for the outlier detection task. Next, we will detail the similarity metrics that we applied to numerical and categorical attributes. Then, we will introduce the way that we implement to obtain data clusters (prototypes). Immediately, we will describe how we conceptualize the explanation facility and finally, we will summarize the complete algorithm.

2.1 Outlier Detection Algorithm

The proposed algorithm was adapted from the ANN paradigm known as Adaptive Resonance Theory or simply ART [12], using ideas from the CBR methodology and Data Mining area.

Although more sophisticated algorithms exist to obtain clusters, like *k-means* or *Expectation Maximization* [1], we choose ART due to their simplicity and acceptable algorithmic complexity. As it will be seen in Section 3, our ART hybrid variation has faster processing time than other well established cluster methods.

At the beginning, ART accepts instances (records or transactions). Depending on the similarity degree between two instances, ART joins them, forming a prototype (a cluster), or separates them, forming two prototypes (two clusters).

The original ART algorithm is described in [12], where we can observe that the user defines a threshold parameter, that we named U: with the U parameter the desired similarity /dissimilarity among instances is controlled.

ART paradigm has certain similarity with the CBR methodology; nevertheless, originally ART was designed only to handle binary data, for image classification, and using an unsupervised learning schema. For our necessities, the ART paradigm was useful for the clusterization task: the cluster that contains the smaller number of instances will be the one considered as the exception group, or outlier data.

2.2 Metrics

To be able to apply ART paradigm to continuous and categorical attributes, we defined the following similarity metrics (adapted from CBR and Data Mining), applicable when an instance is compared against another instance.

Continuous-numerical attributes:

To apply the proposed metric, firstly attributes with numerical values should be normalized among 0 and 1. Once this is done, the distance d_N between two numerical values, of a same attribute A, will be simply its absolute difference:

$$d_N = abs\ (value1 - value2) \tag{1}$$

Categorical attributes:

In this case, a value of 1 is assigned if the categorical attributes are different, and a 0 if they are equal. This approach is often used in the Data Mining area, e.g. [14].

$$d_C = 0 \text{ if } value1 = value2, 1 \text{ in other case} \tag{2}$$

Total similarity metric:

Total similarity metric d_T is the attribute distances average. That is to say, it is calculated with the sum of each one of the distances d of each attribute (attributes can be numerical-continuous d_N, or categorical d_C) divided by the total number of attributes A:

$$d_T = \Sigma\ d\ /\ A \tag{3}$$

Thus, if distance d_T results with a value of 1 or close to 1, the compared instances are different; and if the value is 0 or close to 0, the two compared instances are equal or very similar. Equation (3) easily can be modified to give more weight to an attribute (or more), than others: if we knew beforehand that certain attributes are more important than others, each attribute can be multiplied by a weight factor F assigned by a human expert.

$$d_T = \Sigma\ (d * F)\ /\ A \tag{3'}$$

The weight factors F of all the attributes should sum A units. If we do not want to use the weighted schema, then F will take a value of 1 for each attribute.

2.3 Prototype Construction

In this case, the idea that better work was to apply an incremental prototype construction, following a weighted schema according to the number of instances accumulated in each prototype. For relatively small databases (less than 100 instances or transactions), the weighted schema is not relevant: its importance begins for databases with thousands of instances. For our purposes, each prototype is a cluster, where the cluster with less accumulated instances will be the exceptional cluster or, properly, the outlier data.

Additional to a distance calculation, or total similarity d_T between an instance and a prototype, a user-defined threshold parameter U will determine if an instance will be clustered with some existing prototype or if it will form a new independent prototype, according to:

If $U < d_T$ create a new prototype, else combine the instance to the prototype   (4)

To combine an instance to a prototype we apply the following two weighted schemas:

Continuous-numerical attributes:

$$a_{NP} = fa_{TP} / (fa_{TP} + 1) * a_{NP} + (1 - (fa_{TP} / (fa_{TP} + 1)) * a_{NI} \qquad (5)$$

where fa_{TP} is the total number of instances accumulated in the prototype P for the a attribute; a_{NP} is the numerical value of the prototype for the attribute a; a_{NI} is the actual numerical value of the attribute a which instance I is going to be added to the prototype. With equation (5) we want to consider the contribution of a new instance I to the existing prototype: while more instances will be accumulated already in the prototype, smaller will be the weight that will be assigned to the numerical value of the instance that is going to be added to the prototype. For example, if a numerical attribute from the instance has a value of 0.1 and the prototype has only an instance accumulated and the numerical value for that attribute is of 0.9, then the new a_{NP} value is:

$$a_{NP} = 1 / (1 + 1) * 0.9 + (1 - (1 / (1 + 1)) * 0.1 = 0.5$$

but if the prototype has already accumulated 99 instances, then we have:

$$a_{NP} = 99 / (99 + 1) * 0.9 + (1 - (99 / (99 + 1)) * 0.1 = 0.892$$

Categorical attributes:

In this case, we update the occurrence frequency of the categorical attribute value that is going to be added to the prototype, using:

$$fa_{VC} = fa_{VC} + 1 \qquad (6)$$

where fa_{VC} is the number of instances accumulated in the prototype for a certain value V of a categorical attribute C. Thus, to calculate the distance that exists between a categorical attribute of a prototype and a new instance, we employ a weighted variant idea of the well known schema to assign the value of 1 if the categorical attributes are different, and a 0 if they are equal, described previously in the equation (2). Our distance variant consists of a weighted schema that considers the occurrence frequency of the categorical value:

$$d_{BCP} = (\Sigma \ fa_{VC}) / (fa_C + 1) \qquad\qquad B \neq V \qquad\qquad (7)$$

where fa_C is equal to the total of accumulated instances for a certain categorical attribute, and d_{BCP} is the distance of the prototype categorical attribute respect to the categorical attribute with value B of the new instance. With this, we seek to weigh up the distance of an instance to the prototype: while more instances of certain value of a categorical attribute B are accumulated in the prototype, smaller will be the distance to an instance with that same categorical attribute value.

For example, if the prototype only had accumulated categorical values type B, the distance would be 0, because the sum of instances with different values from B (Σ fa_{VC}) is 0. On the other hand, if there were no one value B, and the prototype already has accumulated 99 instances, the distance would be 99 / 100 = 0.99.

The metrics described in Section 3.2 and in the present Section will be essential to obtain the explanation facility, which is presented in the next Section.

2.4 Explanation Facility

Once all the database instances have been processed and all the prototypes have been created, the prototype with less accumulated instances is selected, because we consider that it represents the outlier data. This relatively small prototype is compared against the other constructed prototypes, and applying the distance equations (1), (7) and (3) or (3`), we can obtain automated explanations, which form the explanation facility, following the ideas from Expert System shells.

An example of the explanation facility would be:

Cluster X is different from cluster Y in a p% due to:

   1.   Attribute g (because they differ in a q_g%)
   2.   Attribute h (because they differ in a q_h%

    ...
   N.   Attribute n (because they differ in a q_n%)

where p% is the value of d_T calculated with equations (3) or (3`) and expressed like a percentage; q_g%, q_h% and q_n% are calculated with equations (1) or (7), depending if the attribute is numerical or categorical, and they are expressed in percentage. Each attribute is listed in importance order, that is to say, beginning with the highest percentage and ending with the lowest one. Additionally, we can show only those attributes that have percentages over certain threshold, for example, only list the attributes greater to 70%.

Following the same idea, we can depict the pair of clusters that were the most distant and the attribute that obtained the greatest distance in general:

Clusters X and Y were the most different, with a p%.
The most distant attribute was g, with an q_g%, .

With this explanation, the end user can improve his problem understanding. For example, is possible to see if an attribute related with the payments is abnormal, or if an attribute related with the hour of the day in which the transaction was done is unusual. Furthermore, we can observe if an related attribute with the number of sales is abnormally rising.

2.5 Proposed Method ART-E

According to the previously exposed, the proposed hybrid method, that we denominate like ART-E (ART with Explanation), can be summarized in the way showed in Fig. 1.

```
Given a dataset with R instances and A attributes, P = 0, and an
user-defined threshold U:

  Normalize among 0 and 1 all the numerical attributes.
  Until finish with all the instances, do:
  a) Take randomly an instance.
  b) Compare it against the P existing prototypes applying equation
     (3).
  c) If the smaller obtained d_T is greater than parameter U,
        a. Create a new prototype P = P + 1.
        b. Otherwise, combine the instance with the most similar
           prototype, using equation (5) for numerical attributes
           and equation (6) for attributes with nominal values.
  d) Increment the number n of instances accumulated in the
  prototype P doing n_P = n_P + 1.

Show results applying the explanation facility described in
Section 2.4.

End of ART-E.
```

Fig. 1. ART-E Algorithm

3 Experiments

We conducted several experiments with academic and real datasets to empirically evaluate if the proposed hybrid intelligence artificial method ART-E performs better in outlier detection than other well-known cluster algorithms, in terms of processing time, explanation facility and optimal instance clusterization. We choose academic datasets in our experiments because they are well known in the Data Mining area.

3.1 Experimentation Details

The experimentation objective is to observe the ART-E behavior related to clusterization quality, response time and explanation facility.

First, we tested our proposed method with two well-known academic datasets obtained from the UCI Machine Learning repository [15]: the Iris data, with 150 instances, 3 classes and 4 features or attributes, and the Wine dataset with 178 samples or instances, 3 classes and 13 attributes. For these academic datasets, the class labels were deleted, thus treating it as an unsupervised learning problem.

Additionally, we experimented with a very large real Mexican financial database, with more than 30 million transactions with numerical and categorical attributes. This database contains confidential information of financial transactions related with the major Mexican electric company, named *Comisión Federal de Electricidad*. With this real database, we observe the method scalability and good performance, not only to detect the outlier data, but also to provide an articulated explanation related with reasoning process applied by ART-E in considering those transactions as exceptional.

In order to compare the results obtained with ART-E, we use Weka [16] implementation of cluster algorithms (version 3.6). These experiments were executed using Weka default values. All the experiments were executed in a personal computer with an Intel Core 2 Duo processor, 2 GHz, and 1 Gbyte RAM. ART-E was coded with the Java JDK 6 Update 7 and NetBeans IDE 6.1. In the following Section, the obtained results are shown.

3.2 Experimental Results

Testing over the UCI academic datasets, we can observe (Table 1) that ART-E outperforms several Weka cluster algorithms, and obtains results very near to the optimal. Additionally, ART-E requires approximately 95% less processing time than the EM algorithm (see Table 2).

Table 1. ART-E results for Iris and Wine UCI academic datasets

Dataset	Method	Threshold U	Clusters	Clusters instances		
Iris	ART-E	0.3	3	50	48	52
	Optimal solution		3	50	50	50
	EM		5	28 35 23 42 22		
	SimpleKMeans		2	100 50		
	Coweb		2	100 50		
	FarthestFirst		2	84 66		
Wine	ART-E	0.25	3	58	73	47
	Optimal solution		3	59	71	48
	EM		4	45 31 52 50		
	SimpleKMeans		2	108 70		
	Coweb		220	Clusters with 1 to 10		
	FarthestFirst		2	108 70		

Table 2. ART-E vs EM comparison results for the Financial Database.

Sub Financial Database	Threshold U	Instances	Processing time (secs)		Clusters		Outlier instances	
			ART-E	EM	ART-E	EM	ART-E	EM
Ware House	0.3	24	0.140	122	6	5	1	1
Fixed assets	0.2	39	0.172	72	6	3	2	2
Net Worth	0.7	55	0.187	76	3	3	5	4
Over Heads	0.8	205	0.765	89	9	7	12	9
Liabilities	0.98	1,746	6.375	729	21	13	28	19
Banks	0.99	104,345	358	N/A	637	N/A	563	N/A

To verify if ART-E is able to effectively manage real data, doing the cluster task in an effective and efficient manner, and offering articulated explanations understandable to the end-user, we executed several experiments with six subsets of the Mexican financial database. In Table 2 we can observe that ART-E obtains similar results compared with EM, but much faster. For the Banks subset, Weka EM was unable to manage this data volume, due to memory problems.

The explanations offered by ART-E were shown to the human expert domain and they considered that, with minor changes, these explanations could be understandable and useful for the end-users. Due to confidentiality reasons, in Fig. 2 we only show results from the ART-E explanation facility with fictitious data.

Outlier cluster is different from cluster 3 in a 68.4 % due to:
1. Attribute society (because they differ in a 100%)
2. Attribute hour (because they differ in a 76.6%)
3. Attribute invoice-amount (because they differ in a 72.4%)
4. Attribute vendors (because they differ in a 70.8%)

Clusters 4 and 8 were the most different, with a 73.7%.
The most distant attribute was society, with a 86.9%,

Fig. 2. ART-E Explanation facility example

4   Conclusions and Future Work

We have presented a new hybrid artificial intelligence method for outlier detection that overcomes some drawbacks found in the area, like excessive processing time and the lack of explanation. The proposed method follows an unsupervised learning schema (similar to the ANN ART), with several novel adaptations and metrics inspired from CBR and Data Mining areas. With the proposed method, we found important reductions in processing time, reasonable cluster likelihood, and we obtained reasoned explanations, following Expert Systems ideas, that help the end user gain a better problem understanding.

With the experiments that we performed, we observed that the proposed method ART-E obtains results comparable to, or better than, well-established cluster methods, like EM, Coweb, k-means and FarthestFirst.

Some future research issues arise with respect to ART-E improvement. For example: experimenting with other real databases; comparing our approach against other similar methods (e.g. Trust-Tech [11]); using other metric variations and more efficient search methods and to investigate the possibility of obtaining an optimal U threshold in an automatic way.

References

1. Tang, J., Chen, Z.: Capabilities of outlier detection schemes in large datasets, framework and methodologies. Knowledge and Information Systems 11(1), 45–84 (2006)
2. Caudil, S., Ayuso, M., Guillen, M.: Fraud detection using a multinomial logit model with missing information. The Journal of Risk and Insurance 72(4), 539–550 (2005)
3. Jurowski, C., Reich, A.Z.: An explanation and illustration of cluster analysis for identifying hospitality market segments. Journal of Hospitality & Tourism Research, 67–91 (2000)
4. Kirkos, E., Spathis, C., Manolopoulos, Y.: Data mining techniques for the detection of fraudulent financial statements. Expert Systems with Applications 32(4), 995–1003 (2007)
5. Ferreira, P., Alves, R.: Establishing Fraud Detection Patterns Based on Signatures. In: Perner, P. (ed.) ICDM 2006. LNCS (LNAI), vol. 4065, pp. 526–538. Springer, Heidelberg (2006)
6. Chen, T., Lin, C.: A new binary support vector system for increasing detection rate of credit card fraud. International Journal of Pattern Recognition and Artificial Intelligence 20(2), 227–239 (2006)
7. Pandit, S., Chau, D., Wang, S., Faloutsos, C.: NetProbe: a fast and Scalable System for Fraud Detection in Online Auction Networks. In: Proceedings of the 16th International World Wide Web Conference Committee, Banff, Alberta, Canada, May 2007, pp. 201–210 (2007)
8. Srivastava, A., Kundu, A., Sural, S., Majumdar: Credit Card Fraud Detection Using Hidden Markov Model. IEEE Transactions on dependable and secure computing 5(1), 37–48 (2008)
9. Fast, A., Friedland, L., Maier, M., Taylor, B., Jensen, D., Goldberg, H.G., Komoroske, J.: Relational data pre-processing techniques for improved securities fraud detection. In: 13th International Conference on Knowledge Discovery and Data Mining, San Jose, California, pp. 941–949 (2007)
10. Padmaja, T., Dhulipalla, N., Bapi, R.S., Krishna, P.R.: Unbalanced data classification using extreme outlier elimination and sampling techniques for fraud detection. In: 15th International Conference on Advanced Computing and Communications, pp. 511–516 (2007)

11. Reddy, C.K., Chiang, H., Rajaratnam, B.: Trust-tech-based Expectation maximization for learning finite mixture models. IEEE Transactions on Pattern Analysis and Machine Intelligence 30(7), 1146–1157 (2008)

12. Carpenter, G., Grossberg, S.: Neural dynamics of category learning and recognition: Attention, memory consolidation and amnesia. In: Davis, J. (ed.) Brain structure, learning and memory. AAAS Symposium Series (1986)

13. Waterman, D.: A guide to Expert Systems. Addison-Wesley, Reading (1986)

14. Mitra, S., et al.: Data mining in soft computing framework: a survey. IEEE Trans. on neural networks 13(1), 3–14 (2002)

15. Blake, C., Merz, C.: UCI repository of Machine Learning databases, Univ. of California, Irvine (1998), http://www.ics.uci.edu/mlearn/MLRepository.html

16. http://www.cs.waikato.ac.nz/ml/weka (2004)

Wind Speed Forecasting Using a Hybrid Neural-Evolutive Approach

Juan J. Flores[1], Roberto Loaeza[1], Héctor Rodríguez[1], and Erasmo Cadenas[2]

[1] Division de Estudios de Posgrado, Facultad de Ingenieria Electrica,
Universidad Michoacana, Mexico
[2] Facultad de Ingenieria Mecanica, Universidad Michoacana, Mexico
juanf@umich.mx, roberto.loaeza@gmail.com,
hector120713@gmail.com, ecadenas@umich.mx

Abstract. The design of models for time series prediction has found a solid foundation on statistics. Recently, artificial neural networks have been a good choice as approximators to model and forecast time series. Designing a neural network that provides a good approximation is an optimization problem. Given the many parameters to choose from in the design of a neural network, the search space in this design task is enormous. When designing a neural network by hand, scientists can only try a few of them, selecting the best one of the set they tested. In this paper we present a hybrid approach that uses evolutionary computation to produce a complete design of a neural network for modeling and forecasting time series. The resulting models have proven to be better than the ARIMA and the hand-made artificial neural network models.

1 Introduction

The design of models for time series prediction has traditionally been done using statistical methods. In modeling time series, we find the ARIMA (Auto-Regressive Integrated Moving Average), ARMA, and AR, among others [15]. These models are defined in terms of past observations and prediction errors. Statistical techniques like auto-correlation and partial auto-correlation help scientists identify which of the past observations and/or errors are significant in the construction of the forecasting models.

In the last decade, artificial neural networks have been used successfully to model and forecast time series. Designing an artificial neural network (ANN) that provides a good approximation is an optimization problem. Given the many parameters to choose from in the design of an ANN, the search space in this design task is enormous. On the other hand, the learning algorithms used to train ANNs are only capable of determining the weights of the synaptic connections, and do not include architectural design issues. So, a scientist in need of an ANN model has to design the network on a trial and error basis. When designing an ANN by hand, scientists can try only a few of them, selecting the best one from the set they tested.

We can approach the optimization task involved in ANN design using evolutionary computation. In this paper we present a hybrid approach that uses evolutionary

A. Hernández Aguirre et al. (Eds.): MICAI 2009, LNAI 5845, pp. 600–609, 2009.

computation to produce a complete design of an ANN for modeling and forecasting time series. The architecture we use in the forecasting models is a multi-layer perceptron (MLP). We chose to try 3-layer models, which include an input layer, a hidden layer, and an output layer.

After an ANN is designed, it needs to be trained. Training is the process of determining the weights of the synaptic connections for a given architecture (which does not change in the learning process). The most known learning algorithm is back-propagation. Back-propagation takes every example in the training set, runs the network, and computes the difference between its output and the expected output. The difference is then used to adjust the weights of the network. The process is repeated until convergence, or a maximum number of iterations (epochs) is reached. Unfortunately, back-propagation is a gradient-based optimization algorithm, and as such, opens the possibility of the optimization process to end up in a local maximum.

We propose to design a forecasting ANN using evolutionary computation in three stages. In the first one, the ANN architecture is designed; the second stage optimizes the weight assignments for the synaptic connections; after a suitable candidate has been determined through the first two stages, the third stage fine tunes the weights of the ANN.

We compared our results with a statistically designed model and an expert's hand-crafted ANN. To compare the forecasting accuracy of the different models, we use the following statistical measures: MSE, MAE, and Theil's U. The best model produced through evolutionary computation has proven to be better than the ARIMA and the hand-made artificial neural network models.

The rest of the paper is organized as follows. Section 2 surveys the state of the art in forecasting with statistical methods and artificial neural networks. Section 3 describes the setting, location, and devices involved in the data acquisition process, used to obtain the wind speed time series used in these experiments. Section 4 proposes the evolutionary computation architecture used in the ANN design. Section 5 discusses the results obtained and compares them with traditional approaches. Finally, Section 6 concludes the work.

2   Related Work

Wind forecasting techniques assume that the time series, taken from measurements, is the sum of different components and a random error. The goal of most forecasting techniques is to separate and identify those components (trend, cyclical, seasonal, and irregular). Recently, several techniques have been used from the fields of statistics and artificial intelligence [1], [2], [3], [4]. Scientists have even combined them in order to reduce the forecasting error and to produce more accurate predictions [5], [6].

There have been various studies of the wind speed behavior at La Venta, Oaxaca [7], [8], [9]. Regarding time series forecasting, Cadenas and Rivera [10], [11], and Flores et. al [12], have made models for this purpose. Cadenas and Rivera [10] discuss ARIMA techniques and Artificial Neural Networks (ANN) and make a comparison between the two through the calculation of statistical errors (MAE, MSE, and Theil's U). The final result shows that ARIMA is the best for this case, however, the authors mention that the factor that limits ANN performance is the size of the training set presented to the network. Cadenas and Rivera [11] present a comparison of Artificial Neural Networks with different configurations, under minimum operating requirements, and suggest the network model with the minimum statistical errors. Flores et al. [12] model wind speed with genetic programming, producing a forecasting model which reduces the statistical errors generated by the technique ARIMA.

According to the development of studies, it would be useful to conduct an exploration of the entire universe of configurations that form the neural networks and realize if there is an optimal configuration that can reduce the errors found with statistical techniques.

The area of combining Evolutionary Computation and Artificial Neural Networks to produce Neural Systems capable of classifying, predicting, or controlling complex systems has been explored. The proposal presented in this paper makes contributions to the area not present in previous work, therefore advancing human knowledge on the deployment of ANNs. This section contrasts our proposal with related research work, highlighting the differences and the advantage of the proposed methodology, presented here.

Yiau and Liu [17] present a scheme based on evolution programming, emphasizing on evoling ANN's behaviors. A mixture of other ideas is incorporated in their proposal: mutations are provided by partial training (a la memetic algorithms) and node splitting. Their work, called EPNet evolves architectures and connection weights; while their approach presents a combination of techniques, ours uses pure evolutionary computation.

The work of Abraham [21], [18] presents several differences with respect to our proposal. He uses evolutionary algorithms to determine the network architecture, connection weight, and learning algorithms. Our approach also designs the inputs to the ANN, but does not determine the learning algorithm. That decision relies on the fact that we are training the ANN through evolutionary computation as well. Another difference is that he uses a binary encoding for the weights, while we use real encoding.

Mayer and Schwaiger [19] present a system that evolves ANNs in an evolutionary scheme. Low complexity ANNs guides the evolution of ANNs of greater generalization ability. Evolution is achieved by GAs, using error back-propagation to train the networks. The evolutionary processes they propose consider ANNs of fixed architecture. They also use co-evolution to determine the training data set. They Mackey-Glass benchmark was used to test their results; given that, the benchmark is well known, the inputs to the ANN are fixed.

In summary, our proposal differs from previous works in different aspects. Some of them do not evolve the ANN architecture at all, others evolve it partially. The proposed scheme and representation allows to design the totality of the ANN architecture. In addition, most of the schemes adopt a hybrid approach, router leaving training (using different learning algorithms) with evolution. Our approach is based on pure evolutionary computation; at the end, though, for the winner ANN, we push it a bit forward using back-propagation [22]. Since we have explored the research space, the winner ANN architecture is expected to reach the local optimum where the GA left it, which most likely will be the global optimum.

3 Data Measurement

The Comision Federal de Electricidad (CFE, the governmental electricity supplier in Mexico) has made wind speed measurements since 1994, through a network of measurement stations located in the places of interest. The sensors were located at different heights in the measurement towers (20, 30, 40 m from ground level). Their characteristics are shown in Table 1.

The information generated by the sensors is accumulated in the data acquisition systems through chips or memory cards that are later downloaded to a computer in order to be processed.

Fig. 1 shows the monthly behavior of the wind speed in La Venta, Oaxaca, for the period from June 1994 to May 2000. A seasonal behavior in the series is observed, the strongest winds appear at the end of every year and are weakest in the middle.

Table 1. Specification of the measurement sensors

Specification	Anemometer	Wind Vane
Measuring rank	0.78-45 m/s	0-360°
Exactness	±5%	±5%
Resolution	0.78m/s	1 m/s

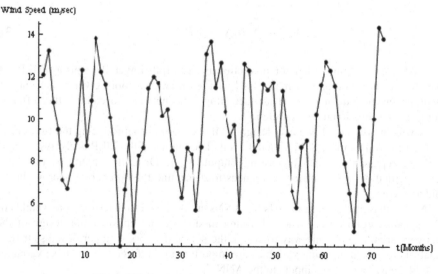

Fig. 1. Wind speed times series in La Venta, Oaxaca

4 Evolving ANNs

Given a time series, we need to provide a neural model capable of producing an acceptable forecasting of the process it represents.

In order to define the term acceptable model, i.e. the fitness measure of a given model, there exist statistical measures that allow us to compare two time series, for instance, the Mean Squared Error (MSE), the Mean of the Absolute Value of the Errors (MAE), etc. [15]. Among these measures, we find Theil's U, which, for an acceptable model, must return a value in the interval [0.5, 1].

An ARIMA model [15] is a statistical model that allows us to model time series, and to predict their behavior. These models have the following form:

$$y_{t+1} = \sum_{k=0}^{w} a_k y_{t-k} + b_k e_{t-k} + \varepsilon_t.$$ (1)

$$e_t = y_t - \hat{y}_t.$$ (2)

Where y_t represents the measurement at time t in the time series and $\hat{y}_t$ is the forecast produced by the ARIMA model; ε_t represents the effects of random factors; w is the window width. The window represents how far behind in time we consider measurements as probably important inputs for the ARIMA model. Outside of the window, observations are not taken into account. Using statistical procedures, the numerical value of the coefficients a_k and b_k are determined.

In the approach presented in this paper, we are using an Auto-Regressive model (AR), which is a reduced version of ARIMA. The AR model does not consider past forecasting errors as forecasting variables. AR has the following form:

$$y_{t+1} = \sum_{k=0}^{w} a_k y_{t-k} + \varepsilon_t.$$ (3)

The ANN architecture used for prediction is the Multi-Layer Perceptron (MLP). A MLP, as a universal approximator [13], can learn any function, given it has enough neurons in the hidden layer. That fact allows the network to capture the different forms of the function to be modeled.

Given an AR model, we can design a MLP capable of reproducing the time series at least as well as the ARIMA model itself. The output of the MLP is always a single neuron, representing the forecasting output, y_{t+1}. Once the inputs to the MLP are specified, the design process reduces to determine the number of neurons in the hidden layer.

Notice that the learning models for ANNs are designed to determine the weights of the synaptic connections. Those learning models do not consider the design of the network architecture. One way to design the neural network is to perform a statistical analysis to determine what variables are important in the forecasting. Those variables will be considered as the inputs to the ANN.

In this work we intend to design the MLP completely, without the need of any statistical analysis. That is, we design the number of input neurons and what they represent, and the number of hidden neurons (the output neuron will always be the same). The design process includes the determination of the weights of the synaptic connections, without the need of a learning algorithm (v.g. back-propagation). The reason to avoid those learning methods is that since they are gradient-based, they are likely to stop at a local optimum. This fact may make a MLP behave badly, even with an adequate architecture.

The proposal is to use Evolutionary Computation to perform the complete design of the ANN used in forecasting. The scheme involves two nested evolutionary processes followed by a third one. The first one designs the network architecture, while the second (inner) one, once determined the architecture, determines the weights of the synaptic connections. A last evolutionary process refines the weights for the winner network of the previous two processes. The proposed architecture of the hybrid ANN-Evolutionary scheme is shown in Fig. 2.

This evolutionary scheme uses two types of chromosomes. The first one, for the outer evolutionary process, contains a bit vector (Vars), whose size is the window size, followed by an integer (NH). A value of 1 in position k of the bit vector indicates that variable y_{t-k} appears as an input variable in the MLP being designed; a 0 indicates that variable is not taken into account in the model. NH indicates the number of neurons included in the MLP's hidden layer. Fig. 3 shows the structure of this chromosome.

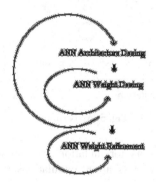

Fig. 2. Hybrid ANN-Evolutionary Forecasting Scheme

For each individual in the outer evolutionary process, we proceed to the inner evolutionary process. The chromosome of this second process contains a vector of real numbers with NC elements. Let us say NV is the number of 1s appearing in Vars. NC is the number of synaptic connections in the neural model, where NC = (NV + 1) NH. Fig. 4 shows the structure of the second chromosome.

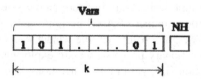

Fig. 3. Structure of the chromosome of the outer evolutionary process

Fig. 4. Structure of the chromosome of the inner evolutionary process

Fig. 5 shows an example of an individual belonging to the inner evolutionary process. The chromosome shows the proposed inputs (y_{t-1}, y_{t-3} and y_{t-4}); the network contains 5 neurons in the hidden layer; the remaining real values are the weights of the synaptic connections.

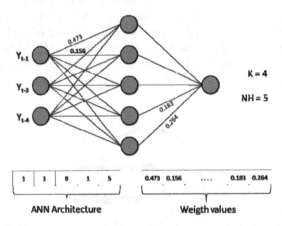

Fig. 5. ANN Model

We provide genetic operators for mutation and crossover for both evolutionary processes. Those genetic operators allow populations to evolve and produce optimized solutions.

Once the first two evolutionary processes are performed, we have the best of the inspected models. At that time a third evolutionary process is performed. This third process is similar to the second one, but we allow a larger population, in order to allow the synaptic weights to be refined.

The search space we are exploring and optimizing in the solution of this problem is huge. That made us play with the different parameters in the evolutionary processes and refine them, to be able to explore the search space more efficiently. For instance, the number of ANNs to be explored is very large, and for each designed architecture, the possibilities for the synaptic weights are just too many. Given that, we decided to let the process explore a good number of ANN designs, and for each design try not too many combinations of weights. After that, the winner ANN is further refined. At that moment (the third evolutionary process), we are exploring a single architecture and give it a larger population size, with more generations, and also, a larger chance of mutations.

5 Results

The experiments performed were divided in ANN Architecture Design, ANN Weight Design, and ANN Weight Refinement processes. The ANN Architecture Design process evaluates about 1,400 architectures. Each architecture was evaluated with about 2040 different combination of weights (ANN Weight Design process). About 2,000,000 evaluations in total.

The ANN Weight Refinement process uses the best Architecture obtained and continues evolving the best combinations of weights. About 200,000 different combination of weights were evaluated.

The final ANN Weight Refinement process uses the best Weight's obtained and continues refinement it. About 1,000,000 iterations were performed.

All the experiments were performed using Genetic Algorithms (GA), Evolutionary Strategies (ES) with Evolvica [14] and Back-propagation[22].

The winner ANN was produced using ES, it took about 60 hours and its characteristics are:

— Window width: 18
— Number of inputs: 13
— Number of neurons in the hidden layer: 30
— Number of outputs: 1
— The output was defined as a function of (yt-1, yt-3, yt-5, yt-6, yt-7, yt-8, yt-9, yt-11, yt-12, yt-14, yt-16, yt-17, yt-18)

Fig. 6. Predicted and observed validation set

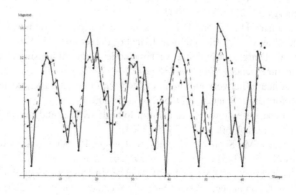

Fig. 7. Observed and predicted data

Table 2. Fitness measure for the different forecasting models

Method/Accuaracy	MAE	MSE	Theil's U
Naïve	2.37	8.28	1
ADALINE	1.8	5.10	0.69
Hybrid	1.72	4.36	0.45

Table 2 shows the results obtained for the statistical measures with the different models obtained by Cadenas and Rivera[2] and our approach. From Table 2 it is clear that the Hybrid model has lower statistical errors than those produced with ADALINE and with the naïve method, Fig. 6 shows the comparison between the observed data and the predicted ones. Continuous lines in Fig. 6 and Fig. 7 are the observed data and the discontinuous lines are the predicted data.

6 Conclusions

We presented a hybrid neural-evolutionary methodology to forecast time-series. The methodology is hybrid because an evolutionary computation-based optimization process is used to produce a complete design of a neural network. The produced neural network, as a model, is then used to forecast the time-series.

Experiments were performed with a real time-series formed by wind speed measurements taken from La Venta, Oaxaca. The forecasts produced with the proposed methodology exhibit a better behavior than the previous ones, produced through statistical methods and hand-crafted ANNs.

The system is fully implemented in Mathematica [16], using Evolvica [14], a Mathematica package developed to perform evolutionary computation.

References

1. Riahy, G.H., Abedi, M.: Short term wind speed forcasting for wind turbine applications using linear prediction method. Renewable energy 33, 35–41 (2008)
2. Cadenas, E., Rivera, W.: Wind speed forecasting in the south coast of Oaxaca, Mexico. Renewable energy 32, 2116–2128 (2007)
3. Ghiassi, M., Saidane, H., Zimbra, D.K.: A dynamic artificial neural network model for forecasting time series events. International Jounal of Forecasting 21, 341–362 (2005)
4. Chen, Y., Bo, Y., Dong, J., Abraham, A.: Time-series forecasting using flexible neural tree model. Informatin Sciences 174, 219–235 (2005)
5. Zhang, G.P.: A neural network ensemble method with jittered training data for time series forecasting. Information Sciences 177, 5329–5346 (2007)
6. Zhang, G.P.: Time series forecasting using a hybrid ARIMA and neural network model. In: Neurocomputing, vol. 50, pp. 159–175 (2003)
7. Elliot, D., Schwartz, M., Scott, G., Haymes, S., George, R.: Wind Energy Rsource Atlas of Oaxaca, NREL/TP-500-34519, http://www.osti.gov/bridge
8. Steenburgh, W.J., Schultz, D.M., Colle, B.A.: The structure and evolution of gap outflow over the Gulf of Tehuantepec, Mexico. Monthly Weather review 126, 2673–2691 (1998)

9. Jaramillo, O.A., Borja, M.A.: Wind Speed Analysis in La Ventosa México: a bimodal probability ditribution case. Renewable Energy 29, 1613–1630 (2004)
10. Cadenas, E., Rivera, W.: Wind speed forecasting in the South Coast of Oaxaca, México. Renewable Energy 32, 2116–2128 (2007)
11. Cadenas, E., Rivera, W.: Short Term Wind Speed Forecasting in La Venta Oaxaca, México. Using Artificial Neural Networks, Renewable Energy
12. Flores, J., Graff, M., Cadenas, E.: Wiind Prediction Using Genetic Algorithms and Gene Expression Programming, Techniques and Methodologies for Modelling and Simulation of Systems. In: International Association for Advanced of Modelling and Simulation, Lyon France – México (AMSE), pp. 34–40 ISBN: 970-703-323-1
13. Haykin, S.: Neural Networks a comprehensive foundation. Prentice Hall press, Englewood Cliffs (1999)
14. Jacob, C.: Illustrating Evolutionary Computation with Mathematica. Morgan Kaufman press, San Francisco (2001)
15. Wheelwright, S., Makridakis, S.: Forecasting Methods for management (1985)
16. Wolfram, S.: The Mathematica Book. Cambridge press (1999) ISBN: 0-521-64314-7
17. Yao, X., Liu, Y.: A new evolutionary system for evolving artificial neural networks. IEEE Transactions on Neural Networks 8(3), 694–713 (1997)
18. Abraham, A.: Optimization of evolutionary neural networks using hybrid learning algorithms. In: Proceedings of the 2002 International Joint Conference on Neural Networks IJCNN apos;2002, vol. 3, pp. 2797–2802 (2002)
19. Mayer, H.A., Schwaiger, R.: Evolutionary and coevolutionary approaches to time series prediction using generalized multi-layer perceptrons. In: Proceedings of the 1999 Congress on Evolutionary Computation, CEC 1999, vol. 1 (1999)
20. Belew Richard, K., John, M., Schraudolph Nicol, N.: Evolving Networks: Using the Genetic Algorithm with Connectionist Learning, Cognitive Computer Science Research Group; Computer Science & Engr. Dept. (C-014); Univ. California at San Diego, CSE Technical Report # CS90-174 (June 1990)
21. Abraham, A.: EvoNF: a framework for optimization of fuzzy inference systems using neural network learning and evolutionary computation. In: Proceedings of the 2002 IEEE International Symposium on Intelligent Control, vol. 2002, pp. 327–332 (2002)
22. Freeman James, A.: Simulating Neural Networks with mathematica. Addison-Wesley Publishing Company, Reading (1994)

Hybridization of Evolutionary Mechanisms for Feature Subset Selection in Unsupervised Learning

Dolores Torres[1], Eunice Ponce-de-León[1], Aurora Torres[1],
Alberto Ochoa[2], and Elva Díaz[1]

[1] Universidad Autónoma de Aguascalientes, Centro de Ciencias Básicas. Avenida Universidad
940 Col Ciudad Universitaria, Aguascalientes, AGS. C.P. 20100. México
{mdtorres,eponce,atorres}@correo.uaa.mx,
elva.diaz@itesm.mx
[2] Universidad Autónoma de Ciudad Juárez, Instituto de Ingeniería y Tecnología, Departamento
de Ingeniería Eléctrica y Computación. Henry Dunant # 4016 Circuito Pronaf Apdo. Postal
1594-D, Juárez, Chihuahua. CP 32310. México
megamax8@hotmail.com

Abstract. Feature subset selection for unsupervised learning, is a very important topic in artificial intelligence because it is the base for saving computational resources. In this implementation we use a typical testor's methodology
in order to incorporate an importance index for each variable. This paper presents the general framework and the way two hybridized meta-heuristics work
in this NP-complete problem. The evolutionary mechanisms are based on the
Univariate Marginal Distribution Algorithm (UMDA) and the Genetic Algorithm (GA). GA and UMDA – Estimation of Distribution Algorithm (EDA) use
a very useful rapid operator implemented for finding typical testors on a very
large dataset and also, both algorithms, have a local search mechanism for improving time and fitness. Experiments show that EDA is faster than GA because
it has a better exploitation performance; nevertheless, GA' solutions are more
consistent.

Keywords: Hybridized Evolutionary Mechanisms, Feature Subset Selection,
Univariate Marginal Distribution Algorithm, Genetic Algorithm, Unsupervised
Learning.

1 Introduction

Even though supervised learning has obtained very good results, this, is not the case
of feature selection in unsupervised learning [22], The main goal of feature subset
selection for unsupervised learning, consists on finding out the subset of features that
better describes an original dataset; taking out noise or distortion for identifying the
class data belong. When a good feature subset selection is done, a very important task
is made because removing irrelevant data, increases accuracy of learning and its comprehension according with Blum and Langley [3]. Reducing data dimensionality has
become very important in machine learning during the past decades. We can see large
datasets with instances described by many features in problems as image processing,

A. Hernández Aguirre et al. (Eds.): MICAI 2009, LNAI 5845, pp. 610–621, 2009.

text mining and bioinformatics. To efficiently deal with those data, dimension reduction techniques emerged as a useful preprocessing step in the task of data analysis. A subset of these techniques makes reference to feature subset selection techniques. The difference between these techniques and other reduction techniques (like projection and compression techniques) is that the first ones do not transform the original input features, but select a subset of them [17]. The work presented here, is located in Automated Learning; that according with Tom Mitchell, refers to the study of computational algorithms that improve themselves with experience [19]. Moreno et al., say that it is the kind of learning a machine can achieve [20]. Nowadays, the efficient use of the resources is not a privilege, but also, it is a necessity. An universal problem, all intelligent agent has, consists on defining where to focus its attention [12], that is why features subset selection has so much importance in data mining and automated learning area.

In recent years, real world and large scale problems have been solved by using combinations of metaheuristics with other techniques and mechanisms; that is because the use of a pure metaheuristic has some restrictions. The idea of hybridization refers to take advantage of the potentialities of some of them and save their weaknesses using mechanisms or techniques from others sources that make them more robust; this way, they can provide a more efficient behavior and a higher flexibility. Some recommendations about how hybridization could be done through of the knowledge about the problem that is treated, can be seen on [8], specifically for GA's. This paper presents two evolutionary metaheuristics hybridized that have their base in the well known genetic algorithm and in the estimation of the distribution algorithm (univariate marginal distribution algorithm).

Genetic algorithm (GA) was presented in the sixties decade, and it has proved be an excellent evolutionary tool for solving different kind of problems; this tool is based on natural evolution [10]. The GA presented, is based on the simple genetic algorithm created by Goldberg [9]. Obviously, many novelty modifications have been incorporated to it. This algorithm has been hybridized for improving exploitation and exploration mechanisms and also testor's theory has been added.

Finding all typical testors is a very expensive procedure; all described algorithms have exponential complexity, and also, they depend on the size of the matrix [24]. That is why, nowadays, this is a current problem.

The problem of features subset selection, has been treated with data mining [22], metaheuristics [11, 12, 23, 30, 31, 33] (actually, Inza et al., were the pioneers using EDAs in feature subset selection), multi-objective point of view [18], etc. Reader could review the state of the art for feature selection in [3, 7, 12]. Nevertheless, results at this time are not conclusive. That is why, we propose two evolutionary algorithms hybridized that use our novelty framework that is adaptable to supervised and unsupervised learning and also we reduced considerably the time used for identifying typical testors by means the hybridization.

In this paper, a comparison between two evolutionary algorithms is presented; both of them: Genetic Algorithm and Estimation Distribution Algorithm use an accelerating mechanism that let them have a pseudo-random behavior that address the search toward a promising area, while the improvement mechanism lead toward a local peak (just like a partial hill-climbing). Both algorithms are focused on finding the feature subset selection that better distinguishes the clusters found. The subsequent sections of this work are organized as follow. In section 2, the overall framework is presented; also some important topics related with the problem and the accelerating and improvements mechanisms are presented here. Section 3 is dedicated to describe the two algorithms to be compared: GA and UMDA. Experiments and results are presented on section 4 and finally, conclusions and future work are discussed on section 5.

2 Framework

In 2005, Peng Liu et al. [22] presented an interesting proposal for unsupervised learning using data mining. The main contribution of Peng Liu research consists on first develop clustering phase, and after the searching phase. This strategy was created for making easier and more understanding wrapper-based methodologies. We use that work as a base for our framework named: "Evolutionary Mechanisms Hybridized for Feature Subset Selection in both: supervised and unsupervised learning". Figure 1 shows the framework we created.

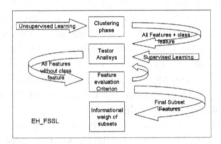

Fig. 1. Framework

An interesting model for intelligent learning environments that also supports supervised an unsupervised learning in other context could be review on [2].

As we can see, our framework is so flexible, that it works in supervised learning as well as in unsupervised learning. We used this framework with a GA and with an UMDA that were hybridized with tools like:

— A logical combinatory approach (typical testors).
— A local search mechanism (improvement mechanism).
— A global search operator (accelerating operator).

In this research, we used k-mean as our clustering algorithm because it is good for finding groups of cases given a specific k (number of classes).

As it can seen in the framework, when unsupervised learning is used:

1. All features are entered to the k-means algorithm.
2. The classification feature is obtained and then, all features plus the classification one are used for starting the testor analysis.
3. If we are working with supervised, then the process starts with the analysis. After that, (in a cyclical process), a subset of features for the clusters given is proposed (as a testor or typical testor).

4. Each subset proposed by the evolutionary algorithm is pondered and a fitness is assigned to each one.
5. Finally, all the typical testors found in the process, are used to calculate the informational weigh of each feature.

The framework presented uses some important concepts that will be presented as follow.

2.1 Typical Testor

The concept of Typical Testor appeared in the middle of the fifties in Russia [4]; it began to be used in failure detection in electrical circuits. Then it was extended to variables selection on geological problems [1]. The pioneer work in the use of typical testors for feature subset selection was from Dmitriev Zhuravlev et al. [6].

This research presents two mechanisms that were applied to both: supervised and unsupervised learning. A testor is a subset of features that distinguishes objects from different classes. Satiesteban and Pons say that a typical testor is the testor which we cannot eliminate any of its features without changing its condition of testor [26]. We could say it in other words: a typical testor is a testor whose redundancy has been eliminated. Let's suppose that U is a collection of objects, and these objects are described by a set of n features; also let's suppose that the objects are grouped into k classes. By comparing each pair of features that belong to different classes, using any criterion, we can obtain the difference matrix DM that is made capturing all the differences between objects from different classes; this difference is coded with 1 if difference exists and 0 if difference does not exist. This DM can be very large when we identify many differences between members of a class with regard to others. Let T to be a subset of the entire set of labels in the columns of DM. We call Basic Matrix BM to a special data set obtained from eliminating all rows belonging to DM that are not basic rows.

Let a and b be two rows from DM, a is sub-row from b if condition presented in ec.1 is satisfied, and also exists at less one that satisfies ec. 2 [15].

$$(\forall_i \mid b_i = 0, a_i = 0).$$ (1)

In other words, a is a basic row from DM, if there is not any other row less than a in DM.

$$(\exists_i \mid b_i = 1 \land a_i = 0).$$ (2)

Given a DM, we create the BM, that is the matrix composed exclusively by the basic rows in DM [15]. BM is composed by 1's and 0's too, because it has the basic differences between classes. We have to pay attention to the three declarations that follow for finding typical testors:

1. T is a testor from a Learning Matrix (LM) if no zero's rows exist in M after eliminating all columns that do not belong to the set T.
2. The set T is typical if by eliminating any feature $j \mid j \in T$, T loses its condition of testor.
3. The set of all typical testors of DM is equal to the set of all typical testors of MB [15].

Some algorithms for determining the complete set of typical testors from a basic matrix have been created. It can be mentioned: BT [28], TB [28], CT-EXT [24], LEX [26] and REC [27] among others as [25]. The framework presented here, use a genetic algorithm on one hand, and on the other, a univariate marginal distribution algorithm to find the entire set of typical testors, to determine how representative is each variable. This way, the informational weigh can be calculated. Interested reader can review the main ideas using testors in [16].

2.2 Feature Subset Selection

Regularly, feature subset selection is used to reduce dimensionality. This task can be done taking out irrelevant or redundant features. This is an important task because reducing the number of features may help to decrease the cost of acquiring data and also make the classification models easier to understand [21]. Also, the number of features could affect the accuracy of classification [32]. Some authors have also studied the bias feature subset selection for classification learning [29, 32]. This justifies the importance of a good feature subset selection for learning.

We can see two groups of authors: who consider that reduced subsets have to conserve features that are relevant and no redundant [5] and who consider the importance of the difference between optimal and relevant features that reduced subset have kept [12]. The two most frequently used feature selection methods, are filter and wrapper [12, 21, 22], Filter methodology selects features based on the general characteristics of the training data, while wrapper methodology uses a learning algorithm to evaluate the accuracy of the potential subsets in predicting a target.

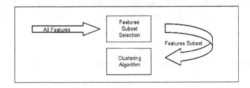

Fig. 2. Filter

Filter methods are independent of the classifier and select features based on properties that a good feature sets are presumed to have, such as class separability or high correlation with the target while wrapper treats the induction algorithm as a black box that is used by the search algorithm to evaluate each candidate [21]. While wrappers methodologies give good results in terms of accuracy of the final classifier are computationally expensive and may be impractical for large datasets [21]. That is why we tried to rescue the better of both methods creating a new framework that is based on the Pen Liu et al. [22]; but ours is more flexible, because it fits to both: supervised and unsupervised learning.

On figure 2, we can see the main components of filter methodology while figure 3 presents wrapper's.

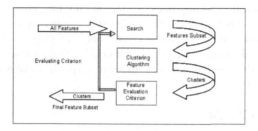

Fig. 3. Wrapper

Even though wrapper methodology has higher computational costs than filter [22], it has received considerable attention [3, 12]. Nevertheless, filter methodologies are faster and also adequate for large dataset [21, 22]. Filter methods can be further categorized into two groups: attribute evaluation algorithms and subset evaluation algorithms, which refer to whether they rate the relevance of individual features or feature subsets. Attribute evaluation algorithms study the features individually, and ponder each one according to each feature's degree of relevance to the target feature. In this research, we focus on feature subset selection (FSS) for both supervised and unsupervised learning.

2.3 Accelerating Operator

The accelerating operator is a very important contribution to our framework because it applies to any metaheuristic we could use. This operator, improves the exploration mechanisms of both: GA and UMDA because it let them to look for in promissory areas. This operator was created for taking the better of exterior and interior scale algorithms for finding typical testors. Exterior scale algorithms, are those that determine typical testors by generating the power set elements of the complete set of columns from BM in a specific order, this way, the algorithm can jump some combinations [27]. While interior scale algorithms, find typical testors studying the internal structure of the matrix, finding conditions which guarantee the condition of testor or typical testor.

Both GA and UMDA have the possibility of exploring different combinations of chains formed by 1's and 0's, for proving if each combination is or not, a typical testor. Nevertheless, we hybridized the two algorithms incorporating an operator that contains the knowledge we obtained from the shallow study of the BM. Such analysis, let us identify atypical cases; which reveal the ownership of certain feature to a typical testor. Once, the entire atypical cases presented in BM have been identified; a special string is built. This string has been called "belonging string", because it is an important little piece of any typical testor. In the core of our metaheuristics, this operator was applied for guiding searching mechanisms. This operator, reduces times in a very considerable way in both metaheuristics.

2.4 Improvement Mechanism

A special mechanism called "improvement mechanism" was used in both metaheuristics. This mechanism acts like a partial hill climbing for improving a particular solution. Obviously, this tool improves the exploitation mechanism. The main idea of using this mechanism, is to increase the probability of obtaining a typical testor from an already obtained testor. This process consists on changing 1's by 0's in the actual chromosome. It was designed based on the knowledge that all testor has inside at least one typical testor. This mechanism, was inspired on the improvement component that scatter search uses [13].

2.5 Informational Weight

The use of the informational weight for feature subset selection in unsupervised learning, is an excellent tool that gives us tangible results [32]. This mechanism consists on determining the most of the typical testor we could; and then, compute a factor of the appearance frequency.

3 Hybridized Evolutionary Algorithms

The framework was used with two algorithms, a GA, and an UMDA; the GA was called "Hybridized Evolutionary Genetic Algorithm for Feature Subset Selection in Learning" (EHGAFSSL) while the UMDA was called "Hybridized Evolutionary Univariate Marginal Distribution Algorithm for Feature Subset Selection in Learning" (EHUMDAFSSL). The two subsections that follow, show each algorithm used.

3.1 HEGAFSSL

The Hybridized Evolutionary Genetic Algorithm for Feature Subset Selection in Learning is shown as follow:

```
Algorithm 1. - HEGAFSSL
begin /* Hybridized Evolutionary GA for FSS in
Learning*/
   Initial pre-processing for taking out some
      features
   If supervised learning then goto ⌂
   else Training clustering mechanism.
   Cluster phase.
⌂ Generating DM phase.
   Generating BM phase.
   Generate initial population (randomly).
   Apply accelerating operator.
   Apply improvement mechanism.
   Compute fitness
   population <- new initial population
   Repeat
      Begin /* New Generation */
         Repeat
            Begin /* reproductive cycle for pairs of
                   individuals */
            Apply selection operator.
            Apply crossover operator.
```

```
            Apply accelerating operator.
            Apply improvement mechanism.
            Compute fitness.
            Count= Count+1;
          End.
          Until Count = (generation size/ 2).
          Order of population by fitness.
          Apply elitism.
          Population <- new population
       End.
    Until (stopping criterion is reached)
    Final set of typical testors analysis.
    Compute informational weight for each feature
    Get final features subset selection.
End /* Hybridized Evolutionary AG for FSS in Learning*
```

Here, we can see that the metaheuristic hybridized with a local mechanism that improves a particular solution looking for a typical testor from a single testor. Besides that, elitism was used for accelerating the typical testors search and also, a global search operator was incorporated. This global operator, was obtained from a simple analysis of the BM. This way, we rescued the best of the interior and exterior scale algorithms, used for finding typical testors. Finally the logical combinatory focus was integrated to our methodology.

3.2 HEUMDAFSSUL

The Hybridized Evolutionary Univariate Marginal Distribution Algorithm for Feature Subset Selection in Learning, was inspired on the EDA called UMDA [14]. And it is shown as follow:

```
Algorithm 2 - HEUMDAFSSL
begin /* Hybridized Evolutionary UMDA for FSS in
Learning*/
   Initial pre-processing for taking out some
      features
   If supervised learning then goto ◊
   else Training clustering mechanism.
   Cluster phase.
◊ Generating DM phase.
   Generating BM phase.
   Generate initial population (randomly).
   Apply accelerating operator.
   Apply improvement mechanism.
   Compute fitness
   population [P(0)] <- new initial population
   Repeat
      Begin /* New Generation */
         Select population of promising solutions
            s(t).
         Build probabilistic model P(t) for s(t).
```

```
          Apply elitism.
          Sample P(t) to complete O(t).
          Apply accelerating operator.
          Apply improvement mechanism.
          Compute fitness.
          Order of P(t) by fitness.
          P(t) <- O(t).
          t=t+1.
     End.
   Until (stopping criterion is reached)
   Final set of typical testors analysis.
   Compute informational weight for each feature
   Get final features subset selection.
End /* Hybridized Evolutionary AG for FSS in Learning*
```

As can be seen, the algorithm includes elitism, an added global search mechanism, and an additional local search mechanism; just like the first algorithm based on GA "HEGAFSSL".

4 Experiments and Results

The main problem, is related with health risk factors in Mexican pregnant women. The cost of taking care of these critical risk factors is insignificant and the benefits are invaluable; so, many researches in health care area, consider it a very important topic. We found that the most important variables, women have to take care of, are 25; but 24 are indispensable while 1 is less vital than the others. The original dataset had 46 features, but we kept 29 after preprocessing phase, after the whole process we kept 25. The experimental results show that even though GA is known as a rapid convergence algorithm, thanks our mechanisms, we had a very stable algorithm that conserved the diversity required for looking in the whole solutions space.

4.1 Accelerating Operator and Improvement Mechanism

We will use AO to refer to the accelerating operator, and IM for the improvement mechanism. Experimental results related with the AO and IM, showed the improvement presented in table 1, for a very large Basic Matrix of 65549 rows and 29 features. An exhaustive algorithm has to do 536 870 911 searches on a basic matrix and this increases exponentially with the size of the matrix. The number of possible combination is 2^n-1, and it depends on the number of cluster we need to look.

Both metaheuristics reported improved in the proportion of table 1. The fitness function (ff) had only three possible values: 5 for no testors, 10 for single testors, and

Table 1. Proportion of testors

Series	Proportion
Without AO-IM	1 of each 100 000
With AO-IM	99 980 of each 100 000

20 for typical testors. We worked with a real-word and benchmark datasets, and here we are reporting here the real-problem results because it was the largest problem we used to probe the algorithms.

4.2 Experimental Results

The results presented in this section, were obtained from 10 executions using both algorithms. Experiments indicate that EDA is faster than GA.

Statistical tests, were done using SPSS software. First of all, a Shapiro Wilk test was selected, because we used the times observed in 10 executions; so, we needed a test for few data. The Shapiro Wilk test threw the following result: p-value is 0.0007434 (less than 0.05), that's why we could not accept that the errors of times had a normal distribution. Since errors have not a normal behavior, a Wilcoxon-Man Whitney test was done, to prove if times of GA and EDA were significantly different. P-value of Wilcoxon-Man Whitney test was 0.002089 (less than 0.05); therefore we can say that times are different. Invariably, EDA was faster; nevertheless, GA was more consistent conserving more diversity of results. EDA found typical testors earlier than GA.

Table 2. Fitness function (ff)

Algorithm	Generation's mean to find any typical Testor	Percentage found of the whole set of typical testors
EDA	0	70
GA	1	100

EDA tended to a single solution while GA showed more diversity even in the last generation. These results are presented in table 2.

GA is more consistent then EDA because the algorithm is able to find more results. EDA find a pattern and it guides strongly its behavior. This research considered a complex problem if the number of cases of the BM is 1 000 000 or more and the number of features is 20 or more; other case is considered as a simple problem.

5 Conclusions and Future Work

It can be concluded that hybridization improves in a very important proportion the results for the problem we attacked. To include the knowledge we have about a particular problem in the algorithm, will help to obtain important improves in our results. Finding the typical testors of a data set is an exponential problem, the accelerator operator had a powerful effect in reducing the time of this process. As long as the string is, is the time for finding the testors. Informational variable weigh is also a very important concept, since it let us to qualify the importance of each variable in a clustering process. In short time, experiments for contrasting the whole set of typical testors in a pareto's front will be done, because we consider that this problem is an inherently multi-objective one. The improvement mechanism will be perfected starting in a random point of our string doing a complete hill-climbing; We hope, applying those strategies, our mechanism will be able to tend to all results and not to find one of them (the EDA). This behavior could be achieved because the accelerating operator guarantees that the common part of any typical testors is considered.

References

1. Alba, C.E., Santana, R., Ochoa, R.A., Lazo, C.M.: Finding Typical Testors By Using an Evolutionary Strategy. In: Proceedings of V Iberoamerican Workshop on Pattern Recognition Lisbon, Portugal, pp. 267–278 (2000)
2. Amershi, S., Conati, C.: Unsupervised and supervised machine learning in user modeling for intelligent learning environments. In: IUI 2007: Proceedings of the 12th international conference on Intelligent user interfaces, pp. 72–81 (2007)
3. Blum, A., Langley, P.: Selection of relevant features and examples in machine learning. In: Artificial Intelligence 1997, pp. 245–271 (1997)
4. Cheguis, I.A., Yablonskii, S.V.: About testors for electrical outlines. Uspieji Matematicheskij Nauk 4(66), 182–184 (1955) (in Russian)
5. Dash, M., Liu, H.: Feature selection for classification. Intelligent Data Analysis 1, 131–156 (1997)
6. Dmitriev, A.N., Zhuravlev, Y. I., Krendeleiev, F.P.: On the mathematical principles of patterns and phenomena classification. Diskretnyi Analiz 7, 3–15 (1966)
7. Dy, J.G., Brodley, C.E.: Feature Selection for Unsupervised Learning. The Journal of Machine Learning Research 5, 845–889 (2004)
8. Glover, F.W., Kochenberger, G.A., Fred, W. (eds.): Handbook of Metaheuristics, pp. 55–82. Springer, Heidelberg (2005)
9. Goldberg, D.E.: Genetic Algorithms in Search. In: Optimization, and Machine Learning, Addison-Wesley, New York (1989)
10. Holland, J.H.: Adaptation in Natural and Artificial Systems: An Introductory Analysis with Applications to Biology, Control, and Artificial Intelligence. In: Ann Arbor, The University of Michigan Press (1975); 2nd edn. The MIT Press, Cambridge (1992)
11. Inza, I., Larrañaga, P., Etxeberria, R., Sierra, B.: Feature Subset Selection by Bayesian networks based optimization. Artificial Intelligence 123(1-2), 157–184 (1999)
12. Kohavi, R.y., Jhon, G.H.: Wrappers for Feature Subset Selection. In: Artificial Intelligence 1997, pp. 273–324 (1997)
13. Laguna, M., Martí, R.: Scatter Search Methodology and Implementations in C. Kluwer Academic Press, Dordrecht (2003)
14. Larrañaga, P., Lozano, J.A.: Estimation of distribution algorithms: a new tool for evolutionary computation. Springer, Heidelberg (2002)
15. Lazo-Cortés, M., Ruiz-Shulcloper, J.: Determining the feature relevance for non classically described objects and a new algorithm to compute typical fuzzy testors. Pattern Recognition Letters 16, 1259–1265 (1995)
16. Lazo-Cortes, M., Ruiz-Shulcloper, J., Alba Cabrera, E.: An Overview of the Evolution of the Concept of Testor. Pattern Recognition 34, 753–762 (2001)
17. Lozano, J.A., Larrañaga, P., Inza, I., Bengoetxea, I.: Towards a new evolutionary computation: advances in the estimation of distribution algorithms, pp. 55–82. Springer, Heidelberg (2006)
18. Mierswa, I., Wurst, M.: Information Preserving Multi-Objective Feature Selection for Unsupervised Learning. In: Proc. of the Genetic and Evolutionary Computation Conference (GECCO 2006), pp. 1545–1552. ACM Press, New York (2006)
19. Tom, M.M.: Machine Learning. Mc.Graw Hill, New York (1997)

20. Moreno, A., Armengol, E., Béjar, J., Belanche, L., Cortés, U., Gavalda, R., Gimeno, J., López, B., Martín, M., Sánchez, M.: Aprendizaje Automático. Ediciones de la Universidad Politécnica de Catalunya. SL (1994)
21. Pelikan, M., Sastry, K., Cantu-Paz, E.: Scalable Optimization via Probabilistic Modeling: From Algorithms to Applications. Springer, Secaucus (2006)
22. Peng L, Jiaxian Z, Lanjuan L, Yanhong L, Xuefeng Z. Data mining application in prosecution committee for unsupervised learning. In: International Conference on Services Systems and Services Management, Proceedings of ICSSSM 2005, 13-15 June 05, Vol. 2, pp 1061 - 1064 (2005)
23. Saeys, Y., Degroeve, S., Van de Peer, Y.: Digging into Acceptor Splice Site Prediction: An Iterative Feature Selection Approach. In: Boulicaut, J.-F., Esposito, F., Giannotti, F., Pedreschi, D. (eds.) PKDD 2004. LNCS (LNAI), vol. 3202, pp. 386–397. Springer, Heidelberg (2004)
24. Sanchez, D.G., Lazo, C.M.: CT-EXT: An Algorithm for Computing Typical Testor Set. In: Rueda, L., Mery, D., Kittler, J. (eds.) CIARP 2007. LNCS, vol. 4756, pp. 506–514. Springer, Heidelberg (2008)
25. Sánchez, D.G., Lazo, C.M.: Modificaciones al Algoritmo BT para Mejorar sus Tiempos de Ejecucion, cuba. Revista Ciencias Matemáticas XX(2), 129–136 (2002)
26. Santiesteban, A.Y., Pons, P.A.: LEX: Un Nuevo Algoritmo para el Cálculo de los Testores Típicos. Revista Ciencias Matemáticas 21(1), 85–95 (2003)
27. Shulcloper, J.R., Alba, C., Lazo, C.: Introducción al reconocimiento de Patrones: Enfoque Lógico Combinatorio, México. Serie Verde, CINVESTAV-IPN 51, 188 (1995)
28. Shulcloper, J.R., Bravo, M.A., Lazo, C.: Algoritmos BT y TB para el cálculo de todos los test típicos, cuba. Revista Ciencias Matemáticas VI(2), 11–18 (1985)
29. Singhi, S.K., Liu, H.: Feature Subset Selection Bias for Classification Learning. In: Proceedings of 23rd International Conference on Machine Learning, Pittsburgh, pp. 849–856 (2006)
30. Torres, M.D., Torres, A., de León, P., Ochoa, A.: Búsqueda Dispersa y Testor Típico. In: XI Simposio de Informática y 6° Mostra de Software Académico SIMS 2006, Uruguaiana, Brasil, Hífen, Noviembre del 2006, vol. 30(58) (2006)
31. Torres, M.D., Ponce, E.E., Torres, A., Luna, F.J.: Selección de Subconjuntos de Características en Aprendizaje no Supervisado utilizando una Estrategia Genética en Combinación con Testores y Testores Típicos. In: Tercer Congreso Internacional de Computación Evolutiva (COMCEV 2007), pp. 57–63 (2007)
32. Torres, M.D., Ponce, E.E., Torres, A., Torres, A., Díaz, E.: Selección de Características Basada en el Peso Informacional de las Variables en Aprendizaje no Supervisado mediante Algoritmos Genéticos. In: Avances en computación Evolutiva. Centro de Investigaciones en Matemáticas, pp. 100–106 (2008)
33. Kim, Y., Street, W.N., Menczer, F.: Feature selection in unsupervised learning via evolutionary search. In: KDD 2000, pp. 365–369 (2000)

A Particle Swarm Optimization Method for Multimodal Optimization Based on Electrostatic Interaction

Julio Barrera and Carlos A. Coello Coello*

CINVESTAV-IPN
Departamento de Computación
Evolutionary Computation Group
Av. IPN No. 2508, Col. San Pedro Zacatenco
México, D.F. 07360, Mexico
julio.barrera@gmail.com, ccoello@cs.cinvestav.mx

Abstract. The problem of finding more than one optimum of a fitness function has been addressed in evolutionary computation using a wide variety of algorithms, including particle swarm optimization (PSO). Several variants of the PSO algorithm have been developed to deal with this sort of problem with different degrees of success, but a common drawback of such approaches is that they normally add new parameters that need to be properly tuned, and whose values usually rely on previous knowledge of the fitness function being analyzed. In this paper, we present a PSO algorithm based on electrostatic interaction, which does not need any additional parameters besides those of the original PSO. We show that our proposed approach is able to converge to all the optima of several test functions commonly adopted in the specialized literature, consuming less evaluations of the fitness function than other previously reported PSO methods.

1 Introduction

There exist certain applications in which the objective function to be optimized has more than one optimum. Such problems are called *multimodal* and may have several local optima and only one global optimum that we want to find, or may have several optima, all of which we aim to locate. This latter problem is probably the most challenging for traditional mathematical programming techniques, which normally are unable to work properly in such problems. When using metaheuristics, multimodal functions are also challenging, because stochastic noise tends to make population-based metaheuristics (e.g., genetic algorithms) to converge to a single solution if run for a sufficiently large number of generations [1]. The currently available metaheuristics designed to deal with multimodal problems, normally require several additional parameters that must be set by the user. More recently, some authors have proposed adaptive procedures for setting

---

* The second author is also affiliated to the UMI-LAFMIA 3175 CNRS.

A. Hernández Aguirre et al. (Eds.): MICAI 2009, LNAI 5845, pp. 622–632, 2009.

up the parameters of some metaheuristics designed for multimodal optimization. However, these adaptive procedures also tend to require additional parameters, and normally do not perform better than other (less elaborate) available methods.

In this paper, we present a method for locating more that one optimum of a fitness function in a given search region. The proposed method does not introduce new parameters into our baseline metaheuristic (particle swarm optimization) other than those originally required by such technique. The method is based on the electrostatic interaction between charged point particles. In our model, we consider the fitness value of a particle to be its charge and then, we compute the magnitude of the force that appears between two charged particles (considering that both have a positive charge). For updating a particle, the particle with the maximum force value in the swarm, replaces the particle with the best global fitness value recorded so far; that is, the particle will follow the particle with which has the maximum electrostatic attraction.

The remainder of the paper is organized as follows. Section 2 reviews some basic notions of the PSO algorithm. In Section 3, we describe some of the available methods for finding more than one optimum using PSO. Section 4 presents our proposed method based on electrostatic interaction. Section 5 shows the test functions used, the parameters setup for the experiments and the results obtained. Finally, in Section 6, we provide our conclusions and some possible paths for future work.

2   Particle Swarm Optimization

The PSO algorithm was originally proposed by Kennedy and Eberhart in the mid-1990s [2] and has been successfully applied to a wide variety of problems. Originally, it was only used for dealing with unimodal problems, but more recently was extended to solve both multimodal and multiobjective optimization problems [3]. One of the reasons for the success of the PSO algorithm is its simplicity and its ease of use. It starts with a population of particles randomly positioned in the search space (such population is called *swarm*). Each particle has a position, a velocity and a fitness function value (evaluated at its current position). The best position obtained by each particle so far (i.e., the position corresponding to the best fitness value obtained so far for that particle), is also stored. At each iteration, the position and velocity of a particle are updated following two simple rules shown in equations (1) and (2).

$$v_{t+1} = v_t + R_1 \cdot C_1 \cdot (g - x_t) + R_2 \cdot C_2 \cdot (p - x_t) \qquad (1)$$
$$x_{t+1} = x_t + v_{t+1} \qquad (2)$$

where R_1 and R_2 are randomly generated numbers in the range $[0, 1]$ using a uniform distribution, C_1 and C_2 are the "learning" constants, p is the position where the particle reached its best fitness value through the iterations, g is the position with the best fitness value of all p positions in the swarm, x_t and v_t are the position and velocity of the particle at iteration t, respectively.

A number of modifications have been proposed to improve the convergence of the PSO algorithm. The most commonly used are the modifications to the velocity update equation (1) proposed by Clerc [4] (called the Constriction Factor model), and by Shi [5] (called the Inertia Weight model). Under the Constriction Factor model, the velocity is updated using equation (3).

$$v_{t+1} = \chi \left[v_t + R_1 \cdot C_1 \cdot (g - x_t) + R_2 \cdot C_2 \cdot (p - x_t) \right] \tag{3}$$

with

$$\chi = \frac{2\kappa}{|2 - \phi - \sqrt{\phi^2 - 4\phi}|} \tag{4}$$

where $\phi = C_1 + C_2$, $\phi > 4$, and κ is an arbitrary value in the range $(0, 1]$. Analogously, the Inertia Weight model uses equation (5) for updating the velocity.

$$v_{t+1} = \omega v_t + R_1 \cdot C_1 \cdot (g - x_t) + R_2 \cdot C_2 \cdot (p - x_t) \tag{5}$$

where ω is an arbitrary value. In contrast with the Constriction factor model, in this case the value of ω does not depend on the value of C_1 nor C_2. The modification of the update equation is complemented confining the particles to the search space using a X_{max} and a X_{min} values for each coordinate. It is also common to limit the velocity of the particles using a value V_{max} for each coordinate, usually $V_{max} = X_{max}$. The Constriction Factor and Inertia Weight model are the two PSO variants most commonly adopted in the literature.

3   Finding More Than One Optimum

Like any other metaheuristic, the PSO algorithm requires certain modifications in order to make it capable of dealing with multimodal problems. In this section, we will review the most relevant previous work in that direction that has appeared in the specialized literature.

Some of the existing proposals borrow ideas from the genetic algorithms literature, such as the species-based PSO introduced by Li [6], and later modified by Iwamatsu [7,8]. This method is a PSO adaptation of the Species Conserving Genetic Algorithm (SCGA), originally proposed by Jiang-Ping Li [9]. It requires the setup of a parameter value called "radius" in order to determine when a particle belongs or not to a certain species. The proper choice of this value is essential for obtaining good results and, in most cases, the optimal value for this parameter depends on the objective function being optimized.

The use of niching (which has been a popular diversity maintenance mechanism in the genetic algorithms literature [10,11,12]) has also been proposed for PSO. Brits et al. [13] proposed a niching PSO algorithm in which the radius of a niche is computed using the mean of the distance among the particles in a sub-swarm. This approach requires the initialization of several parameters. To create a sub-swarm, the variance of the fitness of a particle through the iterations of the algorithm must be measured. To compute the variance, the fitness of the particles is observed during e iterations, and if such fitness variation is less than a δ threshold, a sub-swarm is created. The approach also requires a user-defined threshold μ for preventing the collision of two sub-swarms.

To avoid setting multiple parameters, an adaptive method was introduced by Bird et al. [14]. In this case, the radius is also computed as the mean of the distances among the particles of a swarm, but they use a graph that stores the information of the particles that are at a distance which is smaller than the radius. If two particles are close (according to the computed radius) for a number e of iterations, then a sub-swarm is formed. In this case, there is also a limit m of particles that are allowed per sub-swarm.

Passaro and Sarita [15] present a follow-up of one of Kennedy's papers [16] and use clustering to divide the main swarm into sub-swarms. In order to avoid setting up the number k of clusters, they adopt the x-means algorithm from Pelleg et al. [17] and choose an optimal value for k. A maximum and minimum value for k must be set because the x-means algorithm computes a statistical value for each k in a given range in order to determine the optimum value.

An attempt to develop a PSO method capable of locating more that one optimum, and which does not require any additional parameters was presented by Li [18]. The basis of this work is the proposal of Peram et al. [19] in which a new component is added to the equation for updating the velocity. This component consists of the difference between the position x_t of the particle at iteration t and a computed position p_n for each particle. Each coordinate of the p_n vector is computed separately, and might result in unstable convergence. Li proposed that, instead of computing each coordinate separately at the position p_n, they are selected from the p vectors of the particles in the swarm. Such selection is based on the computation of the ratio of the difference between the fitness value of two particles being compared and their distance. Thus, the p_n position for a particle i is the p vector of the particle j that maximizes the computed ratio. Also, the p_n position, replaces the g position in the update equation of the velocity rather than being added in an additional component. With these modifications, the method performs better, but still shows unstable convergence in some cases.

As we have shown in this section, there are several PSO methods that have been designed to find more than one optimum. Their nature is also varied, but one important problem that is common in almost all of them is that they require additional parameters that have an impact on performance, and whose definition is normally not trivial. In the next section, we describe a PSO method that does not need any additional parameters, besides those of the original PSO algorithm. As we will see later on, the proposed method is also able to find better results than those reported by other PSO methods previously proposed for multimodal optimization.

4   The Electrostatic Interaction Method

As indicated in the previous section, one possible method for finding more than one optimum of a function consists of the selection of a particle's g position from the p positions of the other particles in the swarm. In the case of the FER-PSO of Li [18], the selection method consists of computing and maximizing a ratio for all the particles in the swarm with respect to the particle being updated.

In our proposed method, we follow a similar approach. When the velocity of a particle is being updated, we select the particle's g position by computing and maximizing a value for each particle in the swarm. However, in this case, we take inspiration on electrostatics: according to Coulomb's law, the force between two charged particles can be computed using equation (6).

$$F = \frac{1}{4\pi\varepsilon_0} \frac{Q_1 Q_2}{R^2} \tag{6}$$

where F is the force between the two particles, Q_1 and Q_2 are the charges of the particles, ε_0 is the electric constant (vacuum permitivity), and R is the distance between the two particles. For our proposed method we compute a "force" F_{ij} between a particle being updated i and the rest of the particles in the swarm by replacing the charge Q of a particle by the best fitness value recorded by the particle $f(p)$ so far. Also, we replace the constant $1/(4\pi\varepsilon_0)$ with a scaling constant α computed in the same way as in [18]. Thus, to compute the F_{ij} between two particles i and j, we use equation (7).

$$F_{ij} = \alpha \frac{f(p_i)f(p_j)}{||p_i - p_j||^2} \tag{7}$$

In order to prevent numerical errors, if the distance between the particles i and j is zero, the F_{ij} value is not computed and the particle j is not considered for selection. The best position recorded for the particle j with the maximum value of F_{ij} is used to replace the best position of the particle with the current best global fitness (i.e., the g position). The equations for updating the velocity and position of a particle remain unchanged. For each particle i in a swarm, the values of F_{ij} are computed for the rest of the swarm, that is, this operation has a complexity $O(N^2)$ for each generation, with N being the number of particles in the swarm. In this work, we use a different PSO algorithm that does not modify the update equations like in the Constriction Factor or the Inertia Weight models. Instead, the maximum value of the velocity is decreased in a nonlinear way. For iteration t, the value of the maximum velocity V_{max} is computed according to equation (8).

$$V_{max} = |X_{max} - X_{min}| \cdot \omega^t \tag{8}$$

with ω an arbitrary number in the range $(0, 1)$. The outline of the selection method for the g position is presented in Algorithm 1. In Algorithm 2 we show the procedure to compute the velocity of the particles, and in Algorithm 3, we outline our proposed PSO algorithm for finding more than one optimum: the Electrostatic Particle Swarm Optimization (EPSO).

Algorithm 1. Computeindexmaximum(index) Algorithm for selecting the position p_n with the maximum electrostatic force F for a particle

input : Index of a particle
output: Index j of the recorded position p with the maximum value of F for the particle at the input index

```
1  indexMaximum = 0;
2  FMaximum = 0;
3  for j ← 0 to N-1 do
4      if distance(i,j) > 0 then
5          F = alpha * f(pᵢ) * f(pⱼ)/distance(i, j) * *2;
6          if i=0 then
7              FMaximum = F;
8          end
9          if F > FMaximum then
10             FMaximum = F;
11             indexMaximum = j;
12         end
13     end
14 end
15 return indexMaximum;
```

Algorithm 2. Computevelocities(void) Algorithm to compute the velocity of the particles in the swarm

```
1  indexMaximum = 0;
2  for i ← 0 to N-1 do
3      indexMaximum = computeIndexMaximum(i);
4      velocity = R₁ * C₁ * (position(i) - bestPosition(i)) + R₂ * C₂ * (position(i)
           - bestPosition(indexMaximum);
5      setVelocity(i, velocity);
6  end
```

Algorithm 3. EPSO() Algorithm to find more than one optimum

```
1  generateSwarm();
2  evaluateSwarm();
3  for i ← 0 to iterations do
4      computeVelocities();
5      computeVelocityLimits(i);
6      applyVelocityLimits();
7      updatePositions();
8      applyPositionLimits();
9      evaluateSwarm();
10     updateBestPositions();
11     countOptimaFound();
12 end
```

5   Experiments and Results

For testing and comparison purposes, we use the same set of functions that appeared in the works of Passaro et al. [15] and Bird et al. [14]. We also compare our results with those reported by these authors.

5.1   Test Functions

The set of test functions that we adopted to validate our proposed approach is sumarized in Table 1. For each of them, we show their corresponding equations and the allowable ranges for their decision variables. Next, we describe in more detail each of these test functions.

Table 1. Test functions adopted for our experiments

Function	Equation	Search range
F1	$f(x,y) = \left(y - \frac{5.1x^2}{4\pi} + \frac{5x}{\pi} - 6\right)^2 + 10\left(1 - \frac{1}{8\pi}\right)\cos(x) + 10$	$-5 \leq x \leq 10$ $0 \leq y \leq 15$
F2	$f(x,y) = -4\left[\left(4 - 2.1x^2 + \frac{x^4}{3}\right)x^2 + xy + (-4 + 4y^2)y^2\right]$	$-1.9 \leq x \leq 1.9$ $-1.1 \leq y \leq 1.1$
F3	$f(x) = \sin^6(5\pi x)$	$0 \leq x \leq 1$
F4	$f(x,y) = 200 - (x^2 + y - 11)^2 - (x + y^2 - 7)^2$	$-6 \leq x, y \leq 6$
F5	$f(x,y) = \sum_{i=1}^{5} i \cos[(i+1)x + i] \sum_{i=1}^{5}[(i+1)y + i]$	$-10 \leq x, y \leq 10$

F1 is Branin's RCOS function and has 3 global minima. Its plot is shown in Figure 1(a). F2 is the Six-hump camel back function, which has 2 global and 4 local maxima. Its plot is shown in Figure 1(b). The F3 function is also known as Deb's 1st function. This is the only one-dimensional test function that we adopted. It has 5 global maxima uniformly distributed (see Figure 1(c)). The F4 function is Himmelblau's function with 4 global optima shown in Figure 1(d). The last test function is the two-dimensional Shubert's function. This function is well known for its complexity in the given search range. It has 760 optima, including 18 global minima. The plot of Shubert's function is shown in Figure 2.

In order to allow a fair comparison, we perform experiments similar to those reported by Passaro [15] and Bird [14]. We repeat each experiment 50 times in order to gather statistics and we adopt a maximum of 500 iterations per run. Also, for each test function, we adopt two different sizes for the swarm, 30 and 60 particles. For Shubert's function, we adopt 300 and 500 particles per swarm. In all our experiments, we use a threshold value $\epsilon = 0.00001$ in order to determine if a particle has reached an optimum.

5.2   Results

Table 2 summarizes the results of our experiments. In the first column, we show the name of the test function. The second column indicates the size of the swarm,

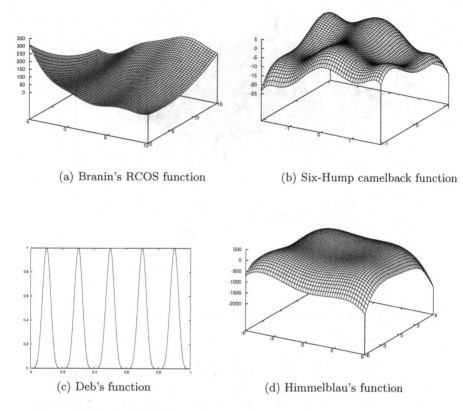

(a) Branin's RCOS function

(b) Six-Hump camelback function

(c) Deb's function

(d) Himmelblau's function

Fig. 1. Test functions used for assessing performance of our proposed electrostatic inspired PSO

and the rest of the columns indicate the mean and standard deviation, of the number of evaluations required to find all the optima of each test function. The last column corresponds to the proposed EPSO method. It is worth mentioning that with the EPSO method, all the global optima are found in all experiments.

From the results shown in Table 2, we can observe that the proposed method needs less function evaluations than any of the other methods with respect to which it was compared. Also, the proposed approach has a smaller standard deviation, which means that it has better stability. We found particularly remarkable the results produced by our proposed approach in the case of the two-dimensional Shubert's function. In that case, our proposed approach requires less than half of the evaluations required by the best approach previously reported for this problem. Additionally, the standard deviation of our proposed approach is, in this case, smaller than that of the best approach previously reported for this problem, by two orders of magnitude in the case of 300 particles, and in one order of magnitude for the case of 500 particles per swarm.

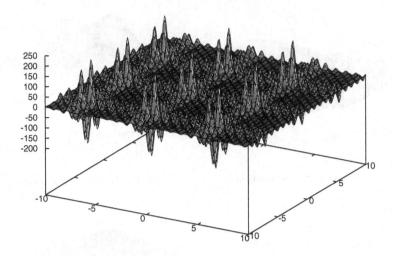

Fig. 2. The two-dimensional Shubert's function

Table 2. Comparison with the results reported for other PSO methods

Function	Particles	SPSO	ANPSO	kPSO	EPSO
F1	30	3169±692	5220±3323	2084±440	**1581±79**
	60	6226±1707	6927±2034	3688±717	**2961±133**
F2	30	2872±827	2798±857	1124±216	**888±78**
	60	5820±1469	4569±1316	2127±341	**1735±128**
F3	30	2007±703	6124±2465	1207±688	**889±93**
	60	4848±2092	8665±2974	1654±705	**1529±174**
F4	30	4096±731	16308±13157	2259±539	**1669±56**
	60	7590±2018	17168±12006	3713±570	**2523±111**
F5	300	166050±42214	82248±10605	81194±45646	**33093±607**
	500	219420±80179	114580±18392	117503±77451	**54010±1111**

6   Conclusions and Future Work

We have presented a new multimodal PSO method that does not require any additional parameters besides those that are inherent to the original PSO algorithm. We argue that our proposed approach is very easy to implement, and we have shown that keeps an $O(N^2)$ complexity (N is the number of particles in the swarm) when computing a numerical value for each pair of particles at each generation. However, in spite of its simplicity, our proposed approach was capable of obtaining better results than other previously reported PSO proposals, in terms of the number of iterations required to locate all the global optima of a function. Our results also show that the standard deviations achieved are small, which is a clear indication of better stability than previously reported PSO methods used for multimodal optimization. Additionally, it is worth mentioning that in the most complex test function adopted (Schubert's function), our proposed approach found all the global optima with less than half of the number of evaluations needed by the best PSO method previously reported for this problem.

As part of our future work, we are interested in applying our proposed approach to problems of higher dimensionality. However, since the current test problems available are normally of very low dimensionality (one or two decision variables), we are currently developing a framework that provides a simple and flexible way to increment the number of variables and optima of the test functions.

The proposed method does not fully model the interaction of a set of charged particles. We only consider the interaction between pairs of particles and select the particle's "best" as the particle with the maximum electrostatic force in the swarm. Thus, a more detailed model of electrostatic interactions would be quite interesting, and is part of our ongoing research. For example, we believe that the use of the electrostatic force to compute the acceleration of the particles might provide an improved accuracy and stability of our proposed approach, when searching for several optima.

References

1. Goldberg, D.E.: Genetic Algorithms in Search, Optimization and Machine Learning. Addison-Wesley Publishing Company, Reading (1989)
2. Kennedy, J., Eberhart, R.: Particle swarm optimization. In: Proceedings of the IEEE International Conference on Neural Networks, December 1995, vol. 4, pp. 1942–1948 (1995)
3. Coello Coello, C.A., Lamont, G.B., Van Veldhuizen, D.A.: Evolutionary Algorithms for Solving Multi-Objective Problems, 2nd edn. Springer, New York (2007)
4. Clerc, M., Kennedy, J.: The particle swarm - explosion, stability, and convergence in a multidimensional complex space. IEEE Transactions on Evolutionary Computation 6(1), 58–73 (2002)
5. Shi, Y., Eberhart, R.: A modified particle swarm optimizer. In: Proceedings of the 1998 IEEE International Conference on Evolutionary Computation. IEEE World Congress on Computational Intelligence, May 1998, pp. 69–73 (1998)
6. Li, X.: Adaptively choosing neighbourhood bests using species in a particle swarm optimizer for multimodal function optimization. In: Deb, K., et al. (eds.) GECCO 2004. LNCS, vol. 3102, pp. 105–116. Springer, Heidelberg (2004)
7. Iwamatsu, M.: Multi-species particle swarm optimizer for multimodal function optimization. Transactions on Information and Systems E89-D(3), 1181–1187 (2006)
8. Iwamatsu, M.: Locating all the global minima using multi-species particle swarm optimizer: The inertia weight and the constriction factor variants. In: Proceedings of the IEEE Congress on Evolutionary Computation (CEC 2006), pp. 816–822 (2006)
9. Li, J.-P., Balazs, M.E., Parks, G.T., Clarkson, P.J.: A species conserving genetic algorithm for multimodal function optimization. Evolutionary Computation 10(3), 207–234 (2002)

10. Goldberg, D.E., Richardson, J.: Genetic algorithm with sharing for multimodal function optimization. In: Grefenstette, J.J. (ed.) Genetic Algorithms and Their Applications: Proceedings of the Second International Conference on Genetic Algorithms, pp. 41–49. Lawrence Erlbaum, Hillsdale (1987)
11. Deb, K., Goldberg, D.E.: An Investigation of Niche and Species Formation in Genetic Function Optimization. In: Schaffer, J.D. (ed.) Proceedings of the Third International Conference on Genetic Algorithms, June 1989, pp. 42–50. Morgan Kaufmann Publishers, San Mateo (1989)
12. Mahfoud, S.W.: Niching methods for genetic algorithms. PhD thesis, University of Illinois at Urbana-Champaign, Champaign, IL, USA (1995)
13. Brits, R., Engelbrecht, A.P., Bergh, F.V.D.: A niching particle swarm optimizer. In: Proceedings of the Conference on Simulated Evolution and Learning, pp. 692–696 (2002)
14. Bird, S., Li, X.: Adaptively choosing niching parameters in a pso. In: Proceedings of the 8th annual conference on Genetic and evolutionary computation (GECCO 2006), pp. 3–10. ACM, New York (2006)
15. Passaro, A., Starita, A.: Particle swarm optimization for multimodal functions: a clustering approach. Journal Artificial Evolution and Applications 8(2), 1–15 (2008)
16. Kennedy, J.: Stereotyping: improving particle swarm performance with cluster analysis. In: Proceedings of the 2000 Congress on Evolutionary Computation, vol. 2, pp. 1507–1512 (2000)
17. Pelleg, D., Moore, A.: X-means: Extending k-means with efficient estimation of the number of clusters. In: Proceedings of the 17th International Conference on Machine Learning, pp. 727–734. Morgan Kaufmann, San Francisco (2000)
18. Li, X.: A multimodal particle swarm optimizer based on fitness euclidean-distance ratio. In: Proceedings of the 9th annual conference on Genetic and evolutionary computation (GECCO 2007), pp. 78–85. ACM, New York (2007)
19. Peram, T., Veeramachaneni, K., Mohan, C.K.: Fitness-distance-ratio based particle swarm optimization. In: Proceedings of the 2003 IEEE Swarm Intelligence Symposium (SIS 2003), April 2003, pp. 174–181 (2003)

Ranking Methods for Many-Objective Optimization

Mario Garza-Fabre[1], Gregorio Toscano Pulido[1], and Carlos A. Coello Coello[2]

[1] CINVESTAV-Tamaulipas. Km. 6 carretera Cd. Victoria-Monterrey
Cd. Victoria, Tamaulipas, 87261, Mexico
mgarza@tamps.cinvestav.mx, gtoscano@cinvestav.mx
[2] CINVESTAV-IPN, Departamento de Computación, Av. IPN No. 2508
Col. San Pedro Zacatenco, México, D.F. 07360, Mexico
ccoello@cs.cinvestav.mx

Abstract. An important issue with Evolutionary Algorithms (EAs) is the way to identify the best solutions in order to guide the search process. Fitness comparisons among solutions in single-objective optimization is straightforward, but when dealing with multiple objectives, it becomes a non-trivial task. Pareto dominance has been the most commonly adopted relation to compare solutions in a multiobjective optimization context. However, it has been shown that as the number of objectives increases, the convergence ability of approaches based on Pareto dominance decreases. In this paper, we propose three novel fitness assignment methods for many-objective optimization. We also perform a comparative study in order to investigate how effective are the proposed approaches to guide the search in high-dimensional objective spaces. Results indicate that our approaches behave better than six state-of-the-art fitness assignment methods.

1 Introduction

Evolutionary algorithms (EAs) are metaheuristics inspired on the mechanism of natural selection, which have been found to be very effective to solve optimization problems. In order to emulate natural selection, it is imperative to implement a function which measures the fitness of the individuals, such that the emulation can favor those individuals that compete most effectively for resources. Usually, when dealing with a single-objective optimization problem such fitness function is directly related to the function to be optimized. However, when the problems involve the simultaneous satisfaction of two or more objectives (the so-called "multiobjective optimization problems", or MOP for short), such relation is not straightforward, since it is required an additional mechanism to map the multiobjective space into a single dimension in order to allow a direct comparison among solutions; this mechanism is known as the *fitness assignment process*[1] [12].

Pareto dominance (PD) has been the most commonly adopted relation to establish preferences among solutions in Multiobjective Optimization (MOO). Multiobjective Evolutionary Algorithms (MOEAs) based on PD have been successfully used to

---

[1] In this study, we will use indistinctly the terms fitness and rank to refer to the value which expresses the quality of solutions and which allows to compare them with respect to each other.

A. Hernández Aguirre et al. (Eds.): MICAI 2009, LNAI 5845, pp. 633–645, 2009.
© Springer-Verlag Berlin Heidelberg 2009

solve problems with two or three objectives. However, when the number of objectives increases, the convergence ability of such approaches can be affected (negatively) [17,13,11]. The main reason is that the proportion of nondominated solutions (*i.e.*, equally good solutions in the multiobjective context) increases exponentially with the number of objectives. As a consequence, it is not possible to impose preferences among individuals for selection purposes and the search process weakens since it is performed practically at random.

Despite the considerable volume of research on evolutionary multiobjective optimization (EMO, for short), most of the MOEAs proposed so far have been validated using only two or three objectives. However, since real-world applications may involve more than 3 objectives, it should be clear how important is to devise alternative approaches to rank solutions when dealing with a higher number of objectives. MOPs having more than 3 objectives are referred to as *many-objective optimization problems* in the specialized literature [8].

In this paper, three novel fitness assignment methods for multiobjective optimization are proposed. We performed a comparative study in order to investigate how effective are the proposed approaches to guide the search process in high dimensional objective spaces. We also included six state-of-the-art methods in such study.

The remainder of this document is structured as follows: the statement of the problem and the PD's drawbacks are described in Section 2. Section 3 describes some state-of-the-art alternatives for fitness assignment in MOO. We present our proposed techniques in Section 4. In Section 5, the results of an empirical study about the convergence properties of the studied approaches are detailed. Finally, Section 6 provides our conclusions as well as some possible directions for future research.

2   Background

In this study, we assume that all the objectives are to be minimized and are equally important. Here, we are interested in solving *many-objective problems* with the following form:

$$\text{Minimize } \mathrm{f}(\mathbf{X}_i) = (f_1(\mathbf{X}_i), f_2(\mathbf{X}_i), \dots, f_M(\mathbf{X}_i))^T$$
$$\text{subject to } \mathbf{X}_i \in \mathcal{F} \tag{1}$$

where $\mathbf{X}_i$ is a *decision vector* (containing our *decision variables*), $\mathrm{f}(\mathbf{X}_i)$ is the M-dimensional *objective vector* ($M > 3$), $f_m(\mathbf{X}_i)$ is the m-th objective function, and $\mathcal{F}$ is the feasible region delimited by the problem's constraints.

2.1   Pareto Dominance (PD)

PD was proposed by Vilfredo Pareto [16] and it is defined as follows: given two solutions $\mathbf{X}_i, \mathbf{X}_j \in \mathcal{F}$, we say that $\mathbf{X}_i$ Pareto-dominates $\mathbf{X}_j$ ($\mathbf{X}_i \prec_P \mathbf{X}_j$) if and only if:

$$\forall m \in \{1,2,\dots,M\} : f_m(\mathbf{X}_i) \leq f_m(\mathbf{X}_j) \; \wedge$$
$$\exists m \in \{1,2,\dots,M\} : f_m(\mathbf{X}_i) < f_m(\mathbf{X}_j) \tag{2}$$

otherwise, we say that $\mathbf{X}_j$ is nondominated with respect to $\mathbf{X}_i$. In this study, we use Goldberg's nondominated sorting method [10] to rank solutions according to PD. The general idea of this approach is the following: first, identify and isolate the nondominated solutions. Then, assign the best rank position to them. Next, from the remainder of the population, identify and isolate the new nondominated solutions and assign the next rank position to them. This process will be repeated until the entire population obtains a rank.

Even though PD has been widely used to develop MOEAs, the performance of such approaches is known to deteriorate as the number of objectives is increased [17,13,11]. The main reason is that the proportion of nondominated solutions increases with the number of objectives. With the aim of clarifying this point, in Figure 1, we show how the number of nondominated solutions grows in a MOEA while the search progresses. This experiment was performed using two well-known scalable test problems, namely DTLZ1 and DTLZ6 [5]. The data in Figure 1 corresponds to the mean of 31 runs of a basic MOEA (described later in Section 5.2) using a population of 100 individuals.

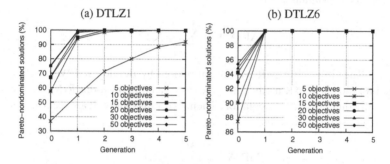

Fig. 1. Proportion of Pareto-nondominated solutions in a randomly generated population

From Figure 1, we can clearly see that an increment of the number of objectives raises the number of nondominated individuals. In other words, PD is useful when there are a few objectives, but it is less suitable as the dimensionality of the objective space increases. For example, let $\mathbf{X}_i$ and $\mathbf{X}_j$ be candidate solutions for a 20-objective MOP. If $\mathbf{X}_i$ is better than $\mathbf{X}_j$ in 19 objectives, but $\mathbf{X}_j$ is superior than $\mathbf{X}_i$ in the remainder objective, both would be equally good solutions if PD is used to compare them. However, a decision maker could say that $\mathbf{X}_i$ has a better performance than $\mathbf{X}_j$. PD does not take into consideration the number of improved (or decreased) objectives nor the magnitude of such improvements (or decreases). Nevertheless, these issues are crucial in the human decision-making process and may lead to several degrees of dominance [9,14].

3   State-of-the-Art Approaches

Weighted Sum Approach (WS). The most natural way to assign fitness in the multiobjective context, is by combining all objectives into a single one using any combination of arithmetical operations. WS is the most commonly used of such approaches. This method expresses the quality of solutions by adding all the objective functions together using weighting coefficients. Thus, the fitness of a solution $\mathbf{X}_i$ is given by:

$$WS(\mathbf{X}_i) = \sum_{m=1}^{M} w_m f_m(\mathbf{X}_i) \tag{3}$$

where w_m is the weighting coefficient denoting the relative importance of the m-th objective[2]. Since WS is a *range-dependent* method, it requires the normalization of the objective values before being applied.

Average Ranking (AR). This method was proposed by Bentley and Wakefield [1]. AR selects one objective and builds a ranking list using the fitness of each solution for such objective. This procedure is performed for all objectives. The ranking positions of a solution $\mathbf{X}_i$ are given by the vector $R(\mathbf{X}_i) = (r_1(\mathbf{X}_i), r_2(\mathbf{X}_i), \ldots, r_M(\mathbf{X}_i))^T$, where $r_m(\mathbf{X}_i)$ is the rank of $\mathbf{X}_i$ for the m-th objective. Once the vector $R(\mathbf{X}_i)$ is calculated for each solution $\mathbf{X}_i$, its global rank is given by:

$$AR(\mathbf{X}_i) = \sum_{m=1}^{M} r_m(\mathbf{X}_i) \tag{4}$$

Maximum Ranking (MR). MR was also proposed by Bentley and Wakefield [1]. As described for the above AR method, the vector of ranking positions $R(\mathbf{X}_i)$ is calculated for each individual $\mathbf{X}_i$ in the population. Then, MR ranks individuals according to their best ranking position. Formally:

$$MR(\mathbf{X}_i) = \min_{m=1}^{M} r_m(\mathbf{X}_i) \tag{5}$$

MR tends to favor extreme solutions, *i.e.*, it prefers solutions with the best performance for some objectives but without taking into account their assessment in the rest of the objectives. This issue is a drawback when searching, for example: if in a 20-objective MOP, $\mathbf{X}_i$ is the solution with the best performance with respect to the first objective, but it is the worst solution for the remainder 19 objectives, $\mathbf{X}_i$ would be classified with the best rank by MR.

Relation (FR). In this alternative dominance relation, proposed by Drechsler et al. [7], a solution $\mathbf{X}_i$ is said to dominate another solution $\mathbf{X}_j$ ($\mathbf{X}_i \prec_f \mathbf{X}_j$) if and only if:

$$|\{m : f_m(\mathbf{X}_i) < f_m(\mathbf{X}_j)\}| > |\{n : f_n(\mathbf{X}_j) < f_n(\mathbf{X}_i)\}|$$
$$\text{for } m, n \in \{1, 2, \ldots, M\} \tag{6}$$

Since FR is not a transitive relation (consider solutions $\mathbf{X}_i$=(8,7,1), $\mathbf{X}_j$=(1,9,6) and $\mathbf{X}_k$=(7,0,9); it is clear that $\mathbf{X}_i \prec_p \mathbf{X}_j \prec_p \mathbf{X}_k \prec_p \mathbf{X}_i$), its authors proposed to rank solutions as follows: to use a graph representation for the relation, where each solution is

---

[2] Since we assume that all objectives are equally important, the weighting coefficients of equation (3) were omitted for this study.

a node and the preferences are given by edges, in order to identify the *Strongly Connected Components* (SCC). An SCC groups all elements which are not comparable to each other (as the cycle of solutions in the above example). A new cycle-free graph is constructed using the obtained SCCs, such that it would be possible to establish an order by assigning the same rank to all solutions that belong to the same SCC.

Preference order ranking (PO). di Pierro et al. [6] proposed a strategy that ranks a population according to the order of efficiency of the solutions. An individual $\mathbf{X}_i$ is considered efficient of order k if it is not Pareto-dominated by any other individual for any of the $\binom{M}{k}$ subspaces where are considered only k objectives at a time. Efficiency of order M for a MOP with exactly M objectives simply corresponds to the original Pareto optimality definition.

If $\mathbf{X}_i$ is efficient of order k, then it is efficient of order $k+1$. Analogously, if $\mathbf{X}_i$ is not efficient of order k, then it is not efficient of order $k-1$. Given these properties, the order of efficiency of a solution $\mathbf{X}_i$ is the minimum k value for which $\mathbf{X}_i$ is efficient. Formally:

$$order(\mathbf{X}_i) = \min_{k=1}^{M}(k \mid \text{isEfficient}(x_i, k)) \tag{7}$$

where isEfficient($\mathbf{X}_i$, k) is to be true if $\mathbf{X}_i$ is efficient of order k. The order or efficiency can be used as the rank of solutions. The smaller the order of efficiency, the better an individual is.

The authors proposed to use this strategy in combination with Goldberg's nondominated sorting [10] described in Section 2.1. The resulting ranking scheme is as follows [6]: first apply nondominated sorting to the population. Next, solutions forming each of the obtained nondominated layers are ranked according to their order of efficiency, in such a way that solutions in the second nondominated layer are penalized by adding the worst obtained rank in the first nondominated layer to their ranks, and so on.

4   Proposed Approaches

Global Detriment (GD). In order to calculate the quality of solutions, we propose a method which takes into account how significant is the difference among solutions. Individuals are compared pairwise in the objective space. The global fitness of a solution is obtained by accumulating the difference by which it is inferior to every other solution, with respect to each objective. The fitness of a solution $\mathbf{X}_i$ is then calculated as follows:

$$GD(\mathbf{X}_i) = \sum_{\mathbf{X}_j \neq \mathbf{X}_i} \sum_{m=1}^{M} \max(f_m(\mathbf{X}_i) - f_m(\mathbf{X}_j), 0) \tag{8}$$

Using GD, $\mathbf{X}_i$ is said to be better than $\mathbf{X}_j$ if it holds that $GD(\mathbf{X}_i) < GD(\mathbf{X}_j)$. This method requires the objective values to be normalized within the same range.

Profit (PF). The proposed PF method expresses the solutions' quality in terms of their profit. Formally, the fitness of a solution $\mathbf{X}_i$ is given by:

$$PF(\mathbf{X}_i) = \max_{\mathbf{X}_j \neq \mathbf{X}_i}\left(gain(\mathbf{X}_i, \mathbf{X}_j)\right) - \max_{\mathbf{X}_j \neq \mathbf{X}_i}\left(gain(\mathbf{X}_j, \mathbf{X}_i)\right) \tag{9}$$

where $gain(\mathbf{X}_i, \mathbf{X}_j)$ is the gain of $\mathbf{X}_i$ with respect to $\mathbf{X}_j$:

$$gain(\mathbf{X}_i, \mathbf{X}_j) = \sum_{m=1}^{M} \max(f_m(\mathbf{X}_j) - f_m(\mathbf{X}_i), 0) \tag{10}$$

PF is a maximization criteria and thus, we say that $\mathbf{X}_i$ outperforms $\mathbf{X}_j$ if and only if $PF(\mathbf{X}_i) > PF(\mathbf{X}_j)$. This method requires the objective values to be normalized within the same range.

Distance To The Best Known Solution (GB). This method uses an ideal reference point in order to rank the population. Such reference point will be referred to as GBEST and it must dominate to the whole population. GBEST is composed by the best known objective value for each dimension. Once that GBEST is calculated, then the fitness of a solution will be assigned with respect to its closeness to the GBEST point (we used the Euclidean distance for this sake). Thus, we will prefer individuals with a small GB fitness value. This method requires the objective values to be into the same range.

5   Experimental Results

In order to investigate the properties of each of the approaches described in Sections 3 and 4, we performed a series of comparisons. Problems DTLZ1, DTLZ3 and DTLZ6 [5] were selected for our experimental study. Due to space limitations we've only included these three test cases. These test functions can be scaled to any number of objectives and decision variables. In these test problems, the total number of variables are $n = M + k - 1$, where M is the number of objectives. The used k values were 5 for DTLZ1, and 10 in the case of DTLZ3 and DTLZ6. In this study, we tested them with 5, 10, 15, 20, 30 and 50 objectives. However, since PO becomes computationally expensive as the number of objectives increases, we only applied it for instances with up to 20 objectives.

5.1   Ranking Distribution

Corne and Knowles [2] proposed to measure the *relative entropy* of the distribution of ranks induced by a method in order to analyze its effectiveness. They described this measure as follows: consider a population of N ranked solutions (there are at most N ranks, and at least 1). The relative entropy of the distribution of ranks D is give by:

$$re(D) = \frac{\sum_r \dfrac{D(r)}{N} \log(\dfrac{D(r)}{N})}{\log(1/N)} \tag{11}$$

where $D(r)$ denotes the number of solutions with rank r. $re(D)$ tends to 1 as we approach to the ideal situation in which each solution has a different rank. On the other hand, when all individuals share the same ranking position, $re(D)$ is equal to zero. Thus, it is supposed that a ranking method providing a richer ordering would lead to a better performing optimization scheme.

López Jaimes et al. [15] performed a similar study. To measure the scalability of ranking methods, they proposed to analyze if the shape and range of the histograms of the ranking distributions is maintained when the number of objectives is increased. As stated by Corne and Knowles, López Jaimes et al. argued that a ranking method will favor the selection process if it is able to generate a richer range of ranks.

Figure 2 shows the relative entropy of the ranking distribution produced by the studied methods. For each experiment, a population of 1000 decision vectors were randomly generated and evaluated for DTLZ1, DTLZ3 and DTLZ6 test problems. Each experiment was run 31 times. Due to space limitations, and the similarity of the results obtained for all test problems, the data was averaged to show the global performance for all test functions using 5, 20 and 50 objectives.

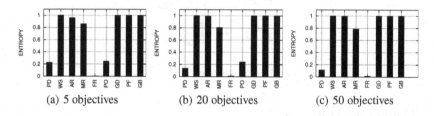

(a) 5 objectives   (b) 20 objectives   (c) 50 objectives

Fig. 2. Average relative entropy of ranks distribution of 1000 randomly generated solutions, for DTLZ1, DTLZ3 and DTLZ6 problems

From Figure 2, we can remark that our proposed methods, as well as the WS and AR approaches, present an stable behavior with the augmentation of the number of optimization criteria. As expected, we can clearly see how PD performance deteriorates with the increment in the number of objectives. Even though we only include PO results for instances with up to 20 objectives, from smaller instances we can see that its entropy results will be (relatively) slightly higher than that of PD. FR performed the worst for this experiment. Its entropy was the lowest in all cases. This property of FR is due to its mechanism, since it is not a transitive relation, for cycles identification. It is usually the case that almost all individuals in the population become part of a cycle, an thus, they share the same ranking position.

5.2   A Generic Multiobjective Evolutionary Algorithm

We implemented a generic multiobjective evolutionary algorithm (MOEA) in order to evaluate the behavior provided by each of the studied ranking mechanisms. The implemented MOEA uses binary tournament selection based on the ranking position of solutions. Simulated binary crossover ($\eta_m = 15$) and polynomial mutation ($\eta_c = 20$) are used as variators. The adopted crossover and mutation rates were set to 1.0 and $1/n$, respectively, where n is the number of decision variables. The combination of parent and children populations is ranked (according to the used method) in order to select the best solutions to form the new parent population (elitism) [3]. If it is the case that in the next ranking position there are more individuals than the available capacity, then, the required individuals are randomly selected. We used a population size of 100 and 300 generations for all our experiments. In order to avoid alterations in the behavior of the studied methods, this MOEA does not incorporate any additional mechanism to maintain diversity in the population. Figure 3 shows the described MOEA's workflow.

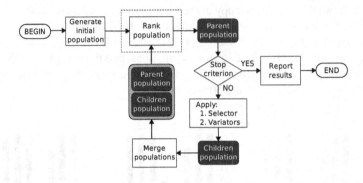

Fig. 3. Base algorithm workflow

5.3   Convergence Properties

All ranking methods were incorporated into the MOEA described in Section 5.2 in order to investigate their convergence ability as the number of objectives increases. The different studied approaches were applied to the normalized objective values: $f'_m(\mathbf{X}_i) = \frac{f_m(\mathbf{X}_i) - GMIN_m}{GMAX_m - GMIN_m}$ for $m = 1, 2, ..., M$, where $GMAX_m$ and $GMIN_m$ are the maximum and minimum known values for the m-th objective. Since we know a priori that for the adopted set of test problems $GMIN_m = 0$, we simply normalized the objectives as follows: $f'_m(\mathbf{X}_i) = \frac{f_m(\mathbf{X}_i)}{GMAX_m}$ for $m = 1, 2, ..., M$. Futhermore, this latter normalization method mantains the proportionality of the data.

Deb et al. [4] proposed a convergence measure in which the average distance of the obtained approximation set from the the Pareto-optimal front is computed. Rather than using the average distance, we computed the distance from the best converged solution (the solution which presents the minimum distance) to the Pareto-optimal frontier. Since equations defining the true Pareto front are known for all the test problems adopted, this measure was determined analytically.

Tables 1, 2 and 3 show the obtained results when the different MOEA's configurations were applied to problems DTLZ1, DTLZ3 and DTLZ6, respectively. The data in these tables corresponds to the average and standard deviation of the convergence measure, at the final of the search, for 31 trials of each experiment.

From Table 1, we can highlight that, for DTLZ1, our proposed GB, GD and PF, and the WS approaches achieved the best results for most instances of this problem. These approaches showed robustness as the number of the objective was increased. AR reached good convergence levels for small instances of this problem. However, its performance was gradually deteriorated as the dimensionality of the problem was increased. The MR results confirm some of its drawbacks mentioned in Section 3, since it was the method which performed the worst for this problem.

Table 1. Final convergence for DTLZ1 problem

	5 Obj.	10 Obj.	15 Obj.	20 Obj.	30 Obj.	50 Obj.
PD	0.637 ± 0.555	3.313 ± 2.545	2.357 ± 1.489	2.147 ± 1.160	1.130 ± 0.789	1.458 ± 0.925
WS	**0.000 ± 0.000**	**0.000 ± 0.000**	0.004 ± 0.023	**0.000 ± 0.000**	0.003 ± 0.016	0.014 ± 0.052
AR	**0.000 ± 0.000**	**0.000 ± 0.000**	0.005 ± 0.023	0.019 ± 0.041	0.107 ± 0.117	0.353 ± 0.259
MR	37.85 ± 14.64	27.10 ± 7.857	22.81 ± 6.854	21.91 ± 7.283	14.28 ± 7.615	10.77 ± 5.471
FR	0.001 ± 0.001	0.016 ± 0.047	0.102 ± 0.145	0.198 ± 0.256	0.634 ± 0.710	2.543 ± 2.713
PO	0.496 ± 0.331	0.824 ± 0.731	0.841 ± 0.755	0.657 ± 0.628	-	-
GD	**0.000 ± 0.000**	**0.000 ± 0.000**	**0.000 ± 0.000**	**0.000 ± 0.000**	0.006 ± 0.023	0.018 ± 0.030
PF	**0.000 ± 0.000**	**0.000 ± 0.000**	**0.000 ± 0.000**	0.000 ± 0.001	0.006 ± 0.022	**0.011 ± 0.026**
GB	**0.000 ± 0.000**	**0.000 ± 0.000**	**0.000 ± 0.000**	**0.000 ± 0.000**	**0.000 ± 0.000**	0.030 ± 0.043

Table 2. Final convergence for DTLZ3 problem

	5 Obj.	10 Obj.	15 Obj.	20 Obj.	30 Obj.	50 Obj.
PD	59.71 ± 19.96	260.6 ± 82.82	244.3 ± 103.2	204.9 ± 86.09	191.7 ± 103.2	164.9 ± 81.59
WS	**0.078 ± 0.245**	0.140 ± 0.333	0.144 ± 0.332	0.239 ± 0.489	0.823 ± 1.277	1.507 ± 1.010
AR	0.920 ± 1.199	3.470 ± 2.344	9.891 ± 6.485	18.21 ± 8.368	50.08 ± 15.78	61.35 ± 22.84
MR	496.1 ± 145.3	290.8 ± 243.5	175.2 ± 189.4	88.15 ± 111.0	45.24 ± 41.57	38.24 ± 11.55
FR	14.11 ± 34.73	25.27 ± 21.41	179.9 ± 85.30	179.6 ± 88.61	328.3 ± 129.2	416.3 ± 133.7
PO	35.77 ± 15.75	33.46 ± 18.47	24.74 ± 10.84	22.50 ± 9.913	-	-
GD	0.171 ± 0.369	0.205 ± 0.471	0.309 ± 0.535	0.336 ± 0.530	0.695 ± 0.686	1.695 ± 1.206
PF	0.094 ± 0.261	0.043 ± 0.176	0.173 ± 0.371	0.236 ± 0.489	**0.340 ± 0.529**	**1.388 ± 1.086**
GB	0.110 ± 0.295	**0.041 ± 0.176**	**0.098 ± 0.265**	**0.151 ± 0.334**	0.693 ± 0.891	1.419 ± 1.300

Table 3. Final convergence for DTLZ6 problem

	5 Obj.	10 Obj.	15 Obj.	20 Obj.	30 Obj.	50 Obj.
PD	5.787 ± 0.503	8.011 ± 0.404	8.036 ± 0.472	8.154 ± 0.483	8.466 ± 0.410	8.529 ± 0.323
WS	0.080 ± 0.030	0.081 ± 0.025	0.089 ± 0.029	0.085 ± 0.031	0.087 ± 0.034	0.091 ± 0.029
AR	0.081 ± 0.032	10.00 ± 0.000	10.00 ± 0.000	10.00 ± 0.000	10.00 ± 0.000	10.00 ± 0.000
MR	6.505 ± 2.207	8.481 ± 0.859	8.494 ± 0.837	8.558 ± 0.832	8.289 ± 0.745	8.074 ± 0.586
FR	5.347 ± 2.238	9.821 ± 0.195	9.380 ± 0.245	9.821 ± 0.084	9.693 ± 0.108	9.573 ± 0.137
PO	4.152 ± 0.631	9.773 ± 0.154	9.939 ± 0.036	9.976 ± 0.018	-	-
GD	0.085 ± 0.027	0.082 ± 0.028	0.089 ± 0.027	0.085 ± 0.029	**0.081 ± 0.033**	0.096 ± 0.027
PF	**0.080 ± 0.029**	**0.076 ± 0.030**	**0.079 ± 0.029**	0.081 ± 0.033	0.089 ± 0.036	**0.089 ± 0.033**
GB	0.086 ± 0.035	0.085 ± 0.027	0.084 ± 0.034	**0.078 ± 0.028**	0.088 ± 0.034	0.090 ± 0.022

Most of the methods failed to converge to the DTLZ3's optimal frontier (Table 2). Our proposed GB and PF methods reached the best results, followed by WS and GD. In this test function MR and FR obtained the worst performing results.

Finally, from Table 3 we can observe that, DTLZ6's optimal frontier was also difficult to reach. Our PF approach showed a remarkable performance for most instances

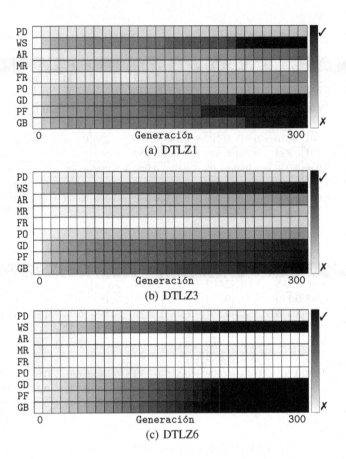

Fig. 4. Online convergence for the 20-objective DTLZ1, DTLZ3 and DTLZ6 problems

(objectives) in this problem, followed by GB, WS and GD. The method which performed the worst for this test problem was AR, since it obtained the poorest convergence value in most cases; this method seems to be the most affected by the problems difficulties.

As general observations, we can highlight that most of the proposed methods were able to guide the search, since they showed robustness as the dimensionality of the different test problems was increased. Results obtained by WS indicates that this simple and efficient approach performs better than some more elaborated ones, since its achieved convergence was nearly the best in most cases. As expected, results confirm that PD's performance deteriorates with the increase in the number of objectives. We consider that MR and FR had the worst performance among the studied approaches.

Additionally, we decided to investigate the convergence speed achieved by the MOEA in Section 5.2 while using the different ranking schemes of our interest. Figure 4 shows the average convergence as the search progresses for the different included

DTLZ problems, considering 31 independent trials of each experiment. Due to space limitations, we only show results for the 20-objective instances, since 20 is the maximum number of objectives for which we performed experiments for all the studied approaches.

The data shown in Figure 4 was plotted in logarithmic scale, in order to highlight the differences in the results obtained using each method. The interpretation of these figures is as follows: each row corresponds to one of the studied methods, and each column represents the search progress in 10-generation steps. The convergence performance is indicated by a grey-scale square, where darker squares refer to the best achieved convergence values.

From Figure 4, we can conclude that the proposed GD, PF and GB, and the WS approaches presented an accelerated convergence, since during the first 150 generations they reached relatively good values for the convergence measure. These results are in agreement with respect to the ranking distributions discussed in Section 5.1, since the methods with the best convergence properties are those which were identified in Section 5.1 to induce a fine-grained discrimination among solutions.

6   Conclusions and Future Work

Since the performance of Pareto-based MOEAs deteriorates as the number of objectives increases, it is necessary to identify alternative approaches to establish preferences among solutions in many-objective scenarios. In this paper, three novel approaches of this sort were proposed. We performed a comparative study in order to investigate how effective are the proposed approaches to guide the search in high dimensional objective spaces. We also included six state-of-the-art methods in such study.

Our experimental results indicate that the proposed methods were able to guide the search process towards the optimum for three well-known scalable test functions. From these results, we can clearly note that our proposed approaches GD, PF and GB were the ones which perform the best. However, it is not possible to select a single approach as the best. Also, one of our main findings is that a simple and efficient aggregation approach was able to perform better than some more sophisticated ranking schemes, since its results were nearly the best in almost all of our experiments.

In this paper, we focused on the convergence properties of different multiobjective ranking approaches. However, an important issue of MOEAs is to converge to a set of well-spread solutions. As part of our future work, we want to extend our experiments to investigate how good is the distribution of the solutions achieved by each of the studied methods. Also, we would like to characterize our methods in terms of the kind of solutions that they tend to prefer, since some methods could favor extreme solutions while some other approaches could focus on solutions which are on the *knee* of the Pareto front.

Acknowledgements

The first author acknowledges support from CONACyT through a scholarship to pursue graduate studies at the Information Technology Laboratory at CINVESTAV-IPN. The second author gratefully acknowledges support from CONACyT through project 90548. Also, this research was partially funded by project number 51623 from "Fondo Mixto Conacyt-Gobierno del Estado de Tamaulipas". Finally, we would like to thank to Fondo Mixto de Fomento a la Investigación científica y Tecnológica CONACyT - Gobierno del Estado de Tamaulipas for their support to publish this paper.

References

1. Bentley, P.J., Wakefield, J.P.: Finding Acceptable Solutions in the Pareto-Optimal Range using Multiobjective Genetic Algorithms. In: Chawdhry, P.K., Roy, R., Pant, R.K. (eds.) Soft Computing in Engineering Design and Manufacturing. Part 5, June 1997, pp. 231–240. Springer, London (1997) (Presented at the 2nd On-line World Conference on Soft Computing in Design and Manufacturing (WSC2))
2. Corne, D., Knowles, J.: Techniques for Highly Multiobjective Optimisation: Some Nondominated Points are Better than Others. In: Thierens, D. (ed.) 2007 Genetic and Evolutionary Computation Conference (GECCO 2007), July 2007, vol. 1, pp. 773–780. ACM Press, London (2007)
3. Deb, K., Agrawal, S., Pratab, A., Meyarivan, T.: A Fast Elitist Non-Dominated Sorting Genetic Algorithm for Multi-Objective Optimization: NSGA-II. KanGAL report 200001, Indian Institute of Technology, Kanpur, India (2000)
4. Deb, K., Mohan, R.S., Mishra, S.K.: Towards a Quick Computation of Well-Spread Pareto-Optimal Solutions. In: Fonseca, C.M., Fleming, P.J., Zitzler, E., Deb, K., Thiele, L. (eds.) EMO 2003. LNCS, vol. 2632, pp. 222–236. Springer, Heidelberg (2003)
5. Deb, K., Thiele, L., Laumanns, M., Zitzler, E.: Scalable Test Problems for Evolutionary Multiobjective Optimization. In: Abraham, A., Jain, L., Goldberg, R. (eds.) Evolutionary Multiobjective Optimization. Theoretical Advances and Applications, pp. 105–145. Springer, USA (2005)
6. di Pierro, F., Khu, S.T., Savić, D.A.: An Investigation on Preference Order Ranking Scheme for Multiobjective Evolutionary Optimization. IEEE Transactions on Evolutionary Computation 11(1), 17–45 (2007)
7. Drechsler, N., Drechsler, R., Becker, B.: Multi-objective Optimisation Based on Relation favour. In: Zitzler, E., Deb, K., Thiele, L., Coello Coello, C.A., Corne, D. (eds.) EMO 2001. LNCS, vol. 1993, pp. 154–166. Springer, Heidelberg (2001)
8. Farina, M., Amato, P.: On the Optimal Solution Definition for Many-criteria Optimization Problems. In: Proceedings of the NAFIPS-FLINT International Conference 2002, June 2002, pp. 233–238. IEEE Service Center, Piscataway (2002)
9. Farina, M., Amato, P.: A fuzzy definition of optimality for many-criteria optimization problems. IEEE Transactions on Systems, Man, and Cybernetics Part A—Systems and Humans 34(3), 315–326 (2004)
10. Goldberg, D.E.: Genetic Algorithms in Search, Optimization and Machine Learning. Addison-Wesley Publishing Company, Reading (1989)
11. Hughes, E.J.: Evolutionary Many-Objective Optimisation: Many Once or One Many? In: 2005 IEEE Congress on Evolutionary Computation (CEC 2005), September 2005, pp. 222–227. IEEE Service Center, Edinburgh (2005)
12. Hughes, E.J.: Fitness Assignment Methods for Many-Objective Problems. In: Knowles, J., Corne, D., Deb, K. (eds.) Multi-Objective Problem Solving from Nature: From Concepts to Applications, pp. 307–329. Springer, Berlin (2008)
13. Khare, V.R., Yao, X., Deb, K.: Performance Scaling of Multi-objective Evolutionary Algorithms. In: Fonseca, C.M., Fleming, P.J., Zitzler, E., Deb, K., Thiele, L. (eds.) EMO 2003. LNCS, vol. 2632, pp. 376–390. Springer, Heidelberg (2003)

14. Köppen, M., Yoshida, K.: Substitute Distance Assignments in NSGA-II for Handling Many-Objective Optimization Problems. In: Obayashi, S., Deb, K., Poloni, C., Hiroyasu, T., Murata, T. (eds.) EMO 2007. LNCS, vol. 4403, pp. 727–741. Springer, Heidelberg (2007)
15. López Jaimes, A., Santana Quintero, L.V., Coello Coello, C.A.: Ranking methods in many-objective evolutionary algorithms. In: Chiong, R. (ed.) Nature-Inspired Algorithms for Optimisation, pp. 413–434. Springer, Berlin (2009)
16. Pareto, V.: Cours d'Economie Politique. Droz, Genève (1896)
17. Purshouse, R.C., Fleming, P.J.: Evolutionary Multi-Objective Optimisation: An Exploratory Analysis. In: Proceedings of the 2003 Congress on Evolutionary Computation (CEC 2003), December 2003, vol. 3, pp. 2066–2073. IEEE Press, Canberra (2003)

Why Unary Quality Indicators Are Not Inferior to Binary Quality Indicators

Giovanni Lizárraga[1], Marco Jimenez Gomez[1], Mauricio Garza Castañon[1], Jorge Acevedo-Davila[1], and Salvador Botello Rionda[2]

[1] Corporación Mexicana de Investigación en Materiales S.A. de C.V. Ciencia y Tecnología 790, Saltillo México
{glizarraga,magarza,marcojimenez,jacevedo}@comimsa.com
[2] Center of Research in Mathematics. Jalisco S/N, Valenciana, Guanajuato, México
botello@cimat.mx

Abstract. When evaluating the quality of non–dominated sets, two families of quality indicators are frequently used: unary quality indicators (UQI) and binary quality indicators (BQI). For several years, UQIs have been considered inferior to BQIs. As a result, the use of UQIs has been discouraged, even when in practice they are easier to use. In this work, we study the reasons why UQIs are considered inferior. We make a detailed analysis of the correctness of these reasons and the implicit assumptions in which they are based. The conclusion is that, contrary to what is widely believed, unary quality indicators are not inferior to binary ones.

1 Introduction

In Multi–objective Optimization using evolutionary algorithms, an open problem is how to evaluate the quality of the non–dominated sets (NSs) that different algorithms generate. The concept of quality with respect to non–dominated sets is difficult to define and several methods to evaluate this quality have been proposed. The most popular approach is to use *quality indicators* (QIs).

A QI is a function that takes one or more non–dominated sets and returns a real number that is related to the quality of the non dominated set(s). The most popular quality indicators are the unary QIs (UQIs) and the binary QIs (BQIs). UQIs take a single NS as argument, and return an evaluation of its quality. A BQI takes two NSs as argument and returns a real number that is the evaluation of the relative quality of one set with respect to the other.

UQIs are easier to use than BQI, so they used to be more popular. But, in 2003, a very influential study of quality indicators were introduced by Ziztler et al. [1]. One of the main conclusions of this study is that UQIs are inferior to BQIs. As a result, UQI are seen as "politicly incorrect" and their use has been discouraged.

In this work we make a revision of the ideas, concepts, assumptions and premises that are used in [1] and we show that some assumptions are not accurate

A. Hernández Aguirre et al. (Eds.): MICAI 2009, LNAI 5845, pp. 646–657, 2009.

and that UQIs are not inferior to BQIs. The rest of this paper is organized as follows: in Section 2, we introduce some basic concepts of optimization. In Section 3, we describe some of the most important concepts when evaluating the quality of a non–dominated set and the desired properties of a quality indicator. We also review the accuracy of this concepts and their limitations. In Section 4, we explain why UQIs are considered as inferior to BQIs and why this is incorrect. In Section 5, we discuss some other topics about quality indicators and the evaluation of non–dominated sets. Finally, in Section 6 we state our conclusions.

2   Basic Concepts

The concepts defined in this section are important to understand the rest of the discussion. We can refer the area we are working on as "problem solving". We have a problem properly stated, and we search for a solution. We want to remark that we track this problem from a practical point of view, and sometimes we appeal to the common sense of when a solution is better than another.

In order to state correctly a problem, there are some important components to consider:

- *A representation of the solutions.* All possible solutions, also known as candidate solutions, must be represented in such a way so we can work with them using mathematical tools and computers. In this work we consider that all solutions are represented as elements of $\mathbf{R}^n$.
- *A search space.* In general, many candidate solutions can not be use to solve a problem. Usually, a solution must not violate a set of constraints in order to be usable. A solution that does not violate any constraint of the problem is called feasible. We call X the set of all feasible candidate solutions, and the final solution to the problem must be an element of X.
- *An objective function.* We want to find the "best" solution for a problem, but in order to decide when one solution is better than another we need a measure of quality. We need a function $f : X \rightarrow \mathbf{R}$, known as objective function, that assigns to each vector in X, a value of how good this solution is. We call x^* to the element(s) of X where $f(x)$ attains its best value.

So, the mathematical statement of an optimization problem is as follows:

$$Minimize \quad f(x) \tag{1}$$
$$subject\ to \quad x \in X \subset \mathbf{R}^n$$

Without lose of generality we only consider problems where the objective function must be minimized. We call *modeling* to the process of constructing the mathematical statement, or model, of the optimization problem, such as the one in Formula 1. We call *optimization* to the process of searching, finding or at least approximating x^*. We consider that a problem is solved when we find a feasible solution with good value in the objective function. For more details about optimization, we recommend reading Nocedal and Wright [2].

There is a special class of problems, known as Multi–objective Problems (MOPs), whose statement is in the following way:

$$Minimize\ F(x) = \langle f_1(x), f_2(x), \ldots, f_m(x) \rangle \qquad (2)$$
$$subject\ to\quad x \in X \subset \mathbf{R}^n$$

In Formula 1, we have a single objective function, while in Formula 2, we have a vector of objective functions. In this work, we consider a MOP as a special case of an optimization problem, where the modeling process is not complete and instead of defining a unique objective function, several objective functions where identified. Each objective function in Formula 2 represent an attribute of interest for the user, but the trade–off between these functions is unknown *a priori*.

Just like in single–objective optimization, a Multi–objective Problem is solved when we find the most satisfactory feasible solution. Unfortunately, this single solution cannot be identified, in general, based in the statement given in Formula 2. In order to identify the "optimal solution" in a Multi–objective Problem we need the help of a *Decision Maker* (DM), the person or group of people who have the last word of when one solution is better than another. We assume that the DM has not clarified his preferences *a priori*, otherwise we could have used those preferences to state the problem in the form of Formula 1. There are three main approaches to solve a Multi–objective Problem based on the interaction with the DM: *a priori* methods, a posteriori methods and iterative methods [3]. In this work we focus on the last two methods.

– *A posteriori method.* Some promising solutions are obtained, so the DM can choose the from them, the one he considers the best.
– *Iterative method.* Some promising solutions are obtained. The DM is presented with these solutions and with this information, the DM clarify preferences. So, a more detailed search can be made and a better set of promising solution is obtained. This process is repeated until the DM is able to find the optimal solution.

In the two methods described above, a promising set of solutions must be obtained. But, how do we define "promising solution" in the context of a Multi–objective Problem, where the evaluation of each solution is based in a vector of objective functions?

A popular approach to compare vectors is the Pareto Optimality Criteria (POC). POC is defined through a binary relation between vectors known as dominance. For two vectors $x, y \in \mathbf{R}^n$, we say that x dominates y, denoted $x\ dom\ y$, if for $i = 1$ to n, $x^{(i)} \leq y^{(i)}$ and $\exists j \in \{1, \ldots, n\}|x^{(j)} < y^{(j)}$. $x^{(i)}$ stands for the i–th component of vector x. When x is not dominated by y and y is not dominated by x, we say that x and y are uncomparable. A set of mutually uncomparable vectors is known as a non–dominated set (NS). For a set of vectors A, we denote by $ND(A)$ the elements of A that are not dominated by any other element of A. We denote an arbitrary NS with script capital letters. For example $\mathscr{A}$, $\mathscr{B}$, $\mathscr{C}$, etc. may represent different non–dominated sets.

The image of X under the vector of objective functions $F(x)$ is called Z. The most promising solutions *a priori* are those vectors in X whose image in Z is not dominated by any other vector in Z. The Pareto Front ($\mathscr{PF}$) is defined as $\mathscr{PF} = ND(Z)$, and is the collection of all promising solutions *a priori*. The sets of vectors whose image are elements of the Pareto Front is known as the Pareto Set (PS). The ideal scenario in an "a posteriori" method is to present the decision maker with the Pareto Front. Unfortunately, the $\mathscr{PF}$ may have an infinite number of elements. So, it is common to present the DM with an approximation of the $\mathscr{PF}$. This approximation consists on a finite non–dominated set that gives enough information of the $\mathscr{PF}$ to make proper decisions. In this work we consider that the objective in Multi–objective Optimization is the following:

Goal of Multi–Objective Optimization 1. *The objective of Multi–objective Optimization is to obtain an approximation of the Pareto Front that contains as much information as possible, so the DM can either choose an element of this set as the final solution, or use this information to specify preferences that allow us to search and find a final solution.*

Multi–objective Evolutionary Algorithms (MOEAs) is a family of "a posteriori" methods that approximate the $\mathscr{PF}$, using concepts from the theory of evolution [4], such as "population", "selection of the fittest", "mutation", "crossover", "fitness" and others. A MOEA takes a Multi–objective Problem and returns a non–dominated set that approximates the $\mathscr{PF}$. From now on, we use the terms approximation set, approximation and non–dominated set as equivalent. For more details about Multi–objective Optimization, we recommend reading Deb [5] and Coello et al. [6].

Many MOEAs have been proposed, for example the ϵ–MOEA [7], NSGA2 [8], SPEA2 [9] and many others. A natural question is which MOEA performs better? Many factors can be considered when evaluating a MOEA, but in this work we only consider one: the approximations that the MOEA produces. So, a very important issue is how to evaluate the quality of an approximation. A good approximation must provide enough information to the DM in order to make good decisions, but this concept must be translated into a numerical value to make the analysis and comparison of MOEAs easier. In the next section we present some published studies related with how to evaluate the quality of an approximation set.

3 Quality Indicators for Non–dominated Sets

Unary and binary quality indicators are often used to evaluate the quality of an approximation. Many quality indicators have been proposed, for example the S–Metric [10] and the Generational Distance [11]. Some important questions are: does these indicators reflect our natural notion of when one non–dominated set is better than another? What QI, or combination of QIs must we use? How can we know if a QI has a good behavior?

One of the most influential studies of quality indicators is a technical report written by Hansen and Jaszkiewicz [12]. In this report, the authors create a framework to analyze quality indicators. Also, they propose some quality indicators and give some directions about how to create QIs. Hansen and Jaszkiewicz [12] rely strongly on what we call the General Assumption 1:

General Assumption 1. *For the Decision Maker, non–dominated solutions are preferred over dominated ones. In other words, for $x, y \in X$, if x dominates y then x is considered a better solution than y.*

This assumption is fundamental, because only when it is true we can justify the use of the POC. From General Assumption 1, Hansen and Jazkciewicz [12] derive the following criterion of when one NS is better than another:

Definition 1. $\mathscr{A}$ *is better than $\mathscr{B}$ if for some DM preferences, $\mathscr{A}$ contains a better solution than $\mathscr{B}$, but there is no DM preference for which $\mathscr{B}$ contains a better solution than $\mathscr{A}$.*

Definition 1 describes a very natural criterion of when one NS is better than another. Actually, this criterion is an extension of the concept of dominance between vectors. Based on Definition 1, Hansen and Jazkiewics [12] derive several binary relations between NSs, for example *weak out–performance* (O_W), *strong out–performance* (O_S) and *complete out–performance* (O_W). Weak out–performance is the most popular out–performance relation, its definition is presented next:

Definition 2. Weak out–performance: $\mathscr{A}$ *weakly outperforms $\mathscr{B}$, denoted by $\mathscr{A} \ O_W \ \mathscr{B}$, when for every point $b \in \mathscr{B}$ there exists a point $a \in \mathscr{A}$ such that a dominates b or $a = b$ and there exists at least a point $c \in \mathscr{A}$ such that $c \notin \mathscr{B}$.*

An out–performance relation O, is a binary relation between two non–dominated sets, so if $\mathscr{A}$ out–performs $\mathscr{B}$ (denoted by $\mathscr{A} \ O \ \mathscr{B}$) then we can conclude that $\mathscr{A}$ is better than $\mathscr{B}$. For the cases where $\neg(\mathscr{A} \ O \ \mathscr{B})$ and $\neg(\mathscr{B} \ O \ \mathscr{A})$ (O may be O_W, O_S, O_C and $\neg$ denotes the logic negation), we say that $\mathscr{A}$ and $\mathscr{B}$ are not comparable under the out–performance relation O.

The out–performance relations represent a partial answer to the question of when one non–dominated set is better than another, because when $\mathscr{A}$ and $\mathscr{B}$ are not comparable, no conclusion can be derived. Actually, Hansen and Jaszkiewicz [12] recommend to use stronger assumptions about DM's preferences for the cases where two non–dominated sets are not comparable. This limitation is very important because it is possible for $\mathscr{A}$ to be better than $\mathscr{B}$ even when $\mathscr{A}$ does not out–perform $\mathscr{B}$. This implies that the out–performance relations have some limitations. Unfortunately, many researchers are not aware of how much limited these out–performance relations can be. To understand better these limitations is useful to study the negations of the out–performance relations. For example, the negation of weak out–performance is as follows:

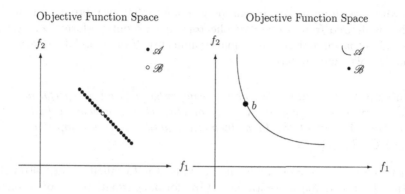

Fig. 1. For $\mathscr{A}$ and $\mathscr{B}$, suppose that $b \in \mathscr{PF}$ and $b \notin \mathscr{A}$. $\mathscr{A}$ and $\mathscr{B}$ are non–comparable under the out–performance relations.

Definition 3. Negation of weak out–performance. *$\mathscr{A}$ does not weakly outperform B when $\mathscr{A} = \mathscr{B}$, or when there exists at least one point $b \in \mathscr{B}$ that is neither contained in $\mathscr{A}$ nor dominated by any point in $\mathscr{A}$.*

We want to remark that a single vector $b \in \mathscr{B}$, that is neither contained in $\mathscr{A}$ nor dominated by elements of $\mathscr{A}$ is enough to conclude that $\mathscr{B}$ is not weakly out–performed by $\mathscr{A}$. Actually, the same condition is enough to consider that $\mathscr{B}$ is not strongly or completely out–performed by $\mathscr{A}$. This leads us to realize how limited is the inference power of the out–performance relations (O_C, O_S and O_W) when the approximations have a similar convergence.

For example, in Figure 1 left panel, we have two NSs $\mathscr{A}$ and $\mathscr{B}$. Suppose that both $\mathscr{A}$ and $\mathscr{B}$ are subsets of $\mathscr{PF}$. It is evident that $\mathscr{A}$ is better than $\mathscr{B}$, because $\mathscr{A}$ gives more information and it is more likely for the DM to find a better solution in $\mathscr{A}$, or to clarify better his preferences from the information in $\mathscr{A}$. Amazingly, according to the out–performance relations mentioned above, $\mathscr{A}$ and $\mathscr{B}$ are not comparable. $\mathscr{B}$ does not outperform $\mathscr{A}$ because there exist many elements in $\mathscr{A}$ that neither are in $\mathscr{B}$ nor are dominated by any element of $\mathscr{B}$. $\mathscr{A}$ does not outperform $\mathscr{B}$ because there exists an element in $\mathscr{B}$, b, that neither is in $\mathscr{A}$ nor is dominated by any element of $\mathscr{A}$. This example can be made more extreme just by adding more elements to $\mathscr{A}$ that neither dominate nor are equal to b. The most extreme example is shown in Figure 1 right panel. $\mathscr{A}$ is represented by the line, and $\mathscr{B}$ only have one element, b. Suppose that b is not in $\mathscr{A}$ and both $\mathscr{A}$ and $\mathscr{B}$ are subsets of $\mathscr{PF}$. Now we have a very significative difference in quality between the NSs, but according to the out–performance relations mentioned above, these NSs are not comparable. This lead us to the following remark:

Remark 1. *There are pairs of approximations with a considerable difference in quality, that are considered, by the out–performance relations, as not comparable.*

Even with these limitations, the out–performance relations represent a minimum of what is desired from a quality indicators. If a $\mathscr{A}$ out–performs $\mathscr{B}$, we want to see this situation reflected in a quality indicator. Hansen and Jaszkiewicz [12] define the following property:

Definition 4. *A quality indicator I is compatible with an out–performance relation O (where O is any of O_W, O_C or O_S) when I always evaluates $\mathscr{A}$ as better than $\mathscr{B}$ if $\mathscr{A}$ O $\mathscr{B}$. In mathematical notation, I is compatible with O when $\mathscr{A}$ O $\mathscr{B}$ $\Rightarrow I(\mathscr{A} > \mathscr{B})$.*

We denote by $I(\mathscr{A} > \mathscr{B})$ that the quality indicator I evaluate $\mathscr{A}$ as better than $\mathscr{B}$. An indicator compatible with the out–performance relations is more robust to misleading cases, and it is desirable to design and use quality indicators that are compatible with these relations. The compatibility with O_W is the most desirable one, because a quality indicator compatible with O_W is also compatible with O_S and O_C. Both UQIs and BQIs can be compatible with the out–performance relations. For example, the S-metric [10], a UQI, is compatible with O_W.

Another influential study on quality indicators was written by Zitzler et al. [1]. In that study, the authors strongly relied on the work of Hansen and Jaszkiewicz [12] and went forward with the objective of analyzing how useful current comparison methods are. They considered the scenario where more than one quality indicator may be used to evaluate or compare the quality of non–dominated sets. Under this scenario, it may be necessary to introduce an interpretation function E. So, when comparing two approximations, we use E to interpret the results of all the quality indicators. The result is a formal definition of a comparison method:

Definition 5. *Let $\mathbf{I} = (I_1, I_2, ..., I_k)$ be a k–tuple of quality indicators. Let $\mathbf{I}(\mathscr{A})$ be a vector of real values $\langle I_1(\mathscr{A}), I_2(\mathscr{A}), ..., I_k(\mathscr{A}) \rangle$ generated by a list $\mathbf{I}$ of UQIs. Let $\mathbf{I}(\mathscr{A}, \mathscr{B})$ be a vector of real values $\langle I_1(\mathscr{A}, \mathscr{B}), I_2(\mathscr{A}, \mathscr{B}), ..., I_k(\mathscr{A}, \mathscr{B}) \rangle$ generated by a list $\mathbf{I}$ of binary quality indicators. Let $E : \mathbf{R}^k \times \mathbf{R}^k \to \{false, true\}$ a Boolean function which takes two real vectors of length k as arguments. When $\mathbf{I}$ consist of unary quality indicators only, a comparison method $C_{\mathbf{I},E}$ is a function in the form:*

$$C_{\mathbf{I},E}(\mathscr{A}, \mathscr{B}) = E(\mathbf{I}(\mathscr{A}), \mathbf{I}(\mathscr{B})) \tag{3}$$

When $\mathbf{I}$ consists of binary quality indicators only, a comparison method $C_{\mathbf{I},E}$ is a function in the form:

$$C_{\mathbf{I},E}(\mathscr{A}, \mathscr{B}) = E(\mathbf{I}(\mathscr{A}, \mathscr{B}), \mathbf{I}(\mathscr{B}, \mathscr{A})). \tag{4}$$

If $C_{\mathbf{I},E}(\mathscr{A}, \mathscr{B})$ is true, then $\mathscr{A}$ is better than $\mathscr{B}$, according to $C_{\mathbf{I},E}$. And if $C_{\mathbf{I},E}(\mathscr{A}, \mathscr{B})$ is false, then $\mathscr{A}$ is not better than $\mathscr{B}$ according to $C_{\mathbf{I},E}$.

Definition 5 establishes a more formal way to define a methodology to compare two non–dominated sets. In [1] is defined a property known as "Compatibility and Completeness"[1], similar to the property of compatibility with the out–performance relations described before.

Definition 6. *A comparison method $C_{I,E}$ is "Compatible and Complete" with respect to an out–performance relation O when $C_{I,E}(\mathscr{A}, \mathscr{B}) = true$ if and only if $\mathscr{A}\ O\ \mathscr{B}$.*

So, a comparison method is "Compatible and Complete" with respect an out–performance relation O[2] if and only if $\forall \mathscr{A}, \mathscr{B}(\mathscr{A}\ O\ \mathscr{B} \Leftrightarrow C_{I,E}(\mathscr{A}, \mathscr{B}) = true)$. "Compatibility and Completeness" transforms the implication of Definition 4 into a co–implication.

In [1] it is stated that a comparison method must be "Compatible and Complete", because only these methods can detect whether one approximation is better than another. Based on the study of Hansen and Jaszkiewicz [12] it is clear that a comparison method must evaluate $\mathscr{A}$ as better than $\mathscr{B}$ if $\mathscr{A}\ O\ \mathscr{B}$, but what is the justification to consider that a comparison method must evaluate $\mathscr{A}$ as better than $\mathscr{B}$ only if $\mathscr{A}\ O\ \mathscr{B}$? The justification also comes from [12], where it is stated that when two non–dominated sets are not comparable under the out–performance relations, it is not possible to decide what NS is better than the other under General Assumption 1. In that case more assumptions about DM's preferences must be used in order to find differences between the sets.

From the framework just described, several important conclusions are derived. Maybe the most important conclusion is the affirmation that unary comparison methods have a limited inference power. This conclusion is derived from the fact that a comparison method based on a finite number of unary quality indicators cannot be "Compatible and Complete" (see the demonstration in [1]). Many of the quality indicators available in the bibliography were revised and it was found that most quality indicators are not "Compatible and Complete".

Unfortunately, "Compatible and Complete" comparison methods (CCCMs) have the same limitations in inference power of the out–performance relations. Recalling the example of Figure 1, where $\mathscr{A}$ and $\mathscr{B}$ do not out–perform each other, a "Compatible and Complete" comparison method, by definition, considers these sets as non–comparable. This is a disadvantage; usually researchers want a comparison method to be able to identify such considerable differences in the quality of the approximations. But, by definition, a CCCM cannot detect any difference when the approximations are not comparable under the out–performance relations.

---

[1] Zitzler et al. [1] define two properties. One property is called "Compatibility" and the other one is called "Completeness". Then, they state that a comparison method must be both compatible and complete. In this work we refer to the property of being both compatible and complete as "compatibility and completeness".

[2] In [1], the property of "Compatibility and Completeness" is defined for any binary relation between approximations. We focus only on the out–performance relations.

Imagine the following scenario. Suppose we need to present the DM with a NS to make important decisions and that we need to choose between the two NSs in Figure 1, left or right panel. A CCCM is suppose to detect whether an approximation is better than another, but in this case a CCCM will consider $\mathscr{A}$ and $\mathscr{B}$ as not comparable. This may lead us to believe that it is indifferent which set we choose and we may end up presenting the DM with $\mathscr{B}$. But, recalling the Goal of Multi–objective Optimization 1, it is evident that $\mathscr{A}$ is much better than $\mathscr{B}$, so a CCCM can be misleading in some cases. As a result, we can state the following remark:

Remark 2. *"Compatibility and Completeness" not only is not necessary but it is an undesirable property for quality indicators. Quality indicators that are "Compatible and Complete" do not reflect properly, in some cases, the natural notion of when one approximation is better than another.*

The problem is that when "Compatibility and Completeness" was designed, it was assumed that the out–performance between two approximation is sufficient and *necessary* to conclude that one approximation is better than another. Based on Remark 1, we realize that this affirmation is incorrect, the out–performance between two approximations is sufficient but not necessary to consider that one approximation is better than another.

4   Why Unary Quality Indicators Are Not Inferior to Binary Quality Indicators

One of the most important conclusions from [1] is that UQIs are inferior to BQIs. First it was demonstrated that UQIs cannot be "Compatible and Complete". Then it was mentioned that BQIs can be "Compatible and Complete". Finally, the chain of reasoning to arrive to that conclusion is the following:

1. A quality indicator must be "Compatible and Complete".
2. Unary Quality Indicators cannot be "Compatible and Complete".
3. Binary Quality Indicators can be "Compatible and Complete".
4. So, Unary Quality Indicators are inferior to Binary Quality Indicators.

Based in the discussion presented in the previous section and in Remark 2, we know that the first statement in the list above is incorrect. "Compatibility and Completeness" is an undesirable property and quality indicators must not be "Compatible and Complete". As a result, we cannot conclude that UQIs are inferior to BQIs. This has very important implications, for several years researches have seen UQIs suspiciously because they are supposed to be inferior. Now we know that they are not.

5   Other Topics of Interest

The reason why Ziztler et al. [1] consider that a quality indicator must be "Compatible and Complete" is that they assume that we cannot consider more preferences than those derived from General Assumption 1. Otherwise, we introduce biases to a quality indicator. We show in this work that "Compatible and Complete" quality indicators have important limitations in their inference power. Something is lacking in a "Compatible and Complete" comparison method. This suggest that General Assumption 1 is not enough to properly model the natural notion of when one approximation is better than another one. The interesting question is: what other information can we consider in a quality indicator to improve its evaluations without introducing important biases? Some interesting directions for future research is to identify and characterize that information. And, to find a proper way to introduce this information into a quality indicator.

The quality of a non–dominated set is usually divided into two characteristics: convergence and diversity. Convergence is related to how near an approximation is (in objective function space) to the Pareto Front. Diversity is related to how well distributed the vectors in the approximation are (in objective function space) among the Pareto Front. General Assumption 1, dominance and the out–performance relations are closely related to convergence. This suggest that the information that we also need to consider in a quality indicator is related to diversity.

There are quality indicators designed to evaluate specific aspects of quality of a non–dominated set. Some quality indicators evaluate convergence, other quality indicators evaluate diversity. When we evaluate convergence and diversity independently, it is important to combine the two evaluations to obtain a general evaluation of an approximation set. There exist one approach that consists on considering convergence and diversity as two independent goals. So we can create a secondary Multi–objective Problem with two objective functions: diversity and convergence. This way, if in the future we obtain the right trade–off between convergence and diversity (maybe from the DM), we can detect the best non–dominated set based on its convergence and diversity. We must be careful when using this approach for the following reasons:

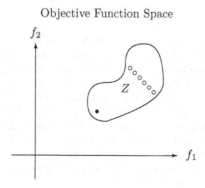

Fig. 2. $\mathscr{A}$ (•) has better convergence but $\mathscr{B}$ (○) has better diversity

1. Considering convergence and diversity as two different goals to optimize does not solve our problem. We want to choose the best one from two sets of vectors. Measuring the quality of an approximation with two characteristics (convergence and diversity) leave us almost at the same point, we still can not decide in all cases which approximation is the best one when they are uncomparable.

2. We, as analysts, are suppose to provide the DM with a good set of solutions so a good decision can be made. Choosing a good approximation is a problem that the analyst must solve, we are not supposed to pass on the problem to the Decision Maker. The DM can tells us what solutions of a set are the best, but not he is not supposed to know about convergence and diversity.

3. If convergence and diversity are considered as two non comparable goals, we can arrive at contradictions. Consider the two NSs in Figure 2. We see that $\mathscr{A}$ has better convergence, while $\mathscr{B}$ has better diversity. So, we have that they are not comparable, but this is a contradiction to General Assumption 1, because all vectors in $\mathscr{B}$ are dominated by vectors in $\mathscr{A}$. If we assume that these sets are not comparable, we are implying that for some DM preferences $\mathscr{B}$ has better solutions than $\mathscr{A}$, but this means that the DM may prefer dominated solutions over non–dominated ones.

6   Conclusions

In this work we demonstrate that Unary Quality Indicators are not inferior to Binary Quality Indicators. First, we show the limitations of the out–performance relations when detecting the differences in quality for some approximation sets. Next, we prove that a property known as "Compatibility and Completeness", contrary to what is widely believed, is an undesirable property for quality indicators. Finally, UQIs are considered inferior because they cannot have the property of "Compatibility and Completeness". But if this property is undesirable, then there is no reason to consider UQIs as inferior.

References

1. Zitzler, E., Thiele, L., Laumanns, M., Fonseca, C.M., da–Fonseca, V.G.: Performance assessment of multiobjective optimizers: An analysis and review. IEEE Transactions on Evolutionary Computation 7(2), 529–533 (2003)
2. Nocedal, J., Wright, S.: Numerical Optimization. Springer, New York (1999)
3. Miettinen, K.: Some methods for nonlinear multi–objective optimization. In: Zitzler, E., Deb, K., Thiele, L., Coello Coello, C.A., Corne, D.W. (eds.) EMO 2001. LNCS, vol. 1993, pp. 1–20. Springer, Heidelberg (2001)
4. Darwin, C.: On the Origin of Species by Means of Natural Selection. John Murray, London (1859)
5. Deb, K.: Multi-objective Optimization Using Evolutionary Algorithms. John Wiley and Sons, Chichester (2001)
6. Coello, C.A., Van–Veldhuizen, D., Lamont, G.B.: Evolutionary Algorithms for Solving Multi-Objective Problems. Kluwer Academic/Plenum Publishers, New York (2002)

7. Laumanns, M., Thiele, L., Deb, K., Zitzler, E.: Combining convergence and diversity in evolutionary multi-objective optimization. Evolutionary Computation 10(3), 263–282 (2002)
8. Kalyanmoy, D., Agrawal, S., Pratab, A., Meyarivan, T.: A Fast Elitist Non-Dominated Sorting Genetic Algorithm for Multi-Objective Optimization: NSGA-II. In: Schoenauer, M., Deb, K., Rudolph, G., Yao, X., Lutton, E., Merelo, J.J., Schwefel, J.P. (eds.) PPSN 2000. LNCS, vol. 1917, pp. 849–858. Springer, Heidelberg (2000)
9. Zitzler, E., Laumanns, M., Thiele, L.: Spea 2: Improving the strength pareto evolutionary algorithm. Technical Report, Technical report 103 (2001)
10. Zitzler, E.: Evolutionary Algorithms Multiobjective Optimization: Methods and Applications. PhD thesis, Swiss Federal Institute of Technology, ETH (1999)
11. Veldhuizen, D.A.: Multiobjective Evolution Algorithms: Classifications, Analyses, and New Innovations. PhD thesis, Force Institute Technology, Wright Patterson AFB (1999)
12. Hansen, M.P., Jaszkiewicz, A.: Evaluating the quality of approximations to the non-dominated set. Technical Report IMM-REP-1998-7 (1998)

Using Copulas in Estimation of Distribution Algorithms

Rogelio Salinas-Gutiérrez, Arturo Hernández-Aguirre,
and Enrique R. Villa-Diharce

Centro de Investigación en Matemáticas, Guanajuato, México
{rsalinas,artha,villadi}@cimat.mx

Abstract. A new way of modeling probabilistic dependencies in Estimation of Distribution Algorithm (EDAs) is presented. By means of copulas it is possible to separate the structure of dependence from marginal distributions in a joint distribution. The use of copulas as a mechanism for modeling joint distributions and its application to EDAs is illustrated on several benchmark examples.

1 Introduction

Estimation of Distribution Algorithms (EDAs) [17] are recognized as a new paradigm in Evolutionary Computation to deal with optimization problems [15]. EDAs are a class of Evolutionary Algorithms (EAs) based on probabilistic models instead of genetic operators such as crossover and mutation. The use of probabilistic models allow us to explicitly represent: 1) dependencies between the decision variables; and 2) their structure. EDAs populate the next generation by simulating individuals from the probabilistic model, therefore, the goal is to transfer both the data dependencies and the structure found in the best individuals into the next population. A pseudocode for EDAs is shown in Algorithm 1.

Algorithm 1. Pseudocode for EDAs

1: assign $t \longleftarrow 0$
   generate the initial population P_0 with M individuals at random
2: select a collection of N solutions S_t, with $N < M$, from P_t
3: estimate a probabilistic model $\mathcal{M}_t$ from S_t
4: generate the new population by sampling from the distribution of S_t.
   assign $t \longleftarrow t + 1$
5: if stopping criterion is not reached go to step 2

As it can be seen in step 3, the interactions among the decision variables are taken into account through the estimated model. The possibility of incorporating the dependencies among the variables into the new population greatly modifies the performance of an EDA. Nowadays, several EDAs have been proposed for

A. Hernández Aguirre et al. (Eds.): MICAI 2009, LNAI 5845, pp. 658–668, 2009.

optimization problems in discrete and continuous domains. They can be grouped by the complexity of the probabilistic model used to learn the interactions between the variables. For instance, the UMDA [18,11,13] can be considered the most simple EDA because it does not take into account dependencies between the variables, while the BMDA [21] and MIMIC [6,11,13] just take dependencies between pairs of variables into account. Probabilistic models such as Bayesian networks and multivariate Gaussian distributions have been used by EDAs for multiple dependencies. Some examples in discrete domain are PADA [24], EBNA [8,12] and BOA [20]. For continuous domain EMNA [14] and EGNA [11,13] are EDAs based on multivariate and bivariate Gaussian distribution respectively. In this paper we deal with continuous optimization problems, and address the use of copulas in EDAs. One motivation of this work is to take advantage of the almost natural capacity of copulas to represent bivariate dependencies through concordance measures, such as Kendall's tau or Pearson's rho. The goal of the paper is to introduce copula functions and implement the copula-based MIMIC; to the best of our knowledge this paper would be one of the first studies on the performance of EDAs based on copulas. Related works have considered EDAs based on multidimensional Gaussian copula with nonparametric marginals [1,3] and EDAs based on two dimensional copulas with Gaussian marginals [27,26].

The structure of the paper is the following: Section 2 is a short introduction to copula functions, Sect. 3 describes the implementation of the MIMIC algorithm with copula functions. Section 4 presents the experimental setting to solve 5 test global optimization problems, and Sect. 5 resumes the conclusions.

2 Copula Functions

The concept of copula was introduced by Sklar [23] to separate the effect of dependence from the effect of marginal distributions in a joint distribution. The separation between marginal distributions and a dependence structure explains the modeling flexibility given by copulas and, for this reason, they have been widely used in many research and application areas such as finance [4,25], climate [22], oceanography [7], hydrology [9], geodesy [2], and reliability [16].

Definition 1. *A copula is a joint distribution function of standard uniform random variables. That is,*

$$C(u_1, \ldots, u_n) = Pr[U_1 \leq u_1, \ldots, U_n \leq u_n] \ ,$$

where $U_i \sim U(0,1)$ for $i = 1, \ldots, n$.

For a more formal definition of copulas, the reader is referred to [10,19]. The following result, known as Sklar's theorem, connects marginal distributions and copula with a joint distribution.

Theorem 1 (Sklar). *Let F be a n-dimensional distribution function with marginals $F_1, F_2, \ldots, F_n$, then there exists a copula C such that for all x in $\overline{\mathbb{R}}^n$,*

$$F(x_1, x_2, \ldots, x_n) = C(F_1(x_1), F_2(x_2), \ldots, F_n(x_n)) \ ,$$

where $\overline{\mathbb{R}}$ denotes the extended real line $[-\infty, \infty]$. If $F_1(x_1), F_2(x_2), \ldots, F_n(x_n)$ are all continuous, then C is unique. Otherwise, C is uniquely determined on $Ran(F_1) \times Ran(F_2) \times \cdots Ran(F_n)$, where Ran stands for the range.

According to Theorem 1, the n-dimensional density f can be represented as

$$f(x_1, x_2, \ldots, x_n) = f_1(x_1) \cdot f_2(x_2) \cdots f_n(x_n) \cdot c(F_1(x_1), F_2(x_2), \ldots, F_n(x_n)) \ ,$$

where $f_i(x_i)$ is the density of variable x_i and c is the density of the copula C. This result allows us to choose different marginals and a dependence structure given by the copula and then merge them to build a multivariate distribution. This contrasts with the usual way to construct multivariate distributions, which suffers from the restriction that the margins are usually of the same type.

There are many families of copulas and each of them is characterized by a parameter or a vector of parameters. These parameters measure dependence between the marginals and are called *dependence parameters* $\boldsymbol{\theta}$. In this paper we use bivariate copulas with one dependence parameter θ. The dependence parameter is related to Kendall's tau through the equation (see [19])

$$\tau(X_1, X_2) = 4 \int_0^1 \int_0^1 C(u_1, u_2; \theta) dC(u_1, u_2; \theta) - 1 \ . \tag{1}$$

Kendall's tau measures the concordance between two continuous random variables X_1 and X_2. Table 1 shows the defining equations of the Frank copula and the Gaussian copula. Observe how the dependence parameter in the copula function is related to the Kendall's tau. The dependence parameter of a bivariate copula can be estimated using the maximum likelihood method. To do so, we need to maximize the *log-likelihood* function given by

$$l(\theta) = \sum_{t=1}^T \ln c(F(x_{1t}), F(x_{2t}); \theta) \ ,$$

where T is the sample size. The value θ which maximizes the log-likelihood is called *maximum likelihood estimator* $\widehat{\theta}_{\text{MLE}}$. Once the value of θ is estimated, the bivariate copula is well defined. For maximizing the likelihood function we use the nonparametric estimation of θ given by Kendall's tau in (1) as an initial approximation to $\widehat{\theta}_{\text{MLE}}$.

3   An EDA Based on Copula Functions

In order to show how a probabilistic model based on copulas can be used in EDAs we proposed an adaptation of the $MIMIC_C^G$ [11,13] with no Gaussian

Table 1. Bivariate copulas used in this paper

Copula's name	Description
Frank	Distribution: $C(u_1, u_2; \theta) = -\dfrac{1}{\theta}\ln\left(1 + \dfrac{(e^{-\theta u_1} - 1)(e^{-\theta u_2} - 1)}{e^{-\theta} - 1}\right)$
	Parameter: $\qquad\qquad\qquad \theta \in (-\infty, \infty)$
	Kendall's tau: $\qquad\qquad \tau = 1 - \dfrac{4}{\theta}[1 - D_1(\theta)],$ where $D_1(\theta) = \dfrac{1}{\theta}\int_0^\theta \dfrac{t}{e^t - 1}dt$
Gaussian	Distribution: $\qquad C(u_1, u_2; \theta) = \Phi_G\left(\Phi^{-1}(u_1), \Phi^{-1}(u_2)\right),$ where Φ_G is the standard bivariate normal distribution with correlation parameter θ
	Parameter: $\qquad\qquad\qquad \theta \in (-1, 1)$
	Kendall's tau: $\qquad\qquad \tau = \dfrac{2}{\pi}\arcsin(\theta)$
	Entropy: $\qquad\qquad H(U_1, U_2) = \dfrac{1}{2}\log(1 - \theta^2)$

assumption over univariate and bivariate density functions. Next, for completeness sake, we describe the principles of the $MIMIC_C^G$ learning algorithm.

Given a permutation of the numbers between 1 and n, $\pi = (i_1, i_2, \ldots, i_n)$ we define a class of density functions, $f_\pi(\boldsymbol{x})$:

$$f_\pi(\boldsymbol{x}) = f(x_{i_1}|x_{i_2}) \cdot f(x_{i_2}|x_{i_3}) \cdots f(x_{i_{n-1}}|x_{i_n}) \cdot f(x_{i_n}) \ . \tag{2}$$

Our goal is to choose the permutation π that minimizes the Kullback-Leibler divergence between the true density function $f(\boldsymbol{x})$ and the proposed density function $f_\pi(\boldsymbol{x})$:

$$D_{KL}\left(f(\boldsymbol{x})||f_\pi(\boldsymbol{x})\right) = E_{f(\boldsymbol{x})}\left[\log\frac{f(\boldsymbol{x})}{f_\pi(\boldsymbol{x})}\right] \ .$$

It is well known that conditional entropy $H(X|Y)$ and mutual information $I(X, Y)$ are related in the following way:

$$H(X|Y) = -I(X, Y) + H(X) \ ,$$

where $H(X) = -E_{f(x)}[\log f(x)]$ denotes the entropy of the continuous random variable X with density $f(x)$. The Kullback-Liebler divergence can be written as:

$$D_{KL}\left(f(\boldsymbol{x})\|f_\pi(\boldsymbol{x})\right) = -H(\boldsymbol{X}) + \sum_{k=1}^{n-1} H(X_{i_k}|X_{i_{k+1}}) + H(X_{i_n})$$

$$= -H(\boldsymbol{X}) + \sum_{k=1}^{n} H(X_{i_k}) - \sum_{k=1}^{n-1} I(X_{i_k}, X_{i_{k+1}}) \ .$$

The first two terms in the divergence do not depend on π. Therefore, minimize the Kullback-Leibler is equivalent to maximize

$$J_\pi(\boldsymbol{X}) = \sum_{k=1}^{n-1} I(X_{i_k}, X_{i_{k+1}}) \ ,$$

where

$$I(X_{i_k}, X_{i_{k+1}}) = E_{f(x_{i_k}, x_{i_{k+1}})}\left[\log \frac{f(x_{i_k}, x_{i_{k+1}})}{f(x_{i_k}) \cdot f(x_{i_{k+1}})}\right] \ .$$

According to [6], the optimal permutation π is the one that equivalently produces the highest pairwise mutual information with respect to the true distribution. But due to computational efficiency reasons we will employ the greedy algorithm originally proposed by [6] and adapted by [11]. Thus, the $MIMIC$ learning algorithm is based on a dependence test, and this is measured through mutual information. In this paper we will use the following fact (see [5]) between copula entropy and mutual information:

$$-E_{c(u_1, u_2)}[\log c(u_1, u_2)] = -E_{f(x_1, x_2)}\left[\log \frac{f(x_1, x_2)}{f(x_1) \cdot f(x_2)}\right]$$

$$H(U_1, U_2) = -I(X_1, X_2) \ ,$$

where $U_1 = F(X_1)$ and $U_2 = F(X_2)$.

Our proposed EDA uses two different dependence functions: a Frank copula and a Gaussian copula. These copulas are chosen because their dependence parameter have associated all range values of Kendall's tau. This means that negative and positive dependence between the marginals are considered in both copulas. However, they differ in the way they model extreme and centered values [25]. For instance, a Frank copula is mostly appropiate for data that exhibit weak dependence between extreme values and strong dependence between centered values. The proposed EDA estimates the copula entropy between each pair of variables in order to calculate the mutual information. The pair of variables with the largest mutual information are selected as the two first variables of the permutation π. The following variables of π are chosen according to their mutual information with respect to the previous variable. Algorithm 2 shows a straightforward greedy algorithm to find a permutation π.

Algorithm 2. Greedy algorithm to pick a permutation π

1: choose $(i_n, i_{n-1}) = \arg\max_{j \neq k} \widehat{I}(X_j, X_k)$, where $\widehat{I}()$ is an estimation of the mutual information between two variables.
2: choose $i_k = \arg\max_j \widehat{I}(X_{i_{k+1}}, X_j)$, where $j \neq i_{k+1}, \ldots, i_n$ and $k = n-1, n-2, \ldots, 2, 1$.

For a Gaussian copula there is a direct way to calculate its entropy and mutual information; for a Frank copula we estimate its entropy with a numerical approximation.

Once a permutation π is found, generating samples follows the order established by (2). In order to do it, we first sample variable $U_{i_n} \sim U(0,1)$ and then we sample variables $U_{i_k} \sim C(U_{i_k}|U_{i_{k+1}} = u_{i_{k+1}})$ from conditional copula of U_{i_k} given the value of $U_{i_{k+1}}$ for $k = n-1, \ldots, 1$. After that, we use values of U_i to find quantiles X_i through expression $X_i = F_{X_i}^{-1}(U_i)$.

It is important to say that, by means of copulas, we can write (2) as

$$f_\pi(\boldsymbol{x}) = \prod_{i=1}^{n} f(x_i) \cdot \prod_{k=1}^{n-1} c(u_{i_k}, u_{i_{k+1}}) \ . \tag{3}$$

where $u_{i_k} = F(x_{i_k})$ and $u_{i_{k+1}} = F(x_{i_{k+1}})$. This means that $MIMIC_C^G$ is a particular copula based EDA with Gaussian copulas $C(u_{i_k}, u_{i_{k+1}})$ and Gaussian marginals $F(x_i)$, for $k = n-1, \ldots, 1$ and $i = 1, \ldots, n$.

In this work we use Beta distributions as marginals. In order to estimate the parameters of the probabilistic model (3), we use the Inference Function for Margins method (IFM) [4]. This method is based on maximum likelihood and estimates first the parameters of marginals and then use them to estimate parameters of copulas. The test problems used in this paper have bounded search space. Each value of variable X_i from search space is transformed to a value in $(0,1)$ through a linear transformation. This explains why we use Beta distributions as marginals.

We summarize in Algorithm 3 the proposed approach. The main aspects, such as the estimation of the probabilistic model and the generation of the new population, are shown.

4   Experiments

We use three algorithms in order to optimize five test problems. The algorithms are $MIMIC_C^G$, copula based EDA using Frank copulas, and copula based EDA using Gaussian copulas. Table 2 shows the definition of the test problems used in the experiments: Ackley, Griewangk, Rastrigin, Rosenbrock, and Sphere functions. We use test problems in 10 dimensions. Each algorithm is run 30 times for each problem. The population size is 100. The maximum number of evaluations is 300,000. However, when convergence to a local minimum is detected the run is stopped. Any improvement less than 1×10^{-6} in 25 iterations is considered convergence. The goal is to reach the optimum with an error less than 1×10^{-6}.

Algorithm 3. Pseudocode for estimating model and generating new population

1: calculate pairwise mutual information using copula entropy
2: use greedy algorithm to pick a permutation (Algorithm 2)
3: calculate concordance measure Kendall's tau between variables in permutation π
4: obtain an initial approximation to the dependence parameter θ_τ using relationship with Kendall's tau (Table 1)
5: estimate marginal and copula parameters using Inference Function for Margins Method with θ_τ as initial approximation
6: simulate U_{i_n} from uniform distribution $U(0, 1)$
7: simulate U_{i_k} from conditional copula $C(U_{i_k}|U_{i_{k+1}})$, $k = n - 1, n - 2, \ldots, 2, 1$
8: determine X_i using quasi-inverse $F_{X_i}^{-1}(U_i)$, $i = 1, \ldots, n$

Table 2. Test functions

Name	Description	
Ackley	Function:	$F(\boldsymbol{x}) = -20 \cdot \exp\left(-0.2\sqrt{\dfrac{1}{n} \cdot \sum_{i=1}^{n} x_i^2}\right)$ $-\exp\left(\dfrac{1}{n} \cdot \sum_{i=1}^{n} \cos(2\pi x_i)\right) + 20 + \exp(1)$
	Search space:	$-10 \le x_i \le 10$, $i = 1, \ldots, 10$
	Minimum value:	$F(\boldsymbol{0}) = 0$
Griewangk	Function:	$F(\boldsymbol{x}) = 1 + \sum_{i=1}^{n} \dfrac{x_1^2}{4000} - \prod_{i=1}^{n} \cos\left(\dfrac{x_i}{\sqrt{i}}\right)$
	Search space:	$-600 \le x_i \le 600$, $i = 1, \ldots, 10$
	Minimum value:	$F(\boldsymbol{0}) = 0$
Rastrigin	Function:	$F(\boldsymbol{x}) = \sum_{i=1}^{n}(x_i^2 - 10\cos(2\pi x_i) + 10)$
	Search space:	$-5.12 \le x_i \le 5.12$, $i = 1, \ldots, 10$
	Minimum value:	$F(\boldsymbol{0}) = 0$
Rosenbrock	Function:	$F(\boldsymbol{x}) = \sum_{i=1}^{n-1}[100 \cdot (x_{i+1} - x_i^2)^2 + (1 - x_i)^2]$
	Search space:	$-10 \le x_i \le 10$, $i = 1, \ldots, 10$
	Minimum value:	$F(\boldsymbol{1}) = 0$
Sphere model	Function:	$F(\boldsymbol{x}) = \sum_{i=1}^{n} x_i^2$
	Search space:	$-600 \le x_i \le 600$, $i = 1, \ldots, 10$
	Minimum value:	$F(\boldsymbol{0}) = 0$

4.1   Numerical Results

In Table 3 we report the fitness value reached by the algorithms in all test functions. The information about the number of evaluations required by each algorithm is reported in Table 4.

To properly compare the performance of the algorithms (using the optimum value reached), we conducted a hypothesis test based on a Bootstrap method for the differences between the means of the three comparison pairs, for all test problems. Table 5 shows the confidence interval for the means, and the corresponding p-value.

Table 3. Descriptive fitness results for all test functions

Algorithm	Best	Median	Mean	Worst	Std. deviation
Ackley					
$MIMIC_C^G$	6.47E-007	8.65E-007	8.62E-007	9.97E-007	1.06E-007
Frank copula	5.79E-007	2.29E-006	3.06E-003	4.71E-002	9.31E-003
Gaussian copula	5.62E-007	9.07E-007	3.64E-006	7.80E-005	1.41E-005
Griewangk					
$MIMIC_C^G$	3.92E-007	8.66E-007	1.30E-003	3.88E-002	7.09E-003
Frank copula	4.30E-007	9.38E-007	2.99E-003	2.90E-002	6.81E-003
Gaussian copula	1.46E-007	8.11E-007	1.81E-002	4.31E-001	7.85E-002
Rastrigin					
$MIMIC_C^G$	4.17E-007	9.96E-001	3.37E+000	2.33E+001	6.24E+000
Frank copula	2.21E+000	4.99E+000	8.05E+000	3.69E+001	9.43E+000
Gaussian copula	7.49E-007	4.00E+000	5.48E+000	2.68E+001	5.35E+000
Rosenbrock					
$MIMIC_C^G$	7.31E+000	8.03E+000	8.89E+000	2.43E+001	3.17E+000
Frank copula	6.87E+000	7.83E+000	7.95E+000	9.69E+000	6.44E-001
Gaussian copula	6.26E+000	8.15E+000	8.53E+000	1.48E+001	1.78E+000
Sphere					
$MIMIC_C^G$	3.55E-007	7.00E-007	7.10E-007	9.86E-007	2.02E-007
Frank copula	3.39E-007	7.40E-007	3.03E-001	8.23E+000	1.50E+000
Gaussian copula	3.42E-007	8.92E-007	4.85E-001	1.22E+001	2.23E+000

Table 4. Descriptive function evaluations for all test functions

Algorithm	Mean	Std. deviation
Ackley		
$MIMIC_C^G$	7660.30	131.24
Frank copula	9310.30	1761.88
Gaussian copula	7657.00	675.63
Griewangk		
$MIMIC_C^G$	6927.70	1581.97
Frank copula	8343.40	2460.13
Gaussian copula	7835.20	2825.34
Rastrigin		
$MIMIC_C^G$	11788.60	3146.69
Frank copula	17055.40	5262.08
Gaussian copula	15408.70	4511.01
Rosenbrock		
$MIMIC_C^G$	12841.30	2665.61
Frank copula	14280.10	1355.85
Gaussian copula	14016.10	1666.29
Sphere		
$MIMIC_C^G$	6175.30	154.87
Frank copula	7069.60	2829.51
Gaussian copula	7874.80	3144.74

Table 5. Results for the difference between fitness means in each problem. A 95% interval confidence and a p-value are obtained through a Bootstrap technique.

Compared algorithms	95% Interval		p-value
Ackley			
$MIMIC_C^G$ vs. Frank copula	-6.15E-03	-7.37E-04	8.13E-02
$MIMIC_C^G$ vs. Gaussian copula	-7.89E-06	1.69E-08	1.94E-01
Frank copula vs. Gaussian copula	7.28E-04	6.14E-03	8.17E-02
Griewangk			
$MIMIC_C^G$ vs. Frank copula	-4.47E-03	1.29E-03	3.26E-01
$MIMIC_C^G$ vs. Gaussian copula	-4.50E-02	-4.33E-04	1.62E-01
Frank copula vs. Gaussian copula	-4.34E-02	1.37E-03	2.60E-01
Rastrigin			
$MIMIC_C^G$ vs. Frank copula	-8.11E+00	-1.48E+00	2.48E-02
$MIMIC_C^G$ vs. Gaussian copula	-4.49E+00	3.20E-01	1.60E-01
Frank copula vs. Gaussian copula	-5.09E-01	5.87E+00	1.89E-01
Rosenbrock			
$MIMIC_C^G$ vs. Frank copula	1.34E-01	2.01E+00	1.12E-01
$MIMIC_C^G$ vs. Gaussian copula	-6.13E-01	1.50E+00	5.68E-01
Frank copula vs. Gaussian copula	-1.18E+00	-6.44E-02	9.48E-02
Sphere			
$MIMIC_C^G$ vs. Frank copula	-8.51E-01	-1.16E-03	1.45E-01
$MIMIC_C^G$ vs. Gaussian copula	-1.28E+00	-1.72E-02	1.42E-01
Frank copula vs. Gaussian copula	-9.84E-01	5.46E-01	6.80E-01

4.2   Discussion

For the Ackley problem, intervals confidence show signicant differences between $MIMIC_C^G$ and Gaussian copula against Frank copula. This means that a dependence structure based on Gaussian copula is more adequate than a dependence structure based on Frank copula.

For the Griewangk problem, the algorithm that shows the best behaviour is $MIMIC_C^G$, closely followed by Frank copula algorithm. In this case, interval confidence between $MIMIC_C^G$ and Gaussian copula shows that better results are obtained using both Gaussian dependence structure and marginals.

The $MIMIC_C^G$ is the algorithm that performed best for the Rastrigin problem. For this problem there is statistically significant difference in the mean fitness between the $MIMIC_C^G$ and the Frank copula algorithms. Although results of Frank copula algorithm are not statistically different of Gaussian copula, is more suitable for this problem to choose a Gaussian structure than a Frank dependence. Respect to marginals distributions we can say something similar between Gaussian copula algorithm and $MIMIC_C^G$ in the sense that is more adequate to choose Gaussian marginals than Beta marginals.

For the Rosenbrock problem, the intervals confidence shows statistical differences between $MIMIC_C^G$ and Gaussian copula against Frank copula. In this case, a Frank dependence between marginals is more adecuate than a Gaussian strucuture between marginals. Fitness results between $MIMIC_C^G$ and Gaussian copula algorithm show no difference between Gaussian or Beta marginals if structure dependence is modeled by a Gaussian copula.

Finally, the fitness results for Sphere problem indicate that $MIMIC_C^G$ obtained the global minimum in all the executions. The selection of Gaussian structure and Gaussian marginals is more adecuate for this problem.

Regarding the number of fitness function evaluations, Table 4, the three algorithms performed in a similar way.

5   Conclusions

In this paper we introduce the use of copulas in EDAs. According to numerical experiments the selection of a copula for modeling structure dependence and the selection of marginals distributions can help achieving better fitness results. Although we use the same structure dependence and the same marginals for each algorithm, it is not necessary to do it. Fitness results are the result of the selected structure dependences and marginals.

The three algorithms performed very similar, however, more experiments are necessary with different probabilistic models in order to identify where the copula functions mean a clear advantage to EDAs.

References

1. Arderí-García, R.J.: Algoritmo con Estimación de Distribuciones con Cópula Gaussiana. Bachelor's thesis, Universidad de La Habana. La Habana, Cuba (2007) (in Spanish)
2. Bacigál, T., Komorníková, M.: Fitting Archimedean copulas to bivariate geodetic data. In: Rizzi, A., Vichi, M. (eds.) Compstat 2006 Proceedings in Computational Statistics, pp. 649–656. Physica-Verlag HD, Heidelberg (2006)
3. Barba-Moreno, S.E.: Una propuesta para EDAs no paramétricos. Master's thesis, Centro de Investigación en Matemáticas. Guanajuato, México (2007) (in Spanish)
4. Cherubini, U., Luciano, E., Vecchiato, W.: Copula Methods in Finance. Wiley, Chichester (2004)
5. Davy, M., Doucet, A.: Copulas: a new insight into positive time-frequency distributions. Signal Processing Letters, IEEE 10(7), 215–218 (2005)
6. De Bonet, J.S., Isbell, C.L., Viola, P.: MIMIC: Finding Optima by Estimating Probability Densities. In: Advances in Neural Information Processing Systems, vol. 9, pp. 424–430. The MIT Press, Cambridge (1997)
7. De-Waal, D.J., Van-Gelder, P.H.A.J.M.: Modelling of extreme wave heights and periods through copulas. Extremes 8(4), 345–356 (2005)
8. Etxeberria, R., Larrañaga, P.: Global optimization with Bayesian networks. In: Ochoa, A., Soto, M., Santana, R. (eds.) Second International Symposium on Artificial Intelligence, Adaptive Systems, CIMAF 1999, Academia, La Habana, pp. 332–339 (1999)
9. Genest, C., Favre, A.C.: Everything You Always Wanted to Know about Copula Modeling but Were Afraid to Ask. Journal of Hydrologic Engineering 12(4), 347–368 (2007)
10. Joe, H.: Multivariate models and dependence concepts. Chapman and Hall, London (1997)

668   R. Salinas-Gutiérrez, A. Hernández-Aguirre, and E.R. Villa-Diharce

11. Larrañaga, P., Etxeberria, R., Lozano, J.A., Peña, J.M.: Optimization by learning and simulation of Bayesian and Gaussian networks. Technical report KZZA-IK-4-99. Department of Computer Science and Artificial Intelligence, University of the Basque Country (1999)
12. Larrañaga, P., Etxeberria, R., Lozano, J.A., Peña, J.M.: Combinatorial optimization by learning and simulation of Bayesian networks. In: Proceedings of the Sixteenth Conference on Uncertainty in Artificial Intelligence, pp. 343–352 (2000)
13. Larrañaga, P., Etxeberria, R., Lozano, J.A., Peña, J.M.: Optimization in continuous domains by learning and simulation of Gaussian networks. In: Wu, A.S. (ed.) Proceedings of the 2000 Genetic and Evolutionary Computation Conference Workshop Program, pp. 201–204 (2000)
14. Larrañaga, P., Lozano, J.A., Bengoetxea, E.: Estimation of Distribution Algorithm based on multivariate normal and Gaussian networks. Technical report KZZA-IK-1-01. Department of Computer Science and Artificial Intelligence, University of the Basque Country (2001)
15. Larrañaga, P., Lozano, J.A.: Estimation of Distribution Algorithms: A New Tool for Evolutionary Computation. Kluwer Academic Publishers, Dordrecht (2002)
16. Monjardin, P.E.: Análisis de dependencia en tiempo de falla. Master's thesis, Centro de Investigación en Matemáticas. Guanajuato, México (2007) (in Spanish)
17. Mühlenbein, H., Paaß, G.: From recombination of genes to the estimation of distributions I. Binary parameters. In: Ebeling, W., Rechenberg, I., Voigt, H.-M., Schwefel, H.-P. (eds.) PPSN 1996. LNCS, vol. 1141, pp. 178–187. Springer, Heidelberg (1996)
18. Mühlenbein, H.: The Equation for Response to Selection and its Use for Prediction. Evolutionary Computation 5(3), 303–346 (1998)
19. Nelsen, R.B.: An Introduction to Copulas. Springer, Heidelberg (2006)
20. Pelikan, M., Goldberg, D.E., Cantú-Paz, E.: BOA: The Bayesian optimization algorithm. In: Banzhaf, W., Daida, J., Eiben, A.E., Garzon, M.H., Honavar, V., Jakiela, M., Smith, R.E. (eds.) Proceedings of the Genetic and Evolutionary Computation Conference GECCO 1999, vol. 1, pp. 525–532. Morgan Kaufmann Publishers, San Francisco (1999)
21. Pelikan, M., Mühlenbein, H.: The Bivariate Marginal Distribution Algorithm. In: Roy, R., Furuhashi, T., Chawdhry, P.K. (eds.) Advances in Soft Computing - Engineering Design and Manufacturing, pp. 521–535. Springer, Heidelberg (1999)
22. Schölzel, C., Friederichs, P.: Multivariate non-normally distributed random variables in climate research – introduction to the copula approach. Nonlinear Processes in Geophysics 15(5), 761–772 (2008)
23. Sklar, A.: Fonctions de répartition à n dimensions et leurs marges. Publications de l'Institut de Statistique de l'Université de Paris 8, 229–231 (1959)
24. Soto, M., Ochoa, A., Acid, S., de Campos, L.M.: Introducing the polytree approximation of distribution algorithm. In: Ochoa, A., Soto, M., Santana, R. (eds.) Second International Symposium on Artificial Intelligence, Adaptive Systems, CIMAF 1999, Academia, La Habana, pp. 360–367 (1999)
25. Trivedi, P.K., Zimmer, D.M.: Copula Modeling: An Introduction for Practitioners. In: vol.1 of Foundations and Trends® in Econometrics Now Publishers (2007)
26. Wang, L.F., Zeng, J.C., Hong, Y.: Estimation of Distribution Algorithm Based on Archimedean Copulas. In: GEC 2009: Proceedings of the first ACM/SIGEVO Summit on Genetic and Evolutionary Computation, pp. 993–996. ACM, New York (2009)
27. Wang, L.F., Zeng, J.C., Hong, Y.: Estimation of Distribution Algorithm Based on Copula Theory. In: Proceedings of the IEEE Congress on Evolutionary Computation, pp. 1057–1063. IEEE Press, Los Alamitos (2009)

Redistricting by Square Cells

Miguel Ángel Gutiérrez Andrade[1] and Eric Alfredo Rincón García[2]

1 Universidad Autónoma Metropolitana, Departamento de Ingeniería Eléctrica, Av. San Rafael
Atlixco 186 Col. Vicentina, 09340 México, D.F.
gamma@xanum.uam.mx
[2] Universidad Nacional Autónoma de México, Facultad de Ingeniería, DIMEI - Departamento
de Sistemas, Ciudad Universitaria, 04510 México, D.F.
caracol_loco@yahoo.com

Abstract. The design of electoral zones is a complex problem in which democracy of the electoral processes is promoted by some constraints such as population balance, contiguity and compactness. In fact, the computational complexity of zone design problems has been shown to be NP-Hard. This paper propose the use of a new measure of compactness, which uses a mesh formed with square cells to measure the quality of the electoral zones. Finally, a practical real case was chosen, which topographical settings causes some traditional measures of compactness to give very poor quality results, and was designed an algorithm based on simulated annealing that realizes a search in the space of feasible solutions. The results show that the new measure favors the creation of zones with straight forms and avoids twisted or dispersed figures, without an important effect to the population balance, which are considered zones of high quality.

Keywords: Redistricting, Compactness, Simulated Annealing, Gerrymandering, GIS.

1 Introduction

The zone design occurs when small areas or geographical basic units (GBU) are aggregated into zones which satisfy the requirements imposed by the studied problem. These requirements can include for example, the construction of connected zones, with the same amount of population, clients, mass media, public services, etc. The zone design is applied in diverse problems like school districting [1], [2], sales territory [3], [4], police district [5], service and maintenance zones [6], [7] and land use [8], [9].

Nevertheless, the design of electoral zones or electoral districting is the most known case, due to its influence in the results of electoral processes. In this case, the geographic units, which are administrative units, in general, are grouped into a predetermined number of zones or districts [10], [11]. In these problems, some restrictions like contiguity, population equality or compactness, must be satisfied to guarantee that elections are fair and competitive.

A. Hernández Aguirre et al. (Eds.): MICAI 2009, LNAI 5845, pp. 669–679, 2009.
© Springer-Verlag Berlin Heidelberg 2009

Thus, the zone design can be modeled as a combinatorial optimization problem, where the objective function measure population equality and compactness while contiguity is guaranteed by restrictions.

This paper proposes a measure of compactness that uses square cells to promote the creation of zones with straight forms. The measure was applied in a difficult real case and it was observed that achieves the proposed objective and has few conflicts with the population equality, resulting in high quality zones, both for their form and for the number of electorate that they contain.

2 Complexity

One of the reasons for which the zone design is a difficult problem, is due to the size of the solution space, which in the real problems, generally makes unfeasible explicitly enumerate all the feasible solutions. Even for a small number of geographical units and zones, the total number of possible arrangements is still enormous, for example, the number of solutions to divide n GBUs in k zones is given by the number of Stirling of the second type:

$$S(n,k) = \frac{1}{k!}\sum_{i=0}^{k}(-1)^i\left(\frac{k!}{(k-i)!i!}\right)(k-i)^n .$$ (1)

In special cases, it is possible to reduce the size of the solution space, for example, if we consider that the GBUs are connected in a chain and each one is contiguous to only two neighbors, the number of solutions is given by:

$$S'(n,k,r) = \frac{(n-1)!}{(k-1)!(n-k)!} .$$ (2)

Even, the following results have been demonstrated in terms of computational complexity [12], [13]:

Proposition 1. Creating equal population zones is NP-Hard.
Proposition 2. Creating minimum cost contiguous zones is NP-Hard.
Proposition 3. Creating an equal population contiguous zones plan is NP-Hard.
Proposition 4. Creating a maximally compact zone plan is NP-Hard.

Therefore, the design of contiguous zones with population equality and compactness will be a problem whose difficulty is at least NP-Hard. Thus, heuristic techniques as tabu search, simulated annealing, evolutionary algorithms or scatter search seems to be the better options to find good solutions in reasonable computational time.

3 Electoral Districting

Electoral districting is a problem that has been analyzed by its influence in the results of these processes. There have been proposed different criteria that regulate the construction of the districts to avoid any political interference, for example, in 1842 the congress of the United States passed the first law related to the construction of electoral districts, which required that all states use contiguous single member districts. Later, in order to guarantee the principle "one person, one vote", was added the concept of population equality. Nevertheless, these two criteria were insufficient to avoid a particular manipulation known as gerrymandering, characterized by the construction of zones that help or hinder particular constituents, such as members of a political, racial, or class group. Therefore, in 1901 was added the concept of geometric compactness to avoid the creation of zones with irregular or confused forms typical of the gerrymandering. At present, contiguity, population equality and compactness are considered essential criteria in the design of democratic electoral zones.

In the sixties, the computers were used to generate electoral zones and it was considered that automation of these processes could be an "antidote" to gerrymandering, [14]. In 1963 [15] and in 1965 [16] were developed the first algorithms and measures capable of performing automated redistricting considering contiguity, population equality and compactness. Since then many algorithms and measures have been proposed to qualify the zoning plans.

At present, there exist different types of algorithms, implemented in software packages, which include the objective function, that maximize the population equality and compactness, the restrictions, like contiguity, and the sequence of steps to get a final plan.

4 Compactness by Cells

As mentioned previously, contiguity, population equality and compactness are essential criteria in electoral districting and there are different proposals to measure them. Nevertheless, the proposed methods to measure population equality and to generate connected zones produce similar results. On the other hand, proposals to measure compactness may produce very different (and inconsistent in some cases) results [17], [18]. In this section we explain the concept of compactness by cells, which will be used in this paper to construct electoral zones.

First, we must see and analyze any figure as a set of square cells of the same size, Fig. 1.

Thus a new area and perimeter can be defined as follow:

Definition 1: The area in cells is the sum of the cells that form the figure.

Definition 2: The perimeter in cells is the sum of the sides that are in the outline of the figure

We observe that a twisted, elongated and dispersed zone will have a smaller area in cells and a larger perimeter in cells than the minimal rectangle, formed by square cells, which inscribes it, Fig. 2. Therefore, the maximum compactness will be obtained when the zone and the rectangle that inscribes it have, approximately, the same area and perimeter in cells. Thus, straight forms will be favored.

Formally, we define compactness as

$$\left(\frac{AC_R}{AC_Z}\right)\left(\frac{PC_Z}{PC_R}\right).\qquad(3)$$

Where:

AC_Z, is the area in cells of the zone

AC_R, is the area in cells of the minimum rectangle that inscribe the zone

PC_Z, is the perimeter in cells of the zone

PC_R, is the perimeter in cells of the minimum rectangle that inscribe the zone

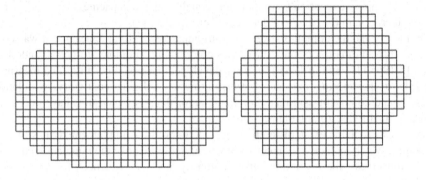

Fig. 1. Ellipse and hexagon formed by 470 and 426 square cells respectively

Thus, equation 3 will measure the difference between the area and the perimeter in cells of the zone and the minimum rectangle that inscribes it.

Finally, we observed that for all zones

$$\frac{PC_Z}{PC_R}\geq 1.\qquad(4)$$

Then the following adjustment is realized.

$$Comp = \left(\frac{AC_R}{AC_Z}\right)\left(\frac{PC_Z}{PC_R}-1\right).\qquad(5)$$

Thus the most compact zones will have values near to zero. We notice that the proposed equation requires few steps to be calculated, which will favor the performance of any algorithm.

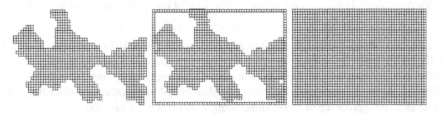

Fig. 2. Analyzed zone and minimal rectangle that inscribes it

5 Model

As mentioned previously, in this study we consider three of the most common criteria, equality, contiguity and compactness, which are included in a model to solve the zones design as a combinatorial optimization problem. In this model, we treat contiguity as a hard constraint and all other criteria through the minimization of a weighted additive multicriteria function.

To measure population equality we apply the definition of the Federal Electoral Institute of México (IFE) used in México's redistricting process in 2006.

$$\sum_{s\in R}\left(\frac{100P_T}{d_a(P_N/300)}\right)^2\left(\frac{P_s}{P_T}-\frac{1}{n}\right)^2 . \tag{6}$$

Where:

P_N, is the electoral population of México.

P_T, is the electoral population of the state.

P_s, is the electoral population of the zone s.

d_a, is the maximum percentage of deviation allowed for the entity.

$R = \{1, 2, 3,..., r\}$, r is the number of electoral zones that must be generated in the entity.

Thus, the lower the cost, the higher the population equality of a solution.

To measure the compactness we used the compactness in cells explained previously.

$$Comp = \left(\frac{AC_R}{AC_Z}\right)\left(\frac{PC_Z}{PC_R}-1\right) . \tag{7}$$

Where:

AC_Z, is the area in cells of the zone

AC_R, is the area in cells of the minimum rectangle that inscribe the zone

PC_Z, is the perimeter in cells of the zone

PC_R, is the perimeter in cells of the minimum rectangle that inscribe the zone

Also in this case, the most compact zones will have smaller values.

To guarantee contiguity of the zones we use the restrictions proposed by Shirabe [19]. These equations formulate contiguity based on a network flows programming. A zone is defined as a sub-network in which one node is like a sink and every other node provides one unit of supply. Thus, a zone is contiguous if the supply sent from every source arrives at the sink, without passing through the outside of the sub-network.

$$\sum_{\{j:(i,j)\in A\}} y_{ijs} - \sum_{\{j:(i,j)\in A\}} y_{jis} \ge x_{is} - Mw_{is} \quad \forall i\in I \;\forall s\in R . \tag{8}$$

$$\sum_{i\in I} w_{is} = 1 \quad \forall s\in R . \tag{9}$$

$$\sum_{\{j:(i,j)\in A\}} y_{jis} \leq (M-1)x_{is} \ \forall i \in I \ \forall s \in R \ . \tag{10}$$

$$x_{is} \in \{0,1\} \ \forall i \in I \ \forall s \in R \ . \tag{11}$$

$$w_{is} \in \{0,1\} \ \forall i \in I \ \forall s \in R \ . \tag{12}$$

$$y_{ijs} \geq 0 \ \forall (i,j) \in A \ . \tag{13}$$

Where:

$I = \{1, 2, 3, \ldots, n\}$, n is the number of GBUs

$R = \{1, 2, 3, \ldots, r\}$, r is the number of zones

$$x_{is} = \begin{cases} 1 & \text{if GBU } i \in s \\ 0 & \text{in other case} \end{cases} \quad i \in I, \ s \in R$$

$A = \{(i,j): \ i, j \text{ are adjacent pairs of GBUs}\}$

$$w_{is} = \begin{cases} 1 & \text{if } i \text{ is a sink in zone } s \\ 0 & \text{in other case} \end{cases} \quad i \in I, \ s \in R$$

y_{ijs}, is a nonnegative continuous decision variable indicating the amount of flow from GBU i to GBU j in zone s

M, is a non-negative integer indicating the maximum allowable number of GBUs that can be chosen for inclusion in a zone

We use the following notation in the model:

$Z_s = \{i : x_{is} = 1\}$, it is the set of GBUs in zone s

$P = \{Z_1, Z_2, Z_3, \ldots, Z_r\}$, is a zoning plan

$C_1(P)$, equality population cost for plan P

$C_2(P)$, compactness cost for plan P

α_1, α_2, weighting factors

Thus, any zone Z_s, is a set of GBUs, x_{is}, and the zoning plan P is a set of zones Z_1, Z_2, Z_3, ... Z_r.

The problem is then modeled as follows:

$$Minimize \ C(P) = \alpha_1 C_1(P) + \alpha_2 C_2(P) \ . \tag{14}$$

Subject to:

$$\sum_{i=1}^{n} x_{is} \geq 1 \ \forall s \in R \ . \tag{15}$$

$$\sum_{s=1}^{r} x_{is} = 1 \ \forall i \in I \ . \tag{16}$$

$$\sum_{\{j:(i,j)\in A\}} y_{ijs} - \sum_{\{j:(i,j)\in A\}} y_{jis} \geq x_{is} - Mw_{is} \ \ \forall i \in I \ \forall s \in R \ . \tag{17}$$

$$\sum_{i\in I} w_{is} = 1 \ \forall s \in R \ . \tag{18}$$

$$\sum_{\{j:(i,j)\in A\}} y_{jis} \leq (M-1)x_{is} \ \ \forall i \in I \ \ \forall s \in R \ . \tag{19}$$

$$x_{is} \in \{0,1\} \ \ \forall i \in I \ \forall s \in R \ . \tag{20}$$

$$w_{is} \in \{0,1\} \ \ \forall i \in I \ \forall s \in R \ . \tag{21}$$

$$y_{ijs} \geq 0 \ \forall (i,j) \in A \ . \tag{22}$$

In this formulation, the objective function measures population equality and compactness. Constraint (15) ensures that the number of zones is equal to R. Constraint (16) ensures that each basic unit is assigned to exactly one zone. Therefore, constraints (15) and (16) guarantee that any feasible plan is formed by the precise number of zones and that all the GBUs are included. Constraints (17), (18) y (19) ensure contiguity.

6   Simulated Annealing

Simulated annealing is a metaheuristic that was introduced in the early 1980s by Kirkpatrick [20]. Originally this concept was inspired on Metropolis's work in the field of statistical thermodynamics [21]. Metropolis modeled the process of annealing of solids by simulating the energy changes in a system of particles while temperature decreases, up to converging to a stable state. The simulated annealing algorithm starts with an initial solution and in each iteration a random neighbor solution is generated. If the neighbor solution improves the current objective's value, it is accepted as the current solution. If the neighbor solution does not improve the objective's value, then it is accepted with a probability given by:

$$p = \exp\left(-\frac{f(P_A)-f(P_B)}{T}\right) \ . \tag{23}$$

Where, $f(P_A)$ is the objective value of the current solution $f(P_B)$ is the objective value of the neighbor solution and T is a control parameter called temperature. These parameters are combined in a cooling schedule that specifies a decrement of the temperature, and a finite number of iterations at each value of the temperature.

At large values of the temperature, virtually all proposed solutions are accepted, and the algorithm can explore the space of solutions without a premature convergence. However, as the algorithm progresses, the temperature and the chance that an inferior solution displaces the current one gradually decreases, and begins a local search to find a superior solution in the neighborhood of the current solution.

The simulated annealing algorithm designed in this paper starts by constructing an initial feasible solution. First the algorithm selects random r GBUs, assigns them to different zones and marks them as not available GBUs. Later, the algorithm selects a random zone and generates an adjacent list of available GBUs. Then, the algorithm selects a random GBU of the list and includes it in the zone. Finally, the selected GBU is marked like not available. These instructions are realized until all the GBU are marked like not available. Thus, the initial solution has r connected zones that include all the GBUs.

A neighbor solution is constructed as follow. A random zone is chosen and a GBU in this zone is moved to a neighbor zone without creating a non-contiguous solution. Thus, a neighbor of a current solution is an identical feasible solution except that one GBU is reassigned from a zone to an adjacent zone.

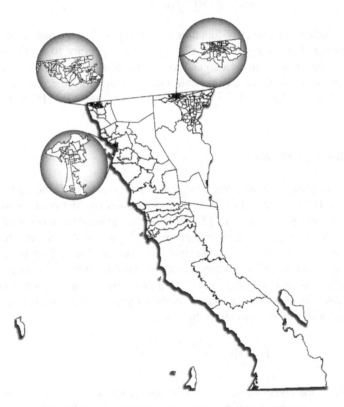

Fig. 3. Geographic basic units of Baja California Norte

7 Baja California Norte Case Study

The model and the compactness measure by cells just described were tested on the state of Baja California Norte, México, with 2,487,367 electors and 319 GBUs from which 8 zones must be created. In accordance with IFE requirements stipulated in Mexico's 2006 elections, we considered a maximum percentage of deviation $d_a = 15\%$

and according to the national population census in 2000, there were 97,483,412 electorates. Finally, we used the weighting factors $\alpha_1= 0.2$ and $\alpha_2= 0.8$.

The 319 GBUs that form the state of Baja California Norte are very different in shapes and sizes, as can be seen in Fig. 3. This variety complicates the design of compact zones.

Table 1. Costs of the zones created using simulated annealing

Zone	Electorates	Population Equality	Compactness
1	305 167	0.00278708	2.101813
2	306 001	0.00203768	0.407717
3	299 633	0.01072637	0.118237
4	302 098	0.00655313	1.894465
5	323 716	0.01378217	0.227546
6	332 889	0.04062698	0.152822
7	312 992	0.00036111	0.421003
8	304 871	0.00308121	0.368051

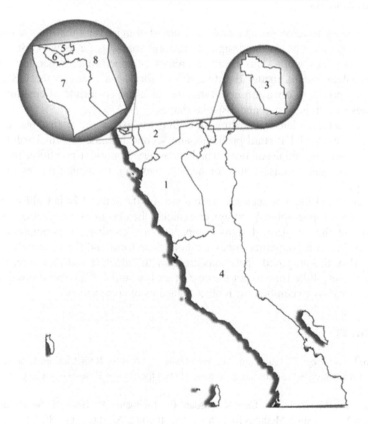

Fig. 4. Zones created using compactness by cells

We use the geographic information system Arcview to divide this state in square cells of 100m^2 and an scrip in avenue to get the data necessary to calculate the area and the perimeter in cells.

The metaheuristic algorithm uses these data to calculate the number of electorates, the area in cells and the perimeter in cells for each zone and the cost of each solution. Then, the algorithm decides if the new solutions are accepted or rejected according to the criteria of simulated annealing.

The results obtained are presented in Table 1. As can be observed the costs of population equality are lower than 1, and the most penalized zones are those which have a quantity of electorate far from the ideal one number, approximately 310,921.

The compactness criteria obtained costs are lower than 1 in the majority of the zones, except in zone 1 and 4, which get high costs due to the priority that is given to the population equality. However, the obtained zones correspond to the proposed objective, since the straight forms are favored avoiding twisted, elongated and dispersed figures, Fig. 4, while population equality is ensured.

8 Conclusions

In this paper we have proposed a new measure of compactness that uses square cells. The use of this measure in the design of electoral zones, as far as we know, does not appear in previous work. This measure creates compact zones regardless the zone design problem and no matter the diversity of forms and sizes of the GBU. In the redistricting problem, it was observed that this measure is capable of creating straight form zones avoiding twisted and disperse shapes.

We used the simulated annealing algorithm for two reasons: first because of its solving efficiency of NP-Hard problems and second because the final solution does not depend on the initial solution, which is a very important condition in electoral processes that seek to ensure the democracy, avoiding manipulations as the Gerrymandering.

The model and the new measure were tested on the state of Baja California Norte, Mexico, which topographical settings complicate the design of compact zones. It was observed that this measure designs compact zones keeping the population equilibrium, resulting in high quality zones for both their form and the electorate that they contain. Also, the proposed operations are easy to calculate and therefore, the algorithm needs very little time to get the cost of each solution. The operations defined in this paper are easy to combine with other measures of compactness.

References

1. Caro, F., Shirabe, T., Guignard, M., Weintraub, A.: School Redistricting: Embedding GIS Tools with Integer Programming. Journal of the Operational Research Society 55, 836–849 (2004)
2. DesJardins, M., Bulka, B., Carr, R., Jordan, E., Rheingans, P.: Heuristic Search and Information Visualization Methods for School Redistricting. AI Magazine 28(3), 59–72 (2006)
3. Ríos-Mercado, R.Z., Fernández, E.: A Reactive GRASP for Comercial Territory Design Problem with Multiple Balancing Requirements. Computers & Operations Research 36, 755–776 (2009)

4. Tavares-Pereira, F., Rui, J., Mousseau, V., Roy, B.: Multiple Criteria Districting Problems: The Public Transportation Network Pricing System of the Paris Region. Ann. Oper. Res. 154, 69–92 (2007)
5. D'Amico, S., Wang, S., Batta, R., Rump, C.: A Simulated Annealing Approach to Police District Design. Computers & Operations Research 29, 667–684 (2002)
6. Blais, M., Lapierre, S.D., Laporte, G.: Solving a Home-Care Districting Problem in an Urban Setting. Journal of the Operational Research Society 54, 1141–1147 (2003)
7. Shortt, N.K., Moore, A., Coombes, M., Wymer, C.: Defining Regions for Locality Health Care Planning: A Multidimensional Approach. Social Science & Medicine 60, 2715–2727 (2005)
8. Macmillan, W.: Redistricting in a GIS environment: An Optimisation Algorithm Using Switching-Points. Journal of geographical systems 3, 167–180 (2001)
9. Williams, J.C.: A Zero-One Programming Model for Contiguous Land Acquisition. Geographical Analysis 34(4), 330–349 (2002)
10. Bong, C.W., Wang, Y.C.: A Multi-Objective Hybrid Metaheuristic for Zone Definition Procedure. Int. J. Services Operations and Informatics. 1(1/2), 146–164 (2006)
11. Bozkaya, B., Erkut, E., Laporte, G.: A Tabu Search Heuristic and Adaptive Memory Procedure for Political Districting. European Journal of Operational Research 44, 12–26 (2003)
12. Altman, M.: Is Automation the Answer: The Computational Complexity of Automated Redistricting. Rutgers Computer and Law Technology Journal 23(1), 81–141 (1997)
13. Gilbert, K.C., Holmes, D.D., Rosenthal, R.E.: A Multiobjective Discrete Optimization Model for Land Allocation. Management Science 31(12), 1509–1522 (1985)
14. Vickrey, W.: On The Prevention of Gerrymandering. Political Science Quarterly 76(1), 105–110 (1961)
15. Weaver, J.B., Hess, S.W.: A Procedure for Nonpartisan Districting: Development of Computer Techniques. The Yale Law Journal 73(2), 288–308 (1963)
16. Hess, S.W., Weaver, J.B., Siegfeldt, H.J., Whelan, J.N., Zitlau, P.A.: Nonpartisan Political Redistricting by Computer. Operations Research 13(6), 998–1006 (1965)
17. Niemi, R.G., Grofman, B., Carlucci, C., Hofeller, T.: Measuring Compactness and the Role of a Compactness Standard in a Test for Partisan and Racial Gerrymandering. Journal of Politics 52(4), 1155–1181 (1990)
18. Young, H.P.: Measuring the Compactness of Legislative Districts. Legislative Studies Quarterly 13(1), 105–115 (1988)
19. Shirabe, T.: A Model of Contiguity for Spatial Unit Allocation. Geographical Analysis 37, 2–16 (2005)
20. Kirkpatrick, S., Gellat, C.D., Vecchi, M.P.: Optimization by Simulated Annealing. Science 220(4598), 671–680 (1983)
21. Metropolis, N., Rosenbluth, A.W., Rosenbluth, M.N., Teller, A.H., Teller, E.: Equation of State Calculation by Fast Computing Machines. Journal of Chemistry Physics 21, 1087–1091 (1953)

Finding Minimal Addition Chains with a Particle Swarm Optimization Algorithm

Alejandro León-Javier, Nareli Cruz-Cortés, Marco A. Moreno-Armendáriz, and Sandra Orantes-Jiménez

Center for Computing Research, National Polytechnic Institute (CIC-IPN), Mexico
leonjavier77@hotmail.com, {nareli,marco_moreno,dinora}@cic.ipn.mx

Abstract. The addition chains with minimal length are the basic block to the optimal computation of finite field exponentiations. It has very important applications in the areas of error-correcting codes and cryptography. However, obtaining the shortest addition chains for a given exponent is a NP-hard problem. In this work we propose the adaptation of a Particle Swarm Optimization algorithm to deal with this problem. Our proposal is tested on several exponents whose addition chains are considered hard to find. We obtained very promising results.

1 Introduction

The finite field exponentiations are the basic operations used in several well known public-key cryptosystems such as RSA, Diffie-Hellman and DSA [6]. Furthermore, they have important applications in the areas of error-correcting codes.

A finite field is a set having finitely many elements in which the usual arithmetic operations (addition, subtraction, multiplication, division by nonzero elements) are well defined.

Field exponentiation can be defined in terms of field multiplication as follows. Let B be an arbitrary element of a finite field F. Let e be defined as an arbitrary positive integer. Then, field exponentiation of an element B raised to the power e is defined as the problem of finding an element $C \in F$ such that,

$$C = B^e mod P \qquad (1)$$

where P is a large prime or an irreducible polynomial.

One manner of reducing the computation of Equation (1) is to reduce the total number of multiplications as much as possible. Thus, for this work, we will assume that arbitrary choices of the base element B are considered but the exponent e has been previously fixed.

Indeed, the problem of computing powers of a base element B is equivalent to an addition calculations. Thus, the field exponentiation problem can be established through the concept of *addition chains*. The concept of an addition chain for a given exponent e can be informally defined in the following manner. An addition chain for e of length l is a sequence U of positive integers,

A. Hernández Aguirre et al. (Eds.): MICAI 2009, LNAI 5845, pp. 680–691, 2009.

$u_0 = 1, u_1, ..., u_l = e$ such that for each $i > 1, u_i = u_j + u_k$ for some j and k with $0 \leq j \leq k < i$.

For example, if we need to find B^{60} one option would be to compute $B \times B \times, ..., \times B$ 60 times. For instance, a more efficient manner to compute this exponentiation is through the addition chain 1-2-4-6-12-24-30-60 that leads to the following scheme:

$$B^1$$
$$B^2 = B^1 B^1$$
$$B^4 = B^2 B^2$$
$$B^6 = B^4 B^2$$
$$B^{12} = B^6 B^6$$
$$B^{24} = B^{12} B^{12}$$
$$B^{30} = B^{24} B^6$$
$$B^{60} = B^{30} B^{30}$$

Notice that the exponents in the left side of each equation conform the addition chains and they determine the manner how the operations are performed.

Another possible addition chain could be 1-2-4-8-10-20-40-60 which leads to the next scheme:

$$B^1$$
$$B^2 = B^1 B^1$$
$$B^4 = B^2 B^2$$
$$B^8 = B^4 B^4$$
$$B^{10} = B^8 B^2$$
$$B^{20} = B^{10} B^{10}$$
$$B^{40} = B^{20} B^{20}$$
$$B^{60} = B^{40} B^{20}$$

However, it has been proved that the problem of finding the shortest addition chain for a given exponent is a NP-hard problem. On the other hand, the Particle Swarm Optimization (PSO) algorithm is a heuristic strategy that has been successfully applied to a wide variety of optimization problems. In this work we propose the adaptation of an PSO algorithm to find short addition chains. We experimented with different exponents obtaining very competitive results.

The rest of this paper is organized as follows: In Section 2 the Problem Statement is provided, in Section 3 some previous Related Work is explained. In Section 4 we give some ideas about the basic Particle Swarm Optimization (PSO) algorithm. Section 5 describes our proposed PSO algorithm to find short addition chains. In Section 6 the Experiments and Results are presented. Finally, in Section 7 some Conclusions and Future work are drawn.

2   Problem Statement

The problem we want to solve is to find the shortest *addition chain* for a given exponent e, that is, the addition chain with the lowest length l.

An addition chain U with length l is defined as a sequence of positive integers $U = u_1, u_2, \ldots, u_i, \ldots, u_l$, with $u_1 = 1$, $u_2 = 2$ and $u_l = e$, and $u_{i-1} < u_i < u_{i+1} < u_l$, where each u_i is obtained by adding two previous elements $u_i = u_j + u_k$ with $j, k < i$ for $i > 2$. Notice that j and k are not necessarily different.

3   Related Work

There exists several deterministic and probabilistic algorithms to find short addition chains, some examples of them were proposed in: [5,9,14,13,11,1,10].

Next, very roughly we will explain some of them classified into deterministic and probabilistic methods:

3.1   Deterministic Methods

Binary Strategies

The binary strategies expand the exponent e to its binary representation with length m. The binary chain is scanned bit by bit (from the left to right or vice versa) and a squaring is performed at each iteration, in addition to that, if the scanned value is *one* then a multiplication is executed. This algorithm requires a total of m-1 squaring operations and $H(e)$-1 field multiplications, where $H(e)$ is the Hamming weight of the binary representation of e.

Window Strategies

The binary method can be generalized by scanning more than one bit at a time. Then, the window method scans k bit at a time. The window method is based on a k-ary expansion of the exponent, where the bits of the exponent e are divided into k-bit words or digits. The resulting words are scanned performing consecutive squarings and a subsequent multiplication as needed. For $k=1, 2, 3$ and 4 the window method is called, binary, quaternary, octary and hexa exponentiation method, respectively. A more detailed explanation can be found in [9].

Adaptive Window Strategy

This strategy is very useful for very large exponents, let us say exponents e with binary representation grater than 128 bits. For this strategy the exponent is partitioned into a series of variable-length zero and non-zero word called windows. This algorithm provides a performance trade-off allowing the processing of zero and non-zero words. The main goal in this strategy is to maximize the number and length of zero words with relatively large values of k. This method significantly outperforms the Window method mentioned before.

3.2   Approaches Based on Evolutionary Heuristics

Very recently the problem of finding short addition chains has attracted the attention of the Evolutionary Computation's researchers.

In the specialized literature we can find the usage of Genetic Algorithms and Artificial Immune Systems to find addition chains for moderated size exponents. For these cases the problem is established as an optimization one, whose goal is to minimize the addition chains's length.

Genetic Algorithms

In [4] it was proposed a Genetic Algorithm to find short addition chains. The individuals represent the addition chains. They are initialized in such manner that they are legal chains (feasible individuals). The genetic operators (crossover and mutation) were specially designed to preserve the individuals's feasibility. It was tested on some small exponents obtaining better results than some deterministic approaches, however it could not find the optimal values in most of the tested exponents.

Artificial Immune System

Another recent approach was presented in [3,2]. Its authors adapted an Artificial Immune System algorithm to optimization to find minimal addition chains. This is a population-based approach. The population is conformed by a set of *antibodies* representing addition chains that are potential solutions to the problem. It is based on the Clonal Selection Principle that establishes that the best antibodies (regarding to a objective function, the addition chains's length in this case) are reproduced by cloning themselves. Then, the new clones are subjected to a mutation process. For this case, the mutation operator maintains the addition chains's legacy. This process is iteratively applied.

This approach was tested on exponents with moderated size obtaining better results than the previously published methods.

4 Particle Swarm Optimization Algorithm

Particle Swarm Optimization (PSO) is a population-based optimization technique inspired by the movements of a flock of birds or fishes. It was proposed by Kennedy and Eberhart in 1995 [8]. The PSO has become very popular mainly because its effectiveness and simplicity. It has been very successful by solving diverse optimization problems such as single [15] and multi-objective numerical [12,16] and combinatorial problems [7].

In the PSO the potential solutions to the problem are called *particles*. The set of all particles is named *swarm* (or population). The particles have a kind of *memory* capable of remembering the particle's best position visited so far (named *pb*), and its neighborhood's best position (known as *lb*). A neighborhood is a set of particles whose "closeness" is defined under certain criterion. In the most extreme case, a neighborhood's size is equal to the whole *swarm*.

The particles modify their position according to a velocity value v. This velocity value is influenced by the particle's best position they have been so far and its best neighbor. For numerical optimization, the velocity v_{id} and position x_{id} of a particle i for the dimension d, are updated according to the following formulae:

$$v_{id} = w \times v_{id} + c_1 r_1 (pb_{id} - x_{id}) + c_2 r_2 (lb_{id} - x_{id}) \tag{2}$$

$$x_{id} = x_{id} + v_{id} \tag{3}$$

where c_1 and c_2 are positive constants, r_1 and r_2 are random numbers with uniform distribution in the range $[0, 1]$ and w is the inertia weight in the same range. The particles's initial position and velocities are randomly determined.

5   A Particle Swarm Optimization Algorithm to Find Short Addition Chains

Our proposal is to adapt a PSO algorithm to find short addition chains.
Thus, for a given exponent e, the algorithm will look for the shortest addition chain. Next, we will explain our algorithm in some detail.

5.1   Particle's Representation

For this algorithm we will only handle feasible addition chains, that is, *legal* addition chains that follow the rules mentioned in Section 2.

Each particle is directly represented by the addition chain. Then, a particle is a chain whose elements correspond to the ones from the addition chain (notice that they must keep the same order).

For example, for the exponent $e = 79$, a possible addition chain is the following:
$1 \rightarrow 2 \rightarrow 4 \rightarrow 6 \rightarrow 12 \rightarrow 24 \rightarrow 30 \rightarrow 60 \rightarrow 72 \rightarrow 78 \rightarrow 79$
then, a particle is represented by the same chain. Because there is no difference between addition chains and particles, in this work we will name them indistinctly.

5.2   Particle's Initialization

The initial particles are randomly generated, however as we mentioned before, they must be *legal* addition chains.

The pseudo-code to generate particles for a given exponent e is shown next:

Let us call $p_n i$ the i-th element in the n-th particle p.

(1)   For $n = 1$ to $n = PopulationSize$ do
(2)   Set $p_{n1} = 1$
(3)   Set $p_{n2} = 2$
(4)   For $i = 3$ to $p_{ni} = e$ do
(5)   Do
(6)   Randomly select k with $1 \leq k < i$
(7) $p_i = p_k + p_{i-1}$
(8)   while $(p_i > e)$
(9)   End-For
(10)   End-For

5.3   Fitness Function

The fitness value for a particle is determined by its length, that is the fitness value is equal to the chain's length. Thus, in our example for the exponent $e = 79$ and a given particle
$$1 \rightarrow 2 \rightarrow 4 \rightarrow 6 \rightarrow 12 \rightarrow 24 \rightarrow 30 \rightarrow 60 \rightarrow 72 \rightarrow 78 \rightarrow 79$$
its fitness value is equal to 10 (notice that we subtract 1 from the length to maintain congruency with previously known works).

As we mentioned before, we are looking for the shortest addition chains, then this is a minimization problem and the fittest particles are those with lower lengths.

5.4   The Velocity Values

For this work, a velocity value is a chain associated with a particle. Both, particles and velocity chains, are the same length. A velocity chain contains the *rules* indicating which elements were added to obtain the current element in the particle.

Let us call p_i and v_i the i-th element in the particle and its velocity, respectively. Then, each element p_i (with $i > 0$) [1] is obtained with $p_i = p_{i-1} + p_{v_i}$.

This idea is illustrated in Figure 1. See for example, for the value in the sixth position $i = 6$ its corresponding velocity value is $v_6 = 3$, then the value in p_6 is computed with $p_{i-1} + p_{v_i} = p_5 + p_3 = 24 + 6 = 30$.

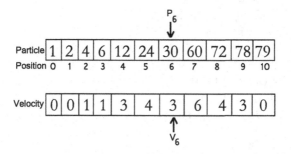

Fig. 1. An example of a particle with its associated velocity chain. The sixth position p_6 is computed with $p_6 = p_{i-1} + p_{v_i} = p_5 + p_3 = 24 + 6 = 30$.

5.5   Particles and Velocities Updating

The particle's position is affected by the particle's best experience (*pBest*) and its neighborhood's best experience (*gBest*). For this case the neighborhood is composed by all the elements in the swarm.

We scan the values in each particle. Then, at each step, we randomly select if the particle is affected by the rule in the velocity chain associated with: (1) the particle's best experience (*pBest*) or, (2) the particle's best neighbor (*gBest*) or, (3) its own velocity.

---

[1] In spite of $v_0 = 0$ the value in p_0 must be 1.

The procedure applied to each particles p with length l, for a given exponent e is as follows:

```
(1) for (i=2 to l) do
(2)    if (2p_{i-1} < e)
(3)         if (flip(Pb)) then /*apply the pBest's velocity rule*/
(4)              if ( pBest's length < i)
(5)                   p_i = p_{i-1} + p_{v_{pBest_i}}
(6)                   v_i = v_{pBest_i}
(7)              else
(8)                   v_i = random(0,i-1)
(9)                   p_i = p_{i-1} + p_{v_i}
(10)        else if (flip(Pg)) then /*apply the gBest's velocity rule*/
(11)             if ( gBest's length < i)
(12)                  p_i = p_{i-1} + p_{v_{gBest_i}}
(13)                  v_i = v_{gBest_i}
(14)             else
(15)                  v_i = random(0, i - 1)
(16)                  p_i = p_{i-1} + p_{v_i}
(17)        else /*apply its own velocity rule*/
(18)             if ( v_i < i)
(19)                  p_i = p_i + p_{v_i}
(20)             else
(21)                  v_i = random(0, i - 1)
(22)                  p_i = p_{i-1} + p_{v_i}
(23)    else
(24)        if (p_{i-1} ≠ e)
(25)           for((m=i-2) to 0)
(26)              if (p_{i-1}+ ≤ e)
(27)                  p_i = p_{i-1} + p_m
(28)                  v_i = m
(29)                  break;
(30)           End-for
(31) End-for
```

The function flip(X) selects "true" with probability X. The parameters Pb and Pg determine the probability that the $pBest$ or $gBest$ been applied to the current step in the particle (lines 3 and 10).

The overall proposed pseudo-code is draw next:

```
Input: An exponent e
(1) Generate an initial set of particles size N.
(2) Repeat
(3)     For each particle n do
(4)         Evaluate the fitness value f(n).
(5)         if (f(n) < pBest) then
```

(6)   pBest = f(n)
(7)   Find the best neighbor and assign its position to gBest.
(8)   Update the position and velocity of the particle n.
(9)   End-For
(10) while a termination criteria is not reached
Output: The shortest particle found.

6   Experiments

The experiments to test our PSO were designed with to main goals: (1) to show that its performance is competitive with those provided by other approaches and (2) to solve more complex instances of the problem. In this way, three experiments were performed. The difference between experiments is the value of the exponent used, which determines the complexity of the problem. Quality (the best solution reached so far) and consistency (better mean, standard, deviation and/or worst values) of the obtained results are considered in the discussion.

All the experiments performed in this work, were obtained by using the following parameter values:

 - Number of particles: 30
 - Number of iterations: 10000
 - Neighborhood configuration: global, i. e. the neighborhood is the total swarm.
 - Probability to apply the pBest's velocities rules (pB): 0.5
 - Probability to apply the gBest's velocities rules (pG): 0.25

The first set of experiments consists on evaluating our algorithm by finding the additions chains for all the exponents into the range from 1 to 1024 ($e = [1, 1024]$). Then, we added all the addition chains's lengths resulting from these exponents. We executed this experiment 30 times with different initial random seeds. The results are compared against the methods Binary and Quaternary and the Artificial Immune System (AIS) published in [2]

The results are presented in Table 1. Note that Binary and Quaternary are deterministic methods, meanwhile AIS and PSO are probabilistic methods. Then, the statistical results for these last two methods are presented (obtained from 30 independent runs).

The second set of experiments consists on finding the optimal addition chains for the exponents $e = 32k$ where k is an integer number from 1 to 28. This is a very important class of exponents for error-correcting code applications. The results are compared against the method presented in [17] and the AIS [2], they are presented in Table 2.

Note that the proposed PSO obtained better results than the other methods in 3 out of the 28 exponents. The addition chains for these three exponents are shown in Table 3.

The last set of experiments consists on finding the addition chains for a especial set of exponents that are considered hard to optimize. Their lengths go from 2 to 30. For all the cases our PSO is capable of finding the minimal addition

Table 1. Results obtained from the addition of the addition chains's lengths for all the exponents into $e = [1, 1024]$

Optimal value=**11115**	
Strategy	**Total length**
Binary	12301
Quaternary	11862

	AIS [2]	**PSO**
Best	11120	11120
Average	11126.433	11122.433
Median	11126.00	11125
Worst	11132	11122
Std. Dev.	3.014	1.951

Table 2. Addition chains's lengths for the exponents $m = 32k - 1$ with $k \in [1, 28]$ obtained by the method propose in [17], the AIS [2], and the proposed PSO

$e = m$	31	63	95	127	159	191	223	255	287	**319**	351	383	415	**447**
Method in [17]	7	8	9	10	10	11	11	10	11	12	11	13	12	12
AIS	7	8	9	10	10	11	11	10	11	12	11	12	11	12
PSO	7	8	9	10	10	11	11	10	11	**11**	11	12	11	**11**

$e = m$	479	511	543	575	607	**639**	671	703	735	767	799	831	863	895
Method in [17]	13	12	12	13	13	13	13	13	12	14	13	13	15	14
AIS	12	12	12	12	13	13	13	13	12	13	13	13	13	13
PSO	12	12	12	12	13	**12**	13	13	12	13	13	13	13	13

Table 3. Addition chains obtained by the PSO for the exponents $e \in [319, 447, 639]$

e	Addition chain	Length
319	1 - 2 - 3 - 6 - 7 - 13 - 26 - 52 - 104 - 208 - 312 - 319	11
447	1 - 2 - 3 - 6 - 12 - 15 - 27 - 54 - 108 - 216 - 432 - 447	11
639	1 - 2 - 3 - 6 - 12 - 24 - 27 - 51 - 102 - 204 - 408 - 612 - 639	12

chains. The AIS [2] also found minimal chains, however the addition chains are different. These addition chains are shown in Table 4.

6.1   Discussion of Results

In the first set of experiments shown in Table 1, we can observe that the deterministic techniques (Binary and Quaternary) obtained the worst results. On the other hand, the AIS and the proposed PSO obtained better results. Notice that from 30 executions, both algorithms obtained the best value with length equal to 11120. However, the PSO seemed to be more robust showing better average, median and standard deviation values.

Table 4. Addition chains obtained by the PSO for some hard exponents

e	Addition chain	length
3	1-2-3	2
5	1-2-4-5	3
7	1-2-4-6-7	4
11	1-2-3-6-9-11	5
19	1-2-4-6-12-18-19	6
29	1-2-3-6-12-13-26-29	7
47	1-2-4-8-9-13-26-39-47	8
71	1-2-4-5-10-20-30-60-70-71	9
127	1-2-3-6-12-24-48-72-120-126-127	10
191	1-2-3-5-10-11-21-31-62-124-186-191	11
379	1-2-3-6-9-18-27-54-108-216-324-378-379	12
607	1-2-4-8-9-18-36-72-144-288-576-594-603-607	13
1087	1-2-4-8-9-18-36-72-144-145-290-580-870-1015-1087	14
1903	1-2-4-5-9-13-26-52-78-156-312- 624-637-1261-1898-1903	15
3583	1-2-3-5-10-20-40-80-160-163- 326-652-1304-1956-3260-3423-3583	16
6271	1-2-3-6-12-18-30-60-120-240-480-960- 1920-3840-5760-6240-6270-6271	17
11231	1-2-3-6-12-24-25-50-100-200-400-800 -1600-3200-4800-8000-11200-11225-11231	18
18287	1-2-3-5-10-20-40-80-160-320-640-1280-2560 -2561-5122-5282-10404-15686-18247-18287	19
34303	1-2-3-6-12-14-28-56-112-224-448-504-1008 -2016-4032-8064-16128-18144-34272-34300-34303	20
65131	1-2-3-6-12-24-36-72-144-288-576- 1152-2304-4608-4611-9222-18444-36888- 55332-64554-65130-65131	21
110591	1-2-4-8-16-32-64-96-192-384-385 -770-1155-2310-2695-5390-10780-21560- 43120-86240-107800-110495-110591	22
196591	1-2-3-6-12-24-30-54-108-216-432- 864-1728-1782-3510-7020-14040-28080- 56160-112320-168480-196560-196591-196591	23
357887	1-2-3-6-12-15-27-29-58-116-232-464- 696-1392-2784-5568-11136-22272-44544-89088- 178176-179568-357744-357860-357887	24
685951	1-2-3-5-10-20-30-60-120-240-480-510 -1020-2040-4080-8160-16320-32640-65280-130560 -261120-522240-652800-685440-685950-685951	25
1176431	1-2-4-5-9-18-27-54-108-216-432-864- 1728-3456-3460-6920-13840-27680-55360- 58820-117640-235280-470560-705840- 1176400-1176427-1176431	26
2211837	1-2-4-8-16-32-34-68-136-272-544-1088- 2176-4352-8704-17408-34816-69632-139264- 139332-148036-296072-592144-1184288-1776432- 2072504-2211836-2211837	27
4169527	1-2-3-6-12-24-48-96-192-384-768-1536- 3072-6144-12288-24576-49152-49344-98688- 148032-296064-592129-592129-1184258-2368516 -3552774-4144903-4169479-4169527	28
7624319	1-2-3-6-12-24-48-96-97-194-206-412-824- 1648-1651-3302-6604-13208-26416-52832-105664- 211328-422656-845312-848614-1693926-3387852- 4236466-7624318-7624319	29
14143037	1-2-3-5-8-13-26-52-104-208-416-832-1664-1667 -3334-6668-13336-26672-26685-53370-106740 -213480-426960-853920-1707840-1711174- 3422348-6844696-7271656-14116352-14143037	30

For the second set of experiments shown in Table 2, the AIS and the PSO obtained similar results. However, in three out of 28 different exponents, the PSO obtained shorter addition chains. The results for this experiment were obtained from 30 independent runs. We show the best result obtained from that.

For the third and last experiment shown in Table 4, both strategies, AIS and PSO, found addition chains with equal lengths. However in most of the cases they obtained different addition chains. The addition chains found by the PSO algorithm are presented there.

In spite that there exists other probabilistic methods based on Genetic Algorithms [4] their results are not presented in this work because they are inferior than those obtained by the AIS.

7 Conclusions and Future Work

We have shown in this paper how a PSO algorithm can be adapted to find short addition chains for a given exponent e. The addition chains are directly represented by the particles. Their velocity values represent the rules indicating which elements must be added to obtain each element in the particles. We experimented with three different sets of exponents with different features and difficulty level. The results demonstrate that this is a very competitive option.

As a future, it is necessary to experiment with higher exponents, furthermore, to improve our strategy and to determine how sensitive is the algorithm to its parameter's values, neighborhood configuration, etc.

Acknowledgment

The authors thank the support of Mexican Government (CONACyT, SIP-IPN, COFAA-IPN, and PIFI-IPN).

References

1. Bergeron, F., Berstel, J., Brlek, S.: Efficient computation of addition chains. Journal de thorie des nombres de Bordeaux 6, 21–38 (1994)
2. Cruz-Cortés, N., Rodríguez-Henríquez, F., Coello, C.A.C.: An Artificial Immune System Heuristic for Generating Short Addition Chains. IEEE Transactions on Evolutionary Computation 12(1), 1–24 (2008)
3. Cruz-Cortés, N., Rodríguez-Henríquez, F., Coello Coello, C.A.: On the optimal computation of finite field exponentiation. In: Lemaître, C., Reyes, C.A., González, J.A. (eds.) IBERAMIA 2004. LNCS (LNAI), vol. 3315, pp. 747–756. Springer, Heidelberg (2004)
4. Cruz-Cortés, N., Rodríguez-Henríquez, F., Juárez-Morales, R., Coello Coello, C.A.: Finding Optimal Addition Chains Using a Genetic Algorithm Approach. In: Hao, Y., Liu, J., Wang, Y.-P., Cheung, Y.-m., Yin, H., Jiao, L., Ma, J., Jiao, Y.-C. (eds.) CIS 2005. LNCS (LNAI), vol. 3801, pp. 208–215. Springer, Heidelberg (2005)

5. Gordon, D.M.: A survey of fast exponentiation methods. Journal of Algorithms 27(1), 129–146 (1998)
6. IEEE P1363: Standard specifications for public-key cryptography, Draft Version D18. IEEE standards documents, http://grouper.ieee.org/groups/1363/ (November 2004)
7. Jarbouia, B., Cheikha, M., Siarryb, P., Rebaic, A.: Combinatorial particle swarm optimization (cpso) for partitional clustering problem. Applied Mathematics and Computation 192(2), 337–345 (2007)
8. Kennedy, J., Eberhart, R.: Particle swarm optimization. In: Proceedings IEEE International Conference on Neural Networks, pp. 1942–1948 (1995)
9. Knuth, D.E.: Art of Computer Programming, Seminumerical Algorithms, vol. 2. Addison-Wesley Professional, Reading (1997)
10. Koç, Ç.K.: High-Speed RSA Implementation. Technical Report TR 201, 71 pages, RSA Laboratories, Redwood City, CA (1994)
11. Koç, Ç.K.: Analysis of sliding window techniques for exponentiation. Computer and Mathematics with Applications 30(10), 17–24 (1995)
12. Koduru, P., Das, S., Welch, S.M.: Multi-objective hybrid pso using -fuzzy dominance. In: Proceedings of the 9th annual conference on Genetic And Evolutionary Computation Conference, pp. 853–860 (2007)
13. Kunihiro, N., Yamamoto, H.: New methods for generating short addition chains. IEICE Trans. Fundamentals E83-A(1), 60–67 (2000)
14. Menezes, A.J., van Oorschot, P.C., Vanstone, S.A.: Handbook of Applied Cryptography. CRC Press, Boca Raton (1996)
15. Poli, R., Kennedy, J., Blackwell, T.: Particle swarm optimization, an overview. Swarm Intelligence 1(1), 33–57 (2007)
16. Sierra, M.R., Coello Coello, C.A.: Improving pso-based multi-objective optimization using crowding, mutation and epsilon-dominance. In: Coello Coello, C.A., Hernández Aguirre, A., Zitzler, E. (eds.) EMO 2005. LNCS, vol. 3410, pp. 505–519. Springer, Heidelberg (2005)
17. Takagi, N., Yoshiki, J., Tagaki, K.: A fast algorithm for multiplicative inversion in $GF(2^m)$ using normal basis. IEEE Transactions on Computers 50(5), 394–398 (2001)

Linear Wind Farm Layout Optimization through Computational Intelligence

José-Francisco Herbert-Acero, Jorge-Rodolfo Franco-Acevedo
Manuel Valenzuela-Rendón*, and Oliver Probst-Oleszewski**

Instituto Tecnológico y de Estudios Superiores de Monterrey, Campus Monterrey
Eugenio Garza Sada 2501 Sur, Monterrey, N.L., México, CP 64849
Fco.Herbert@itesm.mx, jrfa85@hotmail.com,
valenzuela@itesm.mx, oprobst@itesm.mx

Abstract. The optimal positioning of wind turbines, even in one dimension, is a problem with no analytical solution. This article describes the application of computational intelligence techniques to solve this problem. A systematic analysis of the optimal positioning of wind turbines on a straight line, on flat terrain, and considering wake effects has been conducted using both simulated annealing and genetic algorithms. Free parameters were the number of wind turbines, the distances between wind turbines and wind turbine hub heights. Climate and terrain characteristics were varied, like incoming wind speed, wind direction, air density, and surface roughness length, producing different patterns of positioning. Analytical functions were used to model wake effects quantifying the reduction in speed after the wind passes through a wind turbine. Conclusions relevant to the placement of wind turbines for several cases are presented.

Keywords: Optimization, Simulated Annealing, Genetic Algorithms, Sitting, Wind Farm Layout, Wind Turbines, Far Wake Effects, Turbulence and Roughness Length.

1 Introduction

The energy production from wind using wind turbines has become an activity of great importance as it helps to fulfill the increasing energy needs for human activities [1]. Wind energy technology can be applied both on a large scale as well as on a scale suitable for rural communities [2] and it is considered to play a strategic role in the solution of the global energy problem. At present, the use of non renewable energy sources, like oil and fossil fuels, is considered an unsustainable activity because of the limited availability of these resources and the important amount of greenhouse effect gases which are produced from the combustion of these fuels [3].

---

* Professor with Department of Mecatronics and Automatization.
** Professor with the Physics Department.

A. Hernández Aguirre et al. (Eds.): MICAI 2009, LNAI 5845, pp. 692–703, 2009.
© Springer-Verlag Berlin Heidelberg 2009

The power produced by a wind turbine as a function of the incoming free-flow wind speed U_0 can be written as follows:

$$P(U_0) = \frac{1}{2}\rho A C_p(U_0)U_0^3, \tag{1}$$

where ρ is the air density and A the area swept by the rotor. $C_p(U_0)$ is the power coefficient of the turbine, which depends of the wind velocity and which summarizes both the aerodynamic properties of the rotor and the electromechanical and control aspects of the turbine.

A wake effect [4,5] is a phenomenon of interference between wind turbines; this effect is of great importance when deciding the distribution of the turbines that integrate a wind farm. It is said that a wind turbine "b" is located on the wake of another turbine "a" when the wind which initially comes through the turbine "a" reduces its velocity after the collision with the blades, transferring part of the kinetic energy to "a" and eventually reaching "b" with a smaller velocity and, as a consequence, a smaller kinetic energy. As the wind is decelerated by the rotor, mass starts accumulating just before the actuator disk, this generates a volumetric expansion due to the energy loss. The way this effect continues its propagation while moving away from the turbine is often considered linear, forming a cone of wind velocity deficit without appreciable variations of density, as it is shown in figure 1. The arguments sustaining a linear approximation of wake effects are based on continuity. Wake effects can be accumulated when the wind comes across several turbines. In addition, a turbine wake has higher turbulence intensity than ambient turbulence (or added turbulence) which will, however, not be considered in this work for simplicity.

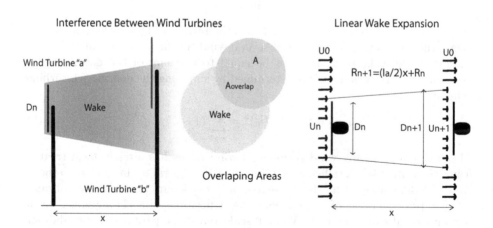

Fig. 1. Schematic of the wake model

Several wake models exists, ranging from models based on simple energy and momentum balances [6,7,8], to more complex models involving the solution of the equations of motion and turbulence models [9].

694   J.-F. Herbert-Acero et al.

Wake models are analytic functions, and the basic structure that defines them is:

$$U(x) = U_0 \left(1 - U_{\text{def}}(x)\right), \tag{2}$$

where U_0 is the velocity of the wind before passing through the actuator disk of the wind turbine, U and U_{def} are the velocity of the wind at a distance x after the wind turbine and the velocity deficit, respectively. The differences between wake models are essentially the way they approximate U_{def}. For the purposes of this study, the Jensen model [6] is used:

$$U_{\text{def}}\left(C_T(v), D, I_a, x\right) = \frac{1 - \sqrt{1 - C_T}}{\left(1 + \dfrac{I_a x}{D_n}\right)^2} \left(\frac{A_{\text{overlap}}}{A}\right)$$

$$= \frac{1 - \sqrt{1 - C_T}}{\left(1 + \dfrac{I_a x}{D_n}\right)^2} \cdot \frac{A_{\text{overlap}}}{\pi \left(\dfrac{D_{n+1}}{2}\right)^2}, \tag{3}$$

where $C_T = C_T(U_0)$ is the thrust coefficient representing a normalized axial force on the rotor by the wind, or conversely, of the rotor on the moving air. C_T depends on the characteristics of the turbine used. D_n is the diameter of rotor (e.g. turbine "a"), D_{n+1} is the diameter of the rotor in the wake (e.g. turbine "b"), x is the distance after the wind turbine on the axis of the rotor, and I_a is the ambient turbulence intensity defined as:

$$I_a = \frac{1}{\ln(z/z_0)}, \tag{4}$$

where z is the height of the rotor and z_0 is the terrain roughness length. To calculate the cumulative wake effect of several wind turbines, the turbulent kinetic energy in the wake is assumed to be equal to the sum of the deficits of kinetic energy. The total deficit is determined by summing the deficit each turbine generates [6]:

$$U_{\text{def,Tot}} = (1 - U/U_0)^2 = \sum_{i=1}^{n} (1 - U_i/U_0)^2. \tag{5}$$

The problem of sitting wind turbines in wind farms has already been treated using computational intelligence algorithms with several optimization methods [10,11,12,13]. Castro et al. [10] presented an evolutionary algorithm to optimize the net present value of a wind farm on a fictitious site characterized by a certain spatial variation of the Weibull scale and shape parameters, but did not give recommendations as to the placement of the turbines. Grady et al. [11] used genetic algorithms to optimize turbine placement patterns on a two-dimensional array, using a variable number of turbines. Marmidis et al. [12] also aimed at the economic optimization of wind farms, using a Monte Carlo approach and presented optimized turbine 2D-placement patterns, improving the total power output over the previous study by Grady et al.

While these and other studies successfully demonstrate the application of methods of artificial intelligence to the problem of wind farm optimization, few general lessons have been learned so far as to the general guidelines for wind turbine placement to be used by the wind farm designer. Such guidelines are important, since generally, due to the complex nature of the wind farm planning process where a number of quantitative and qualitative constraints intervene, it will not be possible to completely rely on an automated optimization process. To obtain some insights into such guidelines, we have gone back to the problem of turbine placement in one dimension, using computational intelligence algorithms as optimization tools. The main objective of this study is to identify general placement strategies which can then be used by the designer in a real-world context.

2 Methodology

In the present work we address the problem of the optimal distribution of a line of turbines lined up in the direction of the wind. The power and thrust for a real wind turbine (Repower MD77) has been used in all simulations; different hub heights have been considered to allow for reduced overlap of the turbine wakes. In order to find the optimal distribution of turbines, simulated annealing [14] and genetic algorithms [15] have been used. Both algorithms are inspired by natural processes and are stochastic searchers. The advantages of using these algorithms are that they can be used in a general way for many different problems; there is no specific knowledge required regarding the mathematical characteristics of the objective function, and almost any kind of restriction can be included. The implementation of these algorithms was done in MATLAB.

Each algorithm requires an objective function; in this study we have chosen the total power (eq. 1) produced by the wind farm. The restrictions are: (1) A minimal distance between turbines of two times the diameter of the turbine; and (2) the geographic limits of the terrain. It is not easy to compare the performance of these two methods, first of all because one can be better in finding the solution of a specific type of problem, but the other can be better dealing with another type. Besides, the performance of the algorithms depends not only on the strategy intrinsic to the algorithm itself, but also on the way the problem is set up and the parameters used for the algorithm, for example the temperature and the neighborhood function for simulated annealing or the length of the chromosomes and number of generations for the genetic algorithms.

In this case even when the genetic algorithm and simulated annealing produced similar problems, the GA required more objective function evaluations. Also, the variation of quality in the solutions obtained from the algorithms is a little more noticeable on the side of the genetic algorithm, giving place for simulated annealing to find slightly better solutions almost all times.

Our implementation of simulated annealing in MATLAB allows for the variation of the temperature inside a Markov chain everytime a transition is accepted. The other parameters of the algorithm are reported in table 1.

Table 1. Parameters used in simulated annealing and genetic algorithms

Simulated Annealing		Genetic Algorithm	
Rate of temperature decrease α	0.9	Population size	50
Temperature variation in Markov chains	On	Generations	600
Markov chain length (accepted transitions)	800	Mutation probability	0.001
Markov chain length (max transitions)	1500	Crossover probability	1.0
Initial temperature	20		

The state is represented by a $4 \times n$ matrix where n is the number of turbines. The first line represents the positions of the turbines, the second line represents heights of the turbines, the third line represents if the turbines are turned on or off (1 or 0, respectively) in the first wind direction (west to east) and the fourth line represents if the turbines are turned on or off in the second wind direction (east to west). In the initial state the n turbines are placed at equal distances. The neighborhood size depends on the temperature of the system, it changes the position and height in the same proportion.

For our genetic algorithm we used the vgGA (virtual gene genetic algorithm) [16] with binary digits and all options set to behave as a traditional genetic algorithm. The parameters are summarized in table 1. Each individual is represented by a vector in which the first 16 bits correspond to the position of a turbine, the following 10 represented the height of the turbine, the following 2 are the on-off bits for the first wind direction (west to east) and the second direction (east to west) respectively, this is repeated in the vector for each turbine.

The implementation of constraints for the genetic algorithms achieved by reinterpretation. At the beginning of the objective function each individual were modified moving the turbines to the right if necessary so the constraint of a minimal space of two diameters between turbines was satisfied. If one or several turbines were out of the valid range of space (length of the terrain) it was ignored at the moment of evaluation. The original individuals remain unchanged for the other processes involved with the GA.

3   Results

Three typical cases have been studied; in all cases a flat terrain of length L [m] has been considered with constant terrain roughness length z_0 [m]; sea level air density of $\rho = 1.2254\,\mathrm{kg/m^3}$ has been used in all calculations. The $C_T(v)$ and $C_p(v)$ curves for the Repower MD77 wind turbine, rated at 1500 kW, with a rotor diameter of 77 m were used. Continuous expressions for the thrust and power coefficient curves were obtained from the adjustment of empirical data to splines with Piecewise Cubic Hermite Interpolating Polynomial's (PCHIP). For comparison of the earlier results in literature the Jensen wake model explained above has been used in all cases.

Case 1. Constant turbine hub height and one-directional wind flow
Figure 2 present the results obtained for this case, assuming an incoming free-stream wind speed of 10 m/s, a roughness length of 0.1 m and L=1550 m. In

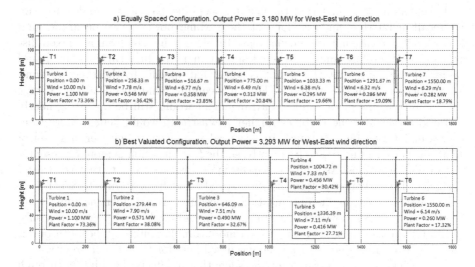

Fig. 2. Results for constant turbine hub height and one-directional wind regime. a) Uniform placement of five turbines. b) Optimized turbine placement.

the initial uniform arrangement, turbines were placed at a constant distance of 260 m (3.35 turbine diameters). It is conspicuous that in this arrangement the wind speed at the downstream turbines quickly drops off, leaving the last turbine with a wind speed of only 6.29 m/s. It is intuitive that this arrangement cannot be optimal. Increasing the turbine distance for those wind turbines operating in the regime of highest variation of turbine power with wind speed (i.e. highest values of the turbine power derivative dP/dv) will clearly increase the overall power production. In the case of the Repower MD77 wind turbine (rated 1500 kW at 14 m/s), the inflexion point of the $P(v)$ curve occurs around 9 m/s; therefore the second turbine in the original uniform arrangement (experiencing a wind speed of 7.78 m/s) will benefit most from an increase in wind speed. Actually, by moving this turbine from its original position at $x = 258$ m to about 280 m, the algorithm achieves an increase in power production from this turbine by about 25 kW (some 5% increase for this turbine). By doing so, the algorithm necessarily pushes the other turbines further downstream. The third turbine experiences a similar situation when is moved from $x = 517$ m to $x = 646$ m, achieving an increase in power production by about 1.32 kW (some 27% increase for this turbine). Interestingly, this increase of power production, by moving subsequent turbines is already higher than the total power production of turbine 7 (producing 262 kW at 6.29 m/s), so this turbine can be removed completely without compromising the overall farm production, allowing for larger distances between the remaining turbines downstream of turbine 1 which provide additional gains in power production. It is therefore completely understood why a six-turbine non-uniform arrangement should produce more power than a seven-turbine uniform turbine row.

To understand the fact that turbine distances decrease as we go downstream we recall again that highest slope of the turbine power curve occurs around 9 m/s, so it is clear that the downstream turbines gains less from a relaxed distance (and, therefore, increased wind speed) than the first turbines. It should be noted, however, that this result is somewhat specific to the range of wind speeds occurring in this example. For an incoming wind speed of, say, 15 m/s the second turbine in the uniform arrangement would still be operating nearly at rated power and the turbines further downstream would benefit more from a greater inter-turbine distance. In practical wind farms, however, a speed of 15 m/s is not encountered so frequently, so the first situation is more representative of real situations. As a final observation of this case we note that the thrust coefficient is nearly constant in the wind speed range from 4.5 m/s to 8.5 m/s, so it has little impact on the optimization procedure. At a higher incoming wind speed (e.g. 15 m/s), however, the turbine is in the pitching regime with a substantially reduced thrust. In this case, a more complex interplay between power and thrust occurs, which will not be discussed here for the sake of brevity.

The variation of the terrain roughness length also affects the solution through the ambient turbulence intensity which promotes wind speed recovery; the higher the roughness length is the higher the number of turbines that form the optimal solution; however, this does not change the general pattern of distribution. Other wake models like Larsen's [7] and Ishihara's [8], which do not necessarily consider a linear wake expansion, produce similar results.

Case 2. Constant turbine hub height and bi-directional wind regime

In this case a bi-directional wind regime is considered with the wind flowing either from west to east with a speed of v_1 and from east to west with a speed of v_2; the probabilities of occurrence for these wind regimes are p_1 and p_2, respectively. Moreover, we have included the option of shutting down turbines when switching from one regime to the other; we assumed that a wind turbine does not generate wake effects while off, that could be convenient specially when wind velocity is low on a regime. The typical solutions are shown on figure 3. The variables used are: $L = 1550$ m, $z_0 = 0.1$ m, $v_1 = 10$ m/s, $v_2 = 10$ m/s, $p_1 = 80\%$, $p_2 = 20\%$, $D = 77$ m, $z = 85$ m, $P = 1500$ kW, $\rho = 1.2254$ kg/m^3, $T = 15\,^{\circ}C$, $p = 1$ atm; again, the curves C_T and C_p correspond to a wind turbine Repower MD77. Jensen model for wake losses is used (eqs. 3-4).

The best performing solutions are similar to those shown in case 1 when v_1 and p_1 are greater than v_2 and p_2, respectively. If the wind speed is equal in both directions and if there is an equal probability of occurrence, then the best valued configuration is the equally spaced one. Symmetrically, the inverse of case 1 occurs when v_2 and p_2 are greater than v_1 and p_1, respectively. This results show that all turbines are turned on all times. Figures (4 and 5) show two different configurations with $v_1 = 10$m/s, $v_2 = 10$ m/s, $p_1 = 50\%$, $p_2 = 50\%$ for the first and $v_1 = 6$ m/s, $v_2 = 10$ m/s, $p_1 = 20\%$, $p_2 = 80\%$ for the second.

The results obtained by shutting down turbines were slightly better, even when the solutions suggest that the optimal positioning in several cases involves

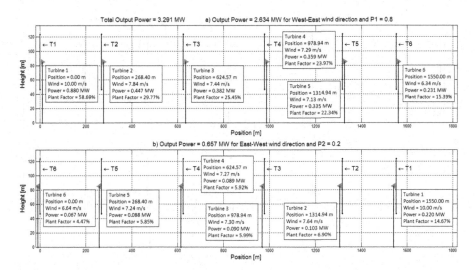

Fig. 3. Results for case 2 (bi-directional wind regime). a) Wind distribution for first wind direction and $p_1 = 80\%$. b) Wind distribution for second wind direction $p_2 = 20\%$.

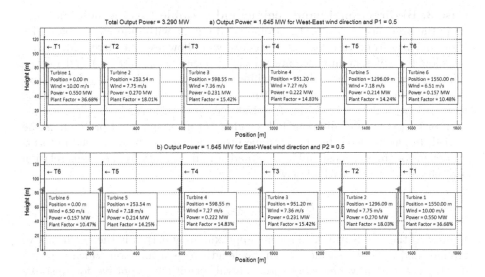

Fig. 4. Results for case 2 (bi-directional wind regime). Optimized solution considering bidirectional wind regime with v1=10 m/s, v2=10 m/s, p1=50%, p2=50%.

turning off some turbines when the wind velocity is low, the overall power produced by the wind farm increases in hardly about 1%, introducing one or even two more turbines than the solution without turning off turbines. Therefore, results will not be reported due to their lack of significance.

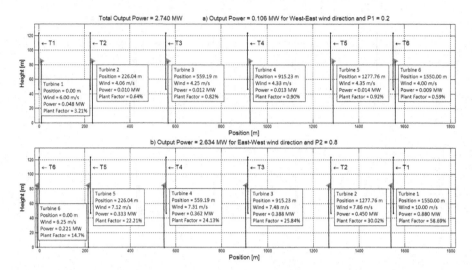

Fig. 5. Results for case 2 (bi-directional wind regime). a) Optimized solution considering bidirectional wind regime with $v_1 = 6\,\mathrm{m/s}$, $v_2 = 10\,\mathrm{m/s}$, p1=20%, p2=80%.

Case 3. Bi-directional wind regime and variable turbine heights

As in the previous case, a bi-directional wind regime with different probabilities of occurrences from west-east and east-west wind flow are considered. Moreover, we allow for different hub heights ($z_{1,\min} = 50\,\mathrm{m}$, $z_{2,\max} = 85\,\mathrm{m}$) and for the shutting down of turbines when switching wind regimes. All other variables have the same values as before: $L = 1550\,\mathrm{m}$, $z_0 = 0.1\,\mathrm{m}$, $v_1 = 10\,\mathrm{m/s}$, $v_2 = 10\,\mathrm{m/s}$, $p_1 = 80\%$, $p_2 = 20\%$, $D = 77\,\mathrm{m}$, $P = 1500\,\mathrm{kW}$, $\rho = 1.2254\,\mathrm{kg/m^3}$.

In order to consider different hub heights, an incoming wind profile must be settled. In nature, wind has logarithmic profile which can be described in terms of terrain roughness, geostrophic wind and stability functions [17]. This relation is often called the logarithmic law:

$$\bar{U}(z) = \frac{u^*}{k}\ln(z/z_0) + \psi, \tag{6}$$

where k is the von Karman constant (approximately 0.4), z is the height above ground, and z_0 is the surface roughness length. ψ is a function which depends on stability and it value is zero for neutral conditions, negative for unstable conditions and positive on stable conditions.

The best performing solutions are similar to those shown in case 1 for turbines at same height and when v_1 and p_1 are greater than v_2 and p_2, respectively, like in case 2. If the wind speed is equal in both directions and there is an equal probability of occurrence, then the best valued configuration is the equally spaced for turbines at same height, conserving a minimal distance of two rotor diameters between wind turbines. Symmetrically, the inverse of case 1 occurs when v_2 and p_2 are greater than v_1 and p_1 respectively. This results show that all turbines are turned on all times.

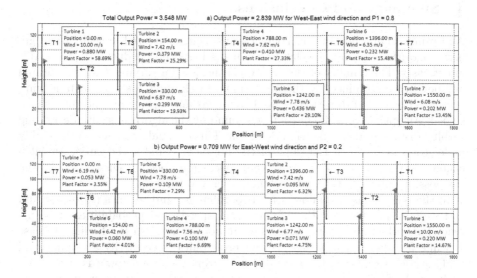

Fig. 6. Results for the variable hub height case with a bi-directional wind regime. a) Solution considering bidirectional wind regime with $v_1 = 10 \, \text{m/s}$, $v_2 = 10 \, \text{m/s}$, $p_1 = 80\%$, $p_2 = 20\%$. b) First wind regime. c) Second wind regime.

Contradicting intuition, there is no symmetry with respect to hub heights. If the arrange of wind turbines changes form taller-smaller-taller (on fig. 6) to smaller-taller-smaller, keeping all the defined variables and positioning, there will be 7.91% less output power. This is because turbulence intensity is different at different heights and therefore the velocity gain is different for each arrangement in each wind turbine. Also, logarithmic profile of wind shows lower wind velocity at lower heights. It is considered that maximum wind speed ($v_1 = 10 \, \text{m/s}$ and $v_2 = 10 \, \text{m/s}$) is reached at $z_{2\,max} = 85 \, \text{m}$ of height and starts to diminish at lower heights (e.g. $v = 8.76 \, \text{m/s}$ at $z_{1\,min} = 50 \, \text{m}$ with $z_0 = 0.1 \, \text{m}$).

Comparing an equally spaced configuration with same conditions as in case 2 (fig. 4), which produces 3.18 MW, with the one in case 3 we find that case 3 produces 10.37% more output power but has one more wind turbine. If in case 3 we vary the heights from 50 m to 120 m and we restrict the number of wind turbines to 6, then the output power will be 4.196 MW, which is 24.21% better than the equally spaced configuration of case 2 and 15.44% better than the original case 3 but also has a wind turbine less than both mentioned cases. Positioning pattern remains placing turbines at heights max-min-max-min... and following patterns of case 1 and 2 for wind turbine distances. Configurations using different wake models (e.g. Larsen and Ishihara) produce similar configurations but with a different number of wind turbines since these models quantify velocity deficit on a different manner. Increasing terrain roughness length produces configurations that have more wind turbines with same positioning patterns. As in case 2, shutting down turbines does not generates important improvements in the results.

4   Summary and Conclusions

An optimization study of linear wind farms using simulated annealing and genetic algorithms considering far wake effects has been conducted in three different cases. The number of turbines was considered a variable and the optimal number was determined by the algorithms. In the first case, a constant wind speed and one wind direction, as well as a constant turbine hub height was assumed. An optimized configuration was found by the algorithms with decreasing turbine distances along the wind direction. This optimal arrangement is readily understood in terms of the slope of the power vs. wind speed curve. Possible changes in the results for different wind speeds have been discussed, as well the the possible interplay between turbine power and thrust.

In the second case we allowed for two different wind regimes (west-east and east-west, respectively) with different wind speeds and probabilities of occurrence for both. The results can be understood in terms of the findings of case 1; while for identical wind speeds and probabilities necessarily a symmetrical configuration results, with largest distances from the center towards either end, an increase of probability in one regime over the other leads to creating larger distances on the side of the prevailing wind. Changing the wind speed of one of regimes has a similar although less pronounced effect; the regime with highest slopes of the power vs. wind speed curve dP/dv and the relevant wind speed range will dominate the distance sequence.

In the third case, a more complex interplay of the variables was observed by allowing the algorithms to vary the turbine heights in order to avoid wake losses. As expected, the algorithms place turbine hubs at very different heights (50m and 120m), thus minimizing wake effects. For each wind regime, a twin row of turbines is obtained as the optimal result, with relative short distances between turbines placed at high and low hub heights and similar distance pattern for each turbine type. When allowing two wind regimes, a combination of the situations presented for cases 1 and 2 was observed.

Algorithms were compared in time, number of evaluations of the objective function (same in both algorithms) and resulting configurations. We found that simulated annealing produces better results and takes less time to achieve them, in contrast genetic algorithms find very acceptable results but takes more time and more evaluations of the objective function. In case 3 both algorithms found the optimal solution were simulated annealing needed to evaluate about 9000 times the objective function while genetic algorithms had to evaluate about 13000 times.

In summary, we have demonstrated the suitability of tools of artificial intelligence such as simulated annealing and genetic algorithms for the optimization of linear wind farms for parallel or anti-parallel wind regimes and have provided a qualitative understanding of the resulting wind farm layouts. Efforts to generalize these results to wind speed distributions and two-dimensional wind regimes are currently under way.

References

1. Burton, T., Sharpe, D., Jenkins, N., Bossanyi, E.: Historical Development. In: Wind Energy Handbook, pp. 1–9. John Wiley, New York (2001)
2. Saheb-Koussa, D., Haddadi, M., Belhamel, M.: Economic and technical study of a hybrid system (wind-photovoltaic-diesel) for rural electrification in Algeria. Applied Energy 86(7–8), 1024–1030 (2009)
3. Graedel, T.E., Crutzen, P.J.: Atmospheric Change: An Earth System Perspective. Freedman, New York (1993)
4. Burton, T., Sharpe, D., Jenkins, N., Bossanyi, E.: Turbulence in Wakes and Wind Farms. In: Wind Energy Handbook, pp. 35–39. John Wiley, New York (2001)
5. Vermeera, L., Sørensen, J., Crespo, A.: Wind turbine wake aerodynamics. Progress in Aerospace Sciences 39, 467–510 (2003)
6. Katic, I., Hojstrup, J., Jensen, N.O.: A simple model for cluster efficiency. In: European Wind Energy Association Conference and Exhibition, Rome, Italy, pp. 407–410 (1986)
7. Larsen, C.G.: A simple wake calculation procedure. Technical Report Risø-M-2760, Risø National Laboratory, Rosklide, Denmark (December 1988)
8. Ishihara, T., Yamaguchi, A., Fujino, Y.: Development of a new wake model based on a wind tunnel experiment. In: Global Wind Power (2004)
9. Crespo, A., Hernandez, J.: Turbulence characteristics in wind-turbine wakes. Journal of Wind Engineering and Industrial Aerodynamics 61, 71–85 (1995)
10. Castro, J., Calero, J.M., Riquelme, J.M., Burgos, M.: An evolutive algorithm for wind farm optimal design. Neurocomputing 70, 2651–2658 (2007)
11. Grady, S.A., Hussaini, M.Y., Abdullah, M.M.: Placement of wind turbines using genetic algorithms. Renewable Energy 30, 259–270 (2005)
12. Marmidis, G., Lazarou, S., Pyrgioti, E.: Optimal placement of wind turbines in a wind park using Monte Carlo simulation. Renewable Energy 33, 1455–1460 (2008)
13. Mosetti, G., Poloni, C., Diviacco, B.: Optimization of wind turbine positioning in large wind farms by means of a genetic algorithm. Journal of Wind Engineering and Industrial Aerodynamics 51(1), 105–116 (1994)
14. Aarts, E., Korst, J.: Simulated Annealing and Boltzman Machines: A Stochastic Approach to Combinatorial Optimization and Neural Computing. John Wiley, New York (1989)
15. Goldberg, D.E.: Genetic Algorithms in Search, Optimization, and Machine Learning. Addison-Wesley, Reading (1989)
16. Valenzuela-Rendón, M.: The virtual gene genetic algorithm. In: Genetic and Evolutionary Computation Conference (GECCO 2003), pp. 1457–1468 (2003)
17. Burton, T., Sharpe, D., Jenkins, N., Bossanyi, E.: The Boundary Layer. In: Wind Energy Handbook, pp. 18–21. John Wiley, New York (2001)

Building Blocks and Search

Alexis Lozano[1,3], Víctor Mireles[1,3], Daniel Monsivais[1,3],
Christopher R. Stephens[1,2], Sergio Antonio Alcalá[1,3], and Francisco Cervantes[4]

[1] C_3 Centro de Ciencias de la Complejidad,
Universidad Nacional Autónoma de México
Torre de Ingeniería, Circuito Exterior s/n Ciudad Universitaria, México D.F. 04510
[2] Instituto de Ciencias Nucleares, Universidad Nacional Autónoma de México
A. Postal 70-543, México D.F. 04510
[3] Posgrado en Ciencia e Ingeniería de la Computación, Universidad Nacional
Autónoma de México Circuito Escolar s/n Ciudad Universitaria, México D.F. 04510
[4] Universidad de las Américas Puebla
Sta. Catarina Mártir, Cholula, Puebla. C.P. 72820. México

Abstract. The concept of "building block" has played an important
role in science and also at a more formal level in the notion of search,
especially in the context of Evolutionary Computation. However, there
is still a great deal that is not understood about why, or even if, building
blocks help in search. In this paper we introduce an elementary search
problem and a class of search algorithms that use building blocks. We
consider how the use of building blocks affects the efficiency of the search
and moreover how the characteristics of the building blocks - the number
of types and their sizes - greatly influences the search.

1 Introduction

The concept of a "building block" is an important one in both science and
engineering, being associated with the idea that more complex systems are constructed by the juxtaposition of simpler units. The idea is indeed an ubiquitous
one. For instance, in physics, molecules have atoms as building blocks, atoms
have nuclei and electrons, nuclei have protons and neutrons and the latter are
composed of quarks as building blocks. Turning to biology, macromolecules such
as proteins, DNA and RNA serve as building blocks for organelles, which in their
turn serve as building blocks for cells, which are blocks for tissues and organs etc.
This concept is also closely related to that of modularity[3], wherein a building
block, thought of as a structure, is associated with a particular function. An
important question is: Why do such hierarchies of building blocks exist and how
are they formed?

In Evolutionary Computation (EC), and especially in Genetic Algorithms
(GAs), the concept of a building block has also played an important role and,
furthermore, it is one that has received a formalization that goes beyond that
of other fields. This formalization is associated with the concept of a schema
and the related idea of a "building block" as a fit, short, low-order schema. The

A. Hernández Aguirre et al. (Eds.): MICAI 2009, LNAI 5845, pp. 704–715, 2009.

Building Block hypothesis [2] and the Schema theorem [1] purportedly describe the evolution of a GA via the juxtaposition of these building blocks.

An important characteristic of building block hierarchies is that the degree of epistasis between the elements of a given building block is higher than that between blocks. Thus, for instance, molecules are less tightly bound than their atomic constituents, while atoms are less tightly bound than nucleons within a nucleus.

Although there have been attempts [4,5] to understand the formation of building block hierarchies in more general terms, and also why they exist and are so common, there is still no general quantitative framework for studying them.

In this paper we will try to understand the formation and utility of building blocks in the process of search. By doing so we will make implicit contact with EC. However, the model we will present, rather than being "GA-like", bears some similarity to physical models that contain building blocks, such as stellar nucleosynthesis.

Within the confines of this model we wish to determine: Firstly, if the inclusion of building blocks aids in the search process; and then, whether search is enhanced by using various different types of building block; and, finally, if certain particular types of building block are preferred over others. Additionally, we try to make a first step in the direction of explaining the ubiquity of hierarchical building blocks in nature.

2   A Simple Hierarchical Model of Search

We consider a set P of n particles and a set B of m lattice sites. Both lattice sites and particles are indistinguishable, and the adjacency of the former is irrelevant in the present model. The search problem consists of constructing states, where a certain, fixed number of particles are found at the same lattice site, this state being the objective of the search. The question then is what type of search algorithm most favors the construction of the objective?

Here, we consider a class of search algorithms all implementable as Markov processes, each of which approaches the goal state by randomly permuting particles or combinations thereof between lattice points, building and then binding together smaller blocks. In general they all proceed as follows: The initial state is that in which all particles are unbound. In each iteration, particles are redistributed among the lattice sites via a random permutation and then blocks of particles are formed at those lattice sites which have an appropriate number of particles. The formation of a block is the binding of the constituent particles so that they will all move together in the next iteration. See figure 1 for an example.

The question then becomes whether this binding of particles makes the algorithm more efficient and, if so, how does the nature of the building blocks - number and type - affect the efficiency of search?

Formally, we define a set $V = \{v_1, v_2 \ldots, v_{k-1}, v_k\} \subset \{1 \ldots n\}$ of valid building block (or subassembly) sizes, so that $v_1 = 1$, $v_k = n$ and $v_i < v_j \ \forall i < j$. We call particles building blocks of size 1.

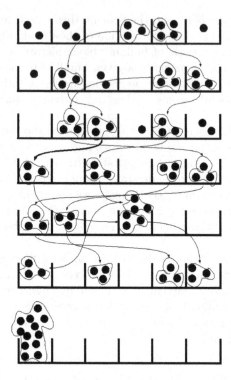

Fig. 1. A sample evolution of the presented model, with $n = 12$ particles and $m = 6$ lattice sites. Evolution is from top to bottom. The valid block sizes are $\{1, 3, 12\}$. Note how, when four particles fall on the same site, only a group of three is formed. Note also, how pairs of groups of three don't bind together, for 6 is not a valid block size.

The discrete-time dynamics of this system consist of synchronously randomly updating the positions of every element of $A(t)$ and subsequently aggregating all building blocks located at the same lattice site that can jointly form a group of valid group size, according to set V.

In this first stage of this work, we abide strictly to the idea of building block hierarchy, and therefore do not allow any already made block to be dismantled. Furthermore, we restrict the dynamics so that a new block can only be formed from blocks all of the same size. Since this restriction can potentially make the objective state an inaccessible one (consider for example $n = m = 6$ and $V = \{1, 3, 4, 6\}$), we relax it in the case of the greatest block-size so that it can be indeed formed by any combination of block sizes.

A first intuitive way to represent the state of the system at any given time would be to specify the position of every particle. This could be done via an n-dimensional vector $(x_1, x_2, \ldots, x_n)$, where $x_n \in B$. This representation is inadequate for two reasons. On the one hand, it fails to represent which particles are bound together in blocks, and on the other, it ignores the fact that both particles and lattice sites are indistinguishable.

Taking into account the following three facts:

1. Building blocks of the same size are indistinguishable, and so are lattice sites.
2. Blocks of non-valid sizes will *decompose* into blocks of valid sizes.
3. The dynamics are defined in terms of blocks and not of particles

we put forth a more adequate representation:

At any given time t, the state of the system can be represented by a k-dimensional vector

$$w(t) = [w_1(t), w_2(t), \ldots, w_k(t)] \tag{1}$$

where $w_i(t) \in \{1, \ldots, [\frac{n}{v_i}]\}$ represents the number of blocks of size v_i that are present at time t. We call W the set of all such possible vectors. Note that the following will be true for all t

$$\sum_{i=1}^{k} v_i \, w_i(t) = n \tag{2}$$

Note that the set W is the set of all integer partitions of n made up only of elements of V and therfore its size, $|W|$, can greatly vary depending on the choice (not only the size) of V.

As mentioned above, the dynamics of the system consist of randomly distributing the blocks present at any given time t, and forming new blocks that will constitute the state at time $t + 1$. This means that the state of the system at time $t + 1$ depends solely on the state at time t, thus making this a Markov process, whose dynamics can be specified by

$$\mathbf{P}(t + 1) = \mathbf{M}\mathbf{P}(t) \tag{3}$$

where $\mathbf{P}(t)$ is a probability vector whose entries denote the probability of the system being in each state $w \in W$ and $\mathbf{M}$ is a left Markov matrix. The dimensionality of both depends on our choice of V, the set of valid block sizes. The matrix $\mathbf{M}$ for the model represents the probability of the system going from any one state to any other. Thus entry $M_{i,j}$ constitutes the probability of going from state i to state j. This matrix was determined computationally with a program developed specifically for it.

Since we are interested in a hierarchy of building blocks, not all states are accessible from all others, since states which contain, for example, only large blocks cannot yield states that contain small blocks. This constitutes in turn a sort of irreversibility of the process, which makes the Markov matrix a triangular one. This implies that the system's fixed point, as predicted by the Perron-Frobenius theorem is the one corresponding to all particles being aggregated in a single block of size v_k.

Table 1 shows the Markov matrix for $n = m = 6$ and $V = \{1, 2, 3, 4, 6\}$, where rows represent states ordered as follows [0,0,0,0,1] [0,1,0,1,0] [2,0,0,1,0] [0,0,2,0,0] [1,1,1,0,0] [3,0,1,0,0] [0,3,0,0,0] [2,2,0,0,0] [4,1,0,0,0] [6,0,0,0,0], i.e., the 10 possible configurations that correspond to the different possible partitions of the six particles at the different lattice sites. Thus, for example, [0,1,0,1,0] corresponds to one block of four particles and one block of two particles, while [0,0,2,0,0] corresponds to two blocks of three particles. On the other hand, the matrix for a system in which no blocks are formed will be have all identical columns, since the configuration in a given step is irrelevant to the configuration in the next step, for all particles remain for ever unbound.

Table 1. Transition matrix for $n = m = 6$ and $V = \{1, 2, 3, 4, 6\}$. See text for ordering of rows.

$$
\begin{pmatrix}
1.000 & 0.167 & 0.028 & 0.167 & 0.028 & 0.005 & 0.028 & 0.005 & 0.001 & 0.000 \\
0.000 & 0.833 & 0.139 & 0.000 & 0.000 & 0.000 & 0.417 & 0.023 & 0.004 & 0.010 \\
0.000 & 0.000 & 0.833 & 0.000 & 0.000 & 0.000 & 0.000 & 0.139 & 0.000 & 0.042 \\
0.000 & 0.000 & 0.000 & 0.833 & 0.000 & 0.023 & 0.000 & 0.000 & 0.000 & 0.006 \\
0.000 & 0.000 & 0.000 & 0.000 & 0.972 & 0.417 & 0.000 & 0.000 & 0.093 & 0.154 \\
0.000 & 0.000 & 0.000 & 0.000 & 0.000 & 0.556 & 0.000 & 0.000 & 0.000 & 0.154 \\
0.000 & 0.000 & 0.000 & 0.000 & 0.000 & 0.000 & 0.556 & 0.139 & 0.069 & 0.039 \\
0.000 & 0.000 & 0.000 & 0.000 & 0.000 & 0.000 & 0.000 & 0.694 & 0.556 & 0.347 \\
0.000 & 0.000 & 0.000 & 0.000 & 0.000 & 0.000 & 0.000 & 0.000 & 0.278 & 0.231 \\
0.000 & 0.000 & 0.000 & 0.000 & 0.000 & 0.000 & 0.000 & 0.000 & 0.000 & 0.015
\end{pmatrix}
$$

Iterating the Markov matrix shown in table 1 one can see how the probabilities to be in the different states evolve as a function of time, with the system eventually converging to the objective state. The initial probability vector has 0's corresponding to all states except the initial one, $[6, 0, 0, 0, 0]$, in which all particles are unbound, and which has a probability 1. As can be seen in fig. 2, this state rapidly loses the status of most probable one due to the fact that the individual particles can join together to form blocks and therefore are a loss term for this state. There is no source term for this state as it cannot be created from other states. For other intermediate building block types there are both loss and gain terms, the latter being associated with the formation of building blocks of the given size from smaller precursors meanwhile the loss term is associated with the joining together of the given size blocks into even larger successors.

Given that the probability to find n particles at a given site starting with a random initial configuration is m^n we see that the probability for forming blocks of larger size is exponentially suppressed relative to that of smaller blocks. It is for this reason that there is a natural hierarchy where one by one states containing building blocks of larger and larger sizes appear, using as material building blocks of smaller size (see fig. 2). Note how all states except the objective one eventually become highly unlikely.

This is in contrast to a system which lacks building blocks. In figure 2 we can see how the algorithm using building blocks, has a greater probability of reaching the objective state in 160 iterations than an equivalent system without building blocks. In the latter, reaching the objective state (all particles bound together) cannot be done with intermediate steps as there are no corresponding building blocks, instead one has to wait for the rather unlikely event that all particles fall on the same lattice site at the same iteration.

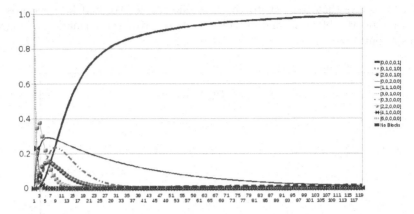

Fig. 2. Evolution of each of the entries of the probability vector for $n = m = 6$ and $V = \{1, 2, 3, 4, 6\}$. The objective state is $[0, 0, 0, 0, 1]$. Also shown is the evolution of the probability of reaching the same objective state (all particles bound together) but using a system without building blocks. The x axis is time and the y axis is probability.

3   The Role of the Number of Different Block Sizes

In this section we examine the effect of changing k the number of intermediate building blocks. Clearly, the presence of intermediate blocks in the dynamic process improves the performance of the search, as we have shown in the previous section, due to the probability difference between a small number of particles to be grouped and any bigger one. However, it's not clear if an increment in the number of intermediate blocks always improves the performance in the search. To clarify this, we simulated the process, fixing $n = m = 27$, and modifying the number and size of the intermediate blocks. Then, we picked out, for each possible number of intermediate blocks, the partition which achieved the best results (less iterations taken to form the final block). The question of which is the best choice for the block's size will be addressed later.

In figure 3 the best results for different numbers of intermediate building ($k-2$, that is, not considering v_1 and v_k) blocks are shown. One can see how, initially, the presence of intermediate blocks reduces drastically the number of iterations needed (for 0 and 1 intermediate blocks the iterations were greater than 200000 steps), but that starting at or above a certain number (6), this improvement in the performance disappears. This stagnation for a large number of intermediate blocks is a finite population effect leading to a lack of diversity in the supply of building blocks. For example, because they tend to form bigger blocks, there may not be any simple particle or small blocks available after a short time and this can hinder search leading to the impossibility of reaching certain block sizes. For example, if we have an intermediate block of size 2 then it will be harder to form blocks with odd sizes, especially if they are big because after a short time is more probable to find the particles be contained in groups of even number of elements. This finite size effect can be ameliorated by having a large population rather than one that is equal to the final block size. We will later propose a relaxation on the block formation rules so as to favor the formation of bigger blocks at any time.

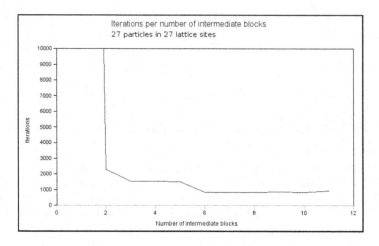

Fig. 3. This figure shows the minimum mean number of iterations per number of valid blocks. For each number of blocks, 1000 iterations were made.

4   The Role of Building Block Sizes

In this section, we investigate the effect of, given a fixed number of building blocks , selecting different block sizes which maximize the improvement of the search speed. Since the first and last block sizes are fixed to 1 and n respectively, we explore the space of combinations of sizes for the remaining blocks. Naive intuition might lead one to expect that equal distances between the sizes would be a good combination. However, this approach ignores the fact that the formation of a second block can be easier using elements of the first block rather than the simple elements when the blocks aren't continuous, because the latter would require more of them to be joined. Therefore one may hypothesize that proportional distances between the sizes would be a better choice, a geometric proportion could be the best option as then the relative number of blocks needed at each level will be about the same, ie, $v_j/v_i = C, \forall i < j < k$. However using the geometric relation limits the size of k ($V = \{1, 3, 9, 27\}$ in this case). Therefore, seems like a good set V should contain some v_i^* which complies a geometric relation, and the other non geometric blocks, should be a multiple of some of this v_i^*. We analyze this question on a system with $n = m = 27$ and we perform, for each choice of building blocks, 100 experiments, noting the average time (number of iterations) for convergence to the objective state. Some results are shown in the graphics of figure 4. Due to the fact that the most interesting results are the ones with a relatively small number of iterations, simulations were truncated. From that figure we glean a first insight as to how block sizes should be chosen. From all the different choices, only those which satisfy certain relations obtain better results. To verify the hypothesis, we fix $n = 27$ as the size of the final block due to the fact that it is a multiple of 3. The better performing runs we obtained have a multiple of 3 as intermediate block size, and sometimes, among them, the block size choice contain just the geometric proportion between blocks, and it's not always the best choice, as one can see in the graphic for three intermediate blocks (a) of the figure 4. The best choice (with intermediate block sizes $v_2 = 3$, $v_3 = 6$, and $v_4 = 12$), doesn't have a geometric proportion between some of their intermediate blocks and achieve the best performance.

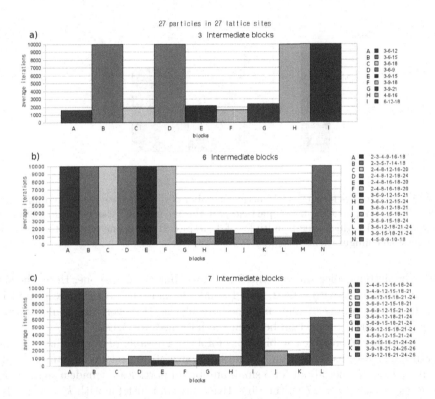

Fig. 4. In the above figures the mean number of iterations to form the final block (size 27), using different choices for the size of the intermediate blocks can be seen. Results are shown for a) 3, b) 6, and c) 7, intermediate $(k-2)$ blocks. Due to the great number of possible choices of the intermediate blocks, only a representative few are shown. The mean number of iterations was truncated above 10000.

Furthermore, we investigate if the relation between blocks that we obtained for $n = m = 27$ also applies for different numbers of elements and blocks. For this we repeat the simulation, using now $n = m = 64$, 81 as the size of the final block. One might expect that choices which include multiples of 2 for the former and of 3 for the latter, would be better performing. This relation can be visualized in the graphics of the figure 5. It is worth to notice, in the 81-size

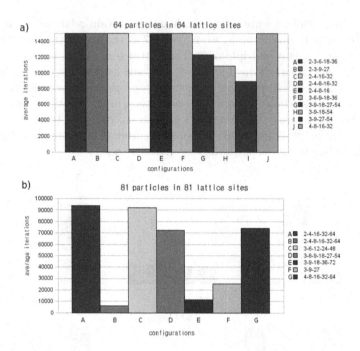

Fig. 5. Some results for runs with distinct number of intermediate blocks and different sizes, for a) 64 particles, b) 81 particles. In the second one, the best run obtained contained blocks which aren't multiples of 3, despite 81 being one (not all the space of the possible configurations was swept).

case, the better choice was not the 3-multiples but the 2-size one. Here, the six intermediate block with sizes 2, 4, 8, 16 and 32, made the best run, despite 81 is not a multiple of any of their intermediate blocks. However, this configuration keeps a geometric proportion between the intermediate sizes.

The number of possible combinations for a fixed number of intermediate blocks number increases as a function of final block size. To better understand how an optimal choice comes about for a small number of particles we made a simulation using again $m = n = 27$ as the objective state and just for with $k = 4$, reads, only 2 intermediate blocks. The results of this simulation are shown in figure 6. We can see that the speed of the process is dominated by slowest step associated with the largest number of particles or blocks that must be combined in order to form the next block in the hierarchy. An element which also affects the optimal choice is that of finite population size.

5   Relaxing the Block Formation Rule

If we relax the rule which determines what blocks will be formed when a group of blocks is present at the same site, turning it into a more greedy one, we actually obtain even better search performance. We briefly studied the evolution of the system if one adopts the following rule for block-formation: Consider the set of

Size of the second block	Size of the first block			
	2	3	4	5
2	>200000			
3	>200000	>200000		
4	>200000	>200000	>200000	
5	>200000	>200000	>200000	>200000
6	191940	165934	>200000	>200000
7	>200000	>200000	>200000	>200000
8	169771	>200000	192291	>200000
9	>200000	2249	>200000	>200000
10	>200000	>200000	>200000	169021
11	>200000	>200000	>200000	>200000
12	197198	5191	170147	>200000
13	>200000	>200000	>200000	>200000

Fig. 6. Average number of iterations for a simulation with $n = m = 27$ and two variable intermediate building blocks

all the blocks at lattice site k at time t. Let the sizes of these blocks be $s_1 \ldots s_r$. Using this rule, we first add up all block sizes, $s = \sum_{i=1}^{r} s_i$ and then form new blocks in a greedy manner. That is, we first try to form as many blocks of size v_k as possible, and once we do this, we try to form, with the remaining blocks, as many blocks of size v_{k-1} as possible and so forth.

Note that this greedy technique may violate the hierarchical property associated with the triangularity of the Markov matrix. Consider for example $V = \{1, 2, 3, 5, 6, 100\}$ and suppose that we find in a given lattice site two blocks of sizes 3 and 5. Following this rule, we would replace this two blocks with a new pair, one of size 6 and one of size 2. This is an example in which larger blocks yield smaller ones, thus violating the hierarchical property.

However, as can be seen in 7, this technique is most of the time faster in reaching the goal state: that in which all particles are bound together. The same procedure than in the previous section was followed to determine the average convergence speed given a certain combination of building block sizes, but this time relaxing the block formation rule, setting the first building block at 3 and varying the second one. This difference is due to the possibility of forming bigger blocks even if there doesn't exist a combination of the current elements which could form the block exactly, favoring thus the creation of the biggest blocks, and reducing the number of small ones faster. The disadvantage of this rule appears in certain cases where the formation of a big intermediate block may be an obstacle to form the objective block, e.g., trying to form a block of 16 with $V = \{1, 4, 9, 16\}$, in the greedy way of this process, the block of 9 will be formed even with the combination of three blocks of 4, nevertheless this decreases the probability of forming the biggest block, because its probability is the probability of joining five elements in the same lattice site, whereas only with smaller blocks of 4 particles it could be formed with four of them. Further work is needed in analyzing the behavior of the system after this relaxation.

Size of the second block	Average iterations without allowing decomposition	Average iterations allowing decomposition
4	>200000	>200000
5	>200000	>200000
6	165934	160754
7	>200000	161669
8	>200000	20649
9	2249	2208
10	>200000	167006
11	>200000	169372
12	5191	5059
13	>200000	143660

Fig. 7. Comparison of number of iterations needed to form the biggest block with $'n = m = 27$ and the first intermediate block with size 3 relaxing the block formation rule and with the original rule

6   Conclusions

In this paper we studied the role of building blocks in search using a class of simple search algorithms associated with different numbers and types of building block. The context was a simple model where the desired search goal was associated with combining a certain number of elements - "particles" at a given location. Our first conclusion is that building blocks definitely help in this type of search. The reason why is that once a stable building block has been formed certain processes, such as the decay of the block are now forbidden. This means that the total number of states being searched is effectively smaller. As building block sizes increase the effective size of the search space correspondingly decreases.

Neglecting the effects of finite population the bigger the number of potential intermediate building blocks the more efficient the search. This follows from the search space reduction argument in that at each intermediate level there is a possibility to reduce the search space size. The more intermediate levels the more possibilities for such reduction.

Finally, we also saw that for a given number of intermediate blocks there existed an optimal sizing for each level block. This again ties into the search space reduction point of view. For k blocks one can think of the search space as being divided up into $k - 1$ parts (each associated to forming blocks of certain sizes, except the original size 1). By having intermediate blocks that are related geometrically one is effectively dividing up the search space into "equal" parts.

We believe that the model we have introduced here and variations thereof offer a rich environment for understanding not only how building block hierarchies and their characteristics can aid in search but also why they are so ubiquitous in the first place.

Acknowledgments. This work is supported by DGAPA grant IN120509, and by CONACYT in form of a grant to the Centro de Ciencias de la Complejidad and scholarships.

References

1. Holland, J.H.: Adaptation in Natural and Artificial Systems. MIT Press, Cambridge (1992)
2. Goldberg, D.E.: Genetic Algorithms in Search, Optimization and Machine Learning. Addison Wesley, Reading (1989)
3. Schilling, M.A.: Modularity in multiple disciplines. In: Garud, R., Langlois, R., Kumaraswamy, A. (eds.) Managing in the Modular Age: Architectures, Networks and Organizations, pp. 203–214. Blackwell Publishers, Oxford (2002)
4. Simon, H.A.: The Architecture of Complexity. Proceedings of the American Philosophical Society 106(6), 467–482 (1962)
5. Simon, H.A.: The Sciences of the Artificial, 3rd edn. MIT Press, Boston (1996)

An Analysis of Recombination in Some Simple Landscapes

David A. Rosenblueth[1] and Christopher R. Stephens[2]

[1] C_3 - Centro de Ciencias de la Complejidad
Instituto de Investigaciones en Matemáticas Aplicadas y en Sistemas, UNAM
A. Postal 20-726, México D.F. 01000
[2] C_3 - Centro de Ciencias de la Complejidad
Instituto de Ciencias Nucleares, UNAM
Circuito Exterior, A. Postal 70-543, México D.F. 04510

Abstract. Recombination is an important operator in the evolution of biological organisms and has also played an important role in Evolutionary Computation. In neither field however, is there a clear understanding of why recombination exists and under what circumstances it is useful. In this paper we consider the utility of recombination in the context of a simple Genetic Algorithm (GA). We show how its utility depends on the particular landscape considered. We also show how the facility with which this question may be addressed depends intimately on the particular representation used for the population in the GA, i.e., a representation in terms of genotypes, Building Blocks or Walsh modes. We show how, for non-epistatic landscapes, a description in terms of Building Blocks manifestly shows that recombination is always beneficial, leading to a "royal road" towards the optimum, while the contrary is true for highly epistatic landscapes such as "needle-in-a-haystack".

1 Introduction

Recombination is an important operator in the evolution of biological organisms. It has been seen as an important element also in the context of Evolutionary Computation (EC), especially in Genetic Algorithms (GAs) [4, 5] and to a lesser extent in Genetic Programming (GP) [6]. However, there has been much debate as to its utility [2].

Importantly, when talking about its utility one has to distinguish two different questions — one in which we ask if recombination itself is useful or not; versus another associated with how we apply it. The former is naturally linked to the probability, p_c, of implementing recombination, versus, $p_c(m)$, the conditional probability that, given that recombination is implemented, what recombination mask is used, i.e., the recombination distribution.[1] Importantly, as we will discuss, neither of these questions can be answered in a way that is independent of

---

[1] Here we restrict attention to recombination that is implementable via a binary recombination mask only. This covers all the standard recombination operators but not variants such as multi-parent recombination.

A. Hernández Aguirre et al. (Eds.): MICAI 2009, LNAI 5845, pp. 716–727, 2009.
© Springer-Verlag Berlin Heidelberg 2009

the fitness landscape on which the population moves. This takes us to a level of complexity which is much more than just asking whether recombination itself is useful.

In this paper using a simple diagnostic as to the utility of recombination in a given generation of a simple GA we will examine how the recombination distribution and fitness landscape affect in an inter-dependent way the efficacy of recombination. We will do this in the context of some simple toy landscapes, trying to indicate what lessons may be learned for more realistic problems. We will also consider the problem by exploiting different descriptions of the population — in terms of Building Blocks and Walsh modes — as well as the canonical representation —the genotype. We will see that in terms of these alternative representations the action of recombination and an understanding of when and under what circumstances it is useful are much more amenable to analysis.

2 Δ - The Selection-Weighted Linkage Disequilibrium Coefficient

In this section we introduce the chief diagnostic we will use to examine the utility of recombination. As we are interested in the interaction of selection and recombination we will leave out mutation. The evolution of a population of length ℓ strings is then governed by the equation [8]

$$\langle P_I(t+1) \rangle = P_I'(t) - p_c \sum_m p_c(m) \Delta_I(m, t) \tag{1}$$

where $P_I'(t)$ is the selection probability for the genotype I. For proportional selection, which is the selection mechanism we will consider here, $P_I'(t) = (f(I)/\bar{f}(t)) P_I(t)$, where $f(I)$ is the fitness of string I, $\bar{f}(t)$ is the average population fitness in the tth generation and $P_I(t)$ is the proportion of genotype I in the population. Finally, $\Delta_I(m, t)$, is the Selection-weighted linkage disequilibrium coefficient [1], for the string I and associated with the recombination mask $m = m_1 m_2 \ldots m_\ell$, such that if $m_i = 0$ the ith bit of the offspring is taken from the ith bit of the first parent, while, if $m_i = 1$ it is taken from the ith bit of the second parent. Explicitly,

$$\Delta_I(m, t) = (P_I'(t) - \sum_{JK} \lambda_I{}^{JK}(m) P_J'(t) P_K'(t)) \tag{2}$$

where $\lambda_I{}^{JK}(m) = 0, 1$ is an indicator function associated with whether or not the parental strings J and K can be recombined using the mask m into the offspring I. For example, for $\ell = 2$, $\lambda_{11}{}^{11,00}(01) = 0$, while $\lambda_{11}{}^{10,01}(01) = 1$. The contribution of a particular mask depends, as we can see, on all possible parental combinations. In this sense, $\Delta_I(m, t)$ is an exceedingly complicated function.

From equation (1), we can see that if $\Delta_I(m) > 0$ then recombination leads, on average, to a higher frequency of the string I than in its absence. In other words, in this circumstance, recombination is giving you more of I than you would have otherwise. On the contrary, if $\Delta_I(m) < 0$ then the converse is true, recombination gives you less of the string of interest than would be the case in its absence. The question then is: Can we characterize under what circumstances $\Delta_I(m)$ is positive or negative? This depends on several factors. First of all, $\Delta_I(m)$ depends explicitly on the recombination mask. Secondly, given that what enters in $\Delta_I(m)$ is $P_I'(t)$, i.e., the selection probability, it clearly depends on the fitness landscape as well. Finally, as $P_I'(t) = \sum_J F_I{}^J P_J(t)$ it depends on the population at generation t. Here we will concentrate on the issue of the dependence on the fitness landscape and to a lesser extent the recombination distribution. To eliminate any bias associated with a particular population we will consider a random, infinite population where each string is represented in the proportion $P_I(t) = 1/|\Omega|$, where $|\Omega|$ is the dimension of the search space.

We will consider only two types of fitness landscape — "counting ones" (CO) and "needle-in-a-haystack" (NIAH). These represent the two ends of a continuous spectrum of problems with varying degrees of epistasis, with CO having no epistasis and NIAH with maximal epistasis. CO is defined by a fitness function for binary strings where $f_I = n$ if string I has n ones. This is a landscape without epistasis. NIAH on the other hand is defined by a fitness function $f_I = f_1$ if I is the optimal string and $f_I = f_0$ for any other string. By considering these two problems we will study how different degrees of epistasis in the fitness landscape impact on the characteristics and utility of recombination.

In the space of genotypes the factor $\lambda_I{}^{JK}(m)$ is very complicated. For a given target string and mask, $\lambda_I{}^{JK}(m)$ is a matrix on the indices J and K associated with the parents. For binary strings, for every mask there are $2^\ell \times 2^\ell$ possible combinations of parents that need to be checked to see if they give rise to the offspring I. Only 2^ℓ elements of the matrix are non-zero. The question is: which ones? The complication of $\lambda_I{}^{JK}(m)$ in terms of genotypes is just an indication of the fact that the latter are not a natural basis for describing the action of recombination. A more appropriate basis is the Building Block Basis (BBB) [7], wherein only the building block schemata that contribute to the formation of a string I enter. In this case

$$\Delta_I(m,t) = (P_I'(t) - P_{I_m}'(t)P_{I_{\bar{m}}}'(t)) \tag{3}$$

where $P_{I_m}'(t)$ is the selection probability of the building block I_m and $I_{\bar{m}}$ is the complementary block such that $I_m \cup I_{\bar{m}} = I$. Both blocks are specified by an associated recombination mask, $m = m_1 m_2 \ldots m_\ell$, such that if $m_i = 0$ the ith bit of the offspring is taken from the ith bit of the first parent, while, if $m_i = 1$ it is taken from the ith bit of the second parent. For instance, for $\ell = 3$, if $I = 111$ and $m = 001$ then $I_m = 11*$ and $I_{\bar{m}} = **1$. Note that in this case the structure of $\lambda_I{}^{JK}(m)$ is very simple when both J and K are Building Block schemata. For a given I and m one unique building block, I_m, is picked out. The second Building Block $I_{\bar{m}}$ then enters as the complement of I_m in I. This means that $\lambda_I{}^{JK}(m)$ is skew diagonal with only one non-zero element on that skew diagonal for a given m and I. At a particular locus of the offspring, the associated allele is taken from the first or second parent according to the value of m_i. If it is taken from the first parent then it did not matter what was the corresponding allele in the second. This fact is represented by the normal schema wildcard symbol $*$.

The interpretation of equation (3) is that recombination is favourable if the probability $P'_{I_m}(t)P'_{I_{\bar{m}}}(t)$ to select the Building Blocks I_m and $I_{\bar{m}}$ is greater than the probability to select the target string.

Another basis that has been considered in GAs and elsewhere is the Walsh basis [3, 9–11], which is just the Fourier transform for the case of binary strings. In this basis,

$$\hat{\Delta}_I(m,t) = (\hat{P}'_I(t) - \hat{P}'_{I_m}(t)\hat{P}'_{I_{\bar{m}}}(t)) \tag{4}$$

where the component I now refers to a Walsh mode rather than a string and $\hat{P}'_I(t)$ is the Walsh transform of the selection probability and therefore is not necessarily positive definite. The notation I_m and $I_{\bar{m}}$ now refer to the "building blocks" of a particular Walsh mode rather than the blocks of a genotype. The logic is, however, very similar to that of building blocks for genotypes. A difference is that the wildcard symbol does not enter directly but is represented by a 0 in the Walsh mode. Thus, for example, to construct the Walsh mode 111 with a mask 001, the first parental Walsh mode is 110 while the second is 001. The first mode is determined by the mask m. If $I_i = 1$ and $m_i = 0$ then the ith allele in the first parent is 1 and in the second 0. On the contrary, if $I_i = 0$ then the corresponding allele in both parents is 0.

As $\hat{P}'_I(t)$ can be positive or negative, the interpretation of (4) is somewhat more subtle than for genotypes or building blocks. The question is rather, for a given target string there is a corresponding set of target Walsh modes. If recombination is to be favoured then it must favour those particular Walsh modes. As these can be positive or negative then for a positive mode, i.e., $\hat{P}'_I(t) > 0$, recombination will give better results if $\hat{\Delta}_I(m,t) < 0$, while, for a negative mode, if $\hat{\Delta}_I(m,t) > 0$.

3   Results

We first present the results for $\ell = 4$ and a CO landscape. The optimal string is 1111 and there are $2^4 = 16$ possible recombination masks. As previously mentioned, to calculate $\Delta_I(m)$ we set $P_I(t) = 1/2^4$, equivalent to that of a uniform population. In Fig. 1 we see the results for the 16 different genotypes and the 16 different possible masks. Plotted on the vertical axis is $-100 * \Delta_I(m)$, so that masks that favour the formation of a given genotype will have positive values of $-100 * \Delta_I(m)$. The ordering of both genotypes and masks is the standard lexicographical ordering from 0000 to 1111, with genotypes going from the left corner to the bottom corner, while the masks are coloured with dark-blue to the left corresponding to $m = 0000$ and red-brown corresponding to $m = 1111$. The last dark-brown row represents the values of the fitness function for each genotype.

In Table 1 we see the same results in tabular form. What can be gleaned from these results is the following: That, for CO, any recombination mask is such

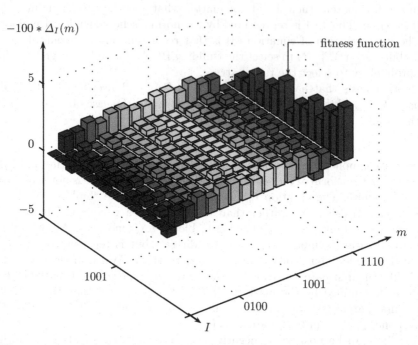

Fig. 1. $-100 * \Delta_I(m)$ as a function of genotype and recombination mask for CO

as to favour the production of the optimal string 1111. Interestingly, the most favourable masks are associated with taking half the alleles from one parent and half from the other as opposed to three from one and one from the other. So, we can see that recombination in a CO landscape is useful in producing the optimal string and that this is true for any mask. However, if we look at these results further we also see that recombination also favours the production of the string 0000 which is the least fit string! What is more, it produces this string in equal proportion to the optimal one. In this sense it is not completely transparent how recombination helps when viewed in terms of genotypes as it helps produce what helps but, at first glance, also produces what hinders.

However, if we turn to a description of what happens in terms of building blocks we see a much more illuminating story. In Fig. 2 we see the values of $\Delta_I(m)$ as a function of mask and building block, as opposed to genotype. Now the lexicographical ordering is from 0000 to 1111 but where each 0 is replaced by a wildcard symbol so that $0000 \rightarrow ****$ for example, with the latter in the left corner and 1111 in bottom corner. Masks go from 0000 (dark blue, to the left) to 1111 (brown, to the right).

The key difference with Fig. 1 and Table 1 is that in terms of building blocks we can see that recombination is *always* beneficial. What this means is that recombination helps in constructing the optimum string while at the same time helping in the construction of the different building blocks of that string. In this sense, there is a "royal road" whereby recombination helps form those building

Table 1. Tabular form of Fig. 1

genotype	0000	0001	0010	0011	0100	0101	0110	0111
mask								
0000	0	0	0	0	0	0	0	0
0001	−0.0117	0.0117	−0.0039	0.0039	−0.0039	0.0039	0.0039	−0.0039
0010	−0.0117	−0.0039	0.0117	0.0039	−0.0039	0.0039	0.0039	−0.0039
0011	−0.0156	0	0	0.0156	0	0	0	0
0100	−0.0117	−0.0039	−0.0039	0.0039	0.0117	0.0039	0.0039	−0.0039
0101	−0.0156	0	0	0	0	0.0156	0	0
0110	−0.0156	0	0	0	0	0	0.0156	0
0111	−0.0117	−0.0039	−0.0039	0.0039	−0.0039	0.0039	0.0039	0.0117
1000	−0.0117	−0.0039	−0.0039	0.0039	−0.0039	0.0039	0.0039	0.0117
1001	−0.0156	0	0	0	0	0	0.0156	0
1010	−0.0156	0	0	0	0	0.0156	0	0
1011	−0.0117	−0.0039	−0.0039	0.0039	0.0117	0.0039	0.0039	−0.0039
1100	−0.0156	0	0	0.0156	0	0	0	0
1101	−0.0117	−0.0039	0.0117	0.0039	−0.0039	0.0039	0.0039	−0.0039
1110	−0.0117	0.0117	−0.0039	0.0039	−0.0039	0.0039	0.0039	−0.0039
1111	0	0	0	0	0	0	0	0

genotype	1000	1001	1010	1011	1100	1101	1110	1111
mask								
0000	0	0	0	0	0	0	0	0
0001	−0.0039	0.0039	0.0039	−0.0039	0.0039	−0.0039	0.0117	−0.0117
0010	−0.0039	0.0039	0.0039	−0.0039	0.0039	0.0117	−0.0039	−0.0117
0011	0	0	0	0	0.0156	0	0	−0.0156
0100	−0.0039	0.0039	0.0039	0.0117	0.0039	−0.0039	−0.0039	−0.0117
0101	0	0	0.0156	0	0	0	0	−0.0156
0110	0	0.0156	0	0	0	0	0	−0.0156
0111	0.0117	0.0039	0.0039	−0.0039	0.0039	−0.0039	−0.0039	−0.0117
1000	0.0117	0.0039	0.0039	−0.0039	0.0039	−0.0039	−0.0039	−0.0117
1001	0	0.0156	0	0	0	0	0	−0.0156
1010	0	0	0.0156	0	0	0	0	−0.0156
1011	−0.0039	0.0039	0.0039	0.0117	0.0039	−0.0039	−0.0039	−0.0117
1100	0	0	0	0	0.0156	0	0	−0.0156
1101	−0.0039	0.0039	0.0039	−0.0039	0.0039	0.0117	−0.0039	−0.0117
1110	−0.0039	0.0039	0.0039	−0.0039	0.0039	−0.0039	0.0117	−0.0117
1111	0	0	0	0	0	0	0	0

blocks. There are no non-beneficial masks, and only a couple, the ones where an entire parent is taken as the building block, that are neutral. The other values of $\Delta_I(m)$ that are zero are those where I represents a schema of order one. As there is no way of cutting such a schema recombination acts trivially.

So, what can we learn from Fig. 2? The most important thing is that $\Delta_{1111}(m)$ is always negative or zero for any mask for the optimum string. For masks that split the components of the optimum asymmetrically, i.e., three and one, as

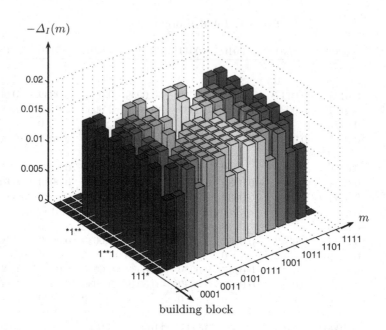

Fig. 2. $-\Delta_I(m)$ for CO in the building block representation

opposed to symmetrically, two and two, $\Delta_{1111}(m)$ is less negative showing that recombination is more effective for symmetric combinations. The most important aspect of these results though is that we can see how recombination of the Building Blocks of the Building Blocks of the optimal string follow the same pattern, i.e., Δ_{I_m} is always negative or zero for the Building Blocks. When Δ_{I_m} is zero it is because the corresponding mask is acting trivially on the Building Block so there is no non-trivial interaction between blocks.

Turning now to what happens in the Walsh basis, in Fig. 3, we see that in contrast with the case of genotypes or Building Blocks, the only non-zero Walsh modes $\tilde{\Delta}_I(m)$ are those Walsh modes with two ones. This is due to the fact that the landscape is counting ones. This means that, given that we start with pure $\tilde{P}_{0000}$, selection can only excite/create Walsh modes of order one, i.e., 0001, 0010, 0100 and 1000. Afterwards, recombination can combine these modes to form an order two mode with two ones. One cannot, however, form an order 3 or order 4 mode in one generation starting only with $\tilde{P}_{0000}$ being non-zero. More generally, if a landscape has epistasis associated with Walsh modes of order b then selection can convert a Walsh mode of $P_I(t)$ of order c and convert it to a Walsh mode of order $d = b+c$. For those Walsh modes of $\tilde{\Delta}_I(m)$ that are of order two we note that $\tilde{\Delta}_I(m)$ is always less than zero. Thus, we see that in the Walsh basis the action of recombination is always manifestly beneficial, producing those "building block" Walsh modes $\tilde{P}'_I(t)$ that are necessary to produce the optimal string 1111.

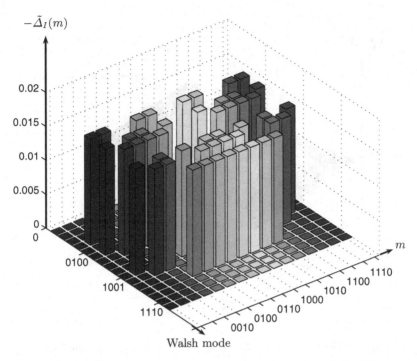

Fig. 3. $-\tilde{\Delta}_I(m)$ as a function of Walsh modes and mask for CO

We now turn to the NIAH landscape. Starting with the genotypic representation as with the case of CO we show the results in Fig. 4. Once again on the vertical axis we have $-100 \times \Delta_I(m)$ with the landscape being the rightmost row (brown). Genotypes run from the left corner, 0000, to the bottom corner, 1111, while masks are denoted by different colours with dark blue corresponding to the mask 0000 and reddish-brown to the mask 1111.

As with the case of the CO landscape, in terms of genotypes it is not clear to what extent recombination is beneficial. Looking at the optimal string, i.e., the "needle" (first row from the bottom), we see that $\Delta_{1111}(m) > 0$ for all masks. In other words, for NIAH, recombination is not beneficial in the production of the optimal string. We do see, however, that there are some other strings for which it is beneficial. For instance, for the string 1110 (second row from the bottom). Overall however, as with CO, in the genotype basis the action and relative merits of recombination are not readily apparent. If we turn to the case of Building Blocks in the NIAH landscape, in Fig. 5 we see the corresponding results. Now, interestingly, we see the complete opposite of what happened with CO. Now, every mask leads to a $\Delta_{I_m}(m') > 0$ for every Building Block I_m of the needle. In other words now there is a "royal road" that leads away from the optimum.

Turning now to the case of NIAH in the Walsh basis for $N = 4$ in Fig. 6 we see the results for $\tilde{\Delta}_I(m)$ for different Walsh modes and different recombination

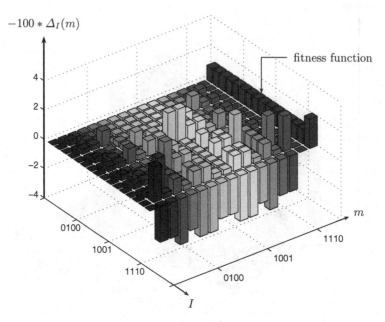

Fig. 4. $-100 * \Delta_I(m)$ as a function of genotype and mask for NIAH

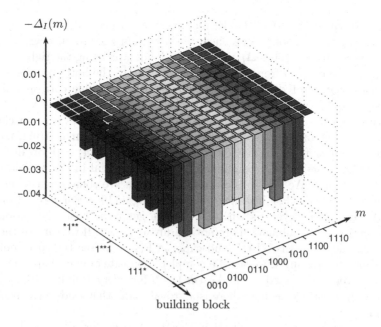

Fig. 5. $-\Delta_I(m)$ for NIAH in the building block representation

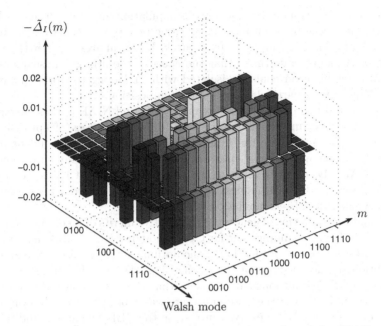

Fig. 6. $-\tilde{\Delta}_I(m)$ as a function of Walsh modes and mask for NIAH

masks. In contrast to the case of the BBB we see that here the values of $\tilde{\Delta}_I(m)$ are both positive and negative. One might interpret this as meaning that recombination is beneficial for some Walsh modes and disadvantageous for others. However, we must remember here that in contrast to the case of Building Blocks $\tilde{P}_I(t)$ does not directly represent a probability but, rather, is a Walsh transform of a probability. In this sense, it is possible for $\tilde{P}_I(t)$ to take on negative as well as positive values. Noting that the Walsh components of $P_I(t)$ are negative when the corresponding Walsh mode has an odd number of ones and positive for an even number we observe that $\tilde{\Delta}_I(m)$ is positive for $\tilde{P}_I(t)$ that are positive and negative for $\tilde{P}_I(t)$ that are negative. Thus, in both cases recombination acts so as to reduce the magnitude of $\tilde{P}_I(t)$ and therefore is disadvantageous irrespective of the sign of $\tilde{P}_I(t)$. This result then concords with the result found for Building Blocks that recombination in a NIAH landscape is uniformly disadvantageous.

4 Conclusion

In this paper we considered the question of under what circumstances recombination is beneficial. To do this we used a theoretical approach, using the exact evolution equations for an infinite population GA. For simplicity, we considered only four-bit strings. However, our general conclusions are valid for any N. We used the Selection-weighted linkage disequilibrium coefficient as a means to measure in a given generation, for a given population and a given fitness landscape whether recombination makes a positive or negative contribution to the production of a given string for a given mask. In our analysis we set the population to be random, uniform so that the effect of the fitness landscape on the effects of recombination could be isolated without having to worry about any bias introduced by a particular population composition.

We considered two particular landscapes —CO and NIAH— these being two extremes of a spectrum of landscapes with different degrees of epistasis, CO having zero epistasis, where the fitness of any given bit is independent of the rest, and NIAH having maximal epistasis, where the fitness of any bit depends on the others. We also chose to analyze $\Delta_I(m)$ in three different bases: genotypes, Building Blocks and Walsh modes; as each one gave a different perspective on the relative benefits of recombination.

We found that in terms of genotypes the relative merits of recombination were not particularly clear, finding that for CO, although recombination made a positive contribution to the production of the optimal string, it also made an equal contribution to the antipode of that string, i.e., the least fit string. Similarly, in the case of NIAH, although it was clear that there was a negative contribution to the production of the optimum there was a positive contribution to the production of other genotypes. By contrast, in the BBB, the action and relative merits of recombination were completely transparent. For CO it was noted that recombination is not only beneficial for the production of the optimum but was also beneficial for the production of any and all building blocks of that string of order greater than one. By contrast, for NIAH, the opposite was true, recombination gave a negative contribution to the production of not only the optimum but also all of its building blocks. We also noted that these results could also be understood using a Walsh mode analysis of $\Delta_I(m)$ with similar conclusions to that with building blocks. In other words, that in CO recombination favoured the production of those Walsh modes which contributed to the production of the optimum, while for NIAH the converse was true. The added benefit of an analysis in terms of Walsh modes is that it is possible to see how different degrees of epistasis in the landscape affect $\Delta_I(m)$. For instance, in CO it was noted that, starting from a random, uniform population corresponding to the presence of $\hat{P}_{0000}$ only in the initial population, the only Walsh modes of the population vector that could be excited by selection were order-one Walsh modes which crossover could then recombine into order-two Walsh modes.

So, what are we to learn from these results? Firstly, that recombination works well with non-epistatic landscapes and badly with very epistatic ones. However, a general landscape is neither CO nor NIAH. What can be said there? Based on the evidence of the current results we can hypothesize that if there are "modules" in the landscape, i.e., epistatically linked groups of bits within a given module or block, that are *not* epistatically linked to bits outside that block, then recombination masks such that all of the block is inherited from one parent will be favoured leading to a positive overall contribution from recombination for that mask. On the contrary, a mask that breaks epistatic links between bits will lead to an overall negative contribution.

Acknowledgements

This work was partially supported by DGAPA grant IN120509 and by a special Conacyt grant to the Centro de Ciencias de la Complejidad. DAR is grateful to the IIMAS, UNAM for use of their facilities and to Carlos Velarde for help with the figures.

References

1. Aguilar, A., Rowe, J., Stephens, C.R.: Coarse graining in genetic algorithms: Some issues and examples. In: Cantú-Paz, E. (ed.) GECCO 2003. Springer, Heidelberg (2003)
2. Aguirre, H., Tanaka, K.: Genetic algorithms on NK-landscapes: Effects of selection, drift, mutation and recombination. In: Raidl, G.R., Cagnoni, S., Cardalda, J.J.R., Corne, D.W., Gottlieb, J., Guillot, A., Hart, E., Johnson, C.G., Marchiori, E., Meyer, J.-A., Middendorf, M. (eds.) EvoIASP 2003, EvoWorkshops 2003, EvoSTIM 2003, EvoROB/EvoRobot 2003, EvoCOP 2003, EvoBIO 2003, and EvoMUSART 2003. LNCS, vol. 2611, pp. 131–142. Springer, Heidelberg (2003)
3. Goldberg, D.E.: Genetic algorithms and Walsh functions: Part I. A gentle introduction. Complex Systems 3, 123–152 (1989)
4. Goldberg, D.E.: Genetic Algorithms in Search, Optimization and Machine Learning. Addison Wesley, Reading (1989)
5. Holland, J.H.: Adaptation in Natural and Artificial Systems. MIT Press, Cambridge (1993)
6. Langdon, W.B., Poli, R.: Foundations of Genetic Programming. Springer, Berlin (2002)
7. Stephens, C.R.: The renormalization group and the dynamics of genetic systems. Acta Phys. Slov. 52, 515–524 (2002)
8. Stephens, C.R., Waelbroeck, H.: Schemata evolution and building blocks. Evol. Comp. 7, 109–124 (1999)
9. Vose, M.D., Wright, A.H.: The simple genetic algorithm and the Walsh transform: Part II, the inverse. Evolutionary Computation 6(3), 275–289 (1998)
10. Weinberger, E.D.: Fourier and Taylor series on fitness landscapes. Biological Cybernetics 65th, 321–330 (1991)
11. Wright, A.H.: The exact schema theorem (January 2000),
    http://www.cs.umt.edu/CS/FAC/WRIGHT/papers/schema.pdf

An Empirical Investigation of
How Degree Neutrality Affects GP Search

Edgar Galván-López[1] and Riccardo Poli[2]

[1] Natural Computing Research & Applications Group, Complex and Adaptive Systems Lab
University College Dublin
edgar.galvan@ucd.ie
[2] University of Essex, School of Computer Science and Electonic Engineering,
Wivenhoe Park, Colchester, CO4 3SQ, UK
rpoli@essex.ac.uk

Abstract. Over the last years, neutrality has inspired many researchers in the area of Evolutionary Computation (EC) systems in the hope that it can aid evolution. However, there are contradictory results on the effects of neutrality in evolutionary search. The aim of this paper is to understand how neutrality - named in this paper degree neutrality - affects GP search. For analysis purposes, we use a well-defined measure of hardness (i.e., fitness distance correlation) as an indicator of difficulty in the absence and in the presence of neutrality, we propose a novel approach to normalise distances between a pair of trees and finally, we use a problem with deceptive features where GP is well-known to have poor performance and see the effects of neutrality in GP search.

1   Introduction

Despite the proven effectiveness of Evolutionary Computation (EC) systems, there are limitations in such systems and researchers have been interested in making them more powerful by using different elements. One of these elements is *neutrality* (*the neutral theory of molecular evolution* [8]) which the EC community has incorporated in their systems in the hope that it can aid evolution. Briefly, neutrality considers a mutation from one gene to another as neutral if this modification does not affect the fitness of an individual.

EC researchers have tried to incorporate neutrality in their systems in the hope that it can aid evolution. Despite the vast number of publications in this field, there are no general conclusions on the effects of neutrality and in fact, quite often, there is a misconception with regard to what neutrality is. There are also many contradictory results reported by EC researchers on neutrality.

For instance, in *"Finding Needles in Haystacks is not Hard with Neutrality"* [20], Yu and Miller performed runs using the well-known Cartesian GP (CPG) representation [9,10] and also used the even-n-parity Boolean functions with different degrees of difficulty ($n = \{5, 8, 10, 12\}$). They compared performance when neutrality was present and in its absence and reported that the performance of their system was better when neutrality was present.

A. Hernández Aguirre et al. (Eds.): MICAI 2009, LNAI 5845, pp. 728–739, 2009.

A few years later, Collins claimed the opposite and presented the paper entitled *"Finding Needles in Haystacks is Harder with Neutrality"* [2]. He further explored the idea presented by Yu and Miller and explained that the choice of this type of problem is unusual and in fact not suitable for analysing neutrality using CGP. This is because both the landscape and the form of the representation used have a high degree of neutrality and these make the drawing of general conclusions on the effects of neutrality difficult.

These works (both nominated as best papers in their conference tracks!) are just two examples of many publications available in the specialised literature which show controversial results on neutrality.

The aim of this paper is to understand the effects of neutrality in GP search. For this purpose a new form of neutrality, called *degree neutrality*, will be proposed and studied in detail and a problem with deceptive features will be used to see how GP behaves in the absence and in the presence of neutrality.

This paper is structured as follows. In the next section, a new form of neutrality will be introduced. In Section 3, a well-known measure of difficulty, called fitness distance correlation will be described. A novel method to normalise the distance between a pair of trees will be explained in Section 4. Section 5 provides details on the experimental setup used. Analysis and conclusions of the results found using our approach are presented in Section 6.

2 Degree Neutrality

Degree neutrality is a form of neutrality induced by adding 'dummy' arguments (i.e., terminals) to internal nodes (i.e., functions). More specifically, this type of neutrality works as follows:

- Once an individual has been created, the internal nodes that have an arity lower than a given arity are marked (i.e., if an internal node is A and needs to be marked, it is relabelled as NA).
- Dummy terminals are added to the internal nodes (i.e., functions) that have been marked. The number of dummy terminals that will be added is determined by a given arity (i.e., an arity specified by the user). For instance, suppose that the given arity is 5 and a tree has functions of arity 1 then those nodes will be marked and 4 dummy terminals will be added to the marked nodes. These are dummy in the sense that their value is totally ignored when computing the output of the node.

Suppose that one is using the language proposed in [14]. That is, the function set is formed by letters (i.e., $F_{set} = \{A, B, C, \cdots\}$) and the arities of each function are as follows: 1 for function A, 2 for function B, 3 for function C and so forth. The terminal set is defined by a single element $T_{set} = \{X\}$. Now, let us define a function set $F_{set} = \{A, B, C, D, E\}$. This function set has a maximum arity of 5. Let us assume that this is the arity that will be used to add degree neutrality. A typical GP individual using the function set F_{set} is shown at the top of Figure 1 and the same individual with degree neutrality is shown at the bottom of the figure. Notice how all internal nodes of the resulting individual (Figure 1 bottom) have now the same arity (i.e., 5).

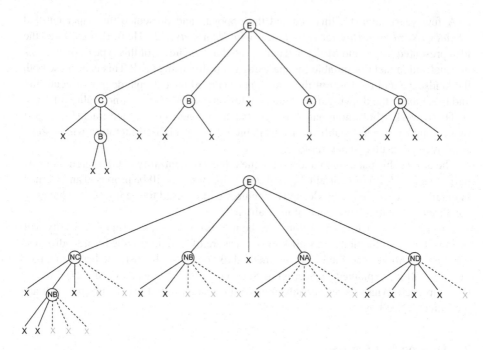

Fig. 1. A typical GP individual created using the language proposed in [14] (top) and the same individual with degree neutrality using a maximum arity of 5 (bottom)

This can easily be extended if one would like to specify, for instance, that the functions defined in the function set are of arities 4 and 5. Then all the function with arities lower than 4 could be marked and extended with dummy arguments. As a result of this, all the functions of the function set would be of arities 4 and 5. The same technique could be applied if the user wanted all the functions defined in the function set to have arities 3, 4 and 5 (i.e., all the internal nodes whose arities are lower than 3 should be "filled" by dummy arguments).

To analyse how degree neutrality will affect the sampling of individuals performed by GP, this form of neutrality will be examined in conjunction with constant neutrality. This form of neutrality was first studied using a binary GA [5,3] and then analysed using GP [4]. Briefly, the idea is that in this approach, neutrality is "plugged" into the traditional GP representation by adding a flag to the representation: when the flag is set, the individual is on the neutral network and its fitness has a pre-fixed value (denoted in this work by f_n). When the flag is not set, the fitness of the individual is determined as usual. See [4] for a full description of its implementation. We decided to use constant neutrality to study the effects of degree neutrality because with many primitive sets, GP has the ability to create a rich and complex set of neutral networks. This may be a useful feature, but it is a feature that is hard to control and analyse. However, using these types of neutrality we are in total control so, we are in position of study its effects in detail.

In the following section, a well-defined measure of hardness will be introduced and this will help us to better understand the effects of neutrality in evolutionary search.

3 Fitness Distance Correlation

In [7], Jones proposed an heuristic called *fitness distance correlation* (fdc) using the typical GA representation (i.e., the bitstring representation) and successfully tested it on several problems.

fdc is an algebraic measure to express the degree to which the fitness function conveys information about distance to the searcher.

The idea of using *fdc* as an heuristic method, as stated in [7], was to create an algebraic metric that can give enough information to determine the difficulty (for a GA) of a given problem when the global optimum is known in advance. To achieve this, Jones explained that it is necessary to consider two main elements:

1. To determine the distance between a potential solution and the global optimum (when using a bitstring representation, this is accomplished using the Hamming distance) and
2. To calculate the fitness of the potential solution.

With these elements in hand, one can easily compute the *fdc* coefficient using Jones' calculation [7] thereby, in principle, being able to determine in advance the hardness of a problem.

The definition of *fdc* is quite simple: given a set $F = \{f_1, f_2, ..., f_n\}$ of fitness values of n individuals and the corresponding set $D = \{d_1, d_2, ..., d_n\}$ of distances of such individuals from the nearest optimum, *fdc* is given by the following correlation coefficient:

$$fdc = \frac{C_{FD}}{\sigma_F \sigma_D},$$

where:

$$C_{FD} = \frac{1}{n} \sum_{i=1}^{n} (f_i - \overline{f})(d_i - \overline{d})$$

is the covariance of F and D, and σ_F, σ_D, $\overline{f}$ and $\overline{d}$ are the standard deviations and means of F and D, respectively. The n individuals used to compute *fdc* are obtained via some form of random sampling.

According to [7] a problem can be classified in one of three classes, depending on the value of fdc:

- *misleading* ($fdc \geq 0.15$), in which fitness tends to increase with the distance from the global optimum,
- *difficult* ($-0.15 < fdc < 0.15$), for which there is no correlation between fitness and distance; and
- *easy* ($fdc \leq -0.15$), in which fitness increases as the global optimum approaches.

There are some known weaknesses with *fdc* as a measure of problem hardness [1,15]. However, it is fair to say that this method has been generally very successful [3,4,5,7,12].

Motivated by the good results found by *fdc*, this measure of hardness has been further explored using tree-like structures. There are some initial works that have attempted calculating the distance between a pair of trees [11,16]. However, these works were

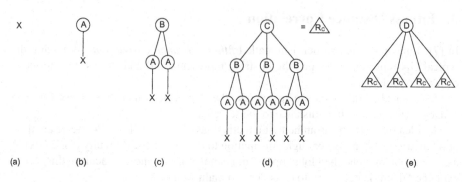

Fig. 2. Distances calculated between different trees. The language that has been used is the one proposed in [14]. $k = 1$ and c has been defined as the arity that each function has (i.e., $c(A) = 1$, $c(B) = 2$, $c(C) = 3$ and so forth). The distance between trees (a) and (b) is denoted by $distance(a,b)$. So, $distance(a,b) = 1.5$, $distance(b,c) = 4.0$, $distance(c,d) = 11.75$ and $distance(d,e) = 36.75$.

limited in the sense that they did not offer a reliable distance. In, [18,19,17] the authors overcame these limitations and computed and defined a distance[1] between a pair of tress.

There are three steps to calculate the distance between tree T_1 and T_2:

- T_1 and T_2 must be aligned to the most-left subtrees,
- For each pair of nodes at matching positions, the difference of their codes c (typically c is the index of an instruction within the primitive set) is calculated, and
- The differences calculated in the previous step are combined into a weighted sum (nodes that are closer to the root have greater weights than nodes that are at lower levels).

Formally, the distance between trees T_1 and T_2 with roots R_1 and R_2, respectively, is defined as follows:

$$dist(T_1, T_2, k) = d(R_1, R_2) + k \sum_{i=1}^{m} dist(child_i(R_1), child_i(R_2), \frac{k}{2}) \qquad (1)$$

where: $d(R_1, R_2) = (|c(R_1) - c(R_2)|)^z$ and; $child_i(Y)$ is the ith of the m possible children of a node Y, if $i < m$, or the empty tree otherwise. Note that c evaluated on the root of an empty tree is 0 by convention. The parameter k is used to give different weights to nodes belonging to different levels in the tree. In Figure 2 using Equation 1, various distances have been calculated using different trees. The distance produced successful results on a wide variety of problems [3,4,17,18,19].

Once a distance has been computed between two trees, it is necessary to normalise it in the range $[0,1]$. In [18], Vanneschi proposed five different methods to normalise the distance. In this work, however, we will introduce a new method that carries out this task more efficiently. This will be presented in the following section.

---

[1] This is the distance used in this work. We will use the code provided in [3, Appendix D] to calculate the distance between a pair of trees.

4 Normalisation by Maximum Distance Using a Fair Sampling

Inspired by the methods proposed in [18], we propose a new method called "Normalisation by Maximum Distance Using a Fair Sampling" which is used to conduct the empirical experiments show in this work. This method works as follows.

1. A sample of n_s individuals is created using the ramped half-and-half method using a maximum depth greater than the maximum depth defined to control bloat,
2. The distance is calculated between each individual belonging to the sample and the global optimum,
3. Once all the distances have been calculated using n_s individuals that belong to the sample, the maximum distance K_s found in the sampling is stored,
4. n_p individuals that belong to the population are created at random,
5. The distance is calculated between each individual belonging to the population and the global optimum,
6. Once all the distances have been calculated using n_p individuals that belong to the population, the maximum distance K_p found in the population is stored,
7. The global maximum distance, K, is the largest distance between K_s and K_p.

At the end of this process, the global maximum distance K is found. Given that the depth d_s used to create a sample of n_s individuals (for our experiments n_s is typically 10 times bigger than the population size) is greater than the depth d_p defined to control bloat (for our experiments $d_s = d_p + 2$), then throughout the evolutionary process, it is highly unlikely we will ever find a higher value for the global maximum distance. In fact, as we mentioned previously, this normalisation method was used to conduct the experiments reported in this work and in none of them was a higher distance found.

The global maximum distance is found after the sampling of individuals and the creation of the population. Clearly, the main advantage of this process is that during the evolution of individuals, the complexity of the process to normalise distances can be reduced substantially compared to Vanneschi's methods (e.g., "constant-normalisation by iterated search").

5 Experimental Setup

5.1 Trap Function

The problem used to analyse the proposed form of neutrality is a Trap function [6]. The fitness of a given individual is calculated taking into account the distance of this individual from the global optimum. Formally, a trap function is defined as:

$$f(\ell) = \begin{cases} 1 - \frac{d(\ell)}{d_{min}} & \text{if } d(\ell) \leq d_{min}, \\ \frac{r(d(\ell) - d_{min})}{1 - d_{min}} & \text{otherwise}, \end{cases}$$

where $d(\ell)$ is the normalised distance between a given individual and the global optimum solution. $d(\ell)$, d_{min} and r are values in the range $[0,1]$. d_{min} is the slope-change

Table 1. Parameters used for the problems used to conduct extensive empirical experiments using degree neutrality

Parameter	Value
Population Size	400
Generations	300
Neutral Mutation Probability (P_{nm})	0.05
Crossover Rate	90%
Tournament group size	10
Independent Runs	100

location for the optima and r sets their relative importance. For this problem, there is only one global optimum and by varying the parameters d_{min} and r, the problem can be made either easier or harder.

For this particular problem, the function and terminal sets are defined using the language proposed in [14] (see Section 2).

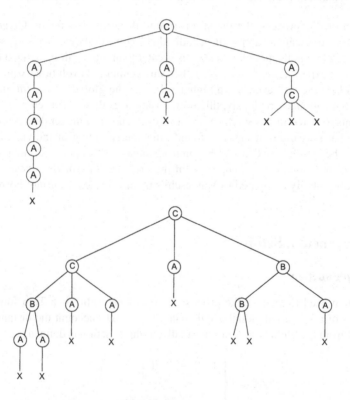

Fig. 3. A global optimum used in our first experiment (top) and a global optimum used in our second experiment (bottom)

Standard crossover was used to conduct our experiments. Tournament selection was used to conduct our experiments. Furthermore, runs were stopped when the maximum number of generations was reached. The parameters used are given in Table 1.

To avoid creating the global optimum during the initialisation of the population for the Trap function, the full initialisation method has been used. Figure 3 shows the global optima used in our experiments.

6   Analysis of Results and Conclusions

The results of *fdc*, the average number of generations required to find these global optima and the percentage of successes are shown in Tables 4 and 5 when constant neutrality and degree neutrality with a maximum arity of 5 are used (i.e., all the functions are of the same arity).

In Tables 2 and 3, we show the results when using constant neutrality (indicated by $\{1,2,3,4,5\}$ which means that there are functions of those arities) and degree neutrality. As we mentioned previously, the performance of the GP system increases when either form of neutrality is added, specifically in the range of $[0.30 - 0.65]$. As can be seen from these results, degree neutrality has almost the same effect as constant neutrality when the functions declared in the function set are of mixed arities.

This, however, is not the case when neutral degree is added and the functions declared in the function set are of the same arity. Under these circumstances, the performance of the GP system is increased in almost all cases, as shown in Tables 4 and 5. As can be seen, the predictions done by *fdc* are roughly correct. That is, in the presence of neutrality, the percentage of successes tends to increase, particularly when the fitness of the constant value lies in the range of $[0.30 - 0.75]$. There is, however, a variation in the percentage of successes when there is a mixture of arities (i.e.$\{1,2,3,4,5\}$) and

Table 2. Performance of GP using swap-crossover. Constant neutrality (i.e., $\{1,2,3,4,5\}$) and degree neutrality with functions of different arities (i.e.,$\{2,3,4,5\}, \{3,4,5\}, \{4,5\}$) were used. f_n stands for the different fixed fitness values used. The global optimum is shown at the top of Figure 3 and setting $B = 0.01$ and $R = 0.8$ (i.e., the problem is considered to be very difficult).

f_n value	$\{1,2,3,4,5\}$		$\{2,3,4,5\}$		$\{3,4,5\}$		$\{4,5\}$	
	Avr. Gen	% Suc.	Avr. Gen	% Suc.	Avr. Gen	% Suc.	Avr. Gen	% Suc.
-	4.50	2%	4.00	1%	4.00	2%	4	1%
0.10	N/A	0%	N/A	0%	N/A	0%	N/A	0%
0.20	6.00	1%	N/A	0%	6.00	1%	N/A	0%
0.30	11.16	6%	15.50	2%	29.66	3%	18.00	9%
0.40	43.75	12%	37.00	9%	65.83	6%	86.57	7%
0.50	76.80	10%	61.30	13%	67.00	10%	80.69	13%
0.60	111.75	8%	97.00	6%	107.85	7%	104.87	8%
0.70	157.25	8%	118.00	3%	124.60	5%	83.00	1%
0.80	266.50	2%	N/A	0%	N/A	0%	N/A	0%
0.90	N/A	0%	N/A	0%	N/A	0%	N/A	0%

Table 3. Performance of GP using swap-crossover. Constant neutrality (i.e., arities of functions = $\{1,2,3,4,5\}$) and degree neutrality with functions of different arities (i.e.,$\{2,3,4,5\},\{3,4,5\},\{4,5\}$) were used. f_n stands for the different fixed fitness values used. The global optimum is shown at the bottom of Figure 3 and setting $B = 0.01$ and $R = 0.8$ (i.e., the problem is considered to be very difficult).

f_n value	$\{1,2,3,4,5\}$		$\{2,3,4,5\}$		$\{3,4,5\}$		$\{4,5\}$	
	Avr. Gen	% Suc.	Avr. Gen	% Suc.	Avr. Gen	% Suc.	Avr. Gen	% Suc.
-	N/A	0%	N/A	0%	N/A	0%	N/A	0%
0.10	N/A	0%	N/A	0%	N/A	0%	N/A	0%
0.20	N/A	0%	N/A	0%	N/A	0%	N/A	0%
0.30	52.20	10%	66.55	20%	62.07	13%	58.50	18%
0.40	79.57	14%	113.69	13%	77.66	9%	86.09	11%
0.50	110.30	10%	109.30	13%	75.07	13%	70.66	3%
0.60	204.00	3%	129.50	6%	82.71	7%	145.00	2%
0.70	128.00	2%	101.00	1%	10.66	3%	161.00	1%
0.80	N/A	0%	N/A	0%	N/A	0%	N/A	0%
0.90	N/A	0%	N/A	0%	N/A	0%	N/A	0%

Table 4. Performance of GP using swap-crossover. Constant neutrality (i.e., arities of functions = $\{1,2,3,4,5\}$) and degree neutrality using the maximum arity (i.e., all function have the same arity) were used. f_n stands for the different fixed fitness values used. The global optimum is shown at the top of Figure 3 and setting $B = 0.01$ and $R = 0.8$ (i.e., the problem is considered to be very difficult).

f_n value	fdc	$\{1,2,3,4,5\}$		$\{5\}$	
		Avr. Gen	% Suc.	Avr. Gen	% Suc.
-	0.9987	4.50	2%	N/A	0%
0.10	0.9623	N/A	0%	N/A	0%
0.20	0.8523	6.00	1%	N/A	0%
0.30	0.7012	11.16	6%	35.60	10%
0.40	0.5472	43.75	12%	49.38	13%
0.50	0.4682	76.80	10%	84.11	18%
0.60	0.3912	111.75	8%	120.40	15%
0.70	0.3298	157.25	8%	100.25	8%
0.80	0.2892	266.50	2%	104.33	3%
0.90	0.2498	N/A	0%	56.00	2%

when all the functions are of the same arity (i.e.,$arity = 5$). So, it is clear that while fdc computes some of the characteristics of a problem in relation to its difficulty, it does not capture all.

By how much neutrality will help the search strongly depends on the constant fitness assigned in the neutral layer, f_n. However, for almost all values of f_n improvements

Table 5. Performance of GP using swap-crossover. Constant neutrality (i.e., arities of functions = $\{1,2,3,4,5\}$) and degree neutrality using the maximum arity (i.e., all function have the same arity) were used. f_n stands for the different fixed fitness values used. The global optimum is shown at the bottom of Figure 3 and setting $B = 0.01$ and $R = 0.8$ (i.e., the problem is considered to be very difficult).

f_n value	fdc	$\{1,2,3,4,5\}$		$\{5\}$	
		Avr. Gen	% Suc.	Avr. Gen	% Suc.
-	0.9991	N/A	0%	N/A	0%
0.10	0.9714	N/A	0%	N/A	0%
0.20	0.8693	N/A	0%	N/A	0%
0.30	0.7023	52.20	10%	82.76	17%
0.40	0.5802	79.57	14%	78.00	14%
0.50	0.4674	110.30	10%	81.40	15%
0.60	0.3879	204.00	3%	147.12	6%
0.70	0.3342	128.00	2%	246.23	2%
0.80	0.2787	N/A	0%	N/A	0%
0.90	0.2467	N/A	0%	N/A	0%

can be seen over the cases where neutrality is absent and when crossover is used. Here there is a rough agreement between fdc and actual performance although as f_n is increased beyond a certain level, fdc continues to decrease (suggesting an easier and easier problem) while in fact the success rate reaches a maximum and then starts decreasing again.

To explain why GP with crossover is able to sample the global optimum in the presence of neutrality (i.e., see for instance Tables 4 and 5 when $f_n = \{0.30, 0.40, 0.50\}$), we need to consider the following elements. Firstly, the flatter the landscape becomes the more GP crossover will be able to approach a Lagrange distribution of the second kind [13]. This distribution samples heavily the short programs. Secondly, since the global optima used in our experiments (i.e., see Figure 3) are all relatively small, this natural bias might be useful.

On the other hand, as can be seen in Tables 4, and 5, when the constant value on the neutral layer is high (i.e., $f_n \geq 0.70$) the perfomance of the GP system tends to decrease. This is easy to explain given that the higher the value of f_n, the flatter the landscape. Thus, flattening completely a landscape (i.e., $f_n \geq 0.80$) makes the search totally undirected, i.e., random. So, there is no guidance towards the optima. The flattening of the landscape also reduces or completely removes bloat. This is a good feature to have in this type of problem and for the chosen global optima because bloat moves the search towards the very large programs, but we know that none of them can be a solution. So, if bloat were to take place during evolution, it would hinder the search. Effectively, we can say that by changing the values of f_n, the user can vary the balance between two countering forces: the sampling of short programs and the guidance coming from fitness.

References

1. Altenberg, L.: Fitness Distance Correlation Analysis: An Instructive Counterexample. In: Back, T. (ed.) Proceedings of the Seventh International Conference on Genetic Algorithms, pp. 57–64. Morgan Kaufmann, San Francisco (1997)
2. Collins, M.: Finding Needles in Haystacks is Harder with Neutrality. In: Beyer, H.-G., O'Reilly, U.-M., Arnold, D.V., Banzhaf, W., Blum, C., Bonabeau, E.W., Cantu-Paz, E., Dasgupta, D., Deb, K., Foster, J.A., de Jong, E.D., Lipson, H., Llora, X., Mancoridis, S., Pelikan, M., Raidl, G.R., Soule, T., Tyrrell, A.M., Watson, J.-P., Zitzler, E. (eds.) GECCO 2005: Proceedings of the 2005 conference on Genetic and evolutionary computation, Washington DC, USA, June 2005, vol. 2, pp. 1613–1618. ACM Press, New York (2005)
3. Galván-López, E.: An Analysis of the Effects of Neutrality on Problem Hardness for Evolutionary Algorithms. PhD thesis, School of Computer Science and Electronic Engineering, University of Essex, United Kingdom (2009)
4. Galván-López, E., Dignum, S., Poli, R.: The Effects of Constant Neutrality on Performance and Problem Hardness in GP. In: O'Neill, M., Vanneschi, L., Gustafson, S., Esparcia Alcázar, A.I., De Falco, I., Della Cioppa, A., Tarantino, E. (eds.) EuroGP 2008. LNCS, vol. 4971, pp. 312–324. Springer, Heidelberg (2008)
5. Galván-López, E., Poli, R.: Some Steps Towards Understanding How Neutrality Affects Evolutionary Search. In: Runarsson, T.P., Beyer, H.-G., Burke, E.K., Merelo-Guervós, J.J., Whitley, L.D., Yao, X. (eds.) PPSN 2006. LNCS, vol. 4193, pp. 778–787. Springer, Heidelberg (2006)
6. Goldberg, D.E., Deb, K., Horn, J.: Massive Multimodality, Deception, and Genetic Algorithms. In: Männer, R., Manderick, B. (eds.) PPSN II: Proceedings of the 2nd International Conference on Parallel Problem Solving from Nature, pp. 37–48. Elsevier Science Publishers, Amsterdam (1992)
7. Jones, T.: Evolutionary Algorithms, Fitness Landscapes and Search. PhD thesis, University of New Mexico, Albuquerque (1995)
8. Kimura, M.: The Neutral Theory of Molecular Evolution. Cambridge University Press, Cambridge (1983)
9. Miller, J.F.: An Empirical Study of the Efficiency of Learning Boolean Functions Using a Cartesian Genetic Approach. In: Banzhaf, W., Daida, J.M., Eiben, A.E., Garzon, M.H., Honavar, V., Jakiela, M.J., Smith, R.E. (eds.) Proceedings of the Genetic and Evolutionary Computation Conference GECCO 1999, Orlando, Florida, vol. 2, pp. 1135–1142. Morgan Kaufmann, San Francisco (1999)
10. Miller, J.F., Thomson, P.: Cartesian genetic programming. In: Poli, R., Banzhaf, W., Langdon, W.B., Miller, J., Nordin, P., Fogarty, T.C. (eds.) EuroGP 2000. LNCS, vol. 1802, pp. 121–132. Springer, Heidelberg (2000)
11. O'Reilly, U.-M.: Using a Distance Metric on Genetic Programs to Understand Genetic Operators. In: IEEE International Conference on Systems, Man, and Cybernetics, Computational Cybernetics and Simulation, Orlando, Florida, USA, vol. 5, pp. 4092–4097. IEEE Press, Los Alamitos (1997)
12. Poli, R., Galván-López, E.: On The Effects of Bit-Wise Neutrality on Fitness Distance Correlation, Phenotypic Mutation Rates and Problem Hardness. In: Stephens, C.R., Toussaint, M., Whitley, D., Stadler, P.F. (eds.) Foundations of Genetic Algorithms IX, Mexico city, Mexico, pp. 138–164. Springer, Heidelberg (2007)
13. Poli, R., Langdon, W.B., Dignum, S.: On the limiting distribution of program sizes in tree-based genetic programming. In: Ebner, M., O'Neill, M., Ekárt, A., Vanneschi, L., Esparcia-Alcázar, A.I. (eds.) EuroGP 2007. LNCS, vol. 4445, pp. 193–204. Springer, Heidelberg (2007)

14. Punch, B., Zongker, D., Godman, E.: The Royal Tree Problem, A Benchmark for Single and Multi-population Genetic Programming. In: Angeline, P., Kinnear, K. (eds.) Advances in Genetic Programming 2, pp. 299–316. The MIT Press, Cambridge (1996)
15. Quick, R.J., Rayward-Smith, V.J., Smith, G.D.: Fitness Distance Correlation and Ridge Functions. In: Eiben, A.E., Bäck, T., Schoenauer, M., Schwefel, H.-P. (eds.) PPSN 1998. LNCS, vol. 1498, pp. 77–86. Springer, Heidelberg (1998)
16. Slavov, V., Nikolaev, N.I.: Fitness Landscapes and Inductive Genetic Programming. In: Smith, G.D., Steele, N.C., Albrecht, R.F. (eds.) Artificial Neural Nets and Genetic Algorithms: Proceedings of the International Conference, ICANNGA 1997, University of East Anglia, Norwich, UK, Springer, Heidelberg (1997)
17. Tomassini, M., Vanneschi, L., Collard, P., Clergue, M.: Study of Fitness Distance Correlation as a Difficulty Measure in Genetic Programming. Evolutionary Computation 13(2), 213–239 (2005)
18. Vanneschi, L.: Theory and Practice for Efficient Genetic Programming. PhD thesis, Faculty of Science, University of Lausanne, Switzerland (2004)
19. Vanneschi, L., Tomassini, M., Collard, P., Clergue, M.: Fitness Distance Correlation in Structural Mutation Genetic Programming. In: Ryan, C., Soule, T., Keijzer, M., Tsang, E.P.K., Poli, R., Costa, E. (eds.) EuroGP 2003. LNCS, vol. 2610, pp. 455–464. Springer, Heidelberg (2003)
20. Yu, T., Miller, J.F.: Finding Needles in Haystacks is not Hard with Neutrality. In: Foster, J.A., Lutton, E., Miller, J., Ryan, C., Tettamanzi, A.G.B. (eds.) EuroGP 2002. LNCS, vol. 2278, pp. 13–25. Springer, Heidelberg (2002)

Author Index